Remote Sensing of the North Eastern Himalayan Ecosystem

Shiv P. Aggarwal · Shyam S. Kundu ·
Bijoy K. Handique
Editors

Remote Sensing of the North Eastern Himalayan Ecosystem

Springer

Editors
Shiv P. Aggarwal
Department of Space, Government of India
North Eastern Space Applications Centre
Umiam, Shillong, Meghalaya, India

Shyam S. Kundu
Department of Space, Government of India
North Eastern Space Applications Centre
Umiam, Shillong, Meghalaya, India

Bijoy K. Handique
Department of Space, Government of India
North Eastern Space Applications Centre
Umiam, Shillong, Meghalaya, India

ISBN 978-981-95-3220-9 ISBN 978-981-95-3221-6 (eBook)
https://doi.org/10.1007/978-981-95-3221-6

This Springer imprint is published by the registered company Springer Nature Singapore Pte Ltd.
The registered company address is: 152 Beach Road, #21-01/04 Gateway East, Singapore 189721, Singapore

Foreword

The Eastern Himalayan region is ecologically diverse and environmentally sensitive. Characterized by complex terrain, dense forests, rich water resources, dynamic weather systems, and biodiversity, this region demands advanced and sustained observation for informed decision-making. Remote sensing, with synoptic coverage, multi-temporal, as well as high spatial and spectral resolution observation capability, is an important tool to monitor, evaluate and manage such ecosystems.

The edited volume, *Remote Sensing of North Eastern Himalayan Ecosystem*, published as part of the Silver Jubilee celebrations of the North Eastern Space Applications Centre (NESAC/DoS), is a significant contribution to this domain. The book presents a comprehensive account of remote sensing applications tailored to the geographical and developmental contexts of the region. A wide spectrum of themes, viz., ecosystem challenges, forest resource management, agricultural monitoring, wetland and water resource assessment, disaster management, infrastructure planning and atmospheric and climate studies are covered in the volume. Methodological aspects of remote sensing applications have been integrated with case studies, demonstrating how space-based observations are translated into region-specific applications.

This publication also highlights the capabilities of Indian space program to address complex mountain-specific challenges, in alignment with sustainable development goals and disaster resilience. It would foster wider awareness, inspire research, and support collaborative efforts in the field of remote sensing studies of mountain ecosystems.

I am sure that this compilation would serve as a valuable reference for researchers, students, professionals and policy makers, engaged in Himalayan ecosystem and highland studies. I congratulate the editorial team and the authors for this scientifically important publication.

Bengaluru, India
September 2025

Dr. V. Narayanan
Chairman, Space Commission
and Secretary, Department of Science

Preface

The North Eastern Himalayan region, encompassing the North Eastern states of India, is one of the most ecologically rich, culturally diverse, and geodynamically complex regions in the world. It hosts some of the highest rainfall zones on Earth, cradles critical river systems, supports unique forest and agricultural ecosystems, and sustains a mosaic of ethnic communities. At the same time, this region faces a wide array of challenges, including natural hazards, environmental degradation, climate variability, and development constraints arising from difficult terrain and limited infrastructure.

In this context, the role of space technology becomes indispensable that provides access to the remotest part of this difficult and complex terrain. Satellite and drone based remote sensing and geospatial analytics powered by artificial intelligence are increasingly transforming how we understand and manage such complex landscapes. As the geospatial ecosystem is rapidly evolving, its application strategy and effectiveness is also going through a rapid progression. It is within this backdrop that this edited book, *Remote Sensing of North Eastern Himalayan Ecosystem*, is conceptualized with an objective of sharing the practical knowledge gained on Earth observation (EO) applications to address real problems on the ground covering all corners of the NE region of India. Brought out as part of the Silver Jubilee celebration of the North Eastern Space Applications Centre (NESAC), this book reflects NESAC's long-standing commitment to applying space-based technologies for the sustainable development of the North Eastern Region of India.

Over the past 25 years, NESAC has played a pioneering role in utilizing satellite remote sensing and GIS to address region-specific challenges. This book aims to encapsulate NESAC's extensive contributions to EO science and its applications, enriched by inputs from several partner organizations, in tackling the diverse environmental and developmental issues of the North Eastern Himalayan region. Building on NESAC's 25 years of experience, the 25 chapters of this book present not only the technological advancements in remote sensing, numerical modeling, and geospatial analytics, but also showcase customized solutions, thematic applications, and case studies demonstrating the use of geospatial technology for the intended user community.

The opening chapter sets the stage by outlining the critical issues, vulnerabilities, and the need for technology interventions in the North Eastern Himalayan ecosystem. The initial chapters focus on agriculture, horticulture, sericulture, and soil resource assessment, demonstrating how remote sensing and geospatial tools support productivity and climate-resilient practices in fragmented and rainfed terrains. This is followed by a set of chapters on forest monitoring, biomass estimation, and forest fire assessment that are critical for biodiversity conservation and carbon accounting. The book also delves into geoscientific applications such as mineral mapping, tectonic studies, and landslide hazard assessment. These are of particular importance given the region's active geodynamics and frequent geological hazards. Urban and rural planning applications are covered through chapters on hill-town development, greenfield initiatives, and multi-hazard risk assessments. Given the centrality of water in the region, a substantial section addresses water resource management, flood early warning systems, and hydropower modeling using integrated spatio-hydrological approaches. Another set of chapters focuses on emerging technologies, including GeoAI, drone-based data analytics, and geospatial governance systems that enhance decision-making and local-level interventions. A very important part of the book highlights atmospheric and climatic studies, including aerosol dynamics, mountain weather forecasting, and aspects related to regional climate change. The final chapters discuss SATCOM applications for societal benefits and the development of a single-window disaster risk reduction platform tailored to the NER.

This book is unique in its regional focus. While there exists extensive literature on remote sensing applications in India and elsewhere, very few publications concentrate on the specific geophysical and socio-economic fabric of the North Eastern Himalayas, particularly the North Eastern region of India. By bringing together case studies, methodologies, and policy-relevant analyses grounded in regional experience, this volume bridges that gap. With the impacts of climate change becoming increasingly evident, particularly in the Himalayan arc, there is a pressing need for decision-makers and planners to have access to actionable, data-driven insights. The volume also aligns with India's broader national goals, such as the Digital India initiative, the Sustainable Development Goals (SDGs), Disaster management act, etc.

For researchers and graduate students, the book provides a solid foundation of applied science, combining Earth observation techniques with domain-specific knowledge. For practitioners, including administrators and planners, the case studies and technology workflows presented herein offer direct replicability. For the broader scientific and development community, this book serves as a reference for the integration of space technology in mountainous and ecologically sensitive regions. As we celebrate NESAC's 25 years of service, we also look to the future. The challenges of the North Eastern Himalayan ecosystem are dynamic and evolving. This book, while comprehensive, is only a step in an ongoing journey of scientific inquiry and technological innovation. The Centre, in collaboration with ISRO and other stakeholders, remains committed to expanding the frontiers of knowledge and delivering impactful solutions to the region and the nation at large.

We hope that this book will contribute meaningfully to the scientific literature on Himalayan ecosystems and inspire further innovations in geospatial science for mountain regions.

Umiam/Shillong, Meghalaya, India

Shiv P. Aggarwal
Shyam S. Kundu
Bijoy K. Handique

Acknowledgements

This book on *Remote Sensing of North Eastern Himalayan Ecosystem* is a testament of the years of hard work and dedication of all the contributing authors. We sincerely acknowledge all authors for writing scholarly and up-to-date chapters aligned with the theme of this book, and for diligently revising their manuscripts based on the valuable feedback from anonymous reviewers. The scientific insights and applications presented in this volume are a reflection of the enormous work carried out by many individuals across various departments.

We sincerely acknowledge the vision and perseverance of its founding leaders, former Directors, scientists, engineers, and all associated professionals, whose tireless efforts, over the decades has helped NESAC to grow into a vibrant institution of excellence under the Department of Space, Government of India. We are deeply grateful to the Department of Space, North Eastern Council (MDoNER), and the state governments of the NE states, for their continued support in fostering research and application of remote sensing technology across the region. Our special thanks are due to all collaborating institutions that have played a key role in enabling this work. We thank Springer for agreeing to publish the volume and giving continued guidance in this journey.

It is our honour and privilege to sincerely thank Hon'ble Union Home Minister Shri Amit Shah Ji, who is the President of NESAC Society, for his visionary leadership towards taking NESAC to a greater height. We sincerely thank Dr. V. Narayanan, Secretary, Department of Space and Chairman, ISRO/NESAC Governing Council, and all former Chairmen of ISRO/NESAC Governing Council for their constant encouragement and guidance towards enabling us to take up and complete this work.

We hope that this volume will serve as a valuable resource for researchers, policy planners, and practitioners working in the fields of Earth observation, environmental studies, and sustainable management of the Himalayan ecosystems.

Umiam/Shillong, Meghalaya, India

Shiv P. Aggarwal
Shyam S. Kundu
Bijoy K. Handique

Contents

Editors and Contributors

About the Editors

Shiv P. Aggarwal is the Director of the North Eastern Space Applications Centre (NESAC), Department of Space, Government of India, Shillong. He holds a B.Tech. degree from Allahabad University, a Postgraduate and Ph.D. degree from the Indian Agricultural Research Institute (IARI), New Delhi. He did his post-doctoral research from ITC/IHE, The Netherlands. A distinguished space applications scientist and visionary leader, Dr. Aggarwal has been pivotal in spearheading transformative initiatives that harness space technology for sustainable development across the North Eastern Region of India. Under his leadership, NESAC has adopted cutting-edge technologies (AI, ML, UAVs), enhanced stakeholder engagement through NeUIM, and earned national accolades for excellence in geospatial applications. His contributions span a wide spectrum of applications, including natural resource management, climate-resilient infrastructure planning, disaster risk reduction, weather forecasting, satellite-based communication, and geo-governance. Earlier, he served as Group Head, Water Resources Department, at the Indian Institute of Remote Sensing, ISRO, Dehradun, where he played a crucial role in advancing remote sensing and GIS-based solutions for water resources management, flood forecasting, disaster management, and climate change impact assessment. He also held the position of Program Coordinator of CSSTEAP, a United Nations-affiliated centre, where he significantly contributed to capacity-building in space science and technology education across the Asia-Pacific region. He has published over 200 scientific articles and book chapters, and has delivered more than 100 invited lectures at reputed national and international forums. He has supervised over 30 M.Tech. and Ph.D. theses, mentoring the next generation of scientists and engineers. In recognition of his exemplary contributions to the field of space applications and geospatial sciences, he has received numerous prestigious awards, including the National Geospatial Award and the National e-Governance Award. He is a life member and honorary fellow of several national and international scientific societies. Currently, he serves as the President of the Indian Society of Remote Sensing (ISRS), and has previously held key editorial and leadership roles, including Executive Editor of the Journal of ISRS.

Shyam S. Kundu did his M.Sc. in Physics from University of North Bengal and Ph.D. in Physics from Gauhati University. He has more than 20 years of professional experience at NESAC and currently working as the Head of the Space and Atmospheric Sciences Division at the centre. He was instrumental in setting up the Space and Atmospheric Science research facility at NESAC with several instruments, a S-band Doppler Weather Radar, largest network of weather stations in NER covering difficult terrains, three aerosol observatories including two in Himalayan region, network of lightning detectors, etc. His research interest include extreme weather forecasting in NER, lightning physics, aerosol-cloud-precipitation interaction studies and its climatic impact, climate change assessment, etc. He took lead role to develop an operational lightning early warning system for the NE region of India. He has more than 50 publications in peer reviewed journals and books. He is making significant contribution to professional bodies like ISRS as its Joint Secretary and ISPRS as the Secretary of the WG-4 under TC V. He also represents ISRO in the Committee on Earth Observation Satellites for Precipitation constellations.

Bijoy K. Handique currently working as the Head, Agriculture and Soil Division and Head, Program Planning and Evaluation Group at North Eastern Space Applications Centre under Department of Space, Government of India, has more than 22 years of experience in space technology applications and research. He has been leading a number of research and applications programs at NESAC particularly in space technology applications in agriculture and allied areas. He is also leading a team at NESAC in using unmanned aerial vehicles for various agricultural applications. He has published about 60 research papers in national and international journals and conference proceedings. He has also published 12 book chapters and guided a number of M.Sc./M.Tech. and Ph.D. students. He has been awarded ISRS-SPECK Award for 2005, ISRO Young Scientist Merit Award for 2008 and ISRO-ASI Young Scientist Award for 2014.

Contributors

Shiv P. Aggarwal North Eastern Space Applications Centre (NESAC), Umiam, Shillong, Meghalaya, India

Ritu Anilkumar University of Bristol, Bristol, United Kingdom

B. M. Arjun National Remote Sensing Centre (NRSC), Balanagar, Hyderabad, Telangana, India

Diganta Barman North Eastern Space Applications Centre (NESAC), Shillong, Meghalaya, India

Nilamoni Barman Central Ground Water Board, Kolkata, West Bengal, India

Dhruval Bhavsar North Eastern Space Applications Centre (NESAC), Umiam, Shillong, Meghalaya, India

Kuntala Bhusan North Eastern Space Applications Centre (NESAC), Umiam, Shillong, Meghalaya, India

Siddhartha Bhuyan North Eastern Space Applications Centre (NESAC), Umiam, Shillong, Meghalaya, India

Prakash Biswakarma North Eastern Space Applications Centre (NESAC), Umiam, Shillong, Meghalaya, India

Arup Borgohain North Eastern Space Applications Centre (NESAC), Umiam, Shillong, Meghalaya, India

Aniket Chakravorty North Eastern Space Applications Centre (NESAC), Umiam, Shillong, Meghalaya, India

Kasturi Chakraborty North Eastern Space Applications Centre (NESAC), Umiam, Shillong, Meghalaya, India

Avinash Chouhan North Eastern Space Applications Centre (NESAC), Umiam, Shillong, Meghalaya, India

Dibyajyoti Chutia North Eastern Space Applications Centre (NESAC), Shillong, Meghalaya, India

Himangshu Jyoti Das North Eastern Space Applications Centre (NESAC), Umiam, Shillong, Meghalaya, India

Pratibha Thakuria Das North Eastern Space Applications Centre (NESAC), Shillong, Meghalaya, India

Ramani Kumar Das North Eastern Space Applications Centre (NESAC), Umiam, Shillong, Meghalaya, India

Ranjit Das North Eastern Space Applications Centre (NESAC), Shillong, Meghalaya, India

Santanu Das North Eastern Space Applications Centre (NESAC), Shillong, Meghalaya, India

Anjan Debnath North Eastern Space Applications Centre (NESAC), Umiam, Shillong, Meghalaya, India

Francis Dutta North Eastern Space Applications Centre (NESAC), Shillong, Meghalaya, India

Manasi Gogoi Department of Physics, Dibrugarh University, Dibrugarh, Assam, India

Mukunda M. Gogoi Space Physics Laboratory (SPL), VSSC, Thiruvananthapuram, Kerala, India

Rekha B. Gogoi North Eastern Space Applications Centre (NESAC), Shillong, Meghalaya, India

Chandan Goswami North Eastern Space Applications Centre (NESAC), Shillong, Meghalaya, India

Jonali Goswami North Eastern Space Applications Centre (NESAC), Shillong, Meghalaya, India

Chirag Gupta North Eastern Space Applications Centre (NESAC), Shillong, Meghalaya, India

Bijoy K. Handique North Eastern Space Applications Centre (NESAC), Umiam, Shillong, Meghalaya, India

Suchitra Haobam North Eastern Space Applications Centre (NESAC), Shillong, Meghalaya, India

Pradesh Jena North Eastern Space Applications Centre (NESAC), Shillong, Meghalaya, India

Golfinstar Kharbani North Eastern Space Applications Centre (NESAC), Umiam, Shillong, Meghalaya, India

Praveen Kumar Euro-Mediterranean Center on Climate Change (CMCC), Lecce, Italy

Arundhati Kundu North Eastern Space Applications Centre (NESAC), Umiam, Shillong, Meghalaya, India

Shyam S. Kundu North Eastern Space Applications Centre (NESAC), Umiam, Shillong, Meghalaya, India

Shanbor Kurbah North Eastern Space Applications Centre (NESAC), Shillong, Meghalaya, India

Rosly B. Lyngdoh North Eastern Space Applications Centre (NESAC), Umiam, Shillong, Meghalaya, India

Shashi Bhusan Nayak North Eastern Space Applications Centre (NESAC), Shillong, Meghalaya, India

Nilay Nishant North Eastern Space Applications Centre (NESAC), Umiam, Shillong, Meghalaya, India

Jenita M. Nongkynrih North Eastern Space Applications Centre (NESAC), Shillong, Meghalaya, India

Sanjay Pandit North Eastern Space Applications Centre (NESAC), Shillong, Meghalaya, India

Binita Pathak Department of Physics, Dibrugarh University, Dibrugarh, Assam, India

Rocky Pebam North Eastern Space Applications Centre (NESAC), Shillong, Meghalaya, India

Rahul Pratap North Eastern Space Applications Centre (NESAC), Umiam, Shillong, Meghalaya, India

Md Abu Sahin North Eastern Space Applications Centre (NESAC), Umiam, Shillong, Meghalaya, India

Victor Saikhom North Eastern Space Applications Centre (NESAC), Shillong, Meghalaya, India

K. K. Sarma North Eastern Space Applications Centre (NESAC), Umiam, Shillong, Meghalaya, India

Gopal Sharma North Eastern Space Applications Centre (NESAC), Shillong, Meghalaya, India

Manbhalang Dakermi Shylla North Eastern Space Applications Centre (NESAC), Shillong, Meghalaya, India

M. Somorjit Singh North Eastern Space Applications Centre (NESAC), Umiam, Shillong, Meghalaya, India

Puyam S. Singh North Eastern Space Applications Centre (NESAC), Umiam, Shillong, Meghalaya, India

Abhay Srivastava North Eastern Space Applications Centre (NESAC), Umiam, Shillong, Meghalaya, India

B. C. Sumanth North Eastern Space Applications Centre (NESAC), Umiam, Shillong, Meghalaya, India

Surajit Sutradhar North Eastern Space Applications Centre (NESAC), Umiam, Shillong, Meghalaya, India

Suraj Kr. Swain North Eastern Space Applications Centre (NESAC), Umiam, Shillong, Meghalaya, India

List of Figures

List of Tables

Chapter 1
North East Himalayan Ecosystem: Issues, Challenges and Need for Technology Interventions

K. K. Sarma

Abstract The North Eastern Himalayan Region (NEHR) of India is a unique and diverse area, recognized as a global biodiversity hotspot with nearly half of India's flowering plant species and endangered wildlife. It is a "melting pot of cultures" with over 200 distinct ethnic communities, many of whom practice animistic beliefs that promote environmental conservation. Geologically, the NEHR is highly seismic due to the convergence of three tectonic plates, resulting in rugged terrain and frequent earthquakes. The region experiences exceptional rainfall, particularly in areas like Mawsynram and Cherrapunji, which fosters lush, unique ecosystems. Additionally, the NEHR is home to distinctive sustainable agricultural practices such as Jhum, Alder-based, and Zabo farming systems, along with ingenious bamboo drip irrigation. Despite its remarkable uniqueness, the NEHR faces significant challenges including environmental degradation, infrastructure deficits, socioeconomic issues, and geological hazards. Advanced technologies, especially geospatial tools (GIS, RS, GNSS, analytics), are key to regional development. They drive economic growth, enhance productivity, attract investment, and improve healthcare, education, and urban living. Technology also promotes inclusion, transparent governance, and environmental sustainability through better resource management and clean energy. In India's North East Region, geospatial technology is vital for managing resources, disaster response, infrastructure, and health, with NESAC playing a pivotal role. Ensuring data access and building skills is crucial for smart, sustainable, and inclusive growth.

Keywords North Eastern Himalaya · Geospatial technology · Natural resource management · Disaster management · Climate resilience · Regional development

K. K. Sarma (✉)
North Eastern Space Applications Centre (NESAC), Umiam, Shillong, Meghalaya, India
e-mail: sarmakk@gmail.com

S. P. Aggarwal et al. (eds.), *Remote Sensing of the North Eastern Himalayan Ecosystem*,
https://doi.org/10.1007/978-981-95-3221-6_1

1.1 Introduction

The Himalayas play a crucial role in global ecosystem by acting as a major barrier against cold air from Central Asia, significantly influencing the monsoon patterns in South Asia (Posey 1994). It also serving as a massive reservoir of freshwater stored in glaciers, often referred to as the "Third Pole" due to its vast ice mass, which impacts water availability for millions of people across the region; essentially regulating the climate of the Indian subcontinent and surrounding areas by blocking cold winds and forcing monsoon winds to release their moisture upon reaching the mountain range (Yao et al. 2012; Rao et al. 2016; Chauhan et al. 2024).

The Himalayas mountain range can be divided in several ways, each offering a unique perspective on its diverse features. The Himalayas consist of four parallel mountain ranges from south to north: the Sivalik Hills on the south; the Lower Himalayas; the Great Himalayas, which is the highest and central range; and the Tibetan Himalayas on the north (Sathyakumar et al. 2020; Anonymous 2025a).

Based on geographical location, the Indian Himalayan region can be broadly divided into three divisions i.e. Western Himalayas, Central Himalayas and Eastern Himalayas (Rao et al. 2016). The Eastern Himalayas stretching across eastern part of Nepal, Sikkim, Bhutan, Arunachal Pradesh and other North Eastern Regions (NER) of India, is characterized by lower elevations compared to the Central Himalayas, but it receives more rainfall and has lush vegetation. (Anonymous 2024).

The North-Eastern Indian part of the Eastern Himalayas which also can refer as Northeastern Himalayas (NEH) covers Karbi Anglong and Cachar Hills in Assam and the entire territories of Arunachal Pradesh, Nagaland, Meghalaya, Manipur, Mizoram, Sikkim and Tripura often referred as North Eastern Region (NER) of India which also covers the plain region of Assam (Anonymous 2025b) (Fig. 1.1).

The NEH region constitutes about 8 percent of India's total geographical area, less than 4 percent of population and around 4 percent of GDP (NEC 2024). The North Eastern Himalayan Region (NEHR) is a fascinating and diverse region of India, characterized by its vibrant cultural heritage, stunning natural landscapes, rich biodiversity hotspot, rolling hills, majestic rivers, and the Himalayan mountain range. Each one of the eight state has its unique identity, shaped by distinct languages, traditions, festivals, and ethnic groups, making Northeast India a melting pot of cultures. The North Eastern Himalayan Region of India, with its diverse topography and unique socio-economic conditions, presents a complex set of challenges and opportunities.

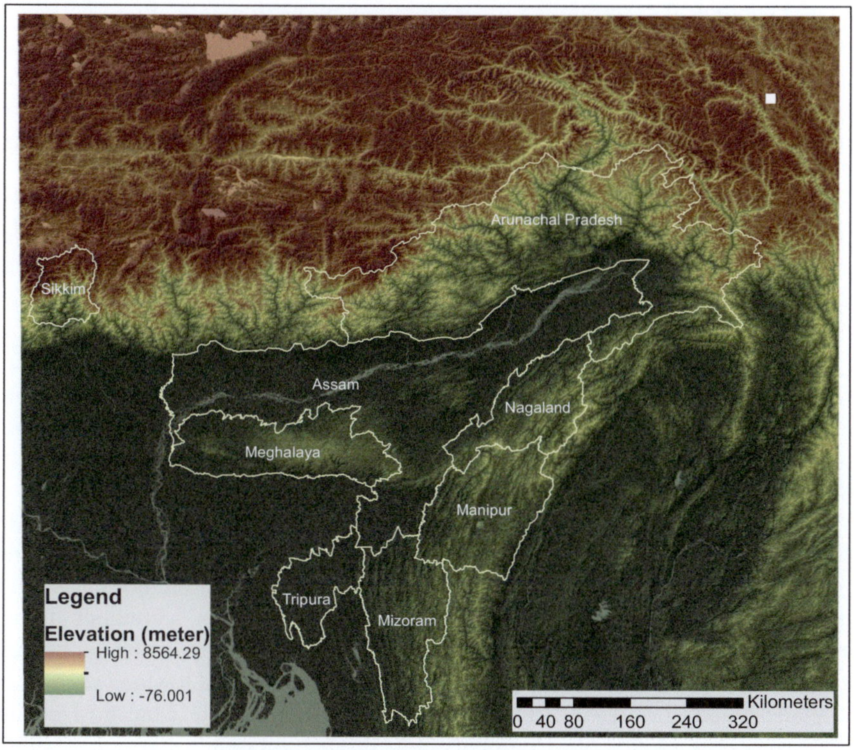

Fig. 1.1 Map showing elevation range over the North Eastern Himalayan Region of India with political boundary of Indian states

1.2 Uniqueness of North Eastern Himalayan Region

The North Eastern Himalayan Region (NEHR) encompassing parts of Northeast India, Bhutan, and the eastern Himalayas, is exceptionally unique with respect to other parts of Himalayas in many areas. A brief summary of its uniqueness is described below.

1.2.1 Rich Biodiversity

The Northeast Himalayan Region (NEHR) stands out as a globally recognized biodiversity hotspot (GIZ 2014), a designation earned through its unparalleled biological richness and the significant threat it faces from habitat loss. The region's extraordinary biodiversity is a direct consequence of its exceptional variety of ecosystems, ranging from the steamy tropical lowland rainforests to the frosty alpine meadows.

This vast ecological spectrum is further amplified by its immense altitudinal range, which creates diverse microclimates and habitats. Crucially, its geographic location acts as a grand crossroads, where the Indo-Malayan, Indo-Chinese, and Himalayan biogeographic zones converge (Rawal et al. 2013). This unique confluence has facilitated a complex evolutionary history, resulting in a remarkable array of life forms (Giz 2022).

A defining characteristic of the NEHR's biodiversity is its high level of endemism, meaning a significant proportion of its plant and animal species are found nowhere else on Earth. This makes the region an irreplaceable natural heritage. Often celebrated as the "Cradle of Flowering Plants," the NEHR astonishingly harbors nearly half of India's total flowering plant species (Malik 2019). This includes an astonishing diversity of orchids, many of which are endemic and highly sought after, along with a rich collection of bamboos that play vital ecological and economic roles, and vibrant rhododendrons that paint the hillsides with color. Furthermore, the region is a crucial genetic reservoir for wild relatives of important crop plants, offering invaluable resources for future agricultural resilience (Samant 2021).

NEH is also a stronghold for numerous charismatic and endangered wildlife species. Among them are the elusive red panda, a reclusive arboreal mammal; the magnificent snow leopard, a master of camouflage in the high altitudes; the enigmatic clouded leopard, with its distinctive coat patterns; and the iconic one-horned rhinoceros, a symbol of conservation success in areas like Kaziranga National Park. The sheer ecological diversity, spanning from its humid tropical lowland rainforests to its bracing alpine meadows, firmly establishes the NEHR as one of the most ecologically diverse regions globally. Complementing this natural wealth, the cultural diversity and traditional conservation practices of indigenous communities further amplify the region's ecological uniqueness (Hazarika 2021, Negi 2023). These communities often possess profound traditional ecological knowledge, employing sustainable practices that have historically contributed to the preservation of the region's rich natural heritage (Rathoure 2024).

1.2.2 Ethnic and Cultural Diversity

The NEHR is a vibrant mosaic of human cultures, embodying a remarkable diversity that parallels its rich biodiversity. It's home to over 200 distinct ethnic communities, each contributing unique threads to the region's cultural tapestry (Mavely 2021). These groups possess their own languages, customs, traditions, and spiritual beliefs, creating a fascinating ethnographic landscape. Among the prominent communities are the Monpa of Arunachal Pradesh, known for their rich Buddhist traditions influenced by Tibetan culture (Lhamu 2023); the Bodo of Assam, a large ethnolinguistic group with a distinct language and vibrant agricultural practices; the Garo and Khasi of Meghalaya, who follow unique matrilineal social systems and have intricate animistic beliefs intertwined with their daily lives; and the Lepcha of

Sikkim, considered the indigenous inhabitants of the region, known for their close ties to nature and distinct language (Choudhury et. al. 2017).

A significant aspect of many of these cultures in the Northeast Himalayas is the prevalence of animistic and nature-worshipping beliefs. This spiritual outlook isn't merely a set of rituals; it deeply reflects the profound connection people have with their natural environment. For centuries, this intrinsic bond has fostered a sense of reverence and responsibility towards the land, helping to conserve the region's natural beauty and resources. For example, sacred groves are maintained by various communities, acting as protected pockets of biodiversity where cutting trees or hunting animals is forbidden due to their spiritual significance. This traditional wisdom and practice, as highlighted by Choudhury et al. (2017), have played a crucial role in safeguarding the ecological integrity of the North Eastern Himalayan Region over generations.

1.2.3 Geological Uniqueness

The NEHR is located at the junction of the Indian, Eurasian, and Burmese plates, and this makes it one of the most seismically active regions on Earth. (Ghosh 2022).

This tectonic activity contributes to the rugged terrain, with breathtaking cliffs, narrow valleys, and fast-flowing rivers like the Brahmaputra (Das and Saraf 2007). This dynamic geological history has also created unique landforms, including fertile valleys and towering mountains that influence the climate and biodiversity.

The Northeastern Himalayan Region is a geological marvel, situated at the critical meeting point of three major tectonic plates makes the region exceptionally dynamic, establishing it as one of the most seismically active regions in the world (Ghosh 2022). The relentless collision and subduction of these colossal landmasses generate immense geological forces, leading to frequent earthquakes and tremors that continually reshape the landscape (Earle 2015). This intense tectonic activity is the primary architect of the region's rugged and dramatic terrain. The colossal pressures and movements have resulted in the uplift of towering mountain ranges, carving out breathtaking cliffs, deep, narrow valleys, and shaping the courses of powerful, fast-flowing rivers like the Brahmaputra. The sheer verticality and intricate geological formations are a direct consequence of this ongoing geological struggle (Das and Saraf 2007).

Beyond just ruggedness, this dynamic geological history has also sculpted a variety of unique and contrasting landforms. It has created not only the formidable, towering mountains but also interspersed them with fertile river valleys and expansive plains. These varied topographical features play a pivotal role in influencing the region's climate and biodiversity. The high mountains act as barriers, dictating monsoon patterns and creating rain shadows, leading to diverse microclimates ranging from tropical to alpine within short distances. This dramatic altitudinal and climatic variation, in turn, provides an extraordinary array of habitats, fostering the region's renowned rich biodiversity and high levels of endemism (Huggett 2007).

The very ground beneath the NEHR is a living testament to the powerful forces that have shaped its unique natural heritage.

1.2.4 Climate and Monsoon Impact

The Northeastern Himalayan Region stands out globally for its extraordinary rainfall, a defining feature that shapes its landscapes, ecosystems, and way of life (Kumar and Khanduri 2024). This climatic phenomenon reaches its zenith in places like Mawsynram and Cherrapunji (Meghalaya), which frequently claim the title of the wettest places on Earth. With annual deluges that defy imagination, these areas epitomize nature's hydrological extremes, fostering lush, rain-drenched ecosystems unlike anywhere else on the planet.

The monsoon season is the dominant force behind this relentless rainfall, sculpting not just the climate but the very essence of NEHR's ecosystems. The deluge acts as the region's lifeblood, sustaining vast, emerald forests that flourish under near-constant saturation. Within these rain-lush realms thrives an astonishing diversity of flora—towering hardwoods draped in moss, tangled undergrowth, delicate ferns, and epiphytes clinging to every available surface—all exquisitely adapted to the rhythm of ceaseless showers and mist-laden air. (Anonymous 2025c).

The NEHR's climatic diversity extends far beyond rainfall alone. The region's dramatic elevation gradients—from subtropical river basins below 300 m to high-altitude peaks exceeding 7,000 m—generate striking microclimatic variability. Studies highlight how this altitudinal zonation (Barry and Chorley 2003) produces sharp ecological transitions, with temperature and moisture regimes shifting profoundly over short distances. For instance, the Brahmaputra Valley's humid warmth contrasts with alpine tundra just 100 km away—a compression of climatic zones typically spanning continents (Rawat 2017).

1.2.5 Unique Agricultural Practices

The Northeast Himalayan region is a testament to ingenious human adaptation to challenging environments, boasting a rich heritage of traditional, sustainable farming methods. These practices, developed over centuries, are deeply intertwined with the local ecosystems and reflect an profound understanding of environmental stewardship.

Traditional agriculture pertaining to shifting cultivation (locally known as Jhum cultivation) is being practiced in Northeast Himalayan region of India since time immemorial (Soni 2021). While often misunderstood, traditional jhum is a sophisticated system that, when practiced sustainably with adequate fallow periods, allows forest regeneration and nutrient cycling. It's an adaptive response to the steep terrains and limited flat land prevalent in the region.

Another remarkable innovation is the Alder-based Farming System, a practice unique to the Angami tribe of Nagaland. This agroforestry system integrates the alder tree (Alnus nepalensis) into agricultural fields. The alder, a fast-growing tree, enriches the soil through nitrogen fixation and provides timber and fuel, demonstrating a symbiotic relationship between forestry and agriculture (Kehie et al. 2017, Steven and Julie 2019).

The Zabo Farming system, practiced by the Chakhesang tribe, also in Nagaland, is an exemplary integrated farming approach. "Zabo" literally means "impounding of water," and this system combines forestry, agriculture, and animal husbandry within a single, interconnected unit (Singh et. al. 2012). It involves constructing intricate water harvesting ponds, often with bunds and channels, to store rainwater which is then used for irrigating rice paddies, feeding livestock, and nurturing forest trees. The system is designed to minimize runoff and maximize resource utilization.

Meghalaya showcases another marvel of indigenous engineering with its Bamboo Drip Irrigation system. Practiced on hill slopes, this ingenious method harnesses gravity to transport water through hollowed bamboo channels, delivering it drop by drop directly to the roots of plants. This efficient system is particularly vital in areas with high rainfall but undulating terrain, preventing soil erosion and ensuring water reaches crops effectively.

Beyond these specific examples, terrace farming is widespread, transforming steep slopes into arable land, preventing soil erosion, and facilitating water management. Agroforestry, the integration of trees and shrubs into agricultural landscapes, is also a common practice, enhancing biodiversity, soil health, and providing diverse products (Selvan and Kumar 2017).

These practices collectively demonstrate a profound ecological mindfulness, reflecting the people's centuries-old expertise in harnessing the natural environment without depleting it.

These unique agricultural practices underscore a deep understanding of local ecosystems and the remarkable resourcefulness of the communities in the NEHR. They serve as living laboratories where traditional knowledge is seamlessly integrated with sustainable methods, offering invaluable lessons for modern agriculture grappling with climate change and resource scarcity. In recent years, this inherent sustainability has found resonance with contemporary movements, leading to a significant surge in organic farming. Notably, states like Sikkim have gone entirely organic, a pioneering achievement that highlights the region's strong emphasis on preserving ecological balance (Dikshit and Dikshit 2014).

1.2.6 Eco-Tourism Potential

The Northeastern Himalayan Region holds immense and largely untapped potential for eco-tourism, positioning itself as an ideal destination for environmentally responsible travel. This potential is rooted in its unparalleled natural beauty, rich biodiversity, and vibrant cultural heritage, promising benefits for both conservation and local communities (Chaudhury et al. 2023).

Northeast India boasts exceptional natural beauty and biodiversity. Its landscapes range from snow-capped peaks and alpine meadows to tropical rainforests and tea gardens. As part of the Eastern Himalayas and Indo-Burma biodiversity hotspots, the region is rich in flora and fauna. The region is ideal for wildlife safaris, bird watching (over 1,000 species), and nature photography. Mighty rivers like the Brahmaputra, along with numerous waterfalls (e.g., Nohkalikai Falls) and serene lakes (e.g., Loktak Lake, Tsomgo Lake), provide opportunities for river rafting, angling, and boating. The region's active tectonic history has created breathtaking cliffs, deep valleys, and unique landforms, including Meghalaya's famous living root bridges, showcasing indigenous engineering. Rich Cultural Heritage and Community Involvement offers immense potential for cultural tourism, allowing visitors to experience authentic local lifestyles, traditional festivals (like Nagaland's Hornbill Festival, which showcases the cultures of 17 major tribes), handicrafts, and cuisine. Many initiatives are focusing on involving local communities directly in eco-tourism, promoting homestays, guided tours by local experts, and showcasing traditional crafts. This ensures that the economic benefits of tourism reach the people who are stewards of the environment, fostering a sense of ownership and encouraging sustainable practices.

1.2.7 Confluence of Historical Trade and Cultural Exchange

Historically, the Northeast Himalayan region served as a vital nexus for ancient trade routes, forging connections between the Indian subcontinent, Southeast Asia, Tibet, and even further afield. These extensive exchanges were far more than just commercial transactions; they were powerful conduits for cultural enrichment. Over centuries, this constant flow of goods, ideas, and people introduced diverse influences that became intricately woven into the local art forms, culinary traditions, and distinctive architectural styles of the region.

The legacy of this historical interconnectedness is strikingly evident today. The Northeast remains a vibrant melting pot of influences, where cross-cultural elements are not just preserved but actively celebrated. This rich tapestry is particularly visible in the region's diverse music, intricate crafts, and colorful festivals. Consider the energetic Wangala Festival of the Garo people, a harvest celebration marked by drums and traditional dances, or the joyous Bihu festival of Assam, which encompasses a series of celebrations reflecting different agricultural cycles. These vibrant events are living testaments to the enduring impact of centuries of cultural exchange, showcasing a unique blend of traditions that makes the Northeast truly exceptional (Post et al. 2015).

1.3 Issues and Challenges of North East Himalayan Region

The NEHR is a biodiversity hotspot and a land of diverse cultures, but it faces a complex web of issues and challenges that hinder its development. Here are some of the key ones.

1.3.1 Environmental Degradation

The NEHR faces severe environmental challenges. Widespread deforestation, driven by practices like jhum cultivation, logging, and agricultural encroachment, is decimating biodiversity and disrupting crucial ecosystem services (Deol 2023). This deforestation, coupled with steep slopes and unsustainable agricultural practices, exacerbates soil erosion, diminishing soil fertility and impacting vital water resources (Hazarika 2021). Furthermore, the region is highly susceptible to the impacts of climate change, experiencing rising temperatures, erratic rainfall patterns, and an increased frequency of extreme weather events such as floods and landslides. Consequently, despite being a biodiversity hotspot, the NEHR is witnessing a decline in species diversity, due to habitat loss, fragmentation, and hunting (Matta et al. 2025).

1.3.2 Infrastructure Deficit

The region suffers from a significant infrastructure deficit that severely affects its development. Poor connectivity, including inadequate road, rail, and air networks, isolates communities, hindering economic growth and limiting access to essential services, particularly in remote and mountainous areas (Koner et al. 2022). Furthermore, an insufficient and unreliable power supply cripples agriculture, industry, and overall economic progress. Compounding these issues, limited irrigation facilities restrict agricultural productivity, especially in the predominantly rain-fed regions, making the NEHR's agricultural sector vulnerable to climate variability (Sharma and Das 2019).

1.3.3 Socioeconomic Challenges

The NEHR grapples with significant socioeconomic challenges. Despite possessing abundant natural resources, a considerable portion of the population lives below the poverty line, facing limited access to essential services like education, healthcare, and employment opportunities. High unemployment rates, particularly among the youth, contribute to social unrest and drive outward migration. Furthermore, the region's rich ethnic diversity, while a cultural asset, is sometimes marked by conflicts over land, resources, and identity, which can disrupt development and create instability. In some areas, ongoing insurgency and militancy further threaten security and severely hinder development activities (Sinha 2015; Anand et al. 2023).

1.3.4 Agricultural Issues

The agricultural sector in the NEHR faces numerous challenges. Low productivity plagues the region due to reliance on traditional farming practices, limited access to improved seeds and modern technologies, and inadequate irrigation facilities. Furthermore, poor connectivity and the absence of well-organized markets hinder farmers' ability to access markets and receive fair prices for their produce. While jhum cultivation has been a traditional practice, increasingly shorter fallow cycles are leading to soil degradation and declining productivity, posing further threats to the region's agricultural sustainability (Barah 2007).

1.3.5 Governance Challenges

The North Eastern Region (NER) of India, despite its rich cultural diversity and natural resources, faces a complex set of governance challenges. These challenges are deeply intertwined with historical, geographical, socio-economic, and political factors.

The rugged terrain and the narrow "Chicken Neck" (Siliguri Corridor) have historically limited accessibility, hindering crucial infrastructure development, with many state capitals still lacking adequate rail and air connectivity. This isolation contributes to the persistent underutilization of the region's abundant natural resources, including water, minerals, forests, and tourism potential, leading to low industrialization. There are issues with significant health and education gaps, evidenced by high infant and maternal mortality rates and inadequate educational facilities, all of which underscore the region's developmental disparities. The overall poor infrastructure, encompassing roadways, railways, and communication networks, severely impedes economic growth and the efficient delivery of public services. Geoinformatics is instrumental in infrastructure development, enabling efficient road and connectivity planning, and ensuring transparent project monitoring. In land administration, it facilitates the modernization of land records which supports the administrators to take informed decisions.

1.3.6 Geological Challenges

The North East Himalayan region is a geologically dynamic and complex area, presenting several significant challenges (Uthaman et al. 2024). It is highly seismically active due to the ongoing collision of the Indian and Eurasian tectonic plates, resulting in frequent and sometimes powerful earthquakes that pose a constant threat to infrastructure and human life (Dasgupta et al. 2021). The complex tectonic setting, including the Indo-Burma ranges, further contributes to diverse earthquake mechanisms. Additionally, the region is highly prone to landslides due to its steep slopes, high rainfall, and unstable geological formations, with deforestation and human activities exacerbating the problem (Shrestha et al. 2024). The presence of numerous faults, thrusts, and folds makes geological assessments and engineering projects particularly challenging, especially in the Eastern Himalayan Syntaxis, where the Himalayan arc meets the Indo-Burma ranges, forming a highly intricate geological zone. Furthermore, the region exhibits varied lithology, encompassing a wide range of rock types, from soft sediments to hard metamorphic rocks, which differ in stability and susceptibility to weathering (Diwakar et al. 2022). These geological challenges have significant implications for infrastructure development, natural hazard management, and the overall sustainability of the region.

1.3.7 Other Challenges

Beyond the core socioeconomic and environmental issues, the North East Himalayan Region (NEHR) faces a range of other pressing challenges. Limited access to quality healthcare services, particularly in remote rural areas, negatively impacts the health and well-being of the population. Inadequate educational infrastructure and a lack of quality education further restrict human capital development, limiting opportunities for individuals and hindering overall progress. Finally, given the region's vulnerability to natural disasters such as floods, landslides, and earthquakes, the development and implementation of effective disaster preparedness and management mechanisms are absolutely crucial for ensuring the safety and resilience of communities.

1.4 Need for Technology Intervention

Advanced technology is crucial for the development of a region for several interconnected reasons. Technology drives regional economic growth by nurturing innovation and creating new industries, which diversifies the economy and enhances its resilience. It also boosts productivity and efficiency through automation and digital tools. Furthermore, a strong tech infrastructure and skilled workforce attract investment, while advanced technology empowers local businesses to compete globally with better offerings (ResearchFDI 2023).

Technology enhances the quality of life in several ways. Healthcare is improved through telemedicine and advanced treatments, especially benefiting remote areas (Haleem et al. 2021). Education is enhanced by e-learning and digital resources, expanding opportunities (Haleem et al. 2022). Smart city technologies and efficient infrastructure improves daily living (ARES 2025). Finally, internet access empowers individuals and businesses with information and online services (Guo et al. 2024).

Technology plays a vital role in enhancing social progress and inclusion. By bridging the digital divide, it extends access to crucial information, education, and economic opportunities to underserved communities, narrowing the gap between urban and rural areas. Furthermore, technology empowers marginalized groups, offering tools and platforms that enable individuals with disabilities and those from disadvantaged backgrounds to participate more fully in both society and the economy. Finally, the adoption of digital platforms in governance enhances transparency, improves efficiency, and fosters greater citizen participation in decision-making processes.

Technological advancements are crucial for achieving environmental sustainability. Sophisticated tools such as precision agriculture, smart grids, and environmental monitoring systems facilitate a more efficient and sustainable management of our planet's natural resources. Furthermore, technology propels the development and wider adoption of clean, renewable energy sources, which is essential for decreasing our dependence on fossil fuels and mitigating the impacts of climate change. Therefore, innovative technologies play a key role in optimizing waste management processes, encouraging recycling initiatives, and ultimately reducing pollution levels.

In today's rapidly evolving world, the integration of advanced technology stands as a crucial imperative for promotion regional development. Acting as a powerful catalyst, technology propels economic growth, fuels social progress, and underpins sustainable development, particularly in regions that are underdeveloped or still developing. The essential role of modern technology in regional advancement is underscored by several key factors. Firstly, sustainable agriculture and food security are enhanced by precision farming techniques utilizing drones and AI, blockchain technology ensuring supply chain transparency, and climate-resilient technologies providing data-driven insights for adaptation. Secondly, the development of smart infrastructure and urbanization is facilitated by IoT-based smart city solutions, renewable energy technologies providing clean power, and AI-driven enhancements in transportation and logistics. Thirdly, environmental sustainability is promoted through green technologies for waste management and disaster prediction and response systems enhanced by AI and satellite technology. Fourthly, healthcare and education witness substantial improvements through telemedicine, which extends healthcare access to remote areas, and e-learning platforms that democratize quality education, bridging regional disparities. Smart healthcare systems, leveraging

IoT and AI, further optimize resource allocation and disease tracking. Fifthly, governance and transparency are strengthened through e-governance platforms streamlining public services and data analytics informing efficient policy making, while blockchain technology ensures the integrity of land records. Sixthly, the economic growth and job creation are significantly boosted through digital infrastructure, such as high-speed internet enabling e-commerce and remote work, and automation and AI enhancing productivity across various sectors. Furthermore, fintech and digital payment systems foster financial inclusion, empowering small businesses and farmers with access to crucial financial resources and broader markets. Finally, bridging the digital divide through affordable internet access and digital literacy programs ensures that the benefits of technological advancements reach even the most marginalized regions. In conclusion we can summarize that the modern technology transcends being merely a luxury; it is an undeniable necessity for meaningful regional development, demanding collaborative efforts from governments, the private sector, and international organizations to ensure equitable access and foster inclusive and sustainable growth.

Geospatial technology, a powerful suite encompassing Geographic Information Systems (GIS), Remote Sensing (RS), Global Navigation Satellite Systems (GNSS), and sophisticated spatial data analytics, is emerging as a transformative force in supporting comprehensive regional development. By providing critical location-based intelligence, this advanced technology empowers stakeholders to make more informed decisions, optimize the management of valuable resources, and cultivate sustainable growth trajectories across diverse regions. Geospatial technology relies on advanced methods for data collection and acquisition. This includes High-Resolution Satellite Imagery, where modern satellites provide incredibly detailed images of the Earth's surface, enabling precise monitoring of changes in landscapes, urban areas, and environmental conditions. Another crucial method is LiDAR (Light Detection and Ranging), a technology that uses lasers to create highly accurate 3D models of surfaces, capturing intricate details of terrain, buildings, and vegetation, which is crucial for applications like autonomous vehicles, forestry management, and urban planning. Furthermore, Drones (UAVs) equipped with various sensors (cameras, multispectral, thermal) can quickly and effectively collect high-resolution data for localized mapping, surveying, and inspection, often reaching areas inaccessible to traditional methods. Lastly, the proliferation of IoT Devices is increasingly generating location-tagged data from countless sensors, contributing to real-time insights essential for smart cities, environmental monitoring, and logistics.

The practical applications of geospatial technology are wide-ranging and deeply impactful across various sectors crucial for regional advancement. In urban planning and the development of smart cities, GIS plays a pivotal role in strategic land use planning, efficient infrastructure development, and the identification of optimal zones for residential, industrial, and green spaces. Furthermore, GNSS and GIS significantly enhance transportation and traffic management by optimizing public transit routes, alleviating congestion, and improving overall mobility. Remote sensing contributes to the creation of disaster-resilient infrastructure by accurately identifying areas prone to floods or earthquakes, facilitating safer construction practices. Within agriculture and the pursuit of food security, precision farming techniques leverage drones and satellite imagery to meticulously monitor soil health, track crop growth patterns, and manage irrigation needs effectively, leading to substantial yield improvements. Hyperspectral imaging enables the early detection of pests and diseases, allowing for timely interventions, while GIS tracks land degradation, including deforestation and soil erosion, supporting sustainable land management practices.

The realm of natural resource management also benefits immensely from geospatial technology. GIS is instrumental in mapping water resources, assessing groundwater potential, monitoring vital river basins, and preventing the detrimental effects of over-extraction. In the crucial area of forest and wildlife conservation, satellite imagery plays a key role in tracking illegal logging activities, monitoring critical wildlife habitats, and supporting afforestation efforts. Moreover, geospatial data aids in the responsible exploration of mineral resources, helping to identify mineral-rich zones while simultaneously minimizing potential environmental damage. During times of crisis, disaster management and the building of climate resilience are significantly enhanced by early warning systems powered by remote sensing, which can predict floods, droughts, cyclones, and wildfires, enabling proactive responses. Post-disaster, GIS facilitates efficient damage assessment and the optimized distribution of essential relief. Looking ahead, spatial analytics plays a vital role in modeling the impacts of climate change, such as sea-level rise and shifts in agricultural zones, informing crucial adaptation strategies.

The benefits of geospatial technology extend to healthcare and epidemic control, where GIS is used to track disease outbreaks, such as malaria and COVID-19, enabling targeted interventions and the optimization of vaccine distribution. Furthermore, mapping healthcare accessibility helps identify underserved regions requiring the establishment of new hospitals or mobile clinics. In the domain of infrastructure and connectivity development, GIS-based route optimization for road and railway planning reduces both costs and environmental impact. Satellite data aids in identifying optimal locations for rural electrification projects utilizing renewable energy sources, while geospatial analysis improves broadband and mobile network coverage in remote and underserved areas. Finally, governance and public service delivery are enhanced through blockchain-integrated GIS for managing property and land records, reducing disputes and promoting transparency. Poverty mapping, combining satellite data with socio-economic indicators, allows for the more effective targeting of welfare schemes, and crime hotspot analysis using predictive policing helps allocate law enforcement resources more efficiently.

The overarching benefits of geospatial technology for regional development are manifold, including data-driven decision making based on accurate spatial analytics, significant cost and time efficiencies through the reduction of manual surveys and optimized resource allocation, enhanced sustainability through support for eco-friendly planning and climate change adaptation, and greater inclusivity by bridging urban–rural gaps and improving access to essential services. Addressing existing challenges, such as ensuring data accessibility through open geospatial data policies, bridging skill gaps through targeted training programs, and fostering the integration of geospatial technology with emerging fields like AI and IoT, is crucial for maximizing its transformative potential. In conclusion, geospatial technology represents a true game-changer for regional development, paving the way for smarter, more sustainable, and equitable growth. To fully harness its potential, governments, non-governmental organizations, and the private sector must prioritize investments in geospatial infrastructure and capacity building initiatives.

Advanced geospatial technology is increasingly becoming a cornerstone for driving multifaceted development across the unique landscape of India's North East Region (NER). Encompassing the power of Geographic Information Systems (GIS), Remote Sensing (RS), Global Navigation Satellite Systems (GNSS), and sophisticated spatial data analytics, these tools offer invaluable insights tailored to the region's distinctive geographical, ecological, and socio-economic fabric. The application of these technologies is fostering transformative advancements across a spectrum of crucial sectors.

In the realm of natural resource management, advanced remote sensing techniques, including high-resolution satellite imagery and LiDAR, are enabling unprecedented precision in monitoring the region's rich forest cover, pinpointing areas of deforestation, and conducting detailed biodiversity assessments. This is particularly significant for states like Arunachal Pradesh and Nagaland, renowned for their extensive forest resources, where near real-time tracking of illegal logging via satellite imagery allows for swift and effective intervention. Furthermore, GIS and remote sensing are indispensable for meticulously mapping and monitoring the region's diverse water bodies, evaluating the health of critical watersheds, and identifying potential groundwater sources, especially in states like Meghalaya, which paradoxically faces water scarcity despite high rainfall. The analysis of time-series satellite data provides crucial insights into the dynamic changes in river courses and water storage levels, underpinning more effective water resource planning and management strategies. In the agricultural sector, precision farming methodologies, leveraging drone imagery and satellite data for in-depth soil health analysis, meticulous crop monitoring, and accurate yield prediction, are gaining traction in states like Assam, a vital agricultural hub. The application of hyperspectral imaging facilitates the early detection of plant stress or disease, enabling timely interventions that minimize potential losses.

Given the North East's vulnerability to natural disasters, particularly the recurrent flooding in the Brahmaputra and Barak river basins, disaster management and the building of climate resilience heavily rely on advanced geospatial technologies. These technologies are fundamental in developing sophisticated flood early warning systems, often integrating real-time satellite data with complex hydrological models, as exemplified by the crucial work of the North Eastern Space Applications Centre (NESAC). This integration provides essential information for timely evacuations and effective disaster response efforts. Moreover, in the hilly terrains of states like Manipur and Mizoram, prone to landslides, GIS-based landslide susceptibility mapping, which considers factors such as slope, geological composition, and rainfall patterns, plays a vital role in identifying high-risk zones and formulating targeted mitigation strategies. Satellite-based monitoring can even detect subtle pre-landslide movements, offering critical early warnings. Looking towards the future, spatial analytics is instrumental in modeling the potential impacts of climate change, such as shifts in temperature and rainfall patterns, on the region's delicate ecosystems and agricultural practices. This foresight enables the development of proactive climate adaptation strategies tailored to the specific vulnerabilities of each state within the NER.

The development of crucial infrastructure and the planning of sustainable urban centers also benefit significantly from advanced geospatial applications. GIS plays a pivotal role in the optimal planning of efficient road and railway networks, carefully considering factors like complex terrain, varying population density, and potential environmental impact. GNSS technology provides the precision necessary for accurate surveying and alignment of infrastructure projects, leading to improved efficiency and reduced construction costs. For burgeoning urban centers like Guwahati and Shillong, sophisticated GIS platforms are being developed to seamlessly integrate diverse city functions, including intelligent traffic management systems, comprehensive utilities mapping, and robust disaster preparedness protocols. This integrated approach fosters smarter governance and ultimately enhances the quality of life for the region's urban residents. The Guwahati Smart City project, for instance, actively utilizes a GIS platform for a multitude of applications, ranging from proactive flood monitoring to enhanced city surveillance.

In the critical domain of healthcare and public health, GIS is proving to be an invaluable tool for mapping and analyzing the spatial distribution of prevalent diseases, such as malaria and Japanese encephalitis, across the NER. This spatial intelligence aids in accurately identifying high-risk areas, understanding intricate disease transmission patterns, and optimizing the strategic allocation of resources for effective disease control and prevention initiatives. Furthermore, geospatial analysis plays a crucial role in identifying underserved areas lacking adequate healthcare facilities, thereby facilitating the strategic placement of new health centers and mobile clinics to improve healthcare accessibility for populations living in remote and often challenging terrains.

Despite the remarkable progress achieved, realizing the full potential of advanced geospatial technology in the NEHR necessitates addressing existing challenges. These include improving data accessibility through the implementation of open data policies, bridging the critical gap in skilled personnel through targeted training and capacity building programs, and fostering the seamless integration of geospatial technologies with other transformative fields such as Artificial Intelligence (AI) and the Internet of Things (IoT). Concerted initiatives focused on enhancing local expertise, promoting transparent data sharing, and encouraging synergistic collaborations between government agencies, academic institutions, and the dynamic private sector are crucial for ensuring the continued and increasingly impactful application of advanced geospatial technologies in driving sustainable and equitable development across the North East Himalayan Region of India.

Several key organizations and collaborative initiatives are at the forefront of driving these advancements in the North East Himalayan region. The North Eastern Space Applications Centre (NESAC), being a pivotal institution under Department of Space, plays a crucial role in providing cutting-edge space technology support tailored for the holistic development of the NER, actively engaging in numerous impactful projects spanning natural resource management, effective disaster management, and crucial capacity building initiatives utilizing the power of remote sensing and GIS. The North Eastern Spatial Data Repository (NeSDR), developed by NESAC, serves as a comprehensive and accessible platform for the seamless discovery, insightful visualization, and powerful analytics of geospatial data relevant to the NER. Its primary aim is to minimize data redundancy and provide robust decision support tools for a wide array of governance applications across the region. Complementing these efforts, each state within the NER has established its own State Remote Sensing Application Centre (SRSAC) as well as departmental RS and GIS Cells in many line departments of the region. These state-level centers work in close collaboration with NESAC and other prominent national agencies to implement geospatial technology applications that directly address the unique needs and priorities of their respective states.

Details of the work on spatial technology applications carried out in different themes have been summarized in various chapters of this volume.

References

S. Anand, M. Das, R. Bhattacharyya, R.B. Singh (Eds.), Sustainable Development Goals in Northeast India Challenges and Achievements (Springer Nature Singapore Pte Ltd.) ISBN 978-981-19-6477-0 (2023)

Anonymous, The Himalayas: formation, divisions, ranges & significance (2024) https://www.nextias.com/blog/the-himalayas/

Anonymous, Himalayas, mountains, Asia (2025a) https://www.britannica.com/place/Himalayas

Anonymous, North Eastern Region of Eastern Himalaya (2025b). https://alliance.indiaagrobdunepproject.cgiar.org/project-overview/target-regions/north-eastern-region-of-eastern-himalaya

Anonymous, Meghalaya subtropical forests (2025c). https://en.wikipedia.org/wiki/Meghalaya_subtropical_forests

ARES, How smart city technology is shaping the future of Urban Development (2025). https://aressecuritycorp.com/2025/01/20/smart-city/

B.C. Barah, Strategies for agricultural development in the North-East India: Challenges and emerging opportunities Indian. J. Agric. Econ. **62**(1),13–31 (2007)

R.G. Barry, R.J. Chorley, *Atmosphere, Weather, and Climate*, 8th ed. Routledge (2003)

S.K. Chaudhury, S. Sarkar, M. Patra, Exploring eco-tourism treasure of north-eastern states of India: a sustainability perspective. BL College J. **5**(2), 55-68 (2023). https://doi.org/10.62106/blc2023v5i2eg1

H.K. Chauhan, D. Gallacher, A. Bhatt, A.K. Bisht, The Himalayas: a climate change laboratory. Environ. Develop. **45**, 100814 (2024). https://doi.org/10.1016/j.envdev.2023.100814

S. Choudhury, B. Bahadur, K.V. Krishnamurthy, S.J. Adams, Ethnic diversity of North-East India, in *Ethnobotany of India: North-East India and Andaman and Nicobar Islands*, 1st ed (2017) (Apple Academic Press)

J.D. Das, A.K. Saraf, Technical Note: remotesensing in the mapping of the Brahmaputra/Jamuna River channel patterns and itsrelation to various landforms and tectonic environment. Int. J. Remote Sens. **28**(16), 3619–3631 (2007)

S. Dasgupta, B. Mukhopadhyay, M. Mukhopadhyay, P. Pande, Geo- and seismo- tectonics of Eastern Himalaya: exploring earthquake source zones from foredeep to Tibetan hinterland. Phy. Chem. Earth, Parts A/B/C **123**, 103013 (2021). https://doi.org/10.1016/j.pce.2021.103013

T. Deol, *Himalayan plunder: ecology changing for the worse due to loss of forest cover, drying springs: States in the northeastern region faring the worst*, Down To Earth, 1–15 February 2023

K.R. Dikshit, J.K. Dikshit, Agriculture in North-East India: past and present, in *North-East India: Land, People and Economy. Advances in Asian Human-Environmental Research. Springer, Dordrecht* (2014). https://doi.org/10.1007/978-94-007-7055-3_16

K.C. Diwakar, K. Gautam, H. Dangi, S. Kadel, L. Hu, Challenges in tunneling in the Himalayas: a survey of several prominent excavation projects in the Himalayan Mountain Range of South Asia. Geotechnics **2**(4), 802–824 (2022). https://doi.org/10.3390/geotechnics2040039

S. Early, *Physical geology*. Ebook ISBN 978-1-989623-71-8. https://opentextbc.ca/geology/chapter/11-2-earthquakes-and-plate-tectonics/

G.K. Ghosh, Delineation of major subsurface structural features and source depth locations using 3-D Euler deconvolution of gravity data at north-eastern part of India. Acta Geophys. **70**(5), 2033–2044 (2022)

Giz, Indo-German Biodiversity Programme (2014). https://indo-germanbiodiversity.com

Giz, Protection and sustainable management of aquatic resources in the North-Eastern Himalayan Region (2022). https://www.giz.de/en/worldwide/117067.html

J. Guo, Z. Cheng, B.Z. Wang, Internet development and entrepreneurship. China Econ. Rev. **88**, 102280 (2024). https://doi.org/10.1016/j.chieco.2024.102280

A. Haleem, M. Javaid, M.A. Qadri, R. Suman, Understanding the role of digital technologies in education: a review. Sustain. Oper. Comput. **3**(2022), 275–285 (2022). https://doi.org/10.1016/j.susoc.2022.05.004

A. Haleem, M. Javaid, R.P. Singh, R. Suman, Telemedicine for healthcare: capabilities, features, barriers, and applications. Sens. Int. (2), 100117 (2021)

A.C. Hazarika, Northeast India and Its Biodiversity. Int. J. Innov. Res. Technol. **8**(5), 304–309. ISSN: 2349-6002 (2021)

R.J. Huggett (2007) Fundamentals of geomorphology. Routledge 2 Park Square, Milton Park, Abingdon, Oxon, OX14 4RN. ISBN 0-203-94711-8 Master e-book

M. Kehie, S. Khamu, P. Kehie, Indigenous alder based farming practices in Nagaland, India: a sustainable agricultural model. J. Trad. Folk Prac. (JTFP) **5**(2) (2017). https://jtfp.jntbgri.res.in/index.php/jtfp/article/view/94

J. Koner, B. Bhadury, A. Purandare, Road infrastructure development and economic growth of North-Eastern States in India: a Revisit. NICMAR-J. Const. Manage. **XXXVII**(4) (Special Issue Part-III) (2022)

S. Kumar, V.P. Khanduri (2024) Impact of climate change on the Himalayan alpine treeline vegetation. Heliyon **10**, e40797 (2024). https://doi.org/10.1016/j.heliyon.2024.e40797

L. McConnell, Exploring geospatial technology: understanding its significance and applications (2023). https://www.aerialsurvey.com/

V. Malik, The cradle of the flowering plants. J. Plant Develop. Sci. **11**(12), 689–696 (2019)

G. Matta, A. Kuma, D.S. Tomar, R. Kuma, Climate change in the Himalayan region: susceptible impacts on environment and human settlements. Front. Environ. Sci. **13** (2025). https://doi.org/10.3389/fenvs.2025.1550843

S. Mavely, Editorial. J. Northeast Indian Cult. **3**(1) (2021)

T. Lhamu, Some elements of animistic beliefs among the Buddhist tribe of Dirang in Arunachal Pradesh. Mizoram Univ. J. Hum.Ities Soc. Sci. **9**(2), 281–290 (2023)

V.S. Negi, R. Pathak, S. Thakur, R.K. Joshi, I.D. Bhatt, R.S. Rawal, Scoping the need of mainstreaming indigenous knowledge for sustainable use of bioresources in the Indian Himalayan Region. Environ. Manage. **72**, 135–146 (2023)

North Eastern Council, NEC year book 2024. Government Printing Press (2024). https://www.nec.gov/yearbook2024

C.A. Posey, The living earth book of wind and weather, Reader's Digest Association (published 1 November 1994). ISBN 978-0-89577-625-9 (1994)

M.W. Post, S. Morey, S. Delancey (Eds.), *Language and Culture in Northeast India and Beyond.* The Australian National University Canberra ACT 2601 Australia (2015)

K.S. Rao, K.G. Saxena, B.K. Tiwari (Eds.), Biodiversity, climate change and socio-economic development in the Indian Himalaya: an overview. (Bishen Singh Mahendrapal Singh). ISBN 978-81-211-0928-1 (2026)

V.B. Rao, S.H. Fran-chito, R.O.P. Gerólamo, E. Giarolla, S.S.V.S. Ramakrishna, B.R.S. Rao, C.V. Nai-du, Himalayan warming and climate change in India. Am. J. Clim. Chang. **5**, 558–574 (2016)

A.K. Rathoure, Cultural practices to protecting biodiversity through cultural heritage: preserving nature, preserving culture. Biodiv. Int. J. **7**(2), 71–75 (2024). https://doi.org/10.15406/bij.2024.07.00213

R.S. Rawal, I.D. Bhatt, K. Chandra Sekar, S.K. Nandi (Eds.), The Himalayan biodiversity: richness, representativeness, uniqueness and life-support values. Almora, Uttarakhand, India. G.B. Pant Institute of Himalayan Environment and Development (GBPIHED). ISBN 81-927373-1-7 (2013)

D.S. Rawat, *Himalayan weather and climate and their impact on the environment* (Springer, 2017)

ResearchFDI, *How technology is changing the landscape of economic development* (2023). https://researchfdi.com/how-technology-is-changing-the-landscape-of-economic-development/

S.S. Samant, Floristic diversity of the Himalaya in relation to climate change: status, values and conservation. J. Graph. Era Univ. **9**(1), 31–54 (2021). https://doi.org/10.13052/jgeu0975-1416.913

S. Sathyakumar, M. Mungee, R. Pal, Biogeography of the mountain ranges of South Asia, in *Encyclopedia of the World's Biomes* (Elsevier, 2020), pp. 543–554. ISBN 978-0-124-09548-9

T. Selvan, S. Kumar, Agroforestry in the North-Eastern Himalayas, in *Agroforestry for Increased Production and Livelihood Security- Chapter: 2* S.K. Gupta, P. Pawar, R. Kaushal (New India Publishing Agency, 2017)

A. Sharma, R. Das, Infrastructure and economic development in Northeast India: challenges and opportunities. J. North East India Stud. **9**(1), 34–56 (2019)

M. Shrestha, S. Sharma, R.P. Shrestha, Landslides in the Himalayas: a comprehensive review of hazards, impacts, and adaptive strategies. Rural. Reg. Dev. **3**(1), 10002 (2024). https://doi.org/10.70322/rrd.2025.10002

R.K. Singh, V. Singh, C. Rajkhowa, Zabo: a traditional way of integrated farming, in *Resilient Shifting Cultivation: Challenges and Opportunities,* ed. by B.C. Deka, M.K. Patra, A. Thirugnanavel, D. Chatterjee, R. Borah Tasvina, S.V. Ngachan, (ICAR Research Complex for NEH Region, Nagaland Centre, Medziphema, Nagaland, 2012). https://doi.org/10.13140/RG.2.1.1773.2647

M. Sinha, Socio-economic problems of Northeast India: an overview. J. Emerg. Technol. Innov. Res. (JETIR) **2**(10), 573–576 (2015)

A. Soni, Traditional agriculture for sustainable development: a Satoyama approach. TERI (2021). https://www.teriin.org/project/traditional-agriculture-sustainable-development-satoyama-approach-1

S.P. Steven, C.P. Julie, Nitrogen-fixing red alder trees tap rock-derived nutrients. Proc. Natl. Acad. Sci. USA **116**(11), 5009–5014 (2019). https://doi.org/10.1073/pnas.1814782116

M. Uthaman, C. Singh, A. Singh, G. Hetényi, A. Dutta, G. Kumar, A.K. Dubey, Complex multi-fault dynamics in Sikkim Himalaya: new insights from local earthquake analysis. Geochem. Geophys. Geosystems **25**(6) (2024). https://doi.org/10.1029/2023GC011363

T. Yao, L.G. Thompson, V. Mosbrugger, F. Zhang, Y. Ma, T. Luo, B. Xu, X. Yang, D.R. Joswiak, W. Wang, M.E. Joswiak, L.P. Devkota, S. Tayal, R. Jilani, R. Fayziev, Third pole environment (TPE). Environ. Develop. **3**, 52–64 (2012). https://doi.org/10.1016/j.envdev.2012.04.002.ISSN2211-4645

K. K. Sarma Ph. D, leads the Remote Sensing Applications Group (RSAG) at the North Eastern Space Applications Centre (NESAC), where he has been a key figure since 2002. Specializing in remote sensing for forestry and biodiversity, his work leverages space technology for critical applications in Biodiversity, Ecosystems, Forestry, and Natural Resource Management. Dr. Sarma has spearheaded over 60 diverse geospatial projects, contributing significantly to natural resource mapping and monitoring across the North Eastern Region (NER). Currently, he coordinates 110 "Plan of Action" projects, focused on using space technology to improve livelihoods in the NER. He also served as the Principal Investigator and NER Coordinator for all ISRO natural resource management projects, including comprehensive Land Use Land Cover Mapping and National Wasteland Updating, Monitoring, and Change Analysis. With over 75 research publications, Dr. Sarma has successfully supervised six Ph.D. students and actively mentors many others across NER Universities.

Chapter 2
Space Technology Applications for Agriculture, Horticulture and Sericulture in NE India

Bijoy K. Handique, Jonali Goswami, Chandan Goswami, Pratibha Thakuria Das, Pradesh Jena, and Francis Dutta

Abstract Geospatial tools and techniques have become critical in addressing agricultural challenges in hill agriculture system like in the north eastern region of India, a diverse agrarian region characterized by complex terrain. Despite hurdles like small landholdings, fragmented farms, and heavy cloud cover, the integration of geospatial technologies with machine learning, deep learning, and artificial intelligence offers crucial spatio-temporal insights for sustainable management of the agricultural resources. These advanced tools support crop acreage estimation using spectral signatures and phenological patterns, enable yield forecasting through vegetation indices and crop simulation models, and improve cropping system analysis and precision agriculture practices. Unmanned aerial vehicles further enhance pest and disease detection and irrigation management. In horticulture, these technologies aid in crop inventory, mapping, and GIS-based site suitability assessments for high-value crops. In sericulture, geospatial tools have been used to identify potential expansion sites. Web-based portals like SILKS have been developed to share spatial data, and GAGAN-enabled mobile apps support asset monitoring. Disease forewarning systems have also been developed using geospatial technologies. Despite challenges,

B. K. Handique (✉) · J. Goswami · C. Goswami · P. T. Das · P. Jena · F. Dutta
North Eastern Space Applications Centre (NESAC), Umiam, Shillong, Meghalaya, India
e-mail: bk.handique@nesac.gov.in

J. Goswami
e-mail: jonali.goswami@nesac.gov.in

C. Goswami
e-mail: chandan.goswami@nesac.gov.in

P. T. Das
e-mail: pt.das@nesac.gov.in

P. Jena
e-mail: pradesh.jena@nesac.gov.in

F. Dutta
e-mail: francis.dutta@nesac.gov.in

S. P. Aggarwal et al. (eds.), *Remote Sensing of the North Eastern Himalayan Ecosystem*,
https://doi.org/10.1007/978-981-95-3221-6_2

geospatial technology is proving to be indispensable, significantly contributing to the sustainable growth and modernization of agriculture and allied sectors.

Keywords Space technology · Agriculture · Horticulture · Sericulture · UAV · Mobile app · Dashboard

2.1 Introduction

The Eastern Himalayan (EH) region, stretching from the Kaligandaki Valley in Nepal to northwest Yunnan in China, are globally recognized for their rich biodiversity, traditional farming practices, and agroforestry systems (Sharma et al. 2007). The North Eastern Region (NER) of India, an integral part of the EH, spans approximately 262,179 square kilometers, accounting for 8% of India's land area. It is characterized by diverse agro-climatic conditions, a rich cultural heritage, and a predominantly agrarian economy. Agriculture is the backbone of the NER, supporting nearly 70% of the population (Vaid 2020; Roy et al. 2014; Seitinthang 2014).

NER constitutes 3.37% of India's total agricultural land and exhibits a complex landscape of hills, mountains, and plains. Mountainous and hilly terrain dominates approximately 70% of the region, characterized by red, lateritic, and mountain soils. These soil types are prevalent in the states of Arunachal Pradesh, Mizoram, Nagaland, Meghalaya, Sikkim, and most of Tripura, with some presence in Assam and Manipur. Fertile alluvial plains are concentrated in the Brahmaputra and Barak Valleys of Assam, the plains of Tripura, and the Imphal Valley of Manipur. This varied landscape has given rise to two predominant agricultural practices. Shifting cultivation, locally known as *Jhum*, is an age-old subsistence practice primarily employed by tribal communities in areas with hilly red and lateritic soils. This practice is found across all districts of Arunachal Pradesh, the hills of southern Assam, the mountainous regions of Tripura, Mizoram, Nagaland, and the hill areas of Manipur, particularly among the Kuki and Naga communities. Settled agriculture, in contrast, flourishes in the fertile alluvial plains, including the Brahmaputra Valley of Assam, southeastern Nagaland, the Brahmaputra plains of southern Arunachal Pradesh, the Barak Valley, parts of Tripura, and Manipur's central Imphal Valley (Dikshit & Dikshit 2014; Punitha et al. 2018; De and Singh 2017).

The NER's agricultural ecosystem is distinguished by its remarkable biodiversity, enabling the cultivation of a wide array of crops. Agricultural practices are deeply rooted in indigenous knowledge systems, characterized by sustainable farming methods refined over generations. These practices reflect local communities' adaptation to challenging geographical conditions and demonstrate a profound understanding of ecological balance and environmental conservation. Climatic variations play a crucial role in determining crop patterns. Diverse microclimatic conditions—from tropical to subtropical and temperate zones—create opportunities for cultivating a wide spectrum of crops. Rainfall patterns, ranging from moderate to extremely high precipitation, coupled with varied temperature ranges, contribute to the region's agricultural versatility. This ecological complexity allows farmers to grow diverse crops, including food grains, horticultural produce, plantation crops, and indigenous varieties.

Rice is the most significant crop in the NER, serving as the primary staple food. The diverse agro-climatic conditions enable the cultivation of multiple rice varieties, including traditional indigenous strains and modern high-yielding varieties (Dikshit & Dikshit 2014). Maize is another important crop, second to rice. Its cultivation is increasing across the north-eastern states due to its growing importance both as a food grain and animal feed component. Farmers cultivate both traditional and hybrid varieties of maize, adapting to different ecological zones and agricultural practices. The region also has suitable microclimates for cultivating pulses, oilseeds, cash crops, spices, fruits, and vegetables. NER is also home to the world's richest diversity of sericigenous fauna (silk producing insects). This is the only region where all four varieties of commercial silk namely Mulberry, Muga, Eri and Oak Tasar are produced. The silk culture has taken deep roots in the socio-cultural milieu of the region as it is entwined with the day-to-day lives of the people especially the tribal communities (Unni et al. 2009).

Despite its agricultural potential, the region faces significant challenges, including small landholdings, low mechanization, inadequate infrastructure, and market access issues. The lack of timely and reliable agricultural data further impairs policymaking. Given these constraints, space-based technologies provide valuable solutions by offering precise, spatio-temporal insights for agricultural planning and resource management (Barah 2007; Gogoi 2019).

The future of the fragile NER is closely tied to the sustainable management of its natural resources. To ensure the long-term viability of agricultural production with minimal negative environmental impacts, advanced technological interventions are crucial. Efforts have been made to promote sustainable development and better resource management using advanced scientific tools and techniques. Geospatial technology, consisting of Remote Sensing (RS), Geographical Information Systems (GIS), and Global Positioning Systems (GPS), has emerged as an effective tool for sustainable development in the agricultural sector. Recent advances in machine learning, deep learning, and artificial intelligence (AI) are helping to improve the accuracy and efficiency of geospatial data, supporting more informed decision-making at the farmer level. Space technology has dramatically transformed agriculture, evolving from early Earth observation to sophisticated precision farming techniques.

2.2 Transformative Role of Space Technology in Agriculture and Allied Sectors

Satellite technology has revolutionized the way we observe, monitor, and understand the Earth, providing unprecedented opportunities across diverse sectors, particularly agriculture. The global proliferation of Earth observation satellites has democratized access to critical data, enabling governments, organizations, and businesses worldwide to leverage this information for informed decision-making (Vali et al. 2020). Major commercial operators like Planet Labs, Maxar Technologies, and Airbus Defence and Space have established extensive constellations offering diverse data products, including high-resolution optical imagery and advanced radar observations (Asner et al. 2017). These innovations have enhanced wide range of applications in agriculture and allied sectors like horticulture, sericulture, fisheries etc. Modern satellites equipped with state-of-the-art sensors such as optical cameras, synthetic aperture radar (SAR), multispectral and hyperspectral imagers, and infrared sensors enable the collection of detailed, high-precision data about Earth's surface and atmosphere, transforming agricultural practices by providing insights into crop health, soil conditions, and water resources. The emergence of small satellite platforms like CubeSats has further reduced the cost of space access, allowing smaller nations and private enterprises to deploy their own systems, fostering competition and innovation in satellite-based services (Liu 2022).

India has achieved remarkable milestones in satellite technology, with the Indian Space Research Organisation (ISRO) spearheading efforts to develop indigenous capabilities. The CARTOSAT series delivers high-resolution imagery vital for farm-level planning and infrastructure development, while the RESOURCESAT series provides crucial data for natural resource monitoring. The RISAT constellation, equipped with SAR technology, ensures all-weather imaging, addressing challenges like crop monitoring during monsoons. Recent reforms in India's space sector, including the establishment of NewSpace India Limited (NSIL) as ISRO's commercial arm, are set to enhance private sector participation, accelerate innovation, and expand the scope of satellite applications. These advancements align with global trends, offering new avenues for addressing societal challenges and supporting sustainable development goals through satellite-based solutions (Guruprasad 2018). The significant missions launched by space agencies around the world and their key agricultural applications are listed in Table 2.1.

2.3 Space Technology Applications in Agriculture

Space technology has become an essential tool for agricultural planning in the complex terrain of the NER. These tools enable comprehensive monitoring of scattered agricultural plots across steep slopes and varying altitudes. It alsohelps in the precise mapping of terraced fields, shifting cultivation areas, and traditional

Table 2.1 Major earth observation satellites and sensors and their applications in agriculture and allied areas

Satellite Name	Sensors/ Instrumentation	Spatial Resolution (m)	Spectral Range (μm)	Applications
Landsat Series	Thematic Mapper (TM)	30 (Visible/ NIR/SWIR), 120 (TIR)	0.45–2.35 (Visible, NIR, SWIR)	Land use/land cover, agriculture, forestry, water resources management
	Operational Land Imager (OLI)	15 (Panchromatic), 30 (Multispectral)	0.43–2.3	
	Thermal Infrared Sensor (TIRS)	100	10.6–12.5	
Radarsat Series	Synthetic Aperture Radar (SAR)	1–100	53,000 (C-Band)	All-weather monitoring, soil moisture, crop mapping
Sentinel 1 & 2	Multispectral Instrument (MSI)	10 (Visible/ NIR), 20 (SWIR), 60 (Coastal)	443–2190 nm	Vegetation monitoring, disaster management, air quality monitoring
	SAR	5–40	53,000 (C-Band)	
Pleiades	High-Resolution Optical Imager	0.5 (Panchromatic), 2 (Multispectral)	0.43–0.90	High-resolution mapping, disaster assessment, urban planning
SMAP	L-band Radar (Active Sensor)	1–3 km	1.26 GHz	Measures soil moisture, freeze/thaw state, climate modelling
	L-band Radiometer (Passive Sensor)	40	1.41 GHz	
PRISMA	Hyperspectral Imager	30	0.4–2.5	Assessment of crop health, productivity, and soil properties
	Panchromatic Camera	5	0.4–0.7	
ALOS	PALSAR (L-band SAR)	10–100	1.2 GHz	Crop monitoring, yield estimation, fallow land analysis

(continued)

Table 2.1 (continued)

Satellite Name	Sensors/ Instrumentation	Spatial Resolution (m)	Spectral Range (μm)	Applications
SPOT	Multispectral Camera	1.5 (Panchromatic), 6 (Multispectral)	0.45–0.89	Land use monitoring, urban planning, natural resources management
PlanetScope&PlanetLab	Multispectral Camera	3	0.43–0.9	Precision agriculture, forestry management, environmental monitoring
IRS Series	Linear Imaging Self Scanner -I (LISS-I)	72	0.52–0.86	Agricultural planning, forestry, land use
	LISS-II	36	0.52–0.86	
Cartosat Series	Panchromatic Camera (PAN)	0.28–2.5	0.45–0.9	Cartography, urban planning, disaster management
	Multispectral (MX)	1–5	0.45–0.86	
RISAT Series	SAR	1–50	53,000 (C-Band)	All-weather surveillance, agriculture, soil moisture studies
Resourcesat Series	LISS-III	23.5	0.52–1.70	Crop monitoring, land use planning
	LISS-IV	5.8	0.52–0.86	
	AWiFS	56	0.62–1.70	

farming systems unique to this region. They assist in tracking seasonal crop patterns, monitoring soil erosion, and analyzing agriculture-forest relationships. This supports climate change adaptation strategies while respecting traditional farming practices of local communities, ultimately ensuring food security and ecological sustainability.

2.3.1 Crop Acreage Estimation

One of the most significant applications of geospatial technology in crop studies is the precise estimation of area under different crops. Accurate crop identification fundamentally relies on the interaction between electromagnetic (EM) radiation and crop canopy. This interaction is governed by the physical and biochemical properties of plants, including their cellular structure, water content, pigmentation, and overall canopy architecture. When solar radiation reaches a crop canopy, it undergoes various processes of reflection, absorption, and transmission, creating distinctive spectral signatures that can be detected by satellite sensors. The spectral response of crops is particularly distinctive in different portions of the EM spectrum. In the visible spectrum (400–700 nm), chlorophyll strongly absorbs blue and red light while reflecting green light, giving healthy vegetation its characteristic green appearance. The near-infrared region (700–1300 nm) is characterized by strong reflection due to the internal cellular structure of leaves, specifically the air-cell wall interfaces within the mesophyll tissue. This creates a dramatic increase in reflectance between the red and near-infrared portions of the spectrum, known as the "red edge," which is a crucial indicator of vegetation health and density. The spectral signature of a few selected crops from NER is depicted in Fig. 2.1.

Crop phenology plays a vital role in identification through remote sensing as different crops exhibit unique temporal patterns in their growth and development cycles, which manifest as changes in their spectral signatures over time. The transition from vegetative to reproductive stages affects the plant's biochemical composition and structural properties, leading to detectable changes in spectral reflectance. This temporal aspect enhances the ability to differentiate between crop types that may appear similar at a single point in time. Additionally, understanding the crop calendar provides an added advantage in accurate crop discrimination, as different crops are cultivated in various seasons based on their specific weather requirements. This knowledge aids in characterizing and quantifying different crops and cropping systems effectively. Crop calendar of major crops grown in the region is highlighted in Fig. 2.2.

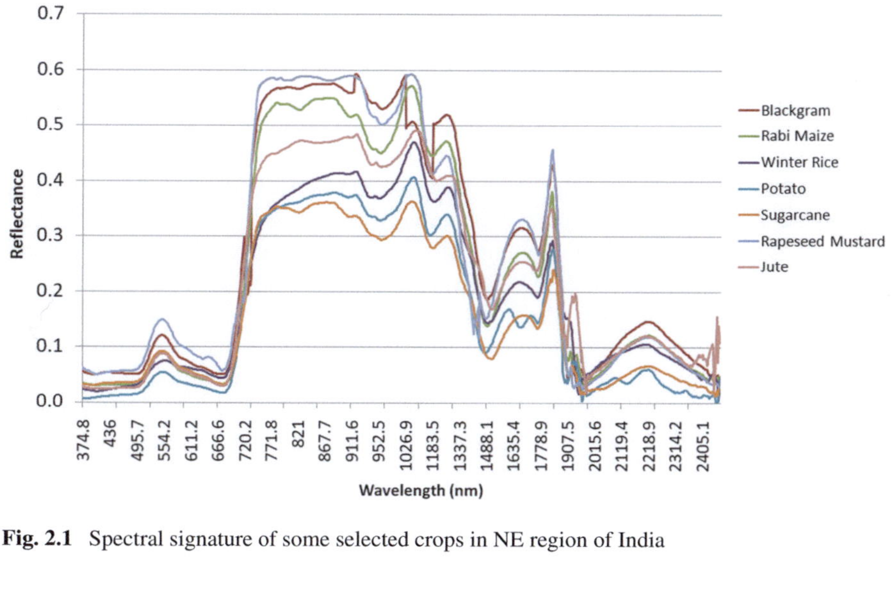

Fig. 2.1 Spectral signature of some selected crops in NE region of India

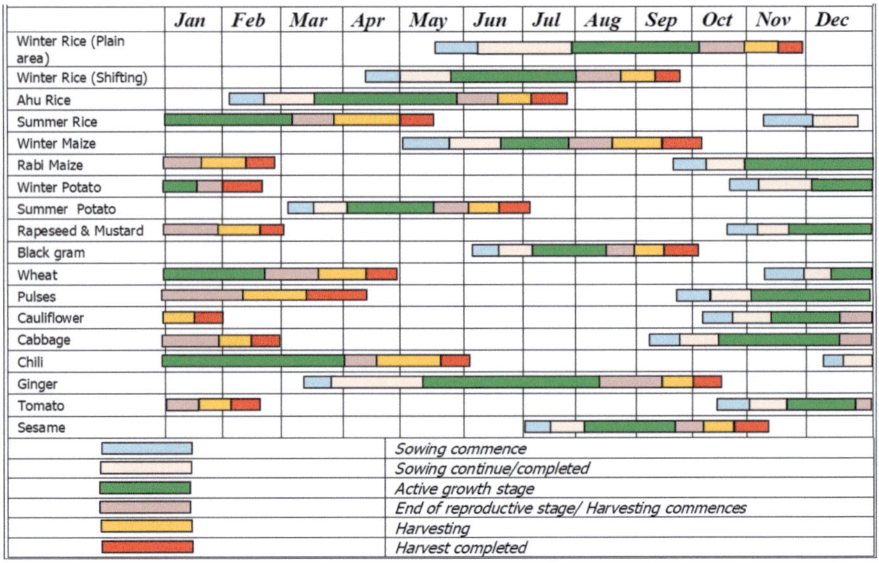

Fig. 2.2 Crop calendar in NER

The evolution of satellite remote sensing applications has shifted from basic land cover classification to more sophisticated analysis of plant biophysical properties through various algorithmic approaches, ranging from traditional supervised classification methods to advanced machine learning techniques. Traditional supervised classification methods, particularly Maximum Likelihood Classification (MLC), have been foundational for basic crop type mapping. MLC, operating on statistical probability principles, has proven effective for distinguishing major crops like rice in valley areas, though its effectiveness is limited in more complex terrain. Advanced machine learning algorithms, such as Support Vector Machines (SVM) and Random Forest (RF) have significantly improved classification accuracy in recent years. SVM has demonstrated exceptional performance in handling non-linear relationships in crop spectral signatures, particularly valuable in areas with mixed cropping patterns. RF classifiers excel in processing high-dimensional time-series data, making them particularly useful for analyzing crop phenology patterns across different altitudes. Deep Learning approaches, including Convolutional Neural Networks (CNNs) and Long Short-Term Memory (LSTM) networks, represent the latest advancement in crop classification. are highly effective in analyzing spatial patterns within satellite imagery, while LSTMs excel in processing time-series data to capture phenological changes. Object-Based Image Analysis (OBIA) has emerged as a powerful tool for high-resolution imagery analysis. This approach considers not just spectral information but also spatial context, texture, and shape parameters, making it particularly valuable in regions with small, irregular field sizes and complex topographic features. Integrating multiple classification approaches through ensemble methods further enhances accuracy in crop identification, as this hybrid strategy combines the strengths of different algorithms to improve classification performance.

Time series multi-spectral and multi-temporal data from MODIS, NOAA-AVHRR, LANDSAT, IRS, SPOT and Chinese HJ-1A/B were used by many researchers at the national and sub-national scale for rice mapping. Singh et al., (2016) used MODIS-NDVI time series data and multi-spectral satellite data (HJ-1A/B) to investigate the combined impact of crop phenology and the Object-based image analysis (OBIA) method for rice mapping in five districts of Assam. Various vegetation indices, i.e. Normalized Difference Vegetation Index (NDVI), Enhanced Vegetation Index (EVI), and Soil Adjusted Vegetation Index (SAVI), derived from different wavelengths, help distinguish between crop types in the region's diverse agricultural landscape (Xue and Su 2017). Multi-temporal optical data from satellites such as Indian Remote Sensing Satellite (IRS), Landsat, MODIS, and SPOT have been effectively utilized for mapping rice areas at national levels. The region's uneven terrain, characterized by small and fragmented croplands, necessitates the integration of satellite data with accurate ground-level information through well-designed sampling strategies. To achieve this, the area can be categorized into homogeneous strata based on key factors such as land-use patterns, soil types, elevation, and agro-climatic conditions. Samples from each stratum can then be systematically collected and analyzed, ensuring a comprehensive and representative assessment of the agricultural landscape. Using this approach, crop acreage of rice has been estimated for the state of Meghalaya. (Sahoo et al. 2006; Handique et al. 2022a). While optical data offer a broad range of spatial and temporal resolutions, their effectiveness is often limited by weather conditions, particularly cloud cover, which can disrupt data acquisition. To address this limitation, weather-independent radar data have been increasingly utilized. Radar technology is highly sensitive to crop structures, alignment, and moisture content, enabling it to provide valuable complementary information to optical datasets. This synergy enhances the overall accuracy and reliability of agricultural monitoring and analysis. Combined information from these two types of sensors increases the information available for distinguishing each target class and its respective signatures, and thus there is a better chance of performing a more accurate classification. Crop identification and mapping benefit from the use of multi-temporal imagery to facilitate classification by taking into account changes in reflectance as a function of plant phenology (stage of growth). This in turn requires calibrated sensors, and frequent repeat imaging throughout the growing seasons (Ghazaryan et al. 2018).

2.3.2 *Yield Estimation*

The interaction of genotype with other factors such as soil characteristics, farming practices, weather conditions, and biotic factors significantly impacts crop growth, development, and yield. Accurate yield prediction is essential for farmers, policymakers, and stakeholders to improve crop management, optimize land use, and

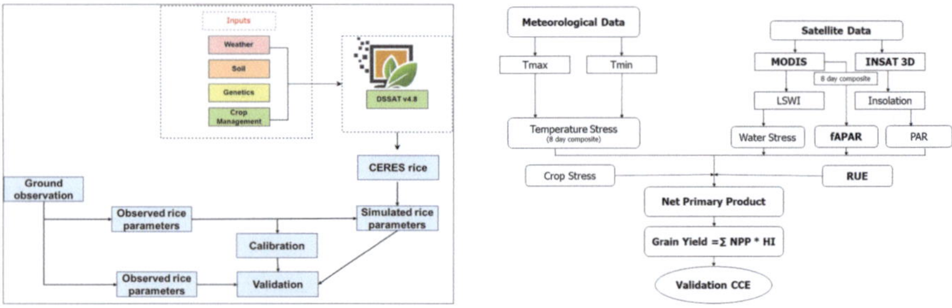

Fig. 2.3 Flowchart for DSSAT (left) and Semi-physical (right) model for crop yield estimation

formulate effective agricultural policies. A variety of remote sensing-based techniques are utilized for yield forecasting. Vegetation indices, which are mathematical formulae derived from remote sensing data, serve as indicators of vegetation health and productivity. Another effective method for yield prediction is Crop Surface Reflectance Modeling (CSRM), which uses remote sensing data to estimate crop surface reflectance at different wavelengths. Crop simulation models play a crucial role in yield estimation through mathematical representations of physiological processes. Models like DSSAT, APSIM, Semi-physical, and WOFOST simulate crop growth and development based on environmental inputs and crop characteristics. Flowchart for DSSAT and semi physical model for yield estimation is given in Fig. 2.3. When these models are integrated with remote sensing data, they become even more powerful tools for yield prediction, allowing real-time calibration based on actual crop conditions. With advancements in technology, machine learning and artificial intelligence models can now analyze remote sensing data to predict agricultural yields. These algorithms are capable of identifying complex patterns in data that are not interpretable easily. The integration of remote sensing data with crop simulation models creates a robust framework for yield estimation. This ensemble approach helps in identifying yield-limiting factors across large areas, providing early warning of potential yield reductions, supporting in-season management decisions, and improving regional crop production estimates. The effectiveness of these approaches relies on incorporating multiple data sources, including ground-based observations for validation and weather data integration (Mamatkulov et al. 2021; Petersen 2018).

2.3.3 *Cropping System Analysis*

Space technology can be used to identify land's potential for growing crops, which is essential for recommending appropriate cropping systems tailored to specific agroclimatic zones. The spatial distribution of crops or fallow lands in a given area over time defines the cropping pattern, while the cropping system encompasses broader aspects including production components and management practices aimed at optimizing resource utilization for sustained food and nutritional security. Remote sensing technology enables comprehensive analysis of agricultural systems, providing insights into cropping patterns, intensity levels, and rotational practices. In the NEHR, cropping systems are shaped by multiple factors including physical parameters such as soil characteristics and elevation, climatic elements like rainfall and temperature patterns, and various socio-economic factors including local food preferences and agricultural policies. The temporal analysis of satellite imagery facilitates detailed monitoring of crop development stages, enabling clear distinction between actively cultivated areas and fallow lands. Crop reflectance characteristics exhibit significant variation based on multiple factors, including crop variety, developmental stage, environmental conditions, and geographical location. These reflectance properties undergo distinct changes as crops progress through various growth phases from initial emergence to maturity and senescence. Satellite-derived vegetation indices provide quantitative measurements of crop health and vigor. The Normalized Difference Vegetation Index (NDVI), particularly, utilizes the distinct spectral responses of vegetation in red and near-infrared wavelengths to assess plant canopy conditions. Multi-temporal NDVI data, when analyzed through knowledge-based decision rules, effectively differentiates between croplands and fallow areas during growing seasons. Actively cultivated fields demonstrate higher NDVI values during growing seasons, indicating healthy vegetation cover. In contrast, fallow lands, which are temporarily uncultivated, show reduced NDVI values due to minimal vegetation coverage as depicted in Fig. 2.4. The seasonal analysis of crop and fallow areas are then can be utilized for the determination of cropping intensity patterns, categorizing lands as single, double, or triple-cropped through spatial integration techniques. This systematic monitoring of cropping systems and intensity variations generates essential data for evaluating the impacts of agricultural practices and developing sustainable agricultural strategies that align with the region's ecological sensitivity.

2.3.4 Precision Agriculture

The modern agricultural system aims to maximize productivity on limited land resources, leading to the development of precision farming—a practice that addresses in-field variability through precise input application using the variable rate method. Remote sensing, particularly high-resolution imagery from satellites, drones, and airplanes, plays a pivotal role in precision farming. These images identify field variations and classify them into management zones to optimize the allocation of inputs such as fertilizers, pesticides, and water. Advancements in sensor technology have further strengthen precision agriculture by providing large-scale, continuous data. Satellites equipped with multi-spectral and hyperspectral sensors offer critical insights into crop health, growth patterns, and stress conditions. Vegetation indices such as the Normalized Difference Vegetation Index (NDVI), Chlorophyll Index (CI) and Enhanced Vegetation Index (EVI) etc. derived from satellite imagery help monitor crop vigor and detect early signs of stress, enabling timely interventions.

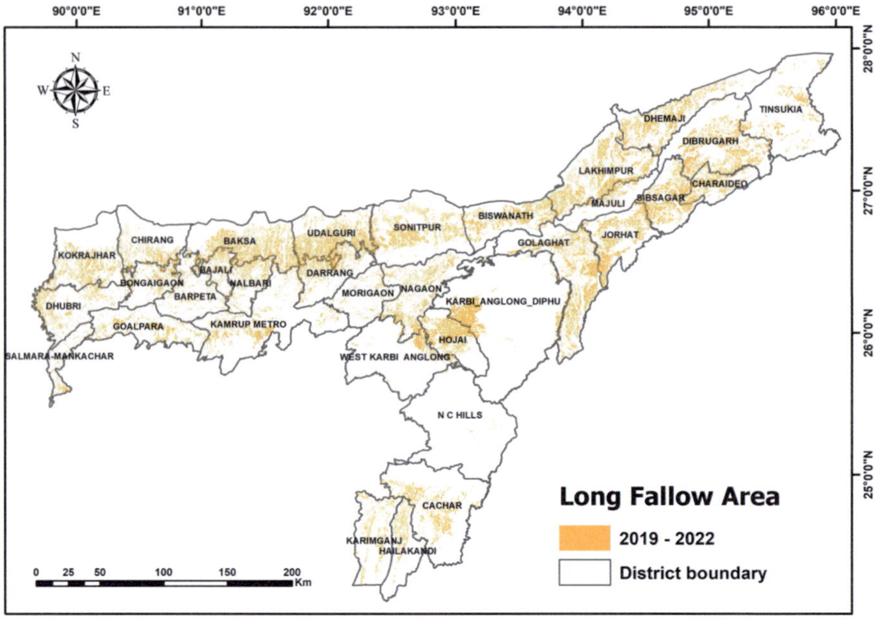

Fig. 2.4 Spatial distribution of agriculture area under long fallow in Rabi season in Assam during 2019–22

Microwave sensors, such as Synthetic Aperture Radar (SAR), are particularly useful for assessing soil moisture, even in cloudy conditions, aiding in efficient irrigation planning. Additionally, satellites contribute to mapping of soil properties, including texture and organic matter content, which are crucial for precision soil management. Moreover, proximal sensors, either handheld or mounted on farm machinery, complement this by collecting detailed environmental data, such as soil moisture, pH, and nutrient levels etc. This data is used to create field maps for optimizing planting schedules, irrigation, and fertilizer application, ensuring better resource utilization and improved crop yields (Schumann 2008).

2.3.5 UAV Applications

Unmanned Aerial Vehicles (UAVs) or drones have brought new dimensions to modern agriculture by offering powerful tools for precision farming and comprehensive field management (Torres et al. 2015). These aerial platforms, equipped with sophisticated sensors and imaging technologies, helps in monitoring agricultural lands in complex terrain with unprecedented efficiency and accuracy. Through multispectral, hyperspectral and thermal imaging capabilities, UAVs can detect subtle variations in crop health, identifying early signs of pest infestations, disease outbreaks, or nutrient deficiencies before they become visible to the naked eye. This early detection system allows farmers to implement targeted interventions, significantly reducing the use of pesticides and fertilizers while maximizing crop yields. Handique et al. (2016a, b) effectively estimated brown plant hopper (*Nilaparvata lugens*, Stal) infestation in Boro paddy in Morigaon district of Assam using UAV-based remote sensing. This timely intervention, helped farmers to implement measures that saved crops from significant losses. UAVs also excel in irrigation management by capturing detailed thermal imagery that reveals moisture stress patterns across fields, enabling precise water distribution and preventing both over and under-watering conditions. The technology's ability to generate high-resolution field maps and 3D models enable farmers to optimize planting patterns, assess crop growth stages, and estimate yields with remarkable precision. Beyond its crop analysis capability, agricultural drones have proven invaluable in precision application of inputs, such as fertilizers and crop protection products, through variable rate application systems that ensure optimal resource utilization while minimizing environmental impact.

2.4 The Significance of Geospatial Technology in Horticulture

Horticulture, derived from the Latin words *hortus* (garden) and *colere* (to cultivate), is a branch of agriculture focused on the cultivation of garden crops such as fruits, vegetables, and ornamental plants (Janick 2024). In recent years, India has experienced a shift towards prioritizing nutritional security over mere caloric intake. This transition is driven by increased health awareness, evolving lifestyles, and a better understanding of how balanced diets can prevent chronic diseases. Consequently, horticulture has become essential for enhancing both the nation's nutritional and economic well-being (Singh et al. 2015). Horticultural crops are pivotal to India's economy, particularly in improving rural incomes. Often referred to as the "Golden Revolution," horticulture offers higher productivity per unit of land compared to other crops, resulting in greater income and employment generation in rural areas. India's diverse agro-climatic conditions make it an ideal location for growing a wide variety of horticultural crops (Peter 2015; Singh 2009). The country is the second-largest producer of fruits and vegetables globally, with the horticulture sector contributing approximately 33% to the Gross Value Added (GVA) of agriculture, thereby having a significant impact on the economy. India's horticulture production is estimated at around 352.23 million tonnes (PIB 2024). As India's population is projected to reach 1.67 billion by 2050, the horticulture sector faces both opportunities and challenges, including rising demand for products, adaptation to climate change, water management, and the integration of advanced technologies. Emerging trends such as urban and vertical farming, improved post-harvest management, and increased export potential will likely shape the sector's future. To support growth, a reliable statistical database is essential. Digital agriculture, regarded as the fourth revolution in farming, has significantly transformed horticultural practices. The Directorate of Economics and Statistics, Ministry of Agriculture (DESMOA), conducts a "Crop Estimation Survey on Fruits and Vegetables" in 11 states using stratified sampling for fruits and a more complex approach for vegetables. The National Horticultural Board (NHB) provides national estimates based on local reports and market data. However, DESMOA's method is time-consuming and limited in scope, while NHB's estimates are often considered less reliable (MoSPI 2024).

Geospatial technologies, offer an effective framework for creating spatial databases of horticultural crops. Remote sensing gathers information about areas or phenomena through sensors mounted on satellites, aircraft, or drones, utilizing either passive energy (from the sun) or active energy (from Radar) (Lillesand et. al 2015). GIS allows for the capturing, storing, and displaying of data related to Earth's surface, enabling spatial pattern analysis. GPS provides precise location data, which is crucial for mapping and surveying. These geospatial technologies can significantly improve the accuracy and efficiency of horticultural data collection and analysis, addressing the limitations of current statistical methods. As India continues to advance its horticultural sector, leveraging technological innovations will be critical to ensuring sustainable growth and meeting the nutritional demands of its expanding population. The horticulture sector has been a pioneer in utilizing remote sensing data. Notably, the inception of remote sensing applications in India began in horticulture with an experiment focused on the early detection of wilt disease in coconut plantations (Dakshinamurty et al. 1971). In the early twenty-first century, the need for detailed spatial databases of horticultural crop distribution led to the use of medium and coarse-resolution satellite data (Dadhwal et al. 2003; Murthy et al. 2007). Institutional efforts began in 2012 with pilot projects using a hybrid approach, combining semi-automatic object-based classification with manual refinement.

GIS, combined with remote sensing and GPS, facilitates comprehensive agricultural data visualization and empowers data-driven decision-making to enhance crop yields and operational efficiency. While GIS has been utilized in agriculture for many years, recent technological advancements have accelerated its adoption in horticulture, allowing for a growing range of innovative applications. GIS-based site suitability analysis for horticultural crops is a powerful approach that uses spatial data to optimize crop production. By integrating various factors such as climate (temperature, rainfall, sunlight, humidity, wind patterns), soil properties (type, texture, pH, organic matter, drainage, nutrient content etc.), topography (slope, elevation, aspect, terrain ruggedness), water availability (surface and groundwater resources, irrigation infrastructure), and accessibility to infrastructure (roads, markets, processing facilities, labour), GIS enables a comprehensive assessment of land suitability. This process involves collecting and processing spatial data from diverse sources like weather stations, soil surveys, digital elevation models, hydrological maps, and transportation networks. Advanced GIS techniques, including spatial interpolation, overlay analysis, and multi-criteria evaluation, are employed to create suitability models within GIS software. These models assign weights to each factor based on its specific importance for the target crop, considering its unique requirements for growth and yield. The resulting site suitability maps delineate areas categorized by their suitability levels (e.g., highly suitable, moderately suitable, marginally suitable, unsuitable), providing a visual representation of optimal planting locations. This approach offers numerous benefits, including improved decision-making for growers in crop selection and land management, maximizing yields by selecting optimal locations that match crop needs, minimizing environmental impact by avoiding sensitive areas prone to erosion or degradation, reducing input costs by optimizing resource use, and facilitating sustainable land use planning for horticulture at local, regional, and national scales. These advancements underscore the growing importance of precise, data-driven methodologies to address evolving challenges and capitalize on emerging opportunities within this critical sector. Below, we explore some of the most significant and emerging uses within this dynamic field.

2.4.1 Horticulture Crop Inventory and Mapping

The ability of spectral reflectance data to differentiate between crops is a cornerstone of accurate crop acreage estimation, production forecasting, and crop health monitoring. Remote sensing utilizes the principle that each crop exhibits a unique spectral signature, shaped by its structural attributes, physiological traits, and growth stages. These spectral signatures, captured across various spectral bands, facilitate the effective identification, classification, and differentiation of crops (Lillesand et al. 2015). However, challenges emerge when crops display similar spectral properties during certain growth phases, complicating accurate classification. In such scenarios, multi-temporal remote sensing data, which captures changes in spectral reflectance over time, becomes indispensable for enhancing precision. For instance, a study on differentiating cotton from other crops at various growth stages highlighted the importance of analyzing reflectance spectra and their derivatives to characterize spectral properties effectively (Baret and Guyot 1991; Zhang et al. 2012). The integration of additional data sources, such as weather patterns, soil characteristics, and crop-specific growth models, further augments the accuracy of remote sensing analyses. These advancements have positioned remote sensing as a vital tool for precision agriculture, supporting informed decision-making in crop management. In orchard management, geospatial technologies have revolutionized practices such as yield prediction, fertilizer and irrigation optimization, and pest control. Remote sensing facilitates large-scale monitoring of orchard health, providing timely insights into the condition of trees and enabling early detection of stress or disease. Digital imaging further supports site-specific interventions by offering high-resolution data for precise management practices. Together, these technologies enhance productivity and sustainability in orchard systems, underscoring their critical role in modern horticulture (Panda et al. 2009).

Crop discrimination relies on the differential spectral response of various crops within a multidimensional feature space generated by distinct spectral bands, temporal domains, or a combination of both. This process is influenced by sensor characteristics and the pattern recognition techniques employed. The interaction of electromagnetic radiation with crops in the optical region is governed by factors such as chlorophyll content and water levels, while in the microwave region, crop geometry and dielectric properties play a critical role. Accurate crop area estimation is often hindered by several complexities, including small field sizes, diverse and scattered cropping patterns, mixed cropping with varying phenological stages, extended sowing periods, short-duration crops, and homestead farming practices. Additional challenges arise from the dynamic nature of cropping patterns, intricate physiography, and seasonal variations. Furthermore, external factors such as weather conditions, pest infestations, and disease outbreaks can significantly alter the sown area. Estimation efforts must account for these variables at critical stages of the crop cycle: pre-sowing (influenced by market forecasts and profit expectations), early sowing, mid-season, and pre-harvest stages (Vogel and Bange 1999). Craig and Atkinson (2013) comprehensively reviewed crop area estimation methodologies, which traditionally involve complete enumeration or sampling techniques. These include Area Frame Sampling (AFS), farm list sampling, and multiple frame sampling—an integrated approach combining the first two methods. Additional data sources include expert opinions, administrative surveys, crop processing unit records, and market data. Final estimates are typically derived through survey data or expert panel evaluations. However, conventional methods are often time-consuming, expensive, biased, and particularly challenging to implement in complex terrains such as hilly regions. Satellite remote sensing has emerged as a transformative tool to overcome these limitations. By providing temporal, synoptic, multispectral, and multiresolution imagery, it enables land use and crop classification either directly or by enhancing existing area sampling schemes. The application of remote sensing in horticultural crop mapping has seen significant progress over the years. One of the early studies by Gupta and Sharma (1990) used Landsat-5 TM data at a 1:50,000 scale to visually delineate mango orchards in Malihabad Tehsil, a prominent mango-growing region in Uttar Pradesh. This pioneering work set the stage for subsequent developments in horticultural crop mapping in the country.

Advancements in remote sensing technology have enabled the use of high-resolution datasets such as Cartosat-1 and LISS-IV for distinguishing and mapping plantation crops like rubber, tea, and coffee. However, conventional per-pixel classification methods have encountered challenges in achieving satisfactory accuracy due to the spatial heterogeneity of horticultural crops (Hebbar and Rao 2002; Srivastava and Gebelein 2006; Yadav et al. 2002). The time-series MODIS NDVI profile has proven effective in differentiating mango-growing regions from other orchards, capturing phenotypic changes during flowering (Nagaraja 2010). Additionally, multispectral Indian Remote Sensing (IRS) satellite data, such as LISS III (23.5 m resolution), has been successfully utilized to map crops like citrus, grapes & other plantation crops (Palaniswami et al. 2006), and apple orchards (Ganie and Nusrath 2014; Sharma and Panigrahy 2007). Ministry of Agriculture & Farmers Welfare (MoAFW) in collaboration with ISRO, Dept. pf Space, Govt. of India integrated remote sensing, GIS, and field data to create spatial inventories for seven horticultural crops (mango, banana, orange, tomato. onion, potato and chilli) in 185 districts representing 12 states of India utilizing data from satellites such as Resourcesat-2, Landsat-8, Sentinel-2A, and IRS-P5, with resolutions ranging from 56 to 10 m. Various classification techniques, including pixel-based methods like K-means and Maximum Likelihood, and object-based methods, were applied depending on the crop type (Sahoo et al. 2005). Spectral mixture analysis (SMA) (Palaniswami et al. 2006), knowledge-based decision tree classification (DTC) (Verma et al. 2020). Deep learning algorithms, particularly Convolutional Neural Networks (CNNs) (Ahmed et al. 2023), have also been successfully used for horticultural crop mapping. Hyperspectral remote sensing, which provides detailed spectral "fingerprints" of materials, has been found to accurately map and estimate the acreage of horticultural crops (Paul et al. 2018).These studies underscore the significant advancements in remote sensing techniques, enhancing the precision and reliability of horticultural crop mapping and estimation.

2.4.2 Site Suitability for Horticultural Crops

The NER represents 8% of country's population and 3.4% of agricultural lands. Therefore, tremendous potential lies in NER for horizontal expansion of crops. Moreover, shifting cultivation is a traditional and popular farming practice of the region and these lands can be brought under plantation crops. Evaluating land suitability is crucial for ensuring the sustainable use of land resources, particularly in light of resource constraints driven by human activities (Joshua et al. 2013; Stefanova et al. 2014). Sustainable agriculture relies on strategic land-use planning that balances economic feasibility with environmental sustainability. Crop suitability analysis plays a key role in identifying land potential for specific crops, facilitating optimal resource use, and supporting horticulture as a viable land-use alternative (Geetha et al. 2019; Rajesh et al. 2019; Karthikeyan et al. 2019).

GIS-based site suitability analysis serves as a powerful tool for enhancing horticultural crop production. This approach integrates spatial data from diverse sources, including climate parameters, topographic parameters, soil properties, water resources, access to infrastructure, and socio-economic parameters. Several studies in the Eastern Himalayan region have utilized GIS-based analysis and multi-criteria decision-making (MCDM), to assess the site suitability for key horticultural crops. GIS-based methods have proven effective for identifying regions suitable for cultivating high-value crops, thus contributing to optimized land-use planning (Baniya et al. 2009; Manandhar et al. 2014; Thapa et al. 2020). In the NER of India, GIS-based weighted overlay analysis was carried out to identify suitable areas for 23 horticultural crops in 77 districts. Site suitability analysis for expansion of pineapple in Meghalaya is presented in Fig. 2.5. The study assessed suitable areas for horticultural crops using MCDM tools, particularly weighted overlay analysis (WOA) (Handique et al. 2022b). A significant portion of Total Geographical Area (TGA) in hilly terrain of NER states are found to be suitable suitable for expansion of horticultural crops which is 5.6% of TGA of NER.

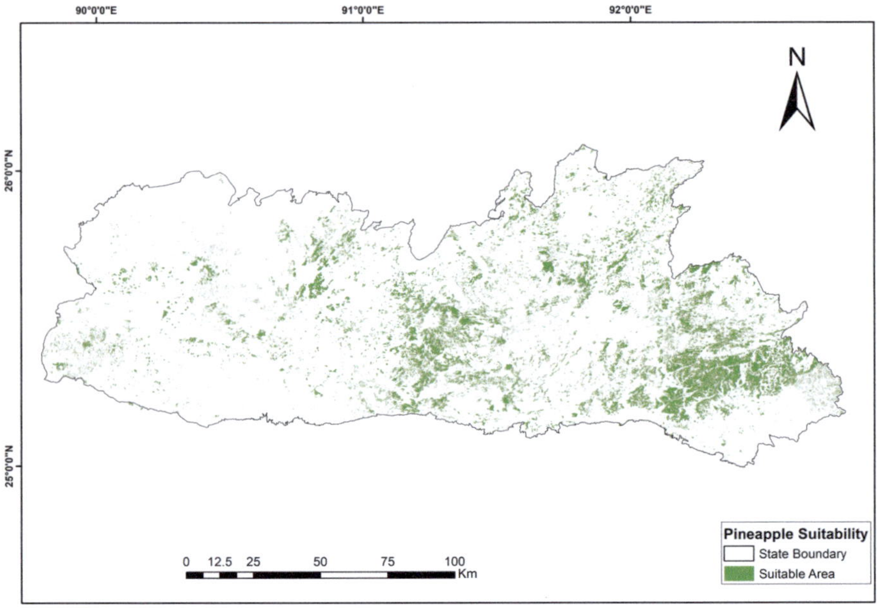

Fig. 2.5 Suitability of pineapple for Meghalaya under SSIDH

2.4.3 Horticultural Resource Mapping and Planning

Mapping horticultural infrastructure is essential for evaluating existing facilities, identifying gaps, and optimizing resource utilization. Key infrastructure components to map include pre-harvest and post-harvest facilities, market infrastructure, and supporting facilities. Mapping of existing horticultural infrastructure is carried out by geo-tagging using mobile application, which is capable of using in-built mobile GPS as well as external dongle (GAGAN dongle) for enhanced accuracy. Despite challenges such as difficult terrain and limited data availability, GIS mapping offers significant benefits, including real-time monitoring, capacity building for stakeholders, and the ability to align with government initiatives aimed at improving horticultural infrastructure. The geo-tagged data collected can be visualized through an interactive dashboard, offering real-time insights into infrastructure distribution, status, and gaps. This visualization aids in informed decision-making and resource allocation.

Again, a site suitability analysis to identify optimal locations for establishment of post-harvest infrastructure is also important for sustainable horticultural development. MCDM approach viz. Analytical Hierarchy process (AHP) is found to be useful in this process. The analysis utilizes inputs such as solar insolation, slope, aspect, existing crop area, potential areas for expansion, proximity to roads, markets, warehouses, and power stations etc.

2.5 Space Technology Applications in Sericulture

Sericulture, the cultivation of silkworms for silk production, originated in ancient China approximately 3000 BC. According to Chinese legend, Empress Leizu (Si-Ling-Chi) discovered the process of unravelling silk threads from silkworm cocoons. By 2000 BC, Chinese artisans had refined techniques for raising silkworms (*Bombyx mori*), feeding them mulberry leaves, and processing silk fibers into fabric. For centuries, China protected sericulture methods as a state secret, which contributed to the development of the Silk Road—an extensive trade network facilitating the exchange of silk, goods, and culture between China and other civilizations. By 200 BC, silk had become a valuable commodity traded with the Roman Empire and other regions. India adopted sericulture around 300 BC, developing regional silk varieties including Banarasi, Kanjivaram, Muga, and Eri silks. Between the fourth and seventh centuries, prominent silk centres emerged in Bengal, Gujarat, and the Deccan Plateau. Asia currently dominates global silk production, accounting for approximately 90% of mulberry silk and nearly 100% of non-mulberry silk. China maintains its position as the world's largest silk producer, with India firmly established as the second-largest producer. Other significant producers include Uzbekistan, Brazil, Japan, South Korea, Thailand, and Vietnam. Today, India's silk industry provides employment to approximately 7.9 million people across rural and urban areas. The sector faces challenges including competition from synthetic fibers, environmental concerns related to mulberry cultivation, fluctuating market demands, and the need for technological modernization. Despite these challenges, India's silk industry continues to thrive due to robust domestic demand, government support programs, and efforts to preserve traditional weaving techniques. As global interest in sustainable and eco-friendly textiles increases, India's diverse silk products maintain their value in both domestic and international markets, reinforcing the country's significant position in the global silk industry.

Space technology has the potential to significantly benefit the field of sericulture, the cultivation of silkworms for silk production, by improving various aspects of silkworm rearing, disease management, and production processes. Space-based satellite systems can monitor climatic factors like temperature, humidity, and rainfall patterns, which are crucial for silkworm health and mulberry cultivation. These systems help farmers determine the best times for planting mulberry trees and rearing silkworms, based on optimal weather conditions. Remote sensing tools can identify suitable regions for mulberry farming by assessing soil quality, water availability, and other environmental parameters. These data help farmers select the best locations for silkworm production. Again, weather satellites can provide accurate weather forecasts, enabling sericulturists to prepare for adverse conditions (like heavy rains, drought, or temperature extremes) that could affect silkworm health and productivity. By analyzing long-term weather data, space technologies help sericulturists adapt to changing climatic conditions and optimize the silkworm rearing process. Using Earth observation satellites, space technology can help monitor the health of mulberry trees, which are the primary food source for silkworms. Early detection of pests, diseases, or stress factors in plants allows for timely intervention. Geospatial technologies can help in accurately mapping and planning sericulture farms, optimizing farm layouts, and tracking production efficiency. These technologies can also support better resource management, including water use and fertilizer application. Space technology can assist in mapping silkworm rearing units, monitoring the health and productivity of different farms, and evaluating the overall performance of the industry in a particular region. Using satellite communication systems and Internet of Things (IoT) devices, farmers can monitor their silkworm rearing units and mulberry plantations remotely. This enables them to track key parameters such as temperature, humidity, and light, all of which are critical to the health of silkworms. Space technology can aid in the detection of insect or pest outbreaks, which can severely affect both mulberry crops and silkworms. Early detection through remote sensing and satellite imagery can help farmers take timely action. Space-based technologies can monitor patterns of disease spread and environmental factors contributing to outbreaks, allowing sericulturists to prevent or manage these issues more effectively.

2.5.1 Identification of Potential Sites for Expansion of Sericulture

Sericulture (silk farming) plays a significant and multifaceted role in contributing to the Sustainable Development Goals (SDGs) of India. As a labour-intensive (especially women), agro-based industry, sericulture supports rural economies, promotes environmental sustainability, and empowers marginalized communities. Thus sericulture contributes to SDG-1 (No Poverty), SDG-2 (Zero Hunger), SDG-5 (Gender Equality), SDG-8 (Economic Growth), SDG-9 (Industry & Infrastructure), SDG-10 (Reduced Inequalities), SDG-13 (Climate Action) and SDG-15 (Life on Land). Therefore, expansion of sericulture for rural livelihood is having great significance.

Space technology applications in the field of sericulture started in India during 1988 when CSB and Indian Space Research Organization (ISRO) joined hands to initiate a project on "Survey of Potential and Actual Areas under Sericulture using Remote Sensing" popularly called as SPAARS. Under the SPAARS project, actual area under mulberry cultivation was estimated at 1:2,50,000 scale. In addition to that, potential areas for expansion of sericultural activities were also identified under the project. The exercise was carried out using satellite imagery captured by Indian Remote Sensing Satellite (IRS) 1A series. In due course of time, with the launch of Resourcesat series of satellites during 2003–2018, CSB and North Eastern Space Applications Centre (NESAC) under Department of Space initiated an ambitious programme to identify potential sites for mulberry and non-mulberry sericulture in 108 districts covering 25 states of India. The exercise was carried out using Linear Imaging Self Scanning Sensor III (LISS-III) of Resourcesat-1 satellite and potential sites were identified at 1:50,000 scale (Handique et al. 2016a, b). After launch of Resourcesat-2 satellite during 2011, the exercise was expanded to another 70

Table 2.2 Suitable sites for expansion of Mulberry and Vanya silk in NER

State*	Mulberry (ha)			Vanya (ha)			Total
	High	Moderate	Marginal	High	Moderate	Marginal	
Arunachal Pradesh (10)	78	2111	39,558	7439	12,331	370,796	432,313
Assam (16)	21,688	108,783	573,139	226,796	369,873	439,557	1,739,835
Manipur (16)	3898	12,884	48,687	7871	13,702	616,417	703,459
Meghalaya (7)	42,344	88,040	131,017	31,760	30,709	748,122	1,071,992
Mizoram (11)	204,060	252,507	111,368	119,345	28,700	267,844	983,824
Nagaland (8)	4145	7180	71,053	18,703	48,468	108,134	257,681
Sikkim (4)	4145	18,604	99,122	6	72	389	122,336
Tripura (4)	219	17,682	41,978	0	0	0	59,879
Total	280,576	507,791	1,115,923	411,919	503,854	2,551,257	5,371,321

* Data in parentheses shows the number of districts covered in the states of NER

districts covering 26 states of India. The potential sites for expansion of mulberry and non-mulberry sericulture was carried out at 1:10,000 scale using LISS-IV sensor of Resourcesat-2 satellite in 76 districts of NER and identified about 53.7 lakhs ha potential areas for expansion of Mulberry and Vanya silk (Table 2.2) (Handique et al. 2018, 2019).

2.5.2 *Dissemination of Spatial Data Using Geo-Portals*

Web based geo-portal is an effective approach to provide geographic information to a wide range of users. It has modified the traditional way of using GIS as a database mapping and spatial analytical tool. To disseminate the large data-base generated for 178 districts covering 26 states; a geo-portal named SILKS (Sericulture Information Linkages and Knowledge System) was developed using open-source software. The SILKS geo-portal has been hosted in the public domain and can be accessed live at http://silks.csb.gov.in. The portal contains interactive maps of suitable areas for silkworm food plants and other relevant spatial and non-spatial information. The contents of the portal is made available in 12 languages/ dialects, viz. English, Hindi, Telugu, Kannada, Assamese, Bengali, Mizo, Manipuri, Khasi, Garo, Ao Naga and Sumi Naga. The spatial and non-spatial data are grouped into 4 categories, namely, planning services, other services, farmer's services and natural resources information. Planning services contain information on silkworm food plants production technologies, techniques of rearing silkworm, diseases and pest management of silkworm food plants, improved varieties of silkworm food plants, species of silkworm, processing of cocoons, infrastructure and equipment and allied sectors and

Fig. 2.6 SILKS Geo-portal

occupations. Other service contains information on micro credit and self-help group, seri marketing, seed distribution centres, weaving reeling centres and schemes and grants for farmers. Farmers' services module consists of weather and weather advisory, disease and pest forewarning and support services (Handique et al. 2016a, b; Singh et al. 2016; Handique et al. 2017). The Silks Geo-portal is illustrated in Fig. 2.6.

2.5.3 Monitoring of Sericultural Assets Using Geo-Spatial Technology

Geospatial technology enables efficient monitoring of host plantations, rearing houses etc. improving efficiency of the system. The monitoring can be carried out using space-based platforms viz. high-resolution satellite sensors; aerial platforms viz. airborne sensors/ drone based sensors and ground based monitoring viz. GPS aided mobile applications. Advantages of space-based monitoring are wide area coverage, cost effective, high repeativity/revisit and non-invasive. Similarly, airborne/ drone-based sensors are also having wider coverage, non-invasive and provide high spatial resolution imagery of the assets. However, airborne monitoring is more costlier than space based monitoring. In the NE hill region, most of the sericultural farms are small & fragmental in nature; mixed with other horticultural and/ or forest plantations and the rearing houses are also very small; made up of bamboo with thatched roof. Therefore, it is difficult to monitor those assets using space-based sensors. In this context, GAGAN (GPS Aided Geo Augmented Navigation) enabled mobile applications have been found to be useful for mapping of sericultural assets in the NE hill region. More than 40,000 NERTPS (North East Region Textile Promotion Scheme) assets were geo-tagged using GAGAN dongle enabled SILKS mobile application in NER (Handique et al. 2022c).

2.5.4 Space Technology Applications in Forewarning of Silkworm Diseases

Silkworms are very delicate and are very sensitive to biotic (pest and diseases) and abiotic (humidity, temperature etc.) stress. Mulberry and Eri silkworms are mostly reared indoor; on the other hand, Muga and Tasar silkworms are reared outdoor. Thus, Muga and Tasar silkworms are popularly known as Vinya Silk. Since, Muga and Tasar silkworms are reared outdoor; they are exposed to extreme weather events viz. heavy rainfall, hail storms, extreme temperatures and humidity etc. (Madhusudhan et al. 2017). Among the biotic stress of Muga silkworms, flacherie, pebrine and muscardine are the major diseases (Subrahmanyam et al. 2019). Utility of space technology for early warning of flacherie disease of Muga silkworm based on Land Use Land Cover (LULC), Land Surface Temperature (LST) and other weather parameters such as temperature and relative humidity has been demonstrated by Goswami et al. (2021).

2.6 Conclusion

Space technology has become an indispensable enabler in addressing the multifaceted challenges of agriculture, horticulture, and sericulture, particularly in the context of the North Eastern Region of the Eastern Himalaya. This region's complex physiography, diverse agro-climatic conditions, and traditional practices pose unique challenges that conventional approaches often fail to address comprehensively. The integration of remote sensing (RS), geographic information systems (GIS), global positioning systems (GPS), and advanced analytics, including machine learning and artificial intelligence, has transformed how resources are managed, monitored, and optimized in this landscape.

In agriculture, space technology has facilitated precise crop acreage estimation, improved yield prediction, and advanced cropping system analysis. By employing multispectral and hyperspectral imagery alongside radar data, it has become possible to overcome the limitations posed by cloud cover and inaccessible terrains. These innovations not only enhance productivity but also contribute significantly to climate-resilient agricultural planning. The advent of precision agriculture, supported by high-resolution satellite and UAV (drone) imagery, highlights the importance of technology in optimizing resource use, minimizing environmental impacts, and increasing farm-level profitability. Horticulture has equally benefitted from these advancements, particularly through site suitability analysis and resource mapping.

GIS-based multi-criteria decision-making tools have been instrumental in identifying optimal locations for cultivating high-value crops, thus unlocking the potential of underutilized lands. Additionally, the creation of spatial database and the use of remote sensing for inventory management have ensured better planning and monitoring of horticultural activities. This is crucial in meeting the growing demand for nutritionally rich crops in a sustainable manner. The sericulture sector, deeply intertwined with the socio-economic fabric of the NER, has witnessed transformative impacts through space technology. Initiatives like the SILKS geo-portal and GAGAN-enabled mobile applications have not only streamlined the identification of potential areas for sericulture expansion but also empowered stakeholders with tools for real-time monitoring and early disease warning systems. By bridging traditional practices with technological advancements, these tools enhance productivity and ensure the economic sustainability of rural livelihoods.

Despite these advances, challenges remain. Crop identification using remote sensing in the North Eastern hilly region of India faces several challenges due to the region's unique geographical, climatic, and socio-economic conditions. Rugged, hilly landscape with steep slopes and varied altitudes complicates field surveys and remote sensing applications. Again, small and fragmented landholdings in the region make it challenging to map and monitor crops using remote sensing technologies. The region's diverse climate leads to a wide variety of crops and cropping patterns which complicates standardized crop classification techniques. Shifting cultivation, practiced widely in the region, involves mixed cropping and short crop cycles followed by fallow periods. This dynamic land use pattern makes it difficult to consistently identify crops over time using remote sensing. The North Eastern region experiences heavy cloud cover, especially during the monsoon season, which limits the effectiveness of optical remote sensing for crop identification.

Addressing these challenges requires a combination of technological innovation, policy support, and community engagement to improve crop identification and support sustainable agriculture in the North Eastern hilly region. Use of radar-based remote sensing to overcome cloud cover issues, availability of space borne hyperspectral data, development of region-specific machine learning models, like random forest algorithms, and advanced AI based models to account for diverse cropping patterns and terrain, capacity Building of local farmers and officials in modern tools and techniques to enhance ground truthing and data validation are some of the potential solutions to address these challenges.

The future of agriculture, horticulture, and sericulture in the region lies in a symbiotic relationship between tradition and innovation. Space technology serves as the bridge, offering unparalleled insights into resource management while respecting the ecological and cultural nuances of the region. By fostering collaboration among scientists, policymakers, and local communities, these technologies can help achieve the twin goals of food security and environmental sustainability.

As we look ahead, it is evident that the role of space technology will continue to expand, driven by advancements in satellite systems, data analytics, and real-time monitoring capabilities. The journey from observation to actionable intelligence marks the evolution of these tools from enablers to indispensable components of agricultural and rural development strategies.

Acknowledgements The authors express their sincere gratitude to the Indian Space Research Organisation (ISRO) for its pivotal role in the development and deployment of advanced satellite technologies essential for Earth observation. We acknowledge the National Remote Sensing Centre (NRSC), Hyderabad, for providing crucial geospatial data, and extend our appreciation to the Space Applications Centre (SAC), Ahmedabad, and the Indian Agricultural Statistics Research Institute (IASRI), New Delhi, for their technical support. We are especially grateful to the North Eastern Council (NEC), Government of India, for the financial support provided towards the implementation of different projects. Our thanks also go to the Central Silk Board for their contributions to the SPAARS & SILKS initiative, and to the Ministry of Agriculture & Farmers Welfare for their continued support. We further acknowledge the collaboration of the State Departments of Agriculture, Horticulture & Sericulture in the North Eastern Region, as well as the State Remote Sensing Applications Centres (SRSACs), for their valuable inputs and cooperation throughout the period.

References

Z. Ahmed, P. Nageswara Rao, P. Srikanth, Area estimation of mango and coconut crops using machine learning in Hesaraghatta Hobli of Bengaluru Urban District, Karnataka. J. Geomat. **17**(1), 79–86 (2023)

G.P. Asner, R.E. Martin, J. Mascaro, Coral reef atoll assessment in the South China Sea using Planet Dove satellites. Remote Sens. Ecol. Conserv. **3**, 57–65 (2017)

N. Baniya, M. Böehme, S. Baniya, Physical land suitability assessment for the large cardamom *Amomum subulatum* Roxb. Cultivation in Hills of Kathmandu Valley. Chin. J. Popul. Resour. Environ. **7**(4), 59–63 (2009)

B.C. Barah, Strategies for agricultural development in the North-East India: challenges and emerging opportunities. Indian J. Agric. Econ. **62**, 13–31 (2007)

F. Baret, G. Guyot, Potential and limits of vegetation indices for LAI and APAR assessment. Remote Sens. Environ. **35**(2–3), 161–173 (1991)

M. Craig, D. Atkinson, *A literature review of crop area estimation* (FAO Publication, Rome, 2013)

V.K. Dadhwal, R.P. Singh, S. Dutta, J.S. Parihar, Remote sensing based crop discrimination and area estimation. A Rev Indian experience. Trop. Ecol. **43**, 107–122 (2003)

C. Dakshinamurty, B. Krishnamurthy, A.S. Summanwar, P. Shanta, P.R. Pisharoty, Remote sensing for coconut wilt, in *Proceedings of the 6th International Symposium on Remote Sensing of Environment* (Ann Arbor, 1971), pp 25–29

L.C. De, D.R. Singh, Natural resources in North East region of India. Int. J. Agric. Sci. Res. **7**(5), 51–66 (2017)

K.R. Dikshit, J.K. Dikshit, Agriculture in North-East India: past and present, in *North-East India: Land People and Economy* (Springer Nature, 2014), pp. 587–637

M.A. Ganie, A. Nusrath, Estimation of apple orchad using remote sensing and agro-metrology land-based observation in Pulwama district of Kashmir Valley. Int. J. Remote. Sens. Geosci. **3**, 21–26 (2014)

G.P. Geetha, G.M. Veerendra Patel, Y. Shruti, V.R. Ramakrishna Parama, A. Sathish, Land evaluation in halayapura1 micro watershed in Tumkur district of Karnataka, India, using remote sensing and geographical information system (GIS) tools. Int. J. Chem. Stud. **7**(1), 84–94 (2019)

G. Ghazaryan, O. Dubovyk, F. Löw, M. Lavreniuk, A. Kolotii, J. Schellberg, N. Kussul, A rule-based approach for crop identification using multi-temporal and multi-sensor phenological metrics. Eur. J. Remote. Sens. **51**, 511–524 (2018)

B. Gogoi, Problem, prospect and role of agriculture in rural development in North-East India. Int. J. Appl. Soc. Sci. **6** (7), 1944–1951 (2019)

J. Goswami, D.K. Gogoi, N. Rasid, B.K. Handique, G. Subhahmanyam, P.P. Bora, R. Das, P.L.N. Raju, Development of a Muga disease early warning system– a mobile-based service for seri farmers. Curr. Sci. **121**(10), 1328–1334 (2021)

P.C. Gupta, R. Sharma, Application of satellite remote sensing technique indelineation and hectarge estimation of mango orchards in parts of Uttar Pradesh, India, in *Proceedings of GIS Development*, pp. 1–3 (1990)

B.R. Guruprasad, Understanding India's international space cooperation endeavour: evolution challenges and accomplishment. India Q. **74**(4), 455–481 (2018)

B.K. Handique, C. Goswami, P.T. Das, J. Goswami, P. Jena, F. Dutta, D.K. Jha, S.P. Aggarwal, Space technology support for development of agriculture in the North Eastern Region of India – scope and challenges. Curr. Sci. **123**(8), 975–986 (2022a)

B.K. Handique, C. Goswami, P. Jena, F. Dutta, R. Samiam, I. Nongrum, D. Jha, P.L.N. Raju, C.R. Deka, R. Sarma, K. Sarmah, H.K. Dutta, S. Deb, B. Yari, S. Maibam, P. Thanpuli, R. Sailo, V. Muansangi, D. Katiry, T. Medo, N.P. Sharma, B. Lepcha, A. Roy, B.K. Bhattacharya, J.G. Patel, C.P. Singh, K.R. Manjunath, M.M. Kimothi, S. Mamatha, P. Kumar, P. Tahlani, S.S. Ray, Applications of advanced geospatial technology for expansion of area under horticultural crops in North Eastern region of India. J. Indian Soc. Remote. Sens. **50**(2), 331–345 (2022b)

B.K. Handique, J. Goswami, A. Qadir, C. Gupta, P.L.N. Raju, Rapid assessment of Boro paddy infestation by brown planthopper in Morigaon district, Assam, India using unmanned aerial vehicle. Curr. Sci. **111**(10), 1604–1606 (2016a)

B.K. Handique et al, Applications of remote sensing & GIS in sericulture development (Phase-II) for NER States. NESAC-SR-190–2018 (2018)

B.K. Handique et al., Applications of remote sensing & GIS in sericulture development Phase-II (18 States). NESAC-SR-198–2019 (2019)

B.K. Handique et al., Geo-tagging of assets created under NERTPS program of CSB in NER. NESAC-SR-279–2022 (2022c)

B.K. Handique, P.T. Das, J. Goswami, C. Goswami, P.S. Singh, C.J. Prabhakar, C.M. Bajpeyi, P.L.N. Raju, Expansion of sericulture in India using geospatial tools and web technology. Curr. Sci. **111**, 1312–1318 (2016b)

B.K. Handique, P.S. Singh, C.J. Prabhakar, P.T. Das, J. Goswami, C. Goswami, D. Chutia, S. Sudhakar, P.P.N. Rao, Sericulture Information Linkages and Knowledge System (SILKS) -a geoportal for sericulture planning and development. Indian Silk **7**(9–11), 8–10 (2017)

K.R. Hebbar, M.V. Krishna Rao, Potential of IKONOS data for horticultural crop inventory, Project report, NRSC, Hyderabad (2002)

J. Janick, R. Perrott, A.C. Geoffrey, P.M. Synge, Horticulture. Encyclopedia Britannica, 29 Aug. 2024

J.K. Joshua, N.C. Anyanwu, A.J. Ahmed, Land suitability analysis for agricultural planning using GIS and multi criteria decision analysis approach in Greater Karu Urban Area, Nasarawa State, Nigeria. Afr. J. Agric. Sci. Technol. **1**(1), 14–23 (2013)

K. Karthikeyan, K. Nirmal, G. Ajit, Pushpanjali, J. Prasad, Assessment of soil site suitability for cotton farming in the semi-arid regions of Central India: an analytic hierarchy process. J. Indian Soc. Soil Sci. **67**(4), 402–410 (2019)

T.M. Lillesand, R.W. Kiefer, J.W. Chipman, *Remote sensing and image interpretation*, 8th ed. (John Wiley & Sons, 2015)

S. Liu, P.I. Theoharis, R. Raad, F. Tubbal, A. Theoharis, S. Iranmanesh, S. Abulgasem, M.U.A. Khan, L. Matekovits, A survey on CubeSat missions and their antenna designs. Electronics **11**(13), 2021 (2022)

K.N. Madhusudhan, M.S.M. Laltlankimi, A.H. Naqvi, V.P. Gupta, A. Sahay, V. Sivaprasad, Impact of varying different abiotic factors on the survivability of Tasar silkworm in outdoor rearing fields. J. Entomol. Zool. Stud. **5**(6), 957–963 (2017)

Z. Mamatkulov, E. Safarov, R. Oymatov, I. Abdurahmanov, M. Rajapbaev, Application of GIS and RS in real time crop monitoring and yield forecasting: a case study of cotton fields in low and high productive farmlands. E3S Web Conf. **227**, 03001 (2021)

S. Manandhar, V.P. Pandey, F. Kazama, Assessing suitability of apple cultivation under climate change in mountainous regions of western Nepal. Reg. Environ. Change **14**, 743–756 (2014)

MoSPI, Production of Horticultural Crops (2024). https://mospi.gov.in/45-production-horticulturalcrops#:~:text=2%20The%20second%20source%20of,Directorates%20of%20Horticulture%20and%20Agriculture.

C.S. Murthy, M.V.R. Sesha Sai, K.V. Bhanuja, P.S. Roy, Agricultural drought assessment at disaggregated level using AWiFS /WiFS data of Indian remote sensing satellites. Geocarto Int. **22**, 127–140 (2007)

A. Nagaraja, Predicting susceptible areas of mango malformation through remote sensing and GIS. Ph. D thesis, IARI, New Delhi (2010)

C. Palaniswami, A.K. Upadhyay, H.P. Maheswarappa, Spectral mixture analysis for sub pixel classification of coconut. Curr. Sci. **91** (12), 1706–1711 (2006)

S.S. Panda, G. Hoogenboom, J. Paz, Distinguishing blueberry bushes from mixed vegetation land-use using high resolution satellite imagery and geospatial techniques. Comput. Electron. Agric. **67**, 51–59 (2009)

N.C. Paul, P.M. Sahoo, T. Ahmad, R. Sahoo, G. Krishna, S. Lal, Acreage estimation of mango orchards using hyperspectral satellite data. Indian J. Hortic. **75**(01), 27–33 (2018)

K.V. Peter, Horticulture for nutrition security. (Daya Publishing House, a division of Astral International Pvt. Ltd, New Delhi, 2015)

L.K. Petersen, Real-time prediction of crop yields m relative vegetation health: a continent-wide analysis of Africa. Remote. Sens. **10**(11), 1726 (2018)

PIB, Second Advance Estimates of 2023–24 of Area and Production of Horticultural Crops (2024). https://pib.gov.in/PressReleseDetailm.aspx?PRID=2022761.

P. Punitha, M.A. Ansari, D.K. Pandey, D. Ram, S. Datt, P.K. Sharma, M. Aheibam, S.S.P. Jyothi, N. Prakash, Shifting cultivation in North East India: Social dimension, cross cultural reflection and strategies for improvement. Indian J. Agric. Sci. **88**(6), 811–819 (2018)

N.L. Rajesh, V. Kirana Kumara, C. Rajesh, S. Ambika Bhandari, B. Deepika, H.V. Ambika, K. Rudramurthy, U.S. Basavaraj, B.K. Desai, Land suitability evaluation for legume crops and horticulture crops in Paman Kallur-1 micro-watershed using geospatial techniques. Int. J. Curr. Microbiol. App. Sci. **8**(12), 1302–1317 (2019)

A. Roy, D.S. Dhar, A.K. Tripathi, N.U. Singh, D. Kumar, S.K. Das, A. Debnath, Growth performance of agriculture and allied sectors in the North East India. Econ. Affairs **59**, 783–795 (2014)

P.M. Sahoo, A. Rai, R. Singh, B.K. Handique, C.S. Rao, Integrated approach based on remote sensing and GIS for estimation of area under paddy crop in North-Eastern hilly region. J. Indian Soc. Agric. Stat. **59**, 151–160 (2005)

P.M. Sahoo, A. Rai, S. Krishnamoorthy, B.K. Handique, P.P.N. Rao, M.P. Oza, J.S. Parihar, Sampling approach for estimation of crop acreage under cloud cover satellite data in hilly regions. Proc. SPIE **64**, 1–9 (2006)

A.W. Schumann, Using precision agriculture technology for precise placement and variable rate fertilizer application. nutrient bmps: keeping water and nutrients in the root zone of Florida's horticultural crops. UF/IFAS, Apopka, FL, p. 13 (2008)

L. Seitinthang, Cropping pattern of North East India: an appraisal. Am. Res. Thoughts **1**, 488–498 (2014)

A. Sharma, S. Panigrahy, Apple orchard characterization using remote sensing and GIS in Shimla district of Himachal Pradesh, in *Proceedings of Remote Sensing and Photogrammetry Annual Conference*, 11–14 September 2007, Newcastle, UK (2007)

R. Sharma, J. Xu, G. Sharma, Traditional agroforestry in the eastern Himalayan region: land management system supporting ecosystem services. Trop. Ecol. **48**(2), 1–12 (2007)

R.J. Shrivastava, J.L. Gebelein, Land cover classification and economic assessment of citrus groves using remote sensing. ISPRS J. Photogramm. Remote. Sens. **61**, 341–353 (2006)

A.K. Singh, A.K. Chaurasiya, S. Mitra, *Role of Horticulture in Agriculture Development and Food Security in India* (2015)
H.P. Singh, Triggering agricultural development through horticultural crops. Indian J. Agric. Econ. **64**(1), 15–39 (2009)
P.S. Singh, B.K. Handique, D. Chutia, P.T. Das, J. Goswami, C. Goswami, C.J. Prabhakar, K. Dhanraj, P.L.N. Raju, Conceptualization and development of sericulture Information linkages and knowledge system (SILKS). J. Geomat. **10**(2), 108–113 (2016)
M. Singha, J. Dong, G. Zhang, X. Xiao, High resolution paddy rice maps in cloud-prone Bangladesh and Northeast India using Sentinel-1 data. Sci. Data **6**, 1–10 (2019)
V. Stefanova, Z. Arnaudova, D. Haytova, T. Bileva, Multi-criteria evaluation for sustainable horticulture. Turk. J. Agric. Nat. Sci. **2**, 1694–1701 (2014)
G. Subrahmanyam, V.G. Esvaran, K.M. Ponnuvel, W. Hassan, M. Chutia, R. Das, Isolation and Molecular Identification of Microsporidian Pathogen Causing Nosemosis in Muga Silkworm, *Antheraea assamensis* Helfer (Lepidoptera: Saturniidae). Indian J. Microbiol. **59**(4), 525–529 (2019)
D.M. Thapa, C.P. Shrivastav, S.C. Shah, K. Sah, Land suitability evaluation using gis for vegetable crops at Sharadanagar, Chitwan, Nepal, Tropical Agrobiodiversity (TRAB), vol. 1, no 1, (Zibeline International Publishing2020), pp. 42–46
J. Torres-Sánchez, F. López-Granados, N. Serrano, O. Arquero, J.M. Peña, High-Throughput 3-D monitoring of agricultural-tree plantations with unmanned aerial vehicle (UAV) Technology. PLoS ONE **2015**(10), e0130479 (2015)
B. Unni, M. Goswami, Y. Kakoty, M. Bhattacarjee, S. Wann, G. Rajkhowa, S. Das, B. Rani, B. Devi, Anamika, D. Chutia, Indigenous knowledge of silkworm cultivation and its utilization in North Eastern region of India. Indian J. Tradit. Knowl. **8**, 70–74 (2009)
S.K. Vaid, An overview of Indian agriculture with focus on challenges and opportunities in North East, in *Microbiological Advancements for Higher Altitude Agro-Ecosystems & Sustainability*, pp. 15–36 (2020)
A. Vali, S. Comai, M. Matteucci, Deep learning for land use and land cover classification based on hyperspectral and multispectral Earth observation data: a review. Remote Sensing **12**, 2495 (2020)
H.C. Verma, S. Rajan, T. Ahmed, Mapping and area estimation of mango orchards of Lucknow region by applying knowledge-based decision tree to Landsat 8 OLI satellite images. Int. J. Innov. Technol. Explor. Eng. (IJITEE)**9**(3). ISSN: 2278–3075 (2020)
F. Vogel, G. Bange, Understanding USDA crop forecasts. Miscellaneous Publication No. 1554. Joint USDA/NASS. (WOAB Publication: Washington, D.C, 1999))
J. Xue, B. Su, Significant remote sensing vegetation indices: a review of developments and applications. J. Sens. **2017**, 17 (2017)
I.S. Yadav, N.K.S. Rao, B.M.C. Reddy, R.D. Rawal, V.R. Srinivasan, N.T. Sujatha, C. Bhattacharya, P.P.N. Rao, K.S. Ramesh, S. Elango, Acreage and production, estimation of mango orchards using Indian Remote Sensing (IRS) satellite data. Sci. Hortic. **93**, 105–123 (2002)
H. Zhang, Y. Lan, C.P. Suh, J.K. Westbrook, R. Lacey, W.C. Hoffmann, Differentiation of cotton from other crops at different growth stages using spectral properties and discriminant analysis. Trans. ASABE **55**(4), 1623–1630 (2012)

Bijoy K. Handique currently working as the Head, Agriculture and Soil Division and Head, Program Planning and Evaluation Group at North Eastern Space Applications Centre under Department of Space, Government of India has more than 22 years of experience in space technology applications and research. He has been leading a number of research and applications programs at NESAC particularly in space technology applications in agriculture and allied areas. He is also leading a team at NESAC in using unmanned aerial vehicles for various agricultural applications. He has published about 60 research papers in national and international journals and conference proceedings. He has also published 12 book chapters and guided a number of M.Sc./M.Tech and Ph.D. students. He has been awarded ISRS-SPECK Award for 2005, ISRO Young Scientist Merit Award for 2008 and ISRO-ASI Young Scientist Award for 2014.

Jonali Goswami currently working as Scientist-SF in Agriculture and Soil Division and Program Coordinator for the Outreach & Capacity Building activities at North Eastern Space Applications Centre under Department of Space, Government of India, has more than 22 years of experience in space technology applications and research. Her research expertise spans Geospatial technology applications in agriculture and allied sectors, disaster management and epidemiological studies. She has developed expertise in hyperspectral and Unmanned Aerial Vehicle (UAV) data analysis and their applications in different fields of research and applications towards precision and climate resilience agriculture. She has published more than 50 research papers in national and international journals and conference proceedings. She has also many book chapters publications and guided a number of M.Sc./M.Tech and Ph.D. students. Applications Centre under Department of Space, Government of India has more than 22 years of experience in space technology applications and research. He has been leading a number of research and applications programs at NESAC particularly in space technology applications in agriculture and allied areas. He is also leading a team at NESAC in using unmanned aerial vehicles for various agricultural applications. He has published about 60 research papers in national and international journals and conference proceedings. He has also published 12 book chapters and guided a number of M.Sc./M.Tech and Ph.D. students. She has been awarded National Award for Women of Excellence in Geospatial Science and Technology by FOSSEE under NMEICT, Ministry of Education, Government of India at IIT Bombay, 2025.

Chandan Goswami received Bachelor Degree in Agriculture and Master Degree in Soil Science from Assam Agricultural University, Jorhat, Assam, India in the year 1999 and 2001, respectively and Ph.D degree in Soil Science & Agricultural Chemistry from Central Agricultural University, Imphal, Manipur India in the year 2020. Dr. Goswami is a Senior Scientist at the North Eastern Space Applications Centre, Shillong, Meghalaya, India. He has specialized in geospatial technology applications in agriculture, soil and natural resources management. His research focuses on soil spectral analysis, soil resources mapping, and land use planning, significantly contributing to sustainable agricultural development in India's North Eastern Region. His work leverages advanced remote sensing and GIS techniques to address challenges like soil health, and crop productivity in the NER's diverse agro-climatic conditions. Dr. Goswami has authored over 84 publications and having more than 600 citations, reflecting his impactful contributions to the scientific community. A recipient of the National e-Governance Award in 2017 for enhancing geospatial data utilization in planning and development.

Pratibha Thakuria Das Scientist/Engineer-SF at the North Eastern Space Applications Centre (NESAC), brings over two decades of expertise in leveraging remote sensing and GIS technologies for soil resource mapping, soil fertility assessment, land degradation analysis, crop suitability evaluation, sericulture development, and land-use planning. Her notable contributions include generating soil maps for several districts across Northeast India, creating fertility maps for Meghalaya, Sikkim, Mizoram, and Arunachal Pradesh, integrating soil data for the Soil Resource Atlas of India, and identifying suitable areas for various crops and silkworm food plants. She has authored several publications in peer-reviewed national and international journals and conferences, focusing on soil fertility mapping, land use planning, and the application of geospatial technologies in agriculture. Her work significantly contributes to sustainable agricultural practices and land management in Northeast India.

Pradesh Jena joined NESAC on 13.05.2019 and has been working extensively in the field of crop acreage and production estimation of major agricultural crops, crop cutting experiments, cropping system analysis, and geo-tagging of sericulture assets in the North Eastern Region using geospatial technologies. He has also gained expertise in applications of UAV technology for precision agriculture. At NESAC, he is a member of the 'Committee for delivery of Data Products & Services from EOS-04 & TRISHNA Mission'. He has published many research papers in national and international journals and conference proceedings. He has also guided a number of M.Sc./M.Tech students.

Francis Dutta joined NESAC in March 2020 and has been actively engaged in the applications of remote sensing and GIS in agriculture and allied sectors. His contributions span multiple projects, including site suitability analysis for horticultural crops, crop acreage estimation, geo-tagging of sericulture assets in the Northeast region, optimization of crop cutting experiments (CCE) using geospatial technology, and UAV-based remote sensing.

Chapter 3
Remote Sensing-Based Soil Resources Assessment

Pratibha Thakuria Das, Chandan Goswami, and Francis Dutta

Abstract Soil resource assessment plays a pivotal role in sustainable land management, particularly in ecologically sensitive and topographically diverse regions such as the North Eastern Region (NER) of India. This chapter highlights the application of remote sensing (RS) and geospatial technologies in soil mapping, fertility evaluation, land degradation monitoring, and land suitability assessment. By integrating multispectral, microwave, thermal, and hyperspectral satellite data with GIS and field-based observations, soil characteristics such as texture, depth, organic carbon, nutrient status, and erosion susceptibility are effectively mapped and analyzed. The use of standard scales (1:250,000 to 1:10,000) across reconnaissance to detailed mapping initiatives enables spatially explicit planning for agriculture, watershed development, and ecological restoration. Programs such as the Soil Health Card Scheme and the National Soil Mapping Programme (NSMP) contribute significantly to data generation and standardization. Spectral indices and soil spectral libraries further enhance precision in soil property estimation. Land degradation, primarily due to erosion, acidification, and anthropogenic pressures is systematically monitored, supporting the goal of land degradation neutrality. The chapter underscores the transformative role of geospatial technologies in building resilient agroecosystems, enabling data-driven decision-making for sustainable agricultural planning and resource conservation in the NER.

Keywords Soil resource mapping · Remote sensing · Soil fertility assessment · Land degradation · Geospatial technology · Sustainable agriculture

P. T. Das (✉) · C. Goswami · F. Dutta
North Eastern Space Applications Centre (NESAC), Umiam, Shillong, Meghalaya, India
e-mail: pt.das@nesac.gov.in

C. Goswami
e-mail: chandan.goswami@nesac.gov.in

F. Dutta
e-mail: francis.dutta@nesac.gov.in

S. P. Aggarwal et al. (eds.), *Remote Sensing of the North Eastern Himalayan Ecosystem*,
https://doi.org/10.1007/978-981-95-3221-6_3

3.1 Introduction

Soil is a fundamental resource that supports plant growth and sustains ecosystems by providing essential nutrients, organic matter, and microbial life (Bridges and Barren 1997). It plays a key role in realizing the UN's Sustainable Development Goals (SDGs), particularly those related to food security, health, water management, sustainable consumption, climate action, and biodiversity conservation (Bouma and Montanarella 2016; Keesstra et al. 2016). As a crucial component of the lithosphere and biosphere, soil supports human needs by providing food, fibre, fodder, and fuelwood. However, its productivity is limited by intrinsic properties, agroecological conditions, and management practices. Increasing human and animal populations have led to greater pressure on soil resources, resulting in declining soil quality and an increased risk of degradation. Studies indicate that a significant portion of India's soil is deteriorating, threatening long-term food security (Sehgal and Abrol 1994; NRSC 2019).

The North Eastern Region (NER) of India includes eight states: Arunachal Pradesh, Assam, Manipur, Meghalaya, Mizoram, Nagaland, Sikkim, and Tripura. This region, known for its cultural and ethnic diversity is home to over 200 distinct ethnic groups, each with unique languages, dialects, and traditions. Covering approximately 7.97% of India's total geographical area, the NER accounts for 3.78% of the national population (Dikshit and Dikshit 2014). Owing to its strategic location, the region borders five countries, Bangladesh, Myanmar, China, Bhutan, and Nepal across a stretch of about 5484 km. While this connectivity presents opportunities for trade, cultural exchange, and regional cooperation, it also poses challenges related to border security, migration, and transnational issues (Chatterjee et al. 2006; Baruah 2020).

The NER's landscape is characterized by hills, dense forests, and river systems, which play a crucial role in its socio-economic development. Rich in natural resources, including biodiversity, fertile land, and hydropower potential, the region faces obstacles such as poor infrastructure and limited connectivity, which hinder economic growth. However, there is immense potential in sectors like tourism, agriculture, renewable energy, and trade (Rizal and Asokan 2013; Roy et al. 2015). The NER features diverse soil types, including alluvial, red loamy, lateritic, and forest soils, which support a variety of crops such as rice, maize, tea, spices, and horticultural produce. However, issues such as soil erosion, shifting cultivation (jhum), and land degradation pose significant challenges to sustainable land use. Proper soil conservation techniques, integrated watershed management, and organic farming practices can restore soil fertility and promote sustainable agricultural growth (Roy et al. 2014).

A thorough understanding of soil resources, including their extent, distribution, characteristics, potential, limitations, and capabilities is crucial for various purposes, such as creating effective land-use plans for agriculture, managing irrigation and drainage, analyzing crop suitability, conserving soil in catchment areas, prioritizing watersheds, and reclaiming degraded lands (Manchanda et al. 2002).

Traditionally soil surveys were conducted through reconnaissance and detailed field observations. Aerial photographs were later introduced for soil mapping but faced limitations, including spectral restrictions, lack of standardization, and inability to penetrate vegetation. Geometric distortions and reliance on visual interpretation further reduced accuracy, while high costs and labor intensity made large-scale application challenging. Satellite-based soil mapping is highly efficient due to multispectral imaging, large-scale coverage, geometric accuracy, and automated processing workflows. The launch of Landsat in 1972 revolutionized soil studies by enabling continuous global monitoring, spectral soil analysis, and digital classification. Its regular data acquisition, cost-effectiveness, and open-access policy made soil mapping more accessible, reliable, and scalable for diverse applications.

Using Remote Sensing (RS) data along with GIS and machine learning helps improve soil type mapping and predict soil health more accurately. These tools enable accurate identification of soil types, monitoring of degradation processes, and forecasting key factors such as soil fertility, compaction, erosion risk etc. The increasing need for sustainable agricultural practices, particularly in regions vulnerable to soil degradation, has underscored the importance of adopting a holistic, climate-smart approach to land use. By incorporating modern geospatial technologies, it is now possible to map and monitor soil health and thereby ensure long-term sustainability.

This chapter will explore the utility of geospatial technology in soil resource mapping, providing spatially explicit information on soil physical and chemical properties, inherent capability of the soil for sustained arable & non arable use, and irrigation, fertility capability and alternative land use plans (e.g., alternate crops, grasses, fruits, vegetables, sericulture, and forest plantations).

3.2 Soil Resource Mapping

A. **Reconnaissance Soil Resource Map**

Soil mapping involves identifying, classifying, and delineating soils based on parent materials/landscapes, climate, relief/topography, and vegetation/organisms, verified through fieldwork and laboratory analysis. Mapped as "map units" with specific symbols and legends, soil maps are produced at various scales for different planning needs. Small-scale maps (1:1 million) support national planning, while 1:250,000 scale maps aid to regional level planning. Standard surveys produce 1:50,000 scale maps for broader land assessments and larger scales (1:10,000 or more) serve specialized purposes like soil conservation, fertility studies, taxation, and crop suitability.

The National Bureau of Soil Survey & Land Use Planning (NBSS&LUP) has carried out systematic soil mapping across India, using a structured three-tier approach. This includes mapping at a 1:1 million scale for the entire country and at a 1:250,000 scale. This involved analyzing Landsat Multi Spectral Sensor (MSS) imagery to delineate landforms, conducting field surveys to study soil profiles and collect samples, and performing laboratory analysis to classify soils based on various properties. The final maps were prepared through cartographic processing and printing as outlined in the Field Manual. This comprehensive methodology ensures accurate and detailed soil mapping, aiding in effective land use planning and sustainable resource management (Sehgal et al. 1987; Velayutham 1999).

The diverse topography and climatic conditions of NER have given rise to a variety of soil types, each playing a crucial role in the region's agricultural productivity and ecosystem sustainability. The classification of these soils provides valuable insights into their formation, distribution, and suitability for different land uses. Inceptisols (recent soils) are the most dominant soil in NER, followed by Entisols (very recent soils), Ultisols (old soils) and other miscellaneous orders occupying 51.7%, 20.4%, 19.2% and 8.7% of the Total Geographical Area (TGA), respectively. Mollisols (soft, dark colour soil) are found only in Sikkim occupying 14.6% of the TGA of the state. The surface soils are mostly loamy followed by clayey in texture occupying more than 60% and 25% of the TGA of NER, respectively. Soils of the region are mostly deep to moderately deep (86% of TGA of NER). Shallow soils are found only in Arunachal Pradesh, Sikkim, Nagaland and hills of Tripura and it covers 11% area of the TGA of the region. The soils of the entire region are acidic in nature. About 63% area is moderately acidic followed by slightly acidic and strongly acidic soils occupy about 25% and 5% of the TGA of NER (Handique et al. 2022a). Figure 3.1 shows the spatial extent of different soil types in NER.

B. **Semi Detailed Soil Resource Map**

Soil resource mapping is essential for sustainable land management, providing critical insights into soil characteristics, fertility, and land suitability. To aid this effort, the Soil and Land Use Survey of India (SLUSI) and the ISRO, in partnership with several institutions, have created district-level soil maps at a 1:50,000 scale. These semi-detailed maps enhance soil classification, land capability assessment, and resource planning, aiding in agricultural development, watershed management, and environmental conservation. North Eastern Space Applications Centre (NESAC) has also contributed in preparation of district wise soil resource maps on1:50 K scale for Assam (Cachar, Dhemaji, Hailakandi, Karimganj, Lakhimpur, Nagaon and Sonitpur district), and Manipur (Bishnupur, Chandel, Churachandpur, Senapati, Tamenglong, Thoubal and Ukhrul district).

District-level soil maps at a 1:50,000 scale are developed using False Colour Composite (FCC) images from Resourcesat-1 and Resourcesat-2, captured by the LISS-III sensor. These images support the creation of landscape, physiographic, and Land Use/Land Cover (LULC) maps through visual interpretation techniques. Cartosat DEM and Topo sheets are used to generate slope map. Each of the above maps represents the factors which influence soil formation. All these maps are overlaid in GIS environment and a composite base map is prepared with tentative soil mapping legends. The detailed methodology is illustrated in Fig. 3.2.

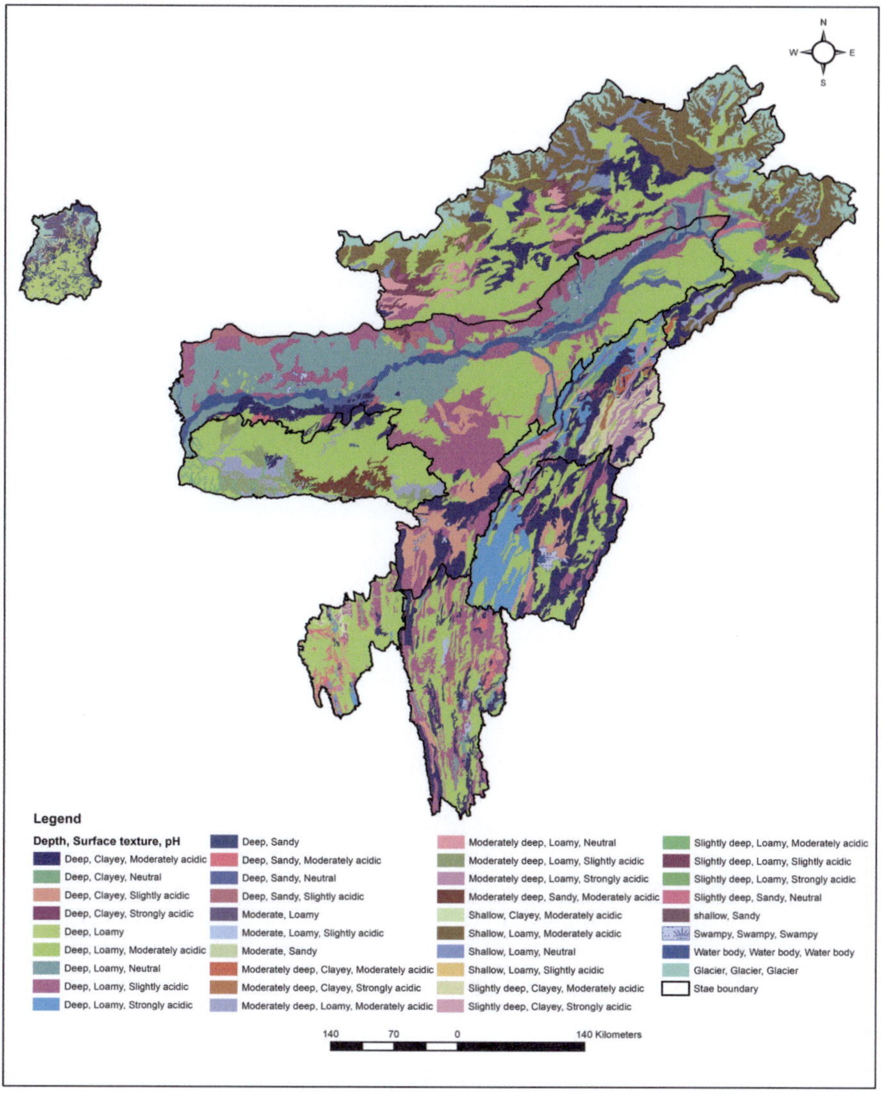

Fig. 3.1 Soils of North Eastern Region of India

National Remote Sensing Centre (NRSC) in collaboration with SLUSI has taken up a project to pool the soil resource maps on 1:50 K scale available with NRSC and SLUSI and standardizing them through common taxonomic classification followed by quality checks, to develop a uniform digital soil data-base for entire India and generating State-wise Soil Resources Atlas of the country that serve the planning needs for various users (NRSC 2021). NESAC, being the coordinator for the Eastern Region, has contributed to this project by bringing

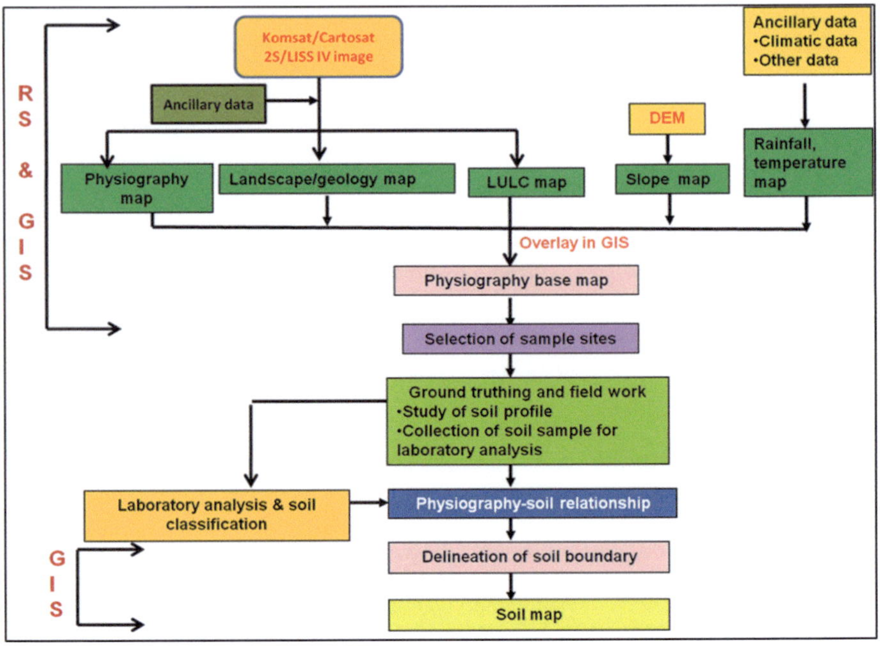

Fig. 3.2 Flowchart of methodology of soil resource mapping using RS and GIS

all soil maps of all the 8 states of NER along with West Bengal and Odisha into a common platform by developing soil legends and taxonomic classification. Figure 3.3 shows the spatial extent of soil taxonomic classes in NER.

C. **Detailed Soil Resource Map**

High-resolution soil mapping at a 1:10,000 scale is essential for managing soil variability at the village and micro-watershed level. However, such detailed data is currently unavailable for the entire country. This highlights the need for large-scale, detailed soil mapping to support agricultural development planning (Srivastava et al. 2004). In view of this, SLUSI has taken up the task of mapping the soils of the country on 1:10,000 (1:10 K) scale under National Soil Mapping Programme (NSMP). Similarly, NESAC has taken up a project on detailed soil mapping for the state of Meghalaya on 1:10 K scale with financial support from ISRO. The work is being carried out in collaboration with SLUSI with logistic support from Directorate of Agriculture and Horticulture, Govt. of Meghalaya. NESAC is also participating in the National Soil Mapping Programme (NSMP) of SLUSI for the remaining 7 states of NER. The Ministry of Agriculture and Farmers Welfare, Government of India, has launched the National One Soil Unified Information System (NOSUIS) to integrate soil-related data from various organizations onto a single digital platform. This initiative aims to make the information accessible to farmers, policymakers, researchers, and other stakeholders,

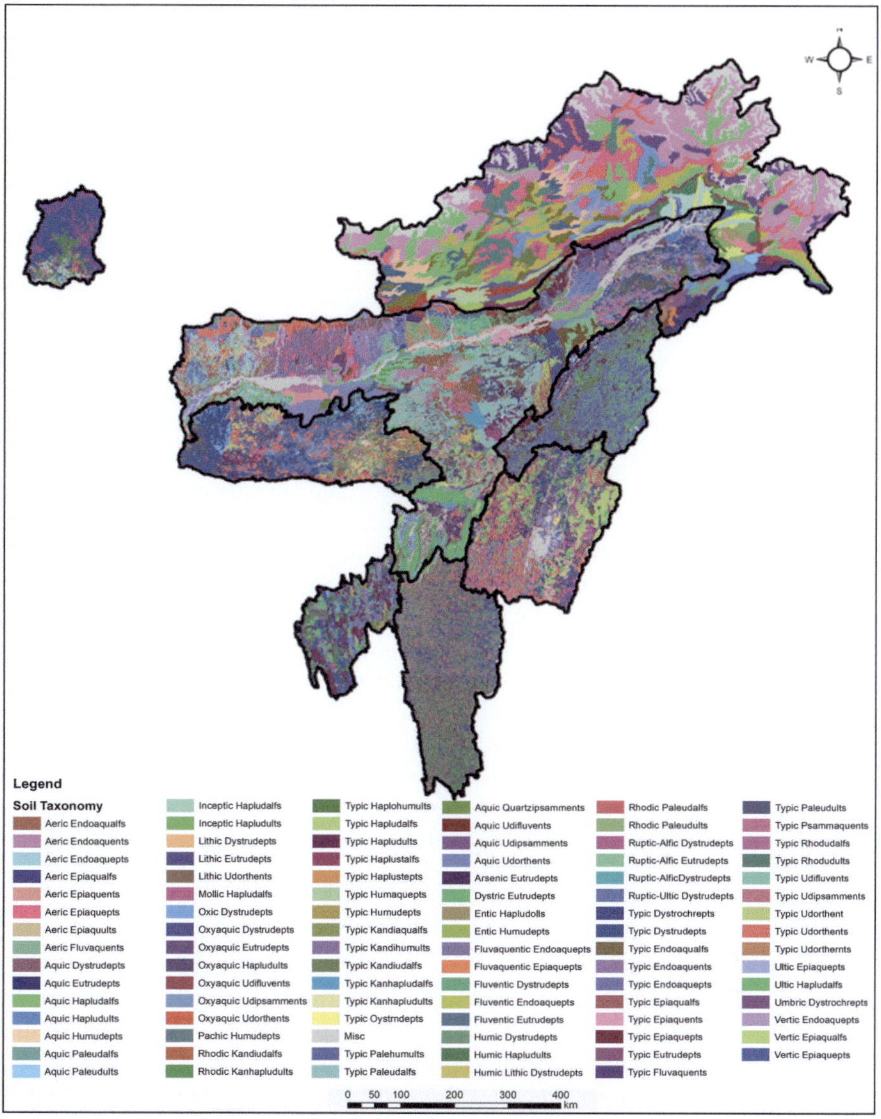

Fig. 3.3 Dominant of soil sub groups of North Eastern Region of India

contributing to the growth of a digital agriculture ecosystem. The soil database generated for Meghalaya state will also be part of NOSUIS.

The National Soil Mapping programme will be carried out by following the standard procedure using satellite data interpretation, soil survey, profile study and soil analysis. Soils will be characterized, classified and mapped based on global standards. The process of detailed soil mapping involves several interconnected steps, including the integration and interpretation of satellite imagery and thematic data, development of alphanumeric legends, soil profile studies, and collection of soil samples. These samples undergo laboratory analysis for taxonomic classification, followed by delineation of soil boundaries, interpretative groupings such as Land Capability Classification (LCC), and finally, the creation of a digital database (SLUSI 2024).

3.3 Soil Fertility Mapping

Maintaining soil fertility is fundamental for ensuring long-term agricultural productivity and environmental sustainability. Efficient soil nutrient management requires a precise understanding of soil health, enabling farmers and policymakers to make informed decisions regarding fertilizer application and land use planning. Advances in soil fertility assessment have significantly improved with the integration of modern technologies, enhancing accuracy and efficiency in monitoring soil conditions. Soil fertility maps play a vital role in promoting sustainable soil and land management by helping assess the current status of soil health. Various methods are used to evaluate soil fertility, such as soil testing, plant tissue analysis, and visual inspection. Among these, soil testing is the most widely adopted and considered the most reliable approach, as it measures the nutrient levels in the soil and helps determine the appropriate fertilizer requirements for achieving optimal crop yields (Khadka et al. 2019). Understanding the spatial variation in soil fertility across agricultural fields was challenging until the advent of technologies like GPS and GIS made it more feasible (Mishra et al. 2013) In general surface soil samples from grid points at regular interval (based on the scale of mapping) are to be collected for preparation different thematic maps (pH, EC, organic carbon, available macro and micro nutrients).

The Soil Health Card (SHC) scheme is a flagship initiative of the Government of India, implemented by the Department of Agriculture & Farmers Welfare under the Ministry of Agriculture and Farmers Welfare. The SHC is a printed report that provides information on the nutrient status of soil based on 12 key parameters. These include pH, electrical conductivity (EC), organic carbon (OC), available nitrogen (N), phosphorus (P), potassium (K), sulphur (S), zinc (Zn), iron (Fe), manganese (Mn), copper (Cu), and boron (B). Under this scheme, surface soil samples are collected from fields. The exact location of each sample is recorded using Global Positioning System (GPS) technology (Bordoloi and Das 2017).these location specific soil analysis data from the SHC Portal have been used for generation of soil fertility maps of Meghalaya, Sikkim, Mizoram and 4 districts of Arunachal Pradesh (Das et al. 2022; NESAC 2022, 2023a, 2023b, 2024). These location specific informations have been interpolated and generated soil fertility maps at block/district level using GIS software (Table 3.1 and Fig. 3.4). The soils in most parts of the study area range from slightly to moderately acidic. Soil organic carbon and available nitrogen levels vary from high to medium. The availability of macronutrients such as phosphorus (P) and potassium (K) is generally medium to low across the region. More than 85% of the area has adequate levels of micronutrients, including zinc (Zn), iron (Fe), copper (Cu), and manganese (Mn).

Table 3.1 Spatial distribution of soil fertility classes in selected states of NER

Parameter	Class	State (% of TGA)			
		Arunachal	Meghalaya	Mizoram	Sikkim
pH	Strongly acidic (<4.5)	3.70	0.09	0.34	0.01
	Moderately acidic (4.5–5.5)	78.69	27.25	35.56	2.83
	Slightly acidic (5.5–6.5)	17.50	69.61	63.93	92.94
	Neutral (6.5–7.5)	0.10	3.05	0.17	4.22
OC (%)	Low (<0.50)	0.00	0.26	31.69	0.03
	Medium (0.50–0.75)	1.09	11.52	55.89	0.16
	High (>0.75)	98.91	88.22	12.44	99.81
P (kg/ha)	Low (<10)	41.09	18.73	4.79	0.06
	Medium (10–25)	56.36	69.89	92.59	41.08
	High (>25)	2.56	11.38	2.64	58.86
K (kg/ha)	Low (<120)	17.06	47.35	0.51	0.13
	Medium (120–280)	64.22	45.54	94.97	73.38
	High (>280)	18.72	7.11	4.54	26.49

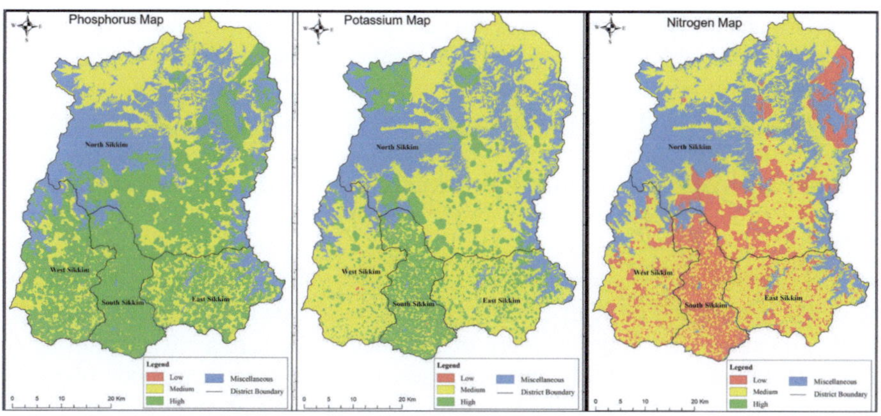

Fig. 3.4 Soil fertility map of Sikkim

3.4 Land Degradation Mapping

Land degradation is a critical environmental concern that affects soil productivity, water resources, and biodiversity, ultimately impacting agricultural sustainability and livelihoods. The NER is home to a rich diversity of plant and animal species. As vegetation is depleting, the habitat for numerous flora and fauna are diminishing, leading to potential extinction or migration of these species.

The region has steep slopes and fragile soils. Deforestation, improper agricultural practices, and overgrazing by livestock contribute to severe soil erosion. This erodes the topsoil, reducing its fertility and leading to less productive agricultural land. Soil erosion also leads to the silting of rivers and streams, causing floods in the downstream. Land degradation disrupts water cycles, reducing groundwater recharge and leading to water scarcity. This is especially concerning in the NER, where communities rely on both surface and groundwater for their daily needs. Reduced water availability impacts agriculture, drinking water supply, and sanitation. The NER is particularly vulnerable to the impacts of climate change. Land degradation, exacerbated by erratic weather patterns is making the region even more vulnerable to environmental stress, such as agricultural droughts, floods, and temperature fluctuations. The Himalayas are vital for maintaining both climatic and hydrological balance. The region's forests, wetlands, and grasslands offer essential ecosystem services. These include carbon sequestration, water purification, and soil stabilization.

Deforestation due to logging, agricultural expansion, and infrastructure development leads to land degradation. Unsustainable agricultural practices such as overcultivation, use of chemical fertilizers, and improper irrigation methods degrade soil quality. Increased human settlement and population growth in the region lead to greater demand for land and resources; further stressing the environment. The shifting climate patterns with erratic rainfall and extreme weather events are increasing land degradation in the NEHR. Efforts to combat land degradation in the NEHR require integrated land management practices, reforestation, sustainable agricultural methods, and community involvement in conservation efforts. Additionally, strengthening policies related to environmental protection and promoting sustainable development will be crucial in mitigating these challenges.

The United Nations Convention to Combat Desertification (UNCCD) recognizes land degradation as a major environmental issue. It has been identified as one of the most pressing and difficult challenges to address. One of the key requirements in this direction is creating inventory of land degradation, periodic monitoring of land degradation, vulnerability and risk assessment and preparing action plans for combating land degradation. India is a signatory to the United Nations Convention to Combat Desertification (UNCCD). The country has pledged to achieve land degradation neutrality by 2030. The Ministry of Environment, Forest and Climate Change (MoEF&CC) serves as India's representative in the UNCCD.

The Space Applications Centre (SAC), ISRO has been monitoring land degradation in the country for more than two decades in collaboration with NESAC, State Remote Sensing Application Centres (SRSACs), Indian Council of Agricultural

Fig. 3.5 Land degradation status of NE states during 2003–05, 2011–13 and 2018–19

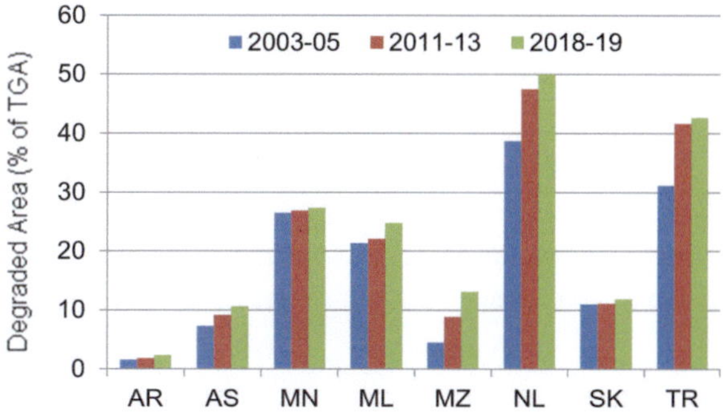

Research (ICAR) Institutes and Academic Institutions. Land degradation mapping has been carried out for the period of 2003–05, 2011–13 and 2018–19 (please refer to Fig. 3.5). Multispectral and multi-temporal satellite imagery of IRS, AWiFS data have been visually interpreted and digitized in GIS environment to prepare land degradation map on 1:5,00,000 (1:500 K) scale (SAC 2021).

An area of 29,248 sq.km have been found to be degraded in the NEHR during 2003–05 representing 17.8% of Total Geographical Area (TGA). The degraded area increased to 34,574 sq. km (21.14% of TGA) and 38,421 sq.km (22.87% of TGA) during 2011–13 and 2018–19, respectively (SAC 2021). Among the 8 states of NER, Nagaland is the most degraded state; 50% of TGA of the state is under degraded land. On the contrary, Arunachal Pradesh is having the lowest area under degraded land (2.4% of TGA).

Land degradation maps prepared at 1:500 K are difficult to use for district level restoration programmes. Therefore, NRSC has initiated land degradation mapping at 1:50 K under Natural Resources Census (NRC) programme of ISRO using temporal Resourcsesat-2 LISS-III data of 2005–06 and 2015–16. NESAC has carried out the project in the NER following standard on-screen visual interpretation keys developed by NRSC for delineating different land degradation classes. The land degradation mapping at 1:50 K addresses 8 major processes and 36 land degradation classes.

The total degraded land in the NER has been mapped as 46,946 sq.km which is 17.91% of TGA of NER during 2005–06. During the ten years (2015–16), there is an overall increase of around 2164 sq.km of land degradation (Table 3.2). Acidification is the most dominant land degradation process followed by water erosion found in the NER. Water logging is found highest in Assam followed by Arunachal Pradesh, Manipur, Meghalaya and Tripura. Glacial are found in Sikkim and Arunachal Pradesh (NRSC 2019) (Table 3.3).

Table 3.2 Spatial distribution of land degradation processes in NER (ha)

Sl. No	Process	Area (sq.km)		% of TGA	
		2005–06	2015–16	2005–06	2015–16
1	Water erosion	12,424	14,246	4.74	5.43
2	Water logging	2779	3056	1.06	1.17
3	Acidification	28,129	28,089	10.73	10.71
4	Glacial	119	119	0.05	0.05
5	Anthropogenic	144	210	0.06	0.08
6	Others	3351	3390	1.28	1.29
	Total	46,946	49,110	17.91	18.73

Table 3.3 State-wise area under land degradation in NER (ha) during 2015–16

Sl No	State	Acidification	Water erosion	Water logging	Anthropogenic	Others	Glacial
1	Arunachal Pradesh	336,881	442,920	35,033	0	167,442	585
2	Assam	73,885	359,168	268,671	11,703	27,990	0
3	Nagaland	743,388	35,843	0	888	0	0
4	Manipur	631,463	221,931	1343	0	261	0
5	Mizoram	469,594	266,577	0	0	0	0
6	Tripura	94,906	6076	298	1050	392	0
7	Meghalaya	458,765	86,767	304	7387	83,264	0
8	Sikkim	0	5362	0	0	59,657	11,271
	Total	2,808,882	1,424,644	305,649	21,028	339,006	11,856

3.5 Land Evaluation for Sustainable Crop Planning

Land evaluation involves determining how suitable a piece of land is for particular uses. This assessment is based on the land's inherent characteristics and how different management practices may affect its performance. It involves analyzing a range of factors, including soil properties, climate, topography, water availability, and ecological conditions, to determine how well land can support activities such as agriculture, forestry, urban development, conservation, and more. The goal of land evaluation is to provide informed recommendations for land use that optimize land productivity, minimize environmental degradation, and align with socio-economic objectives.

The history of land evaluation is a long and evolving process, rooted in the need to understand the capacity and suitability of land for agricultural, environmental, and urban uses. Land evaluation can trace its origins back to ancient civilizations, such as Mesopotamia, Egypt, and the Indus Valley, where societies began to assess the land for agriculture, settlement, and resource management. Early civilizations developed rudimentary methods to evaluate soil fertility, topography, and water availability for crop production. However, these early systems were often based on trial and error rather than scientific principles. During the Middle Ages, land evaluation practices remained somewhat primitive. However, advancements in agricultural techniques were made, and land was still evaluated based on its productivity and the ease of cultivation. The use of the land for grazing, farming, and settlement continued to be critical for societies, though there was limited scientific approach to land suitability.

With the Renaissance, the focus began shifting towards a more scientific approach to land use. The development of agriculture in Europe, along with improvements in cartography, started to provide new insights into the characteristics of different types of land. During this time, the first formalized attempts at mapping land use suitability were made, though the methods were still rudimentary. The nineteenth century was a pivotal time for the field of land evaluation, especially with the onset of the Agricultural Revolution and the growth of soil science. The advent of better tools for measurement and mapping, along with the discovery of key agricultural principles, led to more systematic approaches to land evaluation. Early soil scientists such as Justus von Liebig and others laid the foundations for understanding soil chemistry and fertility, which directly influenced the evaluation of land for agricultural purposes. The twentieth century saw the development of more formalized and comprehensive land evaluation systems. Various methods of classification and evaluation were formalized, and land evaluation became more scientific.

In the early twentieth century, Charles Kellogg developed the concept of LCC, which aimed to categorize land based on its potential for agriculture. The LCC system became a key tool for land evaluation, focusing on factors like soil type, erosion risk, and climate. With the advent of new technologies such as satellite imagery, GIS, remote sensing, and computer modeling, land evaluation became more data-driven and precise. These technologies allowed for more accurate mapping of land attributes and helped to predict land suitability for various uses with greater accuracy. Today, land evaluation continues to evolve, increasingly focusing on sustainability, climate change, and integrated land-use planning. Modern approaches take into account a wide range of factors, including environmental impact, biodiversity, and the long-term productivity of land for various uses. The global focus on food security, climate change adaptation, and urbanization has led to international collaboration on land evaluation, including the creation of global land-use systems and database. Institutions like the Food and Agriculture Organization (FAO) have developed systems to assess land suitability for agricultural and non-agricultural purposes on a global scale.

The history of land evaluation in India has evolved over centuries, influenced by the country's agricultural traditions, technological advancements, and socio-economic changes. From early agricultural practices to modern scientific land evaluation systems, India's approach to land evaluation has been shaped by the need for effective land management, agricultural productivity, and sustainable land use. In ancient India, agriculture was the primary source of livelihood, and land evaluation was done through traditional methods. Farmers would evaluate land based on fertility, water availability, topography, and climate. However, this evaluation was based on experience and local knowledge rather than scientific principles. During the Mauryan Empire (322–185 BCE), King Ashoka is believed to have had a structured approach to land management, focusing on the classification of land and its use for agriculture, conservation, and settlements. Early forms of land taxation and management, though rudimentary, were also in place. During the Mughal period (1526–1857), land revenue systems were established, notably by Akbar. The land was classified into different categories based on its fertility, and land evaluation was tied to the assessment of land revenue, which was based on crop yields and the quality of soil.

The British colonial administration in India introduced more formal systems of land evaluation as part of their revenue collection systems. Key systems like the "Zamindari System" and "Ryotwari System" were implemented, where land was classified based on its productivity, and taxes were levied accordingly. In the nineteenth century, the British government conducted surveys and mapping of land to better understand the agricultural potential and land use. Towards the end of the nineteenth century, scientific agricultural research began to take root in India. The establishment of institutions such as the Indian Agricultural Research Institute (IARI) in 1905 at Pusa, Bihar marked the beginning of more systematic studies of soil types, irrigation, and land productivity. After India gained independence in 1947, the need for systematic land evaluation became more pronounced as the country focused on increasing agricultural productivity and addressing food security. Land use and classification systems were developed to determine which crops should be grown in specific regions. The Soil Survey of India (SSI), established in 1956, played a crucial role in mapping soil types across the country. It conducted detailed surveys to evaluate soil characteristics such as texture, fertility, and suitability for different crops. These surveys laid the foundation for modern land evaluation practices in India. The concept of LCC, developed in the West, was adopted by India during 1970s to 1980s to categorize land based on its suitability for agriculture. This system helped in determining the limitations of land for different crops and the need for soil conservation and sustainable practices. During the 1980s and 1990s, Land Use Planning (LUP) gained importance, with a shift towards balancing agricultural, industrial, urban, and environmental uses. In the 1990s and 2000s, remote sensing technology and GIS revolutionized land evaluation in India. These technologies allowed for more accurate mapping and analysis of land resources, enabling the identification of land suitability for various uses, such as agriculture, urban development, forestry, and conservation.

Effective crop planning requires a thorough understanding of land characteristics to ensure optimal agricultural productivity. Land evaluation serves as a crucial tool in determining the suitability of different areas for specific crops, helping to maximize yields while maintaining soil health and sustainability. By assessing various environmental factors, this process enables informed decision-making, reducing the risk of crop failure and land degradation (Singha and Swain 2016). The suitability of a crop largely depends on various soil characteristics, such as depth, texture, and drainage. These factors are further influenced by the local climate and topography. The FAO (1976) land evaluation framework recommends classifying land into several categories: Orders, Classes, Sub-classes, and Units. Soil and site characteristics are assessed based on degrees of limitation, represented by a scale of 0 to 4. The final evaluation of land suitability is determined by the number and severity of these limitations. The detailed methodology is illustrated in Fig. 3.6. Suitable areas for expansion of important agricultural crops (viz. Rice, Maize, Rapeseed/Mustard, Oil palm etc.), horticultural crops (viz. Apple, Arecanut, Assam Lemon, Banana, Cabbage, Cashew nut, Dragon fruit, Ginger, Grape, Jack fruit, Kiwi fruit, Large cardamom, Marigold, Mousambi, Naga king chilli, Orange, Papaya, Passion fruit, Pineapple, Potato, Tomato, turmeric and walnut) and sericultural host plants (Mulberry, Eri, Muga and Tasar) have been studied adequately (Handique et al. 2016, 2022b) for Oil Palm in Mizoram were identified through soil site suitability evaluation.

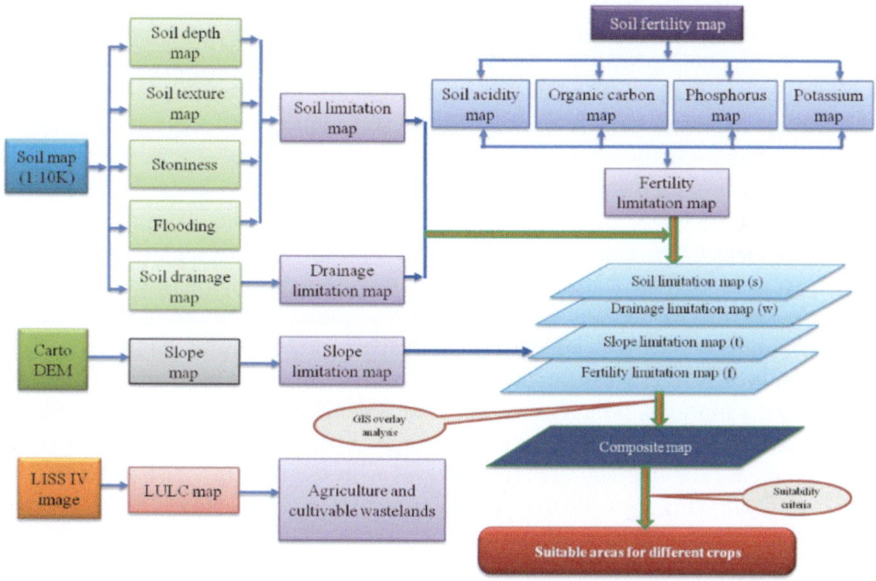

Fig. 3.6 Flowchart of methodology of soil site suitability evaluation for crops

3.6 Application of Advanced Remote Sensing Techniques in Soil Studies

3.6.1 Microwave Remote Sensing Applications

Microwave remote sensing is a valuable tool for monitoring and analyzing soil properties due to its ability to penetrate the soil surface and provide valuable information about soil moisture, texture, temperature, and other characteristics. Synthetic Aperture Radar (SAR) systems, such as Sentinel-1, RADARSAT, RISAT etc., are active sensors and are highly effective for estimating soil moisture. Microwave signals can penetrate the soil surface, and their interaction with the soil surface is strongly influenced by the moisture content. Passive sensors, also referred to as radiometers, detect naturally emitted microwave radiation from the Earth's surface. Examples of such sensors include AMSR-E, Soil Moisture and Ocean Salinity (SMOS), and Soil Moisture Active Passive (SMAP). These instruments do not emit their own signals but measure radiation already present. Since, soil moisture affects the emission of microwave radiation, these sensors provide valuable data for assessing soil moisture at large scales. Soil moisture data from microwave remote sensing can also contribute to predicting crop yields by providing insights into moisture availability.

The interaction between microwave signals and soil depends on the soil's texture (e.g., clay, sand, or silt). Radar backscatter can provide insights into soil texture by assessing how the soil surface reflects microwave radiation. Microwaves are particularly sensitive to clay, which have a higher capacity for retaining moisture. SMAP data can indirectly help assess organic matter content by measuring soil moisture and temperature variations, which correlate with organic matter presence. Microwave remote sensing is a useful tool for monitoring soil salinity, particularly in arid and semi-arid areas. The interaction of microwaves with saline soils is distinct from non-saline soils, allowing for effective salinity detection and mapping. Changes in soil compaction, which affect its ability to hold moisture and support plant growth, can also be detected using microwave remote sensing. The radar backscatter response varies depending on the structure and compaction of the soil surface. Microwave remote sensing helps monitor the effects of erosion on soil, particularly through detecting changes in soil surface roughness. Erosion can alter the surface texture and moisture content, which in turn changes the radar backscatter patterns.

3.6.2 Thermal Remote Sensing Applications

Thermal remote sensing is another powerful tool for studying and monitoring soil properties. By detecting the thermal radiation emitted by the Earth's surface, thermal remote sensing provides insights into various soil characteristics, including moisture content, temperature, and soil health. Thermal infrared sensors can detect soil surface temperatures, which are essential for understanding soil conditions. Soil temperature affects various biological processes, such as microbial activity, plant growth, and nutrient cycling. By monitoring soil temperatures using thermal remote sensing, it is possible to study soil's impact on agricultural productivity and environmental changes. Monitoring the soil's thermal properties helps in detecting frost risks. In colder climates prevailing in NER, this application is particularly important for protecting crops during early planting seasons or late harvesting periods. Soil moisture affects the heat capacity and thermal conductivity of soil. Thermal infrared remote sensing can be used to estimate soil moisture indirectly, as drier soils heat up faster and cool down more slowly compared to moist soils. This is particularly useful for monitoring drought conditions and irrigation scheduling. Evapotranspiration (ET), the process of water transfer from soil to the atmosphere via evaporation and transpiration, influences soil moisture levels. Thermal imagery can be used to estimate ET by capturing the soil surface temperature, which is affected by moisture availability. By using thermal infrared imagery to assess the soil and plant's temperature, it is possible to identify areas that require irrigation and areas that are well-watered. This allows farmers to optimize irrigation systems, reduce water wastage, and improve crop yields.

3.6.3 Hyperspectral Remote Sensing Applications

Hyperspectral remote sensing is a powerful technique that captures detailed spectral information across a wide range of wavelengths, providing high-resolution data about soil properties and conditions. With its ability to detect subtle differences in material composition, hyperspectral sensors can be used to monitor a variety of soil characteristics, including composition, moisture, organic matter content, and even contamination levels. Hyperspectral remote sensing can detect specific absorption features in the electromagnetic spectrum, allowing for the identification of minerals in the soil. This is especially useful in detecting clay minerals, carbonates, and iron oxides, which are critical for understanding soil properties and fertility. For example, specific absorption bands in the SWIR region are sensitive to minerals such as Kaolinite or Illite, which are important for soil texture and classification. Although microwave and thermal sensors are commonly used for soil moisture, hyperspectral remote sensing can also provide information about moisture content. Water has unique spectral characteristics, particularly in the NIR and SWIR regions. Changes in soil moisture affect the reflectance spectra, and hyperspectral sensors can capture these variations to estimate soil moisture content.

Organic matter, a key component of soil fertility, has distinct spectral signatures, particularly in the SWIR range. Hyperspectral sensors can detect organic carbon content in soils, which is important for assessing soil health, fertility, and potential for carbon sequestration. This is crucial for land management, agriculture, and environmental monitoring. Soil salinity is a significant issue, particularly in arid and semi-arid regions. Hyperspectral sensors can help identify the presence of salts in the soil by detecting the specific spectral signatures associated with salinity, such as the absorption features related to sodium and other salts. Hyperspectral remote sensing is valuable for precision agriculture as it allows monitoring of soil properties at a high spectral as well as spatial resolution. Soil microbial communities play a crucial role in soil health and fertility. Hyperspectral remote sensing can help monitor soil health indirectly by detecting spectral changes linked to soil microbial activity, such as organic matter breakdown or nutrient cycling. This is important for understanding the ecological processes occurring within soils.

Hyperspectral remote sensing can help detect soil contamination by heavy metals such as lead (Pb), mercury (Hg), or cadmium (Cd). Certain heavy metals have distinct spectral features in the infrared range, and hyperspectral sensors can identify contaminated areas by analyzing these spectral signatures. Similarly, in areas affected by industrial activities, hyperspectral sensors can detect specific pollutants, such as petroleum hydrocarbons, and map contaminated regions. This is particularly useful for environmental monitoring and remediation efforts.

3.6.3.1 Spectral Signature of Soil

In hyperspectral remote sensing, the spectral characteristics of soils are determined by the interaction of light with the soil surface, including absorption and reflection across different wavelengths, particularly in the visible (400–700 nm), NIR (700–1400) nm, and SWIR (1400–2500 nm) regions. The factors affecting spectral signature of soil include sand, silt, clay, organic matters (peat, muck), salinity, alkalinity, wetness, mineral compositions (e.g., iron oxide, calcium carbonate) etc. Visible spectra are Influenced by soil color, which can be a key indicator of mineral and organic matter content. NIR spectra are sensitive to soil moisture content and organic matter. SWIR provides critical information on minerals, organic matter, and soil moisture. Water Absorption Bands located at around 1400 nm and 1900 nm in the SWIR range, indicate presence of soil moisture. Specific absorption features in and around 2200 nm in the SWIR are related to clay minerals such as kaolinite, illite, and montmorillonite. Spectral features in the visible and NIR regions are influenced by the presence of iron oxides (such as hematite or goethite), giving rise to soil color and reflectance characteristics. Organic matter affects the reflectance in the NIR and SWIR regions, with specific absorption features related to its chemical composition (Goswami 2020a).

A spectral library of soil is a collection of spectral reflectance data that represents the unique spectral characteristics of various soil types under different conditions. These libraries are essential tools for remote sensing applications as they provide reference spectra for identifying and classifying soils based on their spectral signatures. Spectral library is used to create soil maps by comparing the reflectance of soils in the field with the reference spectra in the library. Spectral libraries help in estimating soil properties like moisture content, texture, organic matter, and mineral composition by comparing the observed spectra from remote sensing data with known soil spectral signatures. In precision agriculture, soil spectral libraries help monitor soil variability across fields (nutrient availability, moisture availability etc.) and thus optimize management practices like irrigation, fertilization, and crop rotation to enhance productivity and sustainability.

The United States Department of Agriculture (USDA) has created a spectral library of soils that includes spectral data for different soil types, moisture levels, and conditions. This library is used in various research and practical applications, especially in precision agriculture. The ASTER (Advanced Spaceborne Thermal Emission and Reflection Radiometer) spectral library contains soil reflectance data from various regions worldwide, including spectral signatures of different soil minerals, moisture contents, and textures. The Rochester Institute of Technology (RIT) has developed a comprehensive soil spectral library that is widely used for academic and research purposes. This library includes both field and laboratory measurements. The United States Geological Survey (USGS) has a broad spectral library that includes data on various minerals and soil types, which can be used for geological and agricultural studies.

NESAC has carried out a pilot study to generate the spectral signature of soils of Ri Bhoi, Meghalaya across 3 soil orders (Alfisols, Inceptisols and Ultisols) and 7 LULC classes (kharif crop, double crop, abandoned jhum, current jhum deciduous forest, evergreen forest and wastelands). The soil spectra were collected in-situ and in the laboratory using spectroradiometer. Soil organic carbon, silt and clay content of soil were driving force in influencing the spectral reflectance of laboratory soil spectra irrespective of soil orders and LULC (Goswami 2020b). The spectral signature of soils recorded in the laboratory have been presented in Fig. 3.7.

Spectral indices are key tools in remote sensing for analyzing various properties of soil. These indices utilize the reflectance data from different wavelengths (e.g., visible, NIR and SWIR) to derive meaningful information about soil characteristics. Spectral indices are powerful tools to assess various soil properties (such as moisture, texture, organic content, and clay content) remotely. The choice of index depends on the specific property being studied, the region's vegetation density, and the available spectral bands. By applying these indices, one can obtain insights into soil health and manage agricultural, environmental, and land use practices more effectively. Some of spectral indices used in soil studies are shown in Table 3.4.

3.7 Conclusion

Soil resources form the foundation of agricultural productivity, ecological balance, and sustainable land management in the NER. Given the region's diverse topography, varied climatic conditions, and unique soil characteristics, systematic soil resource assessment is crucial for effective planning and management. Traditional soil mapping methods, while valuable, have limitations in terms of spatial coverage, accuracy, and efficiency. The advent of advanced geospatial technology has significantly improved soil mapping by enabling large-scale, cost-effective, and time-efficient data acquisition.

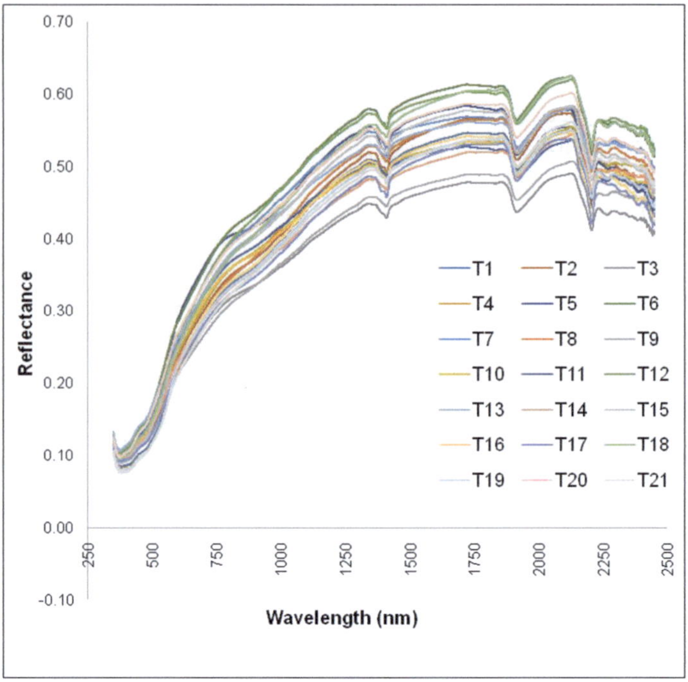

Fig. 3.7 Spectral signature of soils of Ri Bhoi, Meghalaya

Remote sensing, GIS, and allied geospatial technologies have transformed soil resource assessment into a more scientific, scalable, and data-driven process. Satellite-based soil mapping, particularly with multispectral and hyperspectral sensors, has enhanced the ability to classify soils, assess fertility, and monitor degradation trends with greater precision. The integration of high-resolution imagery, digital elevation models (DEMs), and machine learning algorithms has further refined soil classification, allowing for predictive modeling of soil properties such as moisture content, organic matter, and erosion risk. These advancements facilitate the generation of thematic maps that serve as critical decision-support tools for agriculture, watershed management, forestry, and environmental conservation.

In NER, large-scale soil mapping initiatives, such as reconnaissance and semi-detailed soil surveys at 1:1 million to 1:50K scales, have provided valuable insights into soil distribution and suitability. Studies have highlighted that the region is predominantly covered by Inceptisols, Entisols, and Ultisols, with acidic soil conditions affecting agricultural productivity. Mapping at finer scales has enabled the identification of soil constraints, such as erosion-prone areas, waterlogged zones, and degraded lands, aiding in targeted interventions like soil conservation, organic farming, and precision agriculture.

Land evaluation has evolved from early, simplistic methods to highly sophisticated, multi-disciplinary approaches that account for a wide range of environmental,

Table 3.4 Spectral indices used in soil studies

Spectral indices	Equation	Reference
Simple Ratio Index (SRI)	ρ_{NIR}/ρ_{RED}	Rouse et al. (1974)
Relative Vigour Index (RVI)	$\rho_{830\text{ nm}}/\rho_{660\text{ nm}}$	Jordan (1969)
Normalized Difference Vegetation Index (NDVI)	$(\rho_{830\text{ nm}} - \rho_{660\text{ nm}})/(\rho_{830\text{ nm}} + \rho_{660\text{ nm}})$	Rouse et al. (1974)
Renormalized Difference Vegetation Index (RDVI)	$(\rho_{830\text{ nm}} - \rho_{660\text{ nm}})/\sqrt{(\rho_{830\text{ nm}} + \rho_{660\text{ nm}})}$	Rougean and Breon (1995)
Modified Simple Ratio Index (MSRI)	$[(\rho_{830\text{ nm}}/\rho_{660\text{ nm}}) - 1]/\sqrt{[(\rho_{830\text{ nm}}/\rho_{660\text{ nm}}) + 1]}$	Chen (1996)
Soil Adjusted Vegetation Index (SAVI)	$1.5(\rho_{830\text{ nm}} - \rho_{660\text{ nm}})/(\rho_{830\text{ nm}} + \rho_{670\text{ nm}} + 0.5)$	Huete (1988)
Modified Soil Adjusted Vegetation Index (MSAVI)	$2\rho_{830\text{ nm}} + 1 - \sqrt{[\{(2\rho_{830\text{ nm}} + 1)^2 - 8(\rho_{830\text{ nm}} - \rho_{670\text{ nm}})\}/2]}$	Qi et al. (1994)
Photochemical Reflectance Index (PRI)	$(\rho_{531\text{ nm}} - \rho_{570\text{ nm}})/(\rho_{531\text{ nm}} + \rho_{570\text{ nm}})$	Gamon et al. (1997)
Normalized Phaeophytinization Index (NPQI)	$(\rho_{415\text{ nm}} - \rho_{435\text{ nm}})/(\rho_{415\text{ nm}} + \rho_{435\text{ nm}})$	Penuelas et al. (1995a)
Green Ratio Index (GRI)	$\rho_{830\text{ nm}}/\rho_{550\text{ nm}}$	Inoue et al. (1993)
Water Index (WI)	$\rho_{900\text{ nm}}/\rho_{970\text{ nm}}$	Penuelas et al. (1997)
Structure Independent Pigment Index (SIPI)	$(\rho_{800\text{ nm}} - \rho_{445\text{ nm}})/(\rho_{800\text{ nm}} - \rho_{668\text{ nm}})$	Penuelas et al. (1995b)
Short Wave infrared Water Index (SWWI)	$\rho_{1650\text{ nm}}/\rho_{850\text{ nm}}$	Inoue et al. (1993)
Normalized Difference Water Index (NDWI)	$(\rho_{860\text{ nm}} - \rho_{1240\text{ nm}})/(\rho_{860\text{ nm}} + \rho_{1240\text{ nm}})$	Gao (1996)

economic, and social factors. The field continues to advance with new technologies and methodologies, particularly as the world faces challenges like climate change, population growth, and urban expansion.

Microwave remote sensing offers a non-invasive, cost-effective, and large-scale solution to soil monitoring. By utilizing both active and passive microwave sensors, it is possible to gather a wealth of information on soil moisture, texture, organic matter, and more. Similarly, thermal remote sensing plays a vital role in understanding and managing soil properties, particularly in relation to temperature, moisture, and organic content. It aids in optimizing irrigation, managing soil health, detecting soil stress, and supporting sustainable farming practices.

Hyperspectral remote sensing offers a wealth of applications for soil monitoring due to its ability to capture fine spectral details. This capability allows for precise analysis of soil composition, moisture content, fertility, contamination, and many more. With continuous advancements in sensor technology and data processing, hyperspectral remote sensing will likely play an even greater role in soil research and management in the future. The spectral library of soil provides a valuable reference for interpreting and analyzing soil characteristics using remote sensing techniques, particularly hyperspectral imagery. It aids in soil classification, property estimation, and mapping, helping researchers, land managers, and farmers better understand soil health, fertility, and environmental conditions.

The use of advanced remote sensing techniques, coupled with field validation and laboratory analysis, ensures accurate soil classification and sustainable land-use planning. Future research should focus on integrating UAV-based hyperspectral imaging, AI-driven soil health prediction models, and real-time soil monitoring systems to further enhance soil resource management.

Acknowledgements The authors sincerely acknowledge the Soil and Land Use Survey of India (SLUSI) for their collaboration in soil mapping and field survey activities, and for their leadership in the National Soil Mapping Programme (NSMP). We gratefully recognize the contributions of the National Remote Sensing Centre (NRSC), Hyderabad, for access to standardized soil and land degradation datasets and for their role in developing the Soil Resource Atlas and Land Degradation Atlas of India. We also acknowledge the National Bureau of Soil Survey and Land Use Planning (NBSS&LUP) for their foundational work in systematic soil resource mapping across India, which informed much of the regional classification used in this study. Appreciation is extended to the Space Applications Centre (SAC), ISRO, Ahmedabad, for coordinating national-level land degradation monitoring and remote sensing standardization. We further thank the Department of Agriculture and Horticulture, Government of Meghalaya, for field-level logistical assistance, and the Ministry of Agriculture and Farmers Welfare, Government of India, for access to Soil Health Card data used in fertility mapping.

References

S. Baruah, *In the Name of the Nation: India and its Northeast.* (Stanford University Press, 2020)

J. Bordoloi, A.K. Das, *Impact of soil health card scheme on production, productivity and soil health in Assam.* Study No.148. Agro-Economic Research Centre for North-East India, Assam Agricultural University, Jorhat-Assam (2017)

J. Bouma, L. Montanarella, Facing policy challenges with inter-and transdisciplinary soil research focused on the UN sustainable development goals. Soil **2**, 135 (2016)

E. Bridges, J. Van Baren, Soil: an overlooked, undervalued and vital part of the human environment. Environmentalist **17**, 15–20 (1997)

S. Chatterjee, A. Saikia, P. Dutta, D. Ghosh, G. Pangging, A.K. Goswami, *Biodiversity significance of North east India.* (WWF-India, New Delhi, 2006), pp. 1–71

J. Chen, Evaluation of vegetation indices and modified simple ratio for boreal applications. Canadian J. Remote Sens. **22**, 229–242 (1996)

P.T. Das, T. Lakiang, B. Saikia, Soil fertility mapping using GIS in Meghalaya Plateau. Int. J. Curr. Microbiol. Appl. Sci., **11**(3) (2022)

K.R. Dikshit, J.K. Dikshit, *North-east India: Land, people and economy* (Springer, Netherlands, Dordrecht, 2014)

J.A. Gamon, L. Serrano, J.S. Surfus, The photochemical reflectance index: an optical indicator of photosynthetic radiation-use efficiency across species, functional types, and nutrient levels. Oeclogia **112**, 492–501 (1997)

B.C. Gao, NDWI—a normalized difference water index for remote sensing of vegetation liquid water from space. Remote Sens. Environ. **58**, 257–266 (1996)

C. Goswami, Hyperspectral spectroscopic study of soil properties- a review. Int. J. Plant Soil Sci. **32**(7), 14–25 (2020a)

C. Goswami, Hyperspectral spectroscopic study of soil properties in acid soils of north east India. Ph.D. (Agriculture) thesis (Soil Science & Agricultural Chemistry) submitted to Central Agricultural University, Imphal, India, 2020b

B.K. Handique, C. Goswami, P.T. Das, J. Goswami, P. Jena, F. Dutta, D.K. Jha, S.P. Aggarwal, Space technology support for development of agriculture in the North Eastern Region of India—scope and challenges. Curr. Sci. **123**(8), 975–986 (2022a)

B.K. Handique, C. Goswami, P. Jena, F. Dutta, R. Samiam, I. Nongrum, D. Jha, P.L.N. Raju, C.R. Deka, R. Sarma, K. Sarmah, H.K. Dutta, S. Deb, B. Yari, S. Maibam, P. Thanpuli, R. Sailo, V. Muansangi, D. Katiry, T. Medo, N.P. Sharma, B. Lepcha, A. Roy, B.K. Bhattacharya, J.G. Patel, C.P. Singh, K.R. Manjunath, M.M. Kimothi, S. Mamatha, P. Kumar, P. Tahlani, S.S. Ray, Applications of advanced geospatial technology for expansion of area under horticultural crops in North Eastern region of India. J. Indian Soc. Remote. Sens. **50**(2), 331–345 (2022b)

B.K. Handique, P.T. Das, J. Goswami, C. Goswami, P.S. Singh, C.J. Prabhakar, C.M. Bajpeyi, P.L.N. Raju, Expansion of sericulture in India using geospatial tools and web technology. Curr. Sci. **111**, 1312–1318 (2016)

A.R. Huete, A soil-adjusted vegetation index (SAVI). Remote Sens. Environ. **25**, 295–309 (1988)

Y. Inoue, S. Morinanga, M. Shibayama, Non-destructive estimation of water status of intact crop leaves based on spectral reflectance measurements. Jpn. J. Crop Sci. **62**, 462–469 (1993)

C.F. Jordan, Derivation of leaf area index from quality of light on the forest floor. Ecology **50**, 663–666 (1969)

S.D. Keesstra, J. Bouma, J. Wallinga, P. Tittonell, P. Smith, A. Cerdà, The significance of soils and soil science towards realization of the United Nations sustainable development goals. Soil **2**, 111–128 (2016)

D. Khadka, S. Lamichhane, R. Amgain, S. Joshi, S. Vista, K. Sah, N. Ghimire, Soil fertility assessment and mapping spatial distribution of agricultural research station, Bijayanagar, Jumla, Nepal. Eurasian J. Soil Sci. **8**, 237–248 (2019)

M.L. Manchanda, M. Kudrat, A.K. Tiwari, Soil survey and mapping using remote sensing. Trop. Ecol. **43**, 61–74 (2002)

A. Mishra, D. Das, S. Saren, Preparation of GPS and GIS based soil fertility maps for Khurda District. Odisha. **57**, 11–20 (2013)

NESAC, *Soil fertility mapping of Sikkim using geo-spatial technology*. Project Report No. NESAC-SR-286–2022 (North Eastern Space Applications Centre, Umiam, Meghalaya, 2022)

NESAC, *Soil fertility mapping of Arunachal Pradesh using geo-spatial technology*. Project Report No. NESAC-SR-292–2023 (North Eastern Space Applications Centre, Umiam, Meghalaya, 2023a)

NESAC, *Soil fertility mapping of Arunachal Pradesh using geo-spatial technology*. Project Report No. NESAC-SR-292–2023 (North Eastern Space Applications Centre, Umiam, Meghalaya, 2023b)

NESAC, *Soil fertility mapping of Mizoram using geo-spatial technology*. Project Report No. NESAC-SR-320–2024 (North Eastern Space Applications Centre, Umiam, Meghalaya, 2024)

NRSC, *Status of land degradation in India: 2015–16*. (National Remote Sensing Centre, Hyderabad, India, 2019)

NRSC, *Soil Resources Atlas of India*. NRSC-RSA-LRUMG-SLRAD-Oct2021-TD-0001929-V1.0. (National Remote Sensing Centre, Hyderabad, India, 2021)

J. Penuelas, F. Baret, I. Filella, Semiempirical indices to assess carotenoids/chlorophyll a ratio from leaf spectral reflectance. Photosynthetica **31**, 221–230 (1995a)

J. Penuelas, I. Filella, P. Loret, F. Munoz, M. Vilajeliu, Reflectance assessment of mite effects on apple trees. Int. J. Remote Sens. **16**, 2727–2733 (1995b)

J. Penuelas, J. Pinol, R. Ogaya, I. Filella, Estimation of plant water concentration by the reflectance water index (R900/R700). Int. J. Remote Sens. **18**(13), 2869–2875 (1997)

J. Qi, A.L. Chehbouni, A.R. Huete, Y.H. Kerr, S. Sorooshian, A modified soil adjusted vegetation index (MSAVI), *Remote Sens. Environ.* **48**, 119–126 (1994)

P. Rizal, R. Asokan, A comparative study of tourism industry in north-eastern states of India. IOSR J. Bus. Manag. (IOSR-JBM), **12**(4), 56–62

J.L. Rougean, F.M. Breon, Estimating PAR absorbed by vegetation from bidirectional reflectance measurements. Remote Sens. Environ. **51**, 375–384 (1995)

J. Rouse, R.H. Haas, J.A. Schell, D.W. Deering, Monitoring vegetation systems in the Great Plains with ERTS, in *Proceedings of the 3rd Earth resources technology satellite-1 symposium*, Greenbelt, USA. pp. 301–317, 10–14 December 1973

A. Roy, S.K. Das, A.K. Tripathi, N.U. Singh, H.K. Barman, Biodiversity in North East India and their conservation. Progress. Agric. **15**(2), 182–189 (2015)

A. Roy, D.S. Dkhar, A.K. Tripathi, N.U. Singh, D. Kumar, S.K. Das, A. Debnath, Growth performance of agriculture and allied sectors in the North East India. Econ. Aff. **59**, 783 (2014)

SAC,*Desertification and Land Degradation Atlas of India* (Assessment and analysis of changes over 15 years based on remote sensing). (Space Applications Centre, Ahmedabad, India, 2021)

J.L. Sehgal, Soil resource map of India (1:1 M scale) - meeting the challenge. Soil Surv. Land Eval. **7**(2), 123–125

J. Sehgal, I.P. Abrol, *Soil degradation in India: status and impact* (Oxford and IBH Publishing Co., New Delhi, 1994), p.80

C. Singha, K.C. Swain, Land suitability evaluation criteria for agricultural crop selection: za review. Agric. Rev. **37**(2), 125–132 (2016)

SLUSI, *Concept note on National Soil Mapping Programme (NSMP), 1:10K*. (Soil and Land Use Survey of India, New Delhi, India, 2024)

R. Srivastava, A.K. Saxena, Technique of large-scale soil mapping in basaltic terrain using satellite remote sensing data. Int. J. Remote Sens. **25**, 679–688 (2004)

M. Velayutham, D.K. Mandal, C. Mandal, J. Sehgal, *Agro-ecological subregions of India for development and planning*. (NBSS & LUP Publ. 35, NBSS & LUP, Nagpur, India, 1999), p. 452

Pratibha Thakuria Das Scientist/Engineer-SF at the North Eastern Space Applications Centre (NESAC), brings over 22 years of expertise in leveraging remote sensing and GIS technologies for soil resource mapping, soil fertility assessment, land degradation analysis, crop suitability evaluation, sericulture development, and land-use planning. Her notable contributions include generating soil maps for several districts across Northeast India, creating fertility maps for Meghalaya, Sikkim, Mizoram, and Arunachal Pradesh, integrating soil data for the Soil Resource Atlas of India, and identifying suitable areas for various crops and silkworm food plants. She has authored more than 70 publications in peer-reviewed national and international journals, conferences and book chapters. She has trained several newly recruited ICAR scientists in the application of Remote Sensing (RS) and Geographic Information System (GIS) techniques for soil studies, guided 20 M.Sc. students, and delivered numerous lectures. Her work significantly contributes to sustainable agricultural practices and land management in Northeast India.

Chandan Goswami was born on 11th January 1977 in Nagaon district of Assam, India. He received Bachelor Degree in Agriculture and Master Degree in Soil Science from Assam Agricultural University, Jorhat, Assam, India in the year 1999 and 2001, respectively and Ph.D degree in Soil Science & Agricultural Chemistry from Central Agricultural University, Imphal, Manipur India in the year 2020. Dr. Goswami is a Senior Scientist at the North Eastern Space Applications Centre, Shillong, Meghalaya, India. He has specialized in geospatial technology applications in agriculture, soil and natural resources management. His research focuses on soil spectral analysis, soil resources mapping, and land use planning, significantly contributing to sustainable agricultural development in India's North Eastern Region. His work leverages advanced remote sensing and GIS techniques to address challenges like soil health, and crop productivity in the NER's diverse agro-climatic conditions. Dr. Goswami has authored over 84 publications and having more than 600 citations, reflecting his impactful contributions to the scientific community. A recipient of the National e-Governance Award in 2017 for enhancing geospatial data utilization in planning and development.

Francis Dutta joined NESAC in March 2020 and has been actively engaged in the applications of remote sensing and GIS in agriculture and allied sectors. His contributions span multiple projects, including site suitability analysis for horticultural crops, crop acreage estimation, geo-tagging of sericulture assets in the Northeast region, optimization of crop cutting experiments (CCE) using geospatial technology, and UAV-based remote sensing.

Chapter 4
Geospatial Technology for Climate Resilient Agriculture

Jonali Goswami, Chandan Goswami, Bijoy K. Handique, and Pradesh Jena

Abstract Climate change presents significant challenges to Indian agriculture, particularly in the North Eastern Himalayan Region (NEHR), which is vulnerable to abiotic and biotic stresses affecting both crop yield and food security. This chapter explores the pivotal role of geospatial technology in advancing climate-resilient agriculture (CRA) through satellite-based realtime monitoring of crop health and growth stages, aiding in the early detection of abiotic and biotic stresses. Case studies from NEHR demonstrate applications in mapping flood inundated and resilient croplands, identifying agricultural drought hazard zones, and optimizing crop diversification through analytical models which can empower farmers to adapt and sustain productivity through localized agromet advisories and precise resource planning. The chapter highlights geospatial technology as a transformative tool for enhancing resilience, food security, and sustainability in Indian agriculture under changing climatic conditions.

Keywords Geospatial technology · Climate-Resilient Agriculture (CRA) · Crop stress · Crop intensification · Land use planning

J. Goswami (✉) · C. Goswami · B. K. Handique · P. Jena
North Eastern Space Applications Centre (NESAC), Umiam, Shillong, Meghalaya, India
e-mail: jonali.goswami@nesac.gov.in

C. Goswami
e-mail: chandan.goswami@nesac.gov.in

B. K. Handique
e-mail: bk.handique@nesac.gov.in

P. Jena
e-mail: pradesh.jena@nesac.gov.in

S. P. Aggarwal et al. (eds.), *Remote Sensing of the North Eastern Himalayan Ecosystem*,
https://doi.org/10.1007/978-981-95-3221-6_4

4.1 Introduction

India, with its extensive agricultural sector, is particularly vulnerable to the impacts of climate change. Agriculture serves as the backbone of India's economy. While historically, it was the largest contributor to India's Gross Domestic Product (GDP), economic liberalization has led to more rapid growth in the service and industrial sectors. As of 2023, agriculture remains a key component of the Indian economy, contributing approximately 18% to 20% of the total GDP (Anonymous 2023). Agriculture is crucial despite its relatively smaller contribution compared to the service and industrial sectors, as it employs about 41% of India's workforce, according to the National Sample Survey Office (NSSO). However, climate change is already impacting Indian agriculture, causing reduced crop yields, altering growing seasons, and increasing vulnerability to extreme weather events. These challenges pose significant threats to food security, farmer livelihoods, and rural development.

Climate-resilient agriculture (CRA) is pivotal in supporting India's GDP, particularly by strengthening the agricultural sector, which is highly susceptible to climate change. CRA refers to farming practices and systems designed to adapt and mitigate the impacts of climate change while ensuring long-term agricultural productivity, food security, and sustainability. By implementing CRA practices, India can promote sustainable agricultural growth, minimize risks from climate-related disruptions, and enhance agriculture's contribution to the national economy. Globally, countries are adopting innovative farming techniques, utilizing stress-tolerant crop varieties, and encouraging sustainable land management to ensure that agriculture remains productive and resilient despite changing climates. Practices such as crop diversification, drought-tolerant planting, and water-efficient irrigation methods help farmers mitigate climate risks, including floods, droughts, and extreme weather. The System of Rice Intensification (SRI) is a climate-resilient rice-growing method that enhances yields while reducing water usage and reliance on inputs like fertilizers. This approach involves planting fewer, younger seedlings and employing Alternate Wetting and Drying (AWD) techniques to conserve water (Neate 2013).

The United Nations' Sustainable Development Goals (SDGs) emphasizes the critical role of climate-adaptive farming systems in reaching global sustainability targets, specifically connecting to three key objectives: eliminating hunger (Goal 2), combating climate change (Goal 13), and protecting terrestrial ecosystems (Goal 15). The second goal promotes agricultural methods that are both environmentally sustainable and capable of withstanding climate variability while maintaining food security and ecosystem health. The thirteenth goal emphasizes the necessity of developing climate resilience and adaptation strategies throughout various sectors, with agriculture being a primary focus area. Similarly, Goal 15 emphasizes the sustainable management of terrestrial ecosystems, which includes promoting agricultural practices that restore soil health and biodiversity. By addressing climate resilience within agriculture, CRA directly supports food security, poverty reduction, and environmental sustainability, all of which are essential for achieving the SDGs. Climate-resilient agricultural practices are, therefore, crucial in advancing global efforts toward sustainable development, ensuring that food systems are better prepared to face future climate uncertainties, and contributing to the realization of these critical SDGs (United Nations 2015).

India, recognizing the urgent need to adapt its agricultural practices to changing climate conditions, has implemented several initiatives to build climate resilience in agriculture, reducing poverty, ensuring food security and uplifting sustainable agricultural practices. In this direction, the Government of India introduced the National Action Plan on Climate Change (NAPCC) in 2008 as a strategic framework for climate adaptation and mitigation. One of the significant missions under the NAPCC is the National Mission for Sustainable Agriculture (NMSA), which focuses on enhancing the resilience of agriculture by implementing various sustainable and adaptive practices. The mission supports various key initiatives such as the Paramparagat Krishi Vikas Yojana, Soil Health Card Scheme, Rainfed Area Development Program, the Mission Organic Value Chain Development for the North-East Region, Sub-Mission on Agroforestry and National Bamboo Mission. Furthermore, the Pradhan Mantri Krishi Sinchayee Yojana (PMKSY), introduced in 2015, promotes water-efficient farming through the principle of "more crop per drop," helping to conserve water in climate-sensitive regions. PMKSY focuses on expanding irrigation, promoting sustainable water management, and reducing reliance on monsoon rains, aligning with the goals of sustainable agriculture (Anonymous 2015). The Soil Health Card Scheme, also introduced in 2015, supports climate resilience by encouraging farmers to monitor soil health, use balanced fertilizers, and adopt practices that preserve soil quality. This initiative helps prevent soil degradation, contributing to sustainable land use and aligning with SDG 15 (Anonymous 2015). Another significant initiative is the National Innovations in Climate Resilient Agriculture (NICRA), initiated by the Indian Council of Agricultural Research (ICAR) in 2011. This network project is designed to strengthen the adaptive capacity of Indian agriculture by conducting targeted research and field-level demonstrations. NICRA also emphasizes grassroots knowledge-sharing to increase awareness and adoption of climate-smart agricultural practices (Anonymous 2011). These initiatives underscore India's strategic approach to addressing climate resilience in agriculture as a pathway to sustainable development. By equipping farmers with tools and knowledge to adapt to climate change, these programs improve food security, maintain ecosystem health, and advance the SDGs.

Amid current climate challenges, agriculture is confronted with erratic weather, declining productivity, more frequent pest and disease outbreaks, and biodiversity loss. Intelligent technologies have become essential for mitigating the impacts of extreme climate conditions. Advances in Earth Observation (EO), through satellite data acquisition, analysis, and geospatial technologies, enhance the ability to monitor climate dynamics and their effects on agriculture at both local and global scales. By delivering real-time, large-scale data on soil health, crop status, water resources, and weather patterns, geospatial tools empower farmers, researchers, and policymakers to make informed decisions in response to climate variability and extreme weather events. Remote sensing plays a crucial role in regularly monitoring crop health and growth stages, aiding in the early detection of drought stress or pest infestations. This proactive approach enables timely interventions, reducing crop loss and ensuring food security. The integration of GIS further aids in mapping vulnerable areas, assessing soil and water resources, and identifying regions at risk from climate-induced challenges like floods or droughts, thereby facilitating adaptive planning and resource allocation (Anonymous 2016, 2017). In India, the Department of Agriculture, Cooperation & Farmers Welfare has introduced the National Programme on the Use of Space Technology for Agriculture (NPSTA), an integrated initiative aimed at utilizing geospatial technologies to enhance agricultural planning, monitoring and management. This initiative incorporates several key projects, including National Agricultural Drought Assessment and Monitoring System (NADAMS), Forecasting Agricultural Output using Space, Agro-meteorology and Land-based Observations (FASAL), Pradhan Mantri Kisan Samman Nidhi (KISAN), Coordinated Horticulture Assessment and Management using Geo-informatics (CHAMAN), and crop intensification strategies.

The North East Himalayan Region (NEHR), which includes the states of Assam, Manipur, Meghalaya, Mizoram, Nagaland, Tripura, Sikkim, and Arunachal Pradesh, is known for its rich biodiversity, varied agro-ecological zones, and traditional farming practices.Despite its potential, agriculture in this region faces a unique set of challenges that stem from both natural and socio-economic factors. The difficult mountainous terrain, high rainfall, and fragile ecosystem contribute to significant soil erosion, especially in areas where shifting cultivation (locally known as *jhum*) is still widely practiced. Land holdings in the region are typically small, scattered, and fragmented, making it difficult for farmers to adopt modern agricultural machinery or achieve economies of scale. Moreover, limited access to irrigation facilities, particularly in hilly areas, means that agriculture remains predominantly rainfed and vulnerable to seasonal fluctuations. The region is highly susceptible to climate variability, with increasing instances of erratic rainfall, floods, landslides, and droughts affecting both crop yield and food security. Another major issue is the lack of access to timely and accurate agricultural information. To address these challenges, several technological interventions are being introduced to support and transform agriculture in the NEHR. The ability to analyze and respond to spatial and temporal variations in climate and agricultural conditions makes geospatial technology an indispensable tool for sustainable and resilient agriculture. As the global climate evolves, integrating geospatial technology with traditional farming knowledge, modern innovations, and supportive policies is essential for building resilient agricultural systems in NEHR.

4.2 The Role of Geospatial Technology in Climate-Resilient Agriculture

4.2.1 Monitoring of Climate and Weather for Effective Agrometeorological Advisories

Monitoring weather and climate patterns is fundamental to developing climate-resilient agriculture, which aims to adapt farming systems to increasing climate variability and extreme weather events. Globally, organizations like the World Meteorological Organization (WMO), Food and Agriculture Organization (FAO), and other national meteorological services are actively involved in collecting, analyzing, and disseminating weather and climate information based on satellite data, ground-based observations, and advanced climate models. Initiatives such as the Global Framework for Climate Services (GFCS) and FAO's Agricultural Stress Index System (ASIS) provide crucial support to farmers and policymakers in making climate-informed agricultural decisions. These global efforts help monitor droughts, floods, temperature anomalies, and other stress factors that impact agriculture.

In India, India Meteorological Department (IMD), National Centre for Medium-Range Weather Forecasting (NCMRWF), and Indian Institute of Tropical Meteorology (IITM) under the Ministry of Earth Sciences (MoES), provide a range of weather and climate forecast products tailored for agricultural applications. These forecasts support farmers in planning and managing various agricultural operations by providing timely and reliable weather information across different time scales.

IMD offers nowcasting (0–3 h) and short-range forecasts (1–3 days) that include alerts for rainfall, thunderstorms, lightning, and wind gusts. These are crucial for immediate decisions such as harvesting, spraying, and protecting livestock. These forecasts are supported by Doppler Weather Radars, satellite observations, and high-resolution numerical models. NCMRWF delivers short-range deterministic and ensemble forecasts using its global and regional models (e.g., NCUM-R, NEPS-R), that utilizes probabilistic guidance for predicting rainfall intensity and distribution.

IMD also provides the medium-range (4–7 days), district and block-level weather forecasts under the Gramin Krishi Mausam Sewa (GKMS) scheme. These forecasts include daily rainfall, maximum and minimum temperature, relative humidity, wind speed and direction, and cloud cover. Agro-advisories are generated twice a week (on Tuesdays and Fridays) using this information in coordination with Krishi Vigyan Kendras (KVKs) and agricultural universities. These advisories guide farmers on irrigation, sowing, pest and disease management, and harvesting. NCMRWF supports this system by providing ensemble weather forecasts which is very useful in decision-making.

At the extended range (2–4 weeks), IMD issues Extended Range Forecasts (ERFS) on a weekly basis. These forecasts include predictions of rainfall and temperature anomalies and are especially useful for farmers in planning operations like staggered sowing, fertilizer application, and scheduling of intercultural operations. These forecasts also support state-level agricultural contingency planning. NCMRWF contributes with its own coupled model system that provides additional support to IMD's ERFS products. Similarly, IITM contributes to the ERFS system by generating sub-seasonal forecasts using its own coupled models, which help in predicting wet/dry spells, heatwaves, or cold waves up to four weeks ahead.

For seasonal time scales (1–3 months), IMD provides Long-Range Forecasts (LRF) for the southwest monsoon and monthly forecasts for temperature and rainfall. These are essential for planning agricultural activities at policy and administrative levels, including the selection of crops, determining sowing windows, input planning, and disaster preparedness. NCMRWF's coupled seasonal forecasting system is used in conjunction with other models like the Monsoon Mission Climate Forecast System (MMCFS) by IITM and global products from ECMWF and NOAA. IITM provides decadal climate variability assessments and future climate projections under various climate change scenarios, which support research in climate-resilient agriculture.

In addition to these, IMD and NCMRWF also generate specialized agrometeorological products such as Quantitative Precipitation Forecasts (QPF), soil moisture and temperature forecasts, root-zone moisture predictions, and heat stress indices. IMD acts as the central hub, integrating multiple model forecast from NCMRWF and IITM to prepare weather-based agro-advisories through the GKMS network. These advisories are disseminated to millions of farmers via SMS, mobile apps like Meghdoot, the UMANG platform, All India Radio, and Doordarshan, thereby ensuring grass root level access. Similarly, to address the unique challenges faced by farmers in the North Eastern Region of India, arising from its complex geography, difficult terrain, and diverse climate, an interactive mobile application called AASMAN (Agromet Advisory Services and Management of Farm Resources) has been developed (Fig. 4.1). This app provides location-specific information by integrating spatial databases on soil health cards, IMD's weather forecasts, pest alerts, fertilizer and pesticide calculators, extreme weather warnings (such as lightning and heavy rainfall), and crop-specific advisories. Additionally, a 'Crop Doctor' feature enables farmers to upload images of affected crops along with queries, which are sent to experts for guidance. The responses and trends can be monitored through a connected dashboard system for further analysis and support.

The specialized location and crop specific agrometeorological advisories play a crucial role in climate-resilient agriculture by helping farmers make informed decisions. From protecting crops against adverse weather to optimizing input use, these forecasts and advisories are indispensable tools for enhancing productivity, reducing risk, and building resilience in Indian agriculture.

4.2.2 Crop Growth Monitoring and Production Estimates

Remote sensing technology is instrumental in monitoring crop growth by providing precise and timely data on various crop health indicators. This information facilitates early identification of abiotic and biotic stresses over large areas, reducing

Fig. 4.1 AASMAN mobile app and its features

costs, time, and labor while enabling timely interventions to enhance crop resilience. Since stresses disrupt photosynthesis and alter the morphological and biochemical parameters of plants, affecting their light energy absorption and reflectance spectrum, remote sensing has proven highly effective in detecting various stress types, particularly through hyperspectral narrow-band techniques. Hyperspectral remote sensing, an advanced technology, enables the identification and assessment of stress-tolerant crop genotypes by capturing subtle spectral differences associated with plant stress responses. Hyperspectral sensors collect reflectance data across a broad range of wavelengths, allowing for the detection of unique spectral signatures linked to different stress types such as elevated CO_2 levels, temperature, drought, light, and nutrient deficiencies without disturbing the plants. However, to develop effective management strategies, it is essential to accurately correlate spectral characteristics with stress symptoms through laboratory analyses and in-situ observations. Stress-tolerant crop genotypes often exhibit distinct physiological traits influencing their light absorption and reflectance properties under stress conditions. Hyperspectral imaging allows researchers to detect these subtle variations, as stress-tolerant genotypes typically display different spectral patterns compared to susceptible ones. This technology helps identify traits like strong photosynthetic capacity, morpho-physiological changes, and biochemical responses, which are vital for stress tolerance. Offering a non-destructive, efficient, and precise approach for evaluating and selecting stress-tolerant genotypes, hyperspectral imaging supports improved crop productivity and sustainability in the face of climate variability.

Fig. 4.2 Experiment under Carbon Dioxide Temperature Gradient Tunnel (CTGT)

In a study evaluating the response of local rice genotypes (IET 22,238, Inglongkiri, Banglami, and Bash) to elevated carbon dioxide and temperature using a Carbon Dioxide Temperature Gradient Tunnel (CTGT) system (Fig. 4.2), demonstrated that moderate increases in carbon dioxide (550 ppm) and temperature (+4 °C), as simulated in CTGT II, can have a beneficial effect on rice growth and physiology. Under these conditions, several morpho-physiological traits, including the rate of photosynthesis, improved significantly, particularly in the tolerant genotypes i.e. Inglongkiri and Banglami. This suggests that certain local rice varieties are capable of utilizing elevated CO_2 to enhance photosynthetic efficiency (Acock and Allen 1985), supporting the idea that a moderate rise in climate parameters might not always be detrimental and could be optimized under controlled management. However, under more extreme climate scenarios, represented by CTGT III (750 ppm CO_2 with a 6 °C rise), there was a noticeable decline in plant performance. Despite this, tolerant genotypes showed relatively less reduction in photosynthesis and other physiological traits, indicating their potential tolerance to combined heat and CO_2 stress (Fig. 4.3). Their resilience could be attributed to adaptive responses such as increased proline accumulation, better water retention, anatomical changes, and enhanced defence mechanisms, making them promising candidates for climate-resilient agriculture.

The use of hyperspectral remote sensing emerged as a powerful tool in this study. It enabled the identification of varietal differences, nutrient status, and key biochemical traits non-destructively. Spectral indices such as the PSSRa (Penuelas et al. 1995) and NRI1510 (Herrmann et al. 2010) were found to be informative. PSSRa, which reflects chlorophyll a content, showed a strong correlation with photosynthetic activity (Fig. 4.4), while NRI1510 (Fig. 4.5) was associated with proline content and leaf water potential (Curran 1989; Ferwerda et al. 2007) important indicators of stress tolerance. These indices proved effective in tracking plant physiological responses under elevated CO_2 and temperature conditions.

Overall, the investigation highlights that selecting climate-resilient rice genotypes supported by physiological insights and hyperspectral indices, could play a significant role in adapting agriculture to future climate scenarios. Accurate prediction and monitoring of crop responses under such environments can aid both farmers and researchers in making informed varietal choices and management decisions.

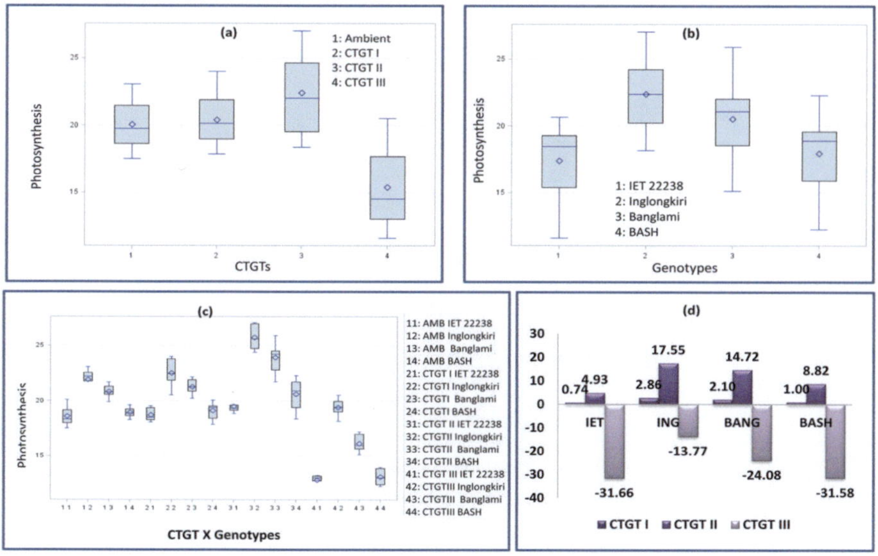

Fig. 4.3 Distribution of photosynthesis (μmol CO_2 m^{-2} s^{-1}) **a** under treatment irrespective of genotype, **b** genotype irrespective of treatment, **c** for Interactive effect of treatment and genotype **d** per cent increase/decrease due to interaction

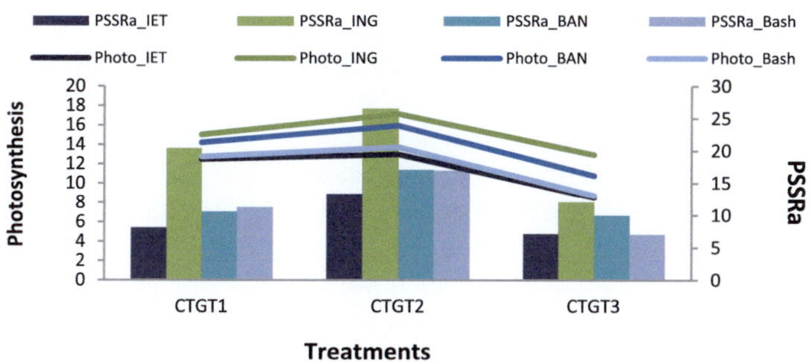

Fig. 4.4 PSSRa and Photosynthesis among genotype under different CTGTs

4.2.3 Agricultural Drought Monitoring and Resilience

Drought is a recurring yet intermittent climate phenomenon. Agricultural drought occurs when soil moisture levels fall below the threshold required to sustain healthy crop growth. Timely detection and early warnings are essential for effectively managing the impacts of such droughts. Satellite technologies play a critical role throughout the drought management process- from tracking the distribution of rainfall over time and space to directly detecting vegetation stress caused by moisture deficits, and facilitating the dissemination of this information. Remote sensing and

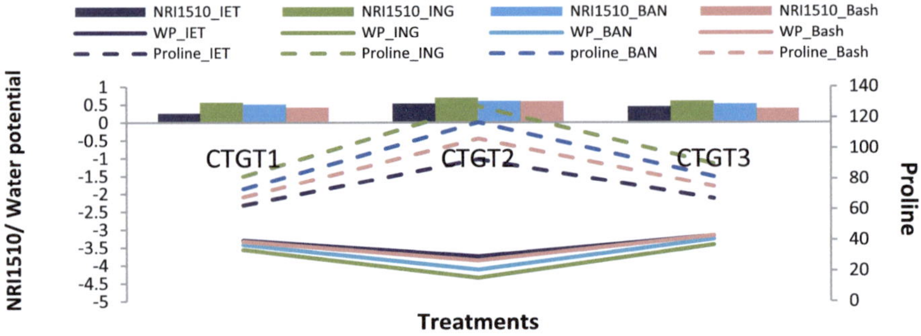

Fig. 4.5 Leaf proline, Water potential and NRI_{1510} variation among genotypes under elevated CO_2 and temperature

GIS models integrate meteorological data to develop early warning systems for impending droughts. Such systems enable farmers to make informed decisions about planting dates, crop varieties, and irrigation strategies. Satellite-based rainfall estimates made using different techniques complement surface data on rainfall measurements. One prominent initiative in this domain is the Global Precipitation Measurement (GPM) mission, a collaborative satellite network that provides advanced global precipitation data. At the heart of this mission is the "Core" satellite, equipped with state-of-the-art radar and radiometer instruments designed to accurately measure rainfall and snowfall, and to standardize data collected from the broader satellite constellation. NASA's Integrated Multi-satellite Retrievals for GPM (IMERG) algorithm processes data from this constellation to deliver near-global precipitation estimates. IMERG is especially useful in data-scarce regions such as oceans and isolated land areas where traditional rain gauges are unavailable.

On the other hand, satellite imagery offers valuable spatial insights into vegetation stress caused by drought, aiding in the evaluation of drought severity and distribution. A variety of vegetation indices have been developed globally to assess plant health, with the Enhanced Vegetation Index (EVI) and Normalized Difference Vegetation Index (NDVI) being among the most widely used. During drought conditions, soil moisture levels drop, leading to noticeable increases in leaf temperatures. Elevated leaf temperature is a strong indicator of plant water stress and can serve as an early warning signal, even before visible symptoms like leaf discoloration appear. This occurs because plants reduce water loss by closing their stomata, which lowers latent heat flux but increases sensible heat flux to maintain energy balance—resulting in warmer leaf surfaces. This physiological change can be effectively monitored using thermal remote sensing techniques. Additionally, the Normalized Difference Water Index (NDWI), (Gao 1996), is a widely adopted remote sensing index for estimating vegetation water content and detecting moisture stress. NDWI is particularly useful for tracking changes in leaf water content, making it a valuable tool for drought surveillance. Since agriculture heavily relies on available soil moisture, it is often the first sector to experience the effects of drought. Prolonged dry spells can quickly exhaust soil water reserves, making efficient water use and management essential for sustaining agriculture in drought-prone and climate-vulnerable regions.

Since 1989, India has been utilizing space-based technologies for agricultural monitoring through the National Agricultural Drought Assessment and Monitoring System (NADAMS). This project provides timely updates on the occurrence, intensity, and duration of agricultural droughts at state, district, and sub-district levels during the kharif season (June to November). However, the states of the NER are not covered under the NADAMS project, primarily because this region is not historically prone to drought. There has been a substantial increase in the frequency and spatial extent of agricultural drought in the NER. Agricultural practices in the NER are mostly rain-fed, and in the absence of proper irrigation systems, even a minor drought can turn into a disaster for poor farmers, directly impacting the economy of the concerned states. Drought events in the NER are more prevalent during the post-monsoon and winter months. To evaluate drought conditions during the Rabi season (December to March), a range of indices were used over a 20-year period. These included vegetation health indices such as NDVI, NDWI, and VCI (Vegetation Condition Index); crop condition indices like LSWI (Land Surface Water Index), SMI (Soil Moisture Index), and SASI (Shortwave Angle Slope Index); and a rainfall-based indicator, SPI (Standardized Precipitation Index). These metrics were employed to characterize drought conditions and assess their severity (Senpakapriya et al. 2025). Ultimately, a Drought Hazard Index (DHI) was generated to map the spatial distribution of agricultural drought risk in the region (Fig. 4.6).

$$DHI = \sum_{i=1}^{n} \sum_{j=1}^{m} \left\{ i * \left(r_j * w_j \right) \right\}$$

where, i = 1 to nrefers to the seven indices used (SPI, NDVI, VCI, NDWI, LSWI, SASI, and SM); j = 1 to m indicates different drought severity levels—mild, moderate, and severe; while r represents the rating assigned to each drought occurrence category (low, medium, high, very high), and www denotes the weight assigned to each severity class.

The agricultural drought hazard map, generated using the Drought Hazard Index (DHI) model, was categorized into four levels of drought risk - medium, high, and very high; based on natural break optimization method. In order to promote climate-resilient agriculture, the findings of this study will assist policymakers in planning for in-season irrigation and reducing the risks related to drought.

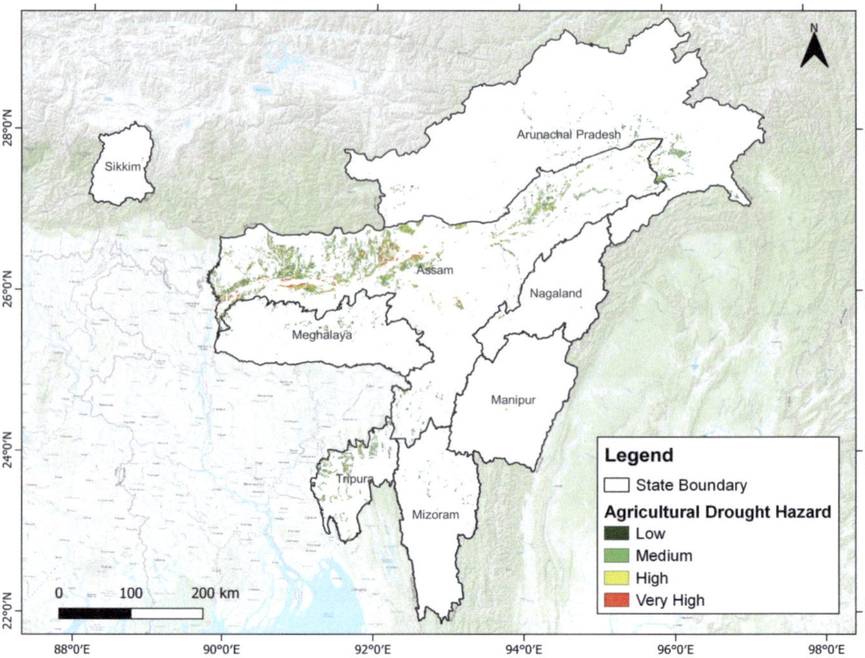

Fig. 4.6 Agricultural drought hazard map of NER during Rabi season

4.2.4 Flood Management and Resilience

4.2.4.1 Flood Inundation Mapping of Agricultural Land

Agriculture is extremely susceptible to floods, not only due to the reduced time frame for sowing and growing but also because of significant risks to good yield production caused by frequent floods of varying intensity. Floods are a major disaster in many regions that not only affect crop production but also crush the zeal of farmers whose families are entirely dependent on agriculture. Damage assessment is now a critical issue for the concerned Government Departments to ensure effective management of climate-resilient activities.

The use of satellite technology, combined with advanced GIS tools for flood monitoring, rapid mapping of inundated areas, and crop damage assessment, can further strengthen decision-making policies in a more effective and transparent manner. The synthetic aperture radar (SAR) data, such as RADARSAT SAR, provides detailed surface information crucial for identifying and mapping flooded areas, even under the cloud cover that typically accompanies severe weather events. SAR sensors operate at microwave frequencies and can function in different polarizations (HH, HV, VV, VH), enhancing the detection and differentiation between water and other surface types. Water surfaces typically produce lower backscatter signals compared to land,

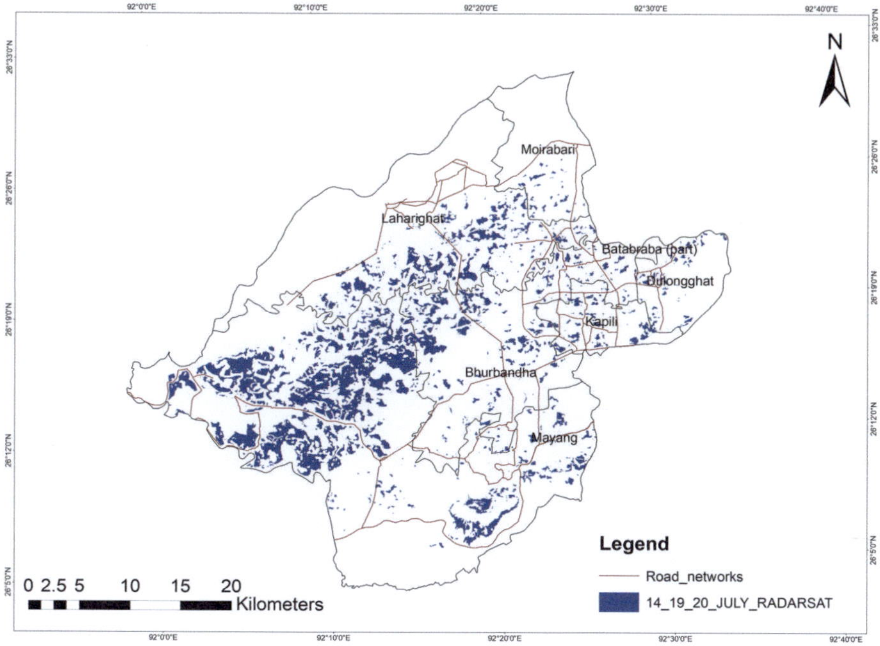

Fig. 4.7 Flood inundation area of Morigaon district of Assam on 14th, 19th and 20th July 2019

making flooded areas easily distinguishable in SAR images.A satellite-derived flood inundation area map in Morigaon district of Assam during July, 2019 is depicted in Fig. 4.7.

4.2.4.2 Assessment of Flood Effected and Resilient Crop Areas

Recurring flood is a major constraint to sustainable agriculture in the NEHR of India. Geospatial technology offer an effective means to assess flood-affected and resilient crop areas towards climate-resilient agricultural planning. Satellite data both optical and microwave data, is employed to monitor flood extent, identify waterlogged fields, and assess crop damage using vegetation indices such as NDVI and SAR backscatter value analysis. Time-series SAR data can be used to monitor crop phenological changes regularly using a rule-based approach, depending on the backscattering coefficients observed during the transplantation of rice. This enables near real-time monitoring of flood incidents and provides an accurate assessment of the extent of floodwater in standing crop areas. Figure 4.8 illustrates the severity of flood damage in agriculture areas during the crop cycle.

Floods are recognized as one of the significant natural hazards occurring almost every year in several parts of the North-Eastern states, most often during the monsoon

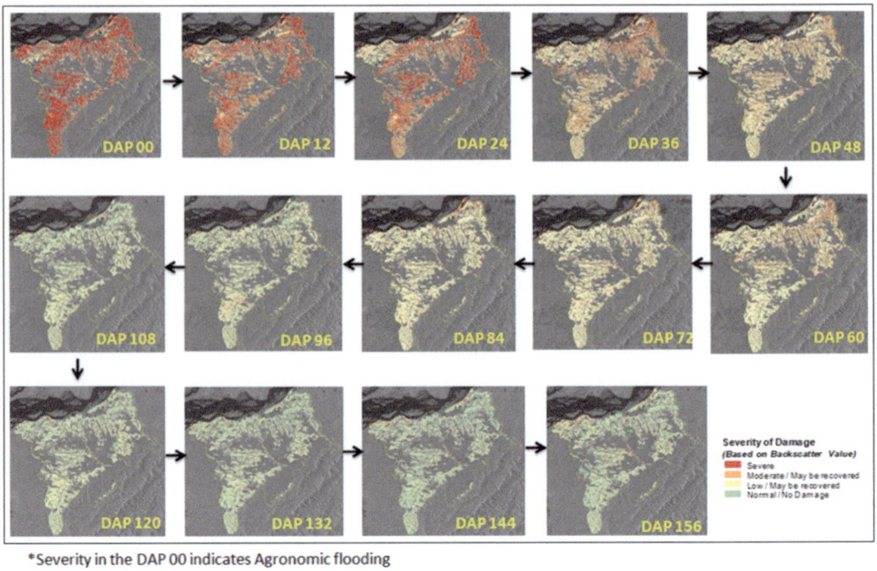

Fig. 4.8 Temporal evolution of flood damage severity in agricultural areas

season, resulting in massive crop losses. The most flood-affected states in North-East India are Assam and Manipur, particularly in the Brahmaputra Valley, Barak Valley, and Manipur Valley. Time series crop phenological stages generated in a scheduled manner from Earth Observation (EO) platforms (Fig. 4.9), along with flood inundation data from SAR data can be used significantly to assess the extent of crop-damaged areas. An extent of crop damage due to flood in Assam during Kharif, 2020 is highlighted in Fig. 4.10.

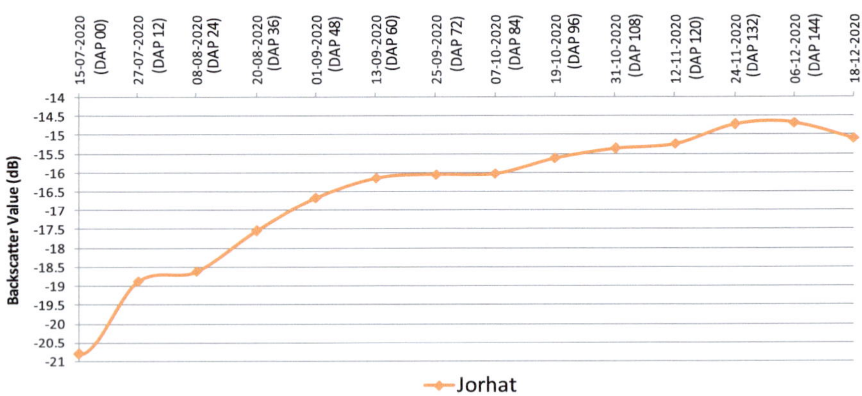

Fig. 4.9 Crop profile based on Sentinel-1 VH polarization for Jorhat district, Assam

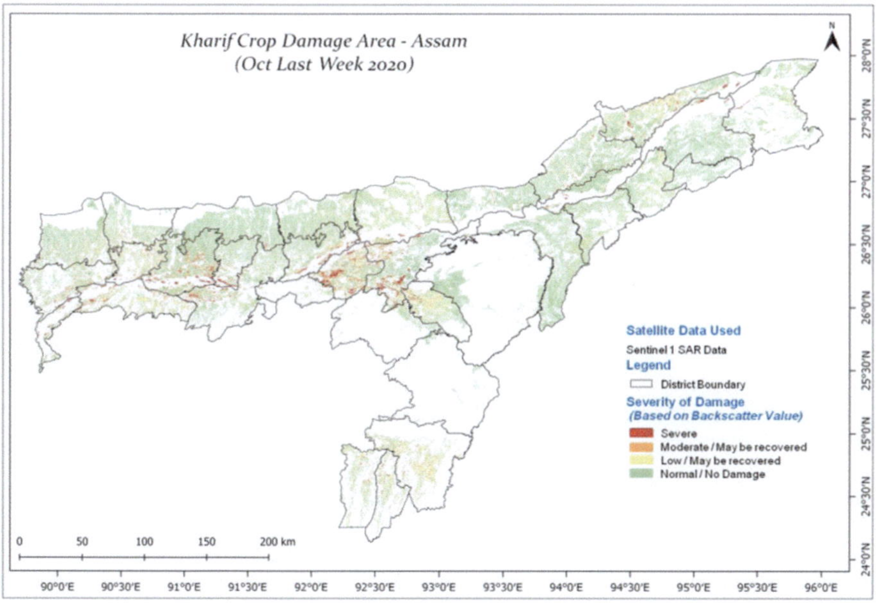

Fig. 4.10 Flood affected crop area in Assam during Kharif, 2020 (Oct last week, 2020)

Flood-resilient cropland is designed to withstand and recover quickly from flood events by cultivating crops that tolerate temporary submersion or recover swiftly after floodwaters recede. Both VV and VH polarization of SAR data are crucial for analysis to eliminate issues caused by taking single polarization. The VV data is more suited for land boundary demarcation but is susceptible to roughening of the water surface, commonly caused by wind, rain or the turbulent flow of the river whereas, VH data is more suitable for vegetation and water-body demarcation. Crop land areas which are not affected by flood and remain above water level can be identified and mapped using the SAR backscattering threshold method.

Assessing site suitability for crop cultivation is essential for improving agricultural productivity and fostering climate-resilient farming in flood resilient crop land. Geospatial technology plays a crucial role in mapping potential sites for introducing flood-tolerant crops or alternative farming practices based on crop-specific requirements. The suitability analysis employs a weighted model, Analytical Hierarchy Process (AHP), integrating the soil and climatic needs of specific crops to assign weights to each factor using Geographic Information System (GIS) techniques (Malczewski 2004). Each parameter is assigned a score ranging from 1 to 5 based on suitability categories—highly suitable, moderately suitable, and marginally suitable. The scores are weighted based on the relative importance of each parameter for crop cultivation (Shaloo et al. 2022).

A study was conducted to identify flood-resilient stable river islands using Sentinel-1 SAR data for the years 2019 to 2022, covering the entire stretch of the

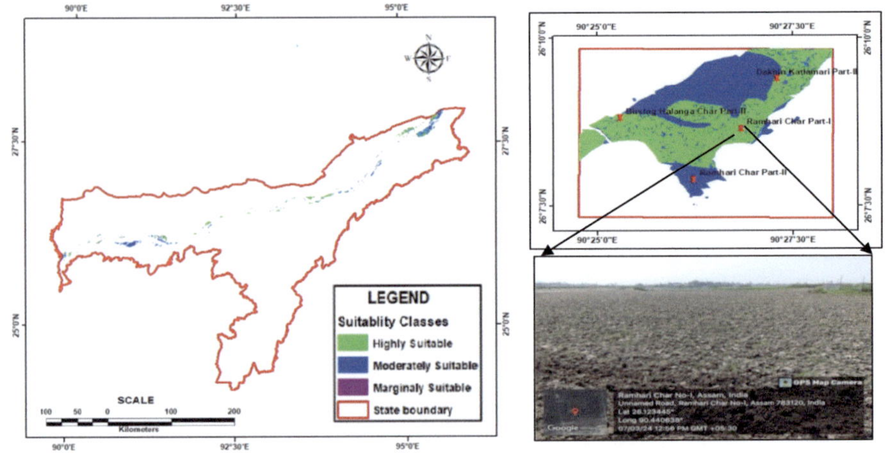

Fig. 4.11 Potential sites for expansion potato crop in the stable island of Brahmaputra river, Assam

Brahmaputra River within Assam for the post-monsoon season activities, such as crop expansion and implementing other government schemes before the next monsoon season. Figure 4.11 shows potential sites for expansion of potato crop in flood resilient river island of Brahmaputra as identified using the AHP model based on soil and climatic parameters.

4.2.5 Climate-Smart Crop Planning

4.2.5.1 Agroclimatic Zone and Crop Planning

Agriculture is both a contributor to and a victim of climate change. Changes in temperature and rainfall patterns are disrupting crop productivity and altering traditional growing seasons across the globe. The increased incidence and severity of extreme weather calamities such as floods droughts, and heat waves are leading to more frequent crop losses. As a result, farmers are being forced to adapt their practices, including shifting planting dates, adopting drought-resistant crop varieties, and improving irrigation systems. Climate-resilient agriculture relies on choosing the right crops and planting schedules based on prevailing and projected climate conditions.

Geospatial technology supports climate-smart crop planning by defining agroclimatic zones through the integration of climate, soil, and physiographic data. Crop yield is determined by weather conditions, whereas climate determines crop suitability and agricultural potential in a region (Pereira 2017).

Precipitation and temperature—both in terms of amount and seasonal variability are key factors used in all agro-climatic classification systems to assess crop suitability (Mavi 2004). When these climate variables are combined with information on soil characteristics and physiography, it becomes possible to identify land areas most suitable for specific agricultural uses, which is essential for adapting to climate change. Selecting crop varieties at the local or regional scale should take into account both the available agro-climatic resources and the climatic requirements of those specific crop varieties. Farmers can use agro-climatic zone information to choose crops best suited for their specific location, reducing vulnerability to climate shocks.

The North Eastern Region has experienced varying levels of change in climate patterns and natural resources across different parts. These changes have been driven by deforestation and over-exploitation under various cropping systems. A wide variety of crops are cultivated traditionallyin the NER, guided by local needs and cultural customs, often with limited consideration of their suitability to the existing climatic conditions. A thorough assessment of agro-climatic resources was conducted in a few selected districts of Mizoram and Nagaland to evaluate and characterize the agro-climate prevalent in these regions. Essential elements in the delineation of agro-climatic zones include bioclimates based on thermal regimes, moisture regimes, and biomass variability (Patel 2000; Goswami 2008). An agro-climatic zone map was prepared by integrating agro-climatic indices such as the Moisture Index (MI) and Thermal Index (TI), which help farmers understand climatic variables relevant to resilient agriculture (Krishnan and Mukhtar 1972). The MI is highly related to annual rainfall and represents the proportion of precipitation that remains available after evaporation. Potential Evapotranspiration (PET) was estimated using the Thornthwaite PET method (Thornthwaite 1948). TI, on the other hand, is extremely useful for determining precipitation effectiveness, which is an important indicator of plant growth. TI is further used to calculate the Length of the Growing Period (LGP) for the study area. The LGP is calculated using a method suggested by the Food and Agriculture Organization of the UN (FAO 1978) as the number of days in a year when precipitation surpasses 50% of the potential evapotranspiration. LGP refers to the duration, measured in days, during which crops can grow under adequate temperature, moisture, and light conditions. It is a critical factor in agricultural planning, determining which crops can be successfully cultivated in a given region and influencing the selection of crop varieties, planting schedules, and overall farm management. A Land Use Land Cover (LULC) map, along with soil parameters, was used to derive Available Water Capacity (AWC). The derived agro-climatic maps, AWC, LGP, soil maps, physiographic maps, and LULC were utilized for land suitability studies and land-use planning (LUP) using weighted overlay analysis (Fig. 4.12). The relative significance of each layer for a specific land use type determines its weight, which ranges from 0 to 100. The attributes are scored from 0 to 9, where 9 indicates the most suited and 0 indicates the least suitable.

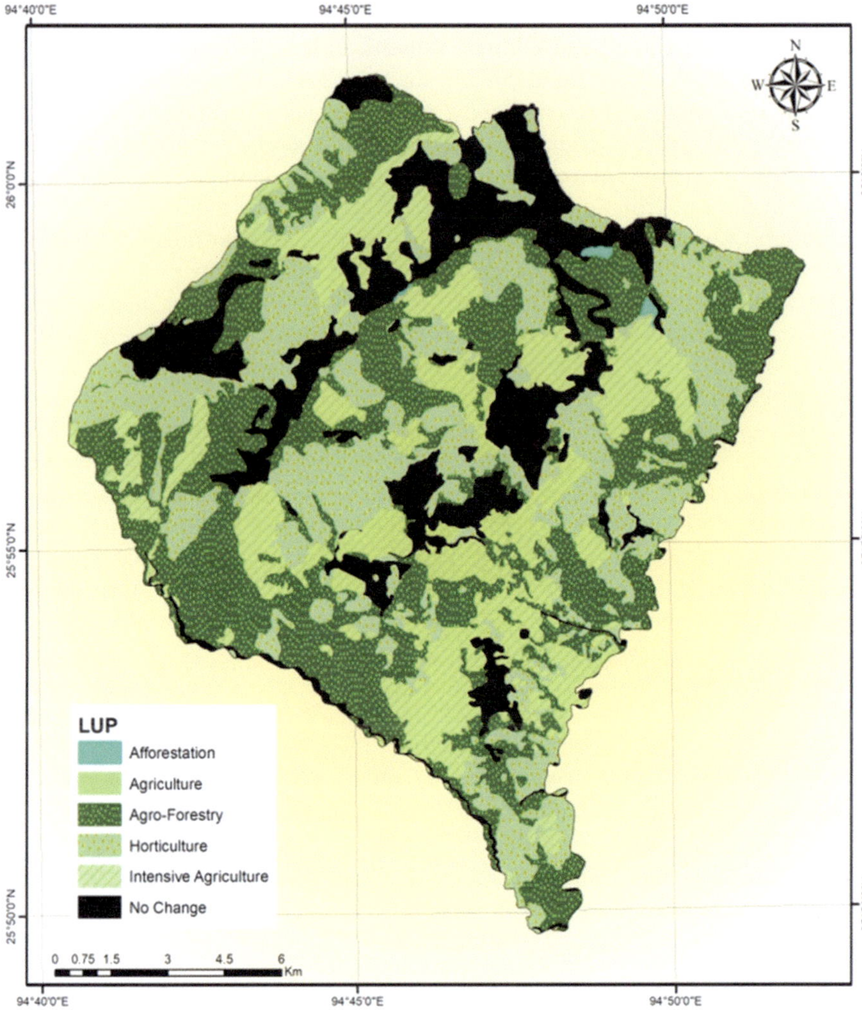

Fig. 4.12 Land use plan (LUP) in Kiphire Sadar community and rural development (C&RD) block of Kiphire district of Nagaland

4.2.5.2 Cropping System Analysis for Crop Diversification

Cropping system refers to the spatial arrangement of crops, cropping patterns, or fallow land in a given area over a period of time. It varies across space and time and reflects the economic, socio-cultural, geo-climatic, historical, and political conditions of an area. It is influenced by physical variables (such as soil type, water bodies, and altitude), climatic factors (such as precipitation and humidity), and socio-cultural elements (such as income, food habits, land size, and government policies). Cropping systems, including intensity and rotation, need to be planned thoroughly for the sustainable development of agricultural land. Before proposing alternate plans to enhance crop productivity, assessing the performance of the existing cropping system in an area is essential. Remote sensing technology plays a crucial role in extracting valuable crop information by utilizing spectral reflectance properties in the electromagnetic spectrum (EMS). The reflectance characteristics of a crop vary significantly depending on various factors, including its type, variety, growth stage, environmental conditions, and geographic location. As crops transition through their growth stages—from emergence to peak growth and senescence—their reflectance properties undergo changes. The NDVI, the most widely used index, captures the unique spectral characteristics of vegetation in the red and near-infrared regions to assess the health and vigor of plant canopies. NDVI effectively captures phenological changes and growth patterns of crops throughout the study period, providing valuable insights into cropping intensity and fallow area identification.

The temporal representation of NDVI values allows for the identification and monitoring of critical phenological events, such as crop emergence, vegetative growth, reproductive stages, and senescence, and quantifies changes in vegetation vigor and biomass over time. A knowledge-based decision tree, when applied to multi-temporal NDVI data, can effectively identify crop and fallow lands during each crop-growing season. During the growing season, actively cultivated croplands exhibit higher NDVI values due to the presence of healthy and dense vegetation cover. Conversely, fallow lands, which are temporarily left uncultivated or in a resting state, display lower NDVI values, reflecting the absence or sparse presence of vegetation. Information on cropping intensity—whether single-cropped, double-cropped, or triple-cropped—can be extracted through spatial integration as depicted in Fig. 4.13, where crop areas for each season are overlaid and intersected. Various indices have been developed to assess and compare the performance of different cropping systems (Panigrahy et al. 2005; Rasid et al. 2019). Some cropping system indices extensively used to characterize cropping patterns, intensity, and rotation within a study area are the Area Diversity Index (ADI), Multiple Cropping Index (MCI), and Cultivated Land Utility Index (CLUI). These indices offer necessary insights into crop growth diversity, land utility, and cropping intensity, which are crucial for formulating effective strategies aimed at enhancing productivity and identifying regions where crop diversification, intensification, or both.MCI measures the intensity of land utilization for crop cultivation by quantifying the extent of multiple cropping practices within a given area.ADI quantifies the diversity of crops grown within a specific region, providing insights into crop diversification and potential risk mitigation strategies. CLUI evaluates the efficiency of land utilization for crop cultivation, taking into account cropping intensity and the potential for multiple cropping cycles. This index helps identify areas where land resources may be underutilized or where there is potential to increase cropping intensity.

There are three cropping seasons in the North-Eastern Region (NER): Kharif (June–November), Rabi (November–January), and Zaid (February–May). Analyzing the cropping system also helps identify fallow areas. Mapping fallow land facilitates the selection of alternate crops, enhancing cropping intensity and crop diversification in the region.

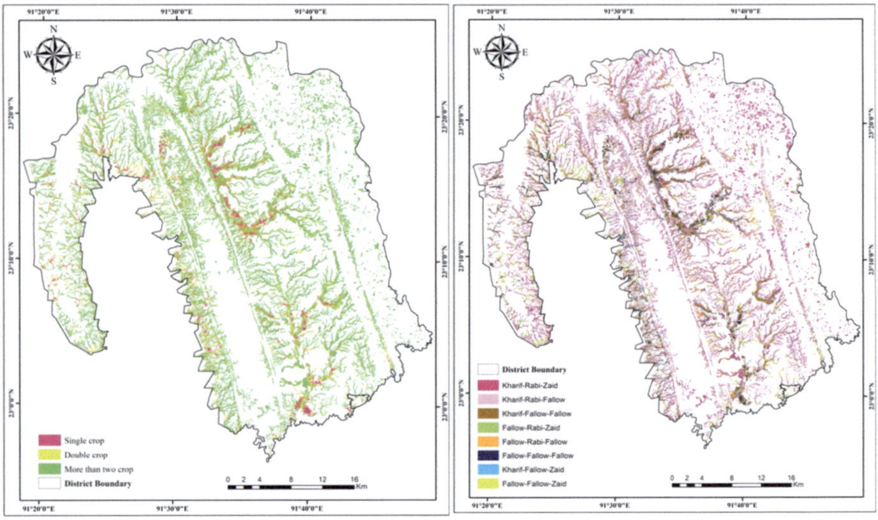

Fig. 4.13 Cropping intensity (left) and Crop rotation (right) of South Tripura District of Tripura

Table 4.1 gives the threshold values for rating these cropping system indices. Areas with low to moderate ADI values generally require crop diversification. Areas with low and medium MCI should prioritize crop intensification, particularly those with low MCI. The Cultivated Land Utility Index (CLUI) offers further insights to determine whether long-duration or short-duration crops would be more suitable substitutes. Thus, these indices hold considerable potential for agricultural planningand monitoring. Rating criteria for different indices are given in Table 4.2.

The System of Crop Intensification (SCI) offers a pathway to more sustainable and climate-resilient agriculture. By integrating traditional knowledge with modern practices, SCI empowers farmers to produce more with fewer resources, conserve natural resources, and build a more secure and sustainable agricultural future. This

Table 4.1 Rating criteria for different indices

Rating	MCI	ADI	CLUI
	Threshold values		
Low	< 130	< 2.0	< 0.5
Medium	130–160	2–5	0.5–0.6
High	> 160	> 5.0	> 0.6

Table 4.2 Planning of cropping system in South Tripura District based on rating

Rating	MCI	ADI	CLUI	Cropping system plan
1	High	Low	High	Crop diversification planning with intensive cropping

approach strengthens the resilience and adaptability of crops. Under SCI, plants typically develop healthier and more robust root systems, which improve their ability to endure heavy rainfall and strong winds (Adhikari et al. 2018). These plants also exhibit a higher tolerance to heat, especially during dry spells. Green gram and black gram are important legume crops grown in India and have emerged as nutritive and remunerative pulse crops. They are capable of providing quick returns in the shortest possible time while contributing to nutritional security for millions of people. These crops can be cultivated by small and marginal farmers as intercrops or during the fallow period following a rice crop. Similarly, The System of Crop Diversification involves the deliberate cultivation of multiple crop types within a farming system instead of depending on a single crop. This approach enhances the resilience of farming systems, optimizes resource use, and improves economic stability for farmers. Crop diversification is especially vital in regions prone to climate variability and economic fluctuations, as it reduces risk and increases overall agricultural productivity (Joshi et al. 2004).

4.3 Challenges and Limitation

The adoption and success of climate-resilient agriculture are often limited by several challenges. One major barrier is the lack of financial resources, as many farmers do not have access to low-cost credit or insurance options needed to invest in adaptive measures like drought-tolerant seeds and water-efficient irrigation technologies. Limited infrastructure, such as inadequate storage and transportation facilities, further impedes resilience efforts. Social and cultural barriers, including resistance to change and gender inequities, also restrict the adoption of innovative practices. Additionally, knowledge gaps, weak policy frameworks, and insufficient institutional support undermine efforts to promote climate-smart strategies. Environmental constraints, such as declining water resources and unpredictable weather patterns, exacerbate these challenges, while poorly developed markets and value chains limit economic incentives. Overcoming these hurdles requires coordinated research, policy, and grassroots implementation efforts to build robust, adaptive, and inclusive agricultural systems.

The application of geospatial technology in climate-resilient agriculture also faces several challenges and limitations that impede its full potential. High costs associated with acquiring and maintaining advanced geospatial tools, such as satellite imagery and Geographic Information Systems (GIS), can be prohibitive for smallholder farmers and resource-constrained regions. Limited technical expertise and inadequate training in interpreting geospatial data hinder its effective use at the grassroots level. Furthermore, inconsistent or incomplete spatial data, especially in remote or underdeveloped areas, reduces the accuracy and reliability of analyses. Infrastructure constraints, such as poor internet connectivity and a lack of electricity, restrict access to geospatial tools in rural areas. Additionally, privacy and data-sharing concerns hinder collaboration and integration across stakeholders. Addressing these limitations requires targeted investments in capacity building, policy support for open data sharing, and developing cost-effective, user-friendly geospatial solutions tailored to local needs.

4.4 Conclusion

Geospatial technology is rapidly transforming agriculture by providing essential tools for climate resilience. By enabling more efficient resource management, enhancing disaster preparedness, and supporting climate-smart decision-making, these technologies are key to ensuring food security in the face of growing climate risks. The future of geospatial technology in climate-resilient agriculture is promising. Improved access to high-resolution and real-time satellite data, coupled with open-source geospatial platforms and artificial intelligence, can enhance the precision and timeliness of climate risk assessments and decision-making. Integrating Internet of Things (IoT) devices with geospatial tools can provide localized, real-time monitoring of soil, weather, and crop health, enabling farmers to adopt more adaptive and sustainable practices. Geospatial technology can play a transformative role in enhancing productivity, improving resource management, and building climate-resilient agricultural systems worldwide by harnessing these advancements.

However, the application of geospatial technology in climate-resilient agriculture also faces several challenges and limitations that impede its full potential. High costs associated with acquiring and maintaining advanced tools, coupled with a lack of technical expertise and insufficient training in data interpretation, pose considerable obstacles, particularly for marginal farmers and resource-constrained regions. Furthermore, the inconsistency and incompleteness of spatial data, especially in remote or underdeveloped areas, compromise the accuracy and reliability of analyses. Addressing these limitations necessitates targeted investments in capacity development, policy frameworks that promote open data sharing, and the creation of cost-effective, user-friendly geospatial solutions tailored to local requirements.

Acknowledgements The authors sincerely acknowledge the support received from Indian Space Research Organisation (ISRO) Department of Space for enabling the research and development through financial as well as technical guidance. We extend our gratitude to the Indian Council of Agricultural Research- Research Complex NEH region, Assam Agricultural Universities (AAU), Jorhat, State agricultural Department, Govt. of Assam & Tripura, State Disaster Management Authorities for providing valuable data and insights. We also appreciate the contributions of the India Meteorological Department (IMD), NCMRWF, and IITM for weather and climate data which has significantly enriched the quality of this work.

References

B. Acock, L.H. Allen Jr, Crop responses to elevated carbon dioxide concentrations, in *Direct effects of increasing carbon dioxide on vegetation,* ed. by B.R. Strain, J.D. Cure. DOE/ER-0238. US Dept. of Energy, Carbon Dioxide Res. Div., Washington DC. pp. 53–97 (1985)

P. Adhikari, H. Araya, G. Aruna, A. Balamatti, S. Banerjee, P. Baskaran, B.C. Barah, D. Behera, T. Berhe, P. Boruah, S. Dhar, System of crop intensification for more productive, resource-conserving, climate-resilient, and sustainable agriculture: experience with diverse crops in varying agroecologies. Int. J. Agric. Sustain. **16**(1), 1–28 (2018)

Anonymous, *National Innovations in Climate Resilient Agriculture (NICRA).* (Indian Council of Agricultural Research, 2011), pp. 1–8

Anonymous, *Annual Report.* Department of Agriculture, Cooperation & Farmers' Welfare, Ministry of Agriculture & Farmers Welfare, Government of India, 2015, pp. 1–158

Anonymous, *Manual for drought assessment using remote sensing and GIS.* Mahalanobis National Crop Forecast Centre, Ministry of Agriculture and Farmers Welfare, Government of India, 2016, pp. 1–160

Anonymous, *National Aeronautics and Space Administration.* (Remote Sensing for Agriculture, 2017)

Anonymous, *Economic survey of India 2022–2023.* (Ministry of Finance, Government of India, New Delhi, 2023)

P.J. Curran, Remote sensing of foliar chemistry. Remote Sens. Environ. **30**(3), 271–278 (1989)

J.G. Ferwerda, A.K. Skidmore, Can nutrient status of four woody plant species be predicted using field spectrometry? ISPRS J. Photogramm. Remote. Sens. **62**, 406–414 (2007)

B.C. Gao, NDWI—a normalized difference water index for remote sensing of vegetation liquid water from space. Remote Sens. Environ. **58**(3), 257–266 (1996)

J. Goswami, Geomatics based agroclimatic characterization of Meghalaya. J. Agrometeorol. **10**(2008), 164–168 (2008)

I. Herrmann, A. Karnieli, D.J. Bonfil, Y. Cohen, V. Alchanatis, SWIR-based spectral indices for assessing nitrogen content in potato fields. Int. J. Remote Sens. **31**, 5127–5143 (2010)

P.K. Joshi, A. Gulati, P.S. Birthal, L. Tewari, Agriculture diversification in South Asia: patterns, determinants and policy implications. Econ. Polit. Wkly., 2457–2467 (2004)

Krishnan, Mukhtar, Soil Climate Zones in relation to cropping patterns, in *Proceedings of the symposium on cropping patterns in India, ICAR,* New Delhi (1972), pp. 172–185

J. Malczewski, GIS-based land-use suitability analysis: a critical overview. Prog. Plan. **62**(1), 3–65 (2004)

H.S. Mavi, G.J. Tupper, *Agrometeorology: principles and applications of climate studies in agriculture.* (CRC Press. New York, NY, USA, 2004). ISBN 9781560229728

P.J. Neate, *Climate-smart agriculture: success stories from farming communities around the world.* (CTA Success Stories, 2013)

S. Panigrahy, K.R. Manjunath, S.S. Ray, Deriving cropping system performance indices using remote sensing data and GIS. Int. J. Remote Sens. **26**(12), 2595–2606 (2005)

N.R. Patel, U.K. Mandal, L.M. Pande, Agro-ecological zoning system: a remote sensing and GIS perspective. J. Agrometeorol. **2**(1), 1–13 (2000)

J. Penuelas, F. Baret, I. Filella, Semi-empirical indices to assess carotenoids/chlorophyll a ratio from leaf spectral reflectance. Photosynthetica **31**(2), 221–230 (1995)

L.S. Pereira, Water, agriculture and food: challenges and issues. Water Resour. Manage **31**(10), 2985–2999 (2017)

N. Rasid, M. Prashnani, J. Goswami, P.L.N. Raju, Cropping pattern analysis of Nagaon district of Assam using geospatial technique. Int. J. Agric. Sci. **4**, 1–17 (2019)

V. Senpakapriya, J. Goswami, D. Battacharjee et al., Geospatial insights for assessing agricultural drought hazards of rabi season in Assam. J. Indian Soc. Remote Sens. (2025). https://doi.org/10.1007/s12524-024-02092-w

Shaloo, R.P. Singh, H. Bisht, R. Jain, T. Suna, R.S. Bana, S. Godara, Y.S. Shivay, N. Singh, J. Bedi, S. Gautam, Crop-suitability analysis using the analytic hierarchy process and geospatial techniques for cereal production in North India. Sustainability **14**(9), 5246 (2022)

Thornthwaite, An approach toward a rational classification of climate, Geogr. Rev., **38**, 55–94 (1948)

United Nations, *Transforming our world: The 2030 agenda for sustainable development*, vol. 1. (United Nations, Department of Economic and Social Affairs, New York, 2015), p. 41

Jonali Goswami currently working as the Scientist/Enginner-SF in Agriculture and Soil Division and Programme Coordinator, Outreach & Capacity Building at North Eastern Space Applications Centre under Department of Space, Government of India, has more than 22 years of experience in space technology applications and research. Her research expertise spans Geospatial technology applications in agriculture and allied sectors, disaster management and epidemiological studies. She has developed expertise in hyperspectral and Unmanned Aerial Vehicle (UAV) data analysis and their applications in different fields of research and applications towards precision and climate resilience agriculture. She has published more than 60 research papers in national and international journals and conference proceedings. She has also many book chapters publications and guided a number of M.Sc./M.Tech and Ph.D. students. She has been awarded National Award for Women of Excellence in Geospatial Science and Technology by FOSSEE under NMEICT, Ministry of Education, Government of India at IIT Bombay, 2025.

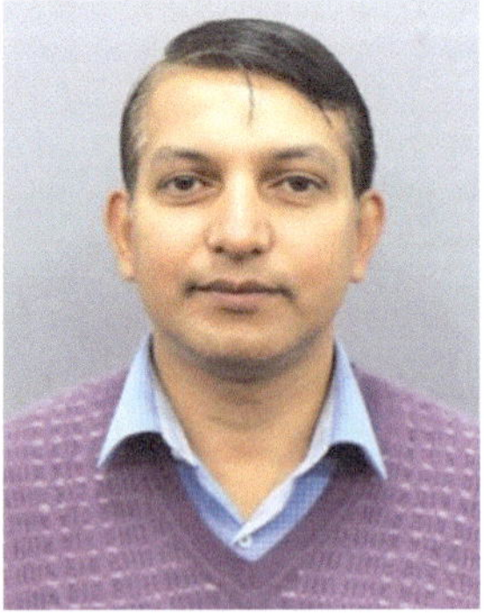

Chandan Goswami was born on 11th January 1977 in Nagaon district of Assam, India. He received Bachelor Degree in Agriculture and Master Degree in Soil Science from Assam Agricultural University, Jorhat, Assam, India in the year 1999 and 2001, respectively and Ph.D degree in Soil Science & Agricultural Chemistry from Central Agricultural University, Imphal, Manipur India in the year 2020. Dr. Goswami is a Senior Scientist at the North Eastern Space Applications Centre, Shillong, Meghalaya, India. He has specialized in geospatial technology applications in agriculture, soil and natural resources management. His research focuses on soil spectral analysis, soil resources mapping, and land use planning, significantly contributing to sustainable agricultural development in India's North Eastern Region. His work leverages advanced remote sensing and GIS techniques to address challenges like soil health, and crop productivity in the NER's diverse agro-climatic conditions. Dr. Goswami has authored over 84 publications and having more than 600 citations, reflecting his impactful contributions to the scientific community. A recipient of the National e-Governance Award in 2017 for enhancing geospatial data utilization in planning and development.

Bijoy K. Handique currently working as the Head, Agriculture and Soil Division and Head, Program Planning and Evaluation Group at North Eastern Space Applications Centre under Department of Space, Government of India has more than 22 years of experience in space technology applications and research. He has been leading a number of research and applications programs at NESAC particularly in space technology applications in agriculture and allied areas. He is also leading a team at NESAC in using unmanned aerial vehicles for various agricultural applications. He has published about 60 research papers in national and international journals and conference proceedings. He has also published 12 book chapters and guided a number of M.Sc./M.Tech and Ph.D. students. He has been awarded ISRS-SPECK Award for 2005, ISRO Young Scientist Merit Award for 2008 and ISRO-ASI Young Scientist Award for 2014.

Pradesh Jena joined NESAC on 13.05.2019 and has been working extensively in the field of crop acreage and production estimation of major agricultural crops, crop cutting experiments, cropping system analysis, and geo-tagging of sericulture assets in the North Eastern Region using geospatial technologies. He has also gained expertise in applications of UAV technology for precision agriculture. At NESAC, he is a member of the 'Committee for delivery of Data Products & Services from EOS-04 & TRISHNA Mission'. He has published many research papers in national and international journals and conference proceedings. He has also guided a number of M.Sc./M.Tech students.

Chapter 5
Remote Sensing Based Forest Resource Assessment and Management in NE India

Rocky Pebam, Suchitra Haobam, K. K. Sarma, and Kasturi Chakraborty

Abstract Forests are one of the major natural resources that support sustenance of the human population. At the same time, they are under the influence of climate change and human—induced pressures affecting sustainability. The working plan is the main instrument through which the scientific management of Forests is being achieved in India. National Working Plan Code which was first adopted in 2004 with a subsequent amendment in 2014 and 2023 brought uniformity and acted as the guiding principle for the preparation of the working plan for scientific management of different forest divisions of the country. NESAC has been supporting the different Forest Departments of the north-eastern states of India in providing geospatial inputs, supplying accurate latest forest canopy density maps and timber growing stock computations using high resolution panchromatic and multispectral satellite data. The improvement in the scale and accuracy of the forest canopy density maps prepared using geospatial technologies could reduce the intensity of sampling that was followed earlier in the older method of estimating growing stock when the remote sensing and GIS techniques were in an early stage. With the geospatial maps of various themes and computational inputs on stock estimates provided to the Forest Departments in the region, forest management plans for many of the divisions in the region have been approved for managing the forest resources. This chapter highlights the techniques and methods that have been used to generate the maps and estimates which have improved over the years following the guidelines of the National Working Plan codes.

Keywords Forest resources · Forest working plans · Growing stock estimates

R. Pebam (✉) · S. Haobam · K. K. Sarma · K. Chakraborty
North Eastern Space Applications Centre (NESAC), Umiam, Shillong, Meghalaya, India
e-mail: rocky.pebam@nesac.gov.in

S. Haobam
e-mail: suchitra.haobam@nesac.gov.in

K. K. Sarma
e-mail: kk.sarma@nesac.gov.in

K. Chakraborty
e-mail: kasturi.c@nesac.gov.in

S. P. Aggarwal et al. (eds.), *Remote Sensing of the North Eastern Himalayan Ecosystem*,
https://doi.org/10.1007/978-981-95-3221-6_5

5.1 Introduction

Forests and other wooded lands have been recognized and valued as important natural resource for centuries and the list of benefits that people derived from forests are numerous. Forest resources play a vital role not only for the economic, social, and cultural well-being of local and regional communities, but also for the maintenance of life on earth as a whole (Korotkov et al. 2004). They are a biodiversity and genetic resource, as well as providers of other environmental services, and a key player in poverty alleviation (Myers 1996; Sellers et al. 1997; FAO 2003). For some forest functions and services, their features and potential still need additional research, but all forest resources need a comprehensive and reliable assessment, as a basis for their proper utilization and management in an efficient and sustainable manner (Korotkov et al. 2004). But this valuable renewable natural resource and repositories of terrestrial biological diversity are in imminent danger due to urban expansion, infrastructural development, agriculture and global warming (Bawa and Dayanandan 1998; Brown and Pearce 1994; Stedman-Edwards 1998). Sustainable forest development, which means, recognizing the limits of forests to withstand environmental change, individually and collectively, and in managing human activities to produce the maximum level of benefits obtainable within the limits (Maini 1992) was influential in the development of the non-binding 'Forest Principles' and Chap. 11 of UNCED Agenda 21 both of which provided an international framework for sustainable forest management. The guiding objective of the Forest Principles is to contribute to the management, conservation and sustainable development of all types of forests and to provide for their multiple and complementary functions and uses (Duinker et al. 1998). The assessment and management of forest resources involves evaluating and managing forest ecosystems to ensure their sustainability and meet the diverse human needs. This includes understanding the extent, condition, and the utilization of forests, as well as implementing practices to conserve and manage them for sustainable development.

Majority of forest functions, both wood and non-wood goods and services, are known to everyone (Korotkov et al. 2004). Besides providing habitats for animals and livelihoods for humans, the forest also offers watershed protection, prevent soil erosion and mitigate climate change (Pokhrel and Gautam 2024). Yet, despite our dependence on forests, we are still allowing them to disappear. To sustainably manage, conserve and utilize the forest resources and to bring uniformity in the forest management planning across the country, Ministry of Environment Forest and Climate Change, Government of India adopted a uniform code, the National Working Plan Code in 2004 (National Working Plan Code 2004). The concept of sustainability in forest management not only implies sustainability of productive function but also environmental functions (soil and water conservation, carbon sequestration, etc.) and socio-economic benefits (meeting livelihood and basic needs) to forest dwellers and other forest-dependent communities. Involvement of forest fringe communities in the protection and management of forest resources through micro-plans and eco-development plans has become imperative and must find linkages to the working plans (National Working Plan Code 2014).

The first planned working of forests in the country was crafted in 1837 by Mr. U. V. Munro, the then Superintendent of Forest in Travancore (Verma et al. 2024). A Forest Working Plan is an essential document for the scientific management of forests. It is a written document for evaluating the status of forest and biodiversity resources of a forest division, assessing the impact of past management practices and deciding on suitable management interventions and silvicultural treatments to manage the forest sustainably in perpetuity (National Working Plan Code 2014). The main task of forest working plan is to prepare a plan for forest division periodically to keep pace with the trends emerging out of the forest-people interface and to address national and international obligations (National Working Plan Code 2014).

The preparation of the working plan is based on stock and vegetation maps which are prepared through a ground survey. The working plan highlights precise strategies for the management of different forest areas, taking into consideration their unique ecological characteristics, socio-economic factors and potential for resource extraction. It is based on broad data collection, including forest inventories, field surveys and remote sensing data and it is designed to achieve multiple objectives such as timber production, watershed protection, wildlife conservation, and soil and water conservation.

In India the Forest Survey of India (FSI) conducts comprehensive assessments of India's forest resources, publishing its findings in the biennial India State of Forest Report (ISFR) since 1987. FSI uses satellite dataand field-based surveys (National Forest Inventory) to map and assess forest resources. In the initial years Landsat Multispectral Scanner (MSS) were used with visual interpretation and later in the 2001 report IRS LISS III data (23.5 m resolution) were used and digital classification techniques were employed to prepare the maps at 1:50,000 scale.

The advent of remote sensing and its rapid technological developments in the scientific field has revolutionised the technological support to forest management globally. Remote Sensing enables the identification of forest cover, biomass estimation (Nandy et al. 2019), detection of changes over time (Kushwaha 2004; Srivastava et al. 2002), distribution of forest ecosystems, fluctuation in plant productivity with season, forest fire monitoring (Chakraborty et al. 2024), etc. It provides a birds-eye view of forest conditions and changes, aiding in decision-making processes across the spectrum of forest management (Ulvdal et al. 2025).

5.2 Remote Sensing and GIS for Forest Resource Assessment

Remote sensing and GIS database forms a platform for monitoring of the implementation of the working plan. It offers very effective tool for an exercise like working plan preparation, which involves analysis of several spatial layers and parameters in relation to the forest for prescribing management practices. The techniques and methods add to the precision, objectivity and scientific analysis into the process and also can lead to significant savings in cost and time (Rao et al. 2007). The Working Plan Code 2014 recommended a grid-based sampling design for the distribution of sampling point and using stratification at the estimation stage. Such a grid-based sampling would require large number of sampling point if the coverage of forest area is huge and also for placing sampling points in centre of grids which are outside the forest area. Fararoda et al. (2021) have demonstrated that by using remote sensing-based forest type and forest canopy density stratification and spatially optimal stratified random sampling the sample size can be reduced with 92% estimation accuracies. Handique and Sudhakar (2014) also reported that employing double sampling design for growing stock estimation in Assam showed a significant reduction in the standard error of the timber volume estimate against estimates brought out by common sampling design.

A spatial database on a forest in a GIS is a set of geo-referenced layers of spatial data of an area in which each layer is devoted to a particular theme or an entity on the ground e.g. administrative boundaries, road network, forest cover, forest types, altitude, etc. (Rao et al. 2007). A spatial database of a forest division with spatial layers gives immense use in the preparation of the working plan and later in day-to-day forest management decisions of the forest division (Kumar et al. 2019).

Although many remote sensing application works have been carried out at different administrative levels ranging from a reserved forest to territorial divisions and whole state, Dutt et al. (2009) opined that the scientific and effective forest management information could be operationalised using the GIS approach with forest division as a functional unit. The organization of data base, data inputting, updating, retrieval and analysis for specific purposes and to obtain outputs apparently is feasible cost effectively at the division level rather than aiming at state level as a central facility.

For a successful implementation of operationalising forest management with GIS, the data base should necessarily precede by generation of reliable spatial maps preferably on 1:10,000 scale to begin with which can be later inputted to GIS for proper data utilization and to obtain outputs with greater degree of accuracy relevant for division level management decision (Dutt et al. 2009). The organization of data feasibility adopting micro scale to macro scale may be a viable approach rather than adopting top to bottom approach starting with coarse resolution data which may fail to reveal any meaningful outputs required at the working level. The organization and aggregation of data base at division level would enhance efficient forest working most needed in the division level. The use of GIS at taluk level for preparation of zonation map revealed micro scale details without sacrificing accuracies for management decisions.

5.2.1 GIS and Remote Sensing Support in Managing Forests in North East India

Forest resources of the north eastern region (NER) of India are diverse owing to its location in the transition zone between the Indian, Indo-Malayan, and Indo-Chinese biogeographic region. The region being remote and a biodiversity hotspot with a wide array of floristic biodiversity recognized as major biome in the world (Mao et al. 2009). However, due to uncontrolled exploitation, these resources are under certain categories of threat and the most challenging regions to govern due to its unique physiographic condition, rich biodiversity, disaster prone and diverse socio-economic characteristics. In this context remote sensing and GIS applications have been taking an important role to assess the status of forests, land use, land cover, wildlife, watershed, wasteland, etc. in a quick and cost-effective way.

The scientific management of forest resources requires assessment and monitoring of forest resources in terms of types and density at a regular interval. A precise database pertaining to forest cover information is a vital input while formulating various management plans. However, the inventory and assessment of forest resources and forest cover assessment / change detection in the rugged topography of the NER is difficult and it is a time-consuming process through the traditional method. Towards this, remote sensing and GIS plays an important role in the scientific and sustainable management of the forest resources.

The research project on 'Biodiversity characterization at landscape level in north east India' which was coordinated and executed by Indian Institute of Remote Sensing (IIRS) Dehradun, was the first comprehensive work on remote sensing applications for spatially assessing the biodiversity and the associated anthropogenic disturbances in north east India in 2002. In this study different biodiversity information layers were generated namely vegetation map, fragmentation map, disturbance map and biological richness map considering the topography, human settlements, biogeographic information, climate, socio-economic and census data etc. (Roy et al. 2002).

The establishment of NESAC (North Eastern Space Applications Centre) in the region has facilitated in making significant progress in natural resources management using space technology in the region. Since its inception, the Centre has been providing remote sensing services in generating geospatial maps for forest inventory and management of forest resources. Different projects like forest cover mapping, estimation of growing stocks, assessment of vegetation and soil carbon, bamboo resource mapping, forest biomass estimations, wetland mapping, shifting cultivation dynamics studies, burnt area assessment, wildlife habitat mapping, etc. has been carried out in the region.

Forest managers, planners and policy makers often need the detailed information about different forestry parameters such as distribution of timber species, volume, biomass, carbon stock, regeneration status, population and structures etc. in different forests of the states for planning and management of the forest resources. Growing stock is one of the most important parameters which quantify forest resources. It is also an indicator of forest productivity. In working plans, growing stock has traditionally been used for calculation of sustainable yield of timber from forests. In the recent times, growing stock estimation has gained importance due to the significant roles of forests in climate change mitigation. The growing stock data forms the basis for calculation of biomass and carbon stock in the forests.

NESAC has provided space-based inputs in the preparation of forest working plans of different NER states. The methodology involves stratification of forest area based on forest type, forest crown density and/or elevation. Over the past one-decade IRS P5 Cartosat-1 data, IRS P6 LISS IV data and carto DEM have been used for stratifying the forest, an improvement in the mapping scale over the IRS 1D LISS III and PAN data generated maps that were used earlier. Based on field enumeration data, timber growing stock information for different girth classes have been generated for each compartment, the details of which are elaborated in the following section.

5.2.2 *Geospatial Inputs for Preparation of Forest Working Plan for Assam*

Forest Working Plans are prepared for duration of ten years and traditionally it takes about 4–5 years to prepare them. The working plans are prepared in two stages. The first stage is to prepare a reconnaissance report highlighting past systems of management, their result, status of vegetation, need for updation and broad future requirements. The second stage is to update the status of vegetation in the form of stock maps, carry out field inventory and assess the quality and quantity of growing stock based on which suitable management prescriptions are made. In a traditional working plan updation exercise, growing stock estimates are made with a sampling intensity 5–10% (sometimes even 100% enumeration in selective areas is also claimed) at the cost of very large time and manpower.

The conventional method of enumeration of forest inventory by a handful of Working Plan staff is very uncertain and highly time consuming. By the time enumeration of the forest stock of one division is completed and the Working Plan is prepared for the next 10 years or so, there may be a sea-change of reality on the ground. Moreover, the forest stock maps are not geo-referenced and hence is not much useful even after digitization of individual compartment under computer-based GIS and hence are not fit for analysis under GIS.

The application of RS & GIS in generation of inputs for the forest working plan preparation for the three divisions under Eastern Assam Forest Circle was carried out in collaboration with Assam Forest Department and National Remote Sensing Agency (NRSA), Hyderabad, during 2002 to 2005. Subsequently, during 2004 to 2007 RS & GIS inputs for preparation of forest working plan for North Cachar Hills (presently Dima Hasao Autonomous Council) was also carried out followed by another 21 divisions of Assam during 2013 to 2017. Brief methods and procedures followed for generating the maps and timber stock estimates are outlined in Fig. 5.1.

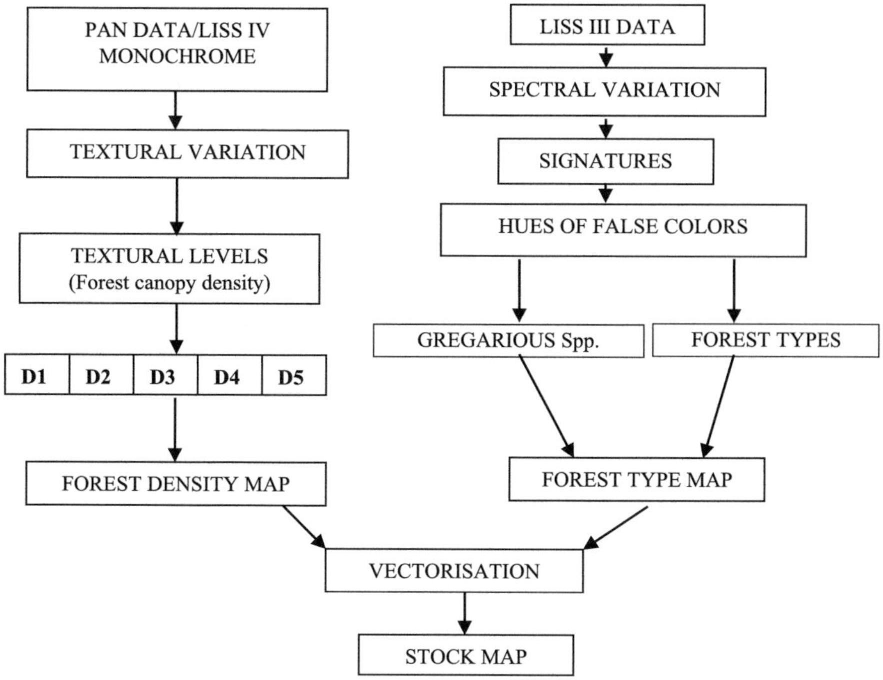

Fig. 5.1 Generalized methodology for stock map preparation

The IRS ID LISS III and PAN data of January-March 2004 were acquired from NRSA Data Center ensuring cloud free coverage. After acquisition of satellite data in the form of a CD-ROM product, it was subjected to various pre-processing techniques in order to obtain geographically referenced data. To prepare the forest type and canopy density maps the multi-spectral images were primarily subjected to digital image classification using both supervised and unsupervised approaches depending upon the land-cover class response in spectral sensing. Supervised classification was operated upon forested as well as agricultural tracts. Forests stand out uniquely due to the spectral distinction because of grades of foliage prevalence. Agricultural tracts consisting of cropped and fallow areas responded spectrally due to their separate class of near infrared/red reflectance relations. Bamboo dominated regions represented another spectral distinction. Other distinguishable categories like water, barren areas, sand could also be recognized spectrally. Based on the interpretation key developed, the IRS 1D PAN data was interpreted for forest crown density (5 categories, D1: < 20%, D2: 20–40%, D3: 40–60%, D4: 60–80%, D5: > 80%) delineation on-screen, based on standardized signatures and correlation from field experience along with ancillary information available. The forest crown density output maps were prepared on 1:25,000 scale. The general methodology flow of forest type and forest density classification is shown in the flow chart (Fig. 5.1).

The determination of appropriate sample size is an important decision in any sampling plan design. The size of sample depends on the variability of characters in the population, allowable error in the estimate and time cost factors (National Working Plan Code 2014). Normally the sample size is estimated using the Eq. 5.1 provided by Chacko (1964):

$$n = \frac{(cv)^2 t_v^2}{(AE)^2} \tag{5.1}$$

where,

n = sample size.

cv = co-efficient of variation of the sample volume.

AE = allowable error (%).

t_v = percent of the *t* distribution with *v* degree of freedom.

Except for the co-efficient of variation, other variables are defined by the user. Co-efficient of variation (Eq. 5.5) which is calculated as the ratio of standard deviation (Eq. 5.4) to the mean (Eq. 5.2) of the sample volume and it indicates the variability within the forest area. Since the variability is require before exercising the final inventory hence it requires advance estimate of the population variance. In order to prepare an advance estimate of population variance a pre inventory survey was conducted by taking a simple random of size *n* for estimating variance (Eq. 5.3) as shown in subsequent equations:

$$\text{Average } (\overline{x}) = \frac{1}{n}\sum_{i=1}^{n} x_i \tag{5.2}$$

$$\text{Variance } \left(s^2\right) = \frac{1}{n}\sum_{i=1}^{n} (x_i - \overline{x})^2 \tag{5.3}$$

$$\text{Standard Deviation(s)} = \sqrt{\frac{1}{n}\sum_{i=1}^{n} (x_i - \overline{x})^2} \tag{5.4}$$

$$\text{Coefficient of Variation (CV)} = \frac{S}{\overline{x}} \times 100 \tag{5.5}$$

Tree enumeration were done from the pre-inventory sample plots (size 0.1 ha) and after analyzing the field data with appropriate volume equations, the plot level timber volume was calculated for deriving the coefficient of variation (*cv*). For allocation of final sampling points, a combination of forest types and canopy densities were performed (‘*union*’ overlay in GIS) to prepare different forest strata and the sampling points were distrusted randomly to represent a stratified random sampling design. Sample points were allocated in proportion of area under different stratum.

The plots belonging to different strata were aggregated within each group to derive the strata wise mean timber volume and per unit area (hectare) estimate of the timber volume were computed as follows:

Mean timber volume for each stratum ȳ (m^3/ha) was calculated using the following formula (Eq. 5.6.):

$$\bar{y} = \frac{1}{n_1}\sum_{i=1}^{n_1} y_i \quad (5.6)$$

where,

n (1,2,3…) = the number of sample plots taken in each stratum

y_i = timber volume in the plot *i*.

For estimation of Reserve Forest wise timber volume estimate, since the compartments and in turn the strata within a compartment are mutually exclusive and independent as per the stratification the total timber volume of the compartment (Eq. 5.7.) and then for the Reserved Forest (Eq. 5.8.) and Range or Division etc. can be estimated as follows:

$$C_{v_i} = \sum\nolimits_{i=1}^{n}\sum\nolimits_{j=1}^{k} A_{ij}m_{ij} \quad (5.7)$$

where,

C_{v_i} is the timber volume of the ith compartment.

A_{ij} is the area under jth stratum of the ith compartment.

m_{ij} is the timber volume per unit area

$$RF_v = \sum_{i=1}^{n} C_{v_i} \quad (5.8)$$

where, C_{v_i} is the total timber volume in ith compartment within the Reserved Forest.

All the timber stock estimates were categories into the prescribed 6 girth classes at an interval of 30 cm *gbh* denoted as G1 (30–59 cm), G2 (60–89 cm), G3 (90–119 cm), G4 (120–149 cm) and G6 (>180 cm). Similarly, the stem densities for each plot were estimated following the procedures as described for timber volume calculation. And the stem densities were also extrapolated as per the values derived for each stratum. The methodology flow for the generation of timber stand and stock tables shown are detailed in Fig. 5.3.

Glimpses of some of the maps and tabular estimates that were generated for the Reserved Forests of North Cachar Hills of Assam are illustrated in the following tables (Tables 5.1, 5.2 and Fig. 5.2).

Table 5.1 Estimated timber volume for Compartment no. 1 of Langting Mupa RF (volume in m^3)

Comp no	Forest type	Crown density				Others	Total
		D1	D2	D3	D4		
1	Semi evergreen	754.90	7131.16	21,083.64	1120.17		30,089.87
	Moist deciduous	1062.30	567.55	3145.86	182.64		4958.35
	Mixed bamboo					1129.28	1129.28
	Plantation					0.00	0.00
	Total	1817.20	7698.71	24,229.50	1302.81	1129.28	36,177.50

D1: Crown density < 10%; D2: Crown density 10–40%; D3: Crown density 40–70%; D4: Crown density > 70%

Table 5.2 Estimated stem density for Compartment no. 1 of Langting Mupa RF (stem numbers)

Comp no	Crown density	Forest type	Girth classes						Total
			G1	G2	G3	G4	G5	G6	
1	D1	Semi evergreen	134	126	267	164	31	15	736
		Moist deciduous	53	0	106	53	53	53	317
	D2	Semi evergreen	188	248	441	163	35	45	1119
		Moist deciduous	0	110	243	177	44	44	618
	D3	Semi evergreen	5272	4313	5935	4829	2248	3280	25,877
		Moist deciduous	0	215	359	143	0	143	861
	D4	Semi evergreen	223	157	142	91	25	51	690
		Moist deciduous	0	0	0	0	0	0	0
	Plantation		0	0	0	0	0	0	0
	Mixed bamboo		121	148	211	124	46	51	703

D1: Crown density < 10%; D2: Crown density 10–40%; D3: Crown density 40–70%; D4: Crown density > 70%

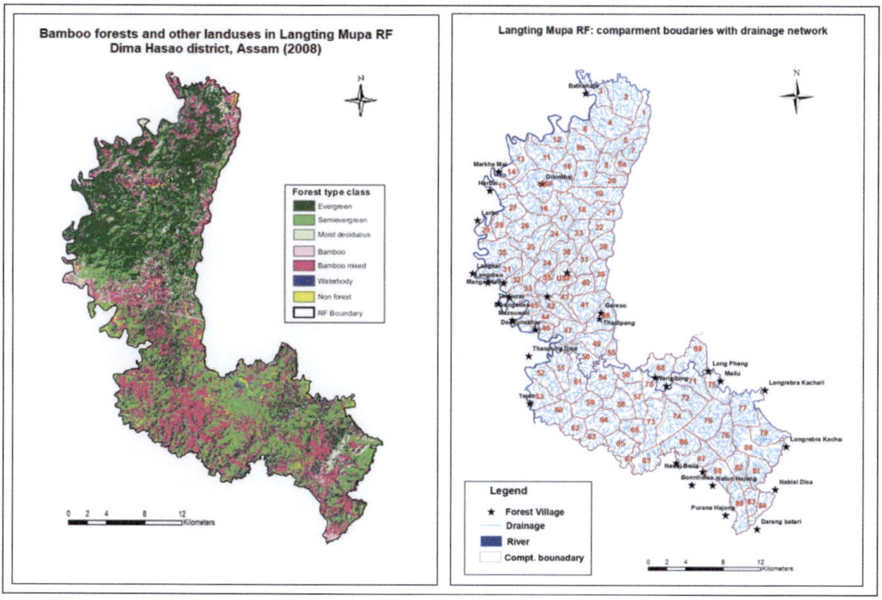

Fig. 5.2 Bamboo bearing area, forests and other landuse in Langting Mupa RF, Assam, India (left) and drainage map overlaid on the compartment map (right)

5.2.3 Geospatial Inputs for Preparation of Forest Working Plan for Mizoram

For the state of Mizoram, the RS & GIS inputs for the preparation of working plan were prepared in collaboration with the Mizoram State Remote Sensing Application Centre and the Mizoram Forest department during 2010 to 2013. The procedures followed were the same as was done for Assam. The only improvement was the use of higher resolution satellite data. IRS P6 LISS IV and IRS P5 Cartosat-1 stereo pair data were used for preparation of forest canopy density maps at 1:10,000 scale. The growing stock estimates and stock maps were prepared for 40 Reserved Forests and Riverine Reserved Forests of Mizoram. The working plans for the different divisions were prepared by the State Forest Department and submitted to the MoEFCC for approval.

Growing stock estimates and stock maps for more forest divisions were prepared during 2018 to 2024 for the divisions of Darlawn, Champhai and Kwarthah using latest satellite data and following the new guidelines as per National Working Plan Code 2014, wherein sampling points have to be located at the center of 25" x 25" grids within the division. Estimates on carbon stock for bamboo as well as timber were also generated and information of phytosociology and diversity of trees were also incorporated (Fig. 5.4).

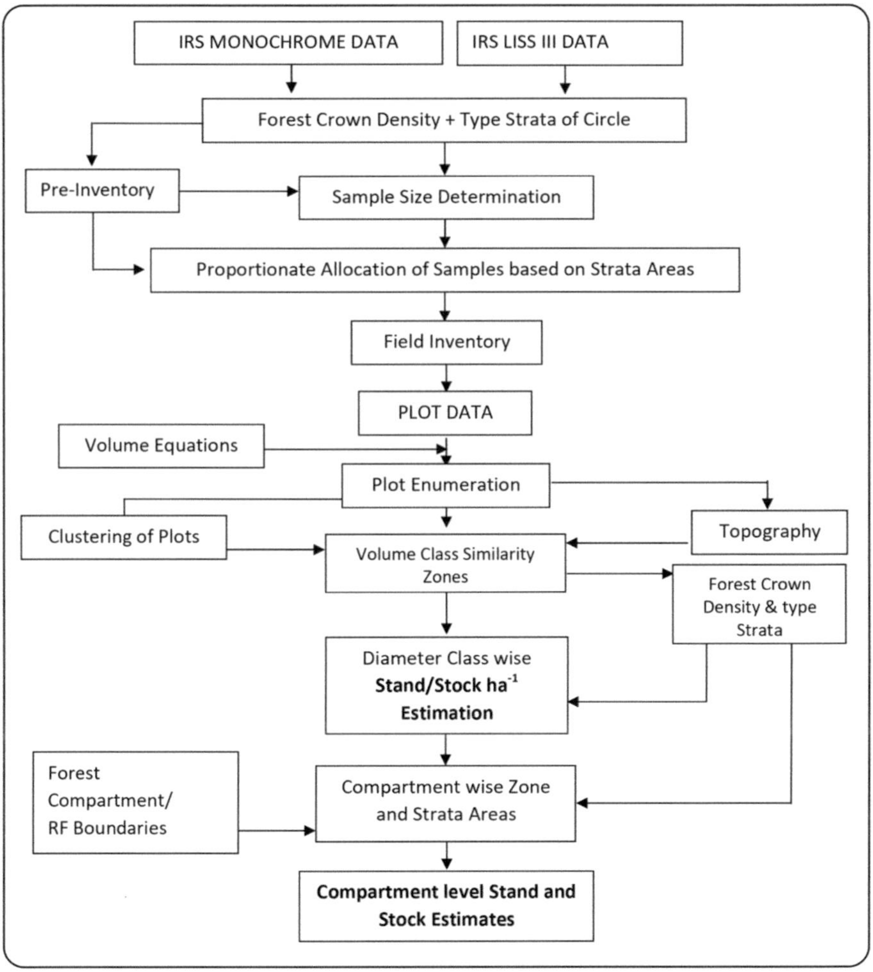

Fig. 5.3 Methodology flow for the generation of forest stand and growing stock tables

5.2.4 Geospatial Inputs for Preparation of Forest Working Plan for Arunachal Pradesh

The preparation of RS & GIS inputs for forest working plan for Arunachal Pradesh has been carried out since 2013. Detailed analysis on timber volume in different canopy density, girth classes at compartment level (wherever boundary available) in spatial domain are being prepared and provided to Department of Environment and Forests as inputs for preparation of e-Working Plan. The state is highly mountainous with dense forests and snow-clad area and is spread over elevations ranging from 100 m to 7,000 m above mean sea level.

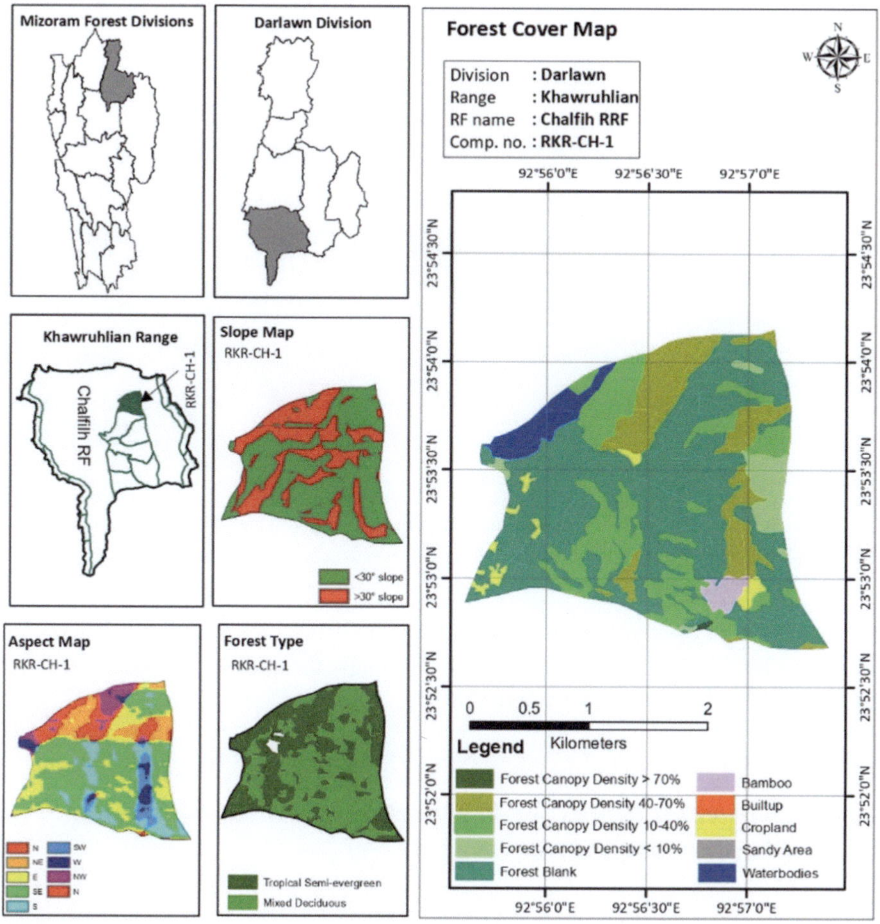

Fig. 5.4 Timber stock map for a compartment in Darlawn Forest division, Mizoram

The forest canopy density maps for the forest were prepared using high resolution satellite data (Cartosat-1 and LISS IV data). Forest canopy densities were categorized into very dense, moderately dense, open and forest scrub based on the signatures developed for visual image interpretation (Fig. 5.5). Forest type maps were also prepared using multi-temporal data. Based on the forest canopy density and forest type overlay, or in the case of hilly mountainous area forest canopy density and elevation were overlaid, to stratify the forest into homogenous classes. Sampling points were distributed within grids of 25" x 25" within and outside the RFs, excluding a buffer area of 200–2000 m from the road network. And field enumeration surveys were carried out by the respective staff of the forest divisions of the states 5.5.

Forest compartments based on micro-watersheds with area ranging from 350–500 ha for whole of the divisions for all the divisions were prepared. Carto DEM and

Fig. 5.5 Forest Canopy density map of Arunachal Pradesh prepared based on Cartosat-1 and LISS IV data pertaining to the period 2009–2012

draining network digitized from the topo maps were used for delineating the micro-watershed boundaries. Timber growing stock maps for each micro-watersheds for the divisions were also prepared (Fig. 5.6).

5.2.5 Geospatial Inputs for Preparation of Forest Working Plan for Meghalaya

For the state of Meghalaya, NESAC have prepared the forest canopy density maps for the entire state using Cartost-1 and LISS IV data pertaining to 2008–2012. Five levels of canopy density maps at 1:10,000 scale were prepared (>70%, 40–70%, 20–40%, 10–20% and < 10%). Forest type map for the entire state was also prepared at 1:50,000 scale. Sampling points were allocated based on stratified random sampling and the state forest department gathered the field enumeration data (Fig. 5.7).

As per the requirement of the department, the growing stock were estimated based on final stratification of forests into 2 categories of forest cover (<40% canopy cover and > 40% canopy cover) and forest types. The compartment boundaries were digitized from the forest maps available with the department.

Based on the field enumeration data provided from the forest department, the growing stocks for all the compartments were estimated. The information on growing stock and stock maps were utilized by the department to prepare the forest working plan for 27 Reserved Forests in the three forest divisions of Jaintia Hills, Khasi Hills and Garo Hills.

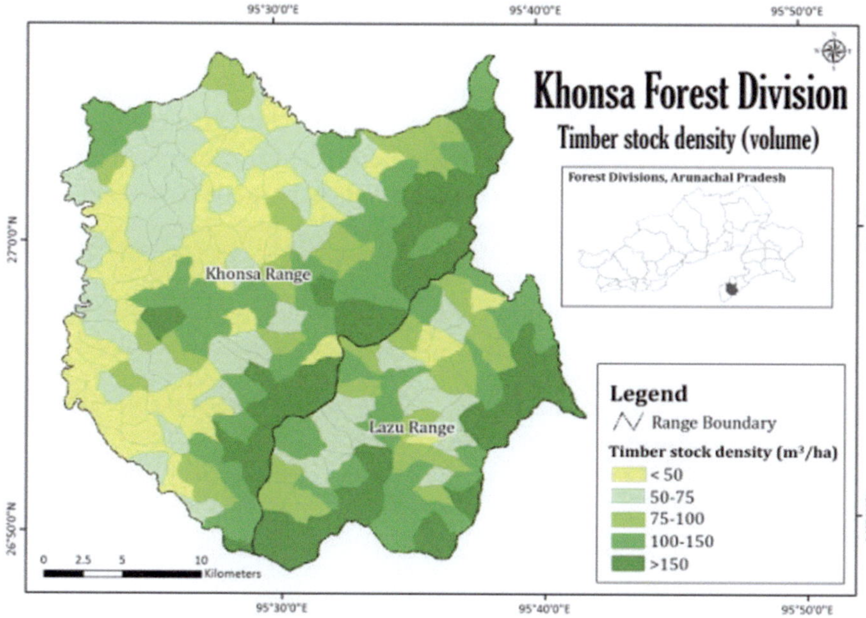

Fig. 5.6 Compartment-wise timber stock density in Khonsa Division, Arunachal Pradesh

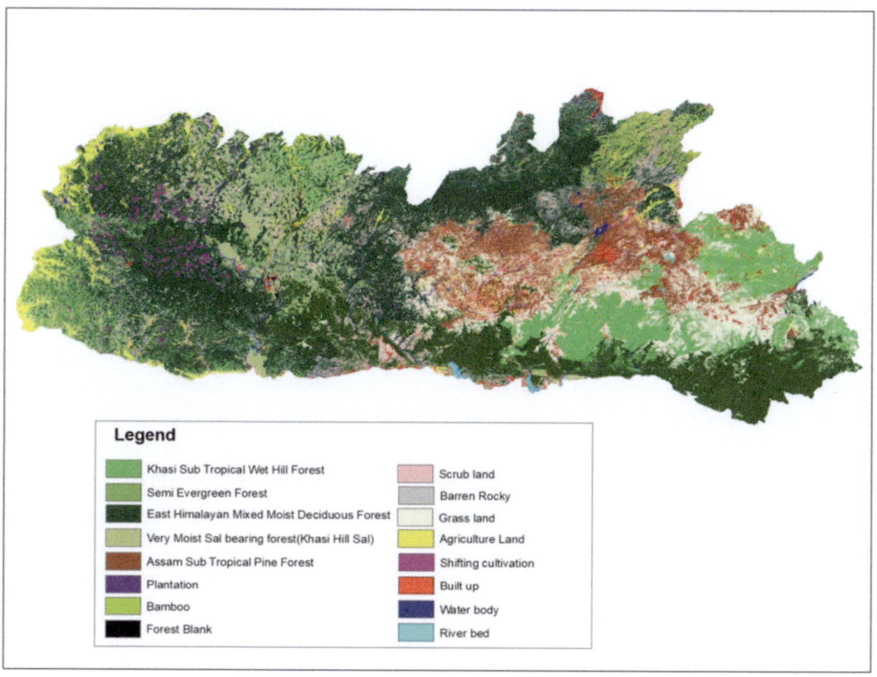

Fig. 5.7 Forest type map of Meghalaya, India

5.3 Assessment of Bamboo Resources in North East India

Inventory and mapping of bamboo resources is being carried out since 2001 in different parts of the North Eastern Region (Talukdar 2001; Lallianthanga et al. 2009; Goswami et al. 2010; Sarma et al. 2010). Based on different vegetation indices, 325.74 km^2 of pure bamboo patches area were mapped in Arunachal Pradesh using space technology. Assessment of the bamboo flowering areas was carried out for Mizoram in collaboration with Mizoram Remote Sensing Application Centre using 2005, 2006 & 2008 satellite data and the maximum loss of bamboo biomass was reported from Mamit district (1539 km^2). Mapping the spatial distribution of bamboo resources in six districts of Nagaland was carried out using high resolution IRS LISS IV imagery to assess the availability of raw materials in Mon, Longleng, Mokokchung, Wokha, Dimapur and Peren districts of Nagaland for the Nagaland Pulp & Paper Company Limited at Tuli, Mokokchung. A total of 41,151.15 ha bamboo growing area that could provide the necessary raw material for the paper mill was estimated. A detailed estimate of bamboo estimates in different bamboo working circles of Dima Hasao district of Assam was also carried out and the Bamboo working scheme for the district was approved by MoEFCC, Govt. of India, based on the inputs provide by NESAC. Spatial distribution of bamboo bearing area and growing stock estimation at village level were also prepared for Mokokchung and Wokha district (Fig. 5.8) in collaboration with Nagaland University. NESAC is currently mapping the bamboo resources for the North Eastern Region using LISS IV satellite data.

5.4 Conclusion

Although most of the states in the northeast part of the country have good forest cover (above 70%) in recent times, it has witnessed drastic changes in the land use practices, resource exploitation and expansion of plantation crops. To manage those forests which are under the control of the Government, every territorial division of a state has to have the working plan with a specific set of silvicultural treatments of working circles. Based on the spatial information on forest cover, growing stock, topography relevant treatments can be prescribed by the forest manager for the following one decade. With the improvement in remote sensing techniques, satellite data resolution, the forest resource assessment has been greatly enhanced. Based on the geo spatial inputs provided by NESAC to the Forest Department of the north-eastern states, Forest Working Plans of different divisions has been approved. Bamboo working schemes in some of the divisions has also been approved. In the forest working plans, the forest managers prescribed different working circles like conservation and improvement working circles in steep slopes and thinly forested areas; selection working circles to enhance regeneration, plantation working circle, soil and water conservation circle or wildlife management and biodiversity working circles where the bio-richness is high. In conclusion, forest resource assessment and management

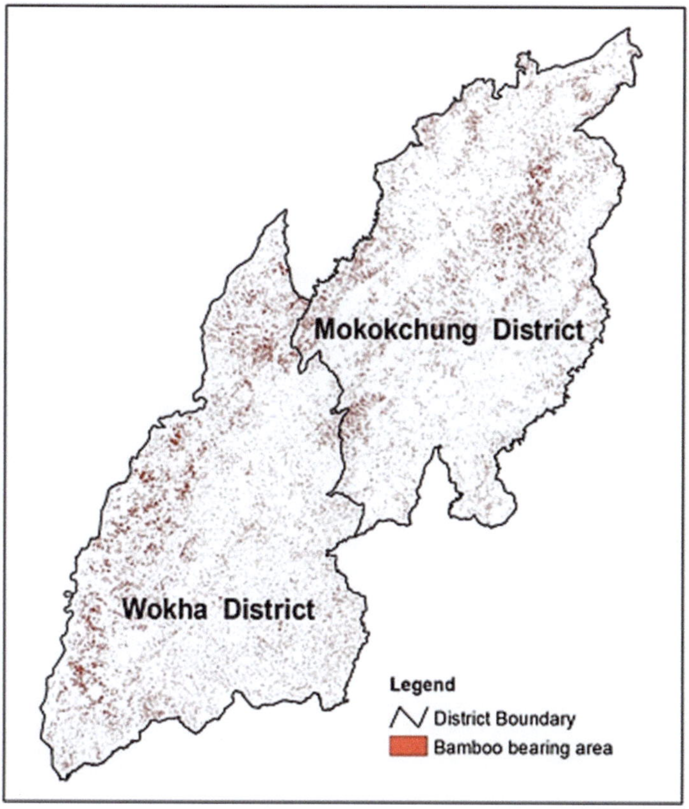

Fig. 5.8 Spatial distribution of bamboo bearing area in Mokokchung and Wokha districts of Nagaland

are crucial for ensuring the sustainability and benefits of forests for current and future generations as the northeast India and many parts of the world are suffering from shortage of forest produce. By understanding the forest ecosystems and status of the forest resources, and implementing responsible management practices with the aid of remote sensing and GIS technology, we can maintain the ecological, economic, and social values of the forests and appreciate the tangible and intangible benefits from the forests.

Acknowledgements The authors wish to acknowledge the support of Dr. B. K. Handique for the statistical design and support to the sampling techniques and growing stock estimation while executing the projects and to all the past and present Directors of NESAC, namely, Shri K. C. Bhattacharya, Dr. P. P. Nageswara Rao, Dr. S. Sudhakar, Dr. P. G. Diwakar, Shri P. L. N. Raju and Dr. S. P. Aggarwal for the support and guidance. The State Forest Departments of Meghalaya, Mizoram, Arunachal Pradesh, Assam and the North Eastern Council, Shillong are also thankfully acknowledged for the financial supports and collaboration for the projects.

References

K.S. Bawa, S. Dayanandan, Climate change and tropical forest genetic resources. Clim. Change **23**, 449–466 (1998)

K. Brown, D.W. Pearce, *The causes of tropical deforestation: the economic and statistical analysis of factors giving rise to the loss of the tropical forests* (University College Press, London, 1994)

V.J. Chacko, *A Manual on Sampling Techniques for Forest Surveys.* (The Manager of Publications, Forest Research Institute, Dehra Dun, 1964), p. 172

K. Chakraborty, D. Bhavsar, S.K. Swain, S. Bhuyan, H. Chaudhary, J. Mohapatra, P. Kumar, B. Lyngdoh, J. Dey, B. Rynjah, N. Nishant, K.K. Sarma, Framework for eForest fire management in the shifting cultivation-dominated landscape of Meghalaya, North East India, using remote sensing and GIS. Curr. Sci. **127**(5), 572–580 (2024)

P.N. Duinker, S. Nilsson, M.E. Chipeta, Forestry for sustainable development and global fibre supply, Unasylva 193, Vol. 49, Global Fibre Supply (1998)

C.B.S. Dutt, V. Udayalakshmt, A.S. Sadhasivaiah, Role of remote sensing in forest management. Geospatial World (2009). https://geospatialworld.net/article/role-of-remote-sensing-in-forest-management/

FAO, *State of the world's forests* (FAO, Rome, 2003)

R. Fararoda, G. Rajashekar, M. Praveen, R.S. Reddy, G. Srivastava, C.S. Jha, S. Tripathi, P.K. Mallick, Geoinformatics in the preparation of forest working plans for Odisha State. Indian For. **147**(9), 813–823 (2021)

J. Goswami, L. Tajo, K.K. Sarma, Bamboo resources mapping using satellite technology. Curr. Sci. **99**(5), 650–653 (2010)

B.K. Handique, S. Sudhakar, Timber volume estimation by double sampling using high resolution satellite data. Indian Forester **140**(2), 154–161 (2014)

https://www.fao.org/4/w7990e/w7990e03.htm

A.V. Korotkov, T. Peck, M. Köhl, Resource assessment- regional and global forest resource assessments, in: *Encyclopedia of Forest Sciences,* ed. by J. Burley, J. Evans, J.A. Youngquist. (Elsevier, Academic Press, 2004), pp 973–982

M. Kumar, M.P. Singh, H. Singh, P.M. Dhakate, N. Ravindranath, Forest working plan for the sustainable management of forest and biodiversity in India. J. Sustain. for. **39**(1), 1–22 (2019)

S.P.S. Kushwaha, Forest-type mapping and change detection from satellite imagery. ISPRS J. Photogramm. Remote. Sens. **45**(1990), 175–181 (2004)

R.K. Lallianthanga, K.K Sarma, Hmingthanpuii, R.L. Sailo, Lalhmanchhuana, *Study of bamboo flowering in Mizoram using remote sensing and GIS.* Scientific Report (2009)

J.S. Maini, *Sustainable development of forests.* Sustainability, 43, 169 (1992). https://www.fao.org/4/u6010e/u6010e03.htm#sustainable%20development%20of%20forests

A.A. Mao, T.M. Hynniewta, M. Sanjappa, Plant wealth of northeast India with reference to ethnobotany. Indian J. Tradit. Knowledge. **8**(1), 96–103 (2009)

N. Myers, The world's forests: problems and potentials. Environ. Conserv. **23**(2), 156–168 (1996)

S. Nandy, S. Ghosh, S.P.S. Kushwaha, A.S. Kumar, Remote sensing-based forest biomass assessment in northwest Himalayan landscape, in *Remote Sensing of Northwest Himalayan Ecosystems,* ed. by R.R. Navalgund, A. Senthil Kumar, S. Nandy (Springer, Singapore, 2019), pp. 285–311

National Working Plan Code, *National Working Plan Code -2004.* Ministry of Environment and Forest, Government of India, New Delhi, India, 2004

National Working Plan Code, *National Working Plan Code* (For Sustainable Management of Forests and Biodiversity in India). Ministry of Environment and Forest, Government of India, New Delhi, 2014, p. 86

R.K. Pokhrel, A.P. Gautam, Community forest, environment conservation and rural livelihood. Geogr. J. Nepal **17**, 151–162 (2024)

M.K. Rao, A.O. Varghese, Y.V.N. Krishnamurthy, Application of remote sensing and GIS in forest working plans. Indian For. **133**(1a), 65–76 (2007)

P.S. Roy, S. Saran, S. Ghosh, N. Prasad, H. Karnatak, G. Talukdar, Development of biodiversity information system for north east India using internet GIS, in *Geospatial Theory, Processing and Applications*, vol. XXXIV Part 4, ed. by C. Armenakis, Y.C. Le ISPRS Symposium, 9–12 July 2002, Ottawa, Canada (2002)

K.K. Sarma, K. Chakraborty, P. Rocky, R. Bharali, *Bamboo resource mapping for six districts of Nagaland using remote sensing and GIS*, Technical Report (NESAC-SR-77–2010) (2010)

P.J. Sellers, F.G. Hall, R.D. Kelly, A. Black, D. Baldocchi, J. Berry, M.M. Ryan, M. Ranson, P.M. Crill, D.P. Lettenmaier, H. Margolis, J. Cihlar, J. Newcomer, D. Fitzjarrald, P.G. Jarvis, S.T. Gower, D. Halliwell, D. Williams, B. Goodison, D.E. Wickland, F.E. Guertin, BOREAS in 1997: experiment overview, scientific results and future directions. J. Geophys. Res. **102**, 731–765 (1997)

S. Srivastava, T.P. Singh, H. Singh, S.P.S. Kushwaha, P.S. Roy, Assessment of large-scale deforestation in Sonitpur district of Assam. Curr. Sci. **82**(12), 1479–1484 (2002)

P. Stedman-Edwards, *Root causes of biodiversity loss: an analytical approach.* (World Wide Fund for Nature, Washington DC, USA, 1998)

G. Talukdar, *Mapping of bamboo brakes and forest area using satellite remote sensing and GIS in Meghalaya*, (Technical Report, NESAC, 2001)

P. Ulvdal, G. Ståhl, L. Sängstuvall, L.O. Eriksson, K. Öhman, Long-term strategic forest planning based on biased remote sensing predictions. For. Monit. **2**(1), 138–175 (2025)

D. Verma, N. Bala, S. Pandey, S.N. Mishra, V.P. Panwar, National working plan code: journey of Indian forests from 2004 to 2023. Indian For. **150**(11), 1027–1032 (2024)

Rocky Pebam obtained his B.Sc. in Forestry from the North Eastern Regional Institute of Science & Technology (NERIST), Arunachal Pradesh in 2000 and an M.Sc. & Ph.D. in Forestry from Mizoram University. He had worked in the field of wildlife ecology, mapping of national parks, socio-economy and ecology of agroforestry systems, etc. under various research projects in different National Institutions before joining the North Eastern Space Applications Centre, Meghalaya, as Scientist in 2009. His area of work in the Forestry & Ecology Division of NESAC includes geospatial applications in forest resource assessment for preparation of forest working plans, wildlife habitat assessment, shifting cultivation monitoring, forest fire assessment, vegetation recovery and time series analyses of *jhum* cycles in the north eastern hilly region of India. He has published about 21 research papers in national/international journals/conference proceedings and 8 book chapters. He has mentored 7 M. Sc. students and successfully co-guided a doctoral student. As Principal Investigator/ Co-Principal Investigator he has completed 12 projects with funding from DST/DOS/ICMR etc. apart from other user agencies.

Suchitra Haobam received her Master's degree in Botany with specialization in Plant Physiology and Biochemistry from Gauhati University, Assam. She joined North Eastern Space Applications Centre, Umiam, as Scientist in the year 2003 and is currently working in Remote Sensing & GIS applications in forestry and her research areas includes wildlife habitat evaluation, species distribution modelling, bamboo resource mapping and time series analysis of vegetation phenology. She is also the Course Director for the regular two week training course on "Remote Sensing & GIS applications in Forestry & Ecology.

K. K. Sarma Ph. D, leads the Remote Sensing Applications Group (RSAG) at the North Eastern Space Applications Centre (NESAC), where he has been a key figure since 2002. Specializing in remote sensing for forestry and biodiversity, his work leverages space technology for critical applications in Biodiversity, Ecosystems, Forestry, and Natural Resource Management. Dr. Sarma has spearheaded over 60 diverse geospatial projects, contributing significantly to natural resource mapping and monitoring across the North Eastern Region (NER). Currently, he coordinates 110 "Plan of Action" projects, focused on using space technology to improve livelihoods in the NER. He also served as the Principal Investigator and NER Coordinator for all ISRO natural resource management projects, including comprehensive Land Use Land Cover Mapping and National Wasteland Updating, Monitoring, and Change Analysis. With over 75 research publications, Dr. Sarma has successfully supervised six Ph.D. students and actively mentors many others across NER Universities.

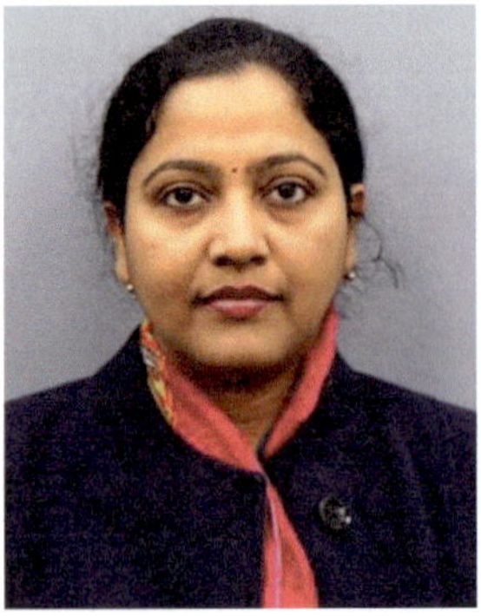

Kasturi Chakraborty completed her M.Sc. and M.Phil in Ecology and completed her Ph.D. on 'Land use/land cover and forest change using Remote sensing and GIS in Barak basin'. She has led project works and contributed as team member for successful completion of important projects at NESAC. Few of the important are in the field of forest fire, forest working plan, forest change analysis at regional scale, land use land cover mapping, wetland conservation and management. She has published research articles on forest and land use change analysis using geospatial techniques. She is a recipient of ISRO, young scientist award (2009) and ISRO project team award (2019).

Chapter 6
Estimation of Forest Above Ground Biomass Using Geospatial Technology

Dhruval Bhavsar, Kasturi Chakraborty, Rocky Pebam, and K. K. Sarma

Abstract Forests play a vital role in global carbon sequestration, with Above Ground Biomass (AGB) serving as a key indicator of forest health and carbon storage capacity. Accurate estimation of AGB is essential, particularly in ecologically sensitive and topographically complex regions like the North Eastern Himalayas of India. In this study, various methods have been presented that integrate field inventory and advanced geospatial technologies for robust AGB estimation. Advanced techniques such as Terrestrial Laser Scanning (TLS), UAV-based LiDAR, and multi-sensor satellite data offer high-resolution measurements of forest structure. Various empirical, nonparametric, and physical modelling approaches enhance AGB prediction using variables like vegetation indices, SAR backscatter coefficients, and canopy metrics. Integrating multisource data significantly improves prediction accuracy, overcoming the limitations of individual sensors. Case studies from Tripura and Meghalaya demonstrate the enhanced accuracy of AGB mapping using dual-polarised SAR and vegetation indices, achieving accuracies above 80%. Robust sampling designs, error metrics (e.g., RMSE, R^2), and uncertainty analyses ensure reliability and compliance with IPCC guidelines. Despite technological advances, challenges remain due to terrain, cloud cover, and data availability. The study recommends harmonising advanced remote sensing with localised models for effective forest carbon monitoring and management in India's northeastern region.

Keywords Above Ground Biomass (AGB) · Field inventory · SAR · Multisource integration · AGB model · Forest carbon

6.1 Introduction

Forests are essential to life on Earth, serving as critical carbon sinks and supporting socio-economic development. They host 80% of the world's land-based animals and plants, making them the most diverse ecosystems on the planet (Saadia 2012).

D. Bhavsar (✉) · K. Chakraborty · R. Pebam · K. K. Sarma
North Eastern Space Applications Centre (NESAC), Umiam, Shillong, Meghalaya, India
e-mail: bhavsardhruval@gmail.com

S. P. Aggarwal et al. (eds.), *Remote Sensing of the North Eastern Himalayan Ecosystem*,
https://doi.org/10.1007/978-981-95-3221-6_6

Covering 31% of the global land area, or 4.06 billion hectares (FAO 2020), forests play a pivotal role in absorbing around 30% of carbon emissions from fossil fuels and deforestation (Le Quere et al. 2018). This creates a significant land residual carbon sink, shaped by emerging environment trends and anthropogenic activities (Ciais et al. 2013). Forests predominantly store carbon as biomass, which is vital not only as a carbon sink during photosynthesis but also for ecosystem productivity and bioenergy generation (FAO 2015).

Globally, tropical rainforests exhibit the highest biomass, compared to extratropical, temperate, and boreal forests (Saatchi 2019). Tropical forests hold approximately two-thirds (262.1 PgC) of the world's terrestrial biomass (Pan et al. 2013). Biomass, the total mass of living and dead organic matter both above and below ground is usually expressed as dry weight per unit area or volume (IPCC 2006; FAO 2015). In forest ecosystems, Above Ground Biomass (AGB) consists mainly of three components:

- **Trees (Woody Biomass)**: Trees dominate AGB, with their trunks, branches, and leaves accounting for the bulk of biomass in forested regions.
- **Shrubs and Understory Vegetation**: Shrubs, saplings, and smaller plants, though contributing less than trees, are significant components, especially in the North Eastern Himalayan region.
- **Dead Biomass**: Dead trees and decaying organic matter contribute to AGB until fully decomposed.

AGB is crucial for understanding forest ecosystems and their role in climate change mitigation, particularly in biodiverse (FAO 2020). As a key metric, AGB aids in evaluating carbon storage potential, forest health, productivity, and sustainable management while supporting climate modelling efforts (Turton et al. 2022). Accurate AGB data are critical for carbon sequestration assessments and programs like REDD + (IPCC 2007). AGB also has applications in forest fire risk assessment, timber management, land-use change monitoring, and quantifying ecosystem services such as biodiversity, food and fibre production, and greenhouse gas accounting. However, despite progress in tracking forest cover changes, understanding terrestrial AGB changes remains limited (Saatchi et al. 2011).

The North Eastern Himalayas, with its diverse vegetation, rugged terrain, and unique climate, is a vital region for biomass studies. It features dense tropical wet evergreen forests, subtropical pine forests, temperate broadleaf and coniferous forests, and alpine meadows, all contributing to varying levels of AGB in eight states of North East India. AGB in tropical forests of Northeast India varies significantly between states due to differences in species composition and forest types. For instance, in Assam, the dense evergreen forests of the Jeypore reserve forest recorded AGB of up to 273 Mg/ha (Hussain et al. 2019). Similarly, the semi-evergreen forests of the Cachar region in Assam exhibited AGB ranging from 32.47 to 261.64 Mg/ha (Borah et al. 2013). In Meghalaya, the tropical moist deciduous forests of Laitkor Reserve Forest had an average AGB of 195.59 ± 49.91 Mg/ha. In Manipur, the tropical dense forests showed an average AGB of 144.2 Mg/ha, whereas the open forests had significantly lower values, averaging 46.5 Mg/ha (Sharma et al. 2020). For

Mizoram, the AGB in tropical forests was estimated to range between 66 and 108 Mg/ha (Hauchhum 2017). In Tripura, the average AGB was found to be 55.82 Mg/ha (Bhavsar et al. 2024), while in Nagaland, the tropical semi-evergreen forests recorded an average AGB of 130 Mg/ha (Sinha and Modak 2023). The temperate forests of Sikkim demonstrated an average AGB of 252.58 ± 72.98 Mg/ha (Prakash et al. 2024). In Arunachal Pradesh, the tropical forests exhibited an average AGB of 163.25 Mg/ha, whereas the temperate forests showed a range of 23.14 to 103.82 Mg/ha (Kashung et al. 2018). The overall forest cover in the states of the North East Region of India is given in Table 6.1 (FSI 2024).

The region's biodiversity and high rainfall result in substantial biomass accumulation, making it a critical global carbon sink. However, the area faces significant threats, including deforestation, shifting agriculture (slash-and-burn), and climate change impacts like extreme weather events (Singh and Chaudhary 2023). These factors can lead to forest degradation, biomass reduction, and the disruption of forest functions. Therefore, accurate AGB estimation in the North Eastern Himalayas is crucial for local forest management and global climate change mitigation. Traditional field methods for AGB measurement involve direct but labour-intensive inventories, impractical for large or inaccessible areas. These methods measure tree height, diameter, and species composition. In contrast, remote sensing (RS) offers a more efficient, broad-scale, and cost-effective approach. RS techniques—using multispectral, hyperspectral, RADAR, and LiDAR data—integrated with geographic information systems (GIS), allow for the mapping, monitoring, and modelling of biomass across large areas (Kumar et al. 2019; Nandy et al. 2019). Unlike field methods, RS enables continuous updates and higher spatial resolution, making it essential for large-scale applications like REDD + and climate modelling (IPCC 2007). However, field data remain essential for refining and confirming RS models to verify their accuracy. (Saatchi 2019). Therefore, combining RS and field data improves AGB estimation and provides a clearer understanding of carbon stocks, forest health, and ecological resilience (Sivasankar et al. 2018; Tian et al. 2023).

Table 6.1 Status of forest cover within reserve forest and tree outside forest in North East Himalayas

Sr. no	State/UT	Forest cover (%) of state's geographical area	Forest cover (%) of India's geographical area	Growing stock Volume (M m^3)
1	Arunachal Pradesh	80.11	2.04	529.45
2	Assam	38.78	0.93	145.21
3	Manipur	75.22	0.51	74.45
4	Meghalaya	78.86	0.54	63.40
5	Mizoram	88.03	0.56	82.01
6	Nagaland	76.10	0.38	55.03
7	Sikkim	48.01	0.10	36.17
8	Tripura	74.69	0.24	33.77

6.2 Field Inventory for Forest Biomass

Field data plays a crucial role in estimating forest biomass, particularly in remote sensing (RS)-based assessments. Serving as the ground truth, field data is essential for calibrating and validating remote sensing models, ensuring that estimates of forest biomass are accurate and reliable (Nandy et al. 2019). This synergy between on-the-ground data and remote sensing technology enhances the precision of biomass assessments, which are key to understanding carbon stocks, forest health, and supporting climate change mitigation efforts. By integrating field measurements with RS-derived data, limitations inherent in each method can be overcome.

6.2.1 Statistical Sampling

Forest inventory measurements involve either the direct measurement of tree biomass through destructive harvesting or indirect estimation using tree size and allometric equations (Chave et al. 2005). Destructive sampling involves cutting down trees to measure their biomass. While it provides accurate data, this method is labour-intensive and harms the environment. It is also not practical for large-scale studies, especially in biodiverse or hard-to-reach areas like the Himalayas. Additionally, it can lead to sampling bias.

Conventional methods for AGB estimating at any location rely on statistical sampling, following guidelines and protocols for greenhouse gas (GHG) inventories in forestlands (IPCC 2006). The two primary sampling methods used are systematic and random sampling, applied either across the entire forest area or in stratified sections. These approaches ensure that biomass estimates are accurate and have low uncertainty. Systematic sampling is commonly used in National Forest Inventory (NFI) systems in developed countries, including the U.S., Finland, Sweden, China and Canada (Saatchi 2019).

In India, the Forest Survey of India (FSI), which conducts the country's forest surveys, also uses a systematic sampling approach for its NFI system. However, in tropical and unmanaged forests, like those in North East India, NFI implementation is challenging due to limited accessibility, deforestation, degradation, and high costs of repeated monitoring. In such cases, stratified random sampling is the most commonly used method. The forest is divided into different strata based on forest type and density, and sample plots are then placed randomly within each stratum. Determining the appropriate number of sample plots is crucial for accurate results. This is usually done through a pilot study or by using previous survey data to estimate sample size using the formula given by Chacko (1965) in Eq. 6.1.

$$N = \frac{t^2 \times CV^2}{(SE\%)^2} \tag{6.1}$$

where N denotes the size of sample, t represents the statistical value at 95% confidence level, CV signifies the coefficient of variation and $SE\%$ is the standard error. The total number of sample plots can be further segregated into different strata using Eq. 6.2 given by Cochran (1963).

$$n_h = \frac{N_h}{N} \times n \tag{6.2}$$

where n_h represent number of samples in h stratum, N_h is the size of h stratum, N is the total size of the population and n represents total number of samples.

6.2.2 *Sample Plot Design*

In forest inventory, sample plot design varies based on plot shape, size, and forest type. Common plot shapes include circular, square, and rectangular, each with its advantages. Circular plots are commonly used in most of the countries in their National Forest Inventory (NFI) systems, including the FSI (Fig. 6.1), because they are easy to set up without needing to mark boundaries. However, while efficient for small-scale studies, circular plots may not align well with remote sensing pixels. Rectangular plots, especially when elongated, can match remote sensing data better, but may introduce edge effects and errors in counting trees along their longer boundaries.

Plot size is crucial and depends largely on the application and the resolution of remote sensing data. For biomass estimation, larger plots (ranging from 0.25 ha to 1 ha) are recommended to reduce errors related to pixel size, especially in studies

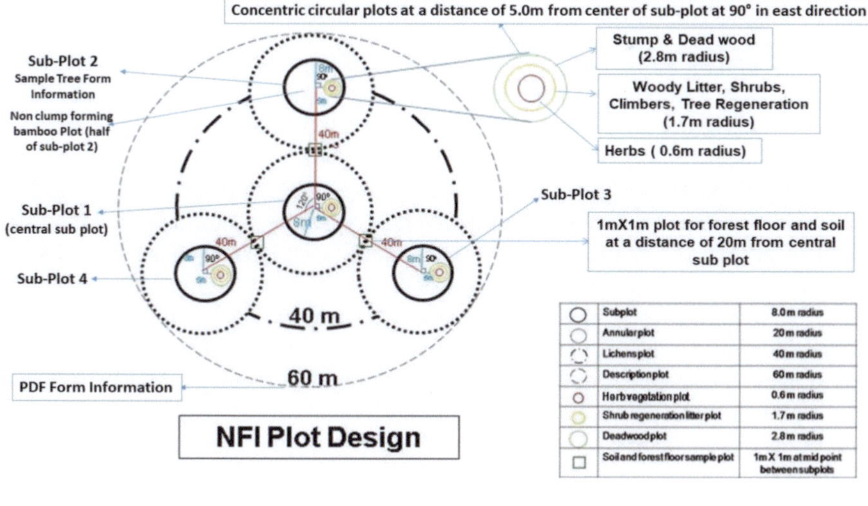

Fig. 6.1 India NFI plot design developed by FSI (FSI 2022)

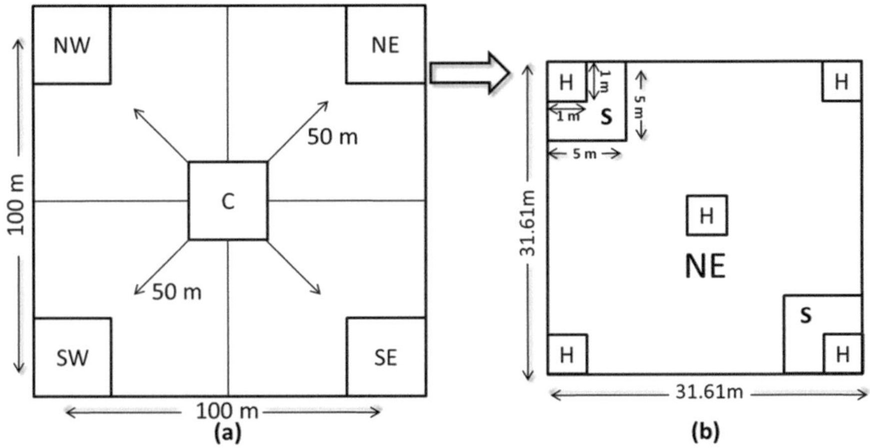

Fig. 6.2 Cluster plot design for biomass estimation of **a** 1 ha plot area having 5 subplots of **b** 0.1 ha of trees with shrubs and herbs design (NE = North East, SW = South West, NW = North West, SE = South East, C = Centre, S = Shrub, H = Herb)

using Synthetic Aperture Radar (SAR) (Saatchi 2019). Cluster plots offer a cost-effective method for collecting tree measurements, particularly in complex ecosystems (as shown in Fig. 6.2). A larger plot of 1 ha (100 m × 100 m) is typically divided into smaller subplots of 0.1 ha, allowing for efficient sampling of trees of varying sizes and densities. The 1 ha plot is divided into five sections—NE, SE, C, SW, and NW. Each 0.1 ha subplot is used for tree sampling, while further subdivisions are made for shrubs and herbs. Shrubs are sampled within 5 m × 5 m plots, and herbs within 1 m × 1 m plots. This plot design is advantageous in the tropical forests of Northeast India, where high biodiversity and challenging terrain make inventory difficult.

The topography in this region often affects plot orientation, requiring careful measurement of slope angles and adjustment for accurate area estimation. When setting up plots using GPS, true or magnetic north should be recorded for accurate comparison with remote sensing data. Tree plot measurements typically include tree height, diameter at breast height (DBH), and species information to assess biomass, wood density, and forest health. These parameters are used for the estimation of AGB. Allometric models are statistical tools that estimate biomass by linking readily measurable tree attributes such as DBH and height, offering a non-destructive approach to biomass assessment (Chave et al. 2005). In the absence of specific allometric equations, volumetric models specific to individual species can be used as an alternative approach. FSI (1996) provide these equations, which can be multiplied by the species-specific wood density and further multiplication by the biomass expansion factor (Nandy et al. 2019). Shrub plots involve recording the total number of individuals by species, with representative samples collected for dry biomass estimation. Herb plots may be harvested and weighed entirely, while litter biomass is measured by weighing material from a 1 m × 1 m plot and taking a 100 g sample for dry biomass calculation. Biomass for the entire forest can be extrapolated based on the plot size and forest area.

6.2.3 *Advanced Remote Sensing Technology for Field Inventory*

While traditional field-based methods like destructive sampling and allometric equations have been effective, they are labour-intensive, time-consuming, and difficult to apply in remote or inaccessible areas, such as the rugged terrains of the North Eastern Himalayas. Advanced technologies like Light Detection and Ranging (LiDAR), specifically Terrestrial Laser Scanners (TLS) and handheld LiDAR sensors, are transforming forest inventories by providing rapid, precise, and non-destructive measurements of forest attributes.

TLS can scan entire plots from a fixed position, providing high-resolution data on tree structure, girth, and height (Nandy et al. 2019), while handheld LiDAR offers flexibility in dense or remote forests where traditional methods are impractical. This ability to cover large areas quickly also reduces human error and sampling bias, making LiDAR an efficient solution for forest measurements in complex terrains. Figure 6.3 presents a three-dimensional point cloud image of a forest stand generated from Terrestrial Laser Scanning (TLS) data. This detailed point cloud visualization captures the structural complexity of the forest, including individual tree trunks, branches, and understory vegetation. Handheld LiDAR is especially valuable for navigating difficult areas and allows for GPS-based geotagging of trees, enhancing spatial data accuracy. Furthermore, Drones equipped with cameras, LiDAR, or multispectral sensors allow for quick and flexible data collection, even in difficult-to-access areas (Brede et al. 2022). They capture high-resolution images and 3D models that can also be used to assess tree height, canopy cover, and spatial structure.

In addition to the precision and efficiency benefits, LiDAR complements conventional field data by enhancing remote sensing (RS) models, which still require ground-based measurements for calibration and validation. This synergy between LiDAR, RS models, and traditional methods ensures robust, large-scale biomass assessments crucial for forest management, carbon accounting, and ecosystem monitoring (Asner and Mascaro 2014). In regions with diverse forests like North East India, LiDAR's advanced capabilities are essential for achieving sustainable forestry and conservation goals.

Fig. 6.3 A 3D point cloud representation of a forest stand captured using Terrestrial Laser Scanning (TLS)

6.3 Role of Geospatial Technology in Biomass Estimation

Geospatial technology has transformed forest biomass estimation by offering efficient, scalable, and non-destructive methods compared to traditional field-based techniques (Brede et al. 2022). This suite of tools includes remote sensing, Geographic Information Systems (GIS), and Global Positioning Systems (GPS). Geospatial technology is particularly beneficial in challenging terrains like the North Eastern Himalayas, where accessibility and environmental concerns often hinder traditional methods. These tools help overcome such limitations by providing accurate data on forest structure, vegetation cover, and landscape features, which are essential for calculating AGB.

Remote sensing involves capturing data about the Earth's surface from a distance using satellite or aerial sensors (Jensen 2009). This technology can collect various types of imagery, including optical, radar, and LiDAR, at different spatial and temporal resolutions, providing critical information on forest biomass and its spatial distribution (Kumar et al. 2019). GIS is a system for capturing, storing, analysing, and visualizing geospatial data, integrating data from multiple sources such as remote sensing and GPS to create detailed maps of forest biomass (San Juan and Domingo-Santos 2018). This enables the analysis of spatial patterns and trends over time. GPS provides precise location information and is often used with remote sensing and GIS to georeferenced field data, ensuring accurate spatial positioning and mapping of biomass. Various remote sensing derived variables can be used for the prediction of biomass distribution, depending on the source or type of remote sensing data.

6.3.1 Optical Remote Sensing

Optical remote sensing sensors detect reflected or emitted electromagnetic radiation from the surface, which can be analysed to derive valuable information about vegetation and other land cover features (Jensen 2009). Two main types of optical remote sensing are multispectral and hyperspectral, providing information on vegetation composition and health. Multispectral imaging captures data in a limited number of broad spectral bands (typically 3 to 10 bands), such as red, green, blue, near-infrared (NIR), and short-wave infrared (SWIR) (Tian et al. 2023). The most commonly used freely available multispectral sensors are presented in Table 6.2.

Hyperspectral imaging captures data in a much larger number of narrow spectral bands (often hundreds or thousands of bands), allowing for a more detailed analysis

Table 6.2 Open source multispectral sensors with specifications

Sr. no	Sensor name	No. of bands	Spatial resolution (m)	Country/agency
1	Landsat MSS (Multispectral Scanner)	4	80	U.S. Geological Survey (USGS) and the National Aeronautics and Space Administration (NASA), USA
2	Landsat TM (Thematic Mapper)	7	30, 120	USGS and NASA, USA
3	Landsat ETM + (Enhanced Thematic Mapper)	8	30,15,60	USGS and NASA, USA
4	Landsat OLI/TIRS (Operational Land Imager/Thermal Infrared Sensor)	11	30,15,100	USGS and NASA, USA
5	Sentinel 2A/B	13	10,20,60	European Space Agency (ESA)
6	AWiFS (Advanced Wide Field Sensor)	4	56	Indian Space Research Organisation (ISRO), India
7	LISS (Linear Imaging Self-Scanning Sensor) III	4	23.5	ISRO, India
8	LISS (Linear Imaging Self-Scanning Sensor) IV	3	5.8	ISRO, India
9	MODIS (Moderate Resolution Imaging Spectroradiometer) Terra/ Aqua	36	250, 500, 1000	NASA, USA
10	ASTER (Advanced Spaceborne Thermal Emission and Reflection Radiometer) Terra	14	15,30,90	USA/Japan

of materials and vegetation health (Thenkabail et al. 2002). The commonly used hyperspectral sensors are presented in Table 6.3.

Various variables derived from optical remote sensing are vital for AGB estimation, including vegetation indices, reflectance, and texture metrics. Different vegetation types and conditions have distinct reflectance patterns, which can be used to estimate biomass. The Vegetation index (VI) is a mathematical combination of different spectral bands designed to highlight specific properties of vegetation (Hemati et al. 2024), a widely used variable for developing a relationship with biomass. Normalized Difference Vegetation Index (NDVI) (Rouse et al. 1974) is the most commonly used vegetation index, which uses the contrast between red and NIR bands to measure vegetation vigour (Hemati et al. 2024). Other commonly used VIs are Enhanced Vegetation Index (EVI) (Huete et al. 2002), Soil Adjusted Vegetation Index (SAVI) (Huete 1988), Green Normalized Difference Vegetation Index (GNDVI) (Gitelson et al. 1996), and Normalized Difference Red Edge (NDRE) (Gitelson and Merzlyak 1994). Texture describes the spatial variation of pixels in an image and can provide canopy roughness and leaf distribution. The statistical texture analysis method includes grey level co-occurrence matrix and semi-variogram analysis. Other techniques include feature extraction, filtering and wavelet decomposition (Nandy et al. 2019). Texture metrics like entropy, contrast, and homogeneity help identify fine-scale structural variations in vegetation, which are valuable for differentiating forest age classes or stand densities that directly relate to biomass.

Table 6.3 The commonly used hyperspectral sensor data with specifications

Sr. No	Sensor	spatial resolution (m)	Spectral range (nm)	Number of bands	Country/Agency
1	PRISMA (PRecursore IperSpettrale della Missione Applicativa)	30	400–2500	239	Italian Space Agency (ASI), Italy
2	EnMAP (Environmental Mapping and Analysis Program)	30	420–2450	242	German Aerospace Center (DLR), Germany
3	DESIS (DLR Earth Sensing Imaging Spectrometer)	30	400–1000	235	DLR, Germany
4	HySIS (Hyperspectral Imaging Satellite)	30	400–2500	55 (VNIR), 60 (SWIR)	ISRO, India

6.3.2 Radar Remote Sensing (Synthetic Aperture Radar—SAR)

Radar remote sensing, particularly Synthetic Aperture Radar (SAR), is vital for estimating AGB because it can penetrate cloud and canopy layers, and detect detailed vegetation structures (Tian et al. 2023). The radar signal's ability to capture information regardless of weather conditions makes it ideal for regions like the North Eastern Himalayas, which experience frequent cloud cover and heavy rainfall. The measurement from radar is strongly dependent on vegetation structure and phenology, soil moisture, and dielectric properties of vegetation components with wavelength and geometry of the sensor (Saatchi 2019).

SAR operates by emitting microwaves, and the wavelengths commonly used in radar remote sensing include X-band (2.4–3.75 cm), C-band (3.75–7.5 cm), L-band (15–30 cm), and P-band (30–100 cm). Each wavelength interacts differently with vegetation. Shorter wavelengths like X- and C-band primarily scatter from leaves and small branches providing detailed information about the canopy. While longer wavelengths such as L- and P-band penetrate deeper and interact with trunks and larger branches, making them effective for estimating AGB in dense forests. The available SAR imagery at different wavelengths is presented in Table 6.4.

Key SAR variables used for biomass estimation include backscatter, polarimetric SAR (PolSAR), and interferometric SAR (InSAR). Backscatter is the intensity of

Table 6.4 List of commonly used SAR imagery with its specifications

Sr. no	SAR Sensor	Wavelength	Polarization	Country/Agency
1	TerraSAR-X	X-band (3.1 cm)	Dual (HH, VV) and Quad (HH, HV, VH, VV)	DLR, Germany
2	COSMO-SkyMed	X-band (3.1 cm)	Quad (HH, HV, VH, VV)	ASI, Italy
3	Sentinel – 1A/B	C-band (5.6 cm)	Dual (VV, VH)	ESA
4	EOS-04 (RISAT 1A)	C-band (5.6 cm)	Dual (HH, HV and RH and RV), Quad (HH, HV VV, VH)	ISRO, India
5	RADARSAT-1/2	C-band (5.6 cm)	Quad (HH, HV, VH, VV)	Canadian Space Agency (CSA), Canada
6	ALOS 2 PALSAR 2	L-band (23.6 cm)	Dual (HH, HV) and Quad (HH, HV, VH, VV)	Japan Aerospace Exploration Agency (JAXA), Japan
7	SAOCOM	L-band (23.6 cm)	Dual (HH, HV)	Argentina National Space Activities Commission (CONAE), Argentina

radar energy reflected back to the sensor, which varies with vegetation density, structure, and moisture content. High backscatter values generally indicate denser or higher biomass vegetation (Saatchi et al. 2011). An advanced SAR technique that uses multiple polarizations (HH, HV, VH, VV) to capture more detailed information about the scattering properties of objects. PolSAR further enhances radar's ability to detect vegetation characterisation and improve AGB estimation accuracy (Sinha et al. 2015). Backscattering coefficients obtained from SAR observations are generally attributed to three main contributors: (1) scattering within the canopy volume, (2) combined volume–surface interactions, and (3) scattering from the ground surface (Saatchi 2019). InSAR technique uses two or more SAR images taken at separate times or from slightly displaced sensor positions to detect surface height changes (Tian et al. 2023). It can help in assessing canopy height and aiding in precise AGB estimation. The Radar Vegetation Index (RVI) as given in Eq. 6.3 is also used as a variable to monitor forest types (Kim and Van Zyl 2009).

$$RVI = \frac{8\gamma_{HV}^{0}}{\left(\gamma_{HH}^{0} + \gamma_{VV}^{0} + 2\gamma_{VH}^{0}\right)} \tag{6.3}$$

where γ^0 is the backscatter coefficient that has been corrected both radiometrically and geometrically, expressed in linear scale for each polarisation mode (m^2/m^2). The value of RVI ranges from 0 and 1 and measures the randomness of scattering. The low value (near 0) of RVI indicates a smooth bare surface, while higher values correspond to increased vegetation cover and hence biomass. Due to its ratio-based calculation, RVI is less sensitive to radar geometry, topography, and calibration errors (Saatchi 2019).

6.3.3 LiDAR (Light Detection and Ranging)

LiDAR is one of the most advanced and accurate remote sensing technologies for biomass estimation. LiDAR sensors use laser pulses emitted toward the ground or object and measure the time it takes for each pulse to return, allowing for precise measurements of vegetation height, canopy cover, and vertical biomass distribution (Brede et al. 2022). Several LiDAR datasets are available for canopy height estimation, each providing various spatial resolutions and coverage scales suited for forest analysis. GEDI (Global Ecosystem Dynamics Investigation) launched by NASA on the International Space Station, provides high-resolution, full-waveform LiDAR data focused specifically on measuring 3D forest structure globally (Duncanson et al. 2022). GEDI's data products include canopy height, vertical structure, and biomass at spatial resolutions of about 25 m. ICESat-2 (Ice, Cloud, and land Elevation Satellite-2) also provides forest canopy height data. Using photon-counting LiDAR, ICESat-2 captures vertical profiles of vegetation with a spatial resolution of around 17 m and is useful for broad-scale canopy height and forest structure assessments. Other airborne

LiDAR data can also be procured from national or local survey agencies, which can give the canopy information at individual tree level.

For LiDAR-based tree volume estimation, the RANSAC method fits cylindrical models to trunks, robustly estimating volume by rejecting outliers (Rodda et al. 2024). Quantitative Structure Models (QSMs) provide highly accurate 3D reconstructions (Raumonen et al. 2013), while Voxelization and Alpha Shapes offer computationally efficient alternatives (Lindberg et al. 2010).

6.3.4 Multi-source Remote Sensing Data

The key parameters for forest AGB estimation are DBH, tree height, density, and canopy structure. However, it is challenging to measure all these parameters using a single remote sensing source. To accurately estimate AGB, multiple remote sensing data sources are integrated, enhancing the precision of AGB predictions.

Combining optical with radar data further improves estimates by merging surface-level details with structural information (Tian et al. 2023). This method proved to be valuable in the reserve forest of Khasi Hills in Meghalaya. The vegetation indices (NDVI, EVI and SAVI) from Landsat 9 and the backscatter coefficient from ALOS-2 PALSAR-2 were combined to estimate AGB. The model accuracy was enhanced to 70%, with an RMSE of 50.34 Mg/ha while combining the dataset compared to a single variable model. The spatial distribution of AGB in mixed deciduous forest in protected forest in Meghalaya is presented in Fig. 6.4. Multi-frequency SAR has also shown significant results in AGB estimation. Bhavsar et al. (2022) integrated dual-pol C-band EOS-04 and L-band ALOS-2 PALSAR-2 to estimate AGB in the tropical forests of Tripura. The study revealed that the C-band was more effective in open and scrub forests, while the L-band was better at predicting high biomass in dense forest regions. However, the saturation was found beyond the AGB value of 120 Mg/ha. The study also found that HV polarized bands from both sensors were more effective than HH polarized bands, while the combined bands increased the AGB mapping accuracy to 81%.

Forest AGB estimation combining LiDAR data with other optical and SAR/InSAR data showed a promising option for accurate measurement in challenging terrains and dense canopies. Nandy et al. (2019) combined terrestrial laser scanning (TLS) with high-resolution optical data in North-Western Himalaya's moist forests and got an accuracy of 87.12%. Ehlers et al. (2022) integrated multi-seasonal Landsat data for species composition, LiDAR for canopy height, image texture for crown size (related to DBH), and SAR for canopy volume to reduce the RMSE of AGB prediction up to 18.8 Mg/ha. In tropical African forests, Laurin et al. (2014) demonstrated that LiDAR metrics and hyperspectral vegetation indices integration effectively captured forest structure for biomass estimation. Duncanson et al. (2022) assessed biomass estimates from NASA's GEDI, ICESAT-2, and NISAR missions in Sonoma County, California, using high-resolution airborne LiDAR maps as a reference to support

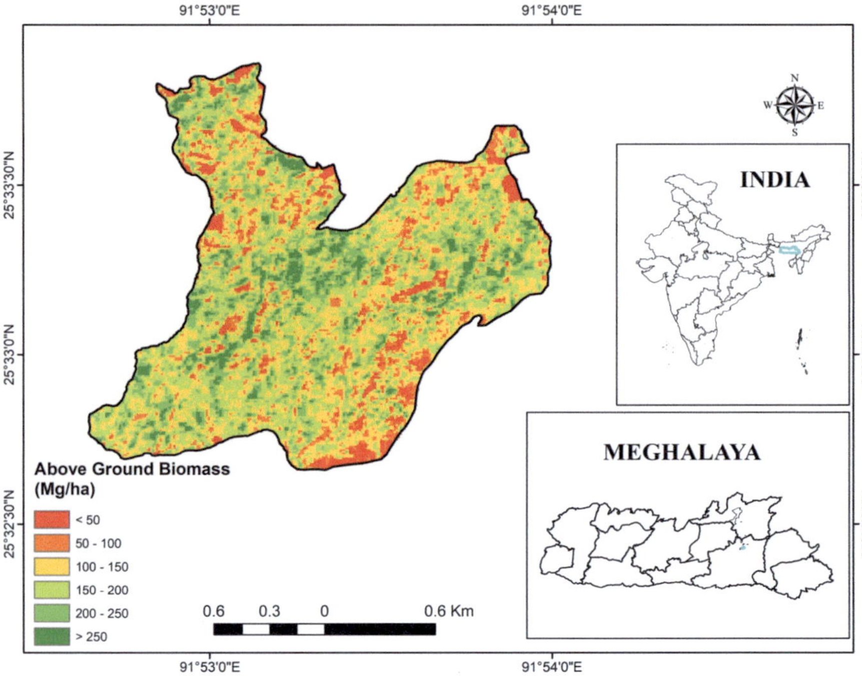

Fig. 6.4 Prediction map of AGB in the Laitkor protected forest of Meghalaya using multisource remote sensing data

inter-comparison and validation of spatial biomass estimates across different satellite missions. These integrated approaches significantly improve the accuracy and reliability of forest AGB estimates.

6.4 Methods for AGB Estimation

Geospatial data generated from various remote sensing imageries can be combined with various modelling approaches to estimate biomass. These models can be empirical, semiempirical, and physical. The empirical model is further divided into parametric and nonparametric models. Each model has its advantages and limitations.

6.4.1 Parametric Model

Parametric models are statistical models that directly relate field measurements of biomass to variables derived from remote sensing data, such as vegetation index or tree height (Nandy et al. 2019). Parametric models, the most commonly used method, are particularly regression approaches which can be linear regression (LR), multiple linear regression (MLR) and non-linear regression (Saatchi 2019; Tian et al. 2023). In regression analysis, the relationship is established between field-measured biomass as a dependent variable and geospatial data as an independent variable such as vegetation index, spectral reflectance, backscatter coefficient etc. In MLR models, multiple independent variables are used to develop the prediction model and estimate biomass. For example, Hussain et al. (2019) used NDVI as an independent variable generated from Resourcesat-2 LISS III satellite data to estimate AGB in the dense Jeypore Reserve Forest of Assam, with values reaching up to 400 Mg/ha. Although vegetation indices are straightforward and easy to understand, their estimation accuracy is typically not very high (Li and Mao 2020).

Radar sensors, due to their penetration ability and scattering from forest stems and large branches, are often used for forest volume and biomass estimation. In the reserve forests of Khasi Hills in Meghalaya, different vegetation indices like NDVI, EVI, and SAVI were used to establish the relationship with AGB, but the coefficient of determination (R^2) was found to be low, ranging from 0.14 to 0.39. However, incorporating the L-band (ALOS-2 PALSAR-2) backscatter coefficient with field inventory data increased the model's prediction accuracy ($R^2 = 0.61$ for the HV polarized band) (Fig. 6.5). Radar backscatter shows a nonlinear relationship with AGB, with increased sensitivity at lower and moderate biomass levels and saturation at higher biomass due to its decreased sensitivity, resulting in a logarithmic or sigmoidal relationship (Saatchi et al. 2011). This saturation threshold differs by radar wavelength, forest type, and structural density, with shorter wavelengths (like C-band) saturating more quickly than longer wavelengths (like L-band). Kumar et al. (2019) applied regression analysis using Sentinel-1 C-band and ALOS PALSAR L-band to estimate AGB in Arunachal Pradesh's temperate forests, while Bhavsar et al. (2024) integrated EOS-04 C-band and ALOS 2 PALSAR 2 L-band for Tripura's tropical forests. The use of varying polarizations, such as VV, VH, HH, and HV, enhances these analyses, as certain polarizations are more effective at detecting specific biomass characteristics, providing a more comprehensive AGB estimate across different forest types.

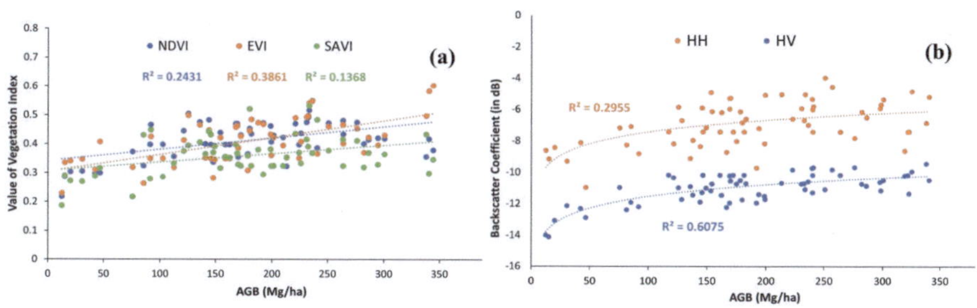

Fig. 6.5 **a** Linear regression analysis between AGB and various vegetation indices **b** Logarithmic regression between AGB and backscatter coefficient of ALOS 2 PALSAR 2 sensors

6.4.2 Nonparametric Model

Parametric models assume an ideal data distribution, typically normal, whereas nonparametric models analyse data directly without relying on assumptions about the underlying distribution (Tian et al. 2023). The AGB maps have also been generated by integrating remote sensing data with climate variables using nonparametric approaches such as interpolation, classification, decision rule technique and machine learning approach (Saatchi 2019). The commonly used non parametric models for forest AGB estimation are artificial neural networks (ANN) (Nandy et al. 2017), k-nearest neighbour (KNN) (Yadav and Nandy 2015), support vector machine (SVM) (Sivasankar et al. 2019), random forest (RF) (Kalita et al. 2022), and gradient boosting (GB). Nonparametric models are beneficial compared to parametric models, especially when there's limited data over a small area or when mapping large regions. These models are less affected by noise from landscape variability, forest structure, moisture changes, and other environmental factors (Saatchi 2019).

Forest structure and its biomass exhibit a complex and nonlinear relationship, which varies across temporal and spatial scales (Saatchi et al. 2011), especially in diverse and heterogenous forests like the North East Himalayas. Nonparametric methods often outperform parametric models here, as shown by Sivasankar et al. (2018), who compared regression and nonparametric SVM models using Sentinel-1 and ALOS PALSAR data in Meghalaya's Nongkhyllem Reserve. The SVM model reduced RMSE from 59 to 20 Mg/ha, achieving greater accuracy. Sivasankar et al. (2019) further improved AGB estimates by including terrain factors (slope, aspect) with L-band SAR, reaching an R^2 of 0.89 and RMSE of 18.21 Mg/ha.

Kalita et al. (2022) mapped AGB in Assam's tea agroforestry systems, using Sentinel-2 data and a random forest (RF) model optimized with spectral indices, achieving high prediction accuracy ($R^2 = 0.86$, RMSE = 7.50 Mg/ha). Roy and Debbarma (2024) found RF to be the best among weighted nonlinear, RF, GB, and SVM models for tree-level AGB estimation, with low RMSE (63.10–132.39 kg) and high modelling efficiency (0.95–0.99). These studies highlight that nonparametric and machine learning approaches offer improved accuracy for AGB estimation in complex forest ecosystems.

6.4.3 Physical Model

Physical models simulate ecological processes governing biomass accumulation, such as photosynthesis, respiration, and water balance, making them more mechanistic compared to empirical models. A physical model is the inversion of remote sensing information, which uses the relationship between biomass and bidirectional vegetation characteristics through remote sensing sensors (Tian et al. 2023). The climate, soil type, and topography variables along with remote sensing data, improve the accuracy of biomass estimates.

Radiative transfer and geometric optical models are commonly used physical modelling methods. These models relate biomass with structural vegetation parameters like canopy cover, LAI, and tree height. LiDAR has been found to enhance the model accuracy by capturing canopy structure and height in the radiative transfer model (Koetz et al. 2007). The most widely used model in microwave remote sensing is the Water Cloud Model, which uses Vegetation Water Content to assess attenuation and scattering effects, focusing on volume and ground scattering (Kumar et al. 2023). This model was originally designed for C-band and shorter wavelengths which does not account for volume-ground interactions. This model is best suited for low density and surface level vegetation biomass estimation (Saatchi 2019). A modified Water Cloud Model (WCM) by Santoro et al. (2002) includes longer wavelength remote sensing data in the model to estimate AGB. Further improvement was carried out by Bouvet et al. (2018) to include volume-surface interaction in the model and designed for L band ALOS PALSAR remote sensing data.

Physical models have shown stability and are effective for localized aboveground biomass (AGB) mapping (Tian et al. 2023). However, their complexity and high data requirements make large-scale AGB estimation challenging, particularly in Northeast India's rugged and remote forest landscapes. Extensive field data collection is difficult in such areas, limiting the application of these models for widespread AGB mapping in this region.

6.5 Accuracy Assessment and Uncertainty Analysis

AGB estimation involves integrating remote sensing data (e.g., optical, radar, or LiDAR data) with field measurements to develop predictive models. However, ensuring the reliability of these models is crucial which requires rigorous accuracy assessment and uncertainty analysis which is also an important ingredient as per IPCC guidelines (IPCC 2006). Accuracy assessment of AGB estimation models is typically conducted by comparing predicted AGB values with ground truth data (reference data collected from field measurements) (Nandy et al. 2019). This is often achieved through statistical metrics that quantify the model's performance. The common steps in the accuracy assessment process include.

6.5.1 Dividing Data into Training and Testing Sets

The field inventory data is usually split into training and testing datasets. The training set is used to build the model, while the testing set (or validation set) assesses the model's accuracy. A common practice is to use 70% of the data for training and 30% for testing (Bhavsar et al. 2024), though this can vary depending on the dataset size and objectives of the study.

6.5.2 Cross-Validation (CV) Techniques

Cross-validation is used to enhance the reliability of accuracy estimates by repeatedly partitioning the data into subsets of training and validation sets. This partitioning help in further performance checks and parameter tuning to prevent overfitting. K-fold and leave-one-out are commonly used in remote sensing to evaluate model accuracy (Saatchi 2019). In K-fold, the dataset is split into k segments (folds). - For each iteration, the model is trained using k–1 folds and evaluated on the remaining fold. This continues until every subset has been used exactly once for validation (Stone 1974). The results are then averaged to provide an overall accuracy estimate. The Leave-One-Out is a special instance of k-fold cross-validation where the number of folds k matches the total number of observations, allowing each data point to act as the validation set once. This can be computationally intensive for large datasets but provides unbiased estimates.

6.5.3 Performance Metrics

The Key metrics used to assess the accuracy of AGB estimation models include the Coefficient of Determination (R^2), Root Mean Squared Error (RMSE), Mean Absolute Error (MAE), and Percent RMSE (%RMSE). The R^2 measures the proportion of variance in observed AGB that the model can predict (Tian et al. 2023). Higher R^2 values imply a well-fitting model, while lower values indicate limited predictive strength, as it reflects how well the model captures variability in the data. RMSE quantifies the average difference between predicted and observed AGB values, providing a measure of error that is sensitive to large deviations (Kalita et al. 2022). RMSE is expressed in the same units as AGB (e.g., Mg ha^{-1}), making it easy to interpret in practical terms (Nandy et al. 2019; Bhavsar et al. 2024). MAE, on the other hand, computes the mean of the absolute differences between predicted outcomes and actual observations (Turton et al. 2022), offering a more intuitive measure of error without placing extra emphasis on large discrepancies, as RMSE does. Percent RMSE represents RMSE as a percentage of the mean observed AGB, facilitating comparisons across different studies and regions with varying AGB ranges. Together, these metrics provide a comprehensive view of model accuracy, understanding the reliability of their AGB estimation models.

Uncertainty analysis in AGB estimation is critical to ensure confidence in reported values, particularly for large-scale applications (e.g., regional or national assessments) that influence policy decisions (Saatchi 2019). Uncertainty analysis involves identifying, minimizing, and quantifying uncertainties from various sources. The key aspects of uncertainty analysis encompass the detection and assessment of contributing factors, efforts to mitigate uncertainty, and approaches for its quantification.

6.5.4 Sources of Uncertainty

In AGB estimation, uncertainties arise from multiple sources that impact the accuracy and reliability of the results. Remote sensing data quality is one of these sources, as variability in sensor resolution, atmospheric effects, and the steps involved in data pre-processing can all introduce uncertainty into AGB estimates derived from remote sensing. Additionally, model errors are a significant source of uncertainty, particularly with statistical and machine learning models, which may have structural limitations, assumptions, or simplifications that prevent them from fully capturing the complex relationships between remote sensing data and AGB (Saatchi 2019). These limitations can lead to inaccuracies in predictions, as models may fail to account for all relevant variables or interactions. Sampling uncertainty in field measurements, which serve as ground truth data, also contributes to overall uncertainty. Field measurements are subject to human error, such as inaccuracies in estimating tree height and DBH, as well as potential lack of representativeness in sample plots (Chave et al. 2005). These sampling errors can introduce biases or inconsistencies when the field data are used to validate or train AGB estimation models. Upon identifying the errors, the uncertainty can be reduced wherever feasible and cost-effective.

6.5.5 Methods for Quantifying Uncertainty

Methods for quantifying uncertainty in AGB estimation include error propagation, Monte Carlo simulation, and confidence intervals, each providing a comprehensive view of the potential error in biomass estimates. Error propagation is a technique that combines uncertainties from multiple sources—such as remote sensing data and field measurements—to calculate an overall uncertainty metric for AGB. By accounting for errors from different sources and propagating them through the model, error propagation produces a single metric that reflects the total uncertainty for biomass estimates at the level of map units or regional averages (Chave et al. 2005; Saatchi 2019). Bhavsar et al. (2024) estimated aboveground biomass (AGB) for Tripura state, with values ranging from less than 30 Mg/ha to over 150 Mg/ha, and provided an uncertainty map for each pixel with propagated uncertainties up to 30 Mg/ha (Fig. 6.6). Monte Carlo simulation is another widely used method, which involves running thousands of simulations by varying input values based on their probability distributions. The resulting spread of simulated AGB values provides insight into the uncertainty range, helping quantify how input variability affects model predictions. Confidence intervals around mean AGB values offer an additional approach to expressing uncertainty, providing a range within which the true AGB value likely lies. For instance, a 95% confidence interval suggests a 95% likelihood that the actual biomass value falls within this specified range. Together, these methods offer robust tools for assessing and reporting uncertainty in AGB estimates, enhancing the reliability of biomass assessments.

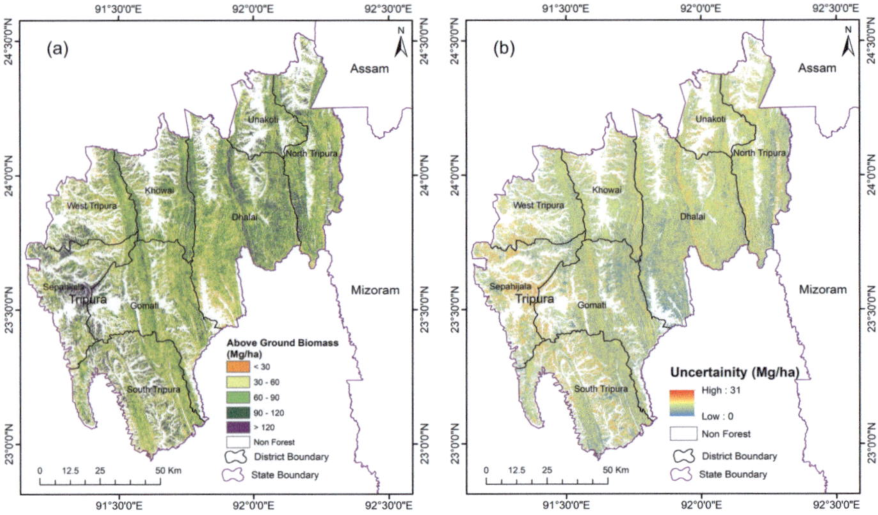

Fig. 6.6 **a** Spatial distribution of AGB of forest in Tripura along with **b** Uncertainty associated with each pixel

The validation of AGB prediction model can also be carried out with independent sample data (Congalton and Green 2019). This helps confirm that the model generalises well beyond the specific dataset used for training. Independent field data collected from different locations or times provides an unbiased test of the model's predictive accuracy. The model accuracy evaluation can be performed using the validation metrics. Accurate, well-quantified AGB estimates with rigorous uncertainty analysis not only improve scientific credibility but also provide stakeholders with reliable information for forest management, carbon accounting, and climate policy.

6.6 Conclusion

AGB estimates provide essential data for monitoring forest health, assessing carbon stocks, and supporting conservation planning, all of which are critical for sustainable forest management. In the North Eastern Himalayan region, with its dense and diverse forests, accurate AGB estimation supports specific applications such as forest degradation monitoring, habitat conservation, and implementing REDD + (Reducing Emissions from Deforestation and Forest Degradation) initiatives. These activities help maintain the region's ecological integrity and contribute to both local and global carbon sequestration goals, underscoring the importance of reliable biomass estimates.

However, several challenges persist in using remote sensing data for biomass estimation, particularly in regions like the North Eastern Himalayas. Frequent cloud cover can hinder optical satellite data acquisition, while limitations in sensor resolution may reduce the accuracy of remote sensing models in capturing fine-scale biomass variations. Additionally, reliable AGB estimation requires extensive ground validation, which is often difficult in the region's rugged and inaccessible areas. Scaling up biomass estimation models from local plots to larger regional landscapes adds another layer of complexity, as it requires careful consideration of environmental heterogeneity and the selection of appropriate remote sensing variables.

Emerging technologies are transforming biomass estimation and may help address some of these challenges. The use of drones and high-resolution imagery offers flexibility in data collection and the ability to capture fine-scale forest details, even in difficult-to-access regions. Artificial intelligence (AI) and deep learning based data analysis are also revolutionising biomass estimation by enabling more precise model calibration and reducing the impact of noise and inconsistencies in large datasets. Additionally, future satellite missions, such as NISAR (NASA-ISRO Synthetic Aperture Radar) and ESA's BIOMASS mission, promise to enhance AGB estimation through improved radar data, capable of penetrating dense canopies and providing detailed structural information. LiDAR technology, which measures forest height and structure with high precision, is also increasingly used in conjunction with other remote sensing data to enhance model accuracy.

Despite advancements, gaps remain in forest biomass data for the North East, where more comprehensive, localized studies are needed to account for the region's unique ecological characteristics. Future research should focus on developing region-specific allometric models, enhancing ground validation networks, and exploring synergies between new technologies and traditional field methods. These steps would improve AGB estimation, providing essential data to inform sustainable forest management and support the region's role in global carbon dynamics.

Acknowledgements The authors gratefully acknowledge the Indian Space Research Organisation (ISRO)/Department of Space (DoS) for funding the study. We sincerely thank the Forest and Environment Department, Meghalaya and Tripura Forest Department for their invaluable field support. We also extend our appreciation to the Forest Survey of India (FSI) for generously providing forest cover and forest type maps. Technical support from the Space Applications Centre (SAC), Ahmedabad, in data processing is deeply appreciated. We are also thankful to the National Remote Sensing Centre (NRSC), Hyderabad, for providing EOS-04 data, which was instrumental in enhancing the accuracy of biomass estimation in this study.

References

G.P. Asner, J. Mascaro, Mapping tropical forest carbon: calibrating remote sensing measurements with ground-based data. Remote Sens. Environ. **145**, 113–127 (2014)

D. Bhavsar, A.K. Das, K. Chakraborty, C. Patnaik, K.K. Sarma, S.P. Aggrawal, Above ground biomass mapping of tropical forest of tripura using EOS-04 and ALOS-2 PALSAR-2 SAR Data. J. Indian Soc. Remote. Sens., 1–11 (2024)

N. Borah, A.J. Nath, A.K. Das, Aboveground biomass and carbon stocks of tree species in tropical forests of Cachar District, Assam, Northeast India. Int. J. Ecol. Environ. Sci. **39**(2), 97–106 (2013)

A. Bouvet, S. Mermoz, T. Le Toan, L. Villard, R. Mathieu, L. Naidoo, G.P. Asner, An above-ground biomass map of African savannahs and woodlands at 25 m resolution derived from ALOS PALSAR. Remote Sens. Environ. **206**, 156–173 (2018)

B. Brede, L. Terryn, N. Barbier, H.M. Bartholomeus, R. Bartolo, K. Calders, M. Herold, Non-destructive estimation of individual tree biomass: allometric models, terrestrial and UAV laser scanning. Remote Sens. Environ. **280**, 113180 (2022)

V.J. Chacko, *A manual on sampling techniques for forest surveys* (Manager of Publications, Government of India, New Delhi, 1965)

J. Chave, C. Andalo, S. Brown, M.A. Cairns, J.Q. Chambers, D. Eamus, T. Yamakura, Tree allometry and improved estimation of carbon stocks and balance in tropical forests. Oecologia **145**(1), 87–99 (2005)

P. Ciais, C. Sabine, G. Bala, L. Bopp, V. Brovkin, J. Canadell, P. Thornton, Carbon and other biogeochemical. Climate change 2013: the physical science basis, in *Contribution of working group I to the fifth assessment report of the intergovernmental panel on climate change.* (Cambridge University Press, Cambridge, 2013)

W.G. Cochran, *Sampling techniques* (John Wiley and Sons Inc., New York, 1963)

R.G. Congalton, K. Green, *Assessing the accuracy of remotely sensed data: principles and practices* (CRC Press, 2019)

L. Duncanson, J.R. Kellner, J. Armston, R. Dubayah, D.M. Minor, S. Hancock, C. Zgraggen, Aboveground biomass density models for NASA's global ecosystem dynamics investigation (GEDI) lidar mission. Remote Sens. Environ. **270**, 112845 (2022)

D. Ehlers, C. Wang, J. Coulston, Y. Zhang, T. Pavelsky, E. Frankenberg, C. Song, Mapping forest aboveground biomass using multisource remotely sensed data. Remote. Sens. **14**(5), 1115 (2022)

FAO, Global Forest Resources Assessment 2015, FAO, Rome, Italy (2015)

FAO, Global Forest Resources Assessment 2020: Main report. Rome (2020)

FSI, Volume equations for forests of India, Nepal and Bhutan. Dehradun: Forest Survey of India, Ministry of Environment and Forests, Dehradun (Uttarakhand), India (1996)

FSI, Manual for Field data collection of National Forest Inventory. Ministry of Environment, Forest and Climate Change Dehradun (Uttarakhand), India (2022)

FSI, India State of Forest Report 2023 (Volume I). Forest Survey of India, Ministry of Environment and Climate Change, Dehradun (Uttarakhand), India (2024). ISBN 978-81-950073-9-4

A.A. Gitelson, Y.J. Kaufman, M.N. Merzlyak, Use of a green channel in remote sensing of global vegetation from EOS-MODIS. Remote Sens. Environ. **58**(3), 289–298 (1996)

A. Gitelson, M.N. Merzlyak, Spectral reflectance changes associated with autumn senescence of Aesculus hippocastanum L. and Acer platanoides L. leaves. Spectral features and relation to chlorophyll estimation. J. Plant Physiol. **143**(3), 286–292 (1994)

R. Hauchhum, Aboveground biomass and carbon stock assessment in forest stands of Gmelinaarborea Roxb. in Mizoram, North-East India. J. Trop. For. Environ. **7**(2) (2017)

M. Hemati, M. Mahdianpari, H. Shiri, F. Mohammadimanesh, Integrating SAR and optical data for aboveground biomass estimation of coastal wetlands using machine learning: multi-scale approach. Remote. Sens. **16**(5), 831 (2024)

A.R. Huete, A soil-adjusted vegetation index (SAVI). Remote Sens. Environ. **25**(3), 295–309 (1988)

A. Huete, K. Didan, T. Miura, E.P. Rodriguez, X. Gao, L.G. Ferreira, Overview of the radiometric and biophysical performance of the MODIS vegetation indices. Remote Sens. Environ. **83**(1–2), 195–213 (2002)

H. Hussain, H.S. Devi, K.K. Sarma, Above ground biomass and carbon stock mapping using NDVI and ecological studies of woody trees of Jeypore Reserve Forest, Assam, India. Indian for. **145**(7), 614–618 (2019)

IPCC, IPCC guidelines for national greenhouse gas inventories, in *National Greenhouse Gas Inventories Programme*, eds. by H.S. Eggleston, L. Buendia, K. Miwa, T. Ngara, K. Tanabe (IGES, Japan, 2006)

IPCC, Climate change 2007: the physical science basis, in *Contribution of Working Group I to the Fourth Assessment Report of the Intergovernmental Panel on Climate Change*. ed. by S. Solomon, D. Qin, M. Manning, Z. Chen, M. Marquis, K.B. Averyt, M. Tignor, H.L. Miller (Cambridge University Press, Cambridge, UK and New York, NY, USA, 2007), pp.1–18

J.R. Jensen, *Remote sensing of the environment: An earth resource perspective 2/e*. Pearson Education India (2009)

R.M. Kalita, S. Nandy, R. Srinet, A.J. Nath, A.K. Das, Mapping the spatial distribution of aboveground biomass of tea agroforestry systems using random forest algorithm in Barak valley, Northeast India. Agrofor. Syst. **96**(8), 1175–1188 (2022)

Y. Kashung, B. Das, S. Deka, R. Bordoloi, A. Paul, O.P. Tripathi, Geospatial technology based diversity and above ground biomass assessment of woody species of West Kameng district of Arunachal Pradesh. For. Sci. Technol. **14**(2), 84–90 (2018)

Y. Kim, J.J. Van Zyl, A time-series approach to estimate soil moisture using polarimetric radar data. IEEE Trans. Geosci. Remote Sens. **47**(8), 2519–2527 (2009)

B. Koetz, G. Sun, F. Morsdorf, K.J. Ranson, M. Kneubühler, K. Itten, B. Allgöwer, Fusion of imaging spectrometer and LIDAR data over combined radiative transfer models for forest canopy characterization. Remote Sens. Environ. **106**(4), 449–459 (2007)

A. Kumar, B.S.P.C. Kishore, P. Saikia, J. Deka, S. Bharali, L.B. Singha, M.L. Khan, Tree diversity assessment and above ground forests biomass estimation using SAR remote sensing: a case study of higher altitude vegetation of North-East Himalayas, India. Phys. Chem. Earth, Parts a/b/c **111**, 53–64 (2019a)

S. Kumar, R.D. Garg, H. Govil, S.P. Kushwaha, PolSAR-decomposition-based extended water cloud modeling for forest aboveground biomass estimation. Remote. Sens. **11**(19), 2287 (2019b)

G.V. Laurin, Q. Chen, J.A. Lindsell, D.A. Coomes, F. Del Frate, L. Guerriero, R. Valentini, Above ground biomass estimation in an African tropical forest with LIDAR and hyperspectral data. ISPRS J. Photogramm. Remote Sens. **89**, 49–58 (2014)

C. Le Quere, R.M. Andrew, P. Friedlingstein, S. Sitch, J. Hauck, B. Zheng, Global carbon budget 2018. Earth System Science Data **10**(4), 2141–2194 (2018)

J. Li, X. Mao, Comparison of canopy closure estimation of plantations using parametric, semi-parametric, and non-parametric models based on GF-1 remote sensing images. Forests **11**(5), 597 (2020)

E. Lindberg, J. Holmgren, K. Olofsson, J. Wallerman, H. Olsson, Estimation of tree lists from airborne laser scanning by combining single-tree and area-based methods. Int. J. Remote Sens. **31**(5), 1175–1192 (2010)

S. Nandy, S. Ghosh, S.P.S. Kushwaha, A.S. Kumar, Remote sensing-based forest biomass assessment in northwest Himalayan landscape, in *Remote Sensing of Northwest Himalayan Ecosystems*, eds by R.R. Navalgund, A. Senthil Kumar, S. Nandy (Springer, Singapore, 2019), pp. 285–311

S. Nandy, R. Singh, S. Ghosh, T. Watham, S.P.S. Kushwaha, A.S. Kumar, V.K. Dadhwal, Neural network-based modelling for forest biomass assessment. Carbon Manag. **8**(4), 305–317 (2017)

Y. Pan, R.A. Birdsey, O.L. Phillips, R.B. Jackson, The structure, distribution, and biomass of the world's forests. Annu. Rev. Ecol. Evol. Syst. **44**(1), 593–622 (2013)

A.J. Prakash, S. Mudi, S. Paramanik, M.D. Behera, S. Shekhar, N. Sharma, B.R. Parida, Dominant expression of SAR backscatter in predicting aboveground biomass: integrating multi-sensor data and machine learning in Sikkim Himalaya. J. Indian Soc. Remote. Sens. pp. 1–13 (2024)

P. Raumonen, M. Kaasalainen, M. Åkerblom, S. Kaasalainen, H. Kaartinen, M. Vastaranta, P. Lewis, Fast automatic precision tree models from terrestrial laser scanner data. Remote. Sens. **5**(2), 491–520 (2013)

S.R. Rodda, R.R. Nidamanuri, T. Mayamanikandan, G. Rajashekar, C.S. Jha, V.K. Dadhwal, Non-destructive allometric modeling for tree volume estimation in tropical dry deciduous forests of India using terrestrial laser scanner. J. Indian Soc. Remote. Sens. **52**(4), 825–839 (2024)

J.W. Rouse, R.H. Haas, J.A. Schell, D.W. Deering, Monitoring vegetation systems in the great plains with ERTS. NASA Spec. Publ **351**(1), 309 (1974)

A.D. Roy, S. Debbarma, Comparing the allometric model to machine learning algorithms for above-ground biomass estimation in tropical forests. Ecol. Front. (2024). https://doi.org/10.1016/j.ecofro.2024.05.010

I. Saadia, Terrestrial biodiversity land ahoy. *Exploring land-based ecosystems, from deep in the soil to high in the mountains*, pp. 70–83 (2012)

S.Saatchi, SAR methods for mapping and monitoring forest biomass. SAR handbook: comprehensive methodologies for forest monitoring and biomass estimation, in eds. by A. Flores, K. Herndon, R. Thapa, E. Cherrington NASA (2019)

S.S. Saatchi, N.L. Harris, S. Brown, M. Lefsky, E.T. Mitchard, W. Salas et al., Benchmark map of forest carbon stocks in tropical regions across three continents. Proc. Natl. Acad. Sci. **108**(24), 9899–9904 (2011)

R.F.D.V. San Juan, J.M. Domingo-Santos, The role of GIS and LIDAR as tools for sustainable forest management. GIS-an Overv. Appl. **1**, 124–148 (2018)

M. Santoro, J. Askne, G. Smith, J.E. Fransson, Stem volume retrieval in boreal forests from ERS-1/2 interferometry. Remote Sens. Environ. **81**(1), 19–35 (2002)

K. Sharma, A. Saikia, S. Goswami, M. Borthakur, Aboveground biomass estimation and carbon stock assessment along a topographical gradient in the forests of Manipur, Northeast India. Arab. J. Geosci. **13**(12), 443 (2020)

V. Singh, N. Chaudhary, Land degradation, desertification, and food security in North-East India: Present and future scenarios. In *Sustainable development goals in Northeast India: Challenges and achievements* (Springer Nature Singapore, Singapore, 2023), pp. 153–166

S. Sinha, S. Modak, Analyzing the spatio-temporal forest aboveground biomass dynamics in Dimapur, Nagaland, India using integrated geospatial techniques. Res. Rev.: J. Space Sci. Technol. **12**(1), 21–32 (2023)

S. Sinha, C. Jeganathan, L.K. Sharma, M.S. Nathawat, A review of radar remote sensing for biomass estimation. Int. J. Environ. Sci. Technol. **12**(5), 1779–1792 (2015)

T. Sivasankar, J.M. Lone, K.K. Sarma, A. Qadir, P.L.N. Raju, The potential of multi-frequency multipolarized ALOS-2/PALSAR-2 and Sentinel-1 SAR data for aboveground forest biomass estimation. Int. J. Eng. Technol. **10**(3), 797–802 (2018)

T. Sivasankar, J.M. Lone, K.K. Sarma, A. Qadir, P.L.N. Raju, Estimation of above ground biomass using support vector. Vietnam. J. Earth Sci. **41**(2), 95–104 (2019)

M. Stone, Cross-validatory choice and assessment of statistical predictions. J. Roy. Stat. Soc.: Ser. B (Methodol.) **36**(2), 111–133 (1974)

P.S. Thenkabail, R.B. Smith, E. De Pauw, Evaluation of narrowband and broadband vegetation indices for determining optimal hyperspectral wavebands for agricultural crop characterization. Photogramm. Eng. Remote Sens. **68**(6), 607–622 (2002)

L. Tian, X. Wu, Y. Tao, M. Li, C. Qian, L. Liao, W. Fu, Review of remote sensing-based methods for forest aboveground biomass estimation: progress, challenges, and prospects. Forests **14**(6), 1086 (2023)

A.E. Turton, N.H. Augustin, E.T. Mitchard, Improving estimates and change detection of forest above-ground biomass using statistical methods. Remote. Sens. **14**(19), 4911 (2022)

B.K. Yadav, S. Nandy, Mapping aboveground woody biomass using forest inventory, remote sensing and geostatistical techniques. Environ. Monit. Assess. **187**, 1–12 (2015)

Dhruval Bhavsar is currently working as scientist in the Forestry and Ecology Division of North Eastern Space Applications Centre (NESAC), Department of Space, Umiam, Meghalaya. He holds M.Sc. and Ph.D. in Forestry from the Forest Research Institute (Deemed to be University), Dehradun. His research expertise includes forest biomass assessment, forest fire monitoring and SAR remote sensing in forestry.With over 10 years of experience in forest ecology, monitoring, and management using remote sensing and GIS, he has contributed to national projects and successfully led research and user-funded projects. He has also authored several publications in peer-reviewed national and international journals and conferences.

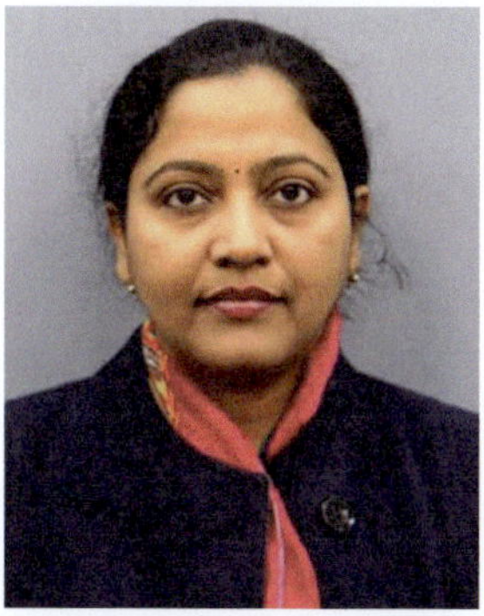

Kasturi Chakraborty completed her M.Sc. and M.Phil in Ecology and completed her Ph.D. on 'Land use/land cover and forest change using Remote sensing and GIS in Barak basin'. She has led project works and contributed as team member for successful completion of important projects at NESAC. Few of the important are in the field of forest fire, forest working plan, forest change analysis at regional scale, land use land cover mapping, wetland conservation and management. She has published research articles on forest and land use change analysis using geospatial techniques. She is a recipient of ISRO, young scientist award (2009) and ISRO project team award (2019)

Rocky Pebam obtained his B.Sc. in Forestry from the North Eastern Regional Institute of Science & Technology (NERIST), Arunachal Pradesh in 2000 and M.Sc. & Ph.D. in Forestry from Mizoram University. He had worked in the field of wildlife ecology, mapping of national parks, socio-economy and ecology of agroforestry systems, etc. under various research projects in different National Institutions before joining North Eastern Space Applications Centre, Meghalaya, as Scientist in 2009. His area of work in the Forestry & Ecology Division of NESAC includes geospatial applications in forest resource assessment for preparation of forest working plans, wildlife habitat assessment, shifting cultivation monitoring, forest fire assessment, vegetation recovery and timer series analyses of *jhum* cycles in the north eastern hilly region of India. He has published about 21 research papers in national/international journals/conference proceedings and 8 book chapters. He has mentored 7 M.Sc. students and successfully guided a Ph.D. candidate. As Principal Investigator/ Co-Principal Investigator he has completed 12 projects with funding from DST/DOS/ICMR apart from other user agencies.

K. K. Sarma Ph. D, leads the Remote Sensing Applications Group (RSAG) at the North Eastern Space Applications Centre (NESAC), where he has been a key figure since 2002. Specializing in remote sensing for forestry and biodiversity, his work leverages space technology for critical applications in Biodiversity, Ecosystems, Forestry, and Natural Resource Management. Dr. Sarma has spearheaded over 60 diverse geospatial projects, contributing significantly to natural resource mapping and monitoring across the North Eastern Region (NER). Currently, he coordinates 110 "Plan of Action" projects, focused on using space technology to improve livelihoods in the NER. He also served as the Principal Investigator and NER Coordinator for all ISRO natural resource management projects, including comprehensive Land Use Land Cover Mapping and National Wasteland Updating, Monitoring, and Change Analysis. With over 75 research publications, Dr. Sarma has successfully supervised six Ph.D. students and actively mentors many others across NER Universities.

Chapter 7
Forest Fire Characterization Using Remote Sensing and GIS: A Case Study from North East Region of India

Kasturi Chakraborty, Dhruval Bhavsar, Suraj Kr. Swain, and K. K. Sarma

Abstract The availability of information on long term forest fire incidents has paved the way for spatial analysis. This chapter presents the characterization of forest fires in the region using historical forest fire incidents from 2003 to 2024. The forest fire incidents have been used to generate the forest fire vulnerability map and the fire prone map. Using spectral index-based analysis, the forest fire burned area map has been prepared. The forest fires in the region show a decrease in the number of fire counts. The Aizawl district of Mizoram has the highest count in NER. The forest fire occurrence in the region spans from the month of January to May, with peak in the number of incidents during the month of March. In this region, the evergreen open forest witness more fire incidents. The forest fire in this region is closely linked to the shifting cultivation practice throughout the region. Remote sensing-based forest fire information has played an important role in the fire assessment. Analysis of the two forest fire incidents in the recent past are presented as case study of major forest fire in this region. The Land use transformation and other ecological implications due to forest fire needs a sustainable outlook to balance the age-old traditional practices of the local communities, integration of local knowledge and framing of strategies based on scientific evidences.

Keywords Forest fire · North East Region (NER) · Forest fire vulnerability · Forest fire prone · Burned area map

7.1 Introduction

The genesis of forest fire is closely tied to the origin of plants. Two of the three necessary components for fire are oxygen and fuel (Pausas and Keeley 2009; Zhao et al. 2016), while the third element is a source of heat obtained from either man-made or natural sources. Around the globe, fire plays a variety of ecological roles.

K. Chakraborty (✉) · D. Bhavsar · S. Kr. Swain · K. K. Sarma
North Eastern Space Applications Centre (NESAC), Umiam, Shillong, Meghalaya, India
e-mail: kasturi.nesac@gmail.com

S. P. Aggarwal et al. (eds.), *Remote Sensing of the North Eastern Himalayan Ecosystem*,
https://doi.org/10.1007/978-981-95-3221-6_7

In ecosystems that rely on fire, it greatly impacts both the structure and function; species have adapted to exist alongside fire, making it vital for the conservation of biodiversity. In contrast, ecosystems that do not depend on fire naturally lack the fuel or ignition sources needed to allow fire to act as an evolutionary agent. New findings about the impact of fire in vegetation distributions are prompting revisions to the theory governing biome distribution (Schimper 1902). Over half (53%) of the earth's ecoregions are thought to be fire-dependent, 22% fire-sensitive, and 15% fire-independent (10% have not been evaluated) (Shlisky et al. 2007). In some ecosystems, fire is used as a land management tool having long cultural significance (Myers 2006; Goldammer et al. 2004). In many places, natural fire regimes with the adaptations of species and natural processes are maintained by periodic burning that help to sustain and rejuvenate ecosystems (Hassan et al. 2005). Forest fire is regarded as a potent ecological and evolutionary force that controls species interactions, community composition, organismal 67 characteristics, population numbers, carbon and 68 nutrient cycling, and ecosystem function. Given that fossil evidence of fire first emerged at the same time as plants began to colonize terrestrial surfaces, the relationship between the earth system and fire has evolved for at least 420 million years (Scott and Glasspool 2006). Rather than being seen as a "threat" to ecosystems as often believed, fire and plants have "co-evolved." Over extensive periods, fire and land plants have influenced one another, with fires being driven by plant growth; however, plants have also changed fire regimes and adapted to changing fire frequency and behaviour (Baker et al. 2022).

However, fire ecologists in the present context, view fires as a dynamic ecological force having evolutionary consequence fundamentally shaped by human actions. (McLauchlan et al. 2020). The evolutionary linkages between fire and Homo sapiens, as well as the long-standing Indigenous ecological knowledge that reflects fire's cultural associations with human cultures, make the connection to humans and human systems inseparable (Kobziar et al. 2024). Research indicates that fires started by humans may hasten the long-term trend of ecological changes brought on by climate change (Kershaw et al. 2002). Fire and humans are inseparable, and how people relate to fire will greatly influence their ability to adapt to continuous and accelerating environmental and climatic change (Kobziar et al. 2024).

7.1.1 Global Forest Fire Scenario

There are no comprehensive national, regional, or worldwide statistics on wildland fires that would enable a trustworthy and accurate comparison of the incidence of fires worldwide in the 1980s and 1990s (FAO 2001). The FAO (2001) study also provides an example of reliable annual area recording that resulted in a wildfire database that showed a concerning trend of larger fires near the end of the twentieth century. According to scientific data, land-use, population, and climate change are all causing changes in fire regimes—the distinctive patterns of fire that have developed over time and space—all across the world. Although the effects of climate change

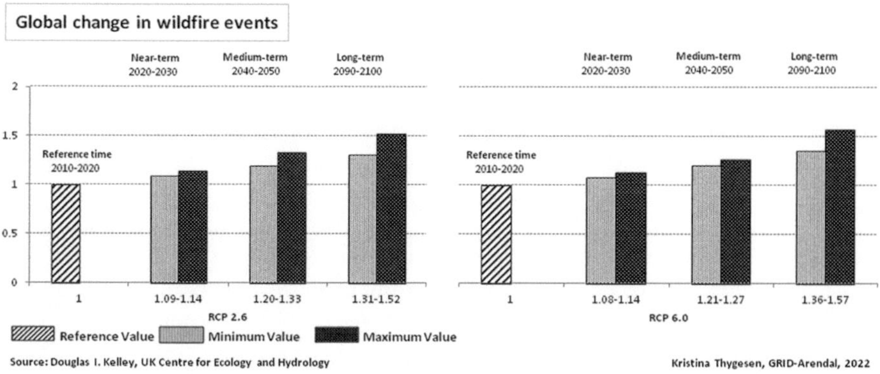

Fig. 7.1 Global change in wildfire events (https://www.grida.no/resources/15602)

on fire behaviour are complicated, current models indicate that by the end of the century, burning is likely to increase significantly in some places, like the Arctic. If greenhouse gas emissions continue at their current rate, there will probably be more fire in areas of tropical forest in the southern Amazon and Indonesia (UNEP 2022). In 2021, Greece, Italy, Algeria, and Turkey saw numerous severe wildfire occurrences that burned over 630 000 ha, making it a year marked by severe wildfires (San-Miguel-Ayanz et al. 2022). In the devastating wildfires of 2021, Algeria experienced a tragic loss of at least 90 lives (ReliefWeb 2021), marking them as the deadliest fires in recent history, following the events in central Portugal (2017) and Mathi, Greece (2018). The projections of the fire events are shown in Fig. 7.1. It is predicted that the number of forest fires will rise by 14% by 2030, 30% by 2050, and 50% by the end of the century. The probability of catastrophic wildfire incidents will rise by a ratio of 1.31 to 1.57 by the end of the century. Wildfire incidents will rise significantly even in the scenario with the lowest emissions (UNEP 2022).

7.1.2 Types of Forest Fire

Ground fire, surface fire, and crown fire are the three main categories of forest fires. During a fire occurrence, all three categories are typically present in variable amounts in a fire pattern. It is discovered that ground fires happen in the rotten roots and debris. Surface fires can start when ground fires break out and ignite surface fuels. Nuts, moss, lichen, herbaceous flora, shrubs, tiny trees, and saplings are all burned by surface fires that travel down the forest floor. The severity of surface fires can be classified as low, medium, or severe. Crown fires move through tree canopies. Crown fires can be active or passive. Crown fires turn active, causing crown-to-crown combustion, whereas passive crown fires entail burning individual trees or clusters of trees. Crown fires are typically severe and stand-replacement and they are greatly impacted by geography, wind, and crown density (Fire Science core curriculum 2017).

7.1.3 Forest Fire Regimes

The pattern of forest fires in both space and time, along with their effects in a particular region over a defined period, is referred to as the fire regime. Frequency, extent, intensity, severity, kind, and purpose of burning are some of the variables that affect wildfire regimes. As a result, a fire regime depends on the frequency of occurrence, fire intensity, fire severity, and amount of fuel burned in a specific area, which is influenced by temperature, wind, humidity, and fuel and igniting sources (Fire Science Core Curriculum 2017). Disruptions to fire regimes have an adverse effect on climate (IPCC 2021), biodiversity (Kelly et al. 2020; Feng et al. 2021), and society (Johnston et al. 2021). According to Sayedi et al. (2024), a location's forest fire regimes are disrupted when fires surpass their historical patterns in terms of intensity, size, severity, seasonality, and frequency. Data on the fire regime is critical for planning, adaption, and mitigation of climate change, as well as for sustainable land management (Cochrane and Bowman 2021; Moritz et al. 2014). According to Whitlock et al. (2010), a significant research problem is separating the impact of human land uses on background fire regimes. This is accomplished by comparing human land use regimes within similar biomes in order to conduct thorough comparative palaeoecological evaluations of the consequences of previous fire activity. An overview of the problem and defines the characteristics of fire regimes as given in the Table 7.1.

Table 7.1 Common fire regime parameters and definitions

Descriptor	Definition
Frequency	The average number of forest fire incidents occurring in a given area per unit time defined by return interval i.e. average number of fires between successive fire events
Magnitude	The magnitude or effect of the disturbance, characterized by intensity, size and impact. The ecological impact assessed by mortality or loss of biomass defines the severity. Magnitude of fire incident is also characterized by flame length and duration
Extent	The total area affected by a single event or over a time period
Fuel characteristics	The fuel type in a region effects the extent of damage
Topography	Determines the spread of fire
Source	The source of the incident, categorized as natural or anthropogenic
Recovery rate	Speed and nature of ecosystem recovery post-fire events, influenced by resilience, resistance, and regeneration capacity
Climate and weather conditions	Determines the ignition characteristics and fire occurrence

7.1.4 Impact of Forest Fire

Wildfire impacts post-fire soil and vegetation processes like soil erosion, debris flow, flooding, and vegetation recovery, as well as the biological functioning of many ecosystems by partially or totally burning the vegetation layers (Morgan et al. 2014). Long-term alterations in vegetation composition, structure, and ecosystem services can result from wildfires. According to the United Nations Environment Programme (2022), wildfires are becoming more intense and sweeping over a variety of habitats, wreaking havoc on infrastructure, wildlife, the environment, and human health. Massive volumes of CO_2 are released into the atmosphere by wildfires in ecosystems that store a lot of irrecoverable terrestrial carbon, such as peatlands and rainforests (UNEP 2022). Nearly 100 million people in five Southeast Asian nations were exposed to elevated air pollution in the second half of 1997 and the first part of 1998 due to a smoky haze that severely affected broad sections of the region and posed an urgent health risk to the general population (WHO 1998). According to van Wees et al. (2022), burning biomass is estimated to release 2069 Tg C year − 1, which is more than 10 Pg of carbon annually. This is the highest rate of carbon release (at least for the dominant fossil fuel emissions) from burning vegetation and fossil fuels in the past 66 million years (Zeebe et al. 2016). Climate change has also exacerbated land use and fuel management practices, leading to modifications in fire patterns that alter species abundance and composition, provoke ecological invasion, and alter carbon and water cycling, wildlife habitat, ecosystem services, and climate resilience (McLauchlan et al. 2020; Rakhmatulina et al. 2021).

7.1.5 The Concept of Forest Fire Hazard, Risk and Vulnerability in the Context of Forest Fire

A popular integrated paradigm for comprehensive fire management combines risk assessment, vulnerability, and wildfire exposure. Any process, phenomena, or human activity that has the potential to result in death, serious injury, negative health effects, property damage, social unrest, economic disruption, or environmental degradation is referred to as a "hazard." The physical and chemical characteristics of fuel, such as loading or biomass, or possible fire behaviour, such as fire line intensity, are typically used to calculate or express wildfire hazard. On the other hand, wildfire risk is the probability that a fire would break out, the behaviour of the fire, and the effects of the fire (Goldammer et al. 2017). The combined impact of wildfire danger (sometimes referred to as fire hazard) and wildfire susceptibility of individuals, ecosystems, and products exposed to wildfires is known as wildfire risk. The likelihood that a wildfire will develop or spread (danger/hazard) and the potential harm it could do (exposure and susceptibility) are the two factors that make up wildfire risk (Scott 2006; Calkin et al. 2011; Chuvieco et al. 2010). On the other hand, vulnerability refers to the traits and conditions of a society, system, or asset that render it vulnerable to the negative consequences of a hazard. Systems, people, property, or other components that are in danger areas and could therefore sustain losses.

7.2 Remote Sensing of Forest Fire

Monitoring wildfires at every stage of their life cycle, that is, determining pre-fire fuel conditions, describing active fire locations and emissions, and analysing post-fire impacts on vegetation, air quality, and climate—requires the use of remote sensing (Chen et al. 2024). Satellite remote sensing offers various advantages like large area coverage, frequent and repetitive coverage, etc. remote sensing sensor detects the heat emitted by active forest fires commonly known as 'hotspot detection' using infrared sensors on board in a satellite. The introduction of the first landsat satellite in the 1970s marked the beginning of space-based fire mapping and detection (Chen et al. 2024). Only since the early 2000s have fire activities been consistently observed, even though remotely sensed observations have been accessible for decades (Vitolo et al. 2020). Since satellite remote sensing technology has advanced so quickly, it is now a commonly utilized instrument for monitoring forest fires (Zheng et al. 2023). A crucial initial step in acquiring prototype baseline data on the size of wildland fires worldwide in the year 2000 is the Global Burned Area Product 2000, which is obtained via a spaceborne sensor system (JRC 2002). The advantages of satellite sensors include a wide monitoring range, quick response times, and strong anti-interference capabilities (Chuvieco et al. 2020). They may also be used to monitor forest fires in large areas in real time (Qin et al. 2015).

7.2.1 Physical Basis for Remote Fire Monitoring

The infrared frequencies of the electromagnetic spectrum are sensitive to the temperature difference. The thermal energy released by ongoing fires is the primary indicator utilized for identifying active fires (Robinson 1991). Therefore, they are used by satellite sensors to detect thermal anomalies as indicators of potential wildfires. This information is used for the spatial and temporal forest fire assessment. Thermal imagers that sense the infrared radiation produced by fires are the main tools employed for real-time surveillance. The monitoring of forest fires utilizing satellite-based remote sensing technology implements the Brightness Temperature (BT) detection approach. This technique identifies BT variations between forest fires and other types of land cover in the Middle Infrared (MIR) and Thermal Infrared (TIR) bands of remote sensing images for fire identification and subsequently uses the reflective characteristics of the visible or Near-infrared (NIR) bands to eliminate false fire detections.

Forest fires can be observed using both polar orbiting satellites and geostationary satellites. Common sensors found on Polar Orbit Satellites include the

Advanced Along-track Scanning Radiometer (ATSR), Advanced Very High-Resolution Radiometer (AVHRR), Moderate-resolution Imaging Spectroradiometer (MODIS), and Visible Infrared Imaging Radiometer (VIIRS), among others (Zheng et al. 2023). NASA's VIIRS and MODIS are frequently utilized sensors for near-real-time wildfire detection. Geostationary Meteorological Satellites employ sensors such as the Advanced Himawari Imager (AHI), Advanced Baseline Imager (ABI), and Advanced Geostationary Radiation Imager (AGRI) for monitoring forest fires. The MODIS is available on both the Terra and Aqua satellites, featuring a wide swath width of 2330 km, which allows it to capture images of the Earth's entire surface every 1 to 2 days. MODIS sensors operate across 36 spectral bands that range from 0.405 to 14.385 μm, delivering data at three different spatial resolutions: 250, 500, and 1000 m. Justice et al. (2002) provide an extensive overview of different MODIS products. Giglio et al. (2003) developed the algorithm for fire detection utilizing MODIS. The spatial resolution for active fire detection pixels obtained from MODIS is 1 km. The AVHRR, which is on board the National Oceanic and Atmospheric Administration's (NOAA) polar orbiting satellites, has been discussed by Flannigan and Vonder Haar (1986) and Kaufman et al. (1990). NOAA/AVHRR presents significant benefits for fire monitoring as it delivers daily global coverage at a moderate resolution of approximately 1 km, which is essential for operational monitoring of global fires, along with a wide range of spectral coverage including visible, near-infrared, mid-infrared, and thermal wavebands. ASTER, or the Advanced Spaceborne Thermal Emission and Reflection Radiometer, is installed on NASA's Terra satellite for wildfire tracking through mapping and monitoring. ASTER's capability to offer extensive spectral coverage is crucial for the understanding and tracking of wildfires. The SEVIRI sensor is part of the Meteosat Second Generation (MSG) system. MSG SEVIRI is one of the geostationary satellites employed to detect and monitor active fires, capturing images of the Earth every 15 min. It supplies image data across four Visible and Near InfraRed (VNIR) channels and eight InfraRed (IR) channels. The VIIRS is installed on the joint NASA-NOAA Suomi National Polar-orbiting Partnership (Suomi NPP) and NOAA-20 satellites. The significance of this is linked to the frequency of the observations. The MODIS instruments on Terra and Aqua, along with the VIIRS instruments on Suomi NPP and NOAA-20, capture images of the entire Earth's surface every 1 to 2 days. These frequent and timely observations yield a wealth of information regarding wildfire occurrences, monitor the progression of known fires in near real-time, and evaluate the effects of significant burns. The VIIRS active fire products enhance the MODIS active fire products by offering better spatial resolution compared to MODIS. The spatial resolution for the active fire detection pixel in VIIRS is 375 m. Moreover, VIIRS can identify smaller fires and assist in outlining the boundaries of ongoing large fires. LANDSAT data offers in-depth insights into forest fires but is limited by its infrequent revisit rate (once every 16 days) and narrow geographic coverage. Since the 1970s, Landsat satellites have been documenting wildfires, aiding land managers and emergency responders in evaluating the effects of fires on ecosystems and communities (USGS). The data from Landsat satellites primarily indicates the location and extent of burned areas, the severity of the burn mosaic, and the subsequent regrowth long after the smoke has dissipated.

7.2.2 Role of Remote Sensing in Forest Fire Management

Traditional practices of forest fire management practices particularly used by Forest departments include methods viz., creations and maintenance of fire lines, controlled burning, fire terracing, rock walls, fire watchers, fire watch and patrol system towers, etc. However, in recent times forest fire management has advanced with the use of remote sensing-based techniques, mobile squads, community participation, etc. Satellite remote sensing offers an important tool for fire monitoring, management and post fire damage assessment. With the help of remote sensing techniques, various types of forest fire can be identified through spectral discrimination. Effective fire control is based on the field survey, delineation and identification risk zone and various methodologies encompassing ground-based systems, aerial surveys etc. Satellite remote sensing plays an important role in prior assessment of the forest fire occurrence (pre-fire), detection of forest fire (during fire) quick and long-term damage assessment (post fire). The damage assessment is done by burnt area assessment using various satellite data. The function of remote sensing in the pre-fire state includes evaluating fire risk, mapping vegetation, analyzing topography, and assessing the likelihood of fire spread, among other factors. While during-fire, remote sensing play's important role by forest fire detection, detection of hotspots and other thermal parameters. The post-fire role of remote sensing includes the burnt area mapping, loss of forest cover etc. The assessment of wildfire exposure, vulnerability, and risk has been widely applied as a comprehensive framework for effective fire management in numerous regions that are prone to wildfires around the globe (Plucinski et al. 2017). By utilizing satellite remote sensing, the affected area from a fire can be identified by analyzing the variations in spectral bands related to vegetation cover and burned scar (Huang 2020).

There are several global burned area datasets available for different time periods and at different resolutions e.g. GBA2000, GLOBSCAR, GFED4S MCD45A1 c51 etc. Satellite based burned area extraction rely on spectral indices generated using spectral reflectance pattern before and after fire. The NIR and SWIR spectral regions are the particularly sensitive to fire effects. There is drastic change in the reflectance pattern in the burned vegetation due to the decline in the NIR and increase in shortwave infrared (SWIR) (Justice et al. 2002). There are several spectral indices derived based on this concept of reflectance pattern e.g. Normalized Difference Vegetation Index (NDVI), Burned Area Index (BAI) (Chuvieco et al. 1988) are based on visible (VIS, 0.4–0.7 μm) and near infrared (NIR, 0.7–1.2 μm) wavelengths. The indices based on visible, near infrared and shortwave infrared (SWIR1, 1.5–1.8 μm) are normalized burn ratio (NBR), improved BAI (BAIM) (Martín et al. 2005) are the commonly used indices. The combination of NIR and SWIR portion of the EMR spectrum is widely used for burned area discrimination.

In addition to the index based approaches, the machine learning techniques such as the random forest algorithm, which relies on a mix of decision tree predictors (Breiman 2001), continue to be widely used and are considered one of the most popular algorithms for mapping burned areas, algorithms like LandTrendR, stochastic continuous change detection (S-CCD), and Continuous Change Detection and Classification (CCDC) are widely used for detecting changes over extended periods and can be accessed on cloud-based geoprocessing platforms (Liu et al. 2023; Pasquarella et al. 2022; Zhang et al. 2023). The application of deep learning algorithms has helped to analyse and map the complex trends of burned area mapping improving overall classification accuracy. The U-Net and its derivatives (including U-Net Red, U-Net Transformer, and U-Net 3D) are among the well-known deep learning algorithms (Nelson et al. 2024).

7.3 Characteristics of Forests and Occurrence of Forest Fire in North Eastern Region (NER) of India

The NER of India (Fig. 7.2) includes eight states: Arunachal Pradesh, Assam, Manipur, Meghalaya, Mizoram, Nagaland, Tripura and Sikkim. The forest and tree cover in these states account for 21.08% of total forest cover of the country. Arunachal Pradesh has 67,083 km^2 forest cover; Mizoram follows with 18,558 km^2, while Meghalaya has 17,687 km^2. Manipur has a forest area of 16,795 km^2 and Nagaland covers 12,616 km^2. Assam reports 30,415 km^2 of forest cover while Tripura has 7,832 km^2 and Sikkim's has 3,406 km^2 are under forest (ISFR 2023). As per Champion and Seth Classification (1968), the state of Assam has 18 different forest types classified into five groups: tropical wet evergreen, tropical semi evergreen, tropical moist deciduous, tropical dry deciduous, and sub tropical pine forests. In Arunachal Pradesh, based on the altitude, precipitation, humidity and species diversity, the vegetation in the state is categorized into three primary forest types (tropical, subtropical, temperate) and ten subtypes (northern tropical semi evergreen forests, northern tropical evergreen forests, northern tropical moist deciduous forests, secondary bamboo brakes, east Himalayan subtropical broad-leaved forests, sub-tropical pine forests, temperate broad-leafed forests, temperate conifer forests, alpine meadows (montane tundra). The state's floral diversity features over 5,000 species of flowering plants, 600 species of orchids, 75 types of rhododendrons, 45 timber-producing tree species, 45 bamboo species, 17 cane varieties, and 500 medicinal plant species. The state has the largest reservoir of orchids and the highest diversity of rhododendrons in the country (https://arunachalforests.gov.in/services/forest-wildlife/forest). The state of Manipur covers an area of 22,327 km^2. The state is recognized for its abundant biodiversity, featuring unique plants and animals. The state's forests comprise six main forest types and ten subtypes. The primary forest types include tropical wet evergreen forest, tropical moist deciduous forest, sub-tropical pine forest, tropical dry deciduous forest, montane wet temperate forest, and sub-alpine forest (https://monitoring.manipurforest.gov.in/Main/AboutUs). Meghalaya's forests are home to over 3,500 flowering plant species, including 352 types of orchids, 40 varieties of bamboo, and approximately 800 medicinal plant resources. There are around 40 species of plants that are endemic to the region, alongside 75 plant species that are threatened. The vegetation varies from tropical rainforest at the foothills to alpine meadows and cold deserts. Due to its diverse plant life, this area is often referred to as a 'Botanist's Paradise.' Hooker described this region as the "Cradle of Angiosperms" (https://www.megforest.gov.in/forest_overview.html).

The forest classifications in Mizoram are primarily based on factors such as altitude, rainfall, and the dominant species present, and can be broadly categorized into tropical wet evergreen forest, montane sub-tropical forest, temperate forests, bamboo forests and *Quercus* forests (https://forest.mizoram.gov.in/page/forest-types). According to Champion and Seth Classification, the forests in this state comprise Cachar tropical evergreen forests, moist mixed deciduous forests, low alluvial savannah woodland, dry bamboo brakes, secondary moist bamboo brakes, and

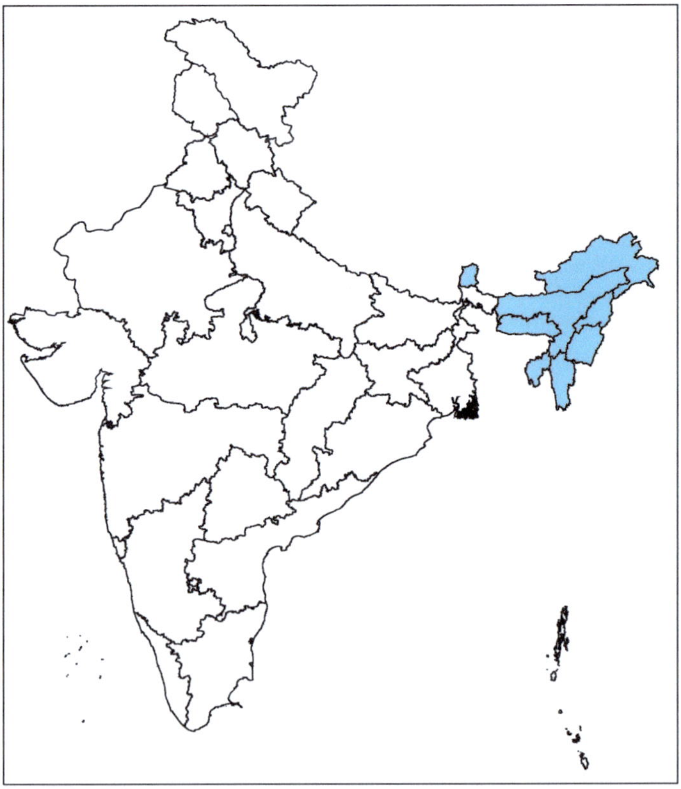

Fig. 7.2 North Eastern Region of India

east Himalayan lower bhabar Sal. A significant portion of the land is occupied by bamboo brakes, which essentially create a "Sub climax" as a result of traditional shifting cultivation practices that have been in place for centuries (https://forest.tripura.gov.in/forest-of-tripura.html). Nagaland possesses a rich variety of forest types due to its distinct geographic setting and diverse physiographic landscape. The forest types identified in this state (following the classification by Champion and Seth 1968) include northern tropical wet evergreen forest, northern tropical semi-evergreen forest, northern sub-tropical broad-leaved wet hill forest, northern sub-tropical pine forest, northern montane wet-temperate forest, and alpine forest (https://forest.nagaland.gov.in/status-of-forests/). Forest fires occur seasonally in this region every year. In northeast India, there has long been a strong connection between forestry and fire practices, which is deeply rooted in the traditional methods of shifting cultivation, making it a seasonal occurrence in the area. Around, 717,536 (MODIS count = 479,262 and VIIRS SNPP = 238,274) forest fire incidences occurred in between 2002 to 2024 (Forest Survey of India).

7.3.1 The Forest Fire Scenario of NER India—An Assessment Under NER DRR Program

NESAC is carrying out forest fire assessment of NER as part of the NER-DRR (North Eastern Regional Node for Disaster Risk Reduction) program. The locations of forest fires are sourced from the Forest Survey of India portal (FSI). A value addition to the forest fire locations is carried out by NESAC during forest fire season for dissemination of information to the respective forest departments and to the state disaster management cells. Such an initiative helps the team at the ground level to plan for the effective strategy for the mitigation of the fire event. The information that are available as attributes with each forest fire event are the coordinate, nearby Land us Land Cover (LULC), slope, aspect, elevation and proximity of the water bodies. Effective forest fire management is possible with the help of adequate temporal and spatial information of any regions. The forest fire occurrence in NER from 2003 to 2024 is shown in the Fig. 7.3. Since the VIIRS SNPP fire locations are available from the year 2018, this leads to increased number of fire incidents in the region, hence the forest fire scenario is presented using MODIS satellite data only. It is found that, over the years northeastern states of India are showing a decreasing tendency of forest fire. In the recent past, majority of fire incidents took place in the year 2021. The comparison of fire events within each state with respective state total events shows almost a cyclic pattern of forest fire incidents in NER states of India. This point towards the anthropogenic nature of forest fire incidents which are mainly caused by shifting cultivation (jhum) agriculture widely practiced in this region. While the comparison of state wise fire events with entire NER fire events, shows (Fig. 7.4) decreasing tendency of forest fire events in each NER states of India.

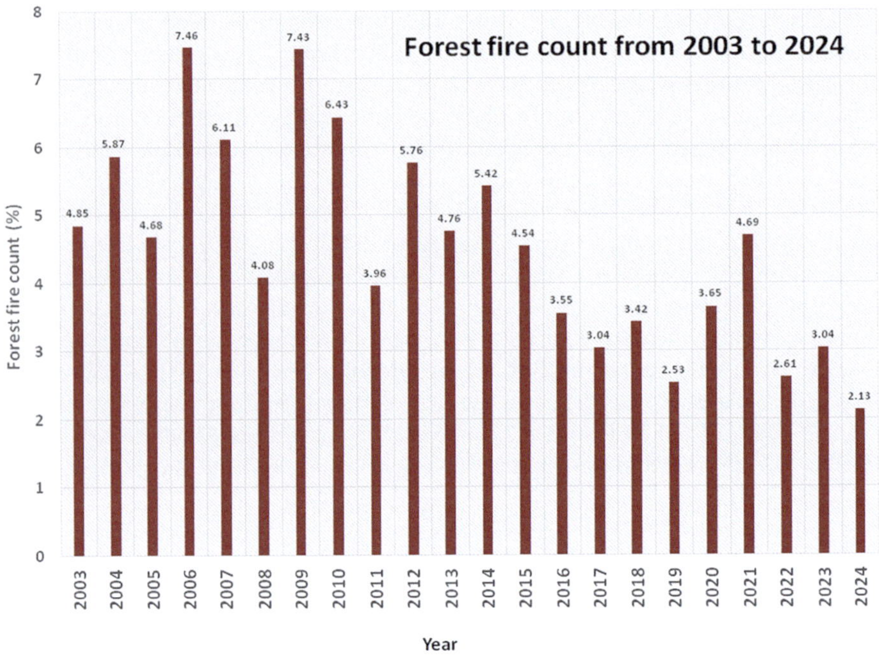

Fig. 7.3 Total forest fire incidents (%) in NER from 2003–2024

For the district wise analysis of the forest fire events the total number of forest fire occurrence from 2003–2024 in the different districts of the respective northeastern states was assessed to understand which region has major fire contribution. It is observed that in Arunachal Pradesh out of the 36,350 number of total fire events, Kameng East and West have contributed to 5481 nos., followed by Siang west (4332.), Tirap (3991), Lohit (3826), Changlang (3479), Dibang valley (2456). In Assam out of total number of fire events (72,501), major contribution is from the districts viz., Karbianglong (25,558), Dima Hasao earlier known as North Cachar Hills (21,114), Cachar (3992), Kamrup (3158), Hailakandi (3048). In Manipur, out the total fire events (106,094), major fire events occurred in Chandel (28,691), Churachandpur (26,908), Tamenglong (22,844), Ukhrul (12,156), Senapati (11,904). In Meghalaya (51,550), RiBhoi district has the maximum number (10,904) followed by West Khasi Hills (10,515), East Garo hills (6232), East Jaintia hills (4381), South Garo Hills (3627), North Garo Hills (3166). In the State of Mizoram (126,183, maximum number of forest fires occurred in Aizwal (48,913), Lunglei (30,237), Lawngtlai (18,208), Champai (13,191), Saiha (9871). In Nagaland out of total number of fire events (46,383), Tuengsang has the highest number of events (10,635), Kohima (9767), Wokha (6100), Mon (5967), Phek (4622). Sikkim although has minimum forest fire incidents with South Sikkim having the maximum fire events. In Tripura, of the total fire events (37,875), North Tripura (21,544) and South Tripura (11,578) have the maximum fires events followed by West Tripura. The top 10 districts of NER

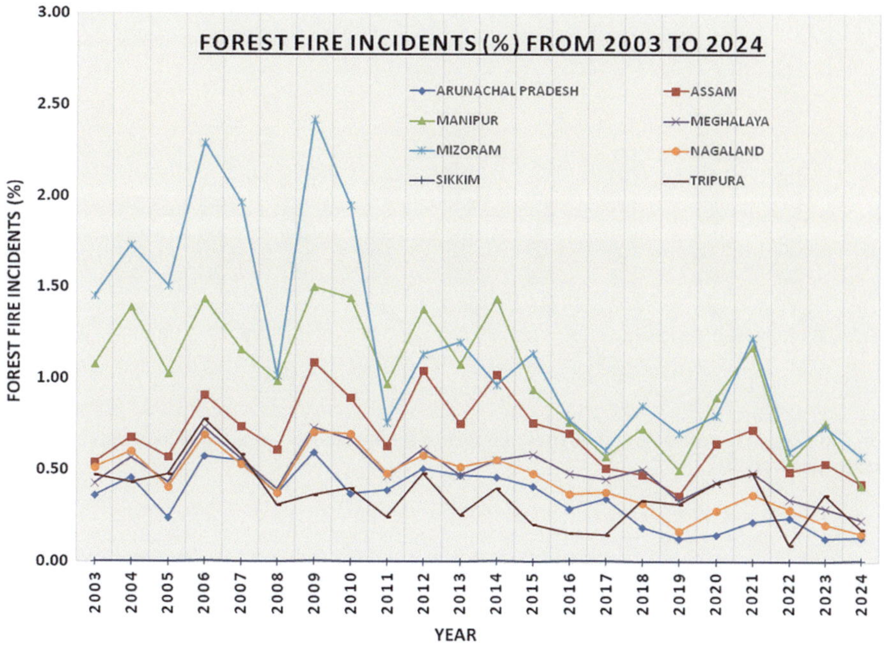

Fig. 7.4 Forest fire incidents (%) from 2003 to 2024 in NER states (based on MODIS) compared with total forest fire events in the region

considering the total fire events in the regions are shown in Fig. 7.5. The forest fire in this region has been compared with the seasonal pattern of occurrence. It is observed that forest fire takes place from January till May with peak in the month of March (Fig. 7.6). Although there are few instances of fire occurrence in the other months also.

For pre-fire preparedness and during fire scenario a forest fire vulnerability map is crucial for making effective and informed decisions. Evaluating fire risk is a vital component of forest fire management to determine when and where fires are likely to ignite and which areas may be impacted. Consequently, a "forest fire vulnerability map" along with a map of fire-prone areas has been created for the entire NER. The vulnerability map has been generated using forest and other land use/land cover categories, topographic features such as elevation, slope, and aspect, in addition to distances from roads and settlements. The vulnerability of each zone has been evaluated by assigning weight based on the factors influencing fire occurrence. The elements considered in the creation of the forest fire vulnerability map for the north-eastern region include fuel characteristics (forest type and density), topographical elements (elevation, slope, and aspect), as well as proximity to roads and settlements. The details of each of the input variable is given in Table 7.2. Each of the factors is observed for their influence using the historical forest fire incidents discussed in the above sections. The assessment of the influencing factors with the historical data is

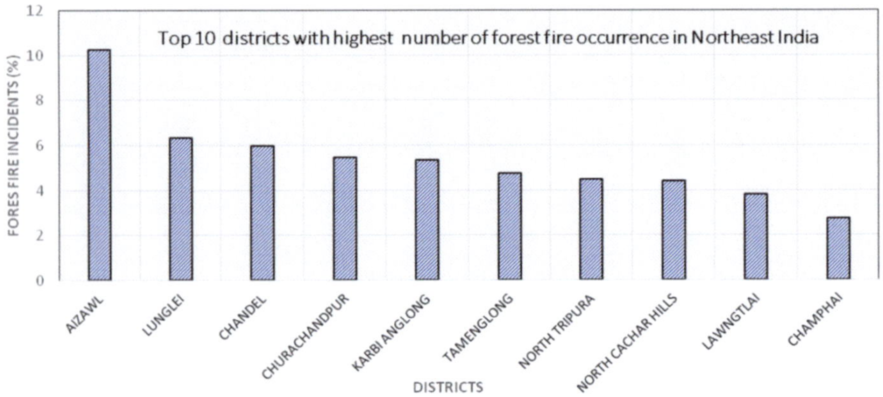

Fig. 7.5 Top 10 districts of with highest number of fire incidents compared to total number of fire occurrence in NER

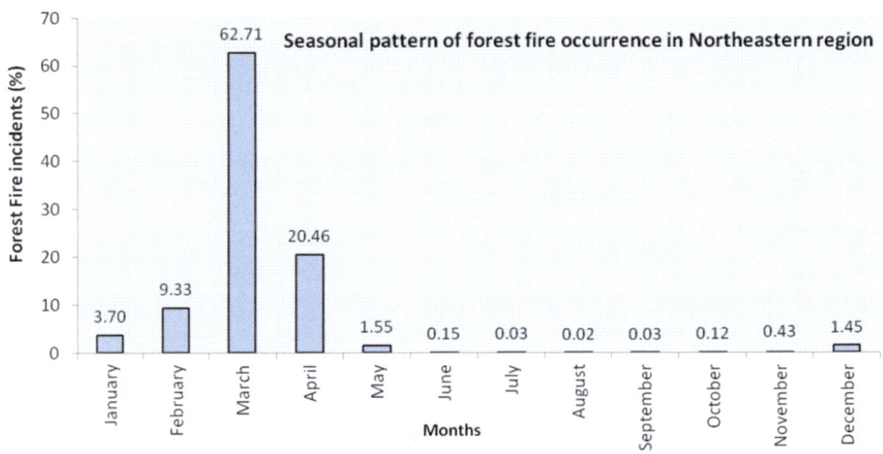

Fig. 7.6 Favourable months of forest fire occurrence in NER

given in Table 7.3. The weightages have been assigned based on decision making process called Analytical Hierarchical Procedure (AHP) so as to determine the influence of each variable on the output. An accuracy rate of 80–85% has been determined based on historical data, and recent fire incidents have shown to be consistent with this assessment. The methodology used to drive forest fire vulnerability map is shown in Fig. 7.7. The map (Fig. 7.8) shows the forest fire vulnerability zones (1:50 K) of northeast India ranging from low to very high-risk zones.

A precedence weighting was derived and combined to determine the global ranking score of each relevant criterion (Saaty 2013). To check the consistency of the decision making and reduce the bias in the process, a consistency ratio (CR) was obtained with Eqs. 7.1 and 7.2

Table 7.2 Input parameters for generation of forest fire vulnerability map

Datasets	Data source	Variables
Topographic	CartoDEM (10 m) (NESAC)	Elevation
		Aspect
		Slope
Forest Fire	FSI (Forest Survey of India)	Historical fire events
		Current fire events
Fuel characteristics	FSI (Forest Survey of India)	Forest type
		Forest density
Base maps for proximity	NeSDR (NESAC)	Distance to road
		Distance to settlement
		Distance waterbody

Table 7.3 Influence of each parameter assessed using historical forest fire incidents

Factors	Forest fire occurrence
Fuel characteristics	
Forest type	Evergreen > Deciduous > Bamboo
Forest density	Open > Moderate > Dense
Topography	
Elevation (m)	200–500 > 500–1000 > 1000–2000
Aspect	All directions
Slope	$15–25^0 > 8–15^0 > 25–35^0 > 3–8^0$
Proximity	
Built up (m)	1000–1500 > 1500–2000 > 500–1000
Road (m)	50–100 > 100–150 > 200–250 > 250–300
Waterbody	1000–3000 > 500–1000 > 300–500

$$CR = \frac{CI}{RI} \tag{7.1}$$

$$CI = \frac{\lambda_{max} - n}{n - 1} \tag{7.2}$$

where, CR is the consistency ratio, CI is the consistency index and RI is the random index in Equation. λmax is the maximum value in the average of dividing the sum of the weights, and n is the number of criteria. CR values of 0.10 and lower are considered tolerable (Malczewski and Rinner 2015).

A total of 35.42% of the NER forest is under high vulnerability zone. The vulnerability of the forest areas under different zones are shown in Fig. 7.9. The area (%) under different categories of vulnerability zones in different northeastern states are shown in Fig. 7.10. Amongst all the states, Tripura has the highest area (35.85%)

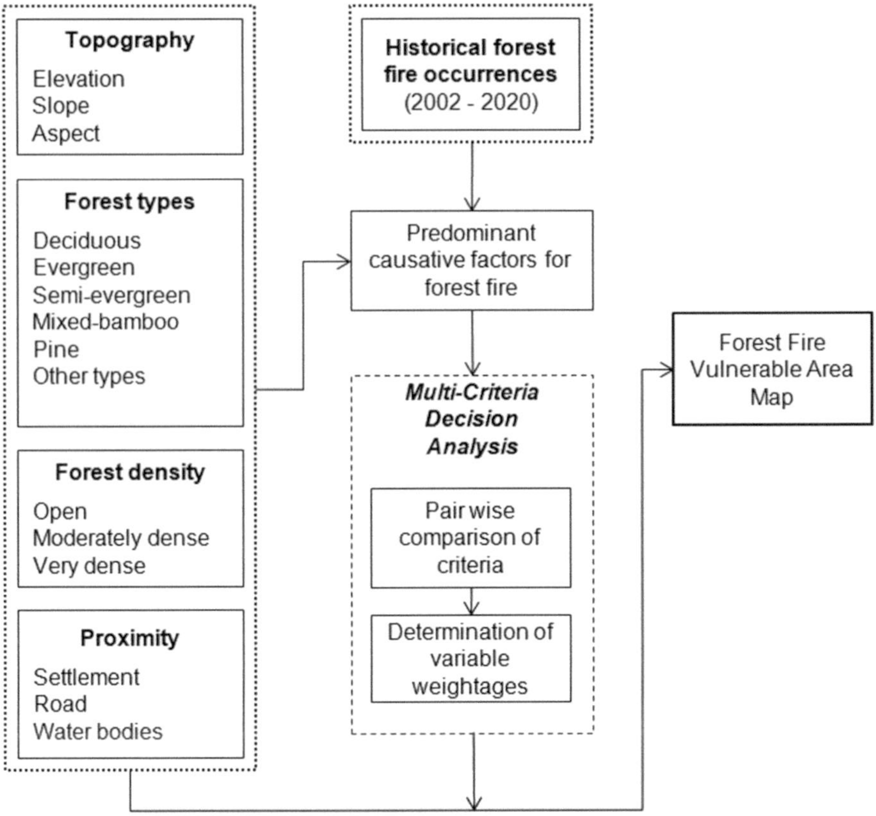

Fig. 7.7 Methodology followed for forest fire vulnerability mapping

under very high category of vulnerability zone. The northeastern states have majority of the forest area under moderately to high vulnerability category.

Besides the forest fire vulnerability information, it is important to generate the forest fire prone areas. Forest fire prone map indicates the frequency of fire occurrence in an area. For generation of forest fire prone map, the frequency of forest fire occurrence has been computed within a grid area of 1 km × 1 km using forest fire points from 2003 to 2024 The analysis using MODIS based historical forest fire incidents reveal that 72.35% forest cover of the region are under very low forest fire prone area category. The total frequency has been divided by the number of years to determine the average annual frequency based on the number of fires recorded in each grid over a span of 22 years of forest fire events and regions at risk of fire have been classified into severity categories (Kumar et al. 2019).

(i) Very low—fire count (0–1)
(ii) Low—fire count (2)
(iii) Moderate—fire count (3–5)

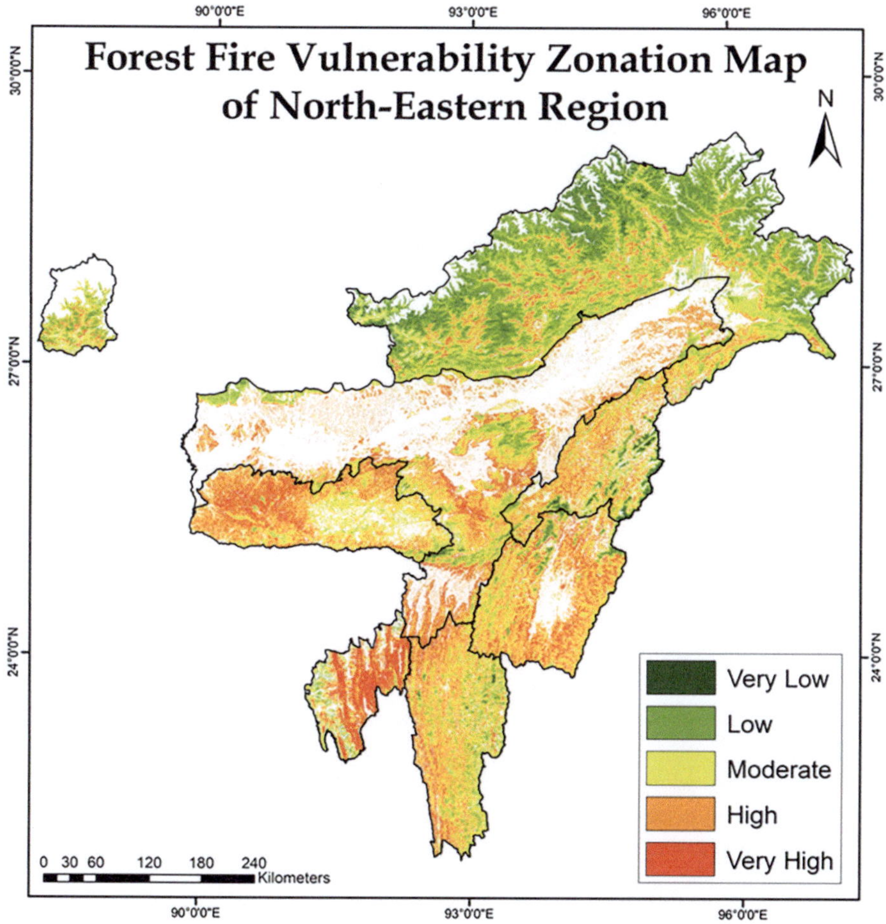

Fig. 7.8 Forest fire vulnerability map of NER

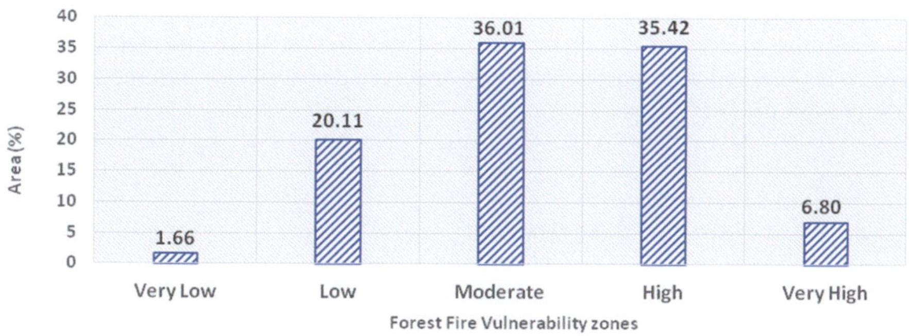

Fig. 7.9 Forest fire (%) vulnerability zones in NER

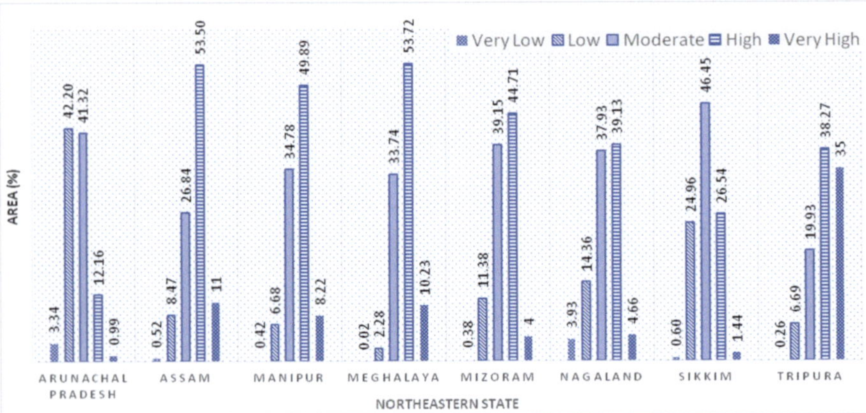

Fig. 7.10 NER state wise (%) forest fire vulnerability

(iv) High—fire count (6–10)
(v) Very High—fire count (>10).

This region has 6.24% area under high to very high prone areas. The prone area has been classified into different risk zones ranging from low to very high. The map of forest fire prone zones of NER is shown in Fig. 7.11. The area under different forest fire prone zones in NER and state wise areas are shown in Figs. 7.12 and 7.13 respectively.

The forest fire prone areas throughout the northeastern states of India, divided into five levels of zones: very low, low, moderate, high, and very high. States like Arunachal Pradesh and Sikkim predominantly fall under the very low risk category, suggesting they face minimal threat from forest fires. Conversely, states such as Tripura, Mizoram, and Manipur have a significant part of their regions classified as moderate to very high fire prone zones, indicating a greater vulnerability to forest fires. Assam, Meghalaya, and Nagaland exhibit a more even spread across the low to moderate fire prone categories. To understand the role of shifting cultivation (jhum) in the forest fire occurrence, the forest fire prone information was compared with jhum area; it is observed that most of the jhum areas are distributed within the moderate to very high fires prone zones.

7.3.2 Fire Damage Assessment

A combination of satellite images like Landsat 8, Sentinel and Resouresat- 2 LISS III has been used for the generation of forest fire burnt area maps and statistics in spatial and temporal context. The burnt area mapping has been carried out by using Normalized Burn Ratio (NBR). NBR highlights the significant decline in NIR caused by damage to the plant canopy (Chuvieco et al. 2018) and a substantial

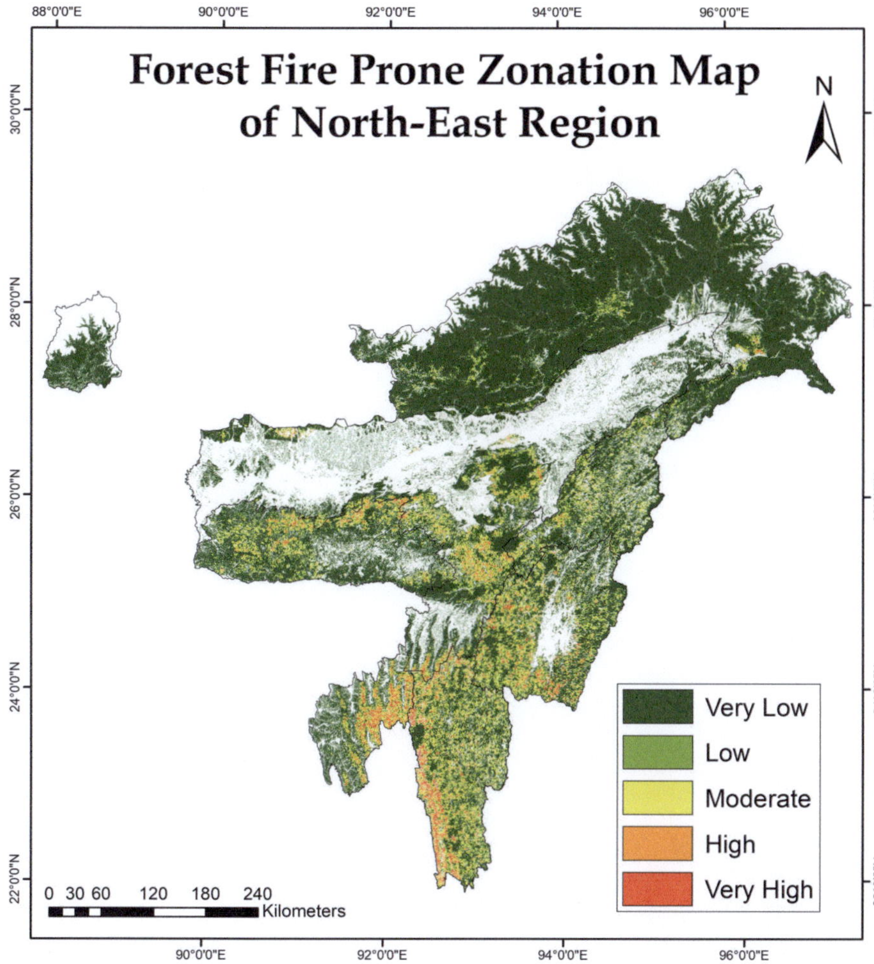

Fig. 7.11 Forest fire prone map of NER of India (based on MODIS data)

increase in SWIR due to consumption by fire (White et al. 1996). NBR is calculated as the normalised ratio between NIR and SWIR bands. The post-fire NBR image was subtracted from a pre-fire NBR image to get the "differenced NBR" (dNBR).

With the above index, it is possible to generate a spatio-temporal map of any area of interest. A map showing extent of forest fire burned area from 2015–2021 available at NESAC from previous study in the state Meghalaya is shown as an example in Fig. 7.14 and area statistics is given in Table 7.4.

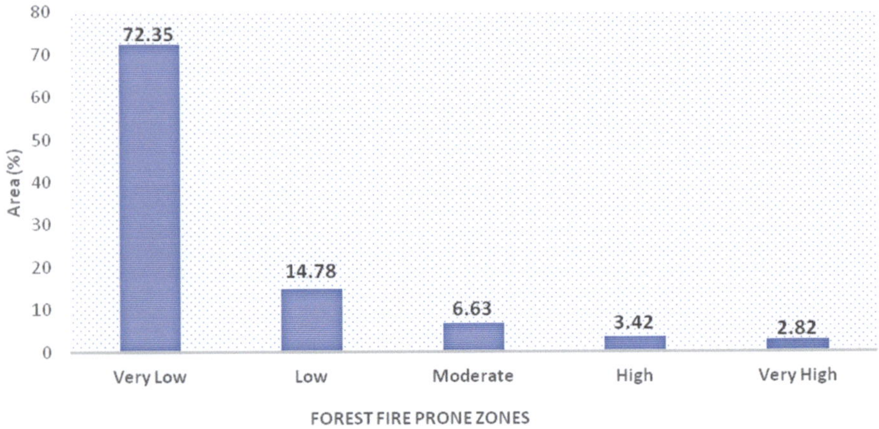

Fig. 7.12 Area (%) under different forest fire prone zones in NER

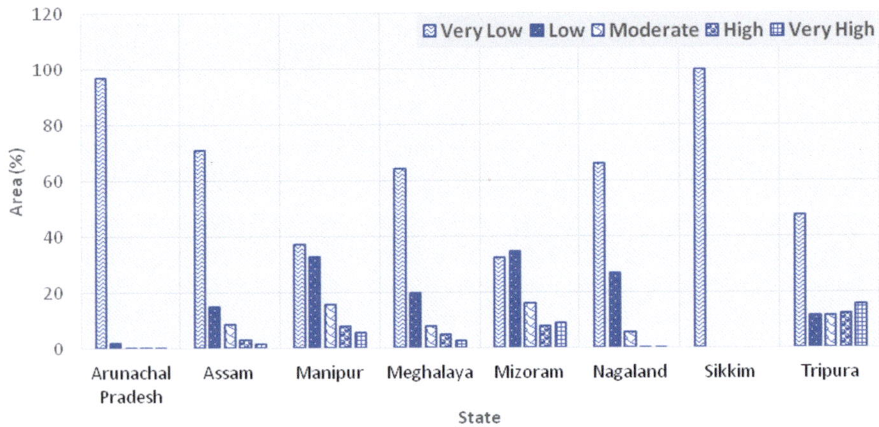

Fig. 7.13 State wise (%) forest fire prone zones in NER

7.3.3 *Monitoring and Mapping of Major Forest Fire Events in NER in Recent Past*

NER of India has witnessed many major forest fire events from time to time which has affected the rich forest cover and biodiversity of the region. Instance of two such fire events is discussed below.

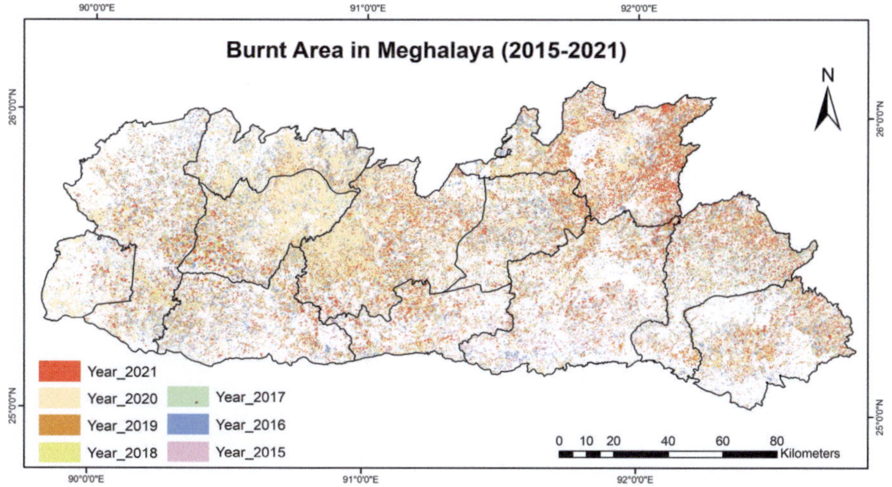

Fig. 7.14 Forest fire burnt Area distribution in Meghalaya (2015–2021)

Table 7.4 Burnt area for the state of Meghalaya (2015–2021)

Year	2015	2016	2017	2018	2019	2020	2021
Burnt area (km^2)	899.61	706.39	483.67	375.64	456.3	1144.35	477.13

7.3.3.1 Dzuko Valley

Dzuko valley is a valley located at the interstate borders of Nagaland and Manipur in NER at an altitude of 2438.4 m. The valley is famous for Dzuko lily found only in this valley and many other rare, and endemic species of flora and fauna, its natural environment and scenic beauty. The rare pink lily *Lilium mackliniae*, *Euphorbia sikkimensis*, Aconitum sp., *Caltha palustris*, Potentilla sp., Ranunculus sp., and hundreds of other species of flowers and plants in varied colour adorn the valley in monsoon (Mao and Rajib 2016). A major for forest fire event occurred in this region between 20th–30th Dec 2020 that took several days to douse the fire event. Huge tracks to the valley and large number of trees were damaged.

The utilization of very high-resolution satellite data helped in the regular monitoring of the fire spread. Further, the burnt area was estimated at NESAC for the Dzuko valley from the available optical satellite imageries of Landsat-7 Enhanced Thematic Mapper Plus (ETM+; spatial resolution of 30 m) 06th January 2021, and Sentinel-2A, -2B MultiSpectral Instrument (MSI, spatial resolution of 10 m) of 03rd January 2021, and 08th January 2021 (Fig. 7.15). Accordingly, the burnt area was estimated in the valley (Fig. 7.16) as 270 ha on 03rd January 2021, 364 ha on 06th January 2021 i.e. 94 ha increase in the burnt area due to continuous forest fire occurrence. While, on the 08th of January 2021 the total burnt area was estimated as 466 ha.

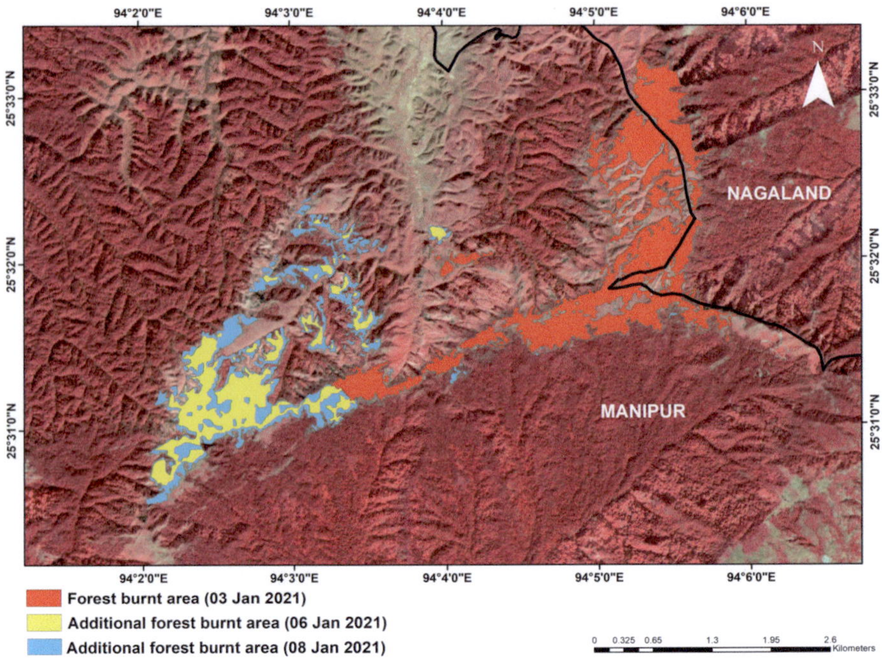

Fig. 7.15 Forest fire burnt area mapping with satellite imagery

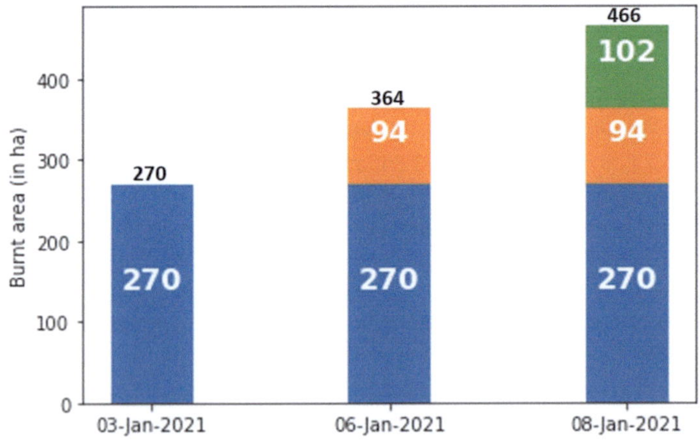

Fig. 7.16 Forest fire burned area statistics

Fig. 7.17 Large number of trees burned due to forest fire in Dzuko valley

The vegetation of the valley comprises of several important tree species (Fig. 7.17) while the major area of the valley has large area under bamboo cover. The dwarf bamboo '*Sinarundinaria rollaona* (Gamble) Chao & Renv.' Found across the entire valley is considered to be an invasion following fire incidents in this region. The dwarf bamboo grows vigorously after the fire resulting in rapid colonization and invasion of the open spaces created by the forest fire (Mao and Gogoi 2010). Hence, it is the fuel availability across this valley that makes it vulnerable to fires leading to serious damage to the tree species available in the valley. The remote sensing-based study supported by ground observations helped to prescribe measures to mitigate the forest fire occurrence in this region. A major suggestion that came out from this study is to involve the local communities in the management of the bamboo and tree species and to monitor the use objects that may cause fire occurrence. A fire line of approximately 500 m was also recommended between the bamboo covered area and the treeline which can greatly help to mitigate the fire spread towards the important tree species.

7.3.3.2 Lunglei, Mizoram

Fire that broke out on 24 April 2021 in Lunglei, Mizoram, raging for over 30 h in the hills of south Mizoram is a major forest fire event in the recent past. A rapid assessment was carried out by NESAC in Lunglei fire incident on the extent of burned area (Fig. 7.18) and the damage caused to the forests of the region. Using the satellite data of two dates, viz. 27 April 2021 (Landsat-8 OLI), and 28 April 2021(Sentinel-2A MSI) the extent of the damage was estimated. The total burnt area

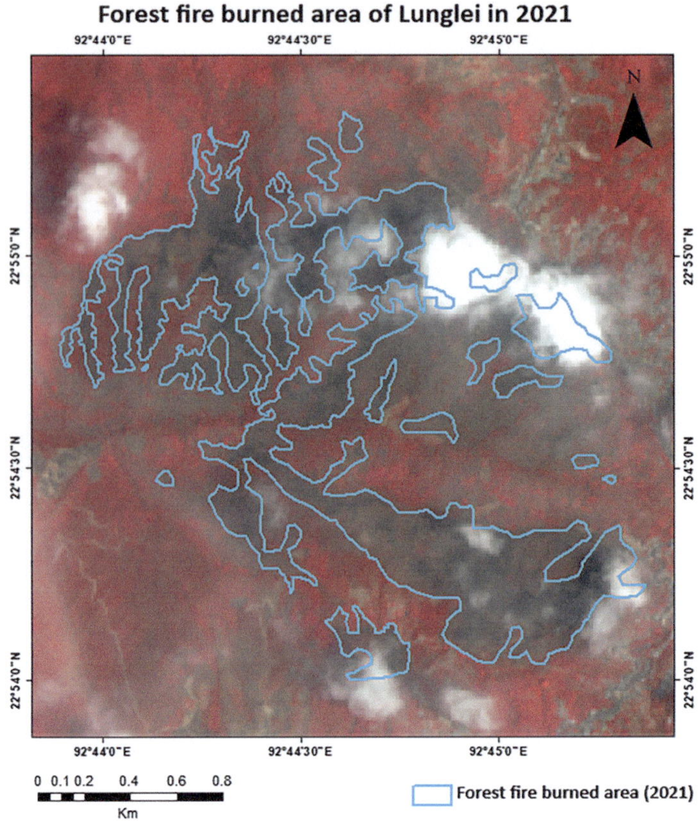

Fig. 7.18 Extent of burnt area due to forest fire in Lunglei

was estimated to be 165.1 ha. In addition to the burned area estimation, another advantage of using remote sensing technology is its utilization in the damage caused to the forest resources of any region. This study found that the fire incident occurred in the open forests of the region. With the help of available forest type map, it was found that the mixed moist deciduous, semi-evergreen and the bamboo areas were mostly impacted.

7.3.4 Relationship Between Climate Variables and Forest Fire Occurrence

A study has been conducted to examine the role of climate factors (such as temperature, rainfall, and humidity) on the incidence of forest fires in Mizoram. Climate data from ERA5 (temperature and precipitation) and FLDAS (specific humidity) at a 0.1° spatial resolution, along with MODIS-based fire occurrence data (2003–2020), were analysed using Python and QGIS to explore trends and correlations. The data were categorized by seasons (winter, pre-monsoon, monsoon, post-monsoon) and districts. The Mann–Kendall (MK) test and Sen's slope were applied for trend analysis and linear correlation coefficients were analysed to understand relationships between variables. A significant increase was found in temperature during the monsoon season (1.97–3.03 Z-score) in districts like Lawngtlai, Lunglei, and Mamit, while precipitation displayed a declining trend in September for Lawngtlai (−2.05) and Mamit (−1.97). Humidity exhibited a positive trend (1.98–3.18 Z-score), particularly in the monsoon and post-monsoon seasons. Forest fire occurrences decreased statewide, except for specific areas in Mamit, Lunglei, and Lawngtlai, which showed an increasing trend. Correlation analysis revealed that temperature had a positive relationship with fire occurrences during winter and pre-monsoon seasons, whereas precipitation and humidity showed a negative correlation with fires. These findings suggest that while climatic conditions contribute to fire susceptibility, the overall decreasing trend in fire occurrences highlights the significant role of anthropogenic factors. The study underscores the importance of understanding localized climate-fire dynamics for effective forest management in Mizoram (Bhavsar et al. 2022).

7.4 Forest Fire Management Through Fire Dashboard

There are various sources of forest fire dashboard worldwide from agencies like The National Aeronautics and Space Administration's "NASA's Fire Information for Resource Management System (FIRMS) (https://firms.modaps.eosdis.nasa.gov/). The FIRMS provides near real time active data using MODIS TERRA, AQUA, VIIRS SNPP, NOAA20 and NOAA 21. The European Forest Fire Information System (EFFIS) provides reliable information on wildfires in Europe (http://effis.jrc.ec.europa.eu). EFFIS gathers and shares data related to forest fires spanning from pre-fire to post-fire stages. The website features interactive maps showing active fires, the locations of fire stations, and fire risk assessments. The Global Wildfire Information System (GWIS) is a collaborative effort between the Group on Earth Observations (GEO) and Copernicus. GWIS offers a thorough overview and analysis of fire patterns and their impacts on a worldwide scale. GWIS also aids in wildfire management operations from local to international levels. The GOFC/GOLD (Global Observations of Forest and Land Cover Dynamics) Fire Monitoring and Mapping Team connects regional networks focused on detecting and monitoring forest fires and collaborates with the Global Fire Monitoring Center (GFMC) and the United Nations International Strategy for Disaster Reduction (UNISDR) Wildland Fire Advisory Group/Global Wildland Fire Network.

The NOAA in the US provides the FIMMA product (Fire Identification, Mapping and Monitoring Algorithm), which utilizes AVHRR-based data. The Canadian Wildland Fire Information System also employs AVHRR and MODIS fire detection techniques. Both Canadian and US services provide extensive information related to forest fires. In addition to monitoring the current conditions and early identification of hot spots, they also provide assessments of the present fire risk and future forecasts to help prepare relevant authorities. Global Forest Watch is an online platform developed by World Resource Institute provides interactive map view of global current and historical forest fire (https://www.globalforestwatch.org). SERVIR-Himalaya at ICIMOD, working closely with Nepal's Department of Forests, has created forest fire detection and monitoring system utilizing MODIS imagery for Nepal.

The Forest Fire Detection and Monitoring System alerts via email if a fire event is identified in a specific district. This information is also made available on ICIMOD's Mountain Geoportal. The Indian Forest Fire Response and Assessment System (INFFRAS) have been established to integrate and disseminate a variety of databases related to pre-fire, during fire, and post-fire situations. INFFRAS is intended to offer services such as fire alerts, tracking fire spread, assessing burnt areas, and developing mitigation strategies. INFFRAS is activated as part of DSC (Disaster Support Centre) operations under NRSC (National Remote Sensing Centre), which combines multi-sensor satellite data (MODIS) and on-ground data using a spatially and temporally detailed GIS analysis framework. However, presently the forest fire service is available in Bhuvan Geo-platform of ISRO on disaster services (https://bhuvan-app1.nrsc.gov.in). The alert system developed by FSI, is a single portal delivery system for forest fire information integration and dissemination entirely for the forest areas only (Forest Fire Alerts System 3.0 | Forest Survey of India, https://fsiforestfire.gov.in). Similarly, under the NER-DRR program, a dashboard has developed for value added product services. Once the forest fire information is received for the northeast region, the associated attributes like the topography, fuel parameters, road connectivity, water availability, nearby settlements etc. are linked to each fire location for effective fire management. The information is disseminated via email services to the respective forest and disaster management officials. A dedicated portal named as Meghalaya Forest Fire Information System (MeFFIS) has been developed for the state of Meghalaya with interactive maps and estimates that can be generated by any forest officials of the department.

7.5 Conclusion

The regional LULC transformations are expected to influence the global climate through biophysical feedback loops, although the potential scale and direction of these long-term modifications remain uncertain (Liu et al. 2019; Walker et al. 2019). To mitigate such effects, there is an urgent necessity to transition from fire-management strategies primarily driven by wildfire control to early warning systems that encompass organized prevention and risk reduction approaches (Trucchia et al. 2023). Focusing solely on ignition sources and weather conditions overlooks the significant opportunities for enhancing knowledge that could lead to identifying priorities for effective wildfire prevention (Trucchia et al. 2023).

Vegetation fires occur frequently across various parts of the world, including South and Southeast Asia (S/SEA). The occurrence and spread of these fires are affected by various factors, such as the type of fuel available, the landscape, climate conditions, weather patterns, and lightning (Prasad et al. 2008; Vadrevu et al. 2008). The impact of fire on ecosystems varies based on the intensity of the fire and the management practices in place (Vadrevu 2011). In terms of mapping and monitoring fires as well as related changes in vegetation, remote sensing technology has been crucial for several decades (Giglio et al. 2016). An analysis of the VIIRS fire data from 2012 to 2016 indicated that India recorded the highest number of annual fire occurrences of 551,649 (Vadrevu et al. 2019).

The difficulties in conducting effective studies on forest fires have been outlined in research by Jones et al. (2024). It has been noted that there is inconsistency in measurable factors and classification processes. Size thresholds differ greatly by region, shaped by local fire dynamics. The variation in criteria such as fire behaviour and its effects reflects differing viewpoints among experts. The subjective nature of what constitutes a significant impact makes it challenging to establish a precise definition. Jones et al. emphasized the necessity for precise, consistent, and scientifically sound definitions to ensure standardized measurements. And, remote sensing based studies in conjunction with GIS have tremendous potential for assessment of forest fire characteristics of any region at spatial and temporal scales.

Acknowledgements The authors would like to thank North Eastern Space Applications Centre, Umiam Meghalaya for supporting this work under the NER-DRR program. We thank the Dr S P Aggarwal, Director, NESAC for the continuous motivation to carry out this work. The authors are grateful to Dr S Sudhakar, Ex. Director, NESAC, under whose guidance the forest fire program was initiated. The authors would like to put on record the support rendered by Dr K K Sarma, Deputy Project Director, NER-DRR program, in the smooth implementation of the forest fire related studies. We would like to acknowledge and thank the Forest Survey of India (FSI), Dehradun for the forest fire incident data provided freely as and when the forest fire incidents are occur in an area. The support and feedback received from the respective State Forest Departments of NER are duly acknowledged. The team would like to thank each and every colleague of NESAC connected directly or indirectly in the execution of this forest fire related study.

References

S.J. Baker, R.A. Dewhirst, J.C. McElwain, M. Haworth, C.M. Belcher, CO2-induced biochemical changes in leaf volatiles decreased fire-intensity in the run-up to the Triassic-Jurassic boundary. New Phytol. **235**(4), 1442–1454 (2022)

L. Breiman, Random forests. Mach. Learn. **45**, 5–32 (2001)

D.C. Calkin, M.A. Finney, A.A. Ager, M.P. Thompson, K.M. Gebert, Progress towards and barriers to implementation of a risk framework for US federal wildland fire policy and decision making. Forest Policy Econ. **13**(5), 378–389 (2011)

H.G. Champion, S.K. Seth, *A Revised Survey of the Forest Types of India* (Manager of publications, 1968)

Y. Chen, D.C. Morton, J.T. Randerson, Remote sensing for wildfire monitoring: insights into burned area, emissions, and fire dynamics. One Earth **7**(6), 1022–1028 (2024)

E. Chuvieco, R.G. Congalton, Mapping and inventory of forest fires from digital processing of TM data. Geocarto. Int. **3**(4), 41–53 (1988)
E. Chuvieco, I. Aguado, M. Yebra, H. Nieto, J. Salas, M.P. Martín, R. Zamora, Development of a framework for fire risk assessment using remote sensing and geographic information system technologies. Ecol. Model. **221**(1), 46–58 (2010)
E. Chuvieco, J. Lizundia-Loiola, M.L. Pettinari, R. Ramo, M. Padilla, K. Tansey, S. Plummer, Generation and analysis of a new global burned area product based on MODIS 250 m reflectance bands and thermal anomalies. Earth Syst. Sci. Data **10**(4), 2015–2031 (2018)
E. Chuvieco, I. Aguado, J. Salas, M. García, M. Yebra, P. Oliva, Satellite remote sensing contributions to wildland fire science and management. Curr. For. Rep. **6**, 81–96 (2020)
M.A. Cochrane, D.M. Bowman, Manage fire regimes, not fires. Nat. Geosci. **14**(7), 455–457 (2021)
Fire Science Core Curriculum. OSU Extension Service, EM 9172: 197 (2017)
Food and Agriculture Organization (FAO). Global Forest Resources Assessment 1990-2000: Working Paper 55. Rome (2001)
Forest Department of Environment, Forest & Climate change, Govt. of Arunachal Pradesh. https://arunachalforests.gov.in/services/forest-wildlife/forest. Accessed 14 Jan 2025
Forest Department of Environment, Forests & Climate change, Govt. of Mizoram. https://forest.mizoram.gov.in/page/forest-types. Accessed 14 Jan 2025
Forest Department of Environment, Forests & Climate change, Govt. of Nagaland. https://forest.nagaland.gov.in/status-of-forests/. Accessed 14 Jan 2025
X. Feng, C. Merow, Z. Liu, D.S. Park, P.R. Roehrdanz, B. Maitner, B.J. Enquist, How deregulation, drought and increasing fire impact Amazonian biodiversity. Nature **597**(7877), 516–521 (2021)
M.D. Flannigan, T.V. Haar, Forest fire monitoring using NOAA satellite AVHRR. Can. J. for. Res. **16**(5), 975–982 (1986)
Forest, Department of Forest, Govt. of Tripura. https://forest.tripura.gov.in/forest-of-tripura.html). Accessed 14 Jan 2025
Forest Department of Forest, Govt. of Manipur. https://monitoring.manipurforest.gov.in/Main/AboutUs. Accessed 14 Jan 2025
Forest Department of Forest & Environment, Govt. of Meghalaya. https://www.megforest.gov.in/forest_overview.html. Accessed 14 Jan 2025
FSI (Forest Survey of India), India State of Forest Report (2021)
L. Giglio, J. Descloitres, C.O. Justice, Y.J. Kaufman, An enhanced contextual fire detection algorithm for MODIS. Remote Sens. Environ. **87**(2–3), 273–282 (2003)
L. Giglio, W. Schroeder, C.O. Justice, The collection 6 MODIS active fire detection algorithm and fire products. Remote Sens. Environ. **178**, 31–41 (2016)
J.G. Goldammer, C. De Ronde, *Wildland fire Management Handbook for Sub-Sahara Africa* (African Minds, 2004)
J. Goldammer, I. Mitsopoulos, G. Mallinis, M. Woolf, *Wildfire Hazard and Risk Assessment* (Words into Action Guidelines National Disaster Risk Assessment; United Nations Office for Disaster Risk Reduction, Geneva, Switzerland, 2017)
D. Havsar, K. Chakraborty, J. Mohapatra, K.K. Sarma, S.P. Aggarwal, Role of Climatic Variables on Forest Fire in the State of Mizoram. In *North-East Research Conclave* (Springer Nature Singapore, Singapore, 2022), pp. 283–309
R. Hassan, R. Scholes, N. Ash (eds.), Findings of the Condition and Trends Working Group of the Millennium Ecosystem Assessment. Ecosystems and Human Well-being: Current State and Trends, Volume 1. University of Pretoria Council for Science and Industrial Research UNEP World Conservation South Africa, South Africa Monitoring Centre, United Kingdom (Island Press Washington, D.C, 2005)
X. Huang, Research on fire point monitoring with high time and high spatial resolution based on multi-source remote sensing data. MS thesis (2020)
India State of Forest Report (ISFR). Forest Survey of India, Dehradun, https://fsi.nic.in (2023)

F.H. Johnston, N. Borchers-Arriagada, G.G. Morgan, B. Jalaludin, A.J. Palmer, G.J. Williamson, D.M. Bowman, Unprecedented health costs of smoke-related PM2. 5 from the 2019–20 Australian megafires. Nat. Sustain. **4**(1), 42–47 (2020)

Joint Research Center of the European Commission (JRC). Global Burnt Area 2000 (GBA2000) dataset: http://www.gvm.jrc.it/fire/gba2000/index.html (2002)

M.W. Jones, S. Veraverbeke, N. Andela, S.H. Doerr, C. Kolden, G. Mataveli, J.T. Abatzoglou, Global rise in forest fire emissions linked to climate change in the extratropics. Science **386**(6719), eadl5889 (2024)

C.O. Justice, J.R.G. Townshend, E.F. Vermote, E. Masuoka, R.E. Wolfe, N. Saleous, J.T. Morisette, An overview of MODIS Land data processing and product status. Remote Sens. Environ. **83**(1–2), 3–15 (2002)

Y.J. Kaufman, A. Setzer, C. Justice, C.J. Tucker, M.C. Pereira, I. Fung, Remote sensing of biomass burning in the tropics. *Fire in the Tropical Biota: Ecosystem Processes and Global Challenges*, 371–399 (1990)

L.T. Kelly, K.M. Giljohann, A. Duane, N. Aquilué, S. Archibald, E. Batllori, L. Brotons, Fire and biodiversity in the Anthropocene. Science **370**(6519), eabb0355 (2020)

A.P. Kershaw, J.S. Clark, A.M. Gill, D.M. D'COSTA, Ahistoryoffire in Australia (2002)

L.N. Kobziar, J.K. Hiers, C.M. Belcher, W.J. Bond, C.A. Enquist, E.L. Loudermilk, A.C. Watts, Principles of fire ecology. Fire Ecol. **20**(1), 39 (2024)

S. Kumar, C. Chaudhary, T. Biswas, S. Ghosh, S. Ashutosh, Identification of fire prone forest areas based on GIS analysis of archived forest fire points detected in last thirteen years. Minist. Environ., For. Clim. Chang. Gov. Ment India Dehradun, India **1** (2019)

T. Kurvits, E. Baker, K. Thygesen, P. Sevaldsen, *Spreading like Wildfire: The Rising Threat of Extraordinary Landscape Fires* (2022)

D. Leavell, *Fire Science core curriculum-Module 2* (2017)

Z. Liu, A.P. Ballantyne, L.A. Cooper, Biophysical feedback of global forest fires on surface temperature. Nat. Commun. **10**(1), 214 (2019)

W. Liu, H. Guan, P.A. Hesp, O. Batelaan, Remote sensing delineation of wildfire spatial extents and post-fire recovery along a semi-arid climate gradient. Eco. Inform. **78**, 102304 (2023)

J. Malczewski, C. Rinner, *Multicriteria Decision Analysis in Geographic Information Science*, vol. 1, pp. 55–77. (Springer, New York, 2015)

A.A. Mao, R. Gogoi, *Flora of Dziiko/Dzukou Valley* (Botanical Survey of India, 2016)

A.A. Mao, R. Gogoooi, Fire-induced invasion of an endemic plant species alters forest structure and diversity: a study from north-east India. Curr. Sci. (00113891) **98**(4) (2010)

M.P. Martín, I. Gómez, E. Chuvieco, Performance of a burned-area index (BAIM) for mapping Mediterranean burned scars from MODIS data. In *Proceedings of the 5th international workshop on remote sensing and GIS applications to forest fire management: fire effects assessment*, pp. 193–198. (Universidad de Zaragoza, GOFC GOLD, EARSeL, Paris, 2005)

V.P. Masson-Delmotte, P. Zhai, S.L. Pirani, C. Connors, S. Péan, N. Berger, P.M. Scheel Monteiro, Ipcc, 2021: Summary for policymakers. in *Climate Change 2021: The Physical Science Basis. Contribution of Working Group i to the Sixth Assessment Report of the Intergovernmental Panel on Climate Change* (2021)

K.K. McLauchlan, P.E. Higuera, J. Miesel, B.M. Rogers, J. Schweitzer, J.K. Shuman, A.C. Watts, Fire as a fundamental ecological process: research advances and frontiers. J. Ecol. **108**(5), 2047–2069 (2020)

Ministry of Environment, F. A. C. C. G. O. I., & World Bank, *Strengthening Forest Fire Management in India* (World Bank, 2018)

P. Morgan, R.E. Keane, G.K. Dillon, T.B. Jain, A.T. Hudak, E.C. Karau, E.K. Strand, Challenges of assessing fire and burn severity using field measures, remote sensing and modelling. Int. J. Wildland Fire **23**(8), 1045–1060 (2014)

M.A. Moritz, E. Batllori, R.A. Bradstock, A.M. Gill, J. Handmer, P.F. Hessburg, A.D. Syphard, Learning to coexist with wildfire. Nature **515**(7525), 58–66 (2014)

R.L. Myers, *Living with Fire: Sustaining Ecosystems and Livelihoods Through Integrated Fire Management. The Nature Conservancy, Arlington, VA, USA* (2006)
D.M. Nelson, Y. He, G.W.K. Moore, Trends and applications in wildfire burned area mapping: remote sensing data, cloud geoprocessing platforms, and emerging algorithms. Geomatica 100008 (2024)
V.J. Pasquarella, P. Arévalo, K.H. Bratley, E.L. Bullock, N. Gorelick, Z. Yang, R.E. Kennedy, Demystifying LandTrendr and CCDC temporal segmentation. Int. J. Appl. Earth Obs. Geoinf. **110**, 102806 (2022)
J.G. Pausas, J.E. Keeley, A burning story: the role of fire in the history of life. Bioscience **59**(7), 593–601 (2009)
M.P. Plucinski, A.L. Sullivan, C.J. Rucinski, M. Prakash, Improving the reliability and utility of operational bushfire behaviour predictions in Australian vegetation. Environ. Model. Softw. **91**, 1–12 (2017)
V.K. Prasad, K.V.S. Badarinath, A. Eaturu, Biophysical and anthropogenic controls of forest fires in the Deccan Plateau, India. J. Environ. Manag. **86**(1), 1–13 (2008)
J. Purkayastha (ed.), *Bioprospecting of Indigenous Bioresources of North-East India* (2016)
X. Qin, X. Chen, X. Zhong, X.F. Zu, G.F. Sun, L.Y. Ying, Development of forest fire early warning and monitoring technique system in China. For. Grassl. Resour. Res **6**, 45–48 (2015)
E. Rakhmatulina, G. Boisramé, S.L. Stephens, S. Thompson, Hydrological benefits of restoring wildfire regimes in the Sierra Nevada persist in a warming climate. J. Hydrol. **593**, 125808 (2021)
ReliefWeb, Algeria: Forest Wildfires-Emergency Appeal №: MDRDZ007, Posted: 2 November 2021, Originally published 30 October 2021 (2021). https://reliefweb.int/report/algeria/algeria-forest-wildfires-emergency-appeal-mdrdz007
J.M. Robinson, Fire from space: global fire evaluation using infrared remote sensing. Int. J. Remote Sens. **12**(1), 3–24 (1991)
T.L. Saaty, Analytic hierarchy process. in *Encyclopedia of Operations Research and Management Science*, pp. 52–64. (Springer, Boston, MA, 2013)
J. San-Miguel-Ayanz, T. Durrant, R. Boca, P. Maianti, G. Libertá, T. Artés- Vivancos, D. Oom, A. Branco, D. de Rigo, D. Ferrari, H. Pfeiffer, R. Grecchi, D. Nuijten, *Advance Report on Forest Fires in Europe, Middle East and North Africa 2021. JRC128678. EUR 31028 EN*. (Publications Office of the European Union, Luxembourg, 2022). https://doi.org/10.2760/039729. ISBN 978–92–76–49633–5
S.S. Sayedi, B.W. Abbott, B. Vannière, B. Leys, D. Colombaroli, G.G. Romera, A.L. Daniau, Assessing changes in global fire regimes. Fire Ecol. **20**(1), 1–22 (2024)
A.F.W. Schimper, *Plant-Geography Upon a Physiological Basis* (Clarendon Press, 1902)
A.C. Scott, I.J. Glasspool, The diversification of Paleozoic fire systems and fluctuations in atmospheric oxygen concentration. Proc. Natl. Acad. Sci. **103**(29), 10861–10865 (2006)
A. Shlisky, J. Waugh, P. Gonzalez, M. Gonzalez, M. Manta, H. Santoso, H, W. Fulks, Fire, ecosystems and people: threats and strategies for global biodiversity conservation. Arlingt. Nat. Conserv. 1–17 (2007)
A. Trucchia, G. Meschi, P. Fiorucci, A. Provenzale, M. Tonini, U. Pernice, Wildfire hazard mapping in the eastern Mediterranean landscape. Int. J. Wildland Fire **32**(3), 417–434 (2023)
United Nations Environment Programme (UNEP). Spreading like Wildfire – The Rising Threat of Extraordinary Landscape Fires. A UNEP Rapid Response Assessment. Nairobi (2022)
United Nations International Strategy for Disaster Reduction, *UNISDR Terminology on Disaster Risk Reduction* (United Nations, Geneva, 2009)
K.P. Vadrevu, C.O. Justice, Vegetation fires in the Asian region: Satellite observational needs and priorities. Glob. Environ. Res. **15**(1), 65–76 (2011)
K.P. Vadrevu, K.V.S. Badarinath, E. Anuradha, Spatial patterns in vegetation fires in the Indian region. Environ. Monit. Assess. **147**, 1–13 (2008)
K.P. Vadrevu, K. Lasko, L. Giglio, W. Schroeder, S. Biswas, C. Justice, Trends in vegetation fires in south and southeast Asian countries. Sci. Rep. **9**(1), 7422 (2019)

D. van Wees, G.R. van der Werf, J.T. Randerson, B.M. Rogers, Y. Chen, S. Veraverbeke, D.C. Morton, Global biomass burning fuel consumption and emissions at 500-m spatial resolution based on the Global Fire Emissions Database (GFED). Geosci. Model Dev. Discuss. **2022**, 1–46 (2022)

C. Vitolo, F. Di Giuseppe, C. Barnard, R. Coughlan, J. San-Miguel-Ayanz, G. Libertá, B. Krzeminski, ERA5-based global meteorological wildfire danger maps. Sci. Data **7**(1), 216 (2020)

X.J. Walker, J.L. Baltzer, S.G. Cumming, N.J. Day, C. Ebert, S. Goetz, M.C. Mack, Increasing wildfires threaten historic carbon sink of boreal forest soils. Nature **572**(7770), 520–523 (2019)

J.D. White, K.C. Ryan, C.C. Key, S.W. Running, Remote sensing of forest fire severity and vegetation recovery. Int. J. Wildland Fire **6**(3), 125–136 (1996)

C. Whitlock, P.E. Higuera, D.B. McWethy, C.E. Briles, Paleoecological perspectives on fire ecology: revisiting the fire-regime concept. Open Ecol. J. **3**(1), 6–23 (2010)

WHO. Report on the biregional workshop on health impacts of haze-related air pollution, Kuala Lumpur, Malaysia, 1–4 June 1998. World Health Organization, Kuala Lumpur, (WP) EHE(O)/ICP/EHH/004-E (1998)

R.E. Zeebe, A. Ridgwell, J.C. Zachos, Anthropogenic carbon release rate unprecedented during the past 66 million years. Nat. Geosci. **9**(4), 325–329 (2016)

Q. Zhang, S. Homayouni, P. Zhao, M. Zhou, Burned vegetation recovery trajectory and its driving factors using satellite remote-sensing datasets in the Great Xing'An forest region of Inner Mongolia. Int. J. Wildland Fire **32**(2), 244–261 (2023)

W. Zhao, W.K. Cornwell, M. van Pomeren, R.S. van Logtestijn, J.H. Cornelissen, Species mixture effects on flammability across plant phylogeny: the importance of litter particle size and the special role for non-Pinus Pinaceae. Ecol. Evol. **6**(22), 8223–8234 (2016)

Y. Zheng, G. Zhang, S. Tan, L. Feng, Research on progress of forest fire monitoring with satellite remote sensing. Agric. Rural Stud. **1**(2) (2023)

Kasturi Chakraborty completed her M.Sc. and M.Phil in Ecology and completed her Ph.D. on 'Land use/land cover and forest change using Remote sensing and GIS in Barak basin'. She has led project works and contributed as team member for successful completion of important projects at NESAC. Few of the important are in the field of forest fire, forest working plan, forest change analysis at regional scale, land use land cover mapping, wetland conservation and management. She has published research articles on forest and land use change analysis using geospatial techniques. She is a recipient of ISRO, young scientist award (2009) and ISRO project team award (2019).

Dhruval Bhavsar is currently working as scientist in the Forestry and Ecology Division of North Eastern Space Applications Centre (NESAC), Department of Space, Umiam, Meghalaya. He holds M.Sc. and Ph.D. in Forestry from the Forest Research Institute (Deemed to be University), Dehradun. His research expertise includes forest biomass assessment, forest fire monitoring and SAR remote sensing in forestry.With over 10 years of experience in forest ecology, monitoring, and management using remote sensing and GIS, he has contributed to national projects and successfully led research and user-funded projects. He has also authored several publications in peer-reviewed national and international journals and conferences.

Suraj Kr. Swain is a dedicated wildlife and forest researcher with a passion to conserve biodiversity and natural habitats. With over six years of professional experience in forest inventory, ecology, and environmental monitoring across various regions of India, his work integrates both field-based ecological research and advanced geospatial analysis. His expertise lies in the application of remote sensing and GIS technologies for forest cover change detection, land use and land cover (LULC) studies, and forest fire assessment and monitoring. He holds a M.Sc. degree in *Biodiversity and Conservation of Natural Resources* from the Central University of Odisha and currently pursuing his Ph.D. at Mizoram University. Presently, He serves as a Research Scientist at the North Eastern Regional Node for Disaster Risk Reduction (NER-DRR), North Eastern Space Applications Centre (NESAC), Department of Space, Government of India.

K. K. Sarma Ph.D., leads the Remote Sensing Applications Group (RSAG) at the North Eastern Space Applications Centre (NESAC), where he has been a key figure since 2002. Specializing in remote sensing for forestry and biodiversity, his work leverages space technology for critical applications in Biodiversity, Ecosystems, Forestry, and Natural Resource Management. Dr. Sarma has spearheaded over 60 diverse geospatial projects, contributing significantly to natural resource mapping and monitoring across the North Eastern Region (NER). Currently, he coordinates 110 "Plan of Action" projects, focused on using space technology to improve livelihoods in the NER. He also served as the Principal Investigator and NER Coordinator for all ISRO natural resource management projects, including comprehensive Land Use Land Cover Mapping and National Wasteland Updating, Monitoring, and Change Analysis. With over 75 research publications, Dr. Sarma has successfully supervised six Ph.D. students and actively mentors many others across NER Universities.

Chapter 8
Geospatial Techniques and Applications for Mineral Mapping in the NE Region of India

Rahul Pratap, Gopal Sharma, M. Somorjit Singh, Golfinstar Kharbani, and K. K. Sarma

Abstract The Northeast Region (NER) of India is endowed with significant mineral resources that are fundamental to its economic development. However, the comprehensive assessment and sustainable management of these resources are impeded by persistent research gaps and the inherent limitations of conventional field surveys in a region characterized by rugged, often inaccessible terrain. This chapter presents a review of modern geospatial technologies, including RS and GIS, as a robust and efficient solution for mineral exploration in the NER. This chapter details a suite of analytical techniques that leverage the diagnostic spectral signatures of minerals for their identification and mapping. It systematically covers multispectral methods, such as band ratioing and Principal Component Analysis, and advanced hyperspectral analyses, including the Spectral Angle Mapper, Matched Filtering, and Linear Spectral Unmixing. The discussion is augmented by an examination of the utility of thermal and microwave remote sensing for geological applications. The practical efficacy and validation of these methodologies are demonstrated through a series of case studies conducted in Meghalaya. These include the successful identification and mapping of calcite, limestone, kaolinite, and ultramafic rock formations utilizing multispectral and hyperspectral satellite data. Collectively, these applications illustrate how integrated geospatial workflows offer a precise, efficient, and scalable approach to identifying unexplored deposits and assessing their economic viability, thereby addressing critical challenges in regional mineral resource management.

R. Pratap (✉) · G. Sharma · M. S. Singh · G. Kharbani · K. K. Sarma
North Eastern Space Applications Centre (NESAC), Umiam, Shillong, Meghalaya, India
e-mail: rahul.pratap@nesac.gov.in

G. Sharma
e-mail: gopal.sharma@nesac.gov.in

M. S. Singh
e-mail: m.somorjit@nesac.gov.in

K. K. Sarma
e-mail: kk.sarma@nesac.gov.in

S. P. Aggarwal et al. (eds.), *Remote Sensing of the North Eastern Himalayan Ecosystem*,
https://doi.org/10.1007/978-981-95-3221-6_8

Keywords Geospatial technologies · Mineral exploration · Northeast India · Remote sensing · Hyperspectral analysis

8.1 Introduction

The mineral-rich Northeast India contribute significantly to the regional economy and local industries. Meghalaya hosts significant mineral resources, particularly the Upper Sylhet Limestone of the Shella formation within the Jaintia group, found in the Larket block (Prasad et al. 2001–2003). Extensive sillimanite deposits occur as strata-bound seams in sillimanite-bearing schist across structural hills and ridges (Prakash 1984), while titaniferous-vanadiferous-magnetite deposits have been identified in Myniar within gneiss and porphyritic granite (Lal 2015–16). In Mizoram, small limestone deposits are scattered across areas such as Muthi, Tuirial, Nisapui, and Daipum, while coal occurs in thin lenses within sandstone near Thinglian, Kolasib, Phuaibuang and Lamzawl. Recent petroleum discoveries have been confirmed by ONGC in Bilkhawthlir, Kolasib District (Kumar et al. 2023). Tripura's 1956–57 geological survey revealed multiple natural gas seepages along four parallel north–south anticlinal ridges (Sen 1956–57). Two distinct types of laterite are found near Agartala: a friable laterite capping hilltops and a loose, unconsolidated, feldspathic sand (Rao 1984). Nagaland's Barail group rocks at Tuensang and Changki contain coal seams (Limaye et al. 1967). Glass sand deposits in the Desai valley are linked to the Girujan clay formation within the Tipam group (Majumdar 1977). A survey in Pukphur, Tuensang district, revealed a nickeliferous magnetite slab within an ultramafic serpentinite belt extending across Manipur and Nagaland (Bhaumik et al. 1973). Sikkim's Reshi Bazar-Tatapani, Mangalbaria Subuk, Kewzing area contains Daling group (Reyang and Buxa formations) and Gondwana supergroup rocks (Damuda formation and Rangit pebble beds). Here, dolomites are massive, compact, and pyrite and chalcopyrite occur sporadically in quartz veins near Lingdum, Rey Khola and Raigaon (Longchari et al. 2014–2015). Arunachal Pradesh hosts refractory-grade quartzite near Kalaktang in West Kameng district, ideal for silica refractory bricks (Srinivasan et al. 1996). White marble is present in the Tezu and Dora river sections near Tidding, Lohit district, alongside graphite deposits and cement-grade limestone. Talc-tremolite-asbestos schist and serpentinite occur in Lalpani and Tidding (Nandy et al. 1967), with Quaternary clays, including whitish grey and yellow–brown varieties, overlaying Neogene sediments in low-level alluvium (Panduranga 1988–89). Assam's Shella formation in the North Boro Hundong Block, Dima Hasao district, holds limestone with high CaO content, suitable for cement grade use due to higher silica content (Thakur et al. 2019–2020).

Extensive clay deposits in Upper Assam, derived from feldspar, are part of the Jaipur measures, along with crystalline rocks suitable for dimension stones. These include equi-granular granite gneiss in the Kukrakata hill range, Phandari parbat, medium-grained porphyritic granite in Seconee and Hatigaon, coarse porphyritic granite in Bura parbat, and greyish granite and gneiss in Karkok (Buragohain 1999).

Coal is also reported in Sonapur in the Jiribam area (Limaye 1966–67), and cement-grade limestone is found in Phungyar, Singkap-Shingoa, and Shangshak blocks in Ukhrul district, Manipur (Khan 1889–90). Numerous brine wells exist around Chingai, with evaporates forming encrustations, dissemination, and thin layers within the shales and sandy shales of the Chingai group (Chattopadhyay et al. 1975).

Despite the abundance of mineral resources in the northeastern states of India, significant research gaps remain regarding mineral mapping and management. The integration of geospatial technology can address these gaps effectively. Accurate mapping and evaluation of mineral resources are made possible by technologies like remote sensing (RS) and geographic information systems (GIS), facilitating the identification of unexplored deposits and assessing their economic viability. Spatial analysis using these technologies aids in planning mining operations, ensuring that decisions are informed by geological, environmental, and socio-economic data to minimise conflicts. Field mapping, while providing in-depth, ground-level data, is slower, more resource-intensive, and less feasible in inaccessible areas. Geospatial technology greatly enhances mineral mapping, offering broad-area coverage, efficiency, and precision over rugged terrains, which would be challenging for traditional field mapping.

This chapter covers the latest geospatial techniques available for mineral mapping, methods and a few case studies in NER (Fig. 8.1).

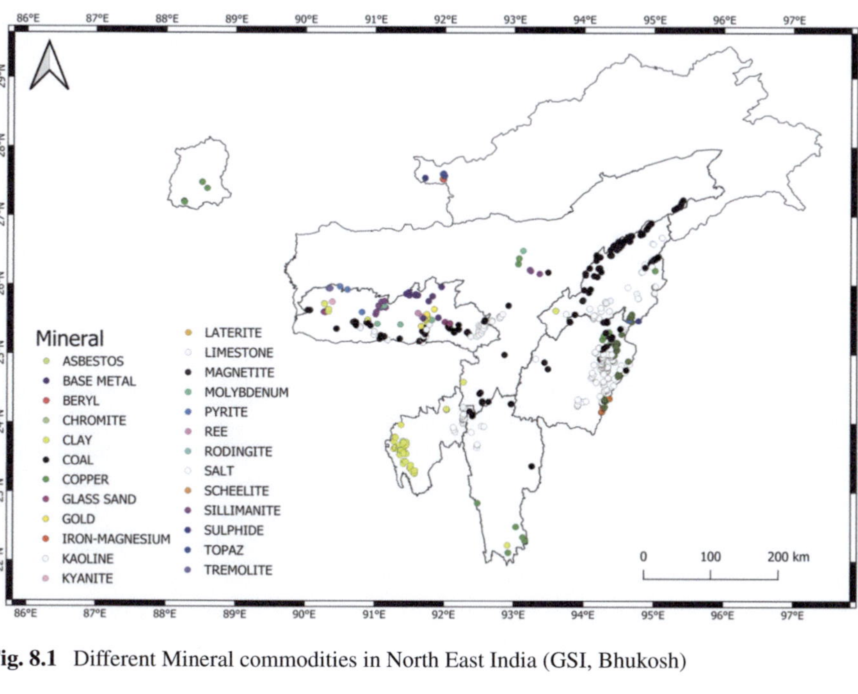

Fig. 8.1 Different Mineral commodities in North East India (GSI, Bhukosh)

8.2 Geospatial Techniques for Mineral Mapping

8.2.1 Multispectral Remote Sensing

A vital tool for mineral mapping is multispectral remote sensing, which records reflectance information in the visible, near-infrared (NIR), and short-wave infrared (SWIR) bands. This information makes it possible to identify minerals using distinctive spectral patterns, which is crucial for mapping mineral compositions, particularly in isolated locations. Satellite-like Landsat enabled the differentiation of surface materials, laying the groundwork for mineral exploration (Sabins 1999). Advances like ASTER have added SWIR bands ideal for distinguishing minerals such as carbonates, clays, and iron oxides, aiding in the identification of hydrothermal alteration zones linked to ore deposits (Rowan and Mars 2003) (Table 8.1).

8.2.1.1 Band Ratio

In remote sensing, a band ratio is a method that determines the proportion of reflectance values from two or more different spectral bands derived from satellite or aerial data (Jensen 1979). This method enhances the contrast of specific features or materials, facilitating better differentiation of various land covers or mineral types. Band ratios are especially effective in highlighting unique spectral characteristics associated with specific minerals or vegetation types, thus improving the interpretability of the imagery. The calculation of band ratios is typically expressed as

$$\text{Band Ratio} = \text{Reflectance in Band A}/\text{Reflectance in Band B} \tag{8.1}$$

where Band A and Band B are selected spectral bands that emphasise specific features of interest (Mather 2010; Lillesand et al. 2015). By calculating the ratio of two different spectral bands, this method can accentuate features or materials that may not be easily identifiable in single-band images (Lillesand et al. 2015). Band ratios utilise the unique spectral signatures of minerals, allowing for better discrimination between various rock types and mineral compositions.

Multispectral remote sensing using Landsat 8 and ASTER offers band ratios to detect various minerals. For iron oxides, Landsat utilises Band 7 divided by Band 5, and ASTER uses Band 4 divided by Band 3 or Band 6 divided by Band 5 for refined detection. Clay minerals are identified with Landsat's Band 4 divided by Band 3 and ASTER's Band 8 divided by Band 3 (Yamaguchi et al. 2006). Carbonates are best detected using Landsat's Band 6 divided by Band 5, while ASTER distinguishes calcite with Band 6 divided by Band 5. For unique minerals, serpentinite and alunite can be detected with ASTER ratios like Band 7 divided by Band 6 and Band 10 divided by Band 9.

Table 8.1 Multispectral sensors and key channels for mineral mapping

Sensor name (Platform)	Key spectral bands (Wavelengths)	Mineral groups mapped/applications	References
ASTER (Terra Satellite)	VNIR (0.52–0.86 μm), SWIR (1.6–2.43 μm), TIR (8.125–11.65 μm)	Hydroxyl-bearing minerals (clays, micas, amphiboles): SWIR bands (e.g., Al–OH at ~2.20 μm, Mg-OH at ~2.32 μm) Carbonates: SWIR bands (e.g., CO3 at ~2.35 μm) Iron Oxides (hematite, goethite, limonite): VNIR bands (e.g., Fe-oxides at ~0.8–1.0 μm) Silicates: TIR bands for emissivity differences	Yamaguchi et al. (1999), Cudahy (2012), Mars and Rowan (2010), Witt et al
Landsat (Various Satellites: TM, ETM +, OLI)	OLI: Deep Blue (0.43–0.45 μm), Blue (0.45–0.51 μm), Green (0.53–0.59 μm), Red (0.64–0.67 μm), NIR (0.85–0.88 μm), SWIR1 (1.57–1.65 μm), SWIR2 (2.11–2.29 μm), Panchromatic (0.50–0.68 μm)	Iron Oxides: Visible and NIR bands, particularly band ratios like Red/Blue, Red/Green, and specific indices related to the 0.8–1.0 μm absorption feature Vegetation (for masking): NIR and Red bands for NDVI General lithological mapping	USGS, MDPI (Geological and Mineralogical Mapping Based on Statistical Methods of Remote Sensing Data Processing of Landsat-8)
Sentinel-2 (Copernicus Programme)	Visible (B2, B3, B4), Red Edge (B5, B6, B7), NIR (B8, B8A), SWIR (B11, B12)	Iron-bearing minerals (e.g., gossans, iron oxides): Red-edge and NIR bands (especially around 0.9 μm absorption) Hydrothermal alteration zones: SWIR bands for hydroxyl and carbonate minerals Vegetation (for masking and exploration in vegetated areas)	Alberta Geological Survey (Sentinel-2 for iron occurrences)

(continued)

Table 8.1 (continued)

Sensor name (Platform)	Key spectral bands (Wavelengths)	Mineral groups mapped/applications	References
WorldView-3 (Maxar)	8 VNIR bands (0.4–1.04 μm), 8 SWIR bands (1.195–2.365 μm)	Direct material identification and chemical measurements: High spectral resolution in VNIR and SWIR allows for more refined mineral identification Clays, Carbonates, Sulfates, Ammonium: Diagnostic absorption features in SWIR Iron-bearing minerals: VNIR bands	Maxar Technologies; AWS (Mineral Identification from Space)
EO-1 ALI (Advanced Land Imager)	Panchromatic (0.48–0.69 μm), Multispectral bands (0.433–0.453 μm to 1.55–1.75 μm, and 1.2–1.3 μm)	General mineral mapping, similar to enhanced Landsat capabilities. Used in studies for simulating hyperspectral data	E3S Web of Conferences (Application of hyperspectral and multispectral datasets for mineral mapping)

Data for band ratio analysis is typically obtained from multispectral satellite imagery, such as Landsat or ASTER. These images are processed to derive reflectance values for each spectral band, which are then used to compute the desired band ratios (Singh et al. 2022). To improve the understanding of mineral properties, the study may additionally incorporate other processing methods as normalised difference indices or principal component analysis. The effectiveness of band ratio approaches in mineral prospecting has been shown in numerous research studies and mapping in the northeastern region of India. For instance, the application of band ratios has been successfully used to identify coal deposits in Assam and limestone formations in Meghalaya, highlighting the method's utility in regional mineral assessments (Bhattacharya et al. 2018). ASTER imagery has also been applied to identify and map hydrothermal alterations associated with mineral deposits, underscoring its versatility in mineral exploration (Niemeyer et al. 2009).

8.2.1.2 Principal Component Analysis (PCA)

This statistical technique preserves the majority of the variance while reducing the dimensionality of huge datasets (Pearson 1901). To do this, the original variables are converted into a group of uncorrelated variables called principal components, which are then rated according to how much variation they can extract from the data (Jolliffe 2002). Typically, a few principal components capture most of the variability, simplifying the data analysis and modelling processes (Wold et al. 1987). The PCA procedure starts with standardising the data to ensure consistency among variables with different units. The data's covariance (or correlation) matrix is then calculated to examine the relationships between the variables. Next, the matrix's eigenvalues and eigenvectors are determined, with eigenvectors serving as principal components ordered by variance captured. Finally, the data is projected into the new component space defined by these principal components (Hotelling 1933; Pearson 1901). The transformation into principal components can be expressed as:

$$\mathrm{Y} = \mathrm{X} \times W \tag{8.2}$$

where X is the data matrix, W is the matrix of eigenvectors, and Y is the matrix of principal components. To maximise variance capture, each principal component Yi is a linear combination of the original variables, weighted by the eigenvectors (Jolliffe and Cadima 2016).

The primary merit of PCA lies in its ability to significantly reduce data dimensionality, which is especially beneficial for high-dimensional hyperspectral datasets, simplifying subsequent analysis and visualisation. By concentrating most of the variance into the leading components, PCA implicitly aids in noise reduction, as noise often contributes less to the overall variance and is relegated to higher-order components. Furthermore, the resulting principal components are uncorrelated, which can be advantageous for various downstream statistical analyses and modelling. PCA also serves as an effective feature extraction method, highlighting the most significant patterns of variance within the data (Abdi and Williams 2010; Ringner 2008). However, PCA has its limitations. The interpretability of principal components can be challenging, as they are abstract linear combinations of original bands, lacking direct physical or geological meaning. While most variance is preserved, some information is inherently lost by discarding lower-variance components; if these components contain subtle yet important mineralogical signals, crucial information might be missed. PCA is also sensitive to the scaling of the original variables, necessitating careful standardisation. Lastly, its effectiveness can be limited for non-linear spectral phenomena due to its underlying assumption of linear relationships between variables. PCA does not directly detect specific minerals but rather excels at identifying and highlighting regions with high spectral variance. This often translates to areas of hydrothermal alteration or distinct lithological boundaries where significant spectral changes occur across multiple bands. Therefore, while not a direct mineral classifier, it is excellent for highlighting general alteration zones characterised by high variance due to the presence of various clays, iron oxides, and carbonates. It also implicitly enhances the detectability of iron-bearing minerals if their features contribute strongly to the overall variance. This is because PCA identifies directions of maximum variance, and minerals with strong, distinct spectral features across multiple bands will contribute significantly to this variance, thus being well-represented in the leading principal components.

8.2.2 *Hyperspectral Remote Sensing*

Typically spanning the visible to infrared portions of the electromagnetic spectrum, hyperspectral remote sensing is a technique that gathers image data across multiple narrow, continuous spectral bands. Hyperspectral sensors gather information in hundreds of narrow bands, in contrast to multispectral sensors, which only record information in a small number of broad bands. For in-depth geological study, this high spectral resolution is crucial because it allows for accurate material identification and classification based on distinct spectral signatures (Goetz et al. 1985).

8.2.2.1 Minimum Noise Fraction (MNF) Transformation

MNF Transformation is a technique in hyperspectral remote sensing that reduces noise in the data, helping to highlight important spectral information for easier analysis. By rearranging spectral bands according to their signal-to-noise ratio, this technique minimises noise in the remaining components and places the most valuable information in the beginning components (Green et al. 1988).

$$Z = P^T * (X - \mu) \tag{8.3}$$

where X is the original data matrix, μ is the original data mean, P is the matrix of the eigenvectors derived from the noise covariance matrix, and Z is the modified data matrix.

The MNF Transformation enhances hyperspectral data analysis by reducing noise and prioritising high signal-to-noise ratio components. It is widely used for preprocessing in mineral mapping and land cover studies, improving classification accuracy and data interpretation. A key merit of MNF is its exceptional effectiveness in isolating and minimising noise, particularly spatially correlated noise, thereby significantly improving the overall quality of the data. By enhancing the SNR of the significant spectral components, MNF makes subtle spectral features more readily detectable. Furthermore, by concentrating the signal into the first few MNF components, it greatly facilitates subsequent analysis, as users can confidently discard the noisy, high-order components without losing significant spectral information. MNF-transformed data often serves as a foundational input for advanced algorithms such as spectral unmixing, classification, and target detection, as it provides cleaner and more informative components. However, the effectiveness of MNF heavily relies on an accurate estimation of the noise covariance matrix; if this estimation is poor, the transformation may not be optimal. While beneficial, the transformation can also be computationally intensive, particularly for very large hyperspectral datasets. Similar to PCA, the MNF components are abstract linear combinations of original bands, which can make their direct physical or geological interpretation challenging. MNF does not directly detect specific minerals but acts as an essential preprocessing step that enhances the detectability of all minerals by substantially improving data quality.

By reducing noise and boosting SNR, MNF makes it considerably easier to discern minerals whose diagnostic spectral features might be subtle or obscured by noise in the raw data. This includes any mineral with subtle spectral features, such as many types of alteration minerals (clays, carbonates, iron oxides), whose absorptions can sometimes be weak. Moreover, when utilised as a precursor to unmixing techniques, MNF helps to refine the endmember extraction process, leading to more accurate abundance maps for minerals in mixed pixels. This is because by cleaning the data and effectively separating signal from noise, MNF ensures that subsequent mineral detection algorithms operate on the highest quality spectral information available, thereby improving the overall accuracy and reliability of mineral mapping.

8.2.2.2 Spectral Angle Mapper (SAM)

SAM is a technique that measures the angle between the spectral signature of a pixel and a known reference spectrum, with smaller angles indicating greater similarity (Kruse et al. 1993). SAM is particularly effective in identifying minerals with distinct spectral features due to its robustness to variations in illumination. Two spectra, one from a pixel and one from a reference material, the spectral angle θ is calculated as:

$$\theta = \cos^{-1}\left(\left(\sum_i {}^n \mathrm{R}_i \mathrm{T}_i\right) / \left(\sqrt{\sum_i {}^n \mathrm{R}_i^2} * \sqrt{\sum_i {}^n \mathrm{T}_i^2}\right)\right) \tag{8.4}$$

where R_i is the reflectance of the reference spectrum at band i, T_i is the reflectance of the target (pixel) spectrum at band i and n is the number of spectral bands.

SAM identifies minerals and materials by comparing pixel spectra to reference spectra, effectively handling illumination variations. It is widely used in hyperspectral remote sensing for mineral exploration and geological mapping due to its accuracy in spectral similarity detection.

The primary merit of SAM is its robustness to variations in illumination and albedo. By focusing on the shape of the spectrum rather than its absolute brightness, SAM remains largely insensitive to differences caused by varying topography or shadows, which is a significant advantage in geological mapping. It is computationally efficient, relatively simple to implement, and provides a direct comparison of image spectra with well-characterised laboratory or field spectra. The output, a spectral angle, offers an intuitive measure of similarity, with lower angles indicating a better match. However, SAM is not without its limitations. While robust to brightness, it is sensitive to noise that can alter the fundamental shape of the spectral curve. The accuracy of SAM is heavily dependent on the quality and representativeness of the reference spectra (endmembers) used for comparison. By focusing solely on the angle, SAM discards information about the overall brightness or magnitude of reflectance, which can sometimes be indicative of material abundance or purity. Furthermore, SAM implicitly assumes that a pixel contains a single, pure material; in reality, many remote sensing pixels are "mixed," containing multiple materials, which can lead to larger angles and less accurate classifications. SAM is particularly effective at detecting minerals that exhibit distinct and unique spectral absorption features (shapes), even if their overall brightness varies. This includes hydroxyl-bearing minerals such as kaolinite, montmorillonite, muscovite, chlorite, and alunite, which possess very characteristic absorption features in the SWIR region (e.g., 2.16–2.20 μm for Al–OH, 2.3–2.35 μm for Mg-OH). Carbonates (e.g., calcite, dolomite) also have well-defined absorption features around 2.3–2.35 μm, producing unique spectral shapes. Similarly, iron oxides (e.g., hematite, goethite) exhibit characteristic spectral shapes due to their absorption features in the visible and NIR regions, which SAM can effectively match. SAM's strength lies in its ability to match spectral *shapes*. Minerals with diagnostic absorption features produce specific and reproducible spectral curve shapes that are relatively independent of illumination changes, making SAM a powerful tool for direct mineral identification when reliable reference spectra are available.

8.2.2.3 Matched Filtering (MF)

Matched Filtering (MF) is a hyperspectral analysis technique that isolates the spectral signature of a target material from background signals, highlighting the target's spectral response while suppressing others. This allows for the identification of specific minerals even when they are present in small amounts within a mixed background (Harsanyi et al. 1994).

$$\mathrm{MF} = \left((\mathrm{x} - \mathrm{m})^{\mathrm{T}} \mathrm{C}^{-1} (\mathrm{s} - \mathrm{m})\right) / \left((\mathrm{s} - \mathrm{m})^{\mathrm{T}} \mathrm{C}^{-1} (\mathrm{s} - \mathrm{m})\right) \tag{8.5}$$

where m is the background's mean spectrum (often taken from the image), x is the spectral vector of the pixel under analysis, C is the covariance matrix of the background data, s is the spectral vector of the target material, and T is the transpose operator. Also (x − m) represents the pixel spectrum after removing the background mean, C^{-1} is the covariance matrix's inverse, which adjusts for the background variability and (s − m) is the target spectrum after removing the background mean (Boardman and Kruse 1994).

MF highlights specific minerals in hyperspectral data by separating their spectral signals from the background. It helps detect even small amounts of target materials, making it useful for mineral exploration and environmental studies.

A significant merit of MF is its excellence in sub-pixel detection, making it highly effective for identifying target materials that occupy only a fraction of a pixel, a common scenario in mineral exploration where target minerals may be disseminated. It explicitly suppresses the spectral contribution of the background, making the target more prominent and easier to detect. MF is also robust to variations in abundance, capable of detecting targets even at low concentrations. Unlike some other unmixing techniques, MF does not require explicit knowledge of all endmembers in the scene, only the target and a statistical model of the background. However, MF has its limitations. Accurate detection heavily relies on a precise and representative target spectral signature. The effectiveness of MF also depends on an accurate statistical model of the background; if the background is highly variable or contains unexpected materials, performance can degrade. There is also a risk of false positives if similar background materials or noise inadvertently mimic the target signature. Furthermore, while it indicates presence, the match score is not a direct quantitative measure of fractional abundance. MF is ideal for detecting specific, known target minerals that are expected to be present in low concentrations or as disseminated occurrences within a more dominant background. This includes trace minerals that serve as indicators of mineralisation but occur in small quantities, such as specific alteration minerals (e.g., alunite, illite, chlorite), or even certain sulfide minerals if they possess detectable spectral features in the VNIR/SWIR. It is also valuable for identifying subtle signatures in gossans and supergene alteration minerals like goethite or jarosite, which can indicate underlying mineralisation. MF's strength lies in its ability to filter out background noise and emphasise the unique spectral response of a specific target mineral, making it particularly useful for "needle in a haystack" scenarios in mineral exploration, where the goal is to locate small occurrences of economically significant minerals or their associated alteration products.

8.2.2.4 Spectral Feature Fitting (SFF)

SFF is a technique that matches the spectral characteristics of each pixel with known mineral spectra, focusing specifically on key absorption features. This method is especially useful for minerals with unique absorptions at certain wavelengths, such as clays or carbonates (Clark et al. 1990a, b). SFF enhances mineral identification by targeting these distinctive absorption characteristics, improving accuracy in mineral mapping. A common expression for SFF is:

$$\mathrm{Fit} = \sum (\mathrm{R}(\lambda) - a * \mathrm{S}(\lambda) - \mathrm{b})^2 \quad (8.6)$$

S(λ) is the reference spectrum at wavelength λ, and R(λ) is the pixel's reflectance spectrum at wavelength λ, a scaling factor for matching amplitudes and b is the offset factor to account for baseline shifts.

Key merit of SFF is its focus on the most informative parts of the spectrum, the absorption features, making it highly specific for accurate mineral identification. By fitting the shape of the feature after continuum removal or by accounting for baseline shifts, SFF is less affected by overall brightness variations or minor changes in the spectral continuum. It can also provide quantitative parameters about the absorption feature, such as depth, width, and symmetry, which can offer insights into mineral chemistry or crystallinity. While not a full unmixing technique, SFF can sometimes identify dominant minerals in relatively simple mixtures by effectively fitting their strong features. However, SFF is most effective for minerals with strong, well-defined, and unique absorption features; minerals without such distinct features are not well-suited for this technique. Noise present within the specific absorption feature can significantly impact the fitting accuracy. Furthermore, it requires prior knowledge or careful definition of the precise spectral regions containing the diagnostic features. SFF is also less effective for minerals characterised by very broad or indistinct absorption features. SFF is excellent for identifying minerals that possess sharp, distinct, and diagnostic absorption features in specific spectral regions. This includes hydroxyl-bearing minerals like kaolinite, dickite, halloysite, illite, smectite, muscovite, and alunite, which are prime targets due to their very characteristic Al–OH and Mg-OH absorption features in the SWIR (e.g., 2.16–2.20 μm, 2.3–2.35 μm). SFF can even differentiate between polymorphs of the same mineral (e.g., kaolinite vs. dickite) based on subtle differences in these features. Carbonates (e.g., calcite, dolomite, magnesite) are also well-suited for SFF due to their distinct CO3 absorption features around 2.3–2.35 μm. Many sulfate minerals, such as gypsum and jarosite, also have diagnostic features that SFF can effectively target. The strength of SFF lies in its direct exploitation of the fundamental spectroscopic principle for mineral identification: specific chemical bonds and crystal structures create characteristic absorption features at particular wavelengths. By precisely focusing on these "fingerprints," SFF provides a robust and accurate method for identifying minerals that exhibit these unique spectral characteristics.

8.2.2.5 Linear Spectral Unmixing (LSU)

A technique that breaks down each pixel's spectrum into a mixture of endmember spectra, which represent pure minerals or materials. By calculating the proportion of each endmember within a pixel, this approach allows for the identification of mineral mixtures, making it highly useful in complex geological environments (Adams et al. 1986).

$$R = \Sigma(f\_i * E\_i) + \varepsilon \tag{8.7}$$

where f_i is the fractional abundance of each endmember i, E_i is the spectrum of each endmember, R is the observed spectrum of a pixel, and ε is the residual error or noise in the data.

A significant merit of LSU is its ability to provide quantitative estimates of the proportion of different materials within a pixel, which is invaluable for accurately mapping mineral distributions in complex geological environments. It directly addresses the common issue of "mixed pixels" in remote sensing imagery, offering a more realistic representation of the Earth's surface composition compared to traditional per-pixel classification methods. By generating abundance maps, LSU can significantly improve the accuracy of geological mapping and precisely delineate areas where mineral concentrations are highest. Furthermore, it can reveal the presence of materials that might not be dominant enough to be classified by conventional per-pixel methods. However, LSU faces a critical limitation in the accurate selection of pure and representative endmembers, which is often a challenging and vital step; incorrect endmembers will inevitably lead to erroneous abundance estimates. LSU's core assumption of a linear mixing model may not always hold true in real-world scenarios, particularly for intimate mixtures or where multiple scattering effects occur. Determining the optimal number of endmembers can also be difficult, with both too many and too few potentially affecting the results. Finally, significant residual error in the unmixing process suggests that the chosen endmembers do not adequately explain the pixel's spectrum, indicating the presence of unmodeled materials or non-linear mixing. LSU is ideal for mapping mineral assemblages and mixtures, especially in geological settings where different alteration minerals or rock-forming minerals coexist within the same pixel. This includes mapping the distribution and relative proportions of various clay minerals (e.g., kaolinite, illite, chlorite), carbonates (e.g., calcite, dolomite), and iron oxides (e.g., hematite, goethite) within hydrothermal alteration halos around mineral deposits. It is also valuable for differentiating and quantifying the presence of different rock-forming minerals (e.g., quartz, feldspars, micas) within various lithological units and for identifying proportions of different detrital minerals in sedimentary environments. LSU provides a more nuanced understanding of mineral distribution by moving beyond simple presence/absence to estimating precise abundances. This is critical in geology because mineral deposits rarely consist of single, pure minerals in large, homogeneous patches; instead, they involve complex mixtures of various alteration minerals and host rocks. LSU helps to decipher these complex spectral mixtures, providing a more detailed and accurate geological interpretation.

8.2.2.6 Spectral Mixture Analysis (SMA)

SMA is an advanced technique that applies linear or non-linear models to estimate the abundance of different materials within a single pixel. This method enables detailed mapping in areas with mixed mineral compositions by determining the fractional presence of each material in a pixel (Adams et al. 1986). SMA highlights its effectiveness in analysing complex geological scenes where multiple minerals coexist within a single pixel.

The primary merit of SMA, especially when incorporating non-linear models, is its potential to better represent complex interactions of light with intimately mixed materials, which can lead to more accurate abundance estimations than purely linear models in certain scenarios. It provides detailed fractional abundance maps, which are invaluable for understanding the spatial distribution and concentration of different minerals. This method is highly effective in improving mapping accuracy in heterogeneous areas, such as complex geological settings where multiple minerals are present within individual pixels. However, SMA has several limitations. If the endmembers themselves exhibit spectral variability due to factors like crystallinity, cation substitutions, or particle size, the accuracy of SMA can be affected. Non-linear SMA models can also be significantly more computationally intensive than their linear counterparts. Selecting the appropriate mixing model (linear versus non-linear, and the specific non-linear model) requires a thorough understanding of the scene characteristics and the physical properties of the materials involved. Furthermore, quantitatively validating the derived abundance maps can be challenging without extensive and precise ground truth data. SMA, particularly in its forms that extend beyond strict linearity, is highly beneficial for situations where the linear mixing assumption might be violated and for mapping complex mineral assemblages with subtle interactions or intimate mixtures. This includes identifying and quantifying minerals in intimately mixed occurrences where different minerals are finely intergrown or layered (e.g., clay mixtures, fine-grained mineral deposits), where light might scatter multiple times between different components, potentially leading to non-linear spectral responses. It is also valuable for mapping alteration zones with variable mineralogy and texture, where the type and proportion of alteration minerals change rapidly over short distances, and where the physical mixing of minerals might be complex. Conceptually, SMA embraces the inherent complexity of real-world spectral mixing. For geological environments where minerals are intimately mixed or possess unique light interaction properties (e.g., dark, opaque minerals), a non-linear SMA approach can offer a more accurate representation of fractional abundances, leading to improved mapping of these intricate mineral compositions.

8.2.3 Thermal Remote Sensing

Using specialised sensors, thermal remote sensing measures the thermal infrared radiation that the Earth's surface emits. By revealing the material composition and physical state, this technology makes it possible to analyse surface temperatures and thermal properties, yielding useful data for uses including mineral mapping, land use classification, and environmental monitoring. Recording surface temperature and emissivity data in the thermal infrared (TIR) spectrum makes it possible to identify minerals according to their distinct thermal emission characteristics. This method is especially useful for mapping minerals, as different minerals emit thermal radiation distinctively due to their composition. The principle of thermal remote sensing relies on detecting blackbody radiation from minerals, which display characteristic emissivity patterns in TIR wavelengths. The thermal remote sensing technique, along with ASTER data, has been employed for mineral mapping in the southern region of Rajasthan, India, specifically between the Udaipur and Dungarpur districts (Jain et al. 2020).

8.2.4 Microwave Remote Sensing

Utilising microwave radiation, microwave remote sensing finds and examines aspects of the Earth's surface. This technique employs sensors that operate in the microwave spectrum, allowing for the measurement of properties like moisture content and vegetation cover, regardless of cloud cover or nighttime conditions (Ulaby et al.

1981). Microwave remote sensing is vital for mineral mapping, utilising the interactions between microwave radiation and mineral materials. This technique effectively distinguishes various minerals based on their dielectric properties, which differ significantly among them (Chakraborty et al. 2008). The interaction of microwaves with surface materials helps detect moisture content and surface roughness, both important indicators in mineral mapping (Njoku and Entekhabi 1996).

8.2.5 Spectral Signature

A spectral signature is a material's unique pattern of reflectance or emittance across different electromagnetic wavelengths, determined by its physical and chemical properties. This allows materials to absorb, reflect, or emit radiation distinctively, making spectral signatures fundamental in remote sensing for identifying minerals, vegetation, water, and structures (Hoffer 1978).

Spectral signature analysis requires high-resolution spectral data from sensors like Landsat or ASTER, ground truth data for context, and environmental data to account for influencing conditions. The process involves acquiring imagery, applying preprocessing steps such as atmospheric and radiometric correction, and extracting spectral signatures from specific regions of interest. These extracted signatures are then compared to reference spectra to identify materials. Because they allow things like minerals, flora, water, and man-made structures to be identified and differentiated based on their distinct reflectance profiles, spectral signatures are fundamental in remote sensing (Lillesand et al. 2015; Jansen et al. 2020).

8.3 Mineral Mapping Case Studies in Meghalaya

This section has information about some of the case studies done in Meghalaya.

8.3.1 Identification and Mapping of Calcite-Bearing Rock in Sung Valley, Meghalaya

The study focuses on identifying calcite-rich rocks in Sung Valley, Meghalaya, using ASTER data and geospatial techniques like band ratio and spectral analysis. The Sung Valley is located in Meghalaya, India, on the border between the districts of East Khasi Hills and West Jaintia Hills. It is a bowl-shaped body that is roughly 35 square kilometres in size. The study's chosen location is in Survey of India Toposheet No. 83C/2. The Sung valley is about 48 km from Shillong and is well connected by road via National Highway 44. The area is characterised by hilly and rugged terrain. The

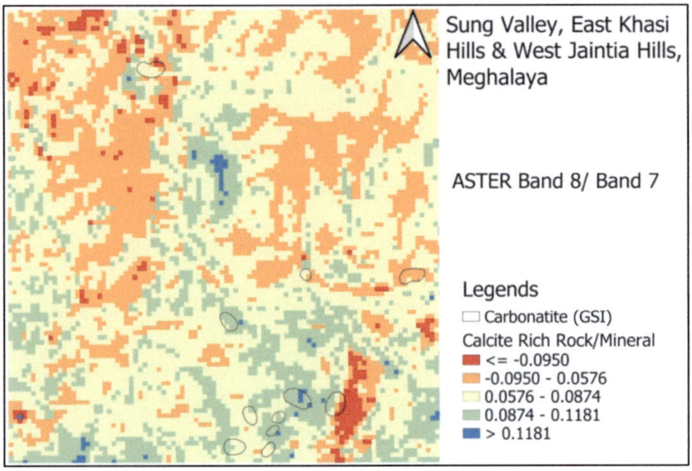

Fig. 8.2 Band ratio for identification of calcite rich rocks shown by blue pixels

complex forms a wide bowl-shaped depression surrounded by ridges composed of quartzites from the Shillong Group. Several seasonal and perennial streams flow into the complex, converging to join the Sung Nala, which runs in a northeast-southwest direction and meets the Um Klen River to the north of the complex.

8.3.1.1 Band Ratio

The band ratios allow for enhanced mapping of calcite-rich areas by highlighting regions where calcite's spectral characteristics dominate. GSI identified and demarcated Carbonatite rock in Sung Valley. Carbonatite consists dominantly of Calcite and Dolomite minerals with minor oxides and phosphates. Calcite mineral identified using Aster Band ratio Band 8/ Band 7 (Fig. 8.2). In areas containing calcite, the reflectance in Band 8 is significantly lower due to absorption, while Bands 7 represent higher reflectance values. This mathematical manipulation results in distinctively high values in the ratio for calcite-rich areas, effectively highlighting carbonate occurrences and suppressing responses from other materials, thereby aiding in the precise mapping of calcite distribution.

In Fig. 8.2, red pixels indicate rocks with the lowest calcium content, while blue pixels represent calcite-rich rocks. The blue pixels align with the carbonatite rock formations shown on the map.

8.3.1.2 Spectral Signature

The spectral signature of a mineral is its unique reflectance or emissivity pattern across different wavelengths of the electromagnetic spectrum. This characteristic pattern acts as a distinct fingerprint, allowing for the identification and differentiation of various materials using remote sensing data. Each mineral possesses a characteristic spectral curve defined by its specific chemical composition and crystal lattice structure, which dictate how it absorbs, reflects, and transmits incident electromagnetic radiation. These interactions result in specific absorption features (dips in reflectance) and reflection peaks (high reflectance). The precise position, depth, and width of these features are highly diagnostic and form the basis for mineral identification.

For calcite, a common carbonate mineral, its spectral signature in the shortwave infrared (SWIR) region is particularly unique and critical for its remote sensing identification (Hunt 1977). Calcite exhibits a strong and characteristic absorption feature specifically at approximately 2.33 μm. This absorption is caused by the overtone and combination vibrational modes of the carbonate ion within its crystal structure. In the context of ASTER data, this diagnostic absorption falls squarely within ASTER's SWIR band 8, which is centered at 2.336 μm (with a bandwidth of 0.081 μm). This precise alignment of the diagnostic feature with an ASTER band is highly advantageous for calcite detection. Furthermore, ASTER's adjacent SWIR bands, such as band 7 (centered at 2.206 μm) and band 9 (centered at 2.400 μm), serve as crucial continuum bands around this diagnostic absorption. The spectral curve of calcite, when plotted using ASTER's SWIR bands, consistently shows a distinct dip at band 8 relative to the higher reflectance values of bands 7 and 9, closely matching reference data from the USGS spectral library (Fig. 8.3).

The utilisation of these specific spectral signatures for mineral identification and mapping involves several key methodologies and insights. Fundamentally, the approach involves comparing the spectral curve of an unknown pixel from the remote sensing image to known, well-characterised spectral signatures stored in geological spectral libraries (such as the USGS Spectral Library). If the pixel's spectral shape, particularly the exact position, depth, and symmetry of its diagnostic absorption features, closely matches that of a known mineral, that mineral can be confidently

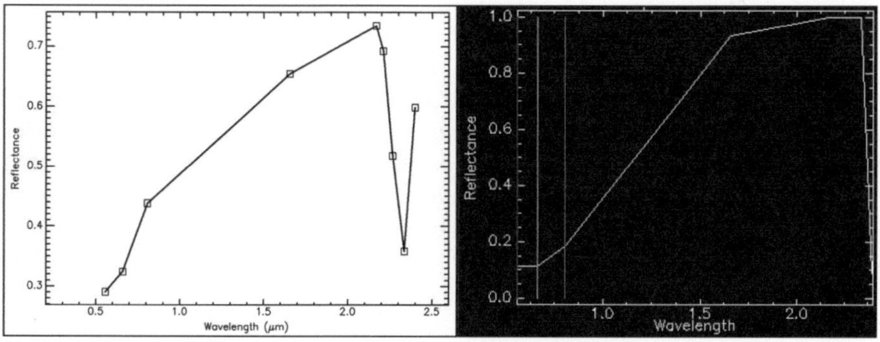

Fig. 8.3 USGS spectral curve (Left) and spectral curve from sung valley (Right)

identified. For calcite, if a pixel exhibits a pronounced absorption feature precisely around ASTER band 8, it provides strong evidence for the presence of calcite.

Moreover, the distinct nature of spectral signatures is paramount for mapping hydrothermal alteration zones, which are critical indicators in mineral exploration. Different types of alteration (e.g., argillic, phyllic, propylitic) are characterised by specific assemblages of minerals, each possessing unique SWIR spectral signatures. For example, while calcite absorbs at 2.33 μm, aluminium-hydroxyl bearing clays like kaolinite absorb around 2.20 μm, and magnesium-hydroxyl minerals like chlorite absorb near 2.30 μm. By analysing the presence, relative intensity, and spatial distribution of these different, yet characteristic, absorption features across an ASTER scene, geologists can accurately delineate and differentiate various alteration types. The spatial pattern of calcite's 2.33 μm absorption can indicate carbonate alteration, which is specifically associated with certain types of economically significant ore deposits. Furthermore, spectral signatures are instrumental in lithological mapping, allowing the differentiation of various rock types based on their constituent minerals. For instance, limestone, being predominantly composed of calcite, will exhibit the characteristic 2.33 μm absorption feature in its spectral signature, clearly distinguishing it from silicate-rich igneous or metamorphic rocks that lack this specific carbonate absorption, facilitating detailed geological mapping from satellite imagery.

8.3.2 Mapping of Ultramafic Rocks Using Thermal Bands of ASTER in Sung Valley, Meghalaya

Ultramafic rocks, distinguished by their exceptionally high content of iron and magnesium-rich minerals (such as olivine and pyroxene), can be effectively identified and mapped using the thermal infrared (TIR) bands of the ASTER sensor. The TIR bands of ASTER, specifically bands 10 through 14, cover a wavelength range between 8.125 and 11.65 μm. This spectral region is crucial because it captures the unique thermal emission and emissivity properties of minerals based on their vibrational fundamental overtones, allowing for their differentiation. Unlike visible and near-infrared (VNIR) or shortwave infrared (SWIR) bands that rely on electronic transitions and overtone/combination vibrations, TIR bands directly measure the energy emitted by the Earth's surface as a function of temperature and emissivity. Mafic minerals, abundant in ultramafic rocks, exhibit distinct emissivity minima in specific TIR wavelengths due to their silicate stretching vibrations, providing a unique spectral signature that can be exploited for mapping.

The utility of ASTER's TIR bands for mapping ultramafic rocks lies in their ability to detect subtle differences in emissivity, which are directly related to mineralogy. For instance, the diagnostic spectral features of mafic minerals, such as olivine and pyroxene, are prominently displayed within these wavelengths. These minerals often show strong absorption features (i.e., low emissivity) in specific TIR bands, particularly in the 9–11 μm range, distinguishing them from felsic minerals (e.g., quartz, feldspars), which have different emissivity patterns. To quantitatively highlight these mafic mineralogical characteristics, indices like the Mafic Bearing Rock Index (MBRI) are employed. As described by Guha and Vinod Kumar (2016), the MBRI, calculated as ((Band12/Band13) × (Band14/Band13)), influences the relative emissivity values across specific ASTER TIR bands to enhance the spectral contrast of mafic-rich rocks. A high value in this index would typically indicate a greater presence of mafic minerals, and thus, ultramafic rock types (Fig. 8.5).

In the context of the Sung Valley in Meghalaya, where ultramafic serpentinite is the predominant rock type (Fig. 8.4), the application of ASTER TIR data and the MBRI is particularly effective. Serpentinite is an alteration product of ultramafic rocks, rich in serpentine group minerals (e.g., antigorite, chrysotile, lizardite), which are hydrated magnesium-iron silicates. These minerals retain the strong mafic spectral characteristics in the TIR region, exhibiting specific emissivity features that are distinct from surrounding lithologies. By calculating the MBRI across the Sung Valley, areas with high index values would directly correspond to serpentinite outcrops, allowing for precise delineation and mapping of these ultramafic bodies. This technique is invaluable for geological mapping, mineral exploration (as ultramafic rocks can host significant deposits of chromite, nickel, platinum group elements, and asbestos), and understanding regional geomorphology and tectonic history by accurately identifying the distribution of these unique rock types.

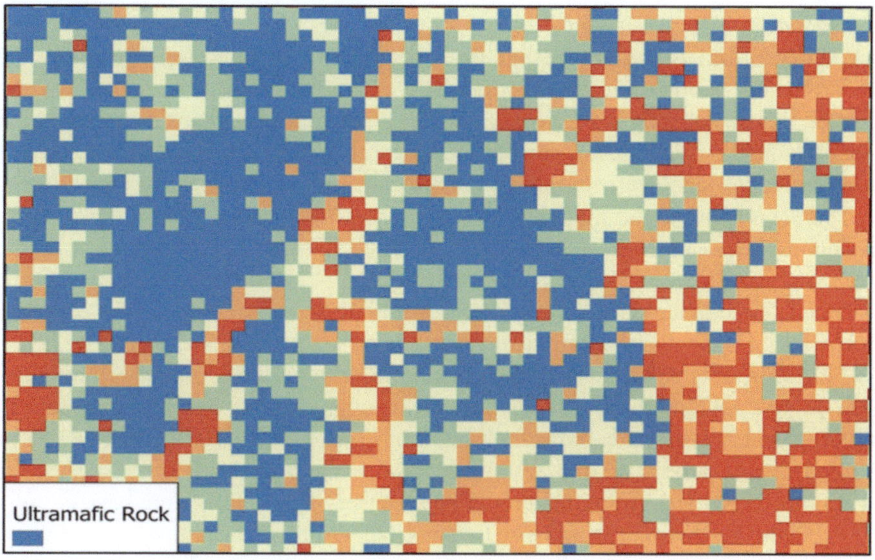

Fig. 8.4 Mapping of ultramafic rock using ASTER (Band12/Band13) × (Band14/Band13) band ratio

Fig. 8.5 Field photo of ultramafic serpentinite rock shown by sparse vegetation in Sung valley

8.3.3 Mapping of Limestone in East Jaintia Hills, Meghalaya

The East Jaintia Hills region of Meghalaya is a geologically significant area, renowned for hosting numerous cement factories. A key factor in this industrial concentration is the abundant presence of limestone, which serves as the primary raw material for cement production. The specific area selected for analysis in the ASTER imagery predominantly features active limestone mines and quarries, highlighting the critical role these geological resources play in supporting the regional cement industry. Accurate mapping of these limestone deposits is therefore essential for resource management, environmental monitoring, and future exploration efforts.

The identification and mapping of limestone in this region heavily relies on its distinct spectral characteristics in the Shortwave Infrared (SWIR) portion of the electromagnetic spectrum, as captured by ASTER. Limestone is primarily composed of calcite, a carbonate mineral, and as previously discussed, calcite exhibits a strong and highly diagnostic absorption feature at approximately 2.33 μm (Hunt 1977). ASTER's SWIR bands are strategically positioned to capture this crucial information. Specifically, ASTER Band 8 (covering wavelengths 2.295–2.365 μm) is precisely centered over this strong carbonate absorption. Conversely, ASTER Band 7 (2.235–2.285 μm) and Band 9 (2.360–2.430 μm) capture the higher reflectance values on either side of this absorption feature, effectively serving as continuum bands.

To effectively distinguish and map carbonate minerals like limestone, a specialised band ratio technique was applied: (Band 7 + Band 9)/Band 8. This method leverages the contrasting reflectance behaviour of carbonate minerals across these specific ASTER SWIR channels. In areas dominated by limestone, the reflectance in Band 8 is significantly lowered due to the strong absorption feature, while the reflectance in the adjacent Bands 7 and 9 remains relatively high. Consequently, when these values are used in the ratio, the denominator (Band 8) is small, and the numerator (sum of

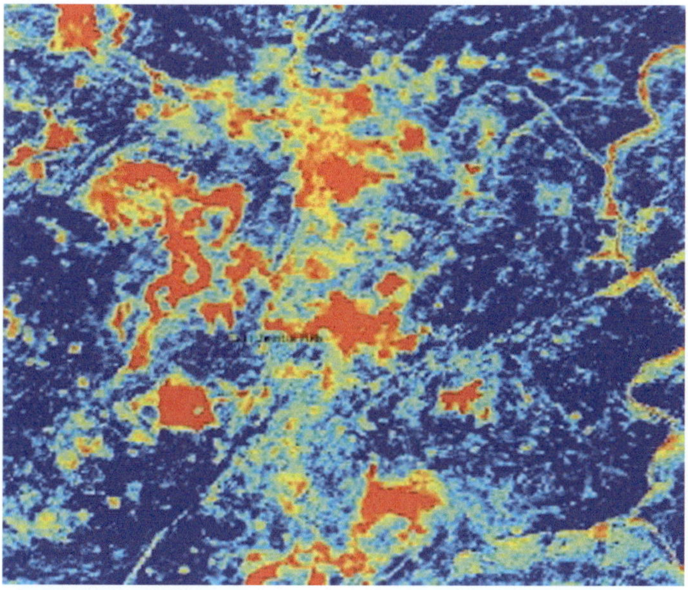

Fig. 8.6 Mapping carbonates minerals shown by red colour using ASTER band ratio technique

Band 7 and Band 9) is large, resulting in a high ratio value for carbonate-rich pixels. This high ratio effectively enhances the spectral signature of carbonates, making them stand out distinctly against other lithologies or vegetation. As a direct result of applying this band ratio, carbonate minerals (limestone) are typically represented as Red pixels in the output image (Fig. 8.6). This clear visual representation allows geologists and resource managers to quickly and accurately delineate the extent of limestone outcrops, active mining areas, and potential new reserves, providing invaluable insights for optimizing quarry operations and planning sustainable resource extraction in the East Jaintia Hills.

8.3.4 Identification of Calcite in Lumshnong, East Jaintia Hills, Meghalaya

Lumshnong is situated in the Khliehriat Block of Jaintia Hills District, Meghalaya, India. It lies approximately 40 kms south of the district headquarters, Jowai, and 16 kms from Khliehriat. The village is located 76 kms from Shillong, the state capital.

Calcite is a carbonate mineral and the most common form of calcium carbonate on Earth, forming a major component of limestone and marble. With a hardness of 3 on the Mohs scale, it crystallises in the hexagonal system with a trigonal crystal structure and displays perfect rhombohedral cleavage at approximately 75°. Calcite occurs in various colours and can also appear opaque. It is widely used in industrial

Fig. 8.7 Satellite image highlighting calcite-bearing lithounits near Lumshnong, Meghalaya. The red box marks the zone of interest with spectral characteristics indicative of calcite presence

processes such as paper and cement production, geochemical engineering for carbon dioxide storage and oil production, biomedical applications including drug delivery, cancer treatment, and bone implants, and as a sorbent material for absorbing sulfur dioxide and other gases from combustion emissions.

Hyperion data consisting of 242 bands was analysed for the Lumshnong region to extract the spectral signatures of Calcite. The spectral signature of calcite, a carbonate-based mineral predominantly found in sedimentary rocks, was identified in this area (Fig. 8.7). In the study, calcite exhibited distinctive absorption bands within the 2300–2400 nm range. Figure 8.8 presents a comparison of the USGS spectral library plot for calcite with the observed mineral spectrum.

8.3.5 Identification of Kaolinite in Brishirnot, East Jaintia Hills, Meghalaya

The study focuses on identifying Kaolinite in Brishirnot, Meghalaya, using Hyperspectral Hyperion data. Brishirnot is situated in the Khliehriat Block of Jaintia Hills District, Meghalaya, India. It lies approximately 68 kms south of the district headquarters, Jowai, and 37 kms from Khliehriat. The village is located 123 kms from Shillong, the state capital, via National Highway 6.

Kaolinite, also known as kaolin clay, is a white clay mineral widely found in nature and used across various industries. Kaolinite is a layered phyllosilicate featuring one silica tetrahedral sheet and one alumina octahedral sheet. Typically white, cream, or pale yellow, it can also appear brown or tan when stained. It has a hardness of 2–2.5

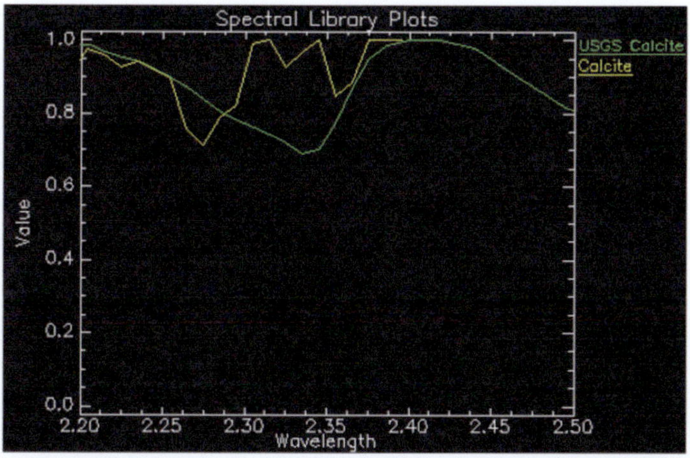

Fig. 8.8 Spectra of calcite showing absorption at 2.33 μm matching with USGS spectra of calcite

Fig. 8.9 Satellite image highlighting Kaolinite mineral near Brishirnot, Meghalaya. The red box marks the zone of interest with spectral characteristics indicative of Kaolinite presence

on the Mohs scale and ranges from translucent to opaque. Kaolinite forms through the weathering or hydrothermal modification of acidic igneous rocks and the chemical breakdown of feldspars in soil. It has diverse applications, including medical uses for treating diarrhoea, soothing upset stomachs, and inducing blood clotting; as an ingredient in skincare barrier creams and toothpaste for its mild abrasiveness; for gastrointestinal detoxification; and in anti-tumour treatments, where its derivatives have demonstrated the ability to reduce tumour mass and metastases in animals (Fig. 8.9).

In the Near Infrared (NIR) region, Kaolinite shows absorption bands at 1410–1430 nm are attributed to hydroxyl (OH) stretching vibrations, while bands at 2200–2220 nm are linked to hydroxyl deformation combined with Si–O stretching (Clark et al. 1990b, a). Figure 8.10 represents a comparison of the spectral library plot for Kaolinite with the observed mineral spectrum.

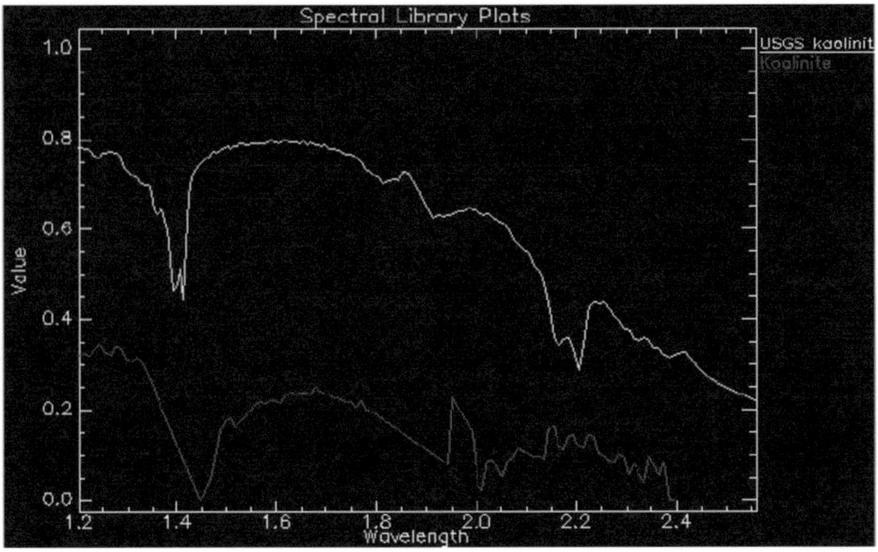

Fig. 8.10 Spectral characteristic of Kaolinite mineral showing absorption at 1.41 and 2.2 μm respectively

Acknowledgements The authors are thankful to the Geological Survey of India (GSI) Bhukosh website for providing different mineral commodity data. Authors also thank NASA Jet Propulsion Laboratory for access to ASTER data and the United States Geological Survey (USGS) for providing HYPERION data. Special acknowledgement to the Indian Space Research Organisation and North Eastern Space Applications Centre for their in-house project funding. This support was crucial, particularly for the foundational work derived from the earlier project, "EO based mineral resource mapping in Sung Valley, Meghalaya," which laid the groundwork for many concepts discussed herein.

References

H. Abdi, L.J. Williams, Principal component analysis. Wiley Interdiscip. Rev. Comput. Stat. **2**(4), 433–459 (2010)

J.B. Adams, M.O. Smith, P.E. Johnson, Spectral mixture modelling: a new analysis of rock and soil types at the Viking Lander 1 site. J. Geophys. Res. **91**(B8), 8098–8112 (1986)

S. Bhattacharya, P. Bandyopadhyay, S. Chakraborty, Application of remote sensing in mineral exploration: a case study of Northeast India. J. Earth Syst. Sci. **127**(3), 1–15 (2018)

N. Bhaumik, M. Majumdar, S.A. Ahmed, *Geological Survey of India Nagaland magnetite a preliminary appraisal. Field season 1972–73, investigation during April 1975* (Geological Survey of India and Directorate of Geology and Mining, Nagaland, 1973)

J.W. Boardman, F.A. Kruse, Automated spectral analysis: a geological example using AVIRIS data, north Grapevine Mountains, Nevada. Proceed. Tenth Themat. Conf. Geol. Remote Sens. **1**, 407–418 (1994)

A.K. Buragohain, *Final report on resource survey for dimension stones in Bura Parbat area of Nagaon and Karbi Anglong districts, Assam (parts of toposheet no. 83f/2) (P-2 stage)* (Geological Survey of India, 1999)

M. Chakraborty, M.K. Sharma, S. Panigrahy, *Dielectric Properties of Minerals and Their Relevance to Microwave Remote Sensing* (2008)

B. Chattopadhyay, R.K. Roy, *Preliminary Report on the Occurrences of Rock Salt in Parts of Manipur East District, Manipur* (Geological Survey of India, 1975)

R.N. Clark, T.V.V. King, M. Klejwa, G.A. Swayze, High-resolution reflectance spectroscopy of minerals. J. Geophys. Res. Solid Earth **95**(B8), 12653–12680 (1990a)

R.N. Clark, G.A. Swayze, A.J. Gallagher, T.V.V. King, W.M. Calvin, Mapping minerals with imaging spectroscopy. U.S. Eol. Surv. **1269**, 1–18 (1990b)

T. Cudahy, *CSIRO, Satellite ASTER Geoscience Product Notes for Australia* (CSIRO Report EP-125895) (2012)

R.O. Duda, P.E. Hart, *Pattern Classification and Scene Analysis* (1973)

A.F.H. Goetz, G. Vane, J.E. Solomon, B.N. Rock, Imaging spectrometry for earth remote sensing. Science **228**(4704), 1147–1153 (1985)

A.A. Green, M. Berman, P. Switzer, M.D. Craig, A transformation for ordering multispectral data in terms of image quality with implications for noise removal. IEEE Trans. Geosci. Remote Sens. **26**(1), 65–74 (1988)

A. Guha, K. Vinod Kumar, New ASTER derived thermal indices to delineate mineralogy of different granitoids of an Archaeancraton and analysis of their potentials with reference to Ninomiya'sindices for delineating quartz and mafic minerals of granitoids–an analysis in Dharwar Craton. Ore Geol. Rev. **2016**(74), 76–87 (2016)

J.C. Harsanyi, C.-I. Chang, Hyperspectral image classification and dimensionality reduction: an orthogonal subspace projection approach. IEEE Trans. Geosci. Remote Sens. **32**(4), 779–785 (1994)

R.M. Hoffer, Biological and Physical Considerations in Applying Computer-Aided Analysis Techniques to Remote Sensor Data, in *Remote Sensing: The Quantitative Approach.* ed. by P.H. Swain, S.M. Davis (McGraw-Hill, New York, 1978), pp.227–289

H. Hotelling, Analysis of a complex of statistical variables into principal components. J. Educ. Psychol. **24**(6), 417–441 (1933)

G.R. Hunt, Spectral signatures of particulate minerals in the visible and near infrared. Geophysics **42**(3), 501–513 (1977)

R. Jain, H. Bhu, R. Purohit, Application of thermal remote sensing technique for mapping of ultramafic, carbonate and siliceous rocks using ASTER Data in Southern Rajasthan, India. Curr. Sci. **119**(6) (2020). https://doi.org/10.18520/cs/v119/i6/954-961

R. Jansen, M P, Advances in spectral remote sensing for monitoring vegetation and soil. J. Appl. Remote Sens. **14**(3), 1–18 (2020)

J.R. Jensen, *Band Ratio Techniques for Remote Sensing* (1979)

I.T. Jolliffe, *Principal Component Analysis*, 2nd edn. (Springer, 2002)

I.T. Jolliffe, J. Cadima, Principal component analysis: a review and recent developments. Philos. Trans. Royal Soc. A Math. Phys. Eng. Sci. **374**(2065), 20150202 (2016)

S.A Khan, *Investigation of cement grade limestone in Phungyar, Singkap-Shingoa and Shangshak blocks, Ukhrul district, Manipur. Progress Report for the field Season* (Geological Survey of India, 1989–90)

F.A. Kruse, A.B. Lefkoff, J.W. Boardman, K.B. Heidebrecht, A.T. Shapiro, P.J. Barloon, A.F.H. Goetz, The spectral image processing system (SIPS)-interactive visualization and analysis of imaging spectrometer data. Remote Sens. Environ. **44**(2–3), 145–163 (1993)

S. Kumar, R. Verma, V.K. Bharti, C. Lalmuankimi, Mineral prospect in Mizoram. *Souvenir Volume of National Seminar on "Geodynamics, Sedimentation and Biotic Response in the Context of India-Asia collision"* (2023)

S. Lal, A. Deka, *Report on Investigation for Titaniferous Vanadiferous Magnetite around Myniar area, West Khasi Hills district, Meghalaya (G-4 stage): field season 2015–16* (2015–16)

T. Lillesand, R.W. Kiefer, J.W. Chipman, *Remote Sensing and Image Interpretation*, 7th ed. (John Wiley & Sons, 2015)

M.D. Limaye, B. Debadhikari, *Geological Survey of India coal and limestone investigations in Nagaland. Progress report for 1966–67* (Geological Survey of India, 1967)

T. Longchari, L.S. Singh, *Final Report on Investigation of the Dolomite and Limestone Around Namchi, Nayabazar and Bijanbari areas of South Sikkim and West Sikkim District, Sikkim (G4), Field Season 2014–15* (Geological Survey of India, 2014–15)

M. Majumdar, *Geological Investigation for Glass Sand near Watiyin New Chungtia Villages in Desai valley, Nagaland and Adjoining Sibsagar District of Assam* (Geological Survey of India, 1977)

P.M. Mather, *Computer Processing of Remotely Sensed Images: An Introduction*, 4th ed. (John Wiley & Sons, 2010)

J.C. Mars, L.C. Rowan, Spectral assessment of new ASTER SWIR surface reflectance data products for spectroscopic mapping of rocks and minerals. Remote Sens. Environ. **114**(9), 2011–2025 (2010)

D.R. Nandy, A. Basak, *Detailed Investigation of the Limestone Deposit at Tidding, Lohit District, N.E.F.A* (Geological Survey of India, 1967)

J. Niemeyer, L. Tedesco, C. Ramo, Mapping hydrothermal alterations in the Brazilian Amazon using ASTER data. Remote Sens. Environ. **113**(4), 867–882 (2009)

E.G. Njoku, D. Entekhabi, Passive microwave remote sensing of soil moisture. J. Hydrol. **184**(1–2), 101–129 (1996)

R. Panduranga, *Report on investigation of plastic clay, around Tezu, Lohit district, Arunachal Pradesh (Part of toposheet no. 83 M/13 & 92 A/1) (Progress Report for F. S. 1988–89)* (Geological Survey of India, 1988–89)

K. Pearson, On lines and planes of closest fit to systems of points in space. Phil. Mag. **2**(11), 559–572 (1901). https://doi.org/10.1080/14786440109462278

P. Prakash, *Final report of Sonapahar Sillimanite Investigation Phase–ii. Report on the Studies on Dispersal and Concentration of Sillimanite and Corundum Float in Sonapahar area, West Khasi Hills District, Meghalaya. Progress report for field seasons 1980–82* (1984)

J. Prasad, S.K. Deb, G.K. Kesari, *Final Report on Investigation for Limestone in Larket Block, East of Litang river, Jaintia Hills District, Meghalaya (E-II stage) F.S.2001–2002&2002–2003* (2001–2003)

B.K.N. Rao, *Progress Report for the Field Season 1981–82 & 1982–83. Report on the Occurrence of Laterite Around Agartala West Tripura district, Tripura* (Geological Survey of India, 1984)

M. Ringnér, What is principal component analysis? Nat. Biotechnol. **26**(3), 303–304 (2008)

L.C. Rowan, J.C. Mars, Mapping hydrothermally altered rocks at Cuprite, Nevada, using the advanced spaceborne thermal emission and reflection radiometer (ASTER), a new satellite-imaging system. Econ. Geol. **98**(5), 1019–1027 (2003)

F.F. Sabins, Remote sensing for mineral exploration. Ore Geol. Rev. **14**(3–4), 157–183 (1999)

S.N. Sen, *A Note on the Occurrence of Natural Gas in Tripura state (with a sketch map)* (Geological Survey of India, 1956–57)

R.P. Singh, A. Kumar, A. Kumar, Utilization of satellite data for mineral resource assessment: a review. Int. J. Remote Sens. **43**(7), 2644–2663 (2022)

V. Srinivasan, S. Kumar, *Preliminary Investigation for Refractory/ Foundry Grade Quartzite Near Kalaktang, West Kameng district. Arunachal Pradesh (progress report for the field season 1995–96)* (Geological Survey of India, 1996)

V. Thakur, S. Jayaswal, *Preliminary Exploration for Limestone in North Boro Hundong Block, Dima Hasao district, Assam (Stage: G-3). Item no.: M2ASMIF-MEP/NC/NER/ASM/2019/25076. Report for field season 2019–20.* (Geological Survey of India, 2019–20)

F.T. Ulaby, R.K. Moore, A.K. Fung, *Microwave Remote Sensing: Active and Passive. Volume I: Fundamentals and Radiometry* (Addison-Wesley, 1981)

S. Wold, K. Esbensen, P. Geladi, Principal component analysis. Chemom. Intell. Lab. Syst. **2**(1–3), 37–52 (1987)

Y. Yamaguchi, S. Aoki, K. Doi, Development of the ASTER mineral mapping algorithm for clay minerals. Remote Sens. Environ. **104**(4), 337–347 (2006)

Y. Yamaguchi, C. Naito, Spectral indices for lithologic discrimination using the ASTER SWIR bands. in *Proceedings of the 13th International Conference on Applied Geologic Remote Sensing* vol. 1, (1999), pp. 151–158

Rahul Pratap holds a M.Sc. Tech degree in Applied Geology from the Indian Institute of Technology (Indian School of Mines), Dhanbad. He is currently serving as a Scientist at the North Eastern Space Applications Centre (NESAC), Shillong, under the Department of Space, Government of India. His research interests lie in mineral exploration, with a particular focus on the integration of hyperspectral remote sensing and geospatial techniques for identifying and mapping mineral resources. His ongoing work involves the application of advanced remote sensing technologies, for the detection and characterization of economically important minerals.

Gopal Sharma has obtained a PhD in Crustal deformation study from IIT Dhanbad. Presently, he is a Senior Scientist at the North Eastern Space Applications Centre (NESAC), Shillong, Department of Space, Government of India. His research interests include Earthquake precursors, Crustal Deformation, and strain accumulation measurements. His ongoing research focuses on large-network GNSS Ionosphere variability for earthquake precursor detection and InSAR-based crustal deformation studies.

M Somorjit Singh is heading the Geoscience Division at the North Eastern Space Applications Centre (NESAC), Shillong, Department of Space, Government of India. In 20 years of service at NESAC, he has contributed to various operational and R&D activities/projects. He has been honored 'ISRO Team Excellence Award for his significant contribution as a team member in 'National Geomorphology and Lineament Mapping (NGLM) project' which was initiated by National Remote Sensing Centre (NRSC), Hyderabad in collaboration with Geological Survey of India (GSI) under Natural Resources Census program. His research interests include Geomorphology & Lineament Studies/Mapping, Infrastructural Planning (Site Suitability/feasibility Studies, road & railway alignment/realignment, Natural/Geo-hazard, Manmade Hazard (Industrial Hazard), Environmental Studies/Pollution (Heavy Metals) etc.

Golfinstar Kharbani holds a M.Sc. in Geology and a PG Diploma in GIS from North-Eastern Hill University (NEHU). He is currently a Research Assistant at NECTAR, Shillong, and has previously worked with NESAC and IIGM-NEGRL. His expertise lies in remote sensing and GIS applications for geoscience, with experience in mineral mapping, landslide assessment, morphometric analysis, and tectonic studies using space geodetic techniques.

K. K. Sarma leads the Remote Sensing Applications Group (RSAG) at the North Eastern Space Applications Centre (NESAC), where he has been a key figure since 2002. Specializing in remote sensing for forestry and biodiversity, his work leverages space technology for critical applications in Biodiversity, Ecosystems, Forestry, and Natural Resource Management. Dr. Sarma has spearheaded over 60 diverse geospatial projects, contributing significantly to natural resource mapping and monitoring across the North Eastern Region (NER). Currently, he coordinates 110 "Plan of Action" projects, focused on using space technology to improve livelihoods in the NER. He also served as the Principal Investigator and NER Coordinator for all ISRO natural resource management projects, including comprehensive Land Use Land Cover Mapping and National Wasteland Updating, Monitoring, and Change Analysis. With over 75 research publications, Dr. Sarma has successfully supervised six Ph.D. students and actively mentors many others across NER Universities.

Chapter 9
Geodynamics of Northeast Region of India: A Geospatial Technology Perspective

Gopal Sharma, M. Somorjit Singh, Prakash Biswakarma, and K. K. Sarma

Abstract Northeastern India lies at the junction of major tectonic regimes, making it one of the most seismically active regions in the Indian subcontinent. This chapter presents an overview of the region's seismotectonic regime and explores the role of advanced geospatial technologies in geodynamic study. Emphasis is placed on Interferometric Synthetic Aperture Radar (InSAR) and its advanced variants, such as Differential InSAR (DInSAR), Persistent Scatterer InSAR (PSInSAR), and Small Baseline Subset (SBAS), which are instrumental in detecting precise ground deformation. The utility of Global Navigation Satellite System (GNSS) technique is discussed for assessing crustal motion, estimating both linear and localised strain, and evaluating regional convergence. A dedicated section addresses the analysis of ionospheric Total Electron Content (TEC) anomalies derived from GNSS observations, highlighting their potential as earthquake precursors. Additionally, the application of satellite-derived Land Surface Temperature (LST) data as a complementary tool for identifying tectonically driven thermal anomalies is discussed. The chapter includes selected case studies such as deformation analysis along the Kopili Fault, detection of pre-earthquake TEC anomalies, and convergence mapping across the Eastern Himalayas. These case studies further highlight the critical value of integrating diverse geospatial datasets for advancing the understanding of geodynamic processes in the region.

Keywords InSAR · GNSS · TEC · Earthquake precursor · Deformation · LST · Geodynamics

G. Sharma (✉) · M. Somorjit Singh · P. Biswakarma · K. K. Sarma
North Eastern Space Applications Centre (NESAC), Umiam, Shillong, Meghalaya, India
e-mail: gops.geo@gmail.com

M. Somorjit Singh
e-mail: m.somorjit@nesac.gov.in

P. Biswakarma
e-mail: p.biswakarma@nic.in

K. K. Sarma
e-mail: kk.sarma@nesac.gov.in

S. P. Aggarwal et al. (eds.), *Remote Sensing of the North Eastern Himalayan Ecosystem*,
https://doi.org/10.1007/978-981-95-3221-6_9

9.1 Introduction

The North Eastern Region (NER) of India, nestled between the mighty Himalayas and the Indian Plate, is known for its scenic landscapes, rich biodiversity, and vibrant cultures (Tshering et al. 2010). Two prevailing tectonic regimes control the geodynamic structure of the NER. (1) Himalayan arc to the north and (2) Indo-Burmese arc to the east. The Himalayan arc is a region of continental-continental collision where the Indian Plate is converging with the Eurasian Plate, leading to crustal shortening, uplift, and high seismic activity (Molnar and Tapponnier 1975; Bilham 2004). The Indo-Burmese Arc is characterized by the oblique subduction of the Indian Plate beneath the Burmese microplate, manifested in a complex system of thrust and strike-slip faults (Steckler et al. 2008). These two tectonic settings converge in the Northeast Region (NER), resulting in a transpressional regime that leads to crustal deformation and significant strain accumulation. The complex interaction between these plates gives rise to a network of active fault lines that traverse the region, leading to significant seismic activity (Lakshmi and Gawali 2022). NER lies in seismic zone V as per IS 1893 (2002) in the country with frequent seismic activity, are the most seismically active zones in India (Longbir and Arjun 2017). The region has experienced several major earthquakes in its geological history, including the 1897 Shillong Earthquake (Mw 8.1), which exhibited significant vertical displacement along the Oldham Fault (Bilham 2004), and the 1950 Assam-Tibet Earthquake (Mw 8.6), one of the largest continental earthquakes of the twentieth century. These events, combined with a network of active faults such as the Kopili Fault, Dauki Fault, and Mishmi Thrust, highlight the region's elevated seismic risk (De and Kayal 2003). These earthquakes have left a lasting impact on the region, shaping its geography and influencing the lives of its residents.

Scientific initiatives such as the Indian Plate Motion (Banerjee and Bürgmann 2002) and regional Continuously Operating Reference Stations (CORS) data have significantly enhanced our understanding of plate kinematics in the Northeast Region (NER). These geodetic models and networks have revealed that the Indo-Burmese wedge accommodates substantial oblique convergence, with GPS-derived velocities ranging from 20 to 50 mm/yr in various directions (Najman et al. 2016; Sahu et al. 2006). The resulting high-resolution data have been instrumental in delineating locked fault segments and evaluating the seismic potential of the region. In addition, geological research like paleoseismological trenching has provided useful information regarding the recurrence intervals of previous earthquakes (Rajendran and Rajendran 2011). This information has contributed to the construction of a longer-term seismic catalogue, which is necessary for probabilistic seismic hazard assessments.

Recently, the use of satellite data and techniques such as Interferometric Synthetic Aperture Radar (InSAR) has helped track ground movement over large areas with very high accuracy. InSAR has been crucial in both post-seismic displacement measurement and interseismic deformation research. Further, thermal Remote Sensing technique, particularly the investigation of Land Surface Temperature (LST) anomalies via has also been utilized to examine pre-seismic thermal signals (Saraf et al. 2008). The fusion of these technologies into Geographic Information System (GIS) platforms enables the overlay of geological, geophysical, and socio-economic information for seismic risk assessment. Integrative strategies of this nature are crucial for successful disaster preparedness and mitigation in the region.

Geospatial technology plays a vital role in understanding the geodynamics of the earth. Utilizing geospatial technologies spanning from GNSS and InSAR to thermal remote sensing and trenching is a potent means to demystify the intricate tectonics and enhance earthquake hazard modelling. Earth scientists have carried out numerous studies on geological and physical processes associated with an earthquake, but early warning of earthquakes is still an elusive goal (Malik and Nakata 2003; Catherine 2004; Sharma and Lindholm 2012). However, the recent advancements in InSAR and Global GNSS technologies have provided significant contributions to seismologists and earth scientists in earthquake precursor and deformation studies. Further, Thermal Remote Sensing techniques offer an effective method for monitoring Land Surface Temperature (LST) associated with earthquakes. This chapter sheds light on some of the recent geospatial techniques, including InSAR, GNSS, and LST, and their applications in understanding the Geodynamics of the NER in India.

9.2 Seismotectonic of North East India

Northeast India (NER) is one of the most seismically active regions in the Indian subcontinent due to the convergence of major tectonic plates and complex geological structures. The major seismotectonic units of the region include the Eastern Himalayas, the Eastern Himalayan Syntaxis, the Indo-Burma Fold Thrust Belt, the Meghalaya or Shillong Plateau, and the Brahmaputra Valley.

The Eastern Himalayas represent the eastward extension of the Himalayan Mountain chain beyond the state of Sikkim and are characterised by several significant thrust systems, including the Main Frontal Thrust (MFT), the Main Boundary Thrust (MBT), and the Main Central Thrust (MCT). These structures accommodate the ongoing convergence between the Indian and Eurasian plates. The Eastern Himalaya are associated with thrust faults that carries the Great Himalayan crystallines and Lesser Himalaya meta-sediments (Nandy and Dasgupta 1991).

One of the most prominent tectonic features in NER, the Shillong Plateau is a unique uplifted block situated between the Himalayan frontal thrust system to the north and the Indo-Burmese subduction zone to the southeast. It is under active compressional stress and is characterized by dominant thrust faulting, as evidenced by the 1897 Shillong earthquake (Kayal and De 1991). The plateau is considered a detached fragment of the ancient Indian Peninsular Shield and is bounded by several significant faults: Dauki Fault to the south, Dudhnoi Fault to the west, Kopili Fault to the east and Oldham Fault to the north (England and Bilham 2015). Other active faults within the Shillong Plateau include the Kulsi Fault and the Bomdila Fault (Nandy 2017). North of the Shillong Plateau lies the Brahmaputra Valley, an alluvium-filled foreland basin of the Eastern Himalayas. It is bounded to the south by the Shillong Plateau and to the north by the Himalayan foothills. The valley is structurally separated from the Indo-Burma Fold Thrust Belt by the Naga and Disang thrusts (Yin et al. 2010). This tectonic zone marks the boundary where the Indian Plate is obliquely subducting beneath the Burmese microplate. The Tripura Folded Belts, which are a part of the Indo-Burma Fold Thrust Belt (IBFTB), have formed as a result of this interaction, with compressional stress oriented westward due to the eastward motion of the Burma Plate.

NER has a long history of intense seismic activity. It has experienced two great earthquakes (Mw > 8.0) and about 20 large earthquakes (Mw 8.0 - 7.0) since 1897 (Kayal et al. 2006). The 1897 Shillong earthquake (Mw 8.1) is one of the most powerful earthquakes recorded in Indian history. It caused widespread devastation across Assam and Meghalaya, including ground ruptures, landslides, and structural collapse (Ambraseys and Bilham 2003). The 1950 Assam-Tibet earthquake (Mw 8.6) was one of the world's largest continental earthquakes. It resulted in severe damage and loss of life across Arunachal Pradesh and neighbouring areas (Jain 1994). The Sikkim Earthquake of 2011 (Mw 6.9), although smaller in magnitude, caused significant infrastructure damage and casualties in Sikkim and parts of eastern Nepal (Dutta et al. 2015). In addition to these major events, many moderate-to-strong earthquakes have occurred in the region, highlighting its persistent seismic threat (Nandy 2001). The spatial distribution of the earthquakes events that triggered in NER and the adjoining areas since 1900 along with major tectonics faults are depicted in the Fig. 9.1. In the year 2023, as per National Centre for Seismology (NCS) data, a total of 450 earthquake have been reported in NER. Out of 450 earthquakes, 35 were reported from Arunachal Pradesh, 68 were reported from Assam, 49 were reported from Manipur, 45 were reported from Meghalaya, 7 from Mizoram, 10 from Nagaland, 3 from Sikkim, 1 from Tripura and 232 events were reported from NER neighborhood.

9.3 Recent Geospatial Technologies for Geodynamic Studies

9.3.1 Interferometric Synthetic Aperture Radar (InSAR)

Synthetic Aperture Radar Interferometry (InSAR) is a tool for measuring topography of the terrain, surface change and surface deformation by utilizing Interference of complex signals from satellite images. The technique utilizes two Radio Detection and Ranging (RADAR) signals to measure phase change that corresponds to high precision topographic changes. The phase change is measured by imaging the earth suface utilizing microwave (radar) signals onboard satellite sensors in nearly, but not exactly, the same positions and/or time. The radar sensor sends its own energy towards the Earth and measures the amount of energy that is reflected back (called backscattered energy) after interacting with the surface. It typically operates in the

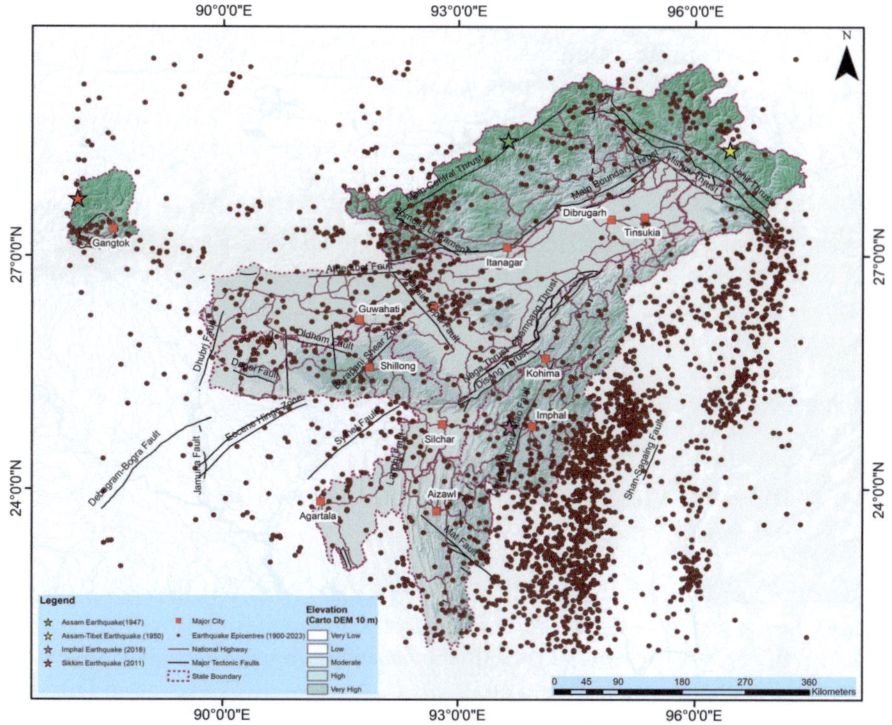

Fig. 9.1 Tectonic setting of NER, India and Historical earthquake events that triggered since 1900. (Base map: ESRI)

microwave region of the electromagnetic spectrum (Dzurisin and Lu 2007). Generally, InSAR functions in the C-band (~5.6 cm wavelength), L-band (~23.5 cm wavelength), or X-band (~3.1 cm wavelength) of the electromagnetic spectrum. P-band is generally not preferred for InSAR analysis because of its longer wavelength, which produces coarser resolution images. K-bands are typically used in an aerial platform, as the atmosphere will interfere with the shorter wavelengths. InSAR uses at least two SAR images of the same area, taken from slightly different positions or times, to measure phase differences. These differences capture both terrain elevation and surface changes in terms of phase difference and can be represented as below (Rosen et al. 2000)

$$\Delta\phi = \frac{4\pi}{\lambda}\Delta r \tag{9.1}$$

where λ is the radar wavelength, $\Delta\phi$ is the interferometric phase difference, and Δr is the change in slant range distance between the radar sensor and the ground target between two SAR acquisitions. The total interferometric phase difference $\Delta\varphi$ in InSAR data generally contains multiple contributions, such as ground deformation, topography, Atmospheric delay and Orbital errors etc. To map only topography using InSAR, retaining the topographic phase component while eliminating or minimizing the effects of deformation, atmosphere, and other noise is essential. This is typically done using single-pass interferometry (example: Shuttle Radar Topography Mission, TanDEM-X/TerraSAR-X etc.) or by careful selection of repeat-pass data.

InSAR can detect millimeter-scale changes in the Earth's surface, providing unparalleled accuracy (Awasthi et al. 2020; Sreejith et al. 2016). InSAR has undergone substantial development since its inception. Initially demonstrated using Seasat data in the 1980s, for topographic mapping (Gabriel et al. 1989). Subsequent missions such as ERS-1/2, ENVISAT, and RADARSAT set the stage for deformation mapping. Sentinel-1 (ESA) and ALOS-2 (JAXA) introduced improved revisit frequencies, spatial resolutions, and global coverage, making InSAR a mandatory tool in geodynamic studies.

9.3.1.1 Differential Interferometric Synthetic Aperture Radar (DInSAR)

Differential Interferometric Synthetic Aperture Radar (DInSAR) was the first radar interferometry method used for ground deformation monitoring. It utilizes two or more SAR images to measure the phase change (interferogram) in a given time with known orbital characteristics (Ullo et al. 2019). A major breakthrough in DInSAR was the mapping of 1992 Landers earthquake deformations (Massonnet et al. 1993). Since then, DInSAR has developed into a stable method of detecting earthquake- and volcanic-induced deformation, land subsidence, glacial motion, as well as human-induced deformation caused by mining and groundwater pumping (Massonnet and Feigl 1995; Peltzer and Rosen 1995; Hanssen 2001). The 1990s saw the emergence of DInSAR, which enabled the detection of fine surface deformations, including those caused by seismic and volcanic activities (Peltzer and Rosen 1995; Massonnet and Feigl 1995, 1998).

The fundamental output of a DInSAR process is an interferogram, a raster image of the phase difference between two SAR acquisitions. In order to isolate deformation, topographic and flat-earth phase contributions need to be removed. In other words, the DInSAR product is the interferograms from which the phase change due to topography has been removed. The differential interferogram thus obtained shows only the surface motion along the radar Line of Sight (LOS). Following the elimination of the flat-earth phase and topographic contribution by a Digital Elevation Model (DEM), the residual phase difference $\Delta\varphi$ in a simplified 2-pass DInSAR can be approximated by Ferretti et al. (2007)

$$\Delta\varphi = -\frac{4\pi}{\lambda}\frac{\mathrm{Bn}\Delta q}{\mathrm{R}\sin\theta} + \frac{4\pi}{\lambda}d \tag{9.2}$$

where the variables 'λ' is the wavelength, 'Bn' is the perpendicular baseline, 's' is the ground range to the targe, 'R' is the radar-target distance, 'θ' is the incidence angle,

'Δq' is the relative elevation (to flat earth) 'ha' is the height of ambiguity and 'd' is the displacement independent of the baseline. Topographic phase is given by the first term in Eq. 9.2, and displacement-induced phase is given by the second term. In DInSAR processing, accurate orbital data and a DEM (such as obtained from SRTM or TanDEM-X) are employed for the removal of topographic contribution and pure deformation extraction.

The major limitation of DInSAR techniques is the decorrelation between radar images due to temporal changes in the physical or geometrical nature of the object on the ground (Zebker and Villasenor 1992). Multi-temporal DInSAR techniques have been put in place to counteract these problems. Specifically, Persistent Scatterer Interferometry (PS-InSAR) and Small Baseline Subset (SBAS), which are discussed in the next section.

9.3.1.2 Persistent Scatterer Interferometric Synthetic Aperture Radar (PSInSAR)

To compensate for decorrelation problems that typically constrain the power of Differential InSAR (DInSAR), Persistent Scatterer Interferometric Synthetic Aperture Radar (PSInSAR) was developed by Ferretti et al. (2002). PSInSAR targets point scatters, also known as persistent scatterers (PS), which exhibit a constant phase response over time, such as human-made objects, rocky outcrops, and urban structures. Utilising the phase stability of these scatterers, PSInSAR facilitates deformation monitoring over long periods of time with high precision. One of the earliest and most highly reported PSInSAR demonstrations was mapping millimeter-scale subsidence in Venice, Italy, from ERS observations (Ferretti et al. 2002). Since then, the technique has been widely used to detect urban subsidence, infrastructure deformation, and geohazards like landslides, tectonic faults, and reservoir-induced seismicity (Hooper et al. 2004; Crosetto et al. 2016). Focusing on stable scatterers such as buildings and rocks for enhanced accuracy in long-term monitoring. PSInSAR has since been widely used in urban areas, such as Venice, for millimeter-scale subsidence monitoring (Ferretti et al. 1999), and for tracking geohazards like landslides and critical infrastructure. PSInSAR technique utilises time series SAR data, co-registered to a common master data and produces a stack of differential interferograms over time. The phase of isolated coherent points is analysed as a function of time. For each Permanent Scatterer (PS), the observed phase is modeled as a combination of

$$\phi(t) = \phi \text{ def} + \phi \text{ atm} + \phi \text{ topo} + \phi \text{ orb} + \phi \text{ noise} \tag{9.3}$$

where Φdef is the phase change caused by deformation, Φatm is the phase due to atmospheric delay, Φorb is the phase due to topography, Φorb is the phase due to orbital errors, and Φnoise is the phase due to noise. By assuming linear deformation over time and slow spatial variation of atmospheric phase, atmospheric noise can be separated using spatial low-pass filtering and Deformation is extracted using temporal linear regression. However, this separation is enabled only by using a large number of SAR images (typically > 20), where statistical methods can be applied effectively.

The key to the successful implementation of PSInSAR is the proper selection of the PS (Permanent Scatterers) pixels, i.e., the radar targets with very high phase stability over time and across different viewing geometries. The PS is generally selected using the amplitude dispersion index.

$$\text{Amplitude dispersion index} = \frac{\sigma A}{\langle A \rangle} \tag{9.4}$$

where σA is the standard deviation of the amplitude over time and ⟨A⟩ is the mean amplitude. PS candidates are typically selected with DA < 0.25. While PSInSAR excels in urban or rocky terrains with abundant reflectors, it suffers from limitations in rural, vegetated, or snow-covered areas where persistent scatterers are sparse or temporally unstable. This limitation hinders its application in areas like the Northeast Indian Region, where dense vegetation and high rainfall promote temporal decorrelation.

9.3.1.3 The Small Baseline Subset (SBAS)

The Small Baseline Subset (SBAS) method, initially suggested by Berardino et al. (2002), was introduced to overcome the shortfalls of Persistent Scatterer Interferometry (PSInSAR) in regions where coherent point targets are limited. Contrary to PSInSAR that depends on the detection of stable, individual scatterers (e.g., buildings, exposed rocks), SBAS can utilize distributed scatterers (DS) over spatially and temporally coherent patches by building up a network of interferograms from SAR image pairs with small spatial and temporal baselines. This approach allows detection of both non-linear and linear surface deformations on large scales, such as forested areas, agricultural land, natural terrain, and rural infrastructure, where permanent scatterers are scarce (Lanari et al. 2004). This technique is useful for measuring deformation over large areas as well as are helpful in forested regions and agricultural fields that lack stable reflectors. As satellite technologies advanced in the 2010s, both PSInSAR and SBAS were refined for broader applications and have become increasingly capable of multi-temporal analysis (Medhat et al. 2022; Alatza et al. 2020), offering more precise deformation monitoring. Advances in atmospheric and ionospheric corrections have further enhanced the accuracy of InSAR, making it a crucial tool in monitoring geohazards, subsidence, and infrastructure.

Unlike PS, in this method given a stack of N + 1 SAR acquisitions at times t0, t1,…,tN, M small-baseline differential interferograms as that

$$\frac{N+1}{2} \leq M \leq \frac{N(N+1)}{2} \tag{9.5}$$

Highly Coherent pixels are selected after topographic phase removal to generate M multi-temporal differential interferogram series (Aobpaet et al. 2013). The deformation phase for a highly coherent pixel in the ith unwrapped differential interferogram for acquisition pair (A, B) can be expressed as follows

$$\Delta\phi i = \phi(t_B) - \phi(t_A) \tag{9.6}$$

where $\phi(t)$ is the deformation phase at time t (small baseline). The primary objective of this technique is to enhance surface deformation estimation by combining multiple differential SAR interferograms with small orbital baselines to minimise decorrelation and improve temporal sampling. This is achieved by utilising singular value decomposition (SVD) to link datasets separated by large baselines. Only interferograms with small spatial/temporal baselines are selected to preserve coherence and reduce spatial decorrelation. Finally, deformation velocity (LOS displacement rate) between multiple acquisitions can be given as

$$vLOS = [-\frac{\lambda}{4\pi}\frac{\phi(t1) - \phi(t0)}{t1 - t0}, -\frac{\lambda}{4\pi}\frac{\phi(t2) - \phi(t1)}{t2 - t1} \ldots n, -\frac{\lambda}{4\pi}\frac{\phi(tN) - \phi(tN - 1)}{tN - tN - 1}] \tag{9.7}$$

SBAS does not rely solely on PS, but PS can still be included in the analysis if they meet coherence criteria. Unlike PS, which are isolated points with stable amplitude and phase, DS represent groups of pixels with similar scattering properties. Selecting DS allows SBAS to increase the density of coherent points, especially in areas where PS are scarce. In SBAS, interferometric pairs are chosen to be such that the perpendicular spatial baseline ($B\perp$) and temporal separation (Δt) between images are kept within predefined limits. This reduces temporal and geometric decorrelation and retains interferometric coherence.

9.3.2 *Global Navigation Satellite System (GNSS)*

One of the reliable methods of survey for geodynamic study is the use of Global Navigation Satellite System (GNSS). It has been used worldwide for its ability to observe crustal motion, interplate and intraplate deformation, and ground displacement associated with earthquakes at high spatial and temporal resolution. GNSS comprises one or more satellite constellations (Such as the Global Positioning System) which broadcast accurate timing and orbit data to receivers on the ground. This technique provides positional accuracy in the order of millimeters and has opened a new avenue for studying the crustal deformation and earthquake mechanisms (Blewitt 2007; Segall and Davis 1997). It is ideal for the measurements of short-term as well as long-term crustal deformation rates. In seismically active region like NER, combined study of seismicity patterns and the GNSS measurement may provide better understanding of earthquake hazards. GNSS technology allows for continuous or episodic monitoring of ground positions over a period of time. By utilising a permanent or semi-permanent GNSS receiver network, scientists are able to monitor horizontal and vertical ground movement to deduce tectonic strain build-up, fault slip rates, postseismic relaxation, and interseismic locking (Bürgmann et al. 2000). In seismically active Northeast India, GNSS is particularly useful. The region is situated at a complex juncture where the Indian, Eurasian, and Burmese tectonic plates converge, resulting in frequent earthquakes due to oblique collision and strain partitioning. A combined interpretation of GNSS data and seismicity characteristics facilitates the identification of fault behaviour, slip distribution, and possible zones of rupture and thus enhances regional earthquake hazard estimation (Najman et al. 2016). The carrier phase measurement from GNSS evolved over time, but its first formal application in satellite-based geodesy can be traced back to the late 1970s and early 1980s, when high-precision GPS techniques were being developed. The most basic observable in GNSS geodesy is the carrier phase measurement, whose equation can be represented as (Blewitt 1989)

$$\Phi = \rho + c(\delta t - \delta T) + \lambda N + \in \tag{9.8}$$

where Φ is the measured carrier phase from satellite to receiver, ρ is the geometric range to the satellite in true sense, c is the velocity of light, δt & δT are the offsets of the satellite and receiver clocks, λ is the carrier wavelength, N is the integer ambiguity and ϵ represent residual errors involving ionospheric/tropospheric delays. Double differencing techniques are applied to eliminate biases from satellite and receiver clocks. High-grade processing (e.g., Precise Point Positioning or network-based Real-Time Kinematic approaches) provides 3D displacement time series, from which scientists can resolve fault motion, block rotation, and elastic strain accumulation.

9.3.2.1 Precise Coordinates and Crustal Deformation Estimation

Precise position and velocity solutions can be obtained using methodology presented by Dong et al. (1998). Post processing of the GNSS data are being done using the software tools such as GAMIT-GLOBK, GIPSY and Bernese etc. Most analysis utilizesL1 and L2 (linear combinations) phase measurements to eliminatefirst order ionospheric delay (Herring et al. 2010). To remove the second and third order ionospheric delays, daily IONEX files from the Centre of Orbit Determination in Europe (CODE) containing vertical Total Electron Content (VTEC) and the latest release of the International Geomagnetic Reference Field (IGRF12) are used (Petrie et al. 2010; Thébault et al. 2015). Vienna Mapping Function (VMF1 grid) is often used to model the propagation delay (hydrostatic and wet zenith delay) due to the atmosphere (Boehm et al. 2007). The IERS 2003 model are used to eliminate site displacement associated with deformation from solid earth tides, while the FES2004 OTL model are used to eliminate the effects of ocean tidal loading. The loosely constrained daily solutions of the GNSS site coordinates are combined for the duration of analysisto define the coordinates and velocity of the GPS sites in the desired reference frame (Altamimi et al. 2011).

9.3.2.2 Linear Strain and Convergence Estimation

The Convergence or shortening between two stations can be computed by the change in their baseline length due to displacements over time (Mukul 2010; Barman et al. 2014). Considering two GPS Stations a and b analysed for time period t years, then Convergence Rate (mm/yr) can be given by

$$Convergence\ Rate_{(a,b)} = \frac{Final\ Baseline_{(a,b)} - Initial\ Baseline_{(a.b)}}{t} \tag{9.9}$$

Linear Strain on the other hand is calculated by dividing the change between the final and initial baseline lengths by the original baseline length (Beer et al. 1992). It is a dimensionless quantity usually reported in Nano-Strain. Therefore, for the same two stations a and b analysed for time period t years, Linear Strain Rate can be written as

$$\begin{matrix} Linear \\ Strain \\ peryear \end{matrix}_{(a,b)} = \frac{(Final\ Baseline_{(a,b)} - Initial\ Baseline_{(a.b)})/Initial\ Baseline_{(a.b)}}{t} \tag{9.10}$$

9.3.2.3 Localised Strain Estimation

Assuming plane strain deformation in the horizontal plane which is homogeneously distributed between GPS stations and ignoring the effect of vertical velocities, we can calculate two-dimensional crustal strain from groups of three or more GPS Stations (Cronin et al. 2012). To better understand the state of localised deformation, one can perform strain analysis using the Inverse Theory based approach (Allmendinger et al. 2011) that provides solutions such as translation, rotation displacement, infinitesimal strain, maximum shear strain and areal strain etc.

9.3.3 Ionosphere Total Electron Content (TEC)

The Total Electron Content (TEC) variations in the ionosphere is a function of phenomena such as geomagnetic storm, solar flare, and seismic activity. The TEC variations due to seismic activity delays and deviates the GNSS signal propagation through the ionosphere. Determination of TEC anomaly in the ionosphere using

GNSS has been proved one of the promising techniques for earthquake precursor studies (Sharma et al. 2017, 2024). Ionospheric TEC has been observed to change within the earthquake preparation zone a few days before an earthquake (Pulinets et al. 2005; Sharma 2021). This phenomenon is widely used to detect potential earthquake precursors. The radius of the preparation zone can be estimated using the method proposed by Dobrovolsky et al. (1979).

$$R = 10^{0.43*M} \tag{9.11}$$

where "M" is the magnitude of the earthquake.

The detailed physical, chemical and geological process associated with TEC anomalies has been explained by the LAIC model presented by Pulinets et al. (2005).

9.3.3.1 TEC Computation from GNSS Data

The Slant Total Electron Content (STEC) between a GNSS satellite and a ground receiver can be estimated by using the difference in pseudorange measurements from two GNSS signal frequencies (Blewitt 1990). The mathematical equation can be represented as

$$STEC = \frac{(P_2 - P_1)}{K(\frac{1}{f1^2} - \frac{1}{f2^2})} \tag{9.12}$$

Here, f1 = 1.57542 GHz and f2 = 1.2276 GHz represent the GPS signal frequencies. The constant K = 80.62 m^3/s^2 is related to the plasma frequency and electron density in the ionosphere. P1andP2 are the pseudorange measurements corresponding to these two frequencies.

For studying earthquake precursors, Slant TEC is converted into Vertical TEC (vTEC), measured in electrons per square meter (Kersley et al. 2004). This is necessary because the satellite signal travels along a slant path, and the actual TEC (STEC) is larger than the vertical TEC. The vertical TEC (vTEC) can be calculated using the following equation

$$\text{vTEC} = (\text{STEC} - \text{bs} - \text{br})\cos z' \tag{9.13}$$

where z' is zenith angle and it is expressed as

$$z\prime = \arcsin\frac{\text{RE}\cos a}{\text{RE} + h} \tag{9.14}$$

where α represents the satellite's elevation angle, RE is the Earth's average radius, and h is the height of the ionospheric layer. The terms bs and br refer to the estimated biases from the satellite and receiver, respectively. These bias values are provided by the CODE analysis center at the University of Bern, while orbit data are sourced from the International GNSS Service (IGS).

9.3.3.2 TEC Anomaly Detection

To detect TEC anomalies that may serve as earthquake precursors, the variation in vertical TEC (vTEC) is continuously monitored using TEC data from the previous 15 days (Sharma et al. 2017, Sharma 2021). The boundary limits are calculated as:

$$\text{Boundary Limit} = \text{X} \pm 1.34\sigma \tag{9.15}$$

where X is the 15-day running median and σ is the standard deviation. Values of vTEC that fall outside these boundary limits are considered anomalies.

Since TEC values are influenced by factors such as solar activity, geomagnetic storms, user location, and pseudorange elevation (Buonsanto and Fuller-Rowell 1997), the anomalies detected each day are compared against geomagnetic and solar activity data from the same period. This comparison helps determine whether the TEC anomaly is related to earthquake activity or other external factors.

9.3.4 Satellite Land Surface Temperature (LST)

The use of Thermal Remote Sensing Satellites to calculate LST has significantly advanced earthquake research by enabling continuous and large-scale monitoring of temperature variations across the Earth's surface. Modern satellite missions equipped with thermal sensors have become essential tools in this area. Notable operational satellites that provide high-quality thermal data for LST estimation include the National Oceanic and Atmospheric Administration's Advanced Very High-Resolution Radiometer (NOAA AVHRR), the Moderate Resolution Imaging Spectroradiometer (MODIS) onboard NASA satellites, the European Space Agency's SENTINEL-2, and the US Geological Survey's LANDSAT 8. These platforms offer diverse temporal and spatial resolutions, enhancing researchers' ability to detect subtle temperature changes potentially related to seismic events.

LST from MODIS are widely analysed before and after the seismic event to identify anomalies (Mishra et al. 2018; Saraf and Choudhury 2005). MODIS MOD11A1 LST products are derived datasets that come with a temperature of each pixel.

The LST in centigrade can be calculated using the given formula

$$\text{LST} = \text{Pixel Value} * 0.02 - 273.15\,^{\circ}\text{C} \tag{9.16}$$

LST derived can be analysed using a statistical and graphical method to identify anomalies due to the earthquake (Saraf and Choudhury 2005).

Several studies have documented that Land Surface Temperature (LST) tends to increase noticeably near earthquake epicenters in the days or weeks leading up to seismic events (Zhang et al. 2009; Tronin 2006). This phenomenon is attributed to a complex interplay of physical and chemical processes occurring within the Earth's crust and the atmosphere during the earthquake preparation phase. Understanding these underlying mechanisms is crucial for effectively using LST as a precursor signal in earthquake monitoring. From a physics perspective, the accumulation of tectonic stress in the crust causes microfracturing and deformation of rock formations. These fractures create pathways for the upward migration of gases such as radon, carbon dioxide, and methane, which originate from deep underground reservoirs. The release and subsequent interaction of these gases with near-surface materials can alter thermal properties, such as emissivity and thermal conductivity, thereby causing localized heating of the ground surface (Tronin 2006).

Simultaneously, the fracturing of rocks influences groundwater dynamics. Changes in groundwater flow and level affect the heat exchange between the subsurface and the surface, contributing to variations in surface temperature. Additionally, chemical reactions involving gas emissions and mineral interactions release latent heat, further amplifying thermal anomalies. Mechanical processes within the fault zone, including friction and deformation, generate heat that may propagate to the surface and be detected via thermal infrared remote sensing (Saraf and Choudhury 2005). Saraf and Choudhury (2005) also emphasize that the energy released during the preparatory stages of a major earthquake is not only mechanical but can manifest as detectable thermal signals on the surface and in the atmosphere above. These thermal anomalies, when observed and analyzed alongside other geophysical data, hold promise for improving the identification of earthquake precursors. As such, satellite-based thermal remote sensing platforms have become essential tools in seismic research, providing continuous, wide-area monitoring capabilities critical for detecting subtle thermal variations linked to seismic activity.

9.4 Selected Applications of Geospatial Technology for Geodynamic Studies in NER

9.4.1 InSAR-Based Crustal Deformation Analysis Along the Tectonically Active Koppili Fault in NER

Ground deformation in tectonically active regions such as Northeast India poses significant challenges for seismic hazard assessment. This region lies at the junction of the Indian Plate, the Eurasian Plate, and the Burmese microplate, making it one of the most geologically complex and seismically active zones in the world. The collision and ongoing convergence of these plates generate numerous active faults and folds, resulting in frequent earthquakes and continuous crustal deformation. One of the prominent faults in this region is the Kopili Fault, which extends across Assam and parts of Arunachal Pradesh. Monitoring ground deformation along such faults is crucial for understanding the accumulation of tectonic strain and for assessing earthquake potential. Figure 9.2 represents the ground deformation in the Kopili Fault region, obtained using the OpenSARLab platform, with the Small Baseline Subset (SBAS) technique. SBAS is an advanced InSAR method that uses multiple satellite images with short time and spatial baselines to generate high-precision maps of surface deformation over time. This technique effectively filters out atmospheric noise and other interferences, allowing for the detection of subtle ground movements often imperceptible by other means.

The dataset analysed spans from 2017 to 2024, capturing both uplift and subsidence phenomena along the fault. Specifically, the northeastern side of the Kopili Fault shows a pattern of subsidence (indicated by the black arrow in Fig. 9.2), while the southwestern side exhibits uplift. These opposite vertical motions reflect differential strain accumulation and release, indicative of complex tectonic processes such as crustal shortening and movements across fault. Such spatial variation in deformation rates emphasizes the dynamic nature of the tectonic environment in Northeast India. These patterns are critical for identifying segments of the fault that may be locked and accumulating stress, versus those that might be slipping aseismically. Continuous monitoring through SBAS-InSAR allows for near real-time assessment of deformation trends, improving the ability to forecast seismic hazards and to design appropriate mitigation strategies.

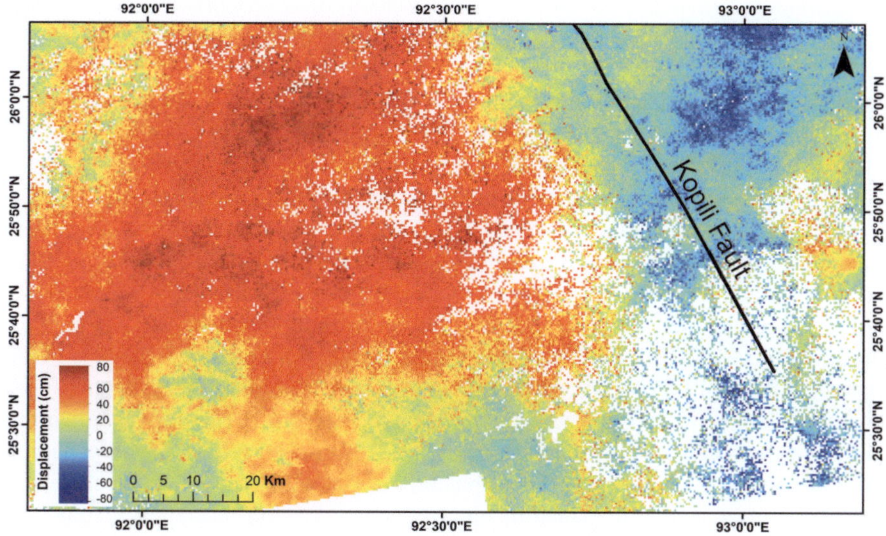

Fig. 9.2 Ground deformation map of the Kopili Fault region (Northeast India) from 2017 to 2024 derived using the Small Baseline Subset (SBAS) InSAR technique on Sentinel-1 SAR data processed through the OpenSARLab platform. The map highlights uplift (redish) on the southwestern side and subsidence (bluish colour) on the northeastern side of the fault (black line). This spatial pattern reflects the dynamic tectonic activity and strain accumulation along the fault

These observations confirm the importance of InSAR technology in detecting regions vulnerable to earthquakes. This comprehensive InSAR analysis, reveals distinct movements on either side of the Kopilli fault. Specifically, the North Eastern block situated to the right of the fault, shows clear signs of subsidence. In contrast, the South Western block, located to the left of the fault, exhibits upliftment. This differential movement suggests active tectonic processes, likely influenced by the regional stress regime and the fault's intrinsic properties. The rates of deformation in this area range from about 4 to 11.5 mm/year. The results showed that the InSAR-derived values closely match with the published GPS values (Line of Sight), demonstrating a high level of accuracy ($R^2 = 92\%$). This research contributes to advancing our understanding of tectonic processes in Northeast India and enhances seismic hazard assessment capabilities. The insights gained from this study add to the scientific understanding of regional tectonic dynamics and their implications for infrastructure development, urban planning, and disaster management.

9.4.2 Application of Ionosphere TEC Anomaly as Precursor to Earthquakes in NER

On April 28, 2021, at 07:51:25 IST, a magnitude 6.4 earthquake struck the Sonitpur district of Assam, India. According to the National Centre for Seismology (NCS), the earthquake's hypocenter was located at a depth of approximately 17 km, with its epicenter situated near two prominent fault systems: the Kopili Fault and the Dighalpani-Kakijan Fault. In an effort to investigate potential precursory signals associated with this seismic event, Sharma (2021) conducted a study focused on the analysis of Total Electron Content (TEC) variations in the ionosphere. The research utilized data from 19 GPS Continuously Operating Reference Stations (CORS) deployed in the region by Earth Scope, formerly known as the University NAVSTAR Consortium (UNAVCO). The study examined TEC time series over a 45-day period leading up to the earthquake, with the aim of detecting any anomalous ionospheric behavior that might correlate with the earthquake preparation phase. Significant anomalies in TEC were identified on April 3, 9, and 10, 2021 (Fig. 9.3). The maximum TEC deviations were recorded at 05.85 UTC on April 3, 08.99 UTC on April 9, and 08.73 UTC on April 10, suggesting localized ionospheric disturbances potentially linked to subsurface geodynamic processes. A particularly noteworthy finding of the study was the spatial pattern of TEC concentration, which appeared to decrease progressively toward the epicenter of the earthquake. This decline was observed within a radial distance of approximately 579 km, closely matching the theoretical earthquake preparation zone of 565 km, as estimated using the empirical relationship proposed by Dobrovolsky et al. (1979). This alignment lends support to the hypothesis that tectonic stress accumulation in the Earth's crust can lead to detectable ionospheric changes prior to a seismic event.

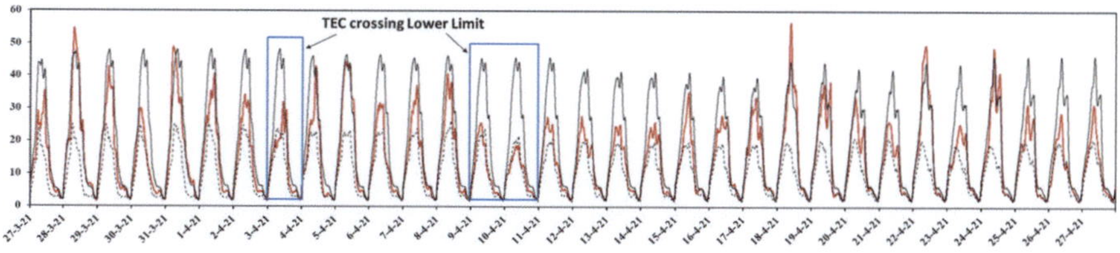

Fig. 9.3 Total Electron Content observations beforeMw 6.4 2021 Sonitpur earthquake. X-axis represent date whereas Y-axis is the TEC observed (Modified after Sharma 2021). Anomalous days are marked by blue rectangle

In another notable case, ionospheric TEC anomalies were identified in connection with the Mw 6.7 Imphal earthquake, which struck in the early hours of January 4, 2016, at 04:35 IST (corresponding to 23:05:21 UTC on January 3, 2016). According to Sharma et al. (2021a, 2017), a detailed analysis of TEC time series data revealed consistent ionospheric disturbances on several days prior to the mainshock—specifically on December 5, 9, 25, and 27, 2015. These anomalies were interpreted as possible precursors associated with crustal stress accumulation and ionospheric coupling phenomena. Extending the scope beyond a single event, Sharma et al. (2021b) conducted a comprehensive statistical analysis of 160 earthquakes that occurred in the North Eastern Region (NER) of India between 2012 and 2018. The study found that detectable ionospheric anomalies were present before 46.5% of earthquakes with magnitudes less than 5 (Mw < 5), and 81.56% of events with magnitudes between 5 and 6 (Mw 5–6).

These findings support the increasing reliability of ionospheric TEC monitoring as a supplementary tool for short-term earthquake research, particularly when integrated with a dense network of GNSS stations. Further spatial analysis by Sharma et al. (2018a, b) demonstrated how GNSS-based TEC data can be used to map the geographical distribution of ionospheric anomalies before earthquakes. These spatial patterns often correlate with known seismotectonic features and potential rupture zones, thereby enhancing our understanding of lithosphere-atmosphere–ionosphere coupling mechanisms.

Recognizing the importance of such capabilities, including deformation monitoring in the country, the Survey of India (SOI) has recently expanded its geodetic infrastructure by deploying over 120 GNSS stations across the eastern Himalayan region and about 1500 across India. This significantly enhanced network provides an opportunity to establish a near real-time TEC monitoring system, which could prove instrumental in developing early warning strategies for seismic hazards. The feasibility and potential of such a system have been demonstrated in preliminary studies by Sharma et al. (2021c). An example of TEC variation captured using this SOI GNSS dataset is illustrated in Fig. 9.4 and is available daily at www.nerdrr.gov.in.

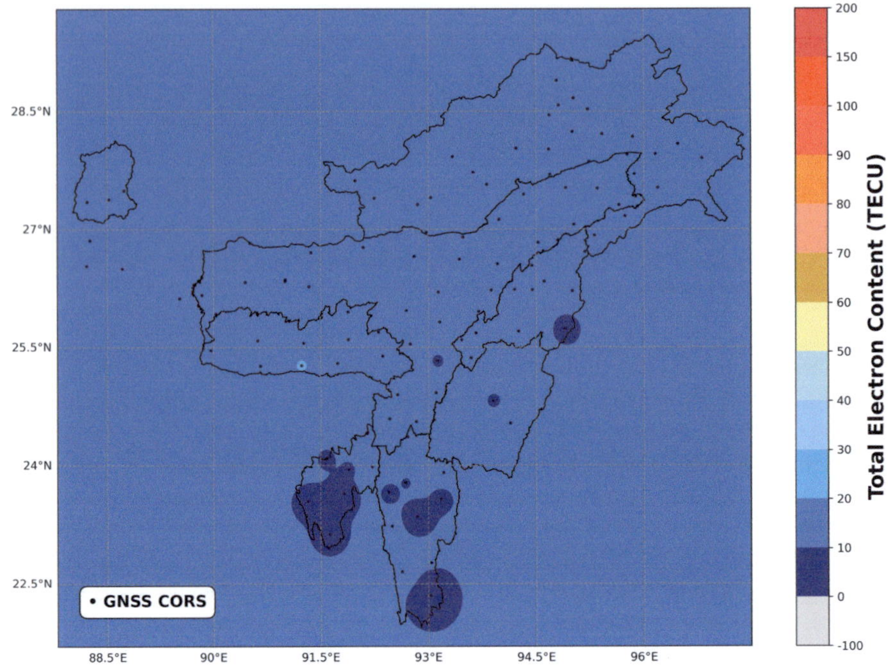

Fig. 9.4 Total Electron Content (TEC) map over the North Eastern Region (NER) of India on July 13, 2025, at 14:00 UTC, derived using data from the Survey of India (SOI) GNSS network. The spatial distribution of TEC values reveals localized ionospheric variations

9.4.3 Crustal Convergence and Deformation in Northeastern India

The eastern segment of the Himalayas is characterized by greater tectonic convergence and higher rates of strain accumulation when compared to the western Himalayan arc. This has been confirmed through several geodetic studies, which highlight the increased geodynamic complexity and seismic potential of the eastern region (Sharma et al. 2021d, 2018b). In the broader Northeastern Region (NER) of India, Global Navigation Satellite System (GNSS) observations during the period 1991 to 1999 indicated a southward velocity of approximately 6.3 $\pm$ 3.8 mm/year for the Shillong region, relative to an India-fixed reference frame (Paul et al. 2001). This early finding suggested the presence of significant crustal motion and initiated a series of subsequent geodetic investigations aimed at refining our understanding of regional deformation patterns. More recent GNSS-based studies have offered improved spatial resolution and longer time series, leading to more accurate estimations of tectonic movements. Notably, measurements from various GPS stations across NER have reported an average arc-normal convergence rate of around 16 mm/year, when measured relative to the Lhasa block, which represents the stable part of the Tibetan Plateau. This convergence rate has been corroborated by multiple independent studies, including those by Barman et al. (2014), Jade et al. (2007, 2017), and Mukul et al. (2010b). Figure 9.5 represents the deformation scenario in the selected locations in NER reported in the literature. These findings highlight the critical role of continuous GNSS monitoring in capturing both horizontal convergence and vertical displacements, which are key indicators of strain accumulation and potential earthquake-generating processes in the seismically active Eastern Himalayas and NER.

A key geodynamic feature within the region is the Shillong Plateau, which plays a crucial role in accommodating tectonic stress. Recent high-precision geodetic observations have documented evidence of vertical uplift, suggesting ongoing crustal thickening or localized fault activity. Vertical displacement rates on the Shillong Plateau have been found to vary spatially, with uplift rates of approximately 5 to 7 mm/year, as reported by Mallick et al. (2020). These uplift trends are significant as they point to long-term crustal deformation that may contribute to future seismic hazard in the region.

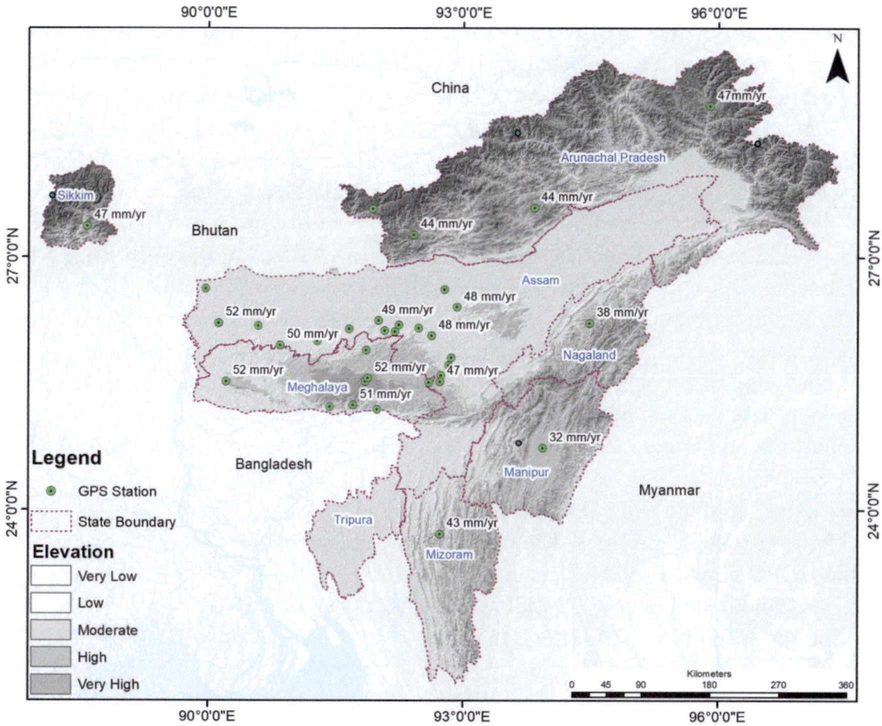

Fig. 9.5 Representation of the GNSS based crustal deformation rates observed at various key locations across the Northeastern Region (NER) of India, as reported in multiple geodetic studies (adopted from Jade et al. 2007, 2017; Mukul et al. 2010; Mahesh et al. 2012; Barman et al. 2017) (Base map: ESRI)

The Kopili Fault, a prominent tectonic structure traversing the state of Assam in Northeast India, is a key locus of active crustal deformation in the region. Geodetic and seismological studies have consistently highlighted the strain accumulation along this fault, pointing to its potential to generate a large-magnitude earthquake in the future. The fault has a documented history of moderate-sized seismic events, reinforcing its classification as a seismically active and hazardous fault system. Its oblique-slip nature and alignment along a major crustal boundary further underline its tectonic significance. In addition to the Kopili Fault, the Indo-Burma Fold and Thrust Belt (IBFTB) located to the east of the Shillong Plateau—exhibits a complex pattern of deformation. This region represents the transitional zone where the Indian Plate subducts obliquely beneath the Indo-Burmese microplate. One of the defining characteristics of this belt is its segmentation into north–south (N-S) trending crustal blocks, which are offset and bounded by east–west (E-W) transverse tectonic zones. These transverse faults accommodate lateral (strike-slip) movement, including Dextral (right-lateral) slip along zones situated between the Nagaland salient and the Manipur recess, and Sinistral (left-lateral) slip in the segment between the Manipur recess and the Tripura-Mizoram salient, as documented by Jade et al. (2007). Such segmentation and the coexistence of thrust and strike-slip mechanisms reflect the highly partitioned nature of deformation in this convergent zone, contributing to a heigh seismic risk. Importantly, recent studies have begun to integrate ionospheric Total Electron Content (TEC) variations derived from GNSS observations with traditional crustal deformation measurements. This multidisciplinary approach offers promising new insights into the coupling between the Earth's surface and the ionosphere. Combining TEC anomaly detection with geodetic data can enhance our understanding of earthquake preparatory processes, particularly in complex and tectonically active regions, such as the NER. Additionally, GNSS has enabled interdisciplinary research when combined with InSAR, opening avenues for 3D deformation mapping (Wright et al. 2004). Such integrated geophysical monitoring frameworks are crucial for advancing early warning systems and improving our knowledge of the spatiotemporal dynamics of seismic hazards in regions with dense populations and vulnerable infrastructure.

India operates a dense network of Continuously Operating Reference Stations (CORS) through the Survey of India (SOI), as detailed in Sect. 4. This advanced geodetic infrastructure plays a crucial role in supporting a wide range of scientific and applied research activities. The continuous, high-precision GNSS data generated by the CORS network enables accurate monitoring of crustal deformation, tectonic plate movement, and strain accumulation across different seismic zones. These data are vital for developing robust geodynamic models that can simulate and predict the behavior of Earth's crust under various stress conditions. In particular, such models contribute to a better understanding of the earthquake cycle, fault interactions, and regional stress fields, which are critical parameters for assessing seismic hazards.

9.5 Conclusions

Geospatial technologies have emerged as indispensable tools for assessing seismic hazards and monitoring tectonic processes, particularly in tectonically complex and seismically active regions like the North Eastern Region (NER) of India. These technologies offer precise, high-resolution spatial and temporal data that enable scientists and disaster management authorities to gain deeper insights into the region's geodynamics.

The use of satellite-based remote sensing tools, such as Interferometric Synthetic Aperture Radar (InSAR) and Global Navigation Satellite Systems (GNSS), has greatly enhanced our capacity to detect and monitor ground deformation associated with strain accumulation along active fault lines. Advanced techniques like PSInSAR and SBAS have enabled millimeter-level ground motion tracking, even in remote or vegetated terrains, while GNSS networks such as CORS offer absolute 3D measurements for calculating strain and plate velocities. The synergy between InSAR and GNSS greatly improves the accuracy of geodynamic models and seismic cycle analysis. These methods provide valuable information on crustal motion and displacement patterns, which are critical for identifying areas of increased seismic risk. Additionally, techniques that observe thermal anomalies and ionospheric Total Electron Content (TEC) variations have shown promising potential in detecting earthquake precursors, adding another layer of early warning capability. The integration of these geospatial techniques with ground-based field observations, geological mapping, and seismic data has led to a more comprehensive and nuanced understanding of the tectonic framework of the NER. This multidisciplinary approach enables more accurate mapping of active faults, identification of seismogenic zones, and improved modelling of strain distribution across the region.

Importantly, the growing network of GNSS-based Continuously Operating Reference Stations (CORS) such as those maintained by the Survey of India provides continuous, real-time positioning data that can be used for long-term monitoring of crustal deformation. This not only supports scientific research but also plays a vital role in disaster risk reduction, infrastructure planning, and earthquake preparedness. Together, these technologies provide critical insights for earthquake preparedness, infrastructure planning, and disaster risk reduction. Their continued development and integration will be vital in building resilient communities and improving seismic hazard management in NER and other vulnerable regions.

Acknowledgements The authors are thankful to Survey of India (SOI) for providing GNSS data. Authors also thank Alaska Satellite Facility (ASF) for providing access to OpenSARLab cloud service. Authors thank Shri. Rahul Kumar, Research Scientist NESAC and Shri. Pritom Pran Dutta, Intern NESAC for their assistance in SAR, GNSS and earthquake data downloading.

References

S. Alatza, I. Papoutsis, D. Paradissis, C. Kontoes, G.A. Papadopoulos, Multi-temporal InSAR analysis for monitoring ground deformation in Amorgos Island, Greece. Sensors **20**(2), 338 (2020)

R.W. Allmendinger, N. Cardozo, D.M. Fisher, *Structural Geology Algorithms: Vectors and Tensors* (Cambridge University Press, 2011)

Z. Altamimi, X. Collilieux, L. Métivier, ITRF2008: an improved solution of the international terrestrial reference frame. J. Geodesy **85**(8), 457–473 (2011)

N. Ambraseys, R. Bilham, Reevaluated intensities for the great Assam earthquake of 12 June 1897, Shillong, India. Bull. Seismol. Soc. Am. **93**(2), 655–673 (2003)

S. Awasthi, K. Jain, V. Mishra, A. Kumar, An approach for multi-dimensional land subsidence velocity estimation using time-series Sentinel-1 SAR datasets by applying persistent scatterer interferometry technique. Geocarto Int. **37**(9), 2647–2678 (2020)

P. Banerjee, R. Bürgmann, Convergence across the northeast Himalaya from GPS measurements. Geophys. Res. Lett. **29**(16), 1555 (2002). https://doi.org/10.1029/2002GL014728

P. Barman, S. Jade, T.S. Shrungeshwara, A. Kumar, S. Bhattacharyya, J.D. Ray, W.M. Jamir, Crustal deformation rates in Assam Valley, Shillong Plateau, Eastern Himalaya, and Indo-Burmese region from 11 years (2002–2013) of GPS measurements. Int. J. Earth Sci. **106**, 2025–2038 (2017)

P. Barman, J.D. Ray, A. Kumar, J.D. Chowdhury, K. Mahanta, Estimation of present-day interseismic deformation in the Kopili fault zone of north-east India using GPS measurements. Geomat. Nat. Haz. Risk **7**(2), 586–599 (2014)

F.P. Beer, E.R. Johnston, J.T. DeWolf, D.F. Mazurek, S. Sanghi, *Mechanics of Materials*, vol. 1 (McGraw-Hill, New York, 1992)

P. Berardino, G. Fornaro, R. Lanari, E. Sansosti, A new algorithm for surface deformation monitoring based on small baseline differential SAR interferograms. IEEE Trans. Geosci. Remote Sens. **40**(11), 2375–2383 (2002)

R. Bilham, Earthquakes in India and the Himalaya: tectonics, geodesy and history. Ann. Geophys. **47**(2–3), 839–858 (2004)

G. Blewitt, Carrier phase ambiguity resolution for the global positioning system applied to geodetic baselines up to 2000 km. J. Geophys. Res. Solid Earth **94**(B8), 10187–10203 (1989)

G. Blewitt, An automatic editing algorithm for GPS data. Geophys. Res. Lett. **17**(3), 199–202 (1990)

G. Blewitt, *GPS and Space-Based Geodetic Methods. In Treatise on Geophysics*, vol. 3, pp. 351–390. (Elsevier, 2007)

J. Boehm, R. Heinkelmann, H. Schuh, Short note: a global model of pressure and temperature for geodetic applications. J. Geodesy **81**, 679–683 (2007)

M.J. Buonsanto, T.J. Fuller-Rowell, Strides made in understanding space weather at Earth. Eos Trans. AG, 78, 1, 0096-394 (1997). https://doi.org/10.1029/97EO00002.

R. Bürgmann, P.A. Rosen, E.J. Fielding, Synthetic aperture radar interferometry to measure Earth's surface topography and its deformation. Annu. Rev. Earth Planet. Sci. **28**, 169–209 (2000)

J.K. Catherine, A preliminary assessment of internal deformation in the Indian Plate from GPS measurements. J. Asian Earth Sci. **23**, 461–465 (2004)

V.S. Cronin, P.G. Resor, W.C. Hammond, C. Kreemer, S.E. Olds, B. Pratt-Sitaula, N.W. West, Developing a curricular module for introductory geophysics or structural geology courses to quantify crustal strain using EarthScope PBO GPS velocities. (Abstract ED41B-0681, Fall Meeting, AGU, San Francisco, 2012)

M. Crosetto, O. Monserrat, M. Cuevas-González, N. Devanthéry, B. Crippa, Persistent scatterer interferometry: a review. ISPRS J. Photogramm. Remote. Sens. **115**, 78–89 (2016). https://doi.org/10.1016/j.isprsjprs.2015.10.011

R. De, J.R. Kayal, Seismotectonic model of the Sikkim Himalaya: constraint from microearthquake surveys. Bull. Seismol. Soc. Am. **96**(6), 2061–2068 (2003)

I.P. Dobrovolsky, S.I. Zubkov, V.I. Miachkin, Estimation of the size of earthquake preparation zones. Pure Appl. Geophys. **117**(5), 1025–1044 (1979)

D. Dong, T.A. Herring, R.W. King, Estimating regional deformation from a combination of space and terrestrial geodetic data. J. Geodesy **72**, 200–214 (1998)

S.C. Dutta, P.S. Mukhopadhyay, R. Saha, S. Nayak, 2011 Sikkim earthquake at Eastern Himalayas: lessons learnt from performance of structures. Soil Dyn. Earthq. Eng. **75**, 121–129 (2015)

D. Dzurisin, Z. Lu, Interferometric synthetic-aperture radar (InSAR). In *Volcano Deformation: Geodetic Monitoring Techniques* (2007), pp. 153–194

P. England, R. Bilham, The Shillong Plateau and the great 1897 Assam earthquake. Tectonics **34**(9), 1792–1812 (2015)

A. Ferretti, A. Monti-Guarnieri, C. Prati, F. Rocca, D. Massonnet, *InSAR Principles: Guidelines for SAR Interferometry Processing and Interpretation*, vol. 19 (2007)

A. Ferretti, C. Prati, F. Rocca, Permanent scatterers in SAR interferometry. IEEE Trans. Geosci. Remote Sens. **39**(1), 8–20 (1999). https://doi.org/10.1109/36.898661

A. Ferretti, C. Prati, F. Rocca, Nonlinear subsidence rate estimation using permanent scatterers in differential SAR interferometry. IEEE Trans. Geosci. Remote Sens. **38**(5), 2202–2212 (2002)

A.K. Gabriel, R.M. Goldstein, H.A. Zebker, Mapping small elevation changes over large areas: differential radar interferometry. J. Geophys. Res. Solid Earth **94**(B7), 9183–9191 (1989)

R.F. Hanssen, *Radar Interferometry: Data Interpretation and Error Analysis* (Springer, 2001)

T.A. Herring, R.W. King, S.C. McClusky, *Introduction to GAMIT/GLOBK, Release 10.4. Department of Earth, Atmospheric and Planetary Sciences* (Massachusetts Institute of Technology, 2010)

A. Hooper, H. Zebker, P. Segall, B. Kampes, A new method for measuring deformation on volcanoes and other natural terrains using InSAR persistent scatterers. Geophys. Res. Lett. **31**(23), L23611 (2004). https://doi.org/10.1029/2004GL021737

S. Jade, M. Mukul, A.K. Bhattacharyya, M.S.M. Vijayan, S. Jaganathan, A. Kumar, R.P. Tiwari, A. Kumar, S. Kalita, S.C. Sahu, A.P. Krishna, S.S. Gupta, M.V.R.L. Murthy, V.K. Gaur, Estimates of interseismic deformation in Northeast India from GPS measurements. Earth Planet. Sci. Lett. **263**(3–4), 221–234 (2007)

S. Jade, T.S. Shrungeshwara, K. Kumar, P. Choudhury, R.K. Dumka, H. Bhu, India plate angular velocity and contemporary deformation rates from continuous GPS measurements from 1996 to 2015. Sci. Rep. **7**(1), 11439 (2017)

J.R. Kayal, R. De, Microseismicity and tectonics in northeast India. Bull. Seismol. Soc. Am. **81**(1), 131–138 (1991)

J.R. Kayal, S.S. Arefiev, S. Barua, D. Hazarika, N. Gogoi, A. Kumar, S. Kalita, Shillong plateau earthquakes in northeast India region: complex tectonic model. Curr. Sci. **90**(1), 109–114 (2006)

L. Kersley, D. Malan, S.E. Pryse, S.M. Radicella, L.R. Cander, A. Belehaki, C.N. Mitchell, Total electron content: a key parameter in propagation—measurement and use in ionospheric imaging. Ann. Geophys. **47**(2/3), 1067–1091 (2004)

B.V. Lakshmi, P.B. Gawali, Soft sediment deformation features in Dauki Fault region: evidence of paleoearthquakes, Shillong Plateau, NE, India. Environ. Earth Sci. **81**(3), 58 (2022)

R. Lanari, O. Mora, M. Manunta, J.J. Mallorquí, P. Berardino, E. Sansosti, A small-baseline approach for investigating deformations on full-resolution differential SAR interferograms. IEEE Trans. Geosci. Remote Sens. **42**(7), 1377–1386 (2004)

S. Longbir, S. Arjun, Assessment of liquefaction potential of Guwahati city based on geotechnical standard penetration test (SPT) data. Disas. Adv. **10**(12), 10–21 (2017)

P. Mahesh, J.K. Catherine, V.K. Gahalaut, B. Kundu, A. Ambikapathy, A. Bansal, S. Kalita, Rigid Indian plate: constraints from GPS measurements. Gondwana Res. **22**(3–4), 1068–1072 (2012)

J.N. Malik, T. Nakata, Active faults and related late Quaternary deformation along the northwestern Himalayan frontal zone, India. Ann. Geophys. **46**, 917–936 (2003). https://doi.org/10.4401/ag-3424

R. Mallick, J.A. Hubbard, E.O. Lindsey, K.E. Bradley, J.D. Moore, A. Ahsan, E.M. Hill, Subduction initiation and the rise of the Shillong Plateau. Earth Planet. Sci. Lett. **543**, 116351 (2020)

D. Massonnet, K.L. Feigl, Discrimination of geophysical phenomena in satellite radar interferograms. Geophys. Res. Lett. **22**(12), 1537–1540 (1995)

D. Massonnet, K.L. Feigl, Radar interferometry and its application to changes in the Earth's surface. Rev. Geophys. **36**(4), 441–500 (1998). https://doi.org/10.1029/97RG03139

D. Massonnet, M. Rossi, C. Carmona, F. Adragna, G. Peltzer, K. Feigl, T. Rabaute, The displacement field of the Landers earthquake mapped by radar interferometry. Nature **364**(6433), 138–142 (1993)

N.I. Medhat, M.Y. Yamamoto, C. Tolomei, A. Harbi, S. Maouche, Multi-temporal InSAR analysis to monitor landslides using the small baseline subset (SBAS) approach in the Mila Basin, Algeria. Terra Nova **34**(5), 407–423 (2022)

V. Mishra, R. Shah, S. Azhar, H. Shah, P. Modi, R. Kumar, Reconstruction of droughts in India using multiple land-surface models (1951–2015). Hydrol. Earth Syst. Sci. **22**(4), 2269–2284 (2018)

P. Molnar, P. Tapponnier, Cenozoic tectonics of Asia: Effects of a continental collision. Science **189**(4201), 419–426 (1975)

M. Mukul, First-order kinematics of wedge-scale active Himalayan deformation: insights from Darjiling–Sikkim–Tibet (DaSiT) wedge. J. Asian Earth Sci. **39**, 645–657 (2010a). https://doi.org/10.1016/j.jseaes.2010.05.005

M. Mukul, S. Jade, A.K. Bhattacharyya, K. Bhusan, Crustal shortening in convergent orogens: Insights from global positioning system (GPS) measurements in northeast India. J. Geol. Soc. India **75**, 302–312 (2010b)

Y. Najman, L. Bracciali, R.R. Parrish, E. Chisty, A. Copley, Evolving strain partitioning in the Eastern Himalaya: the growth of the Shillong Plateau. Earth Planet. Sci. Lett. **433**, 1–9 (2016)

D.R. Nandy, *Geodynamics of North-Eastern India and the Adjoining Region* (ACB Publications, 2001)

D.R. Nandy, *Geodynamics of Northeastern India and the Adjoining Region* (Scientific Book Centre, 2017)

D.R. Nandy, S. Dasgupta, Seismotectonic domains of northeastern India and adjacent areas. Phys. Chem. Earth **18**, 371–384 (1991). https://doi.org/10.1016/0079-1946(91)90006-V

J. Paul, R. Bürgmann, V.K. Gaur, R. Bilham, K.M. Larson, M.B. Ananda, S. Jade, M. Mukal, T.S. Anupama, G. Satyal, D. Kumar, The motion and active deformation of India. Geophys. Res. Lett, 28, 4, 0094-8276 (2001). https://doi.org/10.1029/2000GL011832

G. Peltzer, P. Rosen, Surface displacement of the 17 May 1993 Eureka Valley, California, earthquake observed by SAR interferometry. Science **268**(5215), 1333–1336 (1995)

E.J. Petrie, M.A. King, P. Moore, D.A. Lavallée, Higher-order ionospheric effects on the GPS reference frame and velocities. J. Geophys. Res. Solid Earth **115**(B3) (2010)

S.A. Pulinets, A. Leyva, G. Contreras, G. Bisiacchi, L. Ciraolo, Total electron content variations in the ionosphere before the Colima, Mexico, earthquake of 21 January 2003. GeofísicaInternacional **44**(4), 369–377 (2005)

K. Rajendran, C.P. Rajendran, Revisiting the earthquake sources in the Himalaya: perspectives on past seismicity. Tectonophysics **504**(1–4), 75–88 (2011)

P.A. Rosen, S. Hensley, I.R. Joughin, F.K. Li, S.N. Madsen, E. Rodriguez, R.M. Goldstein, Synthetic aperture radar interferometry. Proc. IEEE **88**(3), 333–382 (2000)

V.K. Sahu, V.K. Gahalaut, S. Rajput, R.K. Chadha, S.S. Laishram, A. Kumar, Crustal deformation in the Indo-Burmese arc region: implications from the Myanmar and Southeast Asia GPS measurements. Curr. Sci. 1688–1693 (2006)

A.K. Saraf, S. Choudhury, NOAA-AVHRR detects thermal anomaly associated with the 26 January 2001 Bhuj earthquake, Gujarat, India. Int. J. Remote Sens. **26**(6), 1065–1073 (2005)

A.K. Saraf, V. Rawat, P. Banerjee, S. Choudhury, S.K. Panda, S. Dasgupta, J.D. Das, Satellite detection of earthquake thermal infrared precursors in Iran. Nat. Hazards **47**, 119–135 (2008)

P. Segall, J.L. Davis, GPS applications for geodynamics and earthquake studies. Annu. Rev. Earth Planet. Sci. **25**, 301–336 (1997)

G. Sharma, Manifestation of earthquake preparation zone in the ionosphere before the earthquake revealed by GPS-TEC data: a case of the 2021 Sonitpur, Assam earthquake. Geodesy Geodyn. (2021)

G. Sharma, P.K. Champatiray, S. Mohanty, Morphotectonic analysis and GNSS observations for assessment of relative tectonic activity in Alaknanda basin of Garhwal Himalaya, India. Geomorphology **301**, 108–120 (2018a)

G. Sharma, P.K. Champatiray, S. Mohanty, S. Kannaujiya, Ionospheric TEC modeling for earthquake precursors from GNSS data. Quatern. Int. **462**, 65–74 (2017)

G. Sharma, G.P.K.R. Kannaujiya, A.K. Taloor, P.K. Champatiray, S. Mohanty, Crustal deformation analysis across Garhwal Himalaya: part of western Himalaya using GPS observations. Quatern. Int. **575**, 153–159 (2021a)

G. Sharma, S. Mohanty, P.K. Champatiray, M.S. Singh, K.K. Sarma, P.L.N. Raju, Ionospheric total electron content and epicentral distance of the 2015 Gorkha earthquake revealed by GNSS observations. Curr. Sci. **115**(1), 27–29 (2018b)

G. Sharma, K. Nayak, R. Romero Andrade, M.A. Mohammed Aslam, K.K. Sarma, S.P. Aggarwal, Low ionosphere density above the earthquake epicenter region of Mw 7.2, El Mayor–Cucapah earthquake evident from dense CORS data. J. Indian Soc. Remote Sens. (2024)

G. Sharma, P. Saikia, D. Walia, P. Banerjee, P.L.N. Raju, TEC anomalies assessment for earthquakes precursors in northeastern India and adjoining region using GPS data acquired during 2012–2018. Quatern. Int. **576**, 120–129 (2021b)

G. Sharma, M. Soubam, D. Walia, N. Nishant, K.K. Sarma, P.L.N. Raju, Development of a monitoring system for ionosphere TEC variability before earthquakes. Appl. Comput. Geosci. **9**, 10052 (2021c)

G. Sharma, D. Walia, P. Banerjee, Statistical behavior of electron density in the ionosphere before the 2016 Manipur earthquake from GPS observation stations at North East India. Geotectonics **55**, 646–653 (2021d)

M.L. Sharma, C. Lindholm, Earthquake hazard assessment for Dehradun, Uttarakhand, India, including a characteristic earthquake recurrence model for the Himalaya Frontal Fault (HFF). Pure Appl. Geophys. **169**(9), 1601–1617 (2012)

K.M. Sreejith, P.S. Sunil, R. Agrawal, A.P. Saji, D.S. Ramesh, A.S. Rajawat, Coseismic and early postseismic deformation due to the 25 April 2015, Mw 7.8 Gorkha, Nepal, earthquake from InSAR and GPS measurements. Geophys. Res. Lett. **43**, 3160–3168 (2016)

M.S. Steckler, S.H. Akhter, L. Seeber, Collision of the Ganges-Brahmaputra delta with the Burma Arc: implications for earthquake hazard. Earth Planet. Sci. Lett. **273**(3–4), 367–378 (2008)

E. Thébault, C.C. Finlay, P. Alken, C.D. Beggan, E. Canet, A. Chulliat, R. Schachtschneider, *Evaluation of Candidate Geomagnetic Field Models for IGRF-12*, vol. 67, pp. 1–23. (Springer Berlin Heidelberg, 2015)

A.A. Tronin, Remote sensing and earthquakes: a review. Phys. Chem. Earth Parts a/b/c **31**(4–9), 138–142 (2006)

K. Tshering, E. Sharma, N. Chettri, A.B. Shrestha, *Climate Change Vulnerability of Mountain Ecosystems in the Eastern Himalayas–Synthesis report* (ICIMOD, Kathmandu, Nepal, 2010). ISBN: 978–92–9115–141–7 (printed), 978–92–9115–142–4 (electronic)

S.L. Ullo, P. Addabbo, D. Di Martire, S. Sica, N. Fiscante, L. Cicala, C.V. Angelino, Application of DInSAR technique to high coherence Sentinel-1 images for dam monitoring and result validation through in situ measurements. IEEE J. Select. Topics Appl. Earth Observ. Remote Sens. **12**(3), 875–890 (2019)

T.J. Wright, B.E. Parsons, Z. Lu, Toward mapping surface deformation in three dimensions using InSAR. Geophys. Res. Lett. **31**(1), L01607 (2004). https://doi.org/10.1029/2003GL018827

A. Yin, C.S. Dubey, A.A.G. Webb, T.K. Kelty, M. Grove, G.E. Gehrels, W.P. Burgess, Geologic correlation of the Himalayan orogen and Indian craton: Part 1. Structural geology, U-Pb zircon geochronology, and tectonic evolution of the Shillong Plateau and its neighboring regions in NE India. GSA Bull. **122**(3–4), 336–359 (2010)

H.A. Zebker, J. Villasenor, Decorrelation in interferometric radar echoes. IEEE Trans. Geosci. Remote Sens. **30**(5), 950–959 (1992)
X. Zhang, X. Shen, J. Liu, X. Ouyang, J. Qian, S. Zhao, Analysis of ionospheric plasma perturbations before the Wenchuan earthquake. Nat. Hazard. **9**(4), 1259–1266 (2009)

Gopal Sharma has obtained a Ph.D. in Crustal deformation study from IIT Dhanbad. Presently, he is a Senior Scientist at the North Eastern Space Applications Centre (NESAC), Shillong, Department of Space, Government of India. His research interests include Earthquake precursors, Crustal Deformation, and strain accumulation measurements. His ongoing research focuses on large-network GNSS and InSAR data for earthquake precursor and deformation studies.

M Somorjit Singh is heading the Geoscience Division at the North Eastern Space Applications Centre (NESAC), Shillong, Department of Space, Government of India. In 20 years of service at NESAC, he has contributed to various operational and R&D activities/projects. He has been honored 'ISRO Team Excellence Award for his significant contribution as a team member in 'National Geomorphology and Lineament Mapping (NGLM) project' which was initiated by National Remote Sensing Centre (NRSC), Hyderabad in collaboration with Geological Survey of India (GSI) under Natural Resources Census program. His research interests include Geomorphology & Lineament Studies/Mapping, Infrastructural Planning (Site Suitability/feasibility Studies, road & railway alignment/realignment, Natural/Geo-hazard, Manmade Hazard (Industrial Hazard), Environmental Studies/Pollution (Heavy Metals) etc.

Dr. Prakash Biswakarma is a Geoscientist specializing in landslides studies with a Ph.D. in Slope Failure Analysis. Currently a Research Scientist at NESAC, Department of Space, he has over 20 peer-reviewed publications and experience across academia, research, and field applications. He is proficient in GIS, remote sensing, SAR tools, and has completed numerous advanced trainings in geospatial technologies.

K. K. Sarma Ph.D., leads the Remote Sensing Applications Group (RSAG) at the North Eastern Space Applications Centre (NESAC), where he has been a key figure since 2002. Specializing in remote sensing for forestry and biodiversity, his work leverages space technology for critical applications in Biodiversity, Ecosystems, Forestry, and Natural Resource Management. Dr. Sarma has spearheaded over 60 diverse geospatial projects, contributing significantly to natural resource mapping and monitoring across the North Eastern Region (NER). Currently, he coordinates 110 "Plan of Action" projects, focused on using space technology to improve livelihoods in the NER. He also served as the Principal Investigator and NER Coordinator for all ISRO natural resource management projects, including comprehensive Land Use Land Cover Mapping and National Wasteland Updating, Monitoring, and Change Analysis. With over 75 research publications, Dr. Sarma has successfully supervised six Ph.D. students and actively mentors many others across NER Universities.

Chapter 10
Understanding the Landslide Hazard Prevailing in North East India

Kuntala Bhusan, Prakash Biswakarma, Aniket Chakravorty, Praveen Kumar, M. Somorjit Singh, and K. K. Sarma

Abstract Landslide is one of the widespread and challenging geo-hazards which occur in the mountainous terrains due to collective effect of several geological factors. All types of landslides of various magnitudes and dimensions either triggered by rainfall, earthquakes or by their combined effect are prevailing in this tectonically active region. Deep seated landslides usually, disrupt transportation network for longer duration while shallow slides, mostly affect the urban life for a short duration. The reported causalities associated to landslide hazards are much high in the North Eastern Region (NER) compared to other parts of India. Fragile, soft and weathered rock formations along the steep slope make the terrain of NER much susceptible to landslides. Further, NER is one of the most seismically active regions in the world, and the region also receives much rainfall compared to other parts of the Country. Geospatial data and tools are extensively being used to study landslide right from inventory mapping to susceptibility-hazard assessment, landslide monitoring and estimating rainfall threshold for development of early warning model. However, in most of the cases selection of Earth Observation (EO) data and technique is governed by the scale and purpose of study apart from location, size, age and exposure condition of existing or fresh landslide. While selecting data emphasis is given on resolution (spatial and temporal), accuracy, swath and revisit time of the sensor. Since it is not possible to

K. Bhusan (✉) · P. Biswakarma · A. Chakravorty · M. Somorjit Singh · K. K. Sarma
North Eastern Space Applications Centre (NESAC), Umiam, Shillong, Meghalaya, India
e-mail: kuntala.bhusan@nesac.gov.in

P. Biswakarma
e-mail: p.biswakarma@nic.in

A. Chakravorty
e-mail: aniket.chakravorty@nesac.gov.in

M. Somorjit Singh
e-mail: m.somorjit@nesac.gov.in

K. K. Sarma
e-mail: kk.sharma@nesac.gov.in

P. Kumar
Euro-Mediterranean Center on Climate Change (CMCC), Lecce, Italy
e-mail: pbr162@gmail.com

S. P. Aggarwal et al. (eds.), *Remote Sensing of the North Eastern Himalayan Ecosystem*,
https://doi.org/10.1007/978-981-95-3221-6_10

avoid developmental activities in the hilly states so proper planning is prerequisite to reduce risk due to impending landslides by understanding the genesis of landslide. Geospatial technology is useful in all the three (pre-, during, post-disaster) phases of Landslide Disaster Management.

Keywords Landslide inventory · Landslide susceptibility mapping · Early warning system · RS and GIS · Rainfall-induced landslides · Earthquake-triggered landslides

10.1 Introduction

A "landslide" is the movement of a mass of rock, debris, soil with or without vegetation, over burden materials, manmade feature, infrastructure etc. present on slopes under the influence of gravity (Cruden and Varnes 1996). Landslides exhibit uncontrolled downward and outward movement of slope materials and are major threat and ranked high among the hydrogeological hazards of mountainous region of India (NDMA 2019). Landslide is highly localized phenomenon and can have longlasting or immediate effect to the existing landscape. Statistics from The Centre for Research on the Epidemiology of Disasters (CRED) show that landslides are responsible for at least 17% of all fatalities from natural hazards worldwide. While about 8% to the total Global death toll by landslide devastations is recorded in India (Froude and Petley 2018). Landslide being most common and widespread geological hazards thus pose considerable threat to human life, infrastructure, dwellable and agriculture land and entire ecosystem in an area. In general, deep-seated landslides are mostly responsible for disrupting transportation network for longer duration while shallow slides are common in urban areas affecting population and settlements.

A combination of multiple terrain factors is responsible to cause landslides while rainfall and earthquake incidences are considered as immediate trigger. Human activities such as deforestation or excavation of slopes for laying road, railway networks and building sites, etc., have become important triggers for landslide too (Dai et al. 2002). The scenario is getting intensified due to unscientific anthropogenic activities coupled with change in rainfall intensity and pattern. The term "landslide" encompasses events such as rock falls, topples, slides, spreads, and flows (Varnes 1958). As explained, the occurrence of landslide in an area is dependent on a combination of predisposing terrain factors i.e., endogenic forces and triggering factors or exogenic forces. Hence to understand type of landslide prevailing in an area, it is important to know landslide classification system and nomenclature.

10.1.1 Types of Landslides

Landslides are classified by the types of material involved and the types of movement executed by the slope material. Additional criteria are the rate of movement and the water content of the material which are characteristically related. Landslide rate of movement ranges from few centimeters over many years to many kilometers per second. Material can be rock or soil or both with or without manmade features, trees etc. Debris contains pieces of rocks and earth, which is primarily composed of sand, silt, and clay. Debris may also include other components, such as trees and construction materials. There are three types of movements, namely fall, in which loose material falls or bounces down a slope; flow, in which material is distorted and moves as a fluid; and sliding, in which the landslide involves relatively undeformed material moving along a discrete surface. Flows can be further classified by their rate of movement. Very rapid flow down a steep slope (usually involving very wet material) is an avalanche, and extremely slow flow deformation is known as creep. Keller (1987) has slightly modified Varnes' classification and has divided landslides into two groups based on the patterns of movements. They are translational and rotational movements. In translational sliding, materials move on a flat surface along geological planes of weakness within rock slopes. Common translational slip planes include fractures, bedding planes, clay partings and foliation planes of insitu rocks. Rotational slides are most common for soil slopes but are also associated with slumping in rock slopes especially where a permeable rock overlies a weak rock (Keller 1987). However, Varnes' classification is still the widely referred landslide classification system (Fig. 10.1) and useful mostly as a tool in evaluating how and why landslides occur in an area. In jointed rock slopes wedge failure is very common. Further, a landslide event may be a sudden movement as a result of brittle deformation post-earthquake or a steady movement over an irregular slip surface due to ductile deformation or combination of both. It is observed that deep and shallow landslides with either rapid or slow movements exhibit different mechanisms. Usually slow/ductile movement of deep-seated landslides have an incubation period, a triggering event resulting in to an initial movement and ultimately leading to the primary failure. However, landslides in nature may be more complex than is understood by theoretical representations. Quantitively, slope stability is evaluated by computing safety factor, defined as the ratio of the resisting forces to the driving forces. Over a period of time the Factor of Safety may change due to various triggers and the initial landslide event may become an episodic one and remain active for few decades.

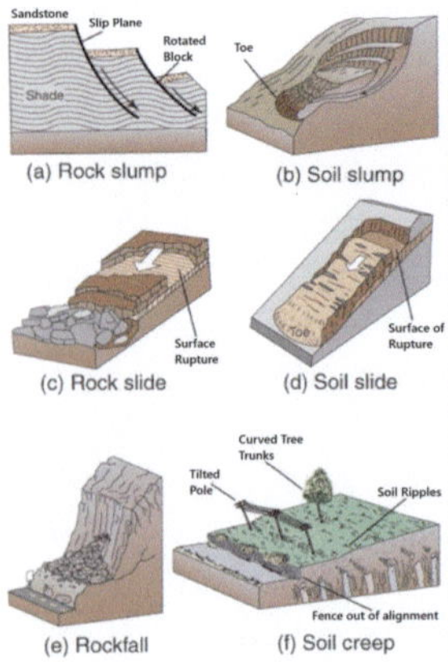

Type of Movement	Materials	
	Rock	Soil
Landslides with variable water content and rate of movement	Rotational	
	Slump(a)	Slump(b)
	Translational	
	Rock slide(c)	Soil slide (slip)(d)
Falls	Rock fall(e)	Soil fall
Flows — Slow	Rock creep	Soil creep(f)
	Unconsolidated rock and soil (saturated)	
Rapid	Earth flow(g)	
	Debris flow / mud flow(h)	
Very rapid	Debris avalanche(i)	
Lateral spread	Rock(j)	Soil
Subsidence	Rock(k)	Soil
Complex	Combination of slides, slumps, and flows(l)	

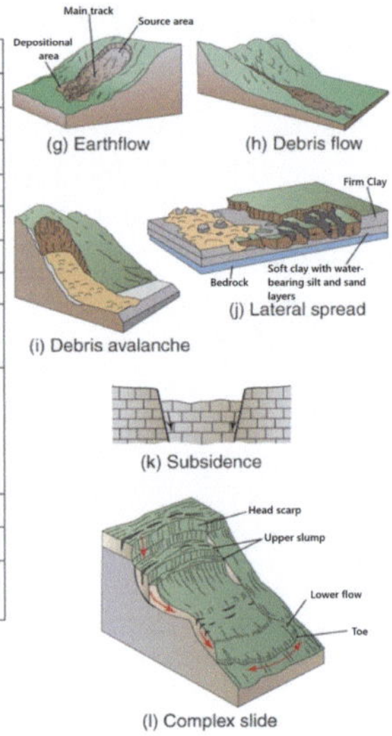

Fig. 10.1 Types of landslides (after Varnes 1985)

10.2 Impacts of Landslides

Landslide being one of the major geo-hazards globally have severe and far-reaching impacts on every sphere of human lives in terms of both direct and indirect consequences and in India have very serious impacts in socio-economic, environmental, and infrastructural sectors every year. Loss of human life and livestock, injuries, psychological trauma to affected people, damage to houses and property as well as habitable areas, damage to infrastructure (roads, bridges, railway lines, etc.), including critical infrastructure, loss of agricultural and farmland, and environmental degradation are some of the direct impacts on human lives, economies, and entire ecosystem. India is one of the most landslide-prone countries in the world about 0.42 million km^2, covering nearly 12.6% of the land area of our country spanning over mountain slopes of the North Western Himalayas and the Sub-Himalayan terrain of the North-East and Western-Eastern Ghats, is prone to landslide hazard, posing a threat to life and livelihood (GSI 2022). Further, in India, the problem of landslide trigger is very complex as landslides in the country are caused due to varied geological set up and hydro-meteorological factors, for example, the entire Himalayan region, Western Ghats, and the North Eastern States have steep slopes, experience heavy rainfall mainly due to monsoon and undergoing active deformation due to tectonic activities. The mountaneous region of India is susceptible to various types and magnitude of landslides and accordingly GSI (2022) and NDMA (2009 and 2019) has categorized different States of India in to various classes as per the landslide vulnerability. In June 2009, NDMA released the National Disaster Management Guideline on Management of Landslides and Snow Avalanches formulated in consultation with the Nodal Ministry/Agency (MoM/GSI) and other core group members from concerned Central, State departments and academia, laying down national policy for the management of landslide related hazards in the country.

Hilly states of North Eastern Region (NER) of India is one of the worst landslides affected regions. Almost entire region spanning over Arunachal Pradesh, Assam, Manipur, Meghalaya, Mizoram, Nagaland, Sikkim and Tripura is susceptible to various degrees and types of landslides except the plain areas of the Brahmaputra and Barak Valleys of Assam and valley portions of Manipur and Tripura. Debris slide and debris flow are among the most critical landslide hazards prevailing in NER especially along the various communication routes. Active Seismo-tectonics, rugged topography, complex lithology, prolonged as well as intense rainfall coupled with unscientific anthropogenic activities at places make NER susceptible to landslides of various magnitudes. Tectonically, the region is one of the seismically active regions in the world and falls in high seismic risk zone, (Zone V), in seismic zoning map of India (BIS 2002) and has 5 distinct seismotectonic zones (Nandy and Dasgupta 1991). Seismically the NER is one of the most active regions in the world due to collision tectonics in the Himalayas to the north and subduction tectonics in the Indo-Burma zone to the east. Physiographically, NER of India is the part of Extra-Peninsula i.e. the mountainous region of the Mighty Himalayan Ranges and its extensions with folded and over thrust mountain chains. Geologically, NER represent complete stratigraphic sequence with distinct geological domains representing rocks from Archaean to Recent (Nandy 2001). North-East India is the rainiest region in the country and receives much higher rain than the average of 1,000 mm for the whole country (Dikshit and Dikshit 2014). In an average the hilly areas of the region receive 2,000–3,000 mm of rain yearly. As a consequence, NER is susceptible to both rainfall- and earthquake-triggered landslides and has a great impact on Developmental Disability Index (DDI) derived by Govt of India (GoI 2013). Though earthquake triggered landslides have been reported in the region (Oldham 1899; Poddar 1950) but rainfall is largely responsible for triggering majority of the landslides (GSI 2022).

Earliest reported landslide study from NER dates back 1941 along NH39 approx. 66 km from Dimapur, Nagaland (Sondhi 1941). Respective State Emergency Operations Centres (SEOC) under State Disaster Management Authority (SDMA) keep the records of such incidences. According to the SEOC, Mizoram, between May and July 2025, a total of 846 landslide events were reported across the state. In Aizawl district alone, 127 landslide events destroyed 22 houses, and road blockages were reported at 119 locations. In August 2024, more than 2,032 landslide events were reported across the state by the SDMA, Tripura, due to an extreme rainfall event. Similarly, extreme rainfall during May 2022 triggered more than 5,000 landslides of various magnitudes across the Dima Hasao district of Assam. On 1st June 2025, intense and continuous rainfall triggered a disastrous landslide near a military camp in Chaten, North Sikkim, killing three army personnel and leaving six others missing. A similar incident was reported on 12 September 2025 from Upper Rimbi, Gyalshing, West Sikkim, killing four people and leaving several others missing. In October 2021, a debris slide triggered by glacier melt (G092671E27916N) led to a change in the course of the Warriyang Bung River by more than 100 m upstream. Downstream, due to the dumping of slide debris, the water of the Kameng River became highly turbid; as a consequence, mass fish mortality was reported from Arunachal Pradesh.

10.3 Space-Based Approaches to Landslide Management

In India, different organizations, and academic institutions are engaged with investigating different aspects of landslide, namely, inventory mapping, susceptibility-hazard-risk mapping and assessment, geo-technical investigation, landslide monitoring and forecasting, capacity building and awareness creation, landslide stabilization and management etc. following conventional or innovative techniques to minimize its devastating effect to the society. Experts from various thematic domains, namely, geologists, civil engineer, remote sensing experts address landslide differently and thus assessment of terrain's susceptibility conditions are represented either quantitively (in terms of Factor of Safety) or qualitatively (in terms of zones, inventory maps etc). In India, landslide studies were initiated in the year 1880, with R.D Oldham's study of the Nainital Landslip. National Disaster Management Authority (NDMA) has identified various Organizations to address different segments of studies using either ground based and/or space-based techniques. The Geological Survey of India (GSI) is identified as the nodal agency for conducting landslide-related investigations in India, while the NDMA has recognized NRSC/ISRO to carry out landslide studies using space-based techniques (NDMA, 2019). The first landslide inventory in the Himalayan terrain was carried out by Lakhera (1982), using RS and GIS techniques. The strength of Earth Observation (EO) data of various spatial, spectral and temporal resolutions for carrying out different aspects of landslide study have been demonstrated by eminent researchers worldwide and India. Various space-based sensors and satellite observations provide probable information on landslides, serving as an alternative to conventional field methods (Table 10.1). Very high-resolution imagery has become the best option now for landslide mapping and number of operational sensors with similar characteristics is growing year by year (van Westen et al. 2008). Champati Ray (2016) evaluated the cost-effective methodology for landslide hazard zonation using RS and GIS. The importance and limitations of optical (both mono and stereoscopic), SAR, LiDAR remote sensing data with varied spatio-temporal resolutions for landslide mapping is elaborately explained by Guzzetti et al. (2012).

The technology is gaining increasing popularity as it is cost effective and less time consuming, in addition availability of Very High Resolution (VHR) images either free or at affordable cost and platform where it is easy to integrate data from various sources. Different techniques are used for (1) detection and classification of landslides basically to prepare landslide inventory maps, (2) monitoring landslide movement and (3) analysis and prediction of slope failure to model landslide susceptibility. However, in most of the cases selection of RS data and technique is determined by the scale and aim of study apart from location, size, age and exposure condition of landslide as each remote sensing techniques has unique advantages and applications.

Table 10.1 Remote Sensing Data for Landslide studies

Applications	Satellites and Sensors		Conventional Methods/ Ground Sensors	Remarks
	Indian	Foreign		
Landslide Inventory	Resourcesat-2/ LISS IV MX data as well as priced and archive data from Cartosat and Resourcesat series.	Sentinel-2 and other high-resolution priced and archive data from WorldView-3/2; GeoEye-1, SPOT 6 / 7, Pleiades etc	Field surveys, GPS mapping, photographic records	Landslide Inventory Maps at various scales.
Landslide Susceptibility/ Hazard Mapping	Resourcesat-(LISS III, IV Mx) data, Cartosat-3 MX data; and CARTO DEM	Sentinel-2, ASTER, Landsat series, DEM from SRTM/ASTER/ ALOS PALSAR Missions and various high-resolution priced data.	Field mapping, Geotechnical investigations, slope stability analysis, soil/rock testing	• RS derived Landslide Suscepti-bility/Hazard Zonation Maps smaller than and up to 1:5000 scale. • Free data are mostly suitable for 1:25,000 and smaller scale of mapping.
Kinematic Mapping/ Landslide Monitoring	NISAR (future)	Sentinel-1 (Free data), ALOS PALSAR, TerraSAR-X etc	Ground-based monitoring, inclinometers, total station measurements, IoT based sensors etc	• Categorisation of stable, slow-moving, and rapidly moving landslides and direction of movement. • Amrita University and IIT Mandi, are leading institutions in India to use ground-based sensors
Rainfall data for Landslide Early Warning System	INSAT 3D/3DR/ 3DS HEM rainfall product, Model derived information from MOSDAC-ISRO, NESAC (only for NER) etc.	Global Precipitation Measurement (GPM)	AWS, ARG or data from IMD	• GSI is currently providing district wise alert. • SILAAS project of NRSC/ISRO.

Remote sensing has emerged as an effective tool for landslide detection, monitoring, and risk assessment due to its ability to provide multi-temporal, and high-resolution data suitable for large-scale study.

10.3.1 Landslide Inventory Mapping

Landslide inventory map refers to the process of systematic identification, mapping, and characterization of past and present landslides in a given area. A landslide inventory map records key information on location, type, date and time of occurrence thus making it easier to identify patterns and trends over time (Guzzetti et al. 2012). An inventory map is prepared by studying details of past and present landslides partly with the aid of EO data, field visit of landslide affected sites and in consultation of records available with District/State Disaster Management Authorities. The final product is landslide map showing the spatial distribution of landslides, represented as point or polygon information at desired scale. Broadly landslide inventory database is categorized as historical and geomorphological inventory (Fig. 10.2). EO data of various spatio-temporal resolutions is found to be very helpful in mapping of Geomorphological Inventory.

A landslide event usually alter landscape and left discernable traces in an area in terms of change in relief, slope gradient, loss of vegetation and exposure of soil and rocks, change in land use pattern etc. By applying appropriate techniques, namely visual interpretation, semi-automated object-based classification or applying deep learning model etc. on pre- and post- event images various characteristic changes in spectral and spatial features are identified and mapped. The morphologic and

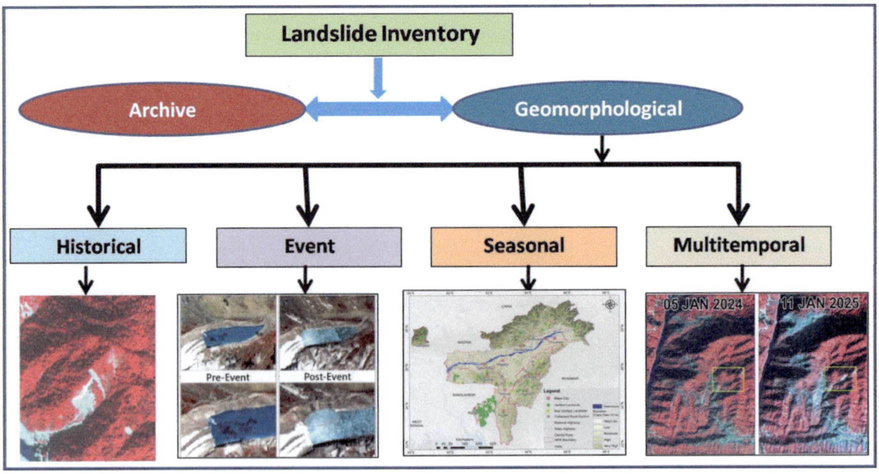

Fig. 10.2 Classification of landslide inventory (After Guzzetti et al. 2012)

radiometric signatures are exploited by geomorphologists to detect, map and classify landslides, both in the field and on remotely sensed imagery (Guzzetti et al. 2012). Landslides of different types and magnitudes can show different radiometric signatures. Deep-seated landslides usually show a more pronounced morphological signature than shallow landslides. An inventory map forms the essential foundation to understand landslide processes active in an area, for landslide susceptibility and hazard mapping, risks assessments, design of appropriate mitigative measures and future developmental activities, developing an active early warning system etc.

10.3.1.1 Case Studies

The availability of date- and time-tagged landslide inventory information for NER is still insufficient as landslide incidents are not instrumentally recorded. Hence, under the ongoing activities of NERDRR program of NESAC, Umiam compilation of reported landslide incidents is being carried out from print and online media immediately after the event is reported. A total 1137 landslide incidents have been documented for the year 2024 from the different states of the NER, mainly triggered by rainfall. For 2025 total 374 events are catalogued. However, these database are not inclusive of all landslide events that might have occurred in a particular year. Subsequently, these locations are verified in terms of precise latitude and longitude either through field survey or from post-dated EO/UAV imagery. This database is available at https://www.nerdrr.gov.in/LandslideDSS/.

On 26 March 2023, a landslide occurred near Sokpay village, Gangtok, Sikkim, destroying four houses, while residents narrowly escaped. Approximately 18 families were relocated, and essential services were disrupted. The run-out distance exceeded 2 km from the crown, with debris deposited into the Teesta V reservoir. Cloud-free pre- and post-event optical data enable effective delineation of the landslide extent and run-out path (Fig. 10.3). During pre-monsoon season of 2022, between May 9th to 18th, Dima Hasao region of Assam has experienced series of landslides across multiple locations. The district has received a rainfall of 672.1 mm (March to May 2022) which is reported to be an above-average record of 414.6 mm (GSI, 2023). While 4 days average rainfall was more than 250 mm that resulted in to a major mudflow event which nearly buried a passenger train and platforms of the New Haflong railway station being specifically reported on May 16th. This incident highlights the effectiveness of UAV imagery for rapid identification of landslide extent, run-out distance, and damage to infrastructure and settlements (Fig. 10.4) in the absence of cloud-free optical data during urgent damage assessment. Although variability in SAR backscatter signatures limits the standalone use of SAR GRD data for landslide inventory mapping, SAR GRD analysis and DInSAR processing of Sentinel-1 SLC data provide effective alternatives for delineating landslide-affected areas in the absence of cloud-free optical imagery. For the Kuliang landslide, which caused significant traffic disruption with two major events reported in 2024, Sentinel-1 GRD data were employed to identify debris accumulation along the roadway where the estimated run-out distance is approximately 440 m from the crown (Fig. 10.5).

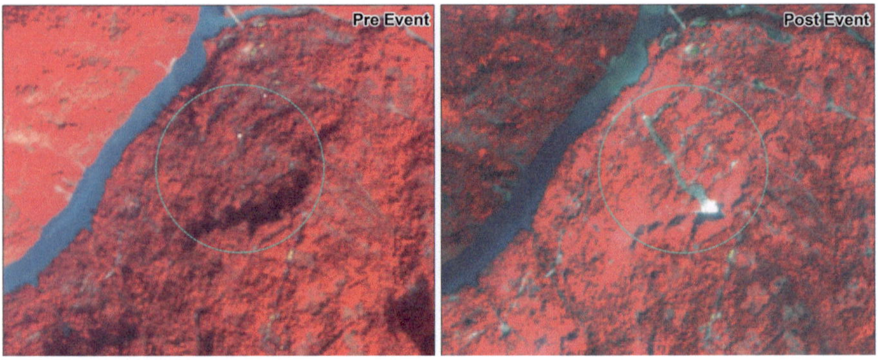

Fig. 10.3 Resoursat 2 (LISS IV) data of pre (19 Dec 2022) and post (7 June 2023) event showing Sokpay landslide in East Sikkim

As part of the national projects executed by NRSC, ISRO, seasonal landslide inventory maps are periodically prepared using pre- and post-monsoon imagery for landslide-affected states of India, including the NER. The database is available at https://bhuvan-app1.nrsc.gov.in/disaster/disaster.php?id=landslide_monitor.

The systematic inventory thus compiled helps in assessing the risk in terms of yearly loss of life and property, disruption of transportation network, frequency and magnitude of landslide incidents in an area and NER as a whole. Immediate/same day cataloguing of landslide incidents enable the inventory maps to be updated in real time and advances the understanding of temporal dynamics of landslides in the region (Sarkar et al. 2020; Martha et al. 2013).

10.3.1.2 Earthquake Induced Landslides

Earthquake-induced landslides are one of the major secondary hazards that not only results in loss of life and property but also significantly affect the post-earthquake rescue process in hilly areas. Earthquake-induced landslides can either be Co-seismic; basically, freshly triggered and/or reactivated landslides or post-seismic-landslides which occur after the earthquake event has passed but owes its origin to fissures, cracks, deformations induced by the earthquake event. According to seismic zoning map of India, major portion of India's highest seismic risk zone, Zone V, falls along Himalayan collision zone to the north and subduction zone along the Indo-Burma border to the east. Thus these regions are much susceptible to both earthquakes and landslides. The 2,500 km long Himalayan collision zone is not uniform in seismicity pattern or in tectonic fabrics (Kayal 2010), so with the geology and slope conditions (Valdiya 1987). As a consequence, landslides (both rainfall and earthquake induced) are major problems for two Eastern Himalayan states, Arunachal Pradesh and Sikkim as well as other hilly states of NER and in the hilly states of Northern and Western Himalaya. In general, research related to the earthquake-induced landslides starts with preparation of co-seismic landslide inventory. Total

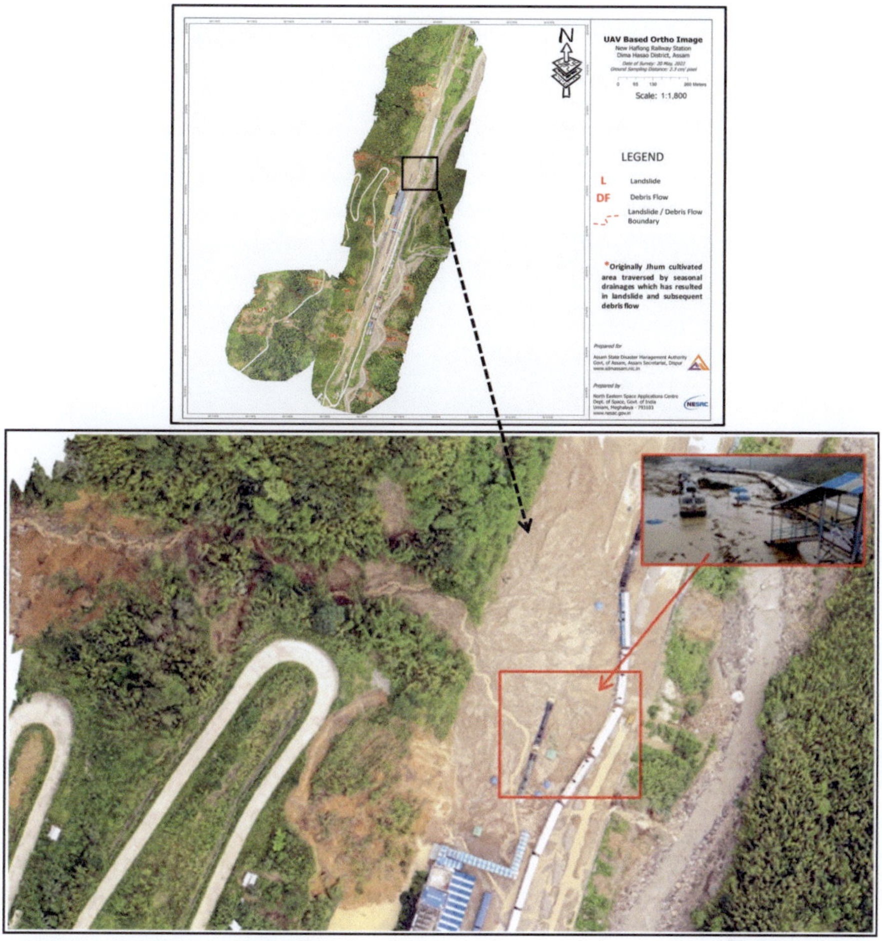

Fig. 10.4 UAV image showing Landslide affected areas near New Halflong Railway Station Dima Hasao District, Assam. Event Date: May 16, 2022. (Inset image source: www.indianexpress.com)

1,196 landslides are mapped within an area of 4,105 km^2 using RS data (Martha et al. 2010) those were triggered by the September 18, 2011 earthquake Mw 6.9 in Sikkim. For preparation of co-seismic landslide inventories differences in colors, textures and terrains between the pre- and post-earthquake images are considered as primary characteristics. In the Fig. 10.6, geo-environmental set up of the NER is shown. NER is crisscrossed by numerous major faults and tectonic features namely, Main Central Thrust (MCT), Main Boundary Thrust (MBT), Himalayan Frontal Thrust (HFT), Indus Tsangpo Suture Zone (ITSZ), Mishmi Trust (MT), Lohit Thrust (LT), Eastern Himalaya Syntaxis (EHS), Jiali Fault (JF), Yemla Fault (YF), Canyon Thrust (CT), Naga Thrust (NT), Disang Thrust (Dis T), Churachandpur Mao Fault (CMF), Mat Fault (MF), Sylhet Fault (Syl F), Kopili Fault (Kop F), Oldham Fault (OlF), Barapani

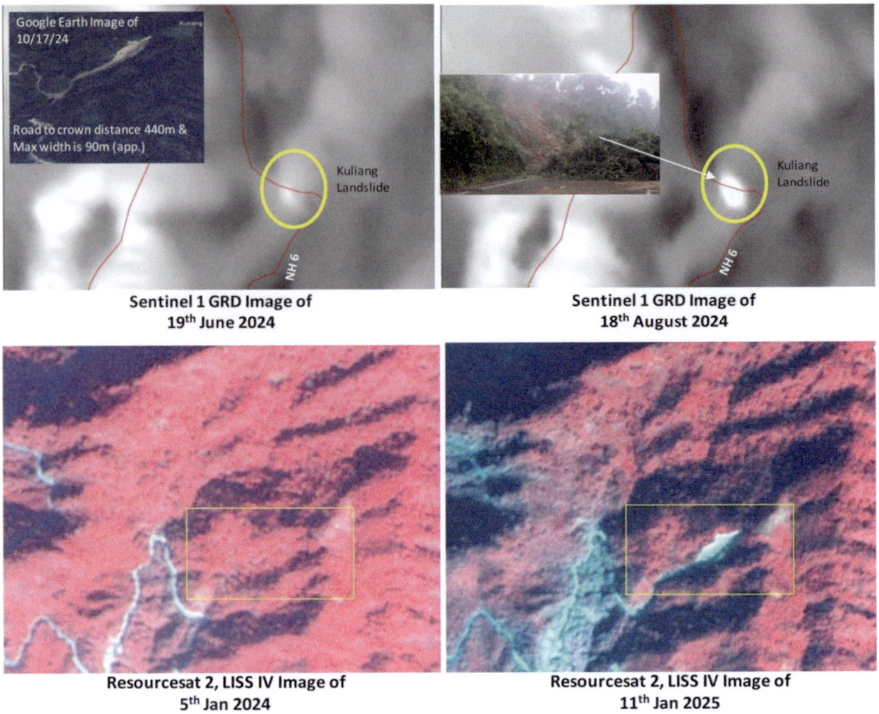

Fig. 10.5 Landslide near Kuliang village, along NH 6, East Jaintia Hills, Meghalaya. Event Date: 6th and 12th August, 2024

Shear Zone (BSZ), Dapsi Thrust (DT), Dauki Fault (DF), Dhubri Fault (Dhb F) etc. The region has experienced two great earthquakes of Mw > 8.0 (largest red stars), large earthquakes of Mw > 7.0 (red circles) and strong earthquakes of Mw 6.0 - < 7.0 (smaller red stars) as shown in the figure.

Post-seismic landslides are observed to be a major cause of recurrent hazards in many regions. Landslides initially triggered by earthquakes often remain active for years and are prone to reactivation, particularly during monsoon seasons. Temporal optical images and time series analysis of SAR data are the best option for regular monitoring of these episodic landslides. Landslide near Sonapur village, East Jaintia Hill, Meghalaya triggered by the 1984 Cachar earthquake of Mw 5.8 (Porwal and Dey 2010) is still an active landslide (Fig. 10.7). The figure shows image view of the slide during March 2005 (A) and December 2022 (B), field view of the slide captured during 2003 (C) and 2024 (D), while at inset 3D view of the terrain is seen (E). More than 10 instances debris slide/flow were reported from this slide during 2023. However, co-seismic landslide also attains stability over a period of time, for example, the first reported seismic induced landslides in Arunachal Himalaya triggered by the 1950 great earthquake (Mw 8.4) being documented by Poddar (1950). 3D Visualization of the old landslide helped in assessing the present condition of the area.

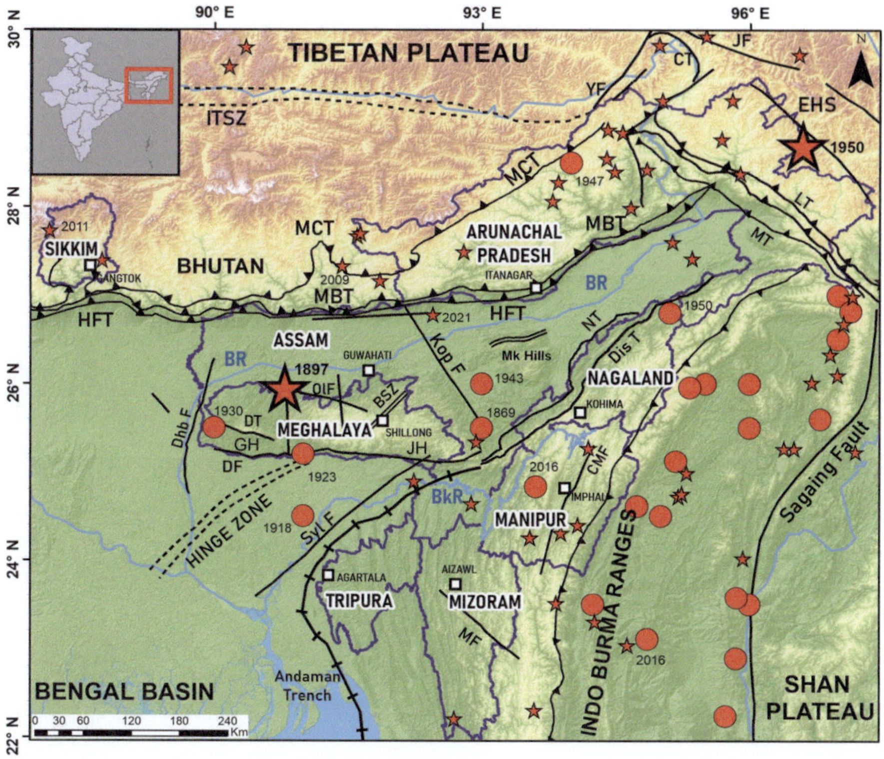

Fig. 10.6 Geo-environmental set up of the NER, prominent reference features are: BR: Brahmaputra River, BkR: Barak River

A seismic-induced landslide can also trigger secondary and cascading effects. One such example is the formation of landslide-induced natural lakes in the Arunachal—EHS zone (Fig. 10.8). Episodes of moderate to strong earthquakes (Mw 5.0–6.0) in this zone frequently trigger landslides, leading to the formation of natural dams and water impoundment. The complex network of faults and thrusts, combined with high-altitude mountainous terrain, makes the region particularly fragile and susceptible to landslides. Sudden breaches of landslide dams located upstream of the international boundary with India have caused massive floods downstream along the Siang River in Arunachal Pradesh, with such events reported since 2000. Earth observation (EO) data remain the primary tool for studying and monitoring these events. Similar incidents have been reported in Sikkim, including the August 2016 landslide near Mantam village in North Sikkim, which blocked the Kanaka River and created a large lake, and the October 2023 glacial-lake outburst at South Lhonak Lake in North Sikkim, triggered by a landslide or collapse of a massive lateral moraine.

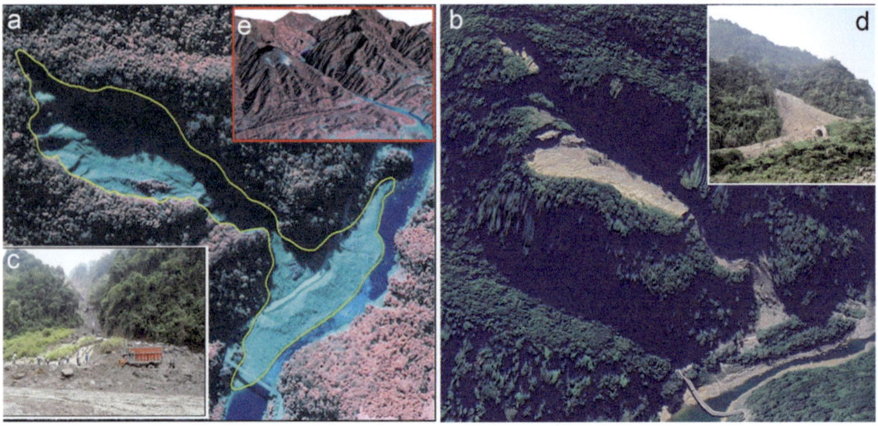

Fig. 10.7 Landslide location near Sonapur village, East Jaintia Hill, Meghalaya

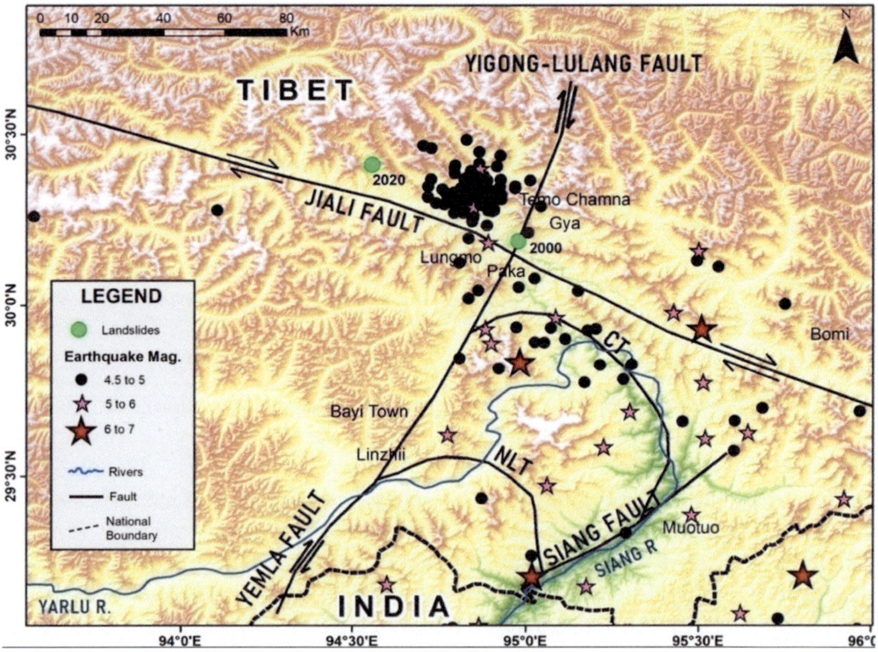

Fig. 10.8 Location of landslides and natural dam in EHS the WNW-ESE right lateral Jiali Fault. Location of clustered earthquake epicenters and geological structures are after B. Mukhopadhyay, and S. Dasgupta, 2015. YLF: YigongeLulang strike slip fault; YF: Yemla Fault; CT: CanyonThrust

10.3.2 Landslide Susceptibility Mapping

Landslide susceptibility mapping is the process of determining the probability of future landslide events in a given area. The susceptibility condition of terrain is evaluated either by Direct Mapping of landslide (active, old, potential areas) and/ or associated mass wasting features. By deriving information on causative factors (various thematic conditions of the terrain) of landslide and their integration at GIS platform, a landslide susceptibility map of an area is carried out using EO data aided by limited field surveys. A 'black-box' model of slope instability is based on the assumption that the factors responsible for past slope failures in a region will also govern future landslide occurrences. A Landslide Susceptibility Map (LSM) assesses only the likelihood of landslide occurrence under given terrain conditions, whereas a Landslide Hazard Zonation (LHZ) map additionally incorporates the probability of triggering events (e.g., rainfall, earthquakes) and establishes relationships between the magnitude and return period of such events and landslide occurrence, either through historical event analysis or physical modeling. Globally, between 1990–2000, focus was on preparation of inventory and Susceptibility/ Hazard zonation maps along with development of various statistical and heuristic models to achieve that. Landslide Susceptibility assessment either may be site specific or regional, accordingly mapping scale also varies. While NDMA has guidelines on scales ranging from < 1:100,000 to > 5000 of LSM depending upon the zoning application (NDMA 2019). Once developed, a Landslide Susceptibility Map (LSM) can support spatially informed developmental planning, quantitative risk assessment by integrating the vulnerability of lifeline infrastructure (Fig 10.9), and the design and implementation of landslide early warning systems.

The purpose of Hazard, Vulnerability and Risk Assessment (HVRA) is to help a community to make risk-based choices to address vulnerabilities, mitigate hazards and prepare for response to and recovery from hazard events. Risk assessment helps in guiding mitigation strategies such as land-use planning, stabilization measures (engineering and bioengineering interventions including both structural and non-structural approaches), and the devlopement of early warning systems. Preparedness plans are also developed, focusing on evacuation procedures, public awareness, and infrastructure resilience (Cao et al. 2021). Landslide risk assessment is the most important way to prevent loss of life and property damage. It helps the government, researchers, and communities in making informed decisions regarding land development, construction, and disaster management. It also helps in resource allocation for mitigation measures in high-risk areas through quantitative risk estimates. Risk assessment map available at GIS environment also helps in validating using post-assessment inventory data and periodic updation. Under the National Landslide Risk Reduction and Mitigation Programme (NLRMP) of NDMA (2025) thrust is also given on holistic landslide risk assessment and mitigation. The current trend is to develop Dynamic Landslide Susceptibility Mapping (DLSM), which evaluates the likelihood of landslide occurrence across space and time by integrating static

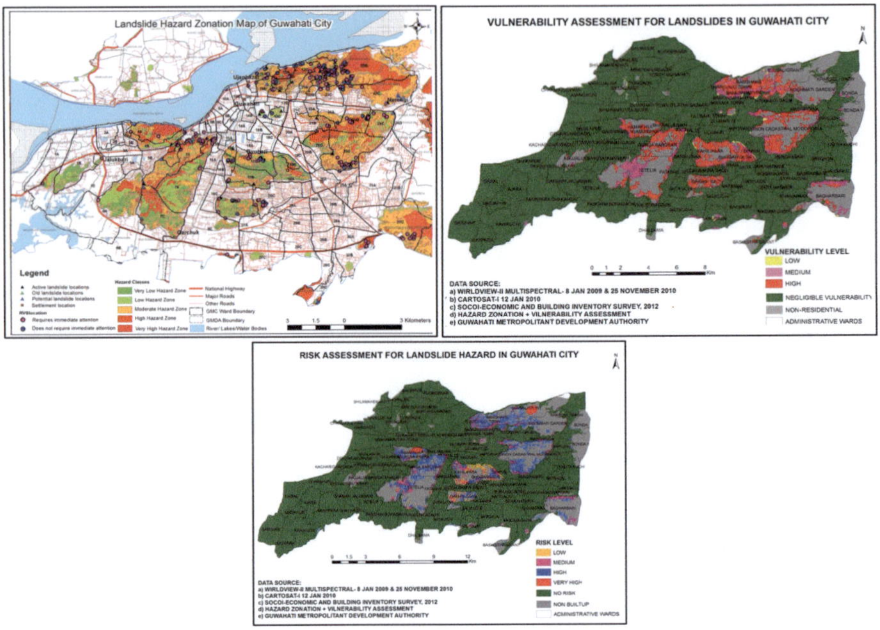

Fig. 10.9 Landslide hazard zonation, risk and vulnerability assessment carried out for Guwahati city, Assam with 2010 landslide database

terrain factors with time-varying triggers—such as rainfall, soil moisture, and ground deformation—using remote sensing-derived inputs and various AI models.

10.3.3 Landslide Monitoring and Early Warning System

Landslide Monitoring and Early Warning System, is ideally designed to detect slope instability, assess potential landslide hazards, and provide timely alerts to reduce risk to life and property. The framework integrates remote sensing and modeling approaches with field measurements to monitor landslide-prone areas continuously. Currently, thrust is to integrate kinematic behaviour of landslides to issue early warning for rainfall triggered landslide(s). To understand kinematic behaviour of landslide continuous monitoring either using ground-based (IoT based hardware and sensors like soil moisture sensor, rain gauge, pore pressure sensor, vibration sensors accelerometer sensor etc.) and/ or space-based (time series analysis of SAR data) system is used. However, the country's current landslide and early warning systems are unable to connect to the last mile (NDMA 2019). Organizations like IIT Mandi, Central Building Research Institute (CBRI), AMRITA University, and Defence Terrain Research Laboratory (DTRL) have taken promising initiatives for developing ground-based technology for landslide monitoring and early warning.

With respect to the space-based approach, only slow-moving landslides can be monitored using Sentinel-1 SAR data, which have a 12-day revisit interval. The combined Line-of-Sight (LOS) measurements from the ascending and descending passes of Sentinel-1 are sensitive to east–west deformation; however, the capability to monitor north–south displacement is limited because landslide movement commonly occurs along the satellite's azimuth direction. However, Synspective Inc. plans to deploy a constellation of 30 high-resolution SAR satellites with daily revisit capability. Such frequent coverage would significantly enhance operational landslide hazard monitoring by enabling near real-time detection and tracking of fast to very fast-moving landslides, supporting timely early warning and risk mitigation.

Early warning of landslides in India requires monitoring of both primary triggers: intense rainfall and seismic activity. In this regard, deterministic seismic hazard and landslide hazard zonation of Arunachal Pradesh is carried out by Anand et al. (2022). For monitoring the earthquake- or vibration triggered landslides, bore hole seismometers can be used as a part of wireless sensor network (WSN) based ground instrumentation and real time monitoring system. However, in India such systems are yet to be implemented operationally. Rainfall Threshold-based Landslide Early Warning Systems (LEWS) assess rainfall conditions relative to past landslide events to determine critical rainfall intensity and duration required to trigger slope failure. The rainfall threshold defines the minimum conditions necessary to initiate a landslide in a given area. Such empirical systems have been implemented in several countries, including Korea, Hong Kong, the United States, Canada, and Norway. LEWS are designed to model landslide occurrence and provide advance warnings, and can vary in scope depending on landslide type, target area, and the population at risk. They may be deployed for individual landslides using movement sensors, for small areas using rainfall thresholds, or for large regions based on weather forecasts. The Geological Survey of India (GSI) has launched a district wise landslide forecasting and early-warning capability, which can be access through https://bhusanket.gsi.gov.in. The Satellite Integrated Landslide Assessment and Alert System (SILAAS) project initiated by NRSC, ISRO, also aims to create a state-of-the-art system for predicting and monitoring landslides to reduce disaster risks by mapping, analyzing terrain, and providing early warnings for landslide-prone areas across India using RS and AI techniques. The system developed for Idukki, Darjeling and Rudraprayag districts is currently under validation. In similar line an experimental alert system is developed for Meghalaya which is also under validation phase by SDMA, Meghalaya. This model takes into account both antecedent rainfall condition and rainfall forecast for the next 24 hours. Landslide susceptibility classes and various terrain parameters, information derived from time-series analysis of Sentinel 1 data are integrated with forecasted rainfall information to delineate probable vulnerable zone(s) where a landslide event is likely to occur.

Despite its potential, several challenges still exist in the practical implementation of an effective EWS for landslide forecasting. However, at many places available models are enhanced by incorporating local factors, such as soil type, soil moisture fluctuation, land use and human activity, seismicity pattern etc which might influence landslide susceptibility along with ground-based monitoring systems.

10.4 Conclusions

The landslide scenario in the North Eastern Region (NER) illustrates the significant challenges to society and developmental activities. Only a few examples are presented here to highlight that each state in NER is susceptible to diverse landslide types of varying intensity, triggered by rainfall, earthquakes, or a combination of both. Landslides in the region disrupt livelihoods, damage infrastructure, and threaten biodiversity. Moreover, landslides can be episodic and remain active for decades specially during monsoon season. Siltation of river due to landslide, as well as damming of rivers resulting impoundment of water and subsequent dam burst also pose indirect threat to the society. Breaching of a landslide dam is time independent phenomenon either occurs immediately or takes year's depending upon the dam material, dam geometry, and characteristics of the upstream catchment area.

For inacessisble terrain of NER, EO data plays a critical role in landslide inventory mapping, susceptibility and hazard assessment and monitoring . Multi-temporal optical and SAR imagery are used to detect slope changes, delineate landslide boundaries, and map debris flow paths. Pre- and post-event imagery allows identification of new or reactivated landslides, while high-resolution data supports analysis of terrain features, slope geometry, and land-cover changes. EO data can also be integrated with rainfall records, seismic information, and topographic models to assess susceptibility, model potential run-out paths, and support early warning systems. UAV surveys and LiDAR data complement satellite imagery for detailed local-scale mapping and monitoring.

A wide variety of landslide types, the lack of date- and time-tagged seamless landslide inventory data, challenges in understanding and modeling the interplay of rainfall in generating pore water pressure at different locations (depending on subsurface rock/soil types and overburden materials), deviations between forecasted (grid-averaged) and observed rainfall at landslide sites, and the revisit period of Sentinel-1 (a freely available SAR sensor) are some of the constraints that limit the development of an effective early warning system for rainfall-triggered landslides. In spite of all the existing challenges, the adoption of geo-spatial technologies in landslide research has significantly improved the understanding of landslide hazards and their effective management.

Acknowledgements The authors sincerely acknowledge the State Disaster Management Authorities (SDMAs) of all North Eastern States for their timely support and for providing essential data for this study. We also extend our gratitude to the Geological Survey of India (GSI), the National Disaster Management Authority (NDMA), and the National Remote Sensing Centre, ISRO (NRSC-ISRO) for the valuable information and resources utilized in this work.

References

G. Anand, A. Rahangdale, S.S. Mantri, S. Singh, S. Kolathayar, Deterministic seismic hazard and landslide hazard zonation of Arunachal Pradesh. J. Earth Syst. Sci. **131**(3), 187 (2022)
BIS, *Seismic Zoning Map of India* (Bureau of Indian Standards, 2002)
Y. Cao, C. Xu, N.M. Aziz, S.N. Kamaruzzaman, BIM–GIS integrated utilization in urban disaster management: the contributions, challenges, and future directions. Remote Sens. **15**(5), 1331 (2023)
R. Varne, Ancient subcontinental mantle: a source for K-rich orogenic volcanics. Geology **13**(6), 405–408 (1985)
D.M. Cruden, D.J. Varnes, Landslide types and processes. Transp. Res. Board Spec. Rep. **247**, 36–75 (1996)
F.C. Dai, C.F. Lee, Y.Y. Ngai, Landslide risk assessment and management: an overview. Eng. Geol. **64**(1), 65–87 (2002)
K.R. Dikshit, J.K. Dikshit, *North-East India: Land* (Springer, People and Economy, 2014)
F. Guzzetti, A.C. Mondini, M. Cardinali, F. Fiorucci, M. Santangelo, K.T. Chang, Landslide inventory maps: new tools for an old problem. Earth Sci. Rev. **112**(1–2), 42–66 (2012)
Geological Survey of India, *Landslide Hazard Zonation Maps* (2022). https://gsi.gov.in
F. Guzzetti, A.C. Mondini, M. Cardinali, F. Fiorucci, M. Santangelo, K.T. Chang, Landslide inventory maps: new tools for an old problem. Earth Sci. Rev. **112**(1–2), 42–66 (2012)
L.F. Hsiao, D.S. Chen, Y.H. Kuo, Y.R. Guo, T.C. Yeh, J.S. Hong, C.T. Fong, C.S. Lee, Application of WRF 3DVAR to operational typhoon prediction in Taiwan. Weather Forecast. **27**(5), 1249–1263 (2012). https://doi.org/10.1175/WAF-D-11-00131.1
J.R. Kayal, Himalayan tectonic model and the great earthquakes: an appraisal. Geomat. Nat. Haz. Risk **1**(1), 51–67 (2010). https://doi.org/10.1080/19475701003625752
E.A. Keller, *Environmental Geology*, 5th ed. (Bell and Howell Company, 1987)
R.C. Lakhera, *Geotechnical Studies of Bowala-Nandprayag Hydel Scheme* (Unpublished Report, IIRS, Dehradun, Chamoli, 1982)
T.R. Martha, N. Kerle, K.V. Kumar, Landslide detection using object-oriented methods. Geomorphology **116**(1–2), 24–36 (2010). https://doi.org/10.1016/j.geomorph.2009.10.004
T.R. Martha, N. Kerle, V. Jetten, C.J. van Westen, K.V. Kumar, Object-oriented landslide detection methods. Geomorphology **116**(1), 24–36 (2013)
D.R. Nandy, S. Dasgupta, Seismotectonic domains of northeastern India and adjacent areas. Phys. Chem. Earth **18**(1), 371–384 (1991)
D.R. Nandy, *Geodynamics of Northeastern India and the Adjoining Region* (Scientific Book Centre, 2001)
National Disaster Management Authority (NDMA), National Disaster Management Guidelines—Management of Landslides and Snow Avalanches (National Disaster Management Authority, Government of India, New Delhi, 2009)
National Disaster Management Authority (NDMA), National Landslide Risk Management Strategy (National Disaster Management Authority, Government of India, New Delhi, 2019)
National Disaster Management Authority (NDMA), National Landslide Risk Mitigation Programme (NLRMP) Phase-I (2025).https://ndma.gov.in
R.D. Oldham, Report on the great earthquake of the 12th June 1897. Memoirs of the Geological Survey of India, 29. Reprinted 1981, (Geological Survey of India, Calcutta, 1899), p. 379
M.C. Poddar, *Macroseismic Survey of the 1950 Assam-Tibet earthquake* (Geological Survey of India, 1950)
A. Porwal, A. Dey, Tunnelling through a highly slide prone area at Meghalaya, India. Geotech. Challenges Megacities, **3**, 1099–1105 (2010). Moscow
S. Sarkar, D.P. Kanungo, K. Krishna, Landslide inventory mapping and its utility. Landslides **17**(3), 689–703 (2020)
V.P. Sondhi, Note on landslips on the Dimapur–Manipur road, Assam. Unpublished report, Geological Survey of India (Acc. No. CHQ-17481) (1941)

K.S. Valdiya, Trans-Himadri Thrust and domal uplifts immediately south of collision zone and tectonic implications. Curr. Sci. **56**, 200–209 (1987)
C.J. van Westen, E. Castellanos, S.L. Kuriakose, Spatial data for landslide susceptibility, hazard, and vulnerability assessment: an overview. Eng. Geol. **102**(3–4), 112–131 (2008)
R. Varne, Ancient subcontinental mantle: a source for K-rich orogenic volcanics. Geology **13**(6), 405–408 (1985)
D.J. Varnes, Landslide types and processes. Landslides Eng. Pract. Highway Res. Board Spec. Rep. **29**, 20–47 (1958)

Kuntala Bhusan is working as a Scientist in North Eastern Space Applications Centre (NESAC), Govt of India, Dept of Space, Umiam, Meghalaya since 2003. She has coordinated several National, User funded, and R&D projects taken up by NESAC related to geoscience and landslide hazard. Currently, she is also contributing as the Project Manager for Landslide Hazard under North Eastern Regional Node for Disaster Risk Reduction (NER-DRR) program of NESAC. She has a couple of publications in national and international journals, conference proceedings and several technical reports in the related subject.

Dr. Prakash Biswakarma is a Geoscientist specializing in landslides studies with a Ph.D. in Slope Failure Analysis. Currently a Research Scientist at NESAC, Dept. of Space, he has over 20 peer-reviewed publications and experience across academia, research, and field applications. He is proficient in GIS, remote sensing, SAR tools, and has completed numerous advanced trainings in geospatial technologies.

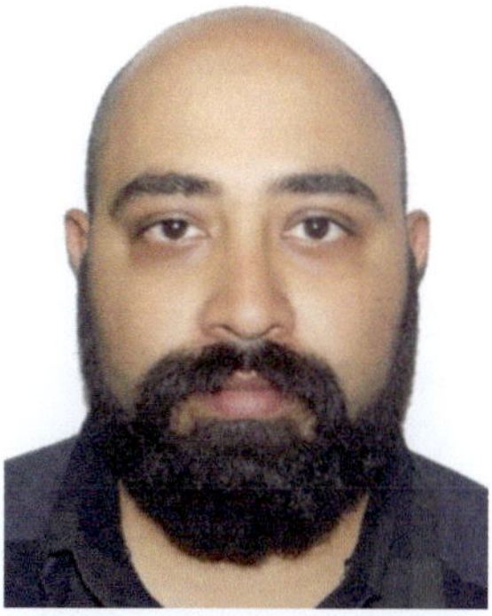

Aniket Chakravorty is a Scientist in the Space and Atmospheric Sciences Division at North Eastern Space Applications Centre, Department of Space, Government of India. He has a doctorate in land atmosphere interactions from the Indian Institute of Technology Delhi. He is working on understanding the changes in the climate of North East India with a special emphasis on extreme precipitation events. He has expertise in uncertainty quantification, time series analysis, and climate downscaling.

Praveen Kumar is currently with the Euro-Mediterranean Center on Climate Change (CMCC), working in the Global Coastal Ocean (GOCO) Division since September 2024. He develops cloud-based geospatial applications and supports decision support systems and oceanographic model monitoring. Previously, he served at NESAC, ISRO from 2019 to 2024, contributing to geospatial portals and disaster management tools. He holds degrees in IT and Computer Science, along with certifications in AI/ML and Geoinformatics from NIT Warangal and the University of Twente.

M Somorjit Singh is heading the Geoscience Division at the North Eastern Space Applications Centre (NESAC), Shillong, Department of Space, Government of India. In 20 years of service at NESAC, he has contributed to various operational and R&D activities/projects. He has been honored 'ISRO Team Excellence Award for his significant contribution as a team member in 'National Geomorphology and Lineament Mapping (NGLM) project' which was initiated by National Remote Sensing Centre (NRSC), Hyderabad in collaboration with Geological Survey of India (GSI) under Natural Resources Census program. His research interests include Geomorphology & Lineament Studies/Mapping, Infrastructural Planning (Site Suitability/feasibility Studies, road & railway alignment/realignment, Natural/Geo-hazard, Manmade Hazard (Industrial Hazard), Environmental Studies/Pollution (Heavy Metals) etc.

K. K. Sarma Ph. D, leads the Remote Sensing Applications Group (RSAG) at the North Eastern Space Applications Centre (NESAC), where he has been a key figure since 2002. Specializing in remote sensing for forestry and biodiversity, his work leverages space technology for critical applications in Biodiversity, Ecosystems, Forestry, and Natural Resource Management. Dr. Sarma has spearheaded over 60 diverse geospatial projects, contributing significantly to natural resource mapping and monitoring across the North Eastern Region (NER). Currently, he coordinates 110 "Plan of Action" projects, focused on using space technology to improve livelihoods in the NER. He also served as the Principal Investigator and NER Coordinator for all ISRO natural resource management projects, including comprehensive Land Use Land Cover Mapping and National Wasteland Updating, Monitoring, and Change Analysis. With over 75 research publications, Dr. Sarma has successfully supervised six Ph.D. students and actively mentors many others across NER Universities.

Chapter 11
Urban Planning in North Eastern Hill Towns of India Through Space Technology

B. C. Sumanth, Santanu Das, Jenita M. Nongkynrih, Surajit Sutradhar, and K. K. Sarma

Abstract The North Eastern Hill Towns of India are experiencing rapid, often unplanned urbanisation, compounded by complex topography and extreme climatic conditions, leading to severe issues such as overcrowding, traffic congestion, and environmental degradation. This paper explores the crucial role of urban planning and the transformative potential of space technology, particularly Remote Sensing and Geographic Information Systems (GIS), in addressing these challenges. These geospatial tools provide data-driven insights for comprehensive urban management, including terrain analysis (slope, aspect, viewshed), urban land use planning and change detection, strategic infrastructure development (water, energy, social services), enhancing traffic management and road safety, facilitating industrial zonation, and accurately identifying hazard-vulnerable zones like landslide and flood-prone areas. Case studies of the Shillong and Itanagar Master Plans highlight the practical application of GIS for sustainable development. The integration of urban planning, zoning, and space technology is vital for fostering resilient urban growth and improving the quality of life in these regions.

Keywords Urban planning · Geospatial technology · North East India · Remote sensing · GIS

B. C. Sumanth (✉) · S. Das · J. M. Nongkynrih · S. Sutradhar · K. K. Sarma
North Eastern Space Applications Centre (NESAC), Shillong, Meghalaya, India
e-mail: sumanth.bc@nesac.gov.in

S. Das
e-mail: santanu.das.nesac@nesac.gov.in

J. M. Nongkynrih
e-mail: jm.nongkynrih@nesac.gov.in

K. K. Sarma
e-mail: kk.sarma@nesac.gov.in

S. P. Aggarwal et al. (eds.), *Remote Sensing of the North Eastern Himalayan Ecosystem*,
https://doi.org/10.1007/978-981-95-3221-6_11

11.1 Introduction

Urbanization is the process of transformation, where the village culture is replaced by the urban culture (Jhawar et al. 2012). It is considered as one of the most important and complex issue in the human civilisation. This process of transformation, the movement among people from rural to urban in search of comparatively better livelihood is so rapid that there is a demand of quick response from the policy makers in the perspective of city and town planning. The migration closely linked to the social process of rationalisation, industrialisation and modernisation. The world is witnessing a rapid urbanization in the 21th century and the UN World Urbanization Prospect report has projected that there will be a rise of 60% of global urban population by 2030 (4.9 billion) from 3.2 billion in 2005 (Jhawar et al. 2012). Although this mode of transformation combined with hapazard planning and lack of proper urban management plan is overshadwoing the positive aspect of urbanization. Pressure on land is high in agro economies, where income is low and population increases in fourfolds unpalnned urbanization is inevitable. Thus it results in creation of bunch of urban problems and the quality of life is detoriarted. Problems like overcrowding, growth of slums, scarcity of water-supply, inadequate public health and sanitation and many more are very common scenerios of the towns and cities in the developing nation (Sudhir 2009).

The growth of urban region is an indicator of a country's overall development but if remain unchecked the outcomes are severe. This is where urban planning and management plays a vital role. According to Levy (2017) "*the need for planning comes down to two words, 'interconnectedness and complexity*'". Urban planning tackles the complex problem of resource allocation and decision making and it uses different methods to make progress in human and environnmental welbeing in an urban area (Jhawar et al. 2012). Thus urban planning is a key approach for transforming the idea of sustainable development into reality where variety of stakeholders need to be engaged. It also assists in allocating and managing the budget of the municipalities by channeling them to the required development of infrastructure and balacing them with the need to protect the environment (Raven et al. 2018).

11.1.1 History of Urbanization in North East India

The process of urbanization in North East India began after the British annexation. Although during the ancient period there existed few urban centres such as Pragjyotishpur and Bhishmak Nagar, but those cities were only for elites, termed as Regal Ritual cities (Dhar 2013). The regional uniqueness in terms of location, tradition and practices, land holding and land tenure pattern, economic system and migration are the reason that the process of urban growth is different from rest of India (Mukherjee et al. 2013). During the British period cities or towns developed within the state capital and district headquarters and notably there were no class 1 cities till 1951 in the region. Introducton of Tea Plantation, coal and oil mining were the determining factors for the development of the urban centres in the region (Singh and Singha 2020). It was because of these reasons there was a growth of small and medium sized town in the region. Most of the hill towns emerged as administrative outpost of British Hill Districts of the Assam Province (Dhar 2013).

The post independence period witnessed some changes although the grwoth of the urban centres was comparativly low in the region. The Northeast region consist of 8 states the overall tempo of urbanisation became distinct by the rural–urban migration after independence. There were severly complementary factors of which the regions urban growth is still below the national growth rate. Although the recent rate of urbanisation in the region is not due to grwoth of industry rather than a shift of individuals from agriculture to another sector which can be marked by low-productivity employment such as handicraft production, retail trading, domestic servives and transport in the urban area (Mukherjee and Biswas 2013).

11.1.2 The Hill Towns of Northeast India and the Issues of Urban Growth

Topographically, the North-Eastern Region (NER) is surrounded by hills and discrete plains, and the land-owning and management system also differs in the hills, as most of the land in the hills is owned by the community, not the state. The region is connected with mainland India through a 22 km Siliguri corridor (chicken neck). The location and topography of the plain have made Guwahati a favourite destination for people from all the parts (Singh and Singha 2020), but the hill towns of Nagaland, Meghalaya, Sikkim, Mizoram, and Arunachal Pradesh face some serious issues, due to which the quality of life is deteriorating. Some major constraints of development in the hill towns are due to sloping terrain, sharp gradient slopes, complex geological structures, and extreme climatic conditions, and these issues are complimented with conflicting goals of environmental protection, economic growth, and sustainable development. The region's location in seismic zone V is another factor in its vulnerability to earthquake hazards.

The rapid process of urbanisation in the hill towns is due to the rural–urban migration in search of better livelihood. It is because of this reason the rural areas in the region is neglected and its leads to the widening socio-economic gap. Rather than the economic advantage in the urban area, the urbanization in the hill towns is highly unplanned. Issues such as slope side construction of buildings, unregulated construction on the fragile hill slopes is a matter of great concern. In areas such as Gangtok, Kohima, Aizawl, Shillong the rampart unplanned construction is causing major disruption to the regions ecology (Singh and Singha 2020). The incessant rainfall in the region not only affects the houses but also disrupts the region's transportation as heavy rainfall is mostly followed by landslides and flash flood, and the road links mostly gets snapped. The flash flood of Sikkim in 2023, not only destructed the built up's but took many lives. Due to the limited availability of land, the scope of road expansion in many hill towns gets restricted and the outcomes are traffic congestion, acute shortage of parking space, poor public transports (Singh and Singha 2020). The limited scope of horizontal buildings in the hill towns is another reason for construction of high rise building to support the growing urban population.

11.1.3 Urban Population North East India

According to census of 2011, each state of north east India shows a wide variation of urban population, Sikkim with the lowest urban population and Assam with the highest urban population as shown in Table 11.1. The urbanization pattern in the NER shows a contradictory situation, as the hilly region experiencing an urban explosion due to its varying economic system, whereas the valleys are experiencing a fading in terms of growth of urban areas (Mandal et al. 2022).

Urbanization in the region is the outcome of demographic explosion induced by rural–urban migration. In the hills several villages re-grouped into one village for administrative convenience. It can be one of the probable reasons for high level of urbanization in Mizoram. In the North-eastern Indian region, the towns are un-uniformly distributed and tends to be concentrated in and around the state and district headquarters. It is because of this reason there is a disparity in the distribution of different size or class of towns (Pautunthang 2018). The state capitals like Guwahati, Itanagar, Shillong and Agartala has already reached the threshold level for sustaining urban population. In the Northeast India region, the present model of urbanization is creating large number of small and middle category town, and these towns qualify the urban status due to population expansion without necessary infrastructure (Mandal et al. 2022).

Table 11.1 Percentage of urban population to total population in north east region

State/India	1971	1981	1991	2001	2011
Arunachal Pradesh	3.7	6.56	12.8	20.41	22.94
Assam	8.82	9.88	11.1	12.72	14.08
Manipur	13.19	26.42	27.52	23.88	21.16
Meghalaya	14.55	18.07	18.6	19.63	15.32
Mizoram	11.36	24.67	46.1	49.5	51.51
Nagaland	9.95	15.52	17.21	17.74	17.32
Sikkim	9.37	16.15	9.1	11.1	11.68
Tripura	10.43	10.99	15.3	17.2	14.87
NER	10.71	16.03	19.72	21.5	18.47

*Source*Urbanization Trend in North East India: An Empirical Study (Mandal et al. 2022)

11.2 Urban Planning Using Geospatial Technology

In the hill towns of Northeast India where urbanization is rapid it is necessary to give immediate attention upon the perspective of city or town planning. With the advent of technology, the utility of Remote Sensing and the incorporation of GIS with the remotely sensed data will help in assessing, understanding, mapping utility and service facility of the complex urban issues. The multi-spectral capabilities of the satellite remote which not only provides a synoptic view of the earth surface but also helps in monitoring the emerging changes in the urban core regions (Kumar and Fidelis 2013). The remote sensing data is suitable for land use mapping and change detection as it provides a spatial–temporal dataset for various regions, whereas GIS as tool is helpful for extracting various useful information like built-up areas, open green spaces and various other attributes. For the hill towns of Northeast India where the topography and climate is complex, these geo-spatial technologies can employ to a good and useful end (Rahman et al. 2011).

Urban planning establishes the vision and framework for growth, space technology provides data-driven insights to inform decision-making and development control regulations or zoning regulations ensure implementation and adherence to established norms.

11.2.1 Understanding the Terrain Using Geospatial Technology

Terrain is one of the utmost vital geographic feature and is the major factor in physical processes (Xiong et al. 2022). The hill towns of Northeast India have a complex topography with rivers and waterfalls dissecting the hill ranges, and the ranges varies from 1000–8000 m mean sea level. Thus for the town planners it is important to have the basic data of geology, topography and landform of the region to determine a suitable location for construction of different urban features. The region is in a high seismic zone, with presence of unstable slope due to landslide or excessive erosion, thus terrain analysis may help to gather some crucial information about the region to help to determine the project design, path selection and construction design of various urban entities (Udin et al. 2022).

The high resolution temporal and spatial data collection capability of Remote Sensing satellites with the wide range of terrain analysis tools of GIS provides researchers and planners to gather some useful insights about the terrain.

The Digital Elevation Model (DEM) can provide some crucial information about the terrain and when interpreted in the GIS environment elevation, slope, aspect, viewshed of a terrain can be analysed.

Slope: It is the most important entity that needs to be considered by a town planner. Regions with slope gradient less than or equal to 30° are most suitable for building houses and other infrastructures. Constructions in slope gradient higher than 30°

makes the buildings susceptible to geo-hazards. Using the DEM data and GIS software, the planners can easily demarcate the zones into most to least suitable for construction using the slope parameter.

Aspect: It is an important parameter to understand the impact of sun upon the local climate of the region, and this can be understood through aspect mapping of the area. The slope which faces the maximum solar insolation can be considered as the most suitable zone for doing agricultural practices and also for constructing residential buildings. The aspect is the orientation of slope, which is measured clockwise in degree from 0 to 360. In the hill towns where space for construction of houses is limited it is very important for the planners to delineate the zones which receives maximum solar insolation, suitable for constructing residential houses and urban green spaces.

ViewShed Analysis: This analysis is generally used to identify locations that are visible from one or more observer points. For urban planner this analysis is useful for model and map green visual exposure from human perspective (Cimburova and Blumentrath 2022). Urban greenery or urban green space is essential for health benefits of urban dwellers also it is important for urban aesthetics.

11.3 Urban Land Use Planning and the Implementation of GIS and Remote Sensing

In the hill towns of North East India urbanization is now becoming an inevitable part of economic development. This process not only possess changes to the demography but results in changes in the urban landscape. It is important of understand and detect the changes in these region where urbanization is rapid, before the mounting impacts of development does not gets too late to mitigate. Integration of Remote Sensing data with GIS not only provides the present land use pattern, it also depicts the growth and trend of transformation. The work of (Amin and Fazal 2012), using GIS and Remote sensing illustrates that in the Srinagar city there has been a significant land use and land cover transformation from 1981–2011. These changes were also validated by estimation of Land Consumption Rate and Land Absorption Coefficient. The study also projected the likely changes that might occur till the year 2021, along with that the study identified the scientifically indifferent pattern of land transformation in Srinagar City, where barren rocky-surface land was brought under residential, parks and gardens, forest were cleared for settlement which is a loss to the ecology. The study also revealed that economic development in the city has triggered a wide social transformation causing changes in land use classes.

In the hill towns where urbanization recently is unplanned, integration of GIS and Remote Sensing with Shannon Entropy Approach can be used as an indicator to recognize and measure the spatial extend of urban land use. In this approach spatial variation and temporal changes of growth areas are statistically measured to understand the urban growth pattern (Deka et al. 2012). It is usefull in addressing the environmental issue of urbanization in developing regions by integrating GIS and Remote Sensing. In this technique if the value of entropy is close to zero—it signifies that urbanization is very compact, and if the value is closer to Log(n), than the urban distribution is dispersed (Sudhira et al. 2004; Deka et al. 2012).

11.4 Urban Infrastructure and Its Planning Using Geospatial Technology

Development of an urban centre can be measured by its infrastructure growth. Many scholars have acknowledged the importance of infrastructure in economic and trade development, as well as provsion of better employment opportunity and reducing the disparity in any region. The states of North East always been a region of poor infrastructure due to issues such undulating topography, extreme climate, insurgency movemnets, its geographical location (Singha 2011; Bhattacharya 2011; Choudhury and Bhuyan 2005).

Urban Infrastructure can broadly be classified into 2 sections.

- Phyicial Infrastructure which includes Water Supple, Drainage, Solid Waste Management and Power,
- Social Infrastructure which includes Health Services and Educational Services.

It is important for the urban planners to focus upon channelising adequate investment on these infrastructure to keep pace with growing demands for these ammenities due to growing population in the urban centres. In almost all the urban hill centres of Northeast India rain water is the major source of water for both domestic and commercial use. To cope up with the demand of regular supply of water, it is important for the planners to trace out the potential zones, which can supply water throughout the year. Remote Sening coupled with GIS technology can be usefull for identifying those potential zones and also be usefull to deliniating proper channels for distribution of water.

Ground water potential assesment can be generated by integrating hydology, land use/ land cover, DEM, using Remote Sensing and GIS techniques. The weighted overlay analysis method applied over the integrated thematic maps, and with manual calculation the potential zones for ground water can be identified. In the Northeastern towns of India where steep terrain previals, ground water is a precious resources with very finite extebt. Thus adaption of this spatial techniques by the urban planners can advantageous for the planners to build sustainable urban centres.

The mapping and analysis capability of the terrain also facilitates in assessing the potential zones for installation of renewable enrgy source. The hilly terrains of Northeastern states which are disected with fast flowing river can be a great source generating hydroelectricty. The hydro electricty potential of the region is 63257 M.W of which only 1911 M.W has been harnessed, it highlights that the region can became, 'The Power House of India' (Das 2013a, b).

Geospatial technology provides a valuable platform for integrating multiple layers, such as land use patterns, population distribution, infrastructure availability and other factors. By exploring this technology, it becomes possible to identify the optimal zones for health and educational services. These selections can be made in accordance with the URDPFI Guidelines and the respective State Byelaws.

11.4.1 Improving Traffic Management

With rapid urbanization transportation is emergening as a complex problem in all urban areas. To manage the problem of traffic congestion or road accidents, the policy makers have to rely upon the traffic management system (Suharyadi et al. 2023). The hill towns of Northeast India are also facing some serious issues of traffic congestion. In towns such as Kohima, Dimapur, and Mokokchung in Nagaland, several critical issues exist. These include unregulated construction of buildings, a lack of parking spaces, and narrow roads that have no room for expansion. These factors contribute to severe traffic congestion, posing significant challenges for residents and authorities alike (Jongkor et al. 2023). Problems like urban sprawling in the sloping areas, in landslide prone areas creates vulnerability to the people during emergency evacuation as those areas are inaccesseble through road networks.

Certain parameteres for road condition can be provided through high resoluation satellite images, and using these data information reagarding road condtion and with, road obstacles information can be gathered (Suharyadi et al. 2023). In town of Northeast where due to heavy rainfall the road condition gets deteriorated and due to less scope of road expansion use of Remote Sensing and GIS can be useful in different ways:

Landslide Monitoring: The region is highly susceptable to landslide which blocks the traffic more hours. Monitoring of those zone using remote sensing data and creation of susceptibilty map using GIS can be used to build a early warning system.

Suitable Road Construction: Using Remote Sensing data and incorporating GIS software suitable areas for road construction by determining the soil quality, drainage and slope of the region.

Geo-spatial technologies can identify the small criticle zones, which are byproduct of poor planning and most common hotspot of traffic congestion. It also highlights the zones which neccessitates the construction of speed breakers, zebra crossing, adequate space for footpath, guard railing to ensure road safety. Also allocation of bus, taxi and rickshaw stands and suitable space for street vendors can also be identified to avoid traffic conegstion.

11.4.2 Industrial Zonation

In the developing hills states of Northeast India, Industries plays a crucial role in improving the state's economy. With a major shift from urban to rural in terms of settlement, and a shift from primary to secondary activities, to sustain the population and providing employment opportunity for the growing urban population, industries have a key role to play. Industrial site selection, whether in plains or hills is a multipart problem which requires a multi-criteria decision analysis which involves economic, environmental and social parameters (Ohri et al. 2010). In the hill regions of Northeast India, where site availability for industrial is already limited due to the limited availability of land and their traditional land holding practices, for industrial development the tasks for planners gets more difficult to select a suitable zone for industrial development.

Industrial development activities always have the potential to cause damage in the environment and thus it possesses major threat, thus while site selection for an industrial zone, some criteria such as proximity from human settlement, distance from road and river, type of soil, type of topography/terrain, nearby forest area needs to be considered. The high resolution spatial data capturing quality of Remote Sensing and the wide range of analytical tools of GIS software can be used to identify the most suitable location for an industry. In GIS the multi-criteria based analysis is among some techniques, which can be used in the site selection process by considering the variety of physical, cultural and economic factors (Reta and Deresso 2021).

This technique is widely used due to its flexibility and can be used to rank options, distinguish acceptable from unacceptable possibilities (Ohri et al. 2010). The output of this technique can be represneted upon a map as the method can seggreagte the region into most suitable to least suitbale zones for an industrial sites based upon the selected criteria and there assigned weightage. It is to have chances of ambiguity due to inconsistent judgement by decisionn makers, and it can be thought as a disadvantage of this technique (Ohri et al. 2010).

11.4.3 Identification of Hazard Vulnerable Zones

With the onset of monsoon, the North Eastern Region receives heavy rainfall and it is a major concern specifically for the hill towns. In most of the hill town the heavy rainfall is followed by landslide and urban flooding due to unplanned construction of various infrastructures, which not only affects the city dwellers but also creates an economic stagnation. Remote Sensing and GIS are important tools to highlight all those hazard vulnerable zones, and provides information about the terrain and landform process active in those areas.

The application of DEM can be observed as a significant data for mapping of water bodies, understanding the flow direction, stream networks, delineation of river basin and creation of various hydrological models (Amatulli et al. 2022; Jiang et al.

2015; Mark 1984; Musa et al. 2015; Sharma and Tiwari 2014; Woodrow et al. 2016; Zhang and Pan 2014; Xionget al. 2022). The dynamic process of urban hydrological system can also be monitored by analysing the DEM data and simulation using GIS software can support the urban developmental process. It is so because these different terrain analyses can be applied to predict various issues in an urban area such as waterlogging, landslides and it opens the door for optimization of various human-made hydrological facilities in an urban area (Xiong et al. 2022). It is important to understand processes such as erosion and landslides are not always anthropogenic in nature, rather these are some important processes which significantly shape the terrain. Thus for urban planners it is important for to map all the zones susceptible to these land degradation processes to properly channelize the land management plan, and reduce the negative effect of these processes.

11.5 Case Study on Formulation of Risk-Informed GIS Based Master Plan Under AMRUT, Shillong Planning Area, Meghalaya

Shillong, situated at an altitude of 1469 m above mean sea level, is the capital of Meghalaya. Known as the "Scotland of the East", the city was founded by Col. Henry Hopkinson in 1864. The city has now become the administrative, commercial and educational hub, but during the colonial period it was planned as a British hill resort. This erstwhile capital of Assam has now overshadowed the hill resort and became a commercial hub, with motor able roads all around the city. Shillong has a sub-tropical climate with very chilly winters and mild summers, and also faces the whim of monsoon. The city is well connected with road networks with other states, have an airport, but does not railway connectivity. The city has seen an increased rural to urban migration in recent years, which has paced the urban growth. It was so to sustain growing urban population and improve the quality of life, the Government of India Launched the Atal Mission for Rejuvenation and Urban Transformation (AMRUT) scheme in the city, with some specific objectives:

- Access of water supply and proper connection to sewage to every household.
- Maintenance of open spaces to increase the amenity value of the city.
- Improvement in public transportation and construction of non-motorized facilities to reduce pollution.
- Improvement in governance through a set of reforms.

For the Formulation of Risk-Informed Master Plan, GIS was the key component used to prepare different thematic map of the Planning area, and also a user-friendly base map at a functional scale of 1:4000 scale was also prepared using GIS (Project Team "Risk-Informed GIS based Master Plan" 2023).

Another key objective was to develop professional expertise in utilizing GIS technology for maintaining and updating databases in the context of urban planning and management.

The demarcated planning area is approximately 288.5 km^2. This included the Shillong urban agglomeration and adjoining 54 villages in the districts of Ri-bhoi and East Khasi hills.

The planners for creation of a sustainable urban area, focused upon 12 important sections, these are:

- Demography
- Urban Land Use
- Land Suitability
- Housing
- Urban Infrastructure
- Traffic and Transportation
- Industry and Commerce
- Ecology & Environment
- Governance & Institutional Arrangement
- Hazard Zonation
- Proposed Land Use & Phasing
- Zoning Regulations.

11.5.1 Demography

It is the study of human composition of a village, town, city, a region or a country. It is important to understand the demography of a region as it helps to plan job opportunity in an economic sector for the upcoming population (Project Team "Risk-Informed GIS based Master Plan" 2023). In a city the growing population also have an impact upon the surrounding village, thus for planners understanding the demography helps to identify the direction in which the city is growing, to demarcate the extent of planning area and the services that need to be provided. In the Shillong Master Plan various entities of a demography were considered such as:

- Existing Demographic Profile
- Population Composition such as towns, village and Ward wise population
- Population Growth rate
- Population Density
- Sex ratio
- Literacy Rate
- Schedule Tribe and Schedule Caste Population
- Work Force Participation Rate
- Population Projection.

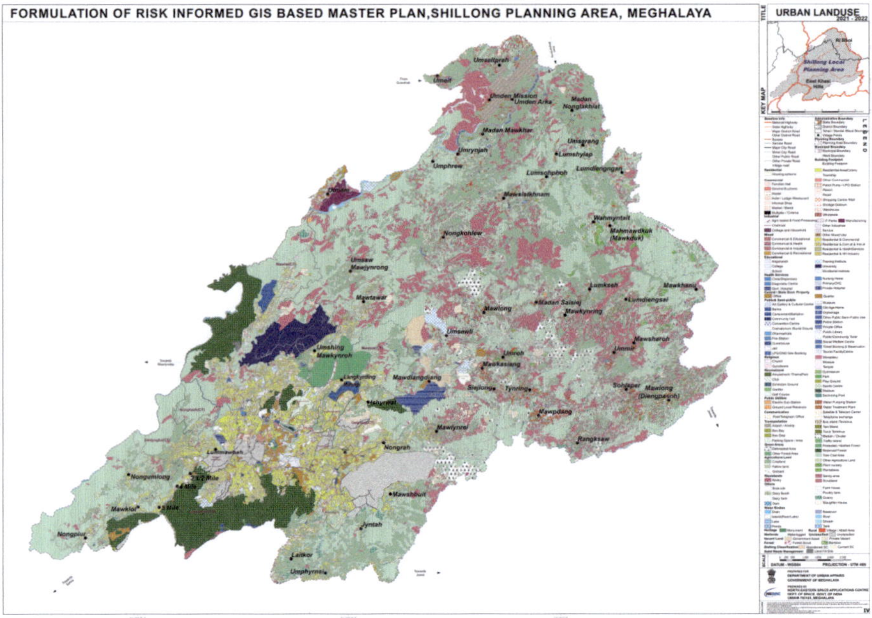

Fig. 11.1 Thematic maps of existing urban land use

11.5.2 Urban Land Use

For sustainable area planning, it is crucial for planners to comprehend the existing land use patterns. In this master plan, the planners have categorized the data into two sections: the existing land use for the entire Shillong Planning Area as shown in Fig. 11.1 and Table 11.2 and district-wise land use classification. To analyse and represent this information, GIS techniques were employed, along with Remote Sensing Satellite data.

11.5.3 Land Suitability

For city planners land suitability analysis is a very important challenge as it is useful in determining the most suitable land for land use in future (Project Team "Risk-Informed GIS based Master Plan" 2023). Land suitability is the ability of a land to accommodate a particular purpose such as residential, commercial, industrial etc. as shown in Fig. 11.2.

The planners here have used multi criteria analysis in a GIS software. The main areas of concern in the GIS suitability analysis were:

- Reserved Forest Area Protection

Table 11.2 Existing urban land use of Shillong Planning Area (2022)

SN	Categories	Area (Ha)	Percentage (%)
1	Residential	1564.74	30.3
2	Commercial	130.46	2.5
3	Industrial	120.99	2.3
4	Mixed	131.51	2.5
5	Educational	652.55	12.6
6	Health Services	160.60	3.1
7	Central Govt. Property	55.06	1.1
8	State Govt. Property	192.64	3.7
10	Public and Semi-public	130.93	2.5
11	Religious	48.16	0.9
12	Recreational	206.03	4.0
13	Eco-Sensitive Areas	1.375	0.0
14	Public Utilities	18.00	0.3
15	Solid Waste Management	6.89	0.1
16	Communication	1.65	0.0
17	Heritage	0.21	0.0
18	Slum	2.54	0.0
19	Vacant Land	579.74	11.2
20	Transportation	840.45	16.3
21	Rural	306.14	5.9
22	Others	8.29	0.2
Total developed area		5158.97	100

- Resource Connectivity
- Economic boost to the Industrial sector
- Proximity to existing developed areas and compatibility with existing land use
- Availability of community facilities

Based on the multi criteria weightage method different suitability thematic maps were prepared, such as:

- Residential Suitability
- Commercial Suitability
- Industrial Suitability
- Logistic Hub Suitability
- IT Hub Suitability
- Animal Husbandry Suitability
- Dairy Suitability.

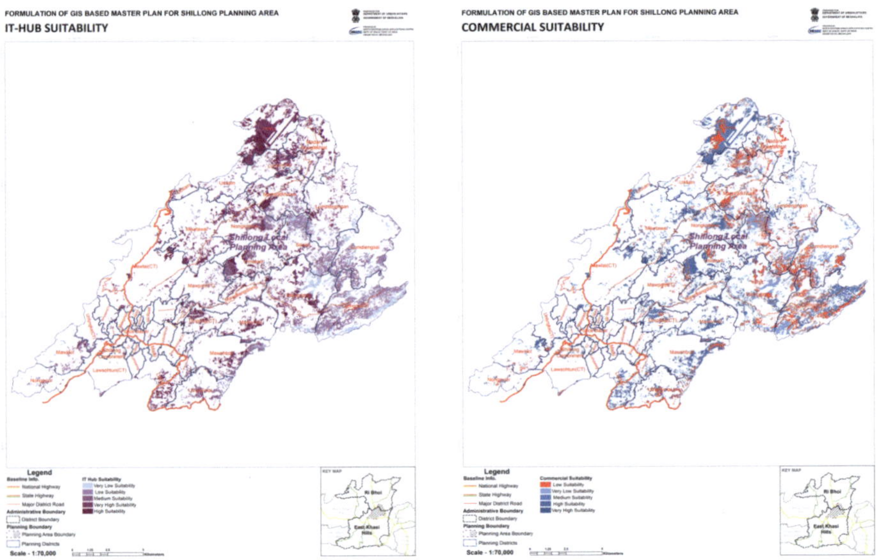

Fig. 11.2 Thematic maps of IT Hub and commercial suitability of SPA

11.5.4 Housing

The planners has analysed the housing scenario in Shillong Planning Area in 2 distinct divisions:

- Shillong Urban Agglomeration Area and
- Adjoining 54 Villages

Also the existing housing status in the Shillong Planning Area is governed by three factors:

- Rapid Urbanization: increasing in the urban population due to migration in search of job opportunities has added demand for housing facility.
- Socio-Economic Diversity and its Influences: variety of social and economic status and homelessness is a major challenge to housing facility.
- Dichotomy in various housing pockets.

These are the reason behind the haphazard growth of urban area and the continuous building pressure to sustain growing urban population is increasing the land prices (Project Team "Risk-Informed GIS based Master Plan" 2023). Several information was gathered to understand the existing housing facility and pattern in the Shillong Planning Area and these are:

- Household Number Information of overall Shillong planning Area
- Household Size Distribution
- Housing Condition and Obsolescence factor

- Housing Stock and Shortage Calculation
- Land Pooling.

11.5.5 Urban Infrastructure

As discussed in the previous sections, development of an urban area depends upon the right balance of urban infrastructure development. The planners have divided the urban infrastructures in two broad categories:

- Physical Infrastructure, which consist of water supply, drainage, Solid waste Management and Power
- Social Infrastructures like educational and health services.

In the planning area rain water may be considered as the original source of all supplied water. By surveying the planning area different sources of drinking water were identified.

Several other primary surveys were conducted to understand the existing condition of other physical and social infrastructures, also using GIS and remote sensing data map of suitable landfill was prepared.

The Multi-criteria weighted analysis was used to locate the location for landfill by considering various criteria as shown in Fig. 11.3.

11.5.6 Transportation Management and Hazard Zonation

For understanding the traffic flow and to facilitate smooth transportation at the first stage primary survey such as Cordon Line Survey and Turning Movement were done. The major objective of the planners for these traffic survey was:

- To get data of traffic flow in the city
- Understand the characteristics of traffic flow and service of the load
- To get the characteristics of on street and off street parking.

The Shillong city is less prone to landslide hazard as compared to its neighbouring area and most of the landslide is experienced during the monsoon season. Although due to its location in very high seismic zone, the earthquake triggered landslide cannot be ruled out. Using GIS planners identified various hazard-vulnerable zones in the region.

By integrating data such as hazard zonation for landslides and floods, land suitability for various uses, existing urban land use, and other relevant factors, an optimal traffic circulation pattern and a proposed land use plan for the horizon year 2041 were developed.

The Risk-Informed GIS Based Master Plan under Atal Mission for Rejuvenation and Urban Transformation (AMRUT), Shillong Planning Area, Meghalaya, covers

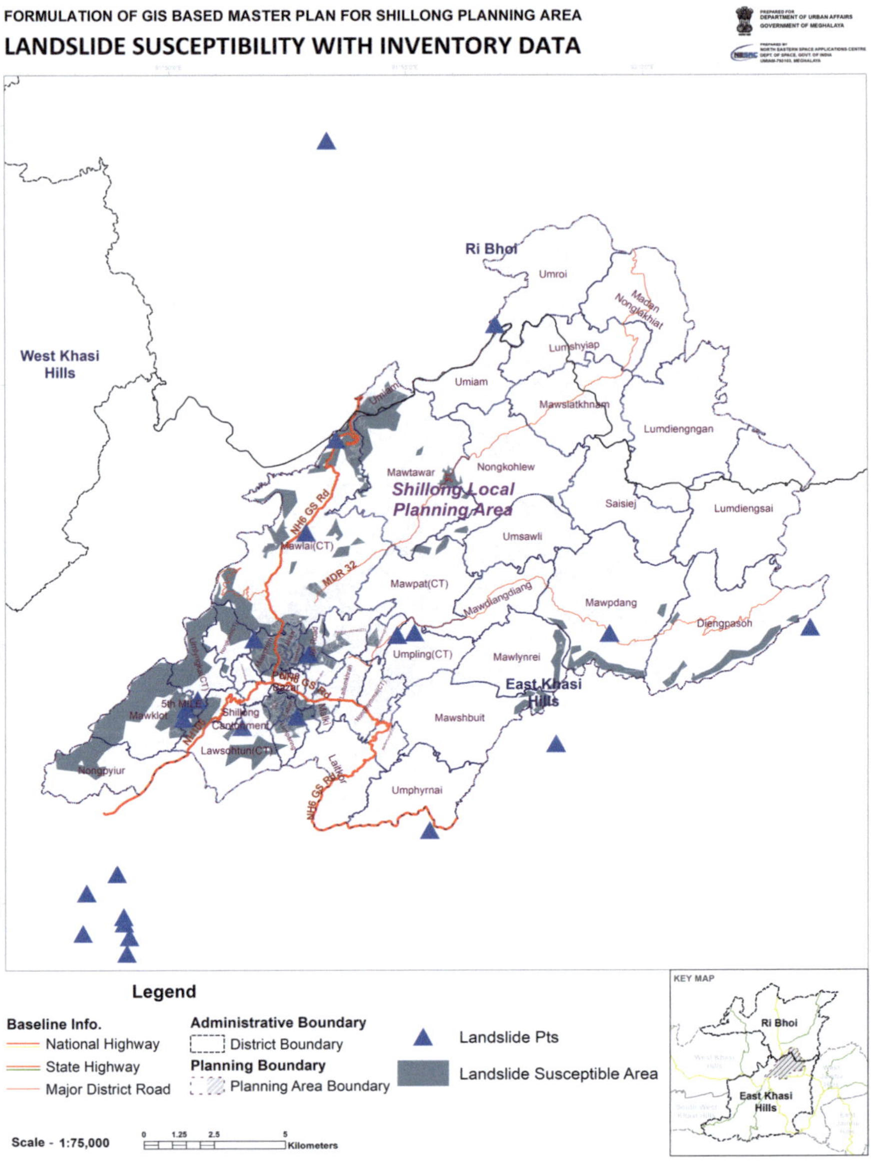

Fig. 11.3 Landslide susceptibility map of Shillong planning area

wider dimension necessary to develop a sustainable urban area, which not only provides a diverse economic prospect, also provides basic amenities to the urban dwellers. The application of GIS and Remote Sensing has improved the horizon of urban planning by providing some crucial data and generating maps for better interpretation of different entities.

11.6 Urban Zoning and Its Importance in Urban Planning

Zoning is an important method of land regulation, where it divides a jurisdiction into geographically contiguous 'zones' and this regulation is undertaken by the local governments. Zoning is useful to determine where it can be built and what can be built in different parcels of land (Lens 2022). Zoning not only allocates land to different kinds of structures in certain areas, also includes restriction in different zoning areas in terms of building height, types of business, use of green space, density of buildings in a particular zone (S et al. 2020). Zoning helps in regulation by preventing the misuse of lands and also prevents encroachment of land, by segregating the land parcels into residential zones, industrial zones, civil zones, recreational zones.

As per S et al. (2020) the advantage of zoning are:

- Zoning helps to regulate the population distribution throughout the town and thus there will be no population concentration in a particular zone.
- Zoning helps to segregate the industrial zones from the residential zone and thus human health is not affected by the industrial effluents and smoke.
- Zoning plays a vital role in promoting the health, safety, prosperity and orderly development of community.
- Height zoning limits the height of building in particular region, by restricting construction of high rise buildings
- Zoning also permits in provisioning of various public utility and services such as water supply, drainage network and proper connectivity.

As discussed in the previous section due to the unplanned growth of the urban centres in the hill towns of North eastern states, which is complemented by the complex topography and climate are creating serious issues to the urban dwellers. Thus a well-structured zoning regulation in these towns is needed for improving the subjective wellbeing.

11.7 Case Study of GIS Based Master Plan of Itanagar Capital Region

Itanagar also known as 'Land of Dawn-lit', the capital city of Arunachal Pradesh. The region is situated in the laps of the Himalayas and houses rich biodiversity. Although the population density of the state is very less as compared to the national data, but its location and is of great strategic imporatance. Under the Atal Mission for Rejuvenation and Urban Transformation, a GIS based Master Plan of the city was prepared to faciliate a well channelised urban development in the Itanagar Capital Region as shown in Fig. 11.4 (Draft Master Plan Report for Itanagar Capital Region" (2022).

The master plan has divided the Itanagar Capital region into 7 major categories, these are.

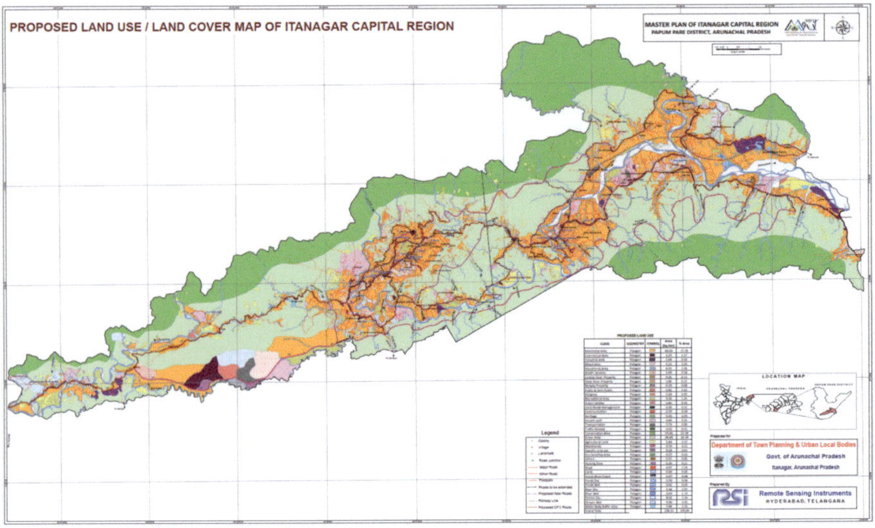

Fig. 11.4 Proposed land use/land cover map of ICR

- Residential Zone
- Commercial Zone
- Mixed-Use Zone
- Public and Semi-Public Zone
- Industrial Zone
- Recreational/ environmental zone
- Agriculture zone.

All these 7 zones were further divided into 3 categories, based upon the permitted utility of the zone, such as.

a. ***Uses permitted***: It consist of the urban structures or entities were permitted to be constructed without as per the guideline in the given zone.
b. ***Conditional Uses***: It refers to the buildings can be constructed in a zone only those fulfilling certain conditions.
c. ***Prohibited Uses***: This are the buildings whose construction is fully prohibited in that particular land parcel.

11.7.1 Residential Zone

In this zone buildings such as plotted, group housing house, residential flats, apartments, research institute, community centres etc. are permitted to be constructed. Buildings such as Dhaemashala, Night Shelter, petrol pumps, moto vehicle repairing garages etc., can be constructed under the conditional uses category.

Residential apartments (whose plots not accesible by 12 m wide road), organised informal market, serivce apartments etc., are strictly prohibited to be constructed in the residential zone.

11.7.2 Commercial Zone

Uses permitted: Shops, shopping mall, super mart, vegetable mart, fish market, commercial centres, research/trainning institute, commercial service centers etc.

Conditional Uses: Non-pollution, non-obnoxious light industries, warehousing, inflammable goods, wood, timber yards, coal, bus and truck depots, hospital and nursing homes.

Prohibited Uses: All types of industrial uses, major infrastructure installations, essential operational, watch and ward personnel, store house for perishable nad inflmmable commodities etc.

11.7.3 Mixed-Use Zone

Uses permitted: 30 to 60% is the maximum gross area allowed with the dominant uses sharing with other uses, such as Multi-family residential apartments above 1st or 2nd floor.

Conditional Uses: 20 to 50% residential use is permitted on commercial and non-polluting industrial land and it depends upon the locational feasibility of the land.

Prohibited Uses: Activities including indutrial which can have an hostile impact on the overall activities of the zones are prohibited.

11.7.4 Public and Semi-Public Zone

Uses Permitted: Central and state government properties, defense copurt, banks, police station, police lines, jails, fire station, post and telegraph. Educational facilities; primary'secondary/high schools, universities and specialized educational institute, polytechnic colleges (not to be situated near hospital or health care facility).

Conditional uses: Single family residents, complementary commercial uses, public swimming pools, burial grounds, cremetion grounds nad cemeteries.

Prohibited Uses: Substantial, widespread and obnoxious, hazardous industries, slaughter houses, junk yards, wholesale mandis, dairy and poultary farm etc.

11.7.5 Open Spaces and Recreational Areas

Uses permitted: Regional parks, district parks, playgrounds, clubs, picnic huts, indoor and outdoor stadium, dedicated recreational spots such as botanical/ zoological garden, bird sanctuary.

Conditional Uses: Building and structures ancillary to use permitted in parks and open spaces such as stand for vehicles on hire, taxis and scooters, commercial use of transit nature like cinema and other shows, public assembly halls, restaurants and caravan parks, sports stadium, open air cinemas.

Prohibited Uses: Buildings which are not required for open air recreation, dwelling unit except for watch and ward personnel and uses not specifically permitted.

11.7.6 Industrial Zone

Uses permitted: Residential buildings for essential staff and for watch and ward personnel, all kind of industries public utilities, parking, loading, unloading spaces, warehousing, storage and depot of non-perishable and non-inflammable commodities etc.

Conditional Uses: Noxious, obnoxious and hazardous industries except storage of perishable and inflammable goods, junkyards, sports facilities, quarrying of gravel, sand clay or stone.

Prohibited Uses: Residential dwellings other than those essential operational, service and watch and ward staff. Schools and colleges, hotels, motels and caravan park etc.

11.7.7 Agricultural Zone

Uses permitted: Rural settlement dwellings, farm-houses and accessory buildings, agriculture, horticulture and forestry, poultry, piggery and dairy farm, cottage industries.

Conditional Uses: Stone, clay and sand quarries, service industries accessory to obnoxious and hazardous industries, school and library, temple, churches, mosque and other religious buildings.

Prohibited Uses: Residential use except those uses permitted in agricultural use zone, heavy extensive, noxious, obnoxious and hazardous industries and any activity creating nuisance and obnoxious in nature.

11.7.8 Transportation and Communication Zone

Uses permitted: Road transport terminals, goods terminals, parking areas, airports-building and infrastructures, truck terminal, motor garage, transmission centre, wireless station.

Conditional Uses: Any other activity subsidiary to transport and communication, residential dwelling units for essential staff and watch personnel.

Prohibited Uses: In vicinity to airport: butcheries, tanneries and solid waste disposal sites shall be prohibited within 10 km from the airdrome reference point.

11.7.9 Solid Waste at Landfill Site

The planning committee has selected landfill sites on the basis of some locational criteria such as:

- The size of the landfill site should be big enough to last for 20–25 years and within 5 km from the city limit.
- The site should at least 500 m away from clustered habitation, forest area, monuments, national parks, wetlands and important cultural, historical and religious interest.
- Landfill should not be constructed within 200 m from any land or pond.
- Prohibition of landfill site within 100 m from navigable water.
- Landfill site should be constructed at least 500 m away from habitation and the 500 m buffer zone should a no development zone.
- Topographical and geographical features such as contour, slope and soil type of the site should be analysed in details to avoid soil contamination and disturbance to drainage pattern.

11.8 Conclusion

The history of urban development has been very critical, the topography, social and economic structure has further deteriorated the urbanization process in the region. Some of the hill towns of north-eastern India are still in the nascent state of urbanization. The region with its rich bio-diversity, culture and heritage are always susceptible to the negative effects of urbanization. Already towns like Kohima, Dimapur, Shillong are facing sever issues of traffic congestion, water scarcity, unchecked urban sprawl, water pollution. Integration of geospatial technology for urban planning will not only eradicate these issues, will widen the scope for further development.

For North Eastern hill towns of India to develop holistically and sustainably, the interplay of urban planning, zoning laws, space technology, and development control laws is essential. Urban planning using space technology for North Eastern

India Hill towns mark a paradigm leap in urban planning and development management. Policymakers may create context-sensitive zoning laws that support orderly growth, safeguard natural resources, and enhance the quality of life for locals by utilizing satellite imagery, remote sensing, and GIS. Resilient and sustainable urban development is encouraged by this integrated strategy.

Acknowledgements The author extends sincerely acknowledges to the Directorate of Town & Country Planning, Shillong, for entrusting NESAC with the project AMRUT 1.0 initiative in 2018 - Formulation of a Risk-Informed GIS-Based Master Plan for the Shillong Planning Area, Meghalaya. In parallel, also grateful for Directorate of Town & Country Planning, Itanagar also making available the report on the GIS-Based Master Plan for the Itanagar Capital Region has provided invaluable insights, serving as classic case studies that illuminate the practical integration of geospatial technologies in urban planning.

References

I. Ahmed, S. Bhatia, GIS-based route optimization for effective traffic management: a case study of Srinagar city. Int. J. Innov. Res. Comput. Sci. Technol. (IJIRCST), **10**(3), 171–175 (2022). https://doi.org/10.55524/ijircst.2022.10.3.28

G. Amatulli, J. Garcia Marquez, T. Sethi, J. Kiesel, A. Grigoropoulou, M.M. Üblacker, L.Q. Shen, S. Domisch, Hydrography90m: a new high-resolution global hydrographic dataset. Earth Syst. Sci. Data **14**(10), 4525–4550 (2022). https://doi.org/10.5194/essd-14-4525-2022

A. Amin, S. Fazal, Quantification of land transformation using remote sensing and GIS techniques. Am. J. Geogr. Inf. Syst. **1**(2), 17–28 (2012). https://doi.org/10.5923/j.ajgis.20120102.01

A. Ashara, S. Salesh, U. Hassan, M.J. Kaura, Traffic management system in Abuja city center using geographic information systems (GIS) and global positioning system (GPS): a case study of FCT, Nigeria. J. Phys. Conf. Ser. **1529**(5), 052068 (2020). https://doi.org/10.1088/1742-6596/1529/5/052068

S.Y. Ayemi, B.K. Kar. Patterns of urbanization and associated infrastructure and socio-economic development in Nagaland, India. Demogr. India, 54–68 (2020)

K.A. Baba, D. Lal, A. Bello, Application of remote sensing and GIS techniques in urban planning, development and management: a case study of Allahabad District, India. Int. J. Sci. Eng. Res. **10**(6), 1127–1134 (2019)

P. Baruah, Y. Millo, Factors affecting development: a study in a state of North Eastern India. Econ. Insights Trends Chall. **3**, 1–10 (2014)

R. Bhattacharya, Development disparities in Northeast India (2011). https://doi.org/10.1017/UPO9788175969193

S. Chakravarty, C.P. Suresh, A. Puri, G. Shukla, North-East India, the geographical gateway of India's phytodiversity. Indian For. 702–709 (2012)

S. Choudhury, R. Bhuyan, Development disparity and North-East Region. Dialogue **7**(2), (2005). Retrieved July 28, 2011, from http://www.asthabharati.org/Dia_Oct%2005/Sas%20&%20Rat.html

Z. Cimburova, S. Blumentrath, Viewshed-based modelling of visual exposure to urban greenery—an efficient GIS tool for practical planning applications. Landsc. Urban Plan. **223**, 104437 (2022). https://doi.org/10.1016/j.landurbplan.2022.104437

P.K. Das, North-East, 'The Power House of India': prospects and problems. IOSR J. Human. Soc. Sci. (IOSR-JHSS) **18**(3), 36–48 (2013a)

P.K. Das, in *Urbanisation: Its Challenges & Environmental Degradation*. Ed. by P.K. Das, Urbanisation: Its challenges & environmental degradation (Eastern Book House, 2013b), p. 278

J. Deka, O.P. Tripathi, M.L. Khan, Urban growth trend analysis using Shannon entropy approach–a case study in North-East India. Int. J. Geomat. Geosci. **2**(4), 1062–1068 (2012)

P.K. Dhar, in *Process of urbanisation and its challenges and quality of urbanisation in India (With special reference to Northeastern and Eastern Regions of India)*. ed. by P.K. Dhar, Urbanisation: Its Challenges & Environmental Degradation. (Eastern Book House, 2013), p. 278

M.J. Fasona, A.S. Omojola, GIS and remote sensing for urban planning: a case of FESTAC Town, Lagos, Nigeria. Geoinformatics **2004**, 451–458 (2004)

D.K. Gogoi, in *Issues of Environmental Degradation in the Context of Urbanisation and Development of Guwahati*. ed. by P.K. Dhar Urbanisation: Its Challenges & Environmental Degradation (Eastern Book House, 2013), p. 278

N. Hoalst-Pullen, M.W. Patterson, Application and trends of remote sensing in professional urban planning. Geogr. Compass **5**(5), 249–261 (2011)

O. Isoiye, I. Enebeli, S.A. Alademomi, A. E, Terrain analysis in GIS and its significance to surface runoff analysis: a study of Basawa Community in Sabon Gari L.G.A. of Kaduna State, Nigeria. IJASETR **1**(1), 72–92 (2012)

M. Jhawar, N. Tyagi, V. Dasgupta, Urban planning using remote sensing. Int. J. Innovat. Res. Sci. Eng. Technol. 42–57 (2012). https://www.ijirset.com/

W. Jiang, L. Zhang, C. Xu, A review of applications of satellite SAR, optical, altimetry and DEM data for surface water modelling, mapping and parameter estimation. Hydrol. Earth Syst. Sci. **19**(8), 3755–3769 (2015). https://doi.org/10.5194/hess-19-3755-2015

T. Jongkor, K.Y. V, L. Longkumer, Urban problems of Kohima, Dimapur, and Mokokchung towns of Nagaland, India. Int. J. Glob. Sci. Res. **10**, 2029–2051 (2023). https://doi.org/10.26540/ijgsr.v10.i1.2023.236

K.S. Krishnaveni, P.P. Anilkumar, Managing urban sprawl using remote sensing and GIS. Int. Archiv. Photogrammetry Remote Sens. Spat. Inf. Sci. 59–66 (2020)

M. Khalesian, P. Pahlavani, M.R. Delavar, A GIS-based traffic control strategy planning at urban intersections. IJCSNS Int. J. Comput. Sci. Netw. Secur. **9**(1), 166–174 (2009)

L. Khamrang, Perceived quality of life in the cities of Northeast India: a welfare geographical perspective. Int. J. Soc. Sci. Tomorrow **3**, 1–10 (2012)

A. Kumar, Impact of building regulations of Indian hill towns. Hous. Build. Nat. Res. Cent. **12**(3), 316–326 (2016). https://doi.org/10.1016/j.hbrcj.2015.02.002

A.K.A. Kumar, S.U. Fidelis, Application of remote sensing, GIS and GPS for efficient urban management plan: a case study of part of Hyderabad city. Novus Int. J. Eng. Technol. **2**(4), 1–14 (2013)

M.C. Lens, Zoning, land use, and the reproduction of urban inequality. Ann. Rev. Sociol. **48**, 421–439 (2022). https://doi.org/10.1146/annurev-soc-030420

J.M. Levy, *Contemporary Urban Planning*, 11th ed. Routledge (2017)

D.R. Mandal, D.A. Alam, D.R. Prasad, D. Kundra, Urbanization trend in North East India: an empirical study. J. Posit. Sch. Psychol. **6**(9), 4641–4653 (2022)

D.M. Mark, Automated detection of drainage networks from digital elevation models. Geogr. Anal. **16**(4), 337–344 (1984). https://doi.org/10.1111/j.1538-4632.1984.tb00225.x

S. Mukherjee, A. Biswas, Complexities in growth of Tura Town in Meghalaya, in *Small Cities and Towns in Global Era: Emerging Changes and Perspectives*. ed. by R.N. Sharma, R.S. Sandhu (Rawat Publications, 2013), pp.295–317

Z.N. Musa, I. Popescu, A. Mynett, A review of applications of satellite SAR, optical, altimetry and DEM data for surface water modelling, mapping and parameter estimation. Hydrol. Earth Syst. Sci. **19**(8), 3755–3769 (2015). https://doi.org/10.5194/hess-19-3755-2015

A. Nath, Cohesive urbanisation in North-East India: issues & concerns. J. Polit. Govern. **2**(3), 66–72 (2013)

A. Ohri, P.K. Singh, P.K. Singh, Spatial multi-criteria analysis for siting industries. Int. J. Indust. Eng. Res. Dev. **1**, 94–114 (2010)

E. Pahl-Weber, F. Schwartze, *Urban Planning* (Academy for Territorial Development, 2018). https://nbn-resolving.org/urn:nbn:de:0156-559923803

C.B. Pande, S.F. Khadri, K.N. Moharir, R.S. Patode, Assessment of groundwater potential zonation of Mahesh River basin, Akola and Buldhana districts, Maharashtra, India using remote sensing and GIS techniques. Sustain. Water Resour. Manag. (2017). https://doi.org/10.1007/s40899-017-0193-5

D. Pandey, H.K. De, L. Geetarani, B.N. Hazarika, Problems and prospects of agricultural diversification in shifting cultivation area of North East India: an empirical study. J. Pharmacogn. Phytochem. **8**(6), 69–73 (2019)

N. Pautunthang, Urbanization in North East India 1991–2011. Int. J. Trend Sci. Res. Dev. (IJTSRD) **2**(6), 711–720 (2018)

Project Team *"Risk-Informed GIS based Master Plan"*, 2023 for Shillong Planning Area, Meghalaya (NESAC and Department of Urban Affairs, Govt. of Meghalaya, 2023)

A. Rahman, Y. Kumar, S. Fazal, S. Bhaskaran, Urbanization and quality of urban environment using remote sensing and GIS techniques in East Delhi, India. J. Geograph. Inf. Syst. **3**(3), 62–84 (2011). https://doi.org/10.4236/jgis.2011.31005

J. Raven, Urban Planning and Urban Design, in *Urban Planning and Urban Design*. ed. by J. Raven (Cambridge University Press, 2018), pp.139–172

Draft Master Plan Report for Itanagar Capital Region–GIS Based Master Plan under AMRUT Scheme, Itanagar: Developed and hosted by National Informatics Centre (Ministry of Electronics & Information Technology, Government of India, 2022)

T. Reta, M. Deresso, Assessment of suitable industrial site using GIS and remote sensing: case study in Assosa Town, Ethiopia. Int. Res. J. Sci. Technol. Educ. Manage. **1**(2), 11–25 (2021)

N.N. S, *Town and Country Planning* (2020). https://www.townandcountryplanninginfo.com/2020/11/zoning.html

A. Sharma, K. Tiwari, A comparative appraisal of hydrological behavior of SRTM DEM at catchment level. J. Hydrol. **519**, 1394–1406 (2014). https://doi.org/10.1016/j.jhydrol.2014.08.029

R. Shaw, A. Das, Identifying peri-urban growth in small and medium towns using GIS and remote sensing technique: a case study of English Bazar urban agglomeration, West Bengal, India. Egypt. J. Remote Sens. Space Sci. **21**(2), 159–172 (2018)

M.A. Singh, K. Singha, in *Understanding Urbanisation in Northeast India: Issues and Challenges*. ed. by M.A. Singh, K. Singha, M.A. Singh, S. Komol, Understanding Urbanisation in Northeast India: Issues and Challenges. (Routledge, 2020), p. 237

K. Singha, Infrastructure and regional disparity: a case of North Eastern Region of India. J. Soc. Dev. Sci. **2**(4), 162–180 (2011). https://doi.org/10.22610/jsds.v2i4.666

H. Sudhir, *Urbanization in Northeast India* (Akansha Publishing House, 2009), p. 128

H.S. Sudhira, T.V. Ramachandra, K. Jagadish, Urban sprawl: Metrics, dynamics and modelling using GIS. Int. J. Appl. Earth Observ. Geoinform. **5**, 29-39 (2004). https://doi.org/10.1016/j.jag.2003.08.002

R. Suharyadi, I.N. Hidayati, M.D. Surya, V. Permatasari, Improving the traffic management system based on remote sensing data imagery. E3S Web Conf. **468**, 03006 (2023). https://doi.org/10.1051/e3sconf/202346803006

A.N. Tiwary, GIS and remote sensing in urban development planning: issues and challenges of developing world. Int. J. Sci. Res. **5**(11), 86–90 (2016)

W.S. Udin, M.N. Aduini, N. Sulaiman, N.S. Shafiee, R.M. Jamil, GIS-based assessment and analysis of terrain variables in Kampung Baru, Jeli, Kelantan. IOP Conf. Ser. Earth Environ. Sci. **1102**, 012023 (2022). https://doi.org/10.1088/1755-1315/1102/1/012023

K. Woodrow, J.B. Lindsay, J. Domingue, Evaluating DEM conditioning techniques, elevation source data, and grid resolution for field-scale hydrological parameter extraction. J. Hydrol. **540**, 1022–1035 (2016). https://doi.org/10.1016/j.jhydrol.2016.06.022

S. Xhafa, A. Kosovrasti, Geographic information system (GIS) in urban planning. Eur. J. Interdiscip. Stud. **1**(1), 74–81 (2015)

L. Xiong, S. Li, G. Tang, J. Strobl, Geomorphometry and terrain analysis: data, methods, platforms and applications. Earth Sci. Rev. **231**, 104191 (2022). https://doi.org/10.1016/j.earscirev.2022.104191

C.-Y. Yiu, K.S. Cheung, Urban zoning for sustainable tourism: a continuum of accommodation to enhance city resilience. Sustainability **13**(13), 7317 (2021). https://doi.org/10.3390/su13137317

A.M. Youssef, B. Pradhan, Integrated evaluation of urban development suitability based on remote sensing and GIS techniques: contribution from the analytic hierarchy process. Arab. J. Geosci. **4**, 463–473 (2010). https://doi.org/10.1007/s12517-009-0118-1

S. Zhang, B. Pan, An urban storm-inundation simulation method based on GIS. J. Hydrol. **517**, 260–268 (2014). https://doi.org/10.1016/j.jhydrol.2014.05.044

B. C. Sumanth is a Scientist/Engineer-SD, at North Eastern Space Applications Centre (NESAC), joined in March 2021. Holds a Bachelor's Degree in Civil Engineering, A Master's in Urban & Regional Planning and a PG Diploma in Urban & Regional Studies. At NESAC, he has contributed to key projects such as the Master Plan for Shillong, Village Level Mapping for Meghalaya, Border Village Planning in Arunachal Pradesh and Hazard Risk Vulnerability Assessments. He is also involved in National initiatives like AMRUT 1.0 and AMRUT 2.0, focusing on sustainable urban & Regional Development.

Santanu Das a Scientist/Engineer-SE at NESAC, has 13 years of Experience across government administration and the private sector. He specializes in leveraging space technology to drive sustainable urban development. His expertise lies in applying geospatial intelligence to urban and regional planning, enabling data-driven decision-making for large-scale operations in planning, scheduling, and implementation. With a strong engineering foundation and a Master's degree in City Planning from the Indian Institute of Technology, Kharagpur, He brings an in-depth understanding of how satellite data and spatial analytics can transform urban systems, support infrastructure development, and enhance governance efficiency.

Jenita M. Nongkynrih is currently Head Urban and Regional Planning Division at North Eastern Space Applications Centre under Department of Space, Government of India. She has more than 21 years of experience in space technology applications for urban & regional planning, natural resource management and disaster management. She has been leading mainly user specific projects relating to Risk-informed Master plan preparations, city disaster management plans and hazard risk vulnerability assessment for urban centers in North Eastern India. She has published research papers in national and international journals. She has guided a number of M.Sc./M.Tech and Ph.D. students. She has been awarded ISRO Young Scientist Merit Award for 2011.

Surajit Sutradhar holds a Master's degree in Geography with a specialization in GIS and Remote Sensing. His academic and professional journey spans multiple research and institutional roles, including GIS Assistant with the Directorate of Land Revenue and Surveys, Assam (Mission Basundhara Project) and Assistant Project Scientist at the Department of Humanities and Social Sciences, IIT Guwahati. He has also contributed to urban and regional planning research as a Junior Research Fellow (JRF) at the Urban and Regional Division (URD), NESAC, Umiam, Meghalaya. Currently, he serves as an Assistant Professor in the Department of Geography at Tihu College, Nalbari, Assam. His core interests lie in the application of geospatial technologies for sustainable development and regional planning.

K. K. Sarma Ph.D, leads the Remote Sensing Applications Group (RSAG) at the North Eastern Space Applications Centre (NESAC), where he has been a key figure since 2002. Specializing in remote sensing for forestry and biodiversity, his work leverages space technology for critical applications in Biodiversity, Ecosystems, Forestry, and Natural Resource Management. Dr. Sarma has spearheaded over 60 diverse geospatial projects, contributing significantly to natural resource mapping and monitoring across the North Eastern Region (NER). Currently, he coordinates 110 "Plan of Action" projects, focused on using space technology to improve livelihoods in the NER. He also served as the Principal Investigator and NER Coordinator for all ISRO natural resource management projects, including comprehensive Land Use Land Cover Mapping and National Wasteland Updating, Monitoring, and Change Analysis. With over 75 research publications, Dr. Sarma has successfully supervised six Ph.D. students and actively mentors many others across NER Universities.

Chapter 12
Geospatial Technology for Greenfield and Rural Development

Santanu Das, Jenita M. Nongkynrih, B. C. Sumanth, and K. K. Sarma

Abstract This chapter explores the unique context of Northeast India and its implications for greenfield and rural development, emphasizing the region's diverse geography, demographic complexity, and socio-cultural richness. It details the significant challenges faced, including infrastructure deficits, environmental vulnerabilities, land tenure complexities, and security concerns. The narrative highlights the transformative role of geospatial technology-encompassing Geographic Information Systems (GIS), remote sensing, and Global Navigation Satellite Systems (GNSS) in overcoming these obstacles. The chapter discusses the use of high-resolution satellite imagery and Unmanned Aerial Vehicles (UAVs) for precise mapping, site suitability analysis, and disaster risk management. It underscores the integration of socio-economic data with spatial information for more effective, targeted interventions and participatory mapping, ensuring inclusivity and accuracy in planning. Case studies from the region illustrate successful community-based greenfield development, sustainable infrastructure design, and socio-economic empowerment. The chapter advocates for leveraging advanced geospatial tools to optimize connectivity, renewable energy site selection, and environmental conservation, thereby fostering inclusive and resilient rural development. It concludes by stressing the importance of collaborative efforts among stakeholders to achieve sustainable growth, cultural preservation, and improved quality of life in the rural landscapes of Northeast India.

Keywords Northeast India · Geospatial technology · Greenfield development · Rural planning · Sustainable development

---

S. Das (✉) · J. M. Nongkynrih · B. C. Sumanth · K. K. Sarma
North Eastern Space Applications Centre (NESAC), Umiam, Shillong, Meghalaya, India
e-mail: santanu.das.nesac@nesac.gov.in

J. M. Nongkynrih
e-mail: jm.nongkynrih@nesac.gov.in

B. C. Sumanth
e-mail: sumanth.bc@nesac.gov.in

K. K. Sarma
e-mail: kk.sarma@nesac.gov.in

S. P. Aggarwal et al. (eds.), *Remote Sensing of the North Eastern Himalayan Ecosystem*,
https://doi.org/10.1007/978-981-95-3221-6_12

12.1 The Unique Context of Northeast India

The eight states of Northeast India—Arunachal Pradesh, Assam, Manipur, Meghalaya, Mizoram, Nagaland, Sikkim, and Tripura—form a region of unparalleled diversity and intricate challenges for sustainable development. This geographical expanse, nestled within the easternmost reaches of the Indian subcontinent, is not merely a collection of states, but a vibrant tapestry woven with distinct geographical, demographic, and socio-cultural threads. These unique characteristics necessitate a departure from conventional development models, demanding bespoke strategies that acknowledge and address the region's specific needs (Karaeva and Mihajlovska 2021).

The very landscape of Northeast India is a study in contrasts, from the soaring peaks of the Himalayas in Arunachal Pradesh and Sikkim to the lush river valleys of Assam and the rolling hills of Meghalaya. This diverse terrain, while breathtakingly beautiful, presents formidable obstacles to infrastructure development and connectivity (Ziipao 2020). The region's demographic profile is equally complex, characterized by a mosaic of ethnic groups and tribal communities, each with its own rich history, language, and cultural practices. These diverse social structures influence land ownership patterns, resource management, and community participation in development initiatives.

Furthermore, the region's socio-cultural fabric is deeply intertwined with its ecological context. Traditional knowledge systems and sustainable practices, often passed down through generations, play a crucial role in shaping livelihoods and resource utilization. However, the pressures of modernization and development can threaten these traditional practices, necessitating a careful balance between progress and preservation (Maldo et al. 2023).

In the context of greenfield and rural development, these unique characteristics demand tailored approaches. Greenfield development, aimed at establishing new infrastructure and industries, must navigate the challenges of difficult terrain, diverse land tenure systems, and environmental sensitivities (Yusoff et al. 2021). Rural development, crucial for improving the livelihoods of the region's predominantly agrarian population, must focus on sustainable agriculture, skill development, and access to essential services.

In this complex landscape, geospatial technology emerges as a powerful tool for informed decision-making. By leveraging Geographic Information Systems (GIS), remote sensing, and Global Positioning System (GPS), planners can gain valuable insights into the region's geography, resources, and socio-economic conditions. These technologies enable precise mapping of land use patterns, monitoring of environmental changes, and efficient planning of infrastructure projects, ultimately facilitating more sustainable and equitable development outcomes. The integration of geospatial information into development strategies allows for more accurate analysis of the region's unique needs, and more effective implementation of programs (Nur et al. 2018).

12.1.1 Geographic and Demographic Overview

Northeast India is characterized by a diverse and challenging geography, dominated by the Himalayas in the north and the fertile Brahmaputra and Barak River valleys. This topographical variation fosters significant ecological diversity, including dense forests, wetlands, and numerous biodiversity hotspots. The region experiences high rainfall, especially during the monsoon, which frequently leads to floods and landslides (Devi 2020).

Demographically, the Northeast is home to a rich tapestry of ethnic groups and tribes, each with distinct languages, cultures, and traditions. This diversity, while a significant strength, also presents challenges for culturally sensitive and inclusive development strategies. Population density varies, with higher concentrations in river valleys and sparser populations in mountainous regions. A significant majority of the population lives in rural areas and primarily relies on agriculture (Izquierdo and Grau 2009).

A defining characteristic of Northeast India's demographic landscape is tribal dominance, with a substantial proportion of the population belonging to Scheduled Tribes. These diverse tribal communities possess unique cultural identities, social structures, and traditional land tenure systems, often communal and deeply rooted in ancestral practices. Understanding and respecting these systems are crucial for equitable resource management and infrastructure development, as ignoring them can lead to land disputes and marginalization. The region also boasts remarkable linguistic diversity, with a multitude of languages and dialects spoken, necessitating tailored communication strategies for effective development initiatives, education, and public service delivery. The significant rural concentration of the population, with a heavy reliance on agriculture and natural resources, underscores the need for development efforts focused on enhancing agricultural productivity, promoting sustainable resource management, and creating alternative rural livelihood opportunities to address poverty and improve food security. Finally, the Northeast's border proximity to China, Myanmar, Bhutan, Bangladesh, and Nepal present both opportunities for cross-border trade and challenges related to security, including illegal immigration, smuggling, and insurgent activities, particularly in remote borderland areas with limited infrastructure (Katara and Pokhriyal 2017).

12.1.2 Challenges in Greenfield and Rural Development in Northeast Region of India

The Northeast region of India, with its unique geographical, cultural, and ecological context, presents a complex landscape for greenfield and rural development. While the region is endowed with immense natural wealth and cultural diversity, it faces significant challenges that hinder its growth potential. Addressing these challenges is critical for achieving sustainable and inclusive development (Cardey et al. 2024).

Infrastructure Deficit: The Northeast's difficult terrain, characterized by hills, forests, and rivers, poses significant challenges to infrastructure development. Poor road connectivity, inadequate railway networks, and limited air transport facilities restrict access to markets, healthcare, education, and other essential services (Kibetu 2020). This infrastructure gap not only hampers economic growth but also exacerbates the isolation of rural communities. Additionally, the region's harsh climatic conditions and frequent natural disasters further strain existing infrastructure, making maintenance and upgrades a persistent challenge.

Environmental Vulnerability: The region is highly susceptible to natural disasters such as floods, landslides, and earthquakes. These events disrupt development efforts, damage infrastructure, and cause significant economic losses. Climate change exacerbates these vulnerabilities, threatening agriculture, livelihoods, and ecosystems. Balancing development with environmental conservation is a critical challenge, as unchecked exploitation of natural resources could lead to long-term ecological degradation (Pech-May et al. 2022).

Land Tenure Issues: Traditional land ownership patterns and unresolved land disputes are significant barriers to development. In many areas, land is communally owned or governed by customary laws, which often conflict with modern legal frameworks. This creates uncertainty for investors and complicates infrastructure projects, industrial development, and urbanization efforts. Resolving land tenure issues requires a sensitive approach that respects local traditions while ensuring clarity and security for development initiatives (Byamugisha and Dubosse 2023; Hilhorst et al. 2021).

Security Concerns: Insurgency and cross-border issues in certain parts of the Northeast create an unstable environment, deterring investment and impeding development activities. Security challenges not only affect economic growth but also disrupt the delivery of public services and infrastructure projects. Addressing these concerns requires a combination of strategic security measures and inclusive development policies that address the root causes of unrest (Azu et al. 2021; Ibaba 2009; Jelilov et al. 2018).

Limited Industrialization: The Northeast's industrial base remains underdeveloped, with limited manufacturing and processing facilities. This lack of industrialization restricts employment opportunities, particularly for the youth, and limits economic diversification. The region's remoteness, coupled with infrastructure deficits, makes it difficult to attract private investment and establish industries that could drive growth (Cao et al. 2014).

Market Access: Poor connectivity and inadequate infrastructure restrict access to domestic and international markets, hindering the growth of agriculture, horticulture, and other sectors. Farmers and small-scale producers often struggle to transport their goods to markets, resulting in post-harvest losses and reduced incomes. Improving market access is essential for unlocking the economic potential of the region's agricultural and artisanal sectors (Mujuka et al. 2019; Sundram 2023).

Skill Gap: The region faces a significant skill gap in key sectors such as technology, construction, and manufacturing. This limits the capacity for industrial development and economic diversification. Additionally, the lack of quality education and vocational training opportunities exacerbates unemployment and underemployment, particularly among rural youth. Bridging this skill gap is crucial for creating a workforce that can drive sustainable development (González-Gil and Virviescas 2024).

12.1.3 Opportunities for Greenfield and Rural Development in NER of India

While facing considerable challenges, Greenfield and rural development present significant opportunities for sustainable economic growth, infrastructure expansion, and improved quality of life in underserved regions. By leveraging these opportunities, governments, investors, and communities can foster inclusive development while addressing environmental and social challenges.

Natural Resource Abundance: The Northeast is rich in natural resources, including forests, minerals, hydropower, and biodiversity. These resources offer immense potential for sustainable development if managed responsibly. For instance, the region's vast forest cover can support eco-friendly industries such as timber, bamboo products, and non-timber forest products. Similarly, its mineral resources can be tapped for industrial development, while its hydropower potential can meet energy needs and support economic growth (Adegboyega and Bolanle 2021; Firmansyah et al. 2019).

Tourism Potential: The Northeast's scenic beauty, cultural diversity, and unique biodiversity make it a promising destination for eco-tourism, adventure tourism, and cultural tourism. Promoting tourism can create employment opportunities, boost local economies, and foster cultural exchange. However, this requires investments in infrastructure, hospitality, and marketing, as well as measures to ensure that tourism development is sustainable and benefits local communities (Bhuiyan et al. 2013).

Organic Agriculture: The region's fertile soil, favourable climate, and low levels of chemical use make it ideal for organic agriculture and horticulture. There is growing demand for organic products in domestic and international markets, providing an opportunity for farmers to access premium prices. Developing value chains for organic products, including certification, processing, and marketing, can enhance rural incomes and promote sustainable farming practices (Peng 2019).

Renewable Energy: The Northeast has significant potential for renewable energy, particularly hydropower and solar energy. Harnessing these resources can provide clean and reliable energy to support industrial and rural development. Additionally, decentralized renewable energy solutions, such as solar microgrids, can improve energy access in remote areas, enhancing the quality of life and enabling economic activities (Islam et al. 2005; Liu et al. 2025).

Connectivity Improvements: Ongoing initiatives to improve connectivity, such as the Northeast Special Infrastructure Development Scheme (NESIDS) and the Bharat Mala Project, are creating new opportunities for development. Enhanced road, rail, and air connectivity will facilitate trade, tourism, and investment, integrating the region with the rest of India and neighbouring countries. Improved digital connectivity can also bridge the gap in access to information, education, and healthcare (Kathuria and Mathur 2019; Sajjad and Dar 2025).

Act East Policy: India's Act East Policy aims to strengthen economic and strategic ties with Southeast Asian countries. The Northeast, as the gateway to Southeast Asia, stands to benefit significantly from this policy. Enhanced cross-border trade, investment, and cultural exchange can drive economic growth and create new opportunities for the region's industries and workforce.

Bamboo Resources: The Northeast is home to abundant bamboo resources, which can be utilized in diverse ways, including construction, textiles, handicrafts, and food products. Bamboo-based industries have the potential to generate employment, particularly in rural areas, and contribute to sustainable development. Promoting bamboo cultivation and processing can also support environmental conservation by reducing pressure on timber resources (Binfield et al. 2025).

12.1.4 The Role of Geospatial Technology in Greenfield Development and Rural Development: An Overview

Geospatial technology is transforming land-use planning in greenfield development and rural areas, enabling data-driven decisions for sustainable growth. In greenfield projects, which involve building on previously undeveloped land, tools like Geographic Information Systems (GIS), remote sensing, and Global Navigation Satellite Systems (GNSS) facilitate comprehensive site suitability analysis. This involves evaluating factors such as topography, soil composition, and flood susceptibility to identify optimal locations for various developments. For instance, remote sensing and digital elevation models are crucial for assessing terrain risks like landslides, while hydrological modeling aids in designing effective drainage systems for flood prevention. Geospatial tools also optimize infrastructure planning by streamlining road networks, utility distribution, and the placement of public services, with 3D urban modeling allowing for visualization of proposed projects.

In rural planning, geospatial technology plays a pivotal role in agricultural development, natural resource management, and infrastructure enhancement. Precision agriculture benefits from satellite imagery and drone surveys for monitoring crop health and optimizing resource use. Governments utilize GIS-based land-use mapping to identify underutilized land and manage land degradation. Geospatial tools are also vital for water resource management, helping to map groundwater potential and design rainwater harvesting systems. Furthermore, spatial accessibility analysis ensures equitable distribution of rural infrastructure like roads and electrification, with least-cost path analysis identifying the most economical routes.

Beyond development, geospatial technology is crucial for disaster risk reduction and climate resilience planning. This includes flood hazard mapping, wildfire risk assessment, and landslide susceptibility modeling, allowing planners to avoid high-risk zones and implement mitigation strategies. LiDAR technology provides high-resolution terrain data for accurate floodplain mapping, while thermal sensors detect forest fire risks. Despite challenges like high initial costs and data interoperability, advancements such as AI-powered geospatial analytics, real-time satellite monitoring, and blockchain-based land records are overcoming these hurdles. The rise of participatory GIS (PGIS) also empowers local communities to contribute their knowledge, leading to more inclusive development. Ultimately, investing in geospatial technologies and building expertise is essential for fostering smarter, more sustainable, and resilient land-use strategies for future generations in both greenfield and rural contexts (Abdullah et al. 2019; Buenemann et al. 2010; Nur et al. 2018).

12.2 Geospatial Data Acquisition and Management in Northeast India

12.2.1 Remote Sensing: Satellite Imagery and Aerial Photography

12.2.1.1 High-Resolution Satellite Data Sources for Greenfield Town Planning and Rural Development

High-resolution satellite imagery has revolutionized urban and rural planning by providing precise, real-time geospatial data essential for sustainable development. With sub-meter resolution capabilities, modern satellites enable detailed mapping, environmental monitoring, and infrastructure planning, making them indispensable for Greenfield town development and rural revitalization. Among the leading optical satellites, Albedo stands out with its ultra-high 10 cm panchromatic and 40 cm multispectral resolution, offering daily revisits for urban cadastral surveys, precision agriculture, and disaster response. Similarly, EROS NG (30–50 cm resolution) excels in rural applications, supporting land-use classification, deforestation tracking, and infrastructure planning with its rapid revisit capability. Pleiades Neo (30 cm multispectral/15 cm HD panchromatic) enhances smart city development with 3D urban modeling and Red Edge band analysis for vegetation health, while Beijing-3 (BJ3N), equipped with AI-driven analytics, aids China's rural revitalization through precision farming, land-use enforcement, and digital twin development (Chen et al. 2020; Yu et al. 2024).

For large-scale rural and Greenfield projects, SuperView-Neo (30 cm) provides daily stereo imaging for DEM generation, flood risk assessment, and cadastral mapping, while WorldView Legion (30 cm multispectral/15 cm HD) offers unmatched spectral diversity-including SWIR and Red Edge bands-for environmental impact assessments, resilient infrastructure planning, and forest health monitoring. Pleiades-1A and 1B (50 cm) support flood-prone region planning with NIR-based water detection, and KOMPSAT-3A (55 cm) enables change detection analysis for illegal land conversion and disaster assessment. Mid-resolution options like Gaofen-2 (80 cm) and TripleSat (80 cm) balance cost and coverage, ideal for agricultural zoning, rural road planning, and large-area monitoring in developing regions. SkySat (50 cm) distinguishes itself with video capabilities, allowing real-time construction tracking and dynamic land-use monitoring-critical for Greenfield project oversight (Tripathy et al. 2024; Yang et al. 2016).

Beyond optical systems, radar satellites like RISAT (1–3 m, C-band) and TerraSAR-X (1 m, X-band) provide all-weather, day-and-night imaging, essential for flood and landslide monitoring in monsoon-prone regions. TerraSAR-X's interferometric deformation analysis ensures stable infrastructure development, while RISAT's soil moisture mapping aids precision agriculture in cloud-covered areas (Rasid et al. 2019).

By integrating these high-resolution satellite datasets, planners can optimize land use, enhance disaster resilience, and ensure sustainable rural–urban transitions. From AI-powered analytics in Beijing-3 to real-time SkySat videos, these technologies empower data-driven decision-making, bridging the gap between geospatial innovation and equitable development. As climate change and urbanization intensify, leveraging these advanced satellite systems will be crucial for building resilient, well-planned communities worldwide.

The Table 12.1 compares leading high-resolution Earth observation satellites by their panchromatic and multispectral resolution, revisit rates, and specialized applications. Frequent revisits and sub-meter clarity enable uses ranging from cadastral mapping and infrastructure monitoring to disaster response and dynamic urban change detection.

12.2.1.2 The Transformative Role of UAV in Rural and Greenfield Planning: Revolutionizing Geospatial Intelligence

Unmanned Aerial Vehicles (UAVs), or drones, have evolved beyond their military origins to become a groundbreaking tool in geospatial data collection and analysis. Equipped with advanced multispectral sensors capable of capturing ultra-high-resolution imagery, down to an astonishing 5 cm. These aerial systems are reshaping rural and greenfield planning by providing unparalleled accuracy in Digital Surface Models (DSMs) and Digital Terrain Models (DTMs). Unlike traditional ground surveys or satellite imagery, UAVs offer a cost-effective, rapid, and dynamic solution for spatial and spectral data acquisition, particularly in challenging environments like India's North Eastern Region (NER), where satellite data is often hindered by

Table 12.1 High-resolution satellite capabilities and applications

Satellite	Resolution (Pan/MS)	Revisit time	Key applications
Albedo	10 cm/40 cm	Daily	Cadastral mapping, precision agriculture
EROS NG	30 cm/50 cm	<1 day	Rapid urban monitoring, disaster response
Pleiades Neo	30 cm/15 cm (HD)	Daily	3D city modelling, infrastructure planning
Beijing-3	30 cm/1.2 m	Daily	Urban expansion, land-use change detection
SuperView-Neo	30 cm/1.2 m	Daily	DEM generation, cadastral surveys
WorldView Legion	30 cm/15 cm (HD)	Multiple/day	Smart cities, precision agriculture
Pleiades-1A/1B	50 cm/2 m	Daily	Stereo mapping, flood risk assessment
KOMPSAT-3A	55 cm/2.2 m	3–4 days	Vegetation health, rural planning
Gaofen-2	80 cm/3.2 m	5 days	Agricultural zoning, regional planning
TripleSat	80 cm/3.2 m	3–4 days	Large-area mapping, rural road planning
SkySat	50 cm/2 m	Multiple/day	Construction monitoring, dynamic changes
RISAT	1–3 m (SAR)	4–5 days	Flood mapping, all-weather imaging
TerraSAR-X	1 m (spotlight)	2–3 days	Subsidence monitoring, urban deformation

dense cloud cover and rugged terrain. The ability to generate precise land-use maps, monitor vegetation health, and assess infrastructure needs makes UAVs indispensable for sustainable rural development and strategic greenfield projects (Bolo et al. 2019; Rokhmana and Utomo 2016).

One of the most significant advantages of UAV-based surveys lies in their capacity to produce high-resolution multispectral imagery, enabling detailed land cover classification and environmental monitoring. These datasets reveal critical insights into agricultural health, soil conditions, and water resources through vegetation indices like the Normalized Difference Vegetation Index (NDVI), facilitating precision farming, forest conservation, and water management. For instance, farmers can identify nutrient-deficient crop areas and apply targeted interventions, while forestry officials can detect deforestation or assess biomass with remarkable accuracy. Additionally, UAV-derived DSMs and DTMs provide essential 3D terrain data, crucial for infrastructure planning, flood mitigation, and ecological assessments. The difference between DSMs and DTMs yields Canopy Height Models (CHMs), which are vital for forestry applications, such as estimating carbon stocks or analysing habitat structures (Vastaranta et al. 2013).

In greenfield planning, UAVs offer a comprehensive baseline assessment of undeveloped land, capturing existing vegetation, topography, and drainage patterns before construction begins. This pre-development data ensures sustainable decision-making by identifying ecologically sensitive zones, optimizing drainage designs, and minimizing environmental disruption. For example, multispectral analysis can pinpoint biodiversity hotspots, allowing planners to avoid or mitigate harm to endangered species. Furthermore, volumetric measurements from UAV-generated 3D models enhance construction efficiency by accurately estimating earthwork requirements, reducing costs and resource waste.

The NER's unique challenges, persistent cloud cover, dense forests, and complex terrain make satellite data unreliable, positioning UAVs as a superior alternative. Drones can operate below cloud layers, ensuring consistent data collection even during monsoons, while their flexibility allows for on-demand, high-frequency monitoring. Though UAVs cover smaller areas per flight than satellites, their ultra-high-resolution data complements broader satellite imagery, creating a more holistic geospatial framework. Advances in AI and machine learning further enhance UAV capabilities by automating feature extraction, such as identifying infrastructure or classifying land use, streamlining data processing for rural planners (Kamarulzaman et al. 2023).

As UAV technology continues to evolve with improvements in sensor miniaturization, battery efficiency, and AI-driven analytics, its role in rural and greenfield planning will only expand. By integrating drone-derived insights with existing geospatial datasets, planners can achieve a more nuanced understanding of landscapes, driving sustainable development in regions where traditional methods fall short. The ascendancy of UAVs marks a paradigm shift in geospatial intelligence, offering precision, adaptability, and efficiency that redefine how we approach rural and environmental planning (Bolo et al. 2019; Ejaz and Choudhury 2024).

12.2.2 Geographic Information Systems (GIS): Data Integration and Analysis

12.2.2.1 Geospatial Data Development and Spatial Data Infrastructure: Foundational Pillars for Comprehensive Rural Planning and Development

Geospatial data development and the establishment of a robust Spatial Data Infrastructure (SDI) represent indispensable foundations for contemporary rural planning and development. They move beyond traditional approaches by enabling data-driven decision-making, fostering equitable resource allocation, and promoting sustainable growth trajectories within rural landscapes. SDI serves as a comprehensive and standardized framework encompassing the intricate processes of collecting, meticulously managing, rigorously analysing, and efficiently disseminating a diverse array of both geospatial and non-spatial data. This framework is paramount in ensuring seamless interoperability and data sharing across disparate sectors, thereby breaking down silos and fostering collaborative efforts. The true power of this integrated approach lies in its capacity to synthesize traditionally separate datasets. By strategically integrating precise topographic and detailed cadastral data with nuanced socio-economic indicators, planners gain unprecedented analytical capabilities. This synthesis allows for the generation of insightful demand-gap analyses for critical infrastructure, the optimization of complex transportation networks to enhance connectivity, the identification of strategically advantageous sites for renewable energy projects promoting energy independence, and the bolstering of overall disaster resilience to safeguard vulnerable communities (Sun et al. 2022). Ultimately, this comprehensive and integrated approach ensures that rural development initiatives are inherently inclusive, proactively addressing existing disparities in access to essential physical and social infrastructure while simultaneously prioritizing the crucial preservation of ecological balance and environmental sustainability.

Geospatial Data Development, at its core, encompasses the multifaceted processes involved in the acquisition, meticulous processing, rigorous analysis, and insightful interpretation of spatial information. This includes a wide spectrum of data sources, ranging from high-resolution satellite imagery providing a synoptic view, detailed surveys conducted using unmanned aerial vehicles (drones) offering granular perspectives, precise location data captured through Global Positioning Systems (GPS), and the wealth of information derived from advanced remote sensing techniques. The true value of this spatially referenced information is unlocked when it is systematically structured and organized within a well-defined **Spatial Data Infrastructure (SDI) framework**. SDI acts as the backbone for making this valuable data readily accessible and usable by a diverse range of stakeholders, fostering crucial collaboration and information sharing among various government agencies operating at different scales, academic and research institutions pushing the boundaries of knowledge, and even private sector entities seeking to contribute to rural development. A well-designed SDI ensures that fundamental datasets, such as comprehensive land use and land cover maps, detailed cadastral records outlining property ownership, and critical environmental data layers, adhere to established standards for data format, metadata, and accessibility. This standardization is crucial for reducing redundancy in data collection efforts, improving the overall efficiency of data management, and facilitating the seamless integration of diverse datasets. The synergistic integration of this rich geospatial data with pertinent socio-economic statistics provides a powerful lens through which to gain a comprehensive and multidimensional understanding of the multifaceted challenges confronting rural areas, thereby enabling the design and implementation of more targeted and effective interventions (Daley 2010).

12.2.2.2 Foundational Datasets: Topography and Cadastre in Rural Development

Topographic Data Management is a cornerstone of informed rural planning, providing essential information about the physical landscape. This includes critical data on terrain elevation, the spatial distribution of water bodies (rivers, lakes, wetlands), and the characterization of various landforms. This information is indispensable for a wide range of applications, including the meticulous planning and design of infrastructure projects (roads, irrigation systems, buildings), the development of accurate flood inundation models to mitigate water-related disasters, and comprehensive hazard assessments to identify areas prone to natural risks. Complementing topographic data is **Cadastral Data**, which provides detailed and legally recognized records of land ownership boundaries, property rights, and land parcel information. This data is fundamental for effective property taxation systems, the establishment and enforcement of coherent land-use zoning regulations, and the efficient resolution of land-related disputes, which can often be significant impediments to rural development. The synergistic combination of these two fundamental datasets forms a robust foundation upon which sustainable and well-planned rural development projects can be built. This ensures that essential infrastructure, such as schools, healthcare facilities, and transportation networks, are strategically located in optimal areas, taking into account both physical suitability and land ownership considerations. Modern technological advancements, such as Light Detection and Ranging (LiDAR), which provides highly accurate three-dimensional terrain models, and high-resolution satellite imagery, significantly enhance the accuracy and detail of topographic mapping. Furthermore, innovative technologies like blockchain are increasingly being explored for establishing secure, transparent, and immutable systems for cadastral data management, promising to improve land tenure security and reduce instances of fraud and disputes (Rashid et al. 2025).

12.2.2.3 Integrating the Human Dimension: Socio-economic Data for Targeted Interventions

The Integration of Socio-Economic Data is crucial for understanding the human context within rural development. This involves the systematic collection and analysis of information pertaining to population demographics (age structure, gender distribution), household income levels and poverty incidence, educational attainment and literacy rates, access to healthcare facilities and health indicators, and employment trends across various sectors. This vital data is gathered through a combination of rigorous primary surveys, including detailed household interviews and comprehensive field studies, and the utilization of reliable secondary sources, such as national census data and relevant government reports. The true power of this socio-economic information is realized when it is effectively assimilated with spatially referenced data within the framework of Geographic Information Systems

(GIS). This integration allows for a nuanced and geographically disaggregated understanding of the social and economic landscape, enabling policymakers to identify underserved regions and vulnerable populations with greater precision. For instance, overlaying spatial poverty maps with the existing road network can visually reveal critical connectivity gaps that hinder economic opportunities (Liu et al. 2023). Similarly, combining spatial data on literacy rates with the geographic locations of schools can highlight areas with significant educational disparities, informing targeted interventions. Advanced analytical techniques, including sophisticated machine learning algorithms, can further refine demand-gap analyses by identifying complex relationships and predicting future needs, ensuring that development initiatives are not only data-driven but also demonstrably equitable and responsive to the specific needs of rural communities.

12.2.3 GPS and GNSS: Field Data Collection and Accuracy

12.2.3.1 The Architecture of Global Navigation: Understanding GPS and GNSS

The Global Positioning System (GPS) and the broader category of Global Navigation Satellite Systems (GNSS) represent humanity's remarkable achievement in leveraging space technology for precise positioning, navigation, and timing (PNT) services anywhere on or near the Earth's surface (Zhang and Sun 2021). These systems are fundamental to countless modern technologies and activities, from simple driving directions on a smartphone to complex military operations and precision agriculture (Ning et al. 2022).

GPS: GPS, originally named NAVSTAR GPS (Navigation System with Timing and Ranging), was developed and is maintained by the United States Department of Defense. Initially conceived for military purposes, it was later made available for civilian use worldwide. GPS operates using a dedicated constellation of satellites orbiting the Earth. While often used colloquially to refer to any satellite navigation, GPS is technically just one component—albeit the most widely recognized—of the larger GNSS ecosystem.

GNSS: GNSS is the umbrella term encompassing all global satellite-based positioning systems. Recognizing the strategic and economic importance of satellite navigation, other nations and blocs developed their own independent systems, creating a multi-constellation environment. This offers users significant advantages, including improved accuracy, availability, and integrity, especially in challenging environments like urban canyons or mountainous terrain where signals from one system might be blocked. The major GNSS constellations include:

- **GPS (United States)**: The pioneering system, currently operating with around 31 active satellites.

- **GLONASS (Russia)**: (Global Navigation Satellite System) Russia's counterpart to GPS, providing global coverage.
- **Galileo (European Union)**: A civilian-controlled system designed for high accuracy and interoperability with GPS and GLONASS.
- **BeiDou (China)**: (BDS) China's system, initially regional but now offering global services (Maciuk and Szombara 2018).

In addition to these global systems, there are also Regional Navigation Satellite Systems (RNSS) like India's NavIC (Navigation with Indian Constellation) and Japan's QZSS (Quasi-Zenith Satellite System), which enhance PNT capabilities over specific geographic areas, often by augmenting the global constellations.

The Fundamental Principle: Measuring Time to Determine Distance

The operational principle behind all GNSS is elegantly simple yet technologically sophisticated: precise timing. Here's a breakdown of the key steps.

Satellite Constellation: Each GNSS maintains a network of satellites (typically 24 or more for global coverage) in specific orbits, usually Medium Earth Orbit (MEO), around 20,000 km altitude. These orbits are carefully designed so that from almost any point on Earth, at least four satellites are "visible" (above the horizon and unobstructed) at any given time.

Signal Transmission: Each satellite continuously broadcasts radio signals (primarily in the L-band frequency range). These signals contain crucial information:

a. **Pseudo-Random Codes (PRN)**: Unique identifying codes for each satellite, allowing the receiver to distinguish between them. These codes are also essential for measuring the signal travel time.
b. **Navigation Message**: Contains data about the satellite's precise orbital position (ephemeris), the general orbital information for all satellites in the constellation (almanac), satellite clock correction data (compared to a master GNSS time standard), and system health status. Highly stable atomic clocks onboard each satellite ensure the extraordinary timing precision required (Li et al. 2020).

Receiver Processing: A GNSS receiver on the ground (in your phone, car, or specialized equipment) detects signals from multiple satellites. It calculates its distance to each satellite by measuring the time it took for the signal to travel from the satellite to the receiver. This is done by comparing the time the signal was transmitted (encoded in the signal) with the time it was received, according to the receiver's internal clock. Since radio waves travel at the speed of light (a known constant), this time delay translates directly into a distance measurement. This calculated distance is called a "pseudorange" because it initially includes errors, most notably the receiver's own clock inaccuracy relative to the highly precise satellite clocks (Liao et al. 2024).

Position Calculation (Multilateration): To determine its precise location, the receiver needs pseudorange measurements from at least four satellites.

- With one satellite, the receiver knows it's somewhere on the surface of a sphere centered on that satellite with a radius equal to the calculated pseudorange.
- With two satellites, the location is narrowed down to the intersection of two spheres—a circle.
- With three satellites, the intersection of three spheres ideally results in two points, one of which is usually easily discarded (e.g., it's far out in space or deep inside the Earth). This gives a 3D position (latitude, longitude, altitude).
- However, the receiver's clock error introduces uncertainty. The signal from a *fourth* satellite provides the necessary additional information to solve for this clock offset, simultaneously refining the latitude, longitude, and altitude calculations, and determining precise time. This process of determining position based on range measurements from multiple points is known as multilateration (Karagiannis et al. 2012).

Achieving Higher Accuracy: Error Correction Techniques

While basic GNSS provides remarkable accuracy (typically within a few meters), various factors introduce errors (atmospheric delays in the ionosphere and troposphere, satellite orbital inaccuracies, clock drift, multipath signal reflection off buildings/terrain, receiver noise). To achieve higher precision required for applications like surveying, precision farming, or autonomous vehicle navigation, correction techniques are employed:

- **Differential GPS (DGPS)/SBAS**: This involves using fixed ground-based reference stations whose exact locations are known. These stations receive the same satellite signals as user receivers, calculate the errors in the signals, and broadcast correction messages. Receivers in the vicinity can apply these corrections to significantly improve accuracy, often to sub-meter levels. Satellite-Based Augmentation Systems (SBAS) like WAAS (USA), EGNOS (Europe), MSAS (Japan), and GAGAN (India) distribute these corrections over wide areas via geostationary satellites (Kim and Kim 2021).
- **Real-Time Kinematic (RTK)**: A more sophisticated technique that also uses a reference station but analyzes the carrier phase of the satellite signal itself, not just the information codes. By resolving ambiguities in the carrier wave cycles, RTK can achieve centimeter-level accuracy in real-time, making it invaluable for high-precision surveying and machine control (Park et al. 2020; Shu et al. 2019).

The GPS and the broader GNSS family represent a constellation of technologies working in concert-precise clocks, orbital mechanics, radio signals, and sophisticated receiver processing—to provide indispensable positioning, navigation, and timing information that underpins a vast array of modern global infrastructure and activities (Koelemeij et al. 2022).

12.2.3.2 Applications of GPS/GNSS in Rural Development and Greenfield Planning

Greenfield Planning and New Township Development

Global Positioning System (GPS) and Global Navigation Satellite System (GNSS) technologies play a pivotal role in the foundational stages of greenfield planning and the development of entirely new townships. A critical component of this process is the use of advanced geospatial technologies, particularly the Global Positioning System (GPS) and Global Navigation Satellite System (GNSS), which provide high-precision data essential for foundational planning. These technologies enable highly accurate topographic and cadastral surveys, forming the basis for land assessment, zoning, and infrastructure design. By leveraging GNSS, urban planners can map existing land use patterns, terrain contours, and natural features such as water bodies, vegetation, and geological formations, ensuring that development aligns with the site's ecological and topographical characteristics.

One of the primary applications of GNSS in greenfield planning is the demarcation of land boundaries, which is crucial for defining residential, commercial, industrial, and recreational zones. Precise boundary delineation helps prevent land disputes, optimizes space utilization, and ensures compliance with regulatory frameworks. Additionally, GNSS-derived terrain data plays a vital role in infrastructure planning, particularly in designing road networks, drainage systems, and utility corridors. By analyzing elevation and slope data, planners can minimize earthwork requirements, reduce construction costs, and enhance the integration of infrastructure with the natural landscape. This approach not only improves efficiency but also reduces environmental disruption.

Another key advantage of GNSS technology is its ability to identify environmental constraints, such as flood-prone areas, unstable slopes, or ecologically sensitive zones. Early detection of these factors allows planners to implement mitigation strategies, such as adjusting building layouts, reinforcing slopes, or preserving natural drainage systems. This proactive approach supports sustainable urban development by minimizing environmental degradation and enhancing resilience to climate-related risks. Furthermore, GNSS facilitates real-time monitoring during the construction phase, ensuring that development adheres to approved designs and regulatory standards.

Beyond infrastructure and environmental considerations, GNSS data supports smart city initiatives by enabling the integration of digital mapping with urban management systems. This includes applications in traffic management, utility distribution, and emergency response planning, ensuring that new townships are future-ready. The use of GNSS in greenfield planning also enhances stakeholder collaboration, as accurate geospatial data can be shared among government agencies, developers, and environmental consultants, fostering coordinated decision-making.

In conclusion, GPS and GNSS technologies are indispensable in greenfield planning and new township development, providing the accuracy and efficiency needed for sustainable urban expansion. From initial land surveys to infrastructure design and environmental risk assessment, these systems ensure that new developments are well-planned, resilient, and adaptable to future growth. As urbanization continues to accelerate globally, the integration of advanced geospatial tools will remain a cornerstone of modern urban planning, enabling the creation of livable, efficient, and environmentally conscious townships (Hameed 2021; Nyakwebara et al. 2023).

Town Planning Schemes

Town Planning Schemes (TPS) serve as a vital instrument for organized urban expansion, particularly in rapidly developing regions like India, where they facilitate the equitable redistribution of land parcels while ensuring the provision of essential infrastructure and public amenities. At the heart of these complex land reorganization efforts lies the indispensable role of Global Navigation Satellite Systems (GNSS), including GPS, GLONASS, Galileo, and BeiDou, which provide the spatial precision required for successful scheme implementation. These technologies revolutionize three critical aspects of TPS execution: land parcel mapping, zoning regulation enforcement, and infrastructure alignment. In land parcel mapping-the foundational stage of any TPS-GNSS enables surveyors to achieve centimetre-level accuracy in demarcating property boundaries, a quantum leap from traditional surveying methods prone to human error and inconsistencies. This geodetic-grade precision is crucial not only for creating legally defensible cadastral records but also for ensuring fair land reallocation, calculating accurate compensation, and preventing the boundary disputes that could otherwise derail entire projects. The technology's reliability transforms abstract planning concepts into physically verifiable markers on the ground, giving both authorities and landowners confidence in the scheme's integrity.

Beyond parcel demarcation, GNSS plays an equally critical role in enforcing zoning regulations-the urban planning mechanism that segregates land into residential, commercial, industrial, and recreational zones to prevent haphazard development. By providing exact georeferenced boundaries, GNSS allows urban authorities to precisely translate zoning maps from planning documents to physical locations, enabling rigorous monitoring of land use compliance. This capability is particularly valuable in preventing unauthorized encroachments and ensuring that development aligns with the TPS's socioeconomic and environmental objectives. Moreover, the spatial clarity offered by GNSS-tagged zoning data enhances public understanding of development regulations, fostering greater community participation and reducing conflicts between stakeholders. When integrated with mobile mapping technologies, GNSS enables real-time verification of construction activities against approved zoning plans, creating a dynamic system of urban governance that can adapt to on-ground realities while maintaining the scheme's planned vision.

The third transformative application of GNSS in TPS lies in infrastructure planning and alignment, where the technology's precision ensures optimal design of roads, utilities, and public services. By providing accurate topographic data-including elevation models and existing feature locations-GNSS allows engineers to design infrastructure networks that harmonize with natural terrain, minimizing costly earthworks and land acquisition. Road alignments can be optimized for gradient and connectivity, drainage systems can leverage natural slopes for gravity flow, and utility corridors can be planned to avoid conflicts with property boundaries. This spatially intelligent approach, often enhanced through integration with Geographic Information Systems (GIS), not only reduces construction costs but also creates more sustainable urban systems. For instance, GNSS-guided alignment of stormwater drains can significantly improve flood resilience, while precise positioning of utility lines simplifies future maintenance and upgrades. The cumulative impact of GNSS in these three domains-parcel mapping, zoning enforcement, and infrastructure design-makes it an irreplaceable tool for executing Town Planning Schemes that are not only legally robust and equitable but also functionally efficient and adaptable to future urban needs. As cities continue to expand, the synergy between TPS methodologies and GNSS technologies will only grow more critical in shaping sustainable, well-planned urban landscapes (Billones et al. 2021; Costa et al. 2024).

The Role of GPS in Infrastructure Development

Infrastructure development is a cornerstone of economic growth and urbanization, demanding precise planning, efficient execution, and sustainable management. Among the most transformative technologies enabling modern infrastructure projects is the Global Positioning System (GPS), which provides unparalleled accuracy in surveying, design, and construction. GPS plays a pivotal role in transportation networks, water supply systems, and sanitation facilities, ensuring efficiency, cost-effectiveness, and environmental sustainability. In road and railway development, GPS eliminates the inefficiencies of traditional surveying by enabling rapid, high-precision terrain mapping. This allows engineers to optimize routes, minimize land conflicts, and ensure safe gradients in challenging terrains. GPS-guided machinery, enhanced by real-time kinematic (RTK) technology, ensures centimetre-level accuracy in earthwork, reducing material waste and accelerating timelines. For water supply and sewerage systems, GPS facilitates detailed 3D utility mapping, preventing clashes with existing infrastructure and ensuring proper slope gradients for gravity-based drainage. In sanitation management, GPS aids in strategically locating waste treatment facilities by analysing terrain, hydrology, and proximity to residential areas, mitigating environmental risks. Beyond construction, GPS supports infrastructure maintenance through real-time structural health monitoring, detecting issues like subsidence in bridges or pipelines. It also enhances disaster resilience by optimizing drainage systems for flood-prone areas and enabling rapid damage assessment post-disasters (Shen et al. 2019). By integrating GPS with geographic information systems (GIS), urban planners can streamline long-term maintenance and upgrades, making cities smarter and more sustainable. Thus, GPS is not just a tool for construction but a critical asset for the entire lifecycle of infrastructure, ensuring resilience, efficiency, and adaptability in an ever-evolving urban landscape.

Applicability of GPS/GNSS in Participatory Mapping

Participatory mapping, empowered by GPS and GNSS technology, has revolutionized rural planning by making it more inclusive, accurate, and community-centric. Unlike traditional top-down approaches, this method actively engages local residents in collecting and interpreting spatial data, ensuring their knowledge and priorities shape development decisions. By equipping communities with GPS-enabled devices, planners can capture precise geospatial information on land use, natural resources, infrastructure, and cultural sites-data critical for effective rural development. A key strength of this approach lies in its ability to integrate indigenous knowledge with technical precision. Rural communities possess intimate understanding of their environment, including undocumented land boundaries, water sources, and hazard-prone areas. When combined with GPS/GNSS, this local expertise translates into highly accurate digital maps, correcting errors from outdated surveys. Farmers, for instance, can map agricultural lands and grazing areas, while community leaders highlight infrastructure gaps like roads or healthcare facilities, revealing real challenges such as land disputes or resource inequities. Beyond improving data quality, participatory mapping builds trust and transparency between communities and authorities. Involving residents fosters ownership of development projects, reducing resistance and empowering marginalized groups-women, indigenous populations, and smallholder farmers-to voice their needs. In land-conflict zones, GPS-verified participatory maps can formalize customary land rights, preventing disputes. Modern tools like handheld GPS devices, smartphone apps, and drones further enhance efficiency, enabling even non-experts to contribute high-precision data for GIS analysis. This integration not only accelerates mapping but also supports scalable, long-term monitoring of rural landscapes. The result is development that is both sustainable and people-driven, ensuring interventions-whether in water management, agriculture, or infrastructure-are practical and culturally grounded. As technology advances, participatory mapping will continue to democratize rural planning, making it more adaptive and equitable, ultimately fostering resilient and inclusive rural futures (Ros-Tonen et al. 2021).

Accuracy Considerations in Mountainous Terrain for GNSS/GPS

GNSS navigation in mountainous regions faces significant challenges due to signal obstruction, multipath errors, and atmospheric disturbances, requiring advanced solutions to maintain positioning accuracy. The rugged topography of mountains often blocks satellite signals, reducing visibility below the critical four satellites needed for 3D positioning, while reflective surfaces like cliffs introduce multipath errors that distort measurements by several meters. Atmospheric variations at high altitudes further degrade signal integrity, compounded by dense forest cover in valleys that attenuates GNSS signals. To overcome these limitations, Differential GPS (DGPS) and Real-Time Kinematic (RTK) GNSS provide sub-meter to centimetre-level precision by leveraging ground reference stations and carrier-phase measurements, respectively-essential for applications like landslide monitoring and infrastructure development. Multi-constellation receivers, which utilize GPS, GLONASS, Galileo, and BeiDou simultaneously, enhance satellite availability and improve geometric strength, reducing position dilution of precision (PDOP) in obstructed environments. Post-processing corrections from Continuously Operating Reference Stations (CORS) refine data accuracy in remote areas, while advanced algorithms mitigate multipath interference by analysing signal patterns. Hybrid solutions, such as GNSS coupled with Inertial Navigation Systems (INS), ensure uninterrupted positioning during signal outages by using accelerometers and gyroscopes for dead reckoning. Augmentation systems like SBAS (WAAS, EGNOS) further enhance accuracy where local base stations are impractical. Emerging technologies, including AI-driven error modeling and LEO satellite augmentation, promise even greater resilience, enabling reliable GNSS performance in mountainous terrain for critical applications in surveying, disaster response, and environmental monitoring. These innovations ensure that despite the inherent difficulties of high-altitude navigation, precise and dependable positioning remains achievable (Yasyukevich et al. 2024).

12.3 Applications of Geospatial Technology in Greenfield Development

12.3.1 Strategic Infrastructure Planning and Informed Site Suitability Analysis

Effective **Infrastructure Planning** in rural areas necessitates a meticulous approach to site selection, aiming to maximize accessibility for the intended beneficiaries while simultaneously minimizing potential negative environmental and social impacts. Geospatial tools provide powerful capabilities for conducting **Multi-Criteria Decision Analysis (MCDA)**, a structured framework that allows planners to systematically evaluate and weigh various relevant factors to determine the most suitable locations. These factors can include proximity to existing settlements and population centers, the inherent stability of the terrain to support construction, the availability of essential resources (water, power), and potential environmental sensitivities of the area (ÖZ 2022). For example, GIS-based suitability modeling can be employed to identify the optimal locations for establishing sustainable water supply systems by analyzing crucial parameters such as groundwater levels, population density patterns, and topographical constraints. Similarly, renewable energy projects, such as small-scale hydropower plants or solar farms, can be strategically located using detailed slope analysis derived from topographic data and accurate river flow data, ensuring both energy generation efficiency and minimal ecological disruption, thereby promoting sustainable energy access for rural communities (Odiji et al. 2021).

12.3.1.1 Optimizing Connectivity: Road and Transportation Network Efficiency

Road and Transportation Network Optimization is a critical catalyst for overall rural development, acting as the vital arteries that connect agricultural producers to markets, facilitate access for students to educational institutions, and ensure that patients can reach essential healthcare facilities. Geospatial data plays a pivotal role in achieving efficient and equitable transportation networks. By analyzing existing traffic patterns and volumes, understanding the spatial distribution of the rural population, and accounting for the inherent challenges posed by the terrain, planners can leverage geospatial tools to optimize existing networks and strategically plan new infrastructure. Network analysis tools within GIS software can identify the shortest, most cost-effective, and most accessible routes for the development of new roads, taking into account factors like slope, soil stability, and existing infrastructure. Furthermore, remote sensing technologies can provide real-time monitoring of road conditions, enabling proactive and timely maintenance interventions, thereby reducing transportation disruptions and costs. Crucially, the integration of transportation data with detailed socio-economic indicators ensures that road expansion projects prioritize connectivity to marginalized communities and underserved areas, actively working to reduce regional inequalities and promote inclusive growth (Sajjad and Dar 2025).

12.3.1.2 Harnessing Renewable Resources: Hydropower and Energy Site Selection

Achieving widespread **Rural Electrification** is a fundamental prerequisite for fostering economic growth and improving the overall quality of life in rural areas. However, many remote and geographically dispersed rural communities continue to lack access to reliable and affordable energy sources. Geospatial technologies offer powerful tools for facilitating the identification of suitable sites for various renewable energy projects, including small-scale hydropower installations, solar energy farms, and wind energy parks. The potential for hydropower generation can be accurately assessed by analysing detailed river flow data, utilizing high-resolution elevation models to determine head and potential energy output, and conducting thorough environmental impact assessments to ensure sustainability. Similarly, the optimal locations for solar energy projects can be determined by analyzing spatial solar irradiance maps, considering land-use suitability, and assessing potential shading from surrounding terrain or vegetation. GIS-based spatial analysis plays a crucial role in ensuring that these renewable energy projects are developed in a sustainable manner, minimizing deforestation, avoiding the displacement of local communities, and mitigating potential environmental consequences. Furthermore, geospatial data is invaluable in the planning and deployment of decentralized microgrids, which can significantly enhance energy access in remote and off-grid villages by efficiently distributing locally generated renewable energy (Caputo et al. 2023).

12.3.1.3 Adapting Urban Innovations: Smart City Concepts for Rural Advancement

While the concept of Smart Cities is often associated with urban environments, the underlying principles of leveraging technology and data for improved efficiency, sustainability, and quality of life can be effectively adapted and applied to the unique context of rural development. By meticulously mapping existing settlement patterns, carefully assessing the availability of local resources, and comprehensively identifying service delivery gaps, rural areas can strategically implement Smart Village Concepts. These concepts can encompass a range of technology-enabled solutions, such as digital governance platforms for improved citizen engagement, the application of Internet of Things (IoT) technologies for precision agriculture to enhance productivity and resource management, and the delivery of remote healthcare services through e-health initiatives. Furthermore, GIS-based urban sprawl analysis, often used in urban planning, can be adapted to rural contexts to better understand and manage rural–urban migration patterns by identifying factors that contribute to out-migration and implementing strategies to improve living conditions and economic opportunities within villages, thereby reducing pressure on urban centers. Geospatial tools also play a vital role in the strategic planning of rural growth centers, ensuring that economic activities are decentralized, sustainably managed, and accessible to a wider rural population.

12.3.2 Environmental Impact Assessment and Monitoring

12.3.2.1 Mitigating Natural Hazards: Landslide Hazard Zonation and Risk Assessment

Mountainous and hilly rural regions are particularly susceptible to the devastating impacts of landslides, which can result in significant loss of life, damage critical infrastructure, and disrupt livelihoods. Geospatial technologies provide essential tools for Landslide Hazard Zonation and comprehensive Risk Assessment. By meticulously analysing a range of contributing factors, including slope steepness and aspect, prevailing rainfall patterns and intensity, underlying soil types and geological formations, and historical records of past landslide events, geospatial models can identify areas with varying degrees of landslide susceptibility. Furthermore, advanced machine learning models, trained on historical landslide data and associated environmental factors, can be employed to predict high-risk zones with increasing accuracy, allowing for the implementation of proactive preventive measures such as terracing and afforestation to stabilize slopes, and the development of effective early warning systems to alert vulnerable communities. The resulting risk assessment maps are invaluable for guiding future infrastructure development, ensuring that the construction of roads, buildings, and other essential facilities avoids unstable areas, thereby significantly enhancing the overall resilience of rural communities to these natural hazards (Padre et al. 2023).

12.3.2.2 Protecting Natural Capital: Deforestation and Biodiversity Conservation

Forests play a critical and multifaceted role in supporting rural livelihoods, providing essential resources such as timber for construction and fuelwood for energy, as well as vital ecosystem services including water regulation, soil conservation, and carbon sequestration. However, unchecked Deforestation poses a significant threat, leading to detrimental consequences such as soil erosion, the alarming loss of biodiversity, and exacerbation of climate change impacts. Geospatial tools, particularly satellite-based forest monitoring systems utilizing data from sensors like MODIS and Sentinel-2, provide the capability to track deforestation activities in near real-time, enabling prompt regulatory action and intervention to curb illegal logging and forest degradation. Furthermore, GIS-based conservation planning allows for the identification and mapping of critical biodiversity hotspots, facilitating the strategic creation and effective management of protected areas and the implementation of community-managed forest initiatives. By integrating socio-economic data with ecological information, policymakers can also design and implement sustainable livelihood programs for forest-dependent communities, reducing their reliance on unsustainable forest exploitation and promoting alternative income-generating activities that are compatible with conservation goals (Njurumana et al. 2020).

12.3.2.3 Ensuring Water Security: Water Resource Management and Flood Monitoring

Water Scarcity and the increasing frequency and intensity of Flooding represent major and growing challenges to sustainable rural development. Geospatial data and technologies provide crucial tools for effective Water Resource Management and proactive Flood Monitoring. By mapping critical hydrological features such as aquifers, river basins, and watershed boundaries, and analysing spatial patterns of rainfall, GIS-based hydrological models can be developed to predict flood risks with greater accuracy, enabling the construction of appropriate mitigation measures such as embankments and efficient drainage systems. Remote sensing technologies can be utilized to monitor crucial parameters like reservoir water levels and rates of groundwater depletion, providing essential information for ensuring sustainable irrigation practices and managing water resources effectively. Additionally, the integration of spatial data on water access (location of wells, distribution networks) with population maps helps to identify villages and communities facing water stress, guiding the strategic deployment of essential infrastructure such as borewells and piped water supply schemes to improve access to clean and reliable water sources (Pires and Martins 2024).

12.3.3 *Disaster Risk Reduction and Management*

12.3.3.1 Building Resilience: Disaster Risk Reduction and Management

Rural communities often experience a disproportionate impact from natural disasters due to factors such as limited infrastructure, inadequate housing, and less developed emergency response mechanisms. Geospatial Technologies play a crucial role in enhancing Disaster Preparedness through the development of detailed hazard maps identifying areas prone to specific risks, comprehensive vulnerability assessments that identify at-risk populations and infrastructure, and the creation of effective evacuation plans based on spatial analysis of population distribution and transportation networks. Advanced flood inundation models, sophisticated cyclone tracking systems utilizing satellite data, and detailed earthquake risk maps enable the implementation of proactive mitigation strategies aimed at reducing the potential impacts of these events. Furthermore, Community-Based Participatory GIS (PGIS) approaches actively engage local stakeholders in the disaster planning process, ensuring that valuable indigenous knowledge and local expertise are effectively integrated with scientific data and geospatial analysis, leading to more relevant and effective disaster risk reduction strategies (Cahyono et al. 2021).

12.3.3.2 Timely Intervention: Early Warning Systems for Natural Hazards

The ability to provide timely and accurate warnings to vulnerable populations can significantly reduce the loss of life and minimize the impact of natural disasters. Geospatial Early Warning Systems integrate real-time data from a network of sensors, such as rainfall gauges, river level monitors, and seismic sensors, with sophisticated predictive modeling capabilities to issue timely alerts and warnings through various communication channels, including SMS messaging, public sirens, and mobile applications. For instance, advanced landslide early warning systems can utilize data from soil moisture sensors and established rainfall thresholds to predict potential slope failures, providing crucial lead time for evacuation. Similarly, flood alerts can be triggered by real-time monitoring of river water levels and integration with hydrological models. These early warning systems are particularly critical in remote rural areas where traditional communication infrastructure may be limited or unreliable, ensuring that even isolated communities receive timely information to take necessary precautions (Smith et al. 2017).

12.3.3.3 Coordinated Response: Emergency Response and Relief Operations

During and immediately following a disaster event, geospatial data becomes an invaluable tool for streamlining Emergency Response and Relief Operations. By rapidly mapping the areas most severely affected, assessing the extent of damage to critical infrastructure (roads, bridges, buildings), and identifying the locations and needs of displaced populations, geospatial information facilitates a more efficient and targeted allocation of resources. Unmanned aerial vehicles (drones) and high-resolution satellite imagery can provide rapid and comprehensive damage assessments, offering a crucial overview of the situation on the ground. GIS platforms are then used to optimize the distribution of essential supplies such as food, medicine, and temporary shelter, ensuring that aid reaches those who need it most quickly and effectively. Furthermore, the integration of social media data and crowdsourced information, when carefully verified, can further enhance situational awareness, providing real-time insights into the evolving needs and challenges faced by affected communities, enabling better coordination between government agencies, non-governmental organizations (NGOs), and volunteer efforts. In the post-disaster recovery phase, geospatial tools play a vital role in rehabilitation planning, ensuring that reconstruction efforts are resilient to future hazards, are implemented in an inclusive manner, and contribute to long-term sustainable development (Hayat et al. 2019; Khadka 2024).

12.3.4 Harnessing Spatial Intelligence for a Prosperous Rural Future

In conclusion, the synergistic integration of robust geospatial data development practices and a well-established Spatial Data Infrastructure (SDI) represents a transformative paradigm shift for rural planning and development. This powerful combination effectively bridges the traditional gap between spatial and socio-economic dimensions, providing policymakers and practitioners with the analytical tools and information necessary to make informed and impactful decisions. By strategically leveraging detailed topographic, cadastral, and comprehensive socio-economic data, it becomes possible to conduct insightful demand-gap analyses for essential infrastructure, optimize complex transportation networks to enhance connectivity and access, and strategically identify suitable sites for renewable energy projects that promote energy independence and sustainability. Furthermore, geospatial technologies are indispensable for enhancing disaster resilience through effective hazard zonation, risk assessment, and the implementation of timely early warning systems. From supporting sustainable renewable energy site selection and optimizing transportation networks to facilitating the adoption of smart village concepts and bolstering environmental conservation efforts, geospatial tools empower rural communities by fostering equitable growth, reducing vulnerabilities to natural hazards, and ultimately improving the overall quality of life for rural populations. Moving forward, sustained investments in the development and maintenance of robust SDI, coupled with targeted capacity building initiatives to enhance geospatial literacy and the promotion of participatory GIS approaches that engage local communities, will be crucial in realizing the full potential of geospatial intelligence for achieving comprehensive, inclusive, and sustainable rural development outcomes.

12.4 Geospatial Technology for Rural Development in Northeast India

12.4.1 Rural Infrastructure Development

Geospatial technology plays a transformative role in rural development, offering innovative solutions to address critical challenges and enhance the quality of life in rural communities. Its application is particularly significant in rural infrastructure development, where it enables efficient planning, implementation, and monitoring of essential services.

Rural Road Connectivity and Accessibility Analysis: Geospatial technology, specifically Remote Sensing and GIS, provides powerful tools for analyzing and improving rural road connectivity. High-resolution satellite imagery can be used to map existing road networks, identify gaps in connectivity, and assess the condition of roads. GIS enables the integration of this spatial data with other relevant information, such as population distribution, land use patterns, and accessibility to essential services like healthcare and education. By analysing these datasets, planners can identify priority areas for road construction and improvement, optimize road alignments to minimize environmental impact and maximize accessibility, and develop efficient transportation networks that connect rural communities to markets, employment opportunities, and social services. Furthermore, GPS technology facilitates accurate surveying and construction of roads, while GIS-based monitoring systems can track road conditions and identify maintenance needs, ensuring the long-term sustainability of rural road infrastructure.

Drinking Water and Sanitation Planning: Geospatial technology is also crucial for effective drinking water and sanitation planning in rural areas. Remote sensing can be used to identify potential water sources, assess water quality, and map groundwater resources. GIS enables the analysis of water availability, demand, and accessibility, helping planners to locate suitable sites for water infrastructure, such as wells, pipelines, and treatment plants. GIS can also be used to map sanitation facilities, identify areas with inadequate sanitation coverage, and plan for the construction of toilets and waste management systems. By integrating geospatial data with public health information, planners can prioritize interventions to improve sanitation and reduce the risk of waterborne diseases. Moreover, GIS-based monitoring systems can track the performance of water and sanitation infrastructure, ensuring its effective operation and maintenance.

Rural Electrification and Energy Access: Geospatial technology plays a vital role in planning and implementing rural electrification projects and promoting access to clean energy. Remote sensing can be used to map existing electricity infrastructure, identify unserved areas, and assess the potential for renewable energy sources, such as solar and wind power. GIS enables the analysis of energy demand, distribution, and accessibility, helping planners to design efficient electricity grids, locate suitable sites for power plants, and develop strategies to extend electricity access to remote villages. GIS can also be used to map households and businesses, assess their energy needs, and plan for the deployment of decentralized renewable energy systems. By integrating geospatial data with socioeconomic information, planners can prioritize interventions to improve energy access for the most vulnerable populations. Furthermore, GIS-based monitoring systems can track energy consumption, identify energy losses, and optimize the operation of electricity infrastructure. In conclusion, geospatial technology is an indispensable tool for rural infrastructure development, enabling data-driven decision-making, efficient resource allocation, and sustainable development outcomes. Its applications in rural road connectivity, drinking water and sanitation planning, and rural electrification and energy access are crucial for improving the quality of life in rural communities and fostering inclusive growth (Jayasree 2022).

12.4.2 Socio-economic Development and Empowerment

Geospatial technology is a powerful catalyst for socio-economic development and empowerment in rural areas, enabling more informed decision-making, targeted interventions, and enhanced community participation. Its applications are crucial in addressing disparities and fostering sustainable progress across various sectors.

Health and Education Planning: Geospatial technology revolutionizes health and education planning in rural areas by providing critical spatial insights. Remote sensing and GIS can map the distribution of healthcare facilities, schools, and educational resources, revealing disparities in access and service coverage. This spatial analysis helps identify underserved populations and prioritize the location of new facilities or the improvement of existing ones. GIS can integrate health data (e.g., disease prevalence, malnutrition rates) with environmental factors (e.g., water quality, sanitation) to understand the spatial patterns of health risks and inform targeted interventions. Similarly, in education, GIS can analyse student demographics, school infrastructure, and teacher availability to optimize resource allocation and improve educational outcomes. Furthermore, mobile GIS applications can empower community health workers and educators to collect and update data in real-time, improving the accuracy and efficiency of service delivery (Gammino et al. 2014).

Poverty Mapping and Resource Allocation: Geospatial technology is instrumental in poverty mapping and more effective resource allocation in rural regions. By integrating data from various sources, including remote sensing, census data, and socioeconomic surveys, GIS can create detailed maps of poverty indicators, such as income levels, access to basic services, and land ownership. These poverty maps help policymakers and development organizations to identify the most vulnerable communities and target interventions where they are most needed. Geospatial analysis can also optimize the allocation of resources, such as food aid, financial assistance, and development projects, ensuring that they reach the intended beneficiaries and maximize their impact. By visualizing the spatial distribution of poverty and related factors, GIS facilitates a more equitable and efficient distribution of resources, contributing to poverty reduction and improved livelihoods (Khasanah and Sapanji 2021).

Participatory Mapping for Community Development: Geospatial technology empowers rural communities through participatory mapping, enabling them to actively engage in development planning and decision-making. Participatory GIS (PGIS) involves the use of GIS tools and techniques by local communities to map their own resources, assets, and challenges. This process allows communities to document their traditional knowledge, land use patterns, and environmental concerns, creating valuable spatial information that can be used to advocate for their rights and needs. PGIS can be used for a variety of community development initiatives, such as land-use planning, disaster management, and natural resource management. By empowering communities to create and control their own spatial data, participatory mapping promotes ownership, accountability, and sustainable development. It ensures that development projects are aligned with local priorities and contribute to the long-term well-being of rural communities. In conclusion, geospatial technology is a vital tool for driving socio-economic development and empowerment in rural areas. Its applications in health and education planning, poverty mapping and resource allocation, and participatory mapping for community development contribute to more equitable, sustainable, and inclusive growth, ultimately improving the lives of millions.

12.5 Case Studies: Successful Geospatial Applications in Northeast India

12.5.1 Community-Based Greenfield Development Projects

North Eastern Space Applications Centre, Umiam, a Department of Space Unit, Government of India has completed a project "setting up a new township Near the Makhru River Bank in Tousem Sub-Division, Tamenglong District, Manipur". The project specifically pertains to development activities which includes identifying the potential of MSME industries, study the impact of the Makhru River cum rail bridge on the region, create a framework for a model township and its feasibility and proposals for other supported infrastructure and other economic boosting proposals.

12.5.1.1 Genesis of the Project

The Tousem subdivision, in Tamenglong district due to extensive practice of Jhum cultivation, has witnessed tremendous forest clearing in the past few decades. Considering, a Household size of five individuals utilises 2 acres of land for the said purpose, thus one thousand Zeme Community households living in that area are consuming 2000 acres of land every year, thus leaving a huge carbon footprint and degrading the environment.

Figure 12.1 depicts the change in forest cover over a decade ranging from the year 2012 and 2023 for the proposed site for township.

The vision of the project was to establish the 1000 Households practicing Shifting Cultivation as their primary activity into the proposed Township. The project also envisages towards developing the proposed township as an eco-hub, establishing MSMEs so as to generate livelihood facilities leading towards a more sustainable habitat. The project aims towards enhancing the standard of living of the pertaining population of the zeme community simultaneously to combat climate change and reduce the impact of deforestation due to shifting cultivation.

Fig. 12.1 Change in forest cover of the A.O.I. from the year 2012 and 2023

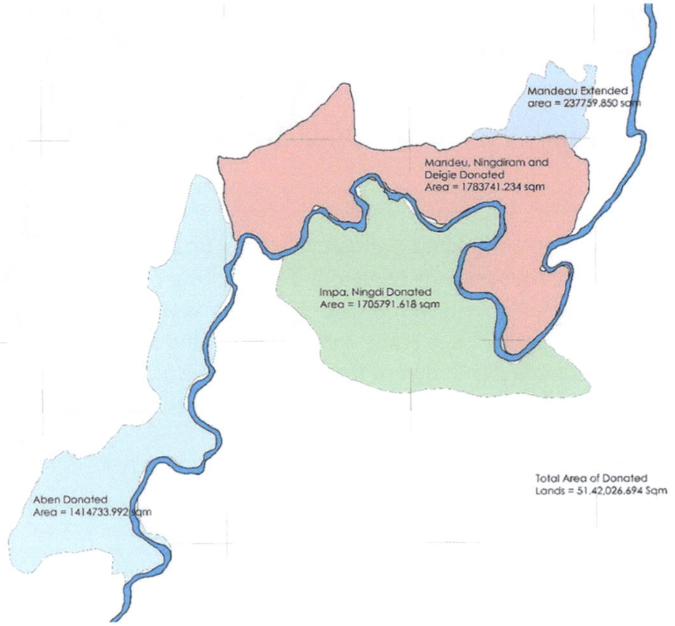

Fig. 12.2 Portion of land donated by the different villages

The development of the township was significantly accelerated by the donation of community-owned lands by the villagers. This act of contribution, following consultations with the Village Council, aimed to boost the development process in an area where shifting cultivation was the predominant agricultural practice. The substantial prevalence of shared land ownership points towards a community-oriented approach to land management, consistent with the customary land tenure systems typical of tribal societies, which emphasize communal rights and responsibilities over individual ownership. The Fig. 12.2 shows proportional contribution of land from participating villages for the project. Landowners from the villages of Mandeu, Ningdiram, Deigie, Aben, and MpaNingdi specifically donated land near the Makru River for the establishment of this new township.

12.5.1.2 Remote Sensing Data Acquisition

UAV data acquisition proved a rapid and cost-effective solution for preliminary site assessment and planning for a new township near the Makhru River Bank, Manipur. The procurement of high-resolution 5 cm spatial resolution Ortho mosaic imagery provided a detailed and geometrically corrected aerial view of the proposed site. This Ortho mosaic served as a fundamental base map, enabling accurate measurements of distances, areas, and identification of existing features like vegetation, water bodies, and existing infrastructure.

Fig. 12.3 Isometric view of the proposed site

Simultaneously, the generation of a Digital Surface Model (DSM) captured the elevation of all visible surfaces, including buildings and vegetation, while the Digital Terrain Model (DTM) represented the bare earth elevation. This bundled DSM and DTM data generation was crucial for detailed topographic analysis, including slope assessment, identification of potential flood-prone zones along the Makhru River, and volumetric calculations for earthwork estimations. The high spatial resolution allowed for precise delineation of land parcels, identification of suitable construction zones, and the planning of road networks and utility corridors with greater accuracy. Furthermore, the UAV data was used to generate contour lines and slope maps, aiding in optimal site layout and minimizing environmental impact. This comprehensive geospatial information derived from UAV data significantly enhanced the planning process, enabling informed decision-making and efficient resource allocation for the development of the new township. The Fig. 12.3 presents a 3D perspective of the development site, where the Ortho-mosaic is draped over a digital surface model. This visualization highlights terrain characteristics, undulations, and site constraints, aiding in assessing the suitability for township planning and development (Tmušić et al. 2020).

12.5.1.3 Land Utilisation Analysis of the Proposed Site

The slope analysis was conducted to determine the most suitable land for infrastructure development within the designated area. This analysis is crucial as it informs us that infrastructure projects can be executed on lands with gentle to moderate slopes, provided that other geographical and environmental factors are favorable. The categorization of land into developable and non-developable zones has been based solely on the slope analysis discussed.

From the total area of 520 hectares, approximately 360 hectares, which accounts for 69.23% of the overall site, have been identified as developable land. These areas feature slopes that are gentle to moderate, making them suitable for various types of infrastructure development. On the other hand, 130.51 hectares, representing 25.09% of the site, have been classified as non-developable. These non-developable areas are largely covered with green vegetation. The steep slopes in these regions make them unsuitable for any form of infrastructure development. The vegetation serves as a crucial environmental component, contributing to the ecological balance by maintaining the stability of the slopes and preventing soil erosion. The preservation of these green covers is not only an environmental necessity but also a legislative mandate in many regions, emphasizing the importance of sustainable development practices. The remaining 29.49 hectares, which constitute 5.68% of the site, are comprised of water bodies. These areas are naturally unsuitable for infrastructure development. Water bodies play a vital role in the local ecosystem, providing habitat for various species of flora and fauna, and contributing to the area's biodiversity. Furthermore, they are critical for groundwater recharge and maintaining the hydrological balance of the region.

The Fig. 12.4 illustrates the slope analysis, and has provided a clear demarcation of the land based on its suitability for infrastructure development. The developable areas with gentle to moderate slopes hold significant potential for future projects, while the non-developable areas with steep slopes and water bodies need to be preserved for environmental sustainability.

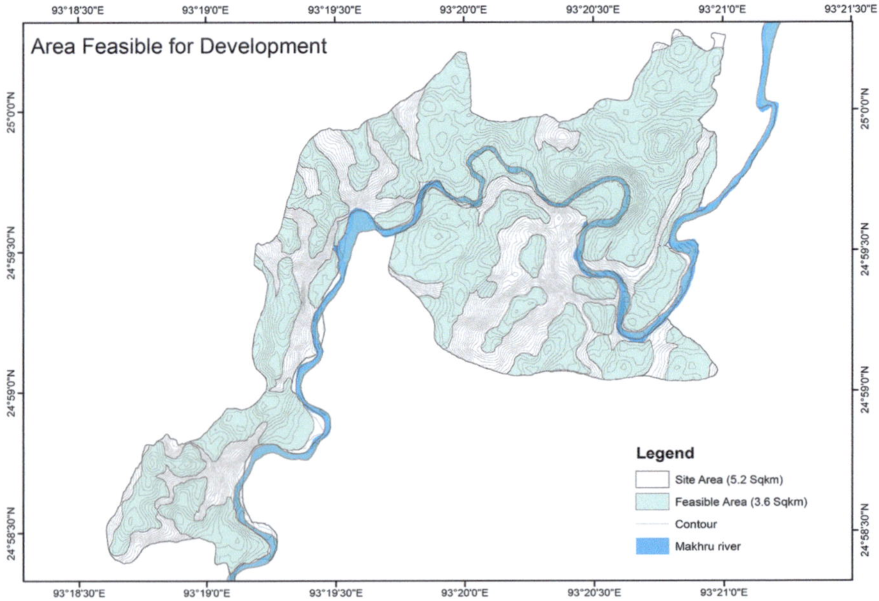

Fig. 12.4 Map showing area feasible for development

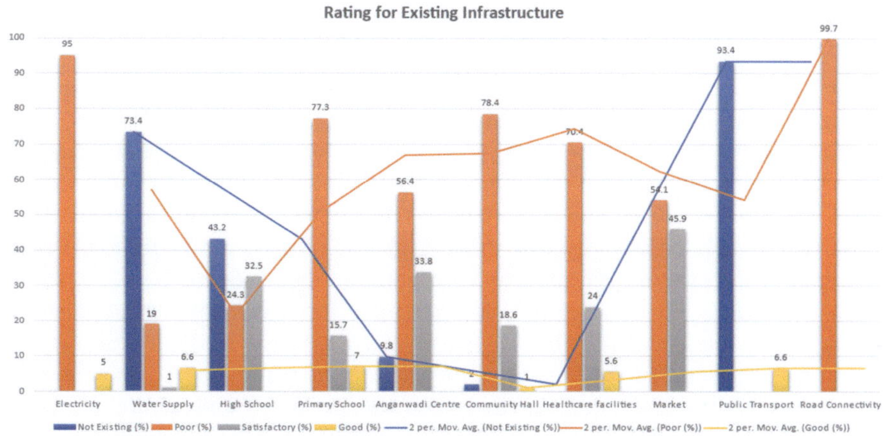

Fig. 12.5 Existing infrastructure rating responses

12.5.1.4 Socio Economic Study

The Study aimed to assess the socio-economic conditions, infrastructure, livelihood sources, skilled labor availability, and local attitudes towards a proposed township near the Makru River bank. It also examined the impact of a rail bridge across the Makru River, the potential for MSME industries, feasible projects that could become entrepreneurial ventures, and the identification of interventions by various stakeholders to participate in the development process in the area.

The Fig. 12.5 provides a comprehensive visualization of stakeholder survey results, assessing the current state and adequacy of key infrastructure elements such as roads, utilities, public facilities, and transport. By graphically representing feedback on the quality and availability of these assets, the data bridges socio-economic realities with spatial analysis. This integrated approach has been a key to identify infrastructure gaps and prioritize improvements based on actual community needs. Furthermore, incorporating participatory mapping ensures that interventions are inclusive and grounded in local experience, enhancing the accuracy and impact of planning efforts and fostering greater community involvement in sustainable development.

12.5.1.5 Residential Zoning

The proposed site is thoughtfully divided into nine distinct sectors, interconnected by a 22 m-wide main road. This arterial thoroughfare serves as the primary circulation route, facilitating movement between sectors and connecting to the National Highway.

To further enhance internal connectivity, a network of 12 m-wide internal roads has been incorporated into the design. This strategic road layout prioritizes safety and smooth traffic flow, adhering to established guidelines and safety measures. The roads are strategically aligned with the site's natural contours, minimizing disruptive interventions and maximizing efficiency.

The design of the road section was guided by standard engineering principles and best practices to ensure durability, safety, and aesthetics.

The proposed design for the new greenfield township offers a comprehensive and sustainable solution for urban development. By carefully considering factors such as connectivity, accessibility, and environmental impact, the township aims to create a vibrant and livable community. The implementation of these design principles will contribute to the overall well-being and prosperity of the residents.

The study has helped develop a clear roadmap for sustainable development in the region. The insights underline the multifaceted nature of development, emphasizing the importance of a holistic approach that considers the socio-economic dynamics, infrastructural needs, environmental sustainability and cultural preservation. The identified challenges-ranging from inadequate road connectivity and electricity supply to educational and healthcare deficiencies-highlight the urgent need for targeted interventions. These challenges, however, also present unique opportunities for transformative change, particularly through the promotion of Micro, Small, and Medium Enterprises (MSMEs) in sectors such as agriculture, tourism, and agroforestry. The potential for economic activities, bolstered by the forthcoming railway bridge and National Highway No. 137 (Mahur-Tamenglong), promises to significantly enhance transportation and market access, creating a ripple effect that could stimulate regional economic growth and improve livelihoods.

Furthermore, the findings underscore the critical role of skilled labour in the region's development trajectory. Addressing skill gaps through vocational training and education is essential for empowering the local workforce and meeting the evolving demands of the economy. Equally important is the need for inclusive community engagement and support mechanisms that ensure all voices are heard and considered in the development process, fostering a sense of ownership and collaboration among residents. Environmental and cultural preservation emerge as critical components within the broader framework of sustainable development, highlighting the importance of a development approach that harmonizes economic growth with the conservation of the region's rich environmental and cultural heritage. By placing a strong emphasis on protecting natural landscapes, biodiversity and the wealth of cultural traditions that define the region, these preservation efforts ensure that development does not come at the cost of losing irreplaceable natural and cultural assets. Initiatives specifically designed to safeguard these resources-ranging from implementing sustainable agricultural practices that protect soil and water quality, to supporting programs that keep alive traditional crafts, languages, and ceremonies-play a crucial role in fostering a development paradigm that prioritizes long-term ecological integrity and the well-being of communities. This approach not only helps in maintaining the unique identity of the region but also strengthens social bonds and cohesion among its inhabitants, creating a shared sense of responsibility towards heritage conservation. In doing so, it lays the foundation for a development model that is not only economically viable but also ecologically sound and culturally rich,

Fig. 12.6 Layout plan for sector D in the proposed township

ensuring that development benefits are enjoyed by present and future generations while preserving the essence of the region's identity.

The proposed design for the new township offers a comprehensive and sustainable solution for Greenfield development. The Fig. 12.6 shows a detailed architectural/engineering plan outlining the spatial arrangement of plots, roads, green spaces, and amenities in Sector D. By carefully considering factors such as connectivity, accessibility and environmental impact, the township aims to create a vibrant and liveable community. The implementation of these design principles will contribute to the overall well-being and prosperity of the residents.

In conclusion, the path forward requires a concerted effort from all stakeholders-government agencies, private institutions, financial entities, NGOs and the community at large. By embracing a comprehensive and integrated approach, the region can navigate the development challenges while capitalizing on its unique strengths and opportunities. This strategy not only would help address immediate needs but also aspires to lay the groundwork for enduring prosperity, environmental sustainability and cultural vibrancy, ensuring a brighter future for current and subsequent generations.

Acknowledgements We gratefully acknowledge North Eastern Development Finance Corporation Ltd. for their opportunity and financial support. We express profound gratitude to Dr. Shiv Prasad Aggarwal (Director, NESAC) for his guidance. Special thanks to the Zeme Committee for their essential local support.

References

R. Abdullah, D. Arief Widianto, I. Muhamad Ilham Kosasih, A. Aditia Pratama, H. Kaulani, M. Ridwan, R. Agung Chrisnawati, R. Aprilia, S. Manurung, Completed participatory mapping of village based on WebGIS. IOP Conf. Ser.: Earth Environ. Sci. **375**, 012004 (2019). https://doi.org/10.1088/1755-1315/375/1/012004

B. Azu, D.M. Uduh, A.I. Mobosi, National security and wealth creation: the Nigerian sustainable growth nightmare. UJAH: Unizik J. Arts Hum. **21**(4), 230–253 (2021). https://doi.org/10.4314/ujah.v21i4.14

Md.A.H. Bhuiyan, C. Siwar, S.M. Ismail, Tourism development in Malaysia from the perspective of development plans. Asian Soc. Sci. **9**(9) (2013). https://doi.org/10.5539/ass.v9n9p11

R.K.C. Billones, M.A. Guillermo, K.C. Lucas, M.D. Era, E.P. Dadios, A.M. Fillone, Smart region mobility framework. Sustainability **13**(11), 6366 (2021). https://doi.org/10.3390/su13116366

L. Binfield, T.L. Britton, C. Dai, J.L. Innes, Evidence on the social, economic, and environmental impact of interventions that facilitate bamboo industry development for sustainable livelihoods: a systematic map. Forests **16**(5), 713 (2025). https://doi.org/10.3390/f16050713

B. Bolo, D. Mpoeleng, I. Zlotnikova, Development of methods acquiring real time very high resolution agricultural spatial information using unmanned aerial vehicle. Agris Online Pap. Econ. Inf. **11**(2), 21–29 (2019). https://doi.org/10.7160/aol.2019.110203

M. Buenemann, C. Martius, J.W. Jones, S.M. Herrmann, D. Klein, M. Mulligan, M.S. Reed, M. Winslow, R.A. Washington-Allen, R. Lal, D. Ojima, Integrative geospatial approaches for the comprehensive monitoring and assessment of land management sustainability: rationale, potentials, and characteristics. Land Degrad. Dev. **22**(2), 226–239 (2010). https://doi.org/10.1002/ldr.1074

F.F.K. Byamugisha, N. Dubosse, The investment case for land tenure security in Sub-Saharan Africa: a cost-benefit analysis. J. Benefit-Cost Anal. **14**(S1), 272–300 (2023). https://doi.org/10.1017/bca.2023.14

S.A. Cahyono, A. Wuryanta, C. Lastiantoro, The local knowledge to mitigate the landslide disaster in Beruk village, Jatiyoso sub-district, Karanganyar regency. IOP Conf. Ser.: Earth Environ. Sci. **874**(1), 012015 (2021). https://doi.org/10.1088/1755-1315/874/1/012015

S. Cao, S. Li, H. Ma, Y. Sun, Escaping the resource curse in China. Ambio **44**(1), 1–6 (2014). https://doi.org/10.1007/s13280-014-0540-9

C. Caputo, M.-A. Cardin, P. Ge, F. Teng, A. Korre, A. del Rio, E. Chanona, Design and planning of flexible mobile micro-grids using deep reinforcement learning. Appl. Energy **335**, 120707 (2023). https://doi.org/10.1016/j.apenergy.2023.120707

S. Cardey, P.J.M. Eleazar, J. Ainomugisha, M. Kalowekamo, Y. Vlasenko, Communication for development: conceptualising changes in communication and inclusive rural transformation in the context of environmental change. Soc. Sci. **13**(6), 324 (2024). https://doi.org/10.3390/socsci13060324

L. Chen, J. Hu, G. Lv, Y. Liu, Space-aerial-ground integrated city dynamic monitoring based on remote sensing. Int. Arch. Photogramm., Remote Sens. Spat. Inf. Sci. **XLIII-B3-2020**, 29–34 (2020). https://doi.org/10.5194/isprs-archives-xliii-b3-2020-29-2020

D.G. Costa, J.C.N. Bittencourt, F. Oliveira, J.P.J. Peixoto, T.C. Jesus, Achieving sustainable smart cities through geospatial data-driven approaches. Sustainability **16**(2), 640 (2024). https://doi.org/10.3390/su16020640

M.R. Daley, A conceptual model for rural social work. Contemp. Rural Soc. Work J. **2**(1) (2010). https://doi.org/10.61611/2165-4611.1011

Devi, R., Geographical background of the Bhabar-Tarai Belt of lower Assam District, India. Int. J. Recent Technol. Eng. (IJRTE) **8**(5), 1299–1304 (2020). https://doi.org/10.35940/ijrte.e4983.018520

N. Ejaz, S. Choudhury, Computer vision in drone imagery for infrastructure management. Autom. Constr. **163**, 105418 (2024). https://doi.org/10.1016/j.autcon.2024.105418

F. Firmansyah, A.S. Kurnia, S. Oktavilia, R. Prayogi, Forestry sector impact on the economy of Central Java, Indonesia. JEJAK **12**(1), 168–189 (2019). https://doi.org/10.15294/jejak.v12i1.18575

V.M. Gammino, A. Nuhu, P. Chenoweth, F. Manneh, R.R. Young, D.E. Sugerman, S. Gerber, E. Abanida, A. Gasasira, Using geographic information systems to track polio vaccination team performance: pilot project report. J. Infect. Dis. **210**(Suppl. 1), S98–S101 (2014). https://doi.org/10.1093/infdis/jit285

N. González-Gil, A.S. Virviescas, *Journey Through the MILtiverse: Media and Information Literacy Toolkit for Youth Organizations*, ed. by S. Acero Pulgarin, M. Martinovic (UNESCO, 2024). https://doi.org/10.58338/apoc5995

A.A.S. Hameed, Urban and regional planning strategies to achieve sustainable urban development: (subject review). Int. J. Adv. Sci. Res. Eng. **07**(03), 22–27 (2021). https://doi.org/10.31695/ijasre.2021.33980

E. Hayat, R. Haigh, D. Amaratunga, A framework for reconstruction of road infrastructure after a disaster. Int. J. Disaster Resil. Built Environ. **10**(2/3), 151–166 (2019). https://doi.org/10.1108/ijdrbe-03-2017-0018

T. Hilhorst, J. Zevenbergen, K. Deininger, Land governance and tenure security at scale: lessons from the field. Land Use Policy **110**, 105451 (2021). https://doi.org/10.1016/j.landusepol.2021.105451

I.S. Ibaba, Violent conflicts and sustainable development in Bayelsa State. Rev. Afr. Polit. Econ. **36**(122) (2009). https://doi.org/10.1080/03056240903346152

M. Islam, A.K.M.S. Islam, M.R. Amin, Small-scale decentralized renewable energy systems for the remote communities of the developing countries, in *ASME 2005 Power Conference* (ASMEDC, 2005), pp. 879–890. https://doi.org/10.1115/pwr2005-50068

A.E. Izquierdo, H.R. Grau, Agriculture adjustment, land-use transition and protected areas in Northwestern Argentina. J. Environ. Manag. **90**(2), 858–865 (2009). https://doi.org/10.1016/j.jenvman.2008.02.013

K. Jayasree, Sustainable rural road network planning with a balance of urban and rural development. IOP Conf. Ser.: Earth Environ. Sci. **982**(1), 012042 (2022). https://doi.org/10.1088/1755-1315/982/1/012042

G. Jelilov, K. Ozden, S.O. Briggs, Impact of insecurity on investment in Nigeria. J. Manag., Econ., Ind. Organ. 41–61 (2018). https://doi.org/10.31039/jomeino.2018.2.3.3

M. Karaeva, V. Goceva Mihajlovska, Policy for balanced regional development: policy settings on regional level—experience and practice, in *International Scientific Conference EMAN—Economics and Management: How to Cope with Disrupted Times, 5th International Scientific Conference—Eman 2021—Economics and Management: How to Cope With Disrupted Times. Association of Economists and Managers of the Balkans* (Belgrade, Serbia, 2021), pp. 21–24. https://doi.org/10.31410/eman.2021.21

M. Karagiannis, I. Chatzigiannakis, J. Rolim, Multilateration: methods for clustering intersection points for wireless sensor networks localization with distance estimation error (version 1). arXiv (2012). https://doi.org/10.48550/ARXIV.1203.3704

S.K. Katara, A. Pokhriyal, Strengthening Indian border security through integrated ICT based framework, in *Proceedings of the Special Collection on eGovernment Innovations in India, ICEGOV '17: 10th International Conference on Theory and Practice of Electronic Governance* (ACM, 2017), pp. 39–44. https://doi.org/10.1145/3055219.3055225

S. Kathuria, P. Mathur (eds.), *Playing to Strengths: A Policy Framework for Mainstreaming Northeast India* (World Bank Publications, 2019)

T.B. Khadka, 4 "R" approach of catastrophic disaster of Melamchi flood: a sustainable urban planning. Proc. IAHS **386**, 327–332 (2024). https://doi.org/10.5194/piahs-386-327-2024

A.D. Khasanah, R.A.E. Virgana Targa Sapanji, Implementation of geographic information system for district and city poverty data distribution using the rational unified process in West Java. J. Comput. Res. Innov. **6**(2), 11–20 (2021). https://doi.org/10.24191/jcrinn.v6i2.185

D.K. Kibetu, GIS analysis of physical accessibility to food markets in Tharaka Region of Kenya. Rev. Agric. Appl. Econ. **23**(2), 36–44 (2020). https://doi.org/10.15414/raae.2020.23.02.36-44
M. Kim, J. Kim, SBAS-aided GPS positioning with an extended ionosphere map at the boundaries of WAAS service area. Remote Sens. **13**(1), 151 (2021). https://doi.org/10.3390/rs13010151
J.C.J. Koelemeij, H. Dun, C.E.V. Diouf, E.F. Dierikx, G.J.M. Janssen, C.C.J.M. Tiberius, A hybrid optical-wireless network for decimetre-level terrestrial positioning. Nature **611**(7936), 473–478 (2022). https://doi.org/10.1038/s41586-022-05315-7
X. Li, Y. Zhu, K. Zheng, Y. Yuan, G. Liu, Y. Xiong, Precise orbit and clock products of Galileo, BDS and QZSS from MGEX since 2018: comparison and PPP validation. Remote Sens. **12**(9), 1415 (2020). https://doi.org/10.3390/rs12091415
M. Liao, C. Tang, L. Zhao, S. Zhou, X. Hu, Y. Chen, K. Li, Y. Gui, Calibration of receiver-dependent pseudorange bias and its impact on BDS augmentation positioning accuracy. Remote Sens. **16**(16), 3022 (2024). https://doi.org/10.3390/rs16163022
M. Liu, M. Liao, R. Zhang, X. Yuan, Z. Zhu, Z. Wu, Quantum computing as a catalyst for microgrid management: enhancing decentralized energy systems through innovative computational techniques. Sustainability **17**(8), 3662 (2025). https://doi.org/10.3390/su17083662
Z. Liu, P. Zhao, Q. Liu, Z. He, T. Kang, Uncovering spatial and social gaps in rural mobility via mobile phone big data. Sci. Rep. **13**(1) (2023). https://doi.org/10.1038/s41598-023-33123-0
K. Maciuk, S. Szombara, Annual crustal deformation based on GNSS observations between 1996 and 2016. Arab. J. Geosci. **11**(21) (2018). https://doi.org/10.1007/s12517-018-4022-4
A.M. Muhmad Kamarulzaman, W.S. Wan Mohd Jaafar, M.N. Mohd Said, S.N.M. Saad, M. Mohan, UAV implementations in urban planning and related sectors of rapidly developing nations: a review and future perspectives for Malaysia. Remote Sens. **15**(11), 2845 (2023). https://doi.org/10.3390/rs15112845
E. Mujuka, J. Mburu, A. Ogutu, J. Ambuko, Returns to investment in postharvest loss reduction technologies among mango farmers in Embu County, Kenya. Food Energy Secur. **9**(1) (2019). https://doi.org/10.1002/fes3.195
J. Ning, Y. Yao, X. Zhang, Overview of the development of global navigation satellite system. J. Geogr. Cartogr. **5**(2), 90 (2022). https://doi.org/10.24294/jgc.v5i2.1676
G.N. Njurumana, K. Ginoga, D. Octavia, Sustaining farmers livelihoods through community forestry in Sikka, East Nusa Tenggara, Indonesia. Biodiversitas J. Biol. Divers. **21**(8) (2020). https://doi.org/10.13057/biodiv/d210846
W.H. Nur, Y. Kumoro, Y. Susilowati, GIS and geodatabase disaster risk for spatial planning. IOP Conf. Ser.: Earth Environ. Sci. **118**, 012046 (2018). https://doi.org/10.1088/1755-1315/118/1/012046
C. Nyakwebara, A.T. Kiggundu, D. Kaijo, E.B. Kaganzi, C. Nakawuki, Harnessing geospatial technologies for sustainable urban planning in the newly created cities in Uganda using rapid physical planning approach (RAPPA): a case study of Hoima City. Curr. Urban Stud. **11**(04), 619–640 (2023). https://doi.org/10.4236/cus.2023.114032
C. Odiji, M. Adepoju, I. Ibrahim, O. Adedeji, I. Nnaemeka, O. Aderoju, Small hydropower dam site suitability modelling in upper Benue river watershed, Nigeria. Appl. Water Sci. **11**(8) (2021). https://doi.org/10.1007/s13201-021-01466-6
İ. ÖZ, Decision making in evaluation and selection of launch site with the best and worst method. Gazi Univ. J. Sci. **35**(4), 1521–1533 (2022). https://doi.org/10.35378/gujs.721823
R.J. Padre, M.A. Baguio, E.B. Panganiban, R.U. Panganiban, C.R. Bautista, J.R.L. Rigates, A.P. Mariano, Risk analysis of urban water infrastructure systems in Cauayan City. Int. J. Adv. Comput. Sci. Appl. **14**(1) (2023). https://doi.org/10.14569/ijacsa.2023.0140160
K.W. Park, J.-I. Park, C. Park, Efficient methods of utilizing multi-SBAS corrections in multi-GNSS positioning. Sensors **20**(1), 256 (2020). https://doi.org/10.3390/s20010256
R.J. Paul Maldo, T. C.F. Castro, B. Alcantara, A. Encinas, R. Galvez, M. Gapayao, Restoring vanishing traditions: "Pabalat" as a heritage tourism product of Bulacan. Glob. Int. J. Manag. IT **14**(2), 36–43 (2023). https://doi.org/10.46360/globus.mgt.120231004

F. Pech-May, R. Aquino-Santos, G. Rios-Toledo, J.P.F. Posadas-Durán, Mapping of land cover with optical images, supervised algorithms, and google earth engine. Sensors **22**(13), 4729 (2022). https://doi.org/10.3390/s22134729

M. Peng, The growing market of organic foods: impact on the US and global economy, in *Safety and Practice for Organic Food* (Elsevier, 2019), pp. 3–22. https://doi.org/10.1016/b978-0-12-812060-6.00001-5

C. Pires, M.V. Martins, Enhancing water management: a comparative analysis of time series prediction models for distributed water flow in supply networks. Water **16**(13), 1827 (2024). https://doi.org/10.3390/w16131827

M.R.A. Rashid, A.A. Rafi, Md.A. Islam, S.U. Sharkar, Z.H. Rafi, M. Hasan, M.S. Ali, M.S.H. Khan, Enhancing land management policy in Bangladesh: a blockchain-based framework for transparent and efficient land management. Land Use Policy **150**, 107436 (2025). https://doi.org/10.1016/j.landusepol.2024.107436

N. Rasid, M. Prashnani, J. Goswami, & P.L.N. Raju, Crop damage assessment in flood inundated area of Morigaon District of Assam. Int. Arch. Photogramm., Remote Sens. Spat. Inf. Sci. **XLII-3/W6**, 489–491 (2019). https://doi.org/10.5194/isprs-archives-xlii-3-w6-489-2019

C.A. Rokhmana, S. Utomo, The low-cost UAV-based remote sensing system capabilities for large scale cadaster mapping. IOP Conf. Ser.: Earth Environ. Sci. **47**, 012005 (2016). https://doi.org/10.1088/1755-1315/47/1/012005

M.A.F. Ros-Tonen, L. Willemen, M.K. McCall, Spatial tools for integrated and inclusive landscape governance: toward a new research agenda. Environ Manag. **68**(5), 611–618 (2021). https://doi.org/10.1007/s00267-021-01547-x

E. Rotimi Adegboyega, A. Oluwakemi Bolanle, Using forest resources for economic development and poverty reduction in Ekiti State. Int. J. Appl. Agric. Sci. **7**(4), 183 (2021). https://doi.org/10.11648/j.ijaas.20210704.17

S. Sajjad, K.H. Dar, Economic infrastructure and its impact on economic growth: a case study of 7 sister states. J. Financ. Manag. Prop. Constr. (2025). https://doi.org/10.1108/jfmpc-03-2024-0016

N. Shen, L. Chen, J. Liu, L. Wang, T. Tao, D. Wu, R. Chen, A review of global navigation satellite system (GNSS)-based dynamic monitoring technologies for structural health monitoring. Remote Sens. **11**(9), 1001 (2019). https://doi.org/10.3390/rs11091001

B. Shu, H. Liu, Y. Feng, L. Xu, C. Qian, Z. Yang, Analysis of factors affecting asynchronous RTK positioning with GNSS signals. Remote Sens. **11**(10), 1256 (2019). https://doi.org/10.3390/rs11101256

P.J. Smith, S. Brown, S. Dugar, Community-based early warning systems for flood risk mitigation in Nepal. Nat. Hazard. **17**(3), 423–437 (2017). https://doi.org/10.5194/nhess-17-423-2017

L. Sun, J. Shawe-Taylor, D. D'Ayala (2022) Artificial intelligence-informed planning for the rapid response of hazard-impacted road networks. Sci. Rep. **12**(1). https://doi.org/10.1038/s41598-022-19637-z

P. Sundram, Food security in ASEAN: progress, challenges and future. Front. Sustain. Food Syst. **7** (2023). https://doi.org/10.3389/fsufs.2023.1260619

G. Tmušić, S. Manfreda, H. Aasen, M.R. James, G. Gonçalves, E. Ben-Dor, A. Brook, M. Polinova, J.J. Arranz, J. Mészáros, R. Zhuang, K. Johansen, Y. Malbeteau, I.P. de Lima, C. Davids, S. Herban, M.F. McCabe, Current practices in UAS-based environmental monitoring. Remote Sens. **12**(6), 1001 (2020). https://doi.org/10.3390/rs12061001

P. Tripathy, K. Baylis, K. Wu, J. Watson, R. Jiang, Investigating the segment anything foundation model for mapping smallholder agriculture field boundaries without training labels (version 1). arXiv. https://doi.org/10.48550/ARXIV.2407.01846

M. Vastaranta, M.A. Wulder, J.C. White, A. Pekkarinen, S. Tuominen, C. Ginzler, V. Kankare, M. Holopainen, J. Hyyppä, H. Hyyppä, Airborne laser scanning and digital stereo imagery measures of forest structure: comparative results and implications to forest mapping and inventory update. Can. J. Remote Sens. **39**(5), 382–395 (2013). https://doi.org/10.5589/m13-046

L. Xin, F. Liu, Modeling urban exodus dynamics considering settlers adaptation time and local authority support. Math. Probl. Eng. **2022**, 1–9 (2022). https://doi.org/10.1155/2022/9954702

T. Yang, X. Wang, B. Yao, J. Li, Y. Zhang, Z. He, W. Duan, Small moving vehicle detection in a satellite video of an urban area. Sensors **16**(9), 1528 (2016). https://doi.org/10.3390/s16091528

Y.V. Yasyukevich, B. Zhang, V.R. Devanaboyina, Advances in GNSS positioning and GNSS remote sensing. Sensors **24**(4), 1200 (2024). https://doi.org/10.3390/s24041200

L. Yu, Z. Du, X. Li, Q. Zhao, H. Wu, D. Weise, X. Yuan, Y. Yang, W. Cai, W. Song, P. Wang, Z. Zhao, Y. Long, Y. Zhang, J. Peng, X. Xin, F. Xu, M. Shen, H. Wang et al., Near surface camera informed agricultural land monitoring for climate smart agriculture. Clim Smart Agric **1**(1), 100008 (2024). https://doi.org/10.1016/j.csag.2024.100008

N.S. Yusoff, M.F. Rashid, N.A. Halim, The impact of regional development towards the well-being of rural community. IOP Conf. Ser.: Earth Environ. Sci. **881**(1), 012056 (2021). https://doi.org/10.1088/1755-1315/881/1/012056

W. Zhang, C. Sun, Progress of surveying and mapping science and technology in the information age. J. Phys: Conf. Ser. **1961**(1), 012042 (2021). https://doi.org/10.1088/1742-6596/1961/1/012042

R.R. Ziipao, Deepening critical infrastructures in Northeast India: people's perspective and policy implications. Strateg. Anal. **44**(3), 208–223 (2020). https://doi.org/10.1080/09700161.2020.1787686

Santanu Das, a Scientist/Engineer-SE at NESAC, has 13 years of Experience across government administration and the private sector. He specializes in leveraging space technology to drive sustainable urban development. His expertise lies in applying geospatial intelligence to urban and regional planning, enabling data-driven decision-making for large-scale operations in planning, scheduling, and implementation. With a strong engineering foundation and a Master's degree in City Planning from the Indian Institute of Technology, Kharagpur, He brings an in-depth understanding of how satellite data and spatial analytics can transform urban systems, support infrastructure development, and enhance governance efficiency.

Jenita M. Nongkynrih is currently Head Urban and Regional Planning Division at North Eastern Space Applications Centre under Department of Space, Government of India. She has more than 21 years of experience in space technology applications for urban and regional planning, natural resource management and disaster management. She has been leading mainly user specific projects relating to Risk-informed Master plan preparations, city disaster management plans and hazard risk vulnerability assessment for urban centers in North Eastern India. She has published research papers in national and international journals. She has guided a number of M.Sc./M.Tech. and Ph.D. students. She has been awarded ISRO Young Scientist Merit Award for 2011.

B. C. Sumanth a Scientist/Engineer-SD, at North Eastern Space Applications Centre (NESAC), joined in March 2021. Holds a Bachelor's Degree in Civil Engineering, A Master's in Urban & Regional Planning and a PG Diploma in Urban & Regional Studies. At NESAC, he has contributed to key projects such as the Master Plan for Shillong, Village Level Mapping for Meghalaya, Border Village Planning in Arunachal Pradesh and Hazard Risk Vulnerability Assessments. He is also involved in National initiatives like AMRUT 1.0 and AMRUT 2.0, focusing on sustainable urban & Regional Development.

K. K. Sarma, Ph.D., leads the Remote Sensing Applications Group (RSAG) at the North Eastern Space Applications Centre (NESAC), where he has been a key figure since 2002. Specializing in remote sensing for forestry and biodiversity, his work leverages space technology for critical applications in Biodiversity, Ecosystems, Forestry, and Natural Resource Management. Dr. Sarma has spearheaded over 60 diverse geospatial projects, contributing significantly to natural resource mapping and monitoring across the North Eastern Region (NER). Currently, he coordinates 110 "Plan of Action" projects, focused on using space technology to improve livelihoods in the NER. He also served as the Principal Investigator and NER Coordinator for all ISRO natural resource management projects, including comprehensive Land Use Land Cover Mapping and National Wasteland Updating, Monitoring, and Change Analysis. With over 75 research publications, Dr. Sarma has successfully supervised six Ph.D. students and actively mentors many others across NER Universities.

Chapter 13
Multi-hazard Risk Assessment Using Geospatial Technology

Jenita M. Nongkynrih, B. C. Sumanth, Santanu Das, and K. K. Sarma

Abstract The North Eastern Region (NER) of India is highly vulnerable to multiple natural hazards, including floods, landslides, and lightning, due to its complex topography, climatic variability, and socio-economic conditions. This chapter presents a comprehensive multi-hazard risk assessment framework utilising geospatial technologies such as remote sensing (RS), geographical information system (GIS), and digital elevation models (DEMs). High-resolution satellite imagery and field-based data collection were employed to map hazard-prone areas, identify vulnerable populations and infrastructure, and assess risks. The study focused on urban centres in Meghalaya and Tripura, considering three key hazards, floods, landslides, and lightning. Vulnerability was analyzed through socio-economic indicators, physical infrastructure assessments, and environmental degradation patterns using Multi-Criteria Decision Analysis (MCDA). Risk indices were developed by integrating hazard zonation and vulnerability layers through GIS-based spatial modelling. The results provide critical inputs for disaster preparedness, urban planning, and risk reduction strategies in the NER. The chapter highlights how an integrated geospatial approach enables better understanding of spatial patterns of risk and supports targeted policy interventions. The findings underscore the importance of multi-hazard assessments for sustainable development and effective implementation of the Sendai Framework for Disaster Risk Reduction.

Keywords HRVA · Vulnerability · Risk · Multi-criteria decision analysis

J. M. Nongkynrih (✉) · B. C. Sumanth · S. Das · K. K. Sarma
North Eastern Space Applications Centre (NESAC), Umiam, Shillong, Meghalaya, India
e-mail: jm.nongkynrih@nesac.gov.in

B. C. Sumanth
e-mail: sumanth.bc@nesac.gov.in

S. Das
e-mail: santanu.das.nesac@nesac.gov.in

K. K. Sarma
e-mail: kk.sarma@nesac.gov.in

S. P. Aggarwal et al. (eds.), *Remote Sensing of the North Eastern Himalayan Ecosystem*,
https://doi.org/10.1007/978-981-95-3221-6_13

13.1 Introduction

The North Eastern Region (NER) of India, comprising eight states, is susceptible to multiple hazards, including earthquakes, riverine floods, urban flash floods, glacial lake outburst floods, landslides, forest fires, thunderstorms, lightning, and urban fires. The region suffers significant loss of life, property, infrastructure damage, and socio-economic losses due to multiple hazards. The devastating earthquakes of 1887 and 1950 hit the region (Oldham 1899; Nakata 1989). Examples of few recent events are the urban flash flood in Guwahati on August 6, 2024 (The Assam Tribune 2024), the South Lhonak glacial lake outburst in Sikkim on October 3, 2023 (The Hindu 2024), the landslide at Nongpriang in Meghalaya on May 30, 2024 (The Hindustan Times 2024) and other disasters have disrupted the socio-economic system, even loss of lives and property losses. The frequency of multi-hazards has recently been witnessed in the region, resulting in massive socio-economic losses. The geographical heterogeneity, restricted accessibility, and haphazard and unplanned urban growth have contributed to the multi-hazards effect on urban areas. In most urban areas, hazards may occur in a cascading manner (Chouhan and Mukherjee 2023). According to the United Nations Office for Disaster Risk Reduction (UNDRR), multi-hazards refer to the recognition of multiple major threats that a country may encounter, along with the specific contexts in which these threats could unfold simultaneously, in sequence, or over time, while considering their potential interlinkages (UNDRR 2015).

The increased vulnerability to natural hazards in the NER region is a major concern. This calls for an inclusive approach to make an assessment of hazard-prone zones and to identify vulnerable and probable risk areas. It is required to adopt a multi-hazard risk approach to reduce disaster risk. An inclusive multi-hazard risk assessment, a step for preparedness towards disaster resilient strategy. The preamble section of the Sendai Framework for Disaster Risk Reduction (2015–2030) emphasises that managing disaster risks effectively is vital to achieving sustainable development. According to the United Nations Office for Disaster Risk Reduction, the concept of multi-hazards involves recognising various major threats that a country may face, along with the particular circumstances that may trigger these hazardous occurrences. These events can happen simultaneously, in succession, or progressively over time, while considering their potential interrelated effects (UNDRR 2019a, b). A multi-hazard risk assessment can serve the purpose of aiding disaster management agencies in addressing vulnerabilities, mitigating hazards, preparing for responses, and recovering. This will enable the community to prioritise efforts in reducing risks. The challenge in multi-hazard risk assessment is estimating the risks of multiple hazards within the same area and understanding how the prevailing hazards interact (Hackl et al. 2015). It is challenging to implement tangible policy actions to mitigate the potential risks posed by various hazards without a comprehensive assessment of multi-hazards (Sekhri et al. 2020). Since the NER is vulnerable to multiple hazards, the only option to get the NER community ready for impending calamities is through risk-resilient development planning. Although there have been fewer assessments in the NER, it is commonly recognised that the area is high-risk for multiple hazards. The North Eastern Space Applications Centre, Umiam, has taken up projects related to multi-hazard risk assessment for various urban centres in Meghalaya and Tripura. Three hazards—Flood, Landslide and Lightning were considered for an integrated multi-hazard risk assessment using various geospatial methods and techniques.

Some concepts like hazard, vulnerability and risk assessment are relevant while considering the aspects of multi-hazard risk assessment. The definitions of these concepts and their relationships are integral to the assessment. A hazard refers to any physical process, event, or human-induced activity that has the potential to cause harm, including loss of life, injuries, damage to property, disruptions to society or the economy, and degradation of the environment. Hazards can arise from various natural causes, including geological, hydro-meteorological, and biological factors, as well as artificial causes like environmental degradation and technological hazards. Additionally, risks can encompass latent conditions that might pose future threats. In contrast, Vulnerability refers to the set of physical, social, economic, and environmental conditions or dynamics that increase the probability of a community suffering negative impacts when exposed to hazardous events (UN/ISDR 2007). However, risk is a result of three main factors: susceptibility, the frequency or severity of the hazard, and the community's exposure to risks (such as those that are impacted by or likely to be affected by natural disasters) (Birkmann 2007). More specifically, it is the potential for harm along with the expected amount of loss from a given hazard magnitude (Alexander 2000).

13.2 Geospatial Technology—A Tool for Multi-hazard Risk Assessment

A multi-hazard risk assessment for the North Eastern Region (NER) of India by geospatial techniques provides an invaluable approach toward better understanding and mitigation of various natural hazards, including floods, landslides, and lightning. The region is generally susceptible due to its complex geography, monsoonal climate, and proximity to tectonic fault lines, which increase the occurrence and intensity of hazards. Generating detailed large-scale hazard maps and risk models utilising geospatial tools, such as remote sensing, GIS and multi-hazard modelling, can help identify vulnerabilities and make data-driven decision-making.

Remote Sensing is an important technology function and has become essential in urban multi-hazard assessment. Very high-spatial-resolution remote sensing data is helpful in capturing fine details on the Earth's surface, especially when the spatial resolution is less than one meter or better. Earth-observing systems regularly image the Earth's land surfaces, thus providing potential inputs at different points of response, recovery, mitigation, and preparedness (Elliott 2020). Different satellite remote sensing (RS) systems have diverse characteristics, viz., spatial, temporal and spectral resolutions. The spatial resolution of satellite images refers to the pixel size that corresponds to covering a portion of the Earth's surface. High spatial resolution ranges from 0.10 to 5 m, whereas low spatial resolution ranges from 6 m to >1000 m. The frequency of a satellite sensor's revisits for a particular location is known as the temporal resolution, and it can vary from less than 24 h to three days for high temporal resolution and from four to sixteen days for medium temporal resolution. The number of spectral bands in which a sensor can gather reflected radiation is determined by its spectral resolution. In contrast to medium spectral resolution, which has three to fifteen bands, and low spectral resolution, which has three bands, high spectral resolution data might have 220 bands or more.

Currently, various remote sensing platforms, such as QuickBird, Worldview-1 and -2, Ikonos II, GeoEye-1, Cartosat-1, and the Earth Remote Observation Satellite (EROS) constellation, are capable of capturing very high-resolution (VHR) optical imagery. Furthermore, faster revisit times offered by future systems, such as the French Pleiades constellation, improve the data timeframes for applications like urban disaster management.

A growing trend in optical imagery technology is providing sub-meter spatial resolution with the 1999 launch of the IKONOS satellite, the first commercial satellite to produce high-resolution images (3.2 m multispectral and 0.80 m panchromatic resolution at nadir) of the Earth. The WorldView Series provides resolutions as fine as 31 cm for multispectral imagery, offering better spatial resolution compared to medium-resolution satellite data (Fig. 13.1). Pleiades is with 0.5 m panchromatic and 2 m in multispectral imagery. QuickBird offered a 61 cm resolution. The CARTOSAT Series provided high-resolution imagery with a resolution of 2.5–0.35 m. In Table 13.1, examples of current, very high-resolution sensors are listed. The high-quality imaging capability with detailed information derived from these satellite data is useful for mapping urban multi-hazard risks. It aids in analysing the spatial distribution and concentration of elements at risks located throughout an area, thereby contributing to disaster preparedness, response, mitigation and prevention strategies. The data has aided in better management of disaster risk at various stages.

Digital Elevation Models (DEMs) provide basic elevation data commonly used to identify hazard-susceptible and vulnerable areas (Fig. 13.2a). The data help determine watershed boundaries, identify geomorphic units, different types of slopes, and elevation (Fig. 13.2b), as well as areas prone to slope instability (Manfreda et al. 2011; Ibrahim et al. 2020; Pourghasemi et al. 2020; Habib et al. 2024). As natural hazards are spatial diverse and complex, there is a requirement for selecting high-quality DEMs, characterised by resolution, accuracy and frequency of updates for practical use in multi-hazard risk assessments. The elevation of an object is the distance and height above sea level. Elevations are usually measured in meters and feet (Fig. 13.2b). Maps that show elevations are called topographic maps.

Derived information from DEMs is useful for many topographic analyses. Some of the derived outputs useful in multi-hazard risk analysis are described in this section. Imaginary lines that connect all sites of equal elevation or altitude above mean sea level are called contours. An isoline map is another name for a contour map. This makes the relief features of any region easier to see. To indicate locations of equivalent height, a contour line is drawn. Typically, two contours with varying altitudes do not

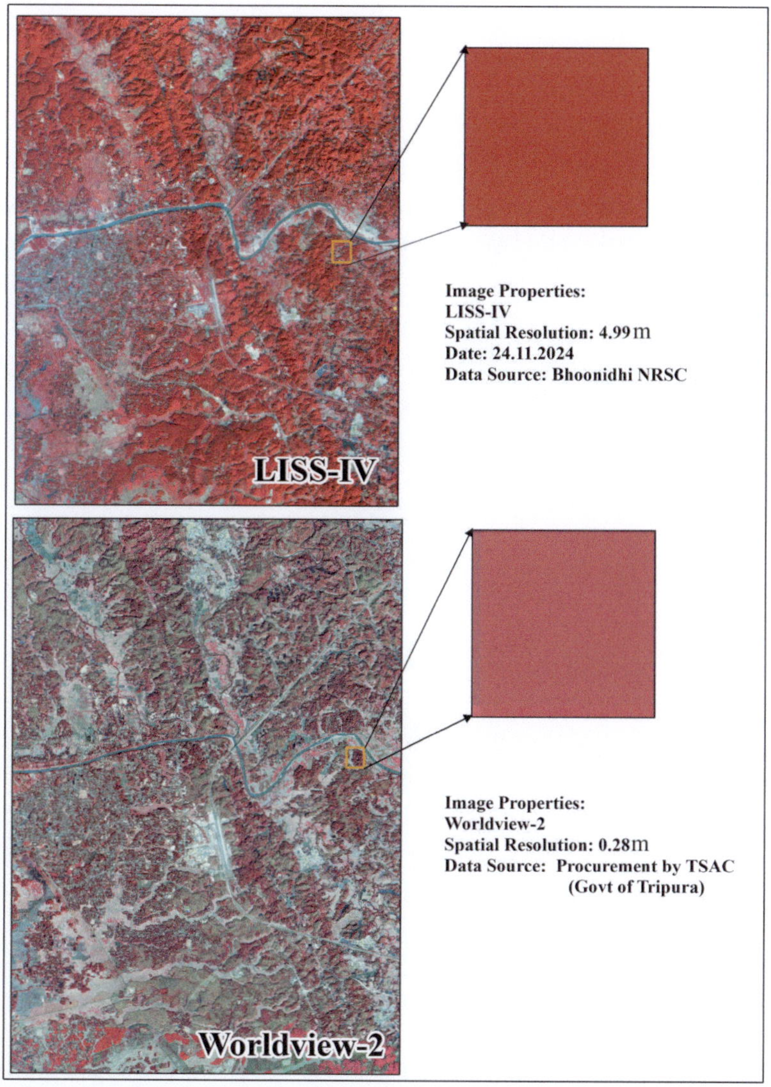

Fig. 13.1 Comparison of spatial resolution-Resourcesat-2 and WorldView-2 (part of Belonia town, Tripura, India)

intersect. A steeper slope is indicated by closer contour lines than by more widely spaced contour lines, which indicate a gentler slope (Fig. 13.3a). Slope describes the shape and relief of the land. The slope measures the changes in that elevation over a certain distance. It is usually measured in degrees and percent. It shows how steep or gentle a slope is at any given point (Fig. 13.3b). A slope Aspect values show which way the actual slopes face. The description direction and slope angle serve as the basis

Table 13.1 Very high-resolution satellite systems (modified from Gamba et al. 2007)

Sl. No.	System	Sensors/geometric resolution		Spectral resolution					
		Panchromatic	Multispectral	Pan	Coastal	Deep blue	Blue	Green	Yellow
1	Albedo	10 cm	40 cm	400–700 nm			450–510 nm	510–580 nm	
2	EROS NG	30 cm	50 cm	*	X	X	*	*	X
3	Pleiades Neo	30 cm	1.2 m	450–800 nm	X	400–450 nm	450–520 nm	530–590 nm	
4	Beijing-3	0.3 m	1.2 m	450–900 nm	X	X	450–520 nm	520–590 nm	
5	Super View Neo2	30 cm	1.2 m	*	X	*	*	*	*
6	Super View Neo1	30 cm	1.2 m	*	X	*	*	*	*
7	World View Legion	34 cm	1.36 m	450–800 nm	400–450 nm	X	450–510 nm	510–580 nm	585–625 nm
8	World View-4	30 cm	1.24 m	450–800 nm	X	X	450–510 nm	510–580 nm	X
9	World View-3	0.31 m	1.24 m	450–800 nm	397–454 nm	X	445–517 nm	507–586 nm	580–629 nm
10	World View-2	0.46 m	1.85 m	400–510 nm	400–450 nm	X	450–510 nm	510–580 nm	585–625 nm
11	World View-1	0.5 m	*	*	X	X	X	X	X
12	GeoEye-1	0.50 m	1.84 m	400–510 nm	X	X	450–510 m	510–580 nm	X
13	Pleiades-1A	50 cm	2 m	480–830 nm	X	X	430–550 nm	490–610 nm	X
14	Pleiades-1B	50 cm	2 m	480–830 nm	X	X	430–550 nm	490–610 nm	X
15	Super View-1	0.5 m	2 m	450–890 nm	X	X	450–520 nm	520–590 nm	X
16	KOMPSAT-3A	0.55 m	2.2 m	450–900 nm	X	X	450–520 nm	520–600 nm	X
17	KOMPSAT-3	0.7 m	2.8 m	450–900 nm	X	X	450–520 nm	520–600 nm	X
18	QuickBird	65 cm	2.62 m	450–900 nm	X	X	450–520 nm	520–600 nm	X
19	Satellogic	X	70 cm	X	X	X	450–510 nm	510–580 nm	X
20	Gaofen-2	0.8 m	3.2 m	450–890 nm	X	X	450–520 nm	520–590 nm	X
21	TripleSat	0.8 m	3.2 m	450–650 nm	X	X	440–510 nm	510–590 nm	X

(continued)

Table 13.1 (continued)

Sl. No.	System	Sensors/geometric resolution		Spectral resolution					
		Panchromatic	Multispectral	Pan	Coastal	Deep blue	Blue	Green	Yellow
22	IKONOS	0.82 m	4 m	526–929 nm	X	X	445–516 nm	506–595 nm	X
23	IKONOS Stereo Imagery	0.82	3.28	526–928 nm	X	X	445–516 nm	506–595 nm	X
24	SkySat	50 cm	1 m	450–900 nm	X	X	450–515 nm	515–595 nm	X
25	Spot-7	1.5 m	6 m	X	X	X	455–525 nm	530–590 nm	X
26	Spot-6	1.5 m	6 m	X	X	X	455–525 nm	530–590 nm	X
27	Spot-5	5 m	10 m	480–710 nm	X	X		500–590 nm	X
28	FORMOSAT-2	2 m	8 m	450–900 nm	X	X	450–520 nm	520–600 nm	X
29	CARTOSAT-1	2.5 M (Stereo)	X	500–850 nm	X	X	X	X	X
30	CARTOSAT-2S (2E + 2F)	65 cm	2 m	0.50–0.85 μm	X	X	430–520 nm	520–610 nm	X
31	CARTOSAT-3	0.28 m	1.12 m	X	X	X	450–520 nm	520–590 nm	X

Sl. No.	System	Spectral resolution						Temporal resolution	Swath width (km)
		Red	Red edge	Red edge 2	NIR	NIR 2	MWIR		
1	Albedo	630–690 nm			770–895 nm	X	X	15 days	35
2	EROS NG	*	X	X	*	X	X	*	12.5
3	Pleiades Neo	620–690 nm	620–690 nm	X	770–880 nm	X	X	26 days	14
4	Beijing-3	630–690 nm		X	770–890 nm	X	X	5 days	11.5
5	Super View Neo2	*	X	X	*	X	X	1 day	15

(continued)

Table 13.1 (continued)

Sl. No.	System	Spectral resolution						Temporal resolution	Swath width (km)
		Red	Red edge	Red edge 2	NIR	NIR 2	MWIR		
6	Super View Neo1	*	X	X	*	X	X	1 day	12
7	World View Legion	630–690 nm	695–715 nm	730–750 nm	770–895 nm	X	X	15 days	10
8	World View-4	655–690 nm	X	X	780–920 nm	X	X	1 day	13
9	World View-3	698–749 nm	698–749 nm	X	765–899 nm	857–1039 nm	X	1 day	13
10	World View-2	705–745 nm	705–745 nm	X	770–895 nm	860–1040 nm	X	1.1 days	16
11	World View-1	X	X	X	X	X	X	*	17
12	GeoEye-1	655–690 nm	X	X	780–920 nm	X	X	16 days	15
13	Pleiades-1A	600–720 nm	X	X	750–950 nm	X	X	1 day	20
14	Pleiades-1B	600–720 nm	X	X	750–950 nm	X	X	2 day	20
15	Super View-1	630–690 nm	X	X	770–890 nm	X	X	2 days	12
16	KOMPSAT-3A	630–690 nm	X	X	760–900 nm	X	3300–5200 nm	*	12
17	KOMPSAT-3	630–690 nm	X	X	760–900 nm	X	X	*	15
18	QuickBird	630–690 nm	X	X	760–900 nm	X	X	1–3.5 days	16.8
19	Satellogic	590–690 nm	X	X	750–900 nm	X	X	*	5
20	Gaofen-2	620–690 nm	X	X	770–890 nm	X	X	5 days	45
21	TripleSat	600–670 nm	X	X	760–910	X	X	*	*
22	IKONOS	632–698 nm	X	X	757–853 nm	X	X	3 days	11.3

(continued)

Table 13.1 (continued)

Sl. No.	System	Spectral resolution						Temporal resolution	Swath width (km)
		Red	Red edge	Red edge 2	NIR	NIR 2	MWIR		
23	IKONOS Stereo Imagery	632–698 nm	X	X	757–853 nm	X	X	*	*
24	SkySat	605–695 nm	X	X	740–900 nm	X	X	6–7 per day	*
25	Spot-7	625–695 nm	X	X	760–890 nm	X	X	*	60
26	Spot-6	625–695 nm	X	X	760–890 nm	X	X	*	60
27	Spot-5	610–680 nm	X	X	780–890 nm	X	X	2–3 days	60
28	FORMOSAT-2	630–690 nm	X	X	760–900 nm	X	X	Daily	24
29	CARTOSAT-1	X	X	X	X	X	X	14 days	29.42
30	CARTOSAT-2S (2E + 2F)	610–690 nm	X	X	760–900 nm	X	X	4/5 days	9.6
31	CARTOSAT-3	620–680 nm	X	X	770–860 nm	X	X	*	17

X—No band available
*—Data not available

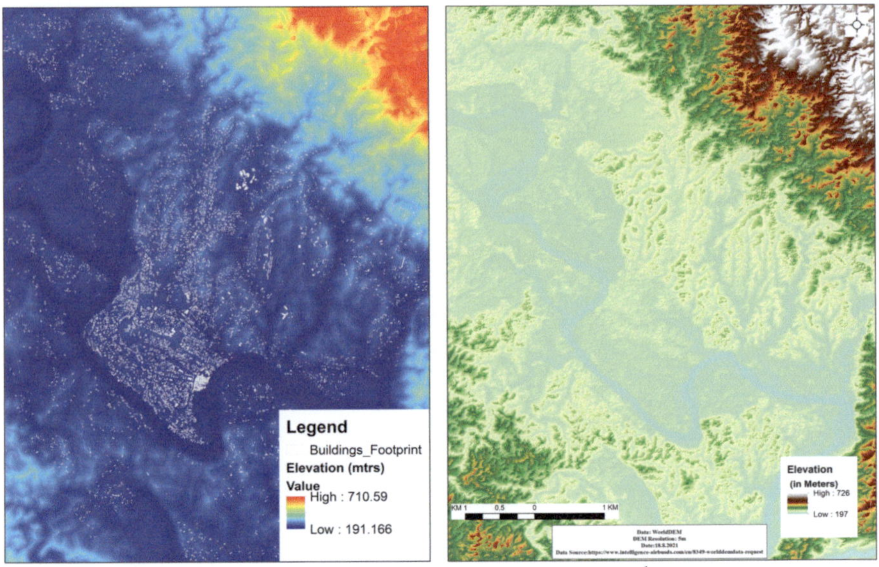

a b

Fig. 13.2 **a** Built-up areas overlaid on CartoDEM (part of Williamnagar town, Meghalaya). **b** Elevation derived form CartoDEM of (part of Williamnagar town, Meghalaya)

for aspect directions. The clockwise measurement begins at 0° north and ends at 360° north. Aspect categories ten different directions (Fig. 13.3c). Hillshade enhances the 3D appearance of the terrain surface based on the sun's relative position, and it is a raster map that assigns a greyscale value to each cell. Hillshade is a technique for visualising terrain determined by the elevation surface's light source and slope aspect.

A grayscale colour ramp is used to display for hillshade elevation model. Hillshade units in degrees from 0 (on the horizon) to 90 (overhead). The default sun altitude for the hillshade is 45° when computing the hillshade (Fig. 13.3d). The application of DEM products in multi-hazard types, including floods, landslides, etc., enables more comprehensive hazard assessments, identifying vulnerable areas for effective risk assessment (Habib et al. 2024).

13.3 Mapping Elements at Risk

Identifying and mapping vulnerable elements is a crucial aspect of multi-hazard risk evaluation. Recognising and mapping the components that may be susceptible to different hazards can facilitate proactive strategies to reduce risks and safeguard individuals, assets, and resources. Elements at risk display both spatial and non-spatial traits.

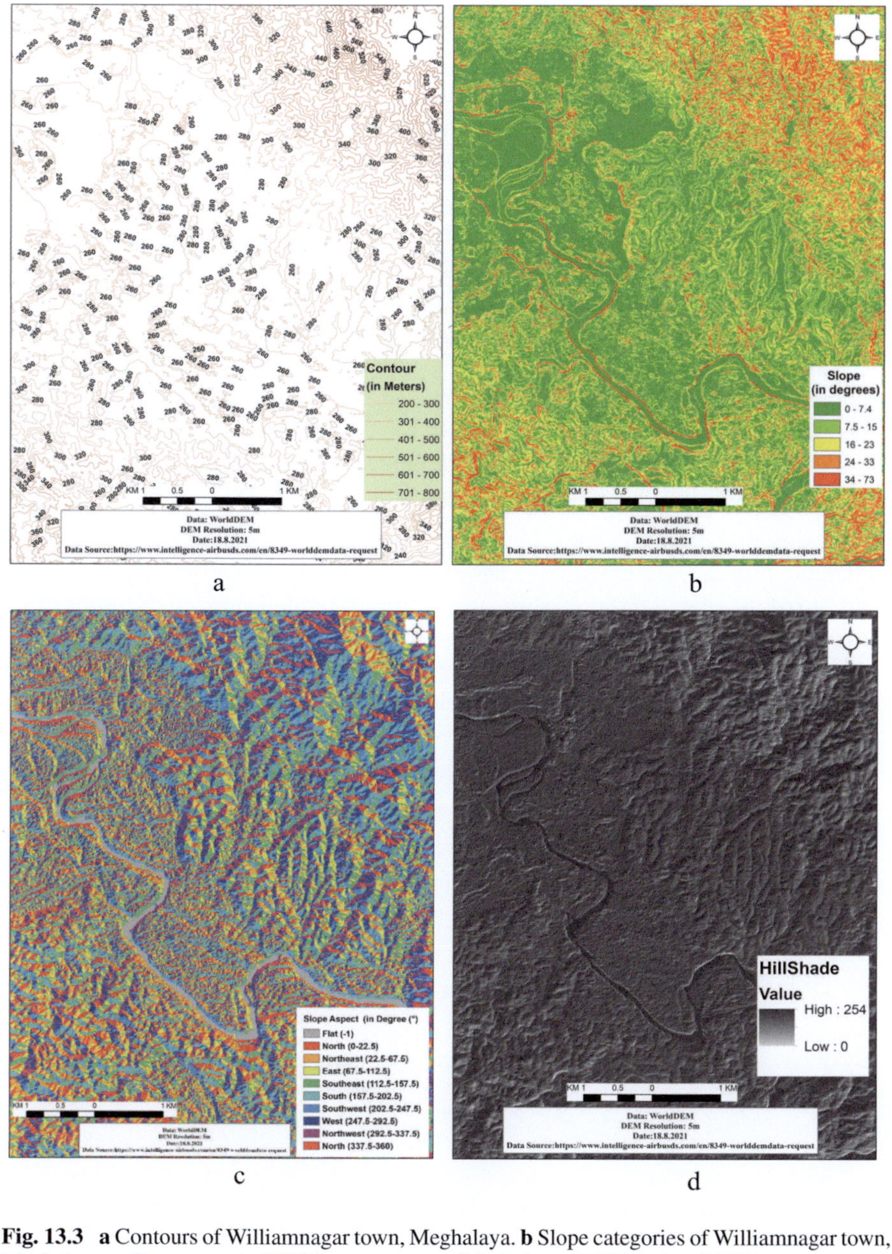

Fig. 13.3 **a** Contours of Williamnagar town, Meghalaya. **b** Slope categories of Williamnagar town, Meghalaya. **c** Slope aspect of Williamnagar town, Meghalaya. **d** Hillshade of Williamnagar town, Meghalaya

13.3.1 What Are Elements at Risk?

Elements at risk refer to people, assets, economic functions, essential services, or other identified values that may be affected by hazards within a given location. These elements can be described through both spatial and non-spatial attributes (Westen et al. 2011). Distinct types of elements at risk can be categorised in several ways in which the amount of elements at risk is characterised:

(a) Human Resources: Number of people, demography (gender and age composition), literacy, number of households, density of population and households.
(b) Infrastructure: Physical assets like roads, buildings (Fig. 13.4), utilities, urban land use, etc.
(c) Economic Assets: Businesses, markets, agricultural sectors, or any industry that may suffer financial loss.
(d) Environmental Resources: Natural resources (forests, water bodies, wildlife, etc.) that may be damaged or endangered.

The way that risky and hazardous factors interact determines how exposed and vulnerable they are. Building a database for assessing the susceptibility of components at risk to certain hazards is the goal of mapping elements-at-risk. Both field-based and remote sensing data can be used to map the elements that are at danger. For example, building footprints, building heights, and urban land use (Fig. 13.5) are mainly obtained from remote sensing data. For administrative units, information

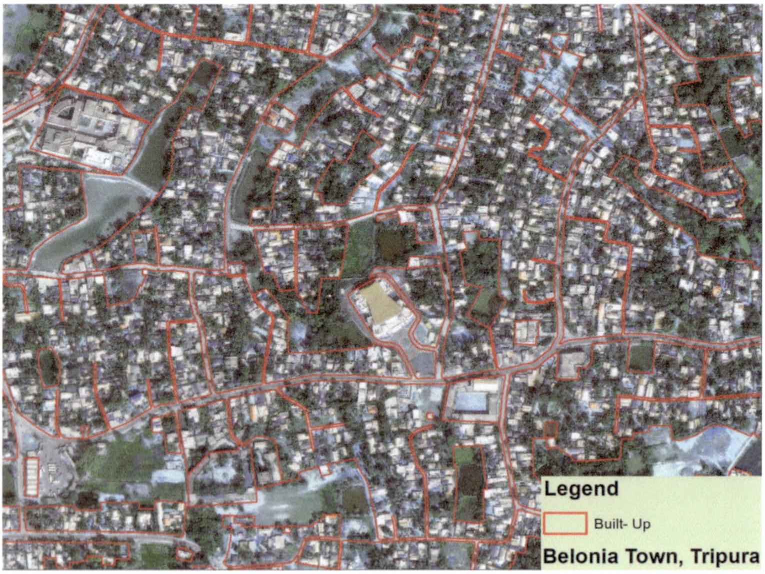

Fig. 13.4 Mapping of element at risk-buildings from WorldView-2 (part of Belonia town, Tripura, India)

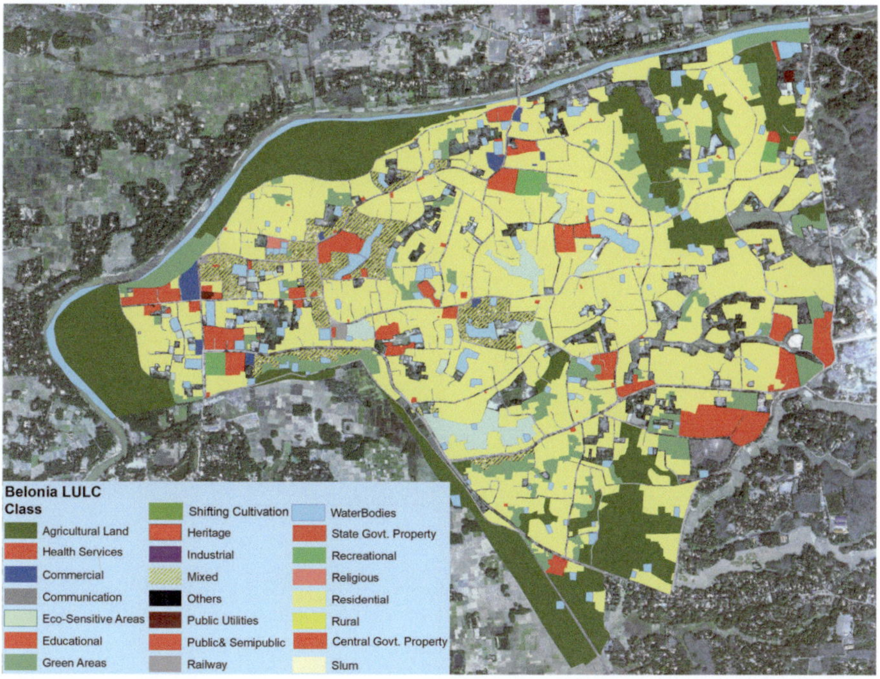

Fig. 13.5 Urban land use as mapped from WorldView-2 (part of Belonia town, Tripura, India)

is available from local governance bodies, viz. municipal ward boundaries. Census data on the population and households for each unit is linked as attribute information to the spatial administrative unit. The basic information on urban land use, building usage and characteristics can be integrated with the survey data using the mobile app (Fig. 13.6). These data can be populated with socio-economic information collected from the field and integrated with the building footprint data (Fig. 13.7).

13.4 Hazard Identification and Mapping

13.4.1 Landslide Susceptibility Mapping

The term "landslide hazard" refers to the probability of an event occurring at a particular location within a defined timeframe. This definition typically addresses the aspects of location and timing. Due to limited data availability, susceptibility maps are often regarded as the primary element in hazard mapping by researchers (Chau et al. 2004). In India, landslides or slope failures are a serious natural hazard that causes annual losses in terms of lives lost, property damage, and communication network interruptions. As they fall into the medium to high and high categories on

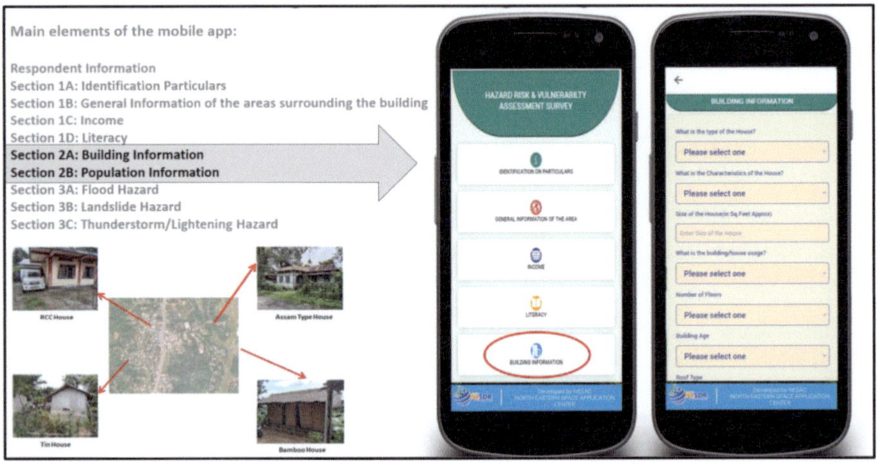

Fig. 13.6 Using the mobile application for field data collection

the Global Landslide Susceptibility Map, they also pose a significant concern in North East India. According to Varnes et al. (1996), the term "landslip" refers to a variety of phenomena, including rock falls, topples, slides, spreads, and flows. These types of events are more common in mountainous areas due to the interplay between geological and human influences. Rainfall, earthquakes, variations in water levels, volcanic activity, forest fires, storm waves, and swift stream erosion are just a few of the triggers that might cause them. Landslide susceptibility maps (Fig. 13.8) can be developed by integrating slope, lithology, land cover, rainfall, and human activity data. Remote sensing data and DEMs are particularly useful for detecting slope instabilities.

Macro-zonation at a scale of 1:50,000 is the main focus of the instructions provided by the Bureau of Indian Standards (BIS) (Ghosh et al. 2009) for the creation of landslip hazard zonation (LHZ) maps in hilly areas. The Landslip Hazard Evaluation Factor (LHEF) rating scheme is used to overlay thematic maps (lithology, structure, slope, relative relief, land use and land cover, and hydrogeology) to create LHZ maps, which are produced following the BIS guideline's knowledge-based rating system. This approach computes the Total Estimated Hazard (TEHD). A slope facet is a portion of a hill slope that consistently exhibits slope direction and inclination, and it is the unit of analysis that the BIS recommends for LHZ. Generally, slope facets in a given area are drawn using SOI Topographical maps, which consider topographical features, including rivers, gullies, ridges, and spurs. The 10-point rating scale is determined by assigning the highest LHEF scores to contributing factors such as lithology, structural features, slope configuration, relative elevation, land use/land cover, and hydrological aspects, with corresponding values of 2, 2, 2, 1, 2, and 1, respectively.

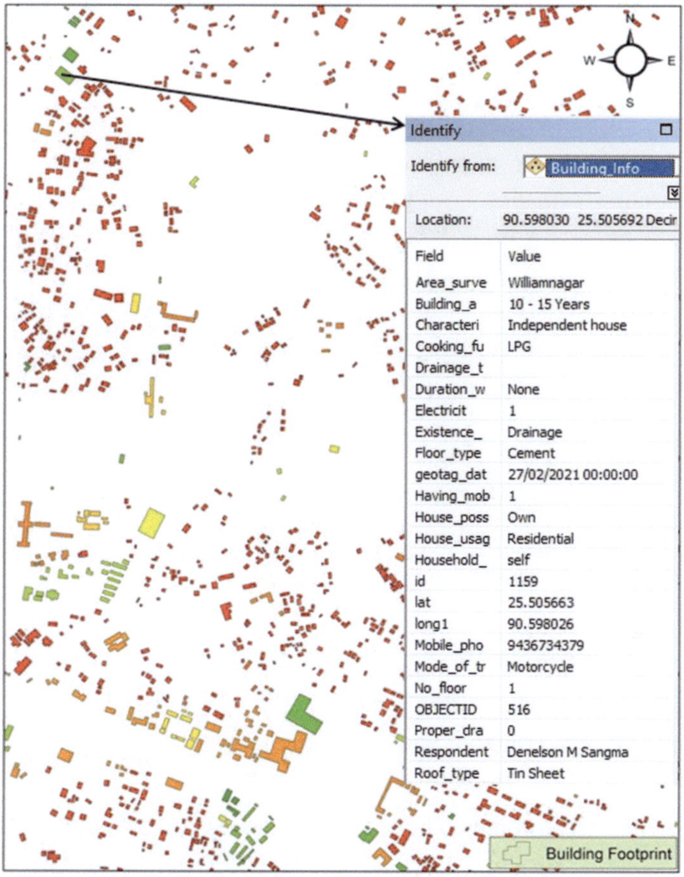

Fig. 13.7 Building information integrated with field information

13.4.2 Flood Hazard Mapping

Flooding is a typical natural disaster that has a severe and widespread impact, resulting in both economic losses and fatalities (Teng et al. 2017). Because of the effects of sudden changes in hydro-climatic conditions and other disruptions, the frequency of climate-related disasters has increased significantly in certain parts of the world (Mihu-Pintilie and Nistor 2019; Dasallas et al. 2019). Land use change and the growth of socioeconomic activity in flood-prone areas are two further possible factors impacting the frequency of flood events (Ramirez et al. 2016). One of the most common hydro-meteorological disasters in the world today is urban flooding. Flooding impacts a significant portion of the global population each year. India, following Bangladesh, holds the second-highest position globally in terms of flood-related fatalities, accounting for approximately 20% of such deaths worldwide,

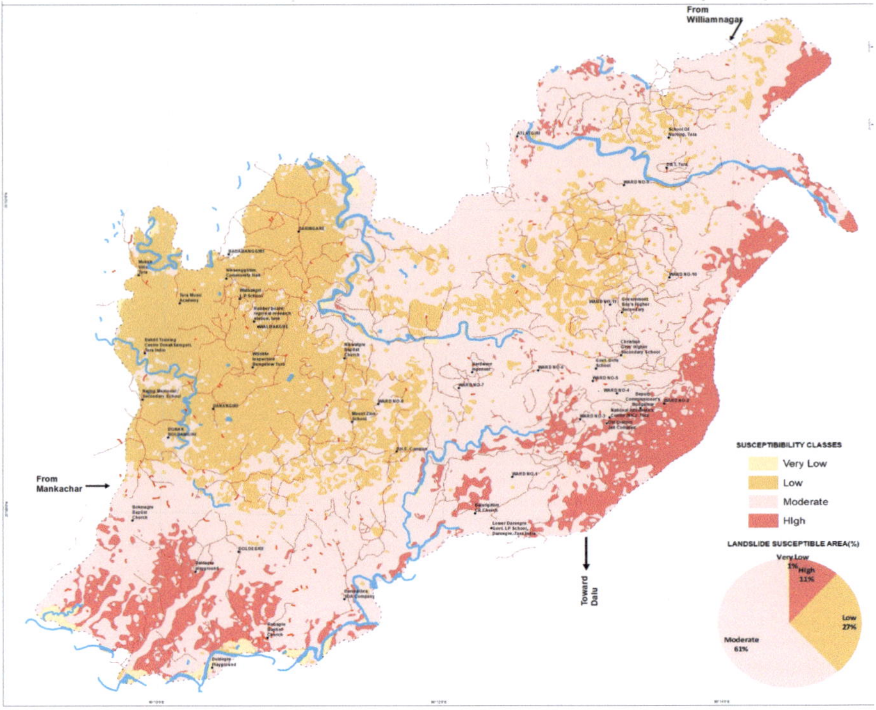

Fig. 13.8 Landslide susceptibility of Tura town, Meghalaya

highlighting its severe vulnerability to flood hazards. Urban flooding is a global phenomenon that is becoming more common and presents a significant challenge to urban planners and city management worldwide. Urban floods can cause cities to be submerged for a few hours or several days, with problems ranging from minor instances to catastrophic ones. Flood hazard maps can be generated using hydrological models such as 'HEC-RAS' or 'MIKE FLOOD', which simulate flood events based on rainfall-runoff relationships, river geometry, and historical flood data (Das 2012). Flood extent can also be mapped using SAR (Synthetic Aperture Radar) images from Sentinel-1A during monsoon seasons (Fig. 13.9).

13.4.3 Lightning Hazard Zonation

A natural electrical discharge with a high voltage and brief duration that occurs between a cloud and the ground, or within a cloud, and is accompanied by a bright flash and usually thunder, is called lightning. "Lightning" as part of a thunderstorm is often accompanied by gusting wind, intense rainfall, and usually hail. Lightning is dangerous when it directly or indirectly strikes humans or animals, leading to loss of

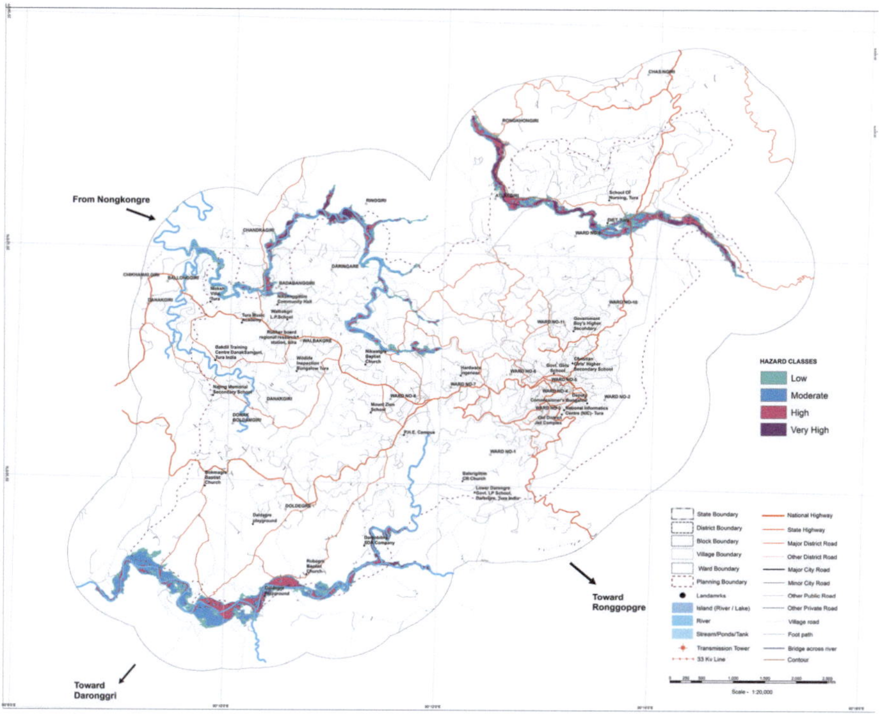

Fig. 13.9 Flood hazard zonation of Tura town, Meghalaya

life or severe injury, and often causing property damage. Recently, the National Crime Report Bureau (NCRB) mentioned more than 2000 fatalities per year due to lightning over India and around 50–60 in the Northeastern region of India. Intra-cloud (IC) lightning flashes occur within clouds and do not reach the Earth's surface, making them generally safe for people. However, some lightning flashes, known as cloud-to-ground (CG) flashes, do reach the ground and can pose a risk. Earth Network Total Lightning Network (ENTLN) provides total lightning data worldwide. The network is expanding annually to various parts of the world, including India. The data discriminate between all types of IC pulses and CG strokes within each lightning flash. NESAC has augmented lightning data for the entire northeastern region from 2017 to mid-2020. In the following report, we have taken the entire CG stroke and IC pulses over the Meghalaya region and used the following steps to prepare the report. First, we extracted the data for the Meghalaya region and then converted it into a density map, using a selected pixel size of 3 square kilometres. The density map counts the number of CG and IC flashes in each district based on the district boundaries. Later, CG strokes were used to show the hazard map (Fig. 13.10) and discuss monthly and annual patterns of lightning flashes.

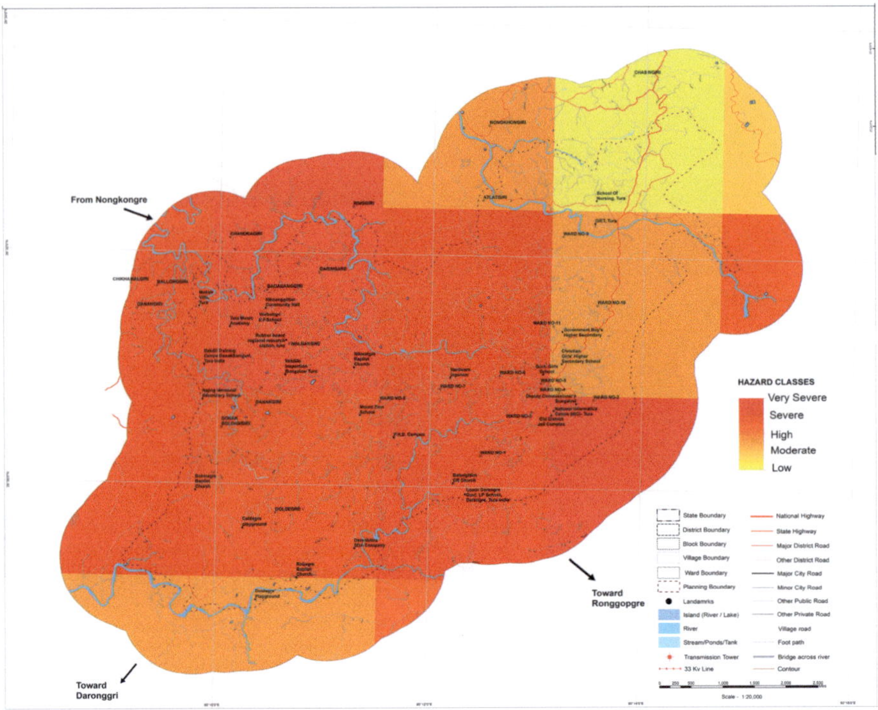

Fig. 13.10 Lightning hazard zonation of Tura town, Meghalaya, India

13.5 Vulnerability Analysis

For vulnerability analysis, socio-economic, physical and environmental parameters are used for further integration in the risk analysis. In the following, the key aspects are discussed:

Socio-Economic Vulnerability: Vulnerability is analysed based on socio-economic parameters, including population density, literacy rates, income levels, and access to critical services such as healthcare and education. Vulnerability indices can be created using multi-criteria decision analysis (MCDA) techniques such as the Analytic Hierarchy Process (AHP) (Villagran de Leon 2023). This analysis helps identify communities most at risk and the factors contributing to their vulnerability.

Physical Vulnerability: The vulnerability of physical infrastructure, such as roads, bridges, and buildings, is assessed through field surveys and GIS-based analysis. Structural vulnerability assessments, particularly in the case of seismic hazards, help determine how infrastructure might perform during disasters.

Environmental Vulnerability: Environmental vulnerability is assessed by analysing land-use changes, deforestation rates, and soil erosion patterns. Remote sensing tools can detect deforestation and degradation, which contribute to increased risks of landslides and floods (Chauhan et al. 2017). Vulnerable ecosystems, such as wetlands and forests, which play a role in hazard mitigation, should also be mapped.

Vulnerability Assessment: MCDA is a decision-making instrument designed to tackle intricate multi-criteria issues that involve both qualitative and quantitative dimensions (Mendoza and Martins 1999). There are two main ways in this methodology: Ranking and Rating. Every decision element in the ranking approach is given a position that represents its relative relevance in the decision-making process. On the other hand, the Rating technique is comparable to ranking but uses numerical scores to indicate the importance of each factor in decision-making (Sharma et al. 2017). Within the framework of multi-hazard risk analysis, the evaluation encompasses the socio-economic susceptibility assessment of buildings and the population within the urbanised area. This evaluation employs the MCDA method to understand the area's susceptibility factors thoroughly. Information about population distribution, building characteristics, general landforms, and previous hazards, as well as existing hazards, is essential for vulnerability assessment. The vulnerability of a population and its buildings to specific hazards varies with the types of hazards and changes over different periods. The primary survey can be carried out using the Mobile Application, following the appropriate sampling method to assess the building, population, and socio-economic vulnerability. Figure 13.11 shows the distribution of socio-economic households surveyed.

The way urban structures have evolved over time has a significant impact on the vulnerability of buildings in metropolitan settings. As a result of the growth of the urban layout that underpins vulnerability, a building aggregate can be thought of as a unit for which it is essential to understand building typology, conservation state, and connecting schemes between buildings. Several criteria were taken into consideration, and ranks and weightages were assigned for assessing building vulnerability based on the type of hazard and the condition of the buildings. The vulnerability zones are categorised into high, moderate, low and very low categories based on the degree of vulnerability. The building's vulnerabilities and other potential risk factors have been assessed and rated by experts. These factors have been assigned scores ranging from 1 to 9 according to their impact on hazardous characteristics. Additionally, weights are assigned to various parameters based on their relative importance. The region is classified into very low to high hazard zones based on these vulnerabilities (Jaman et al. 2025). The built-up area considers various socioeconomic data and risks, and assessments of building vulnerabilities have been conducted.

13.5.1 Building Vulnerability for Landslide

Landslides are geological phenomena where rock, soil, and debris undergo downward movement on a slope. These events can pose significant risks to human life and property, particularly when buildings are located in vulnerable areas. Buildings

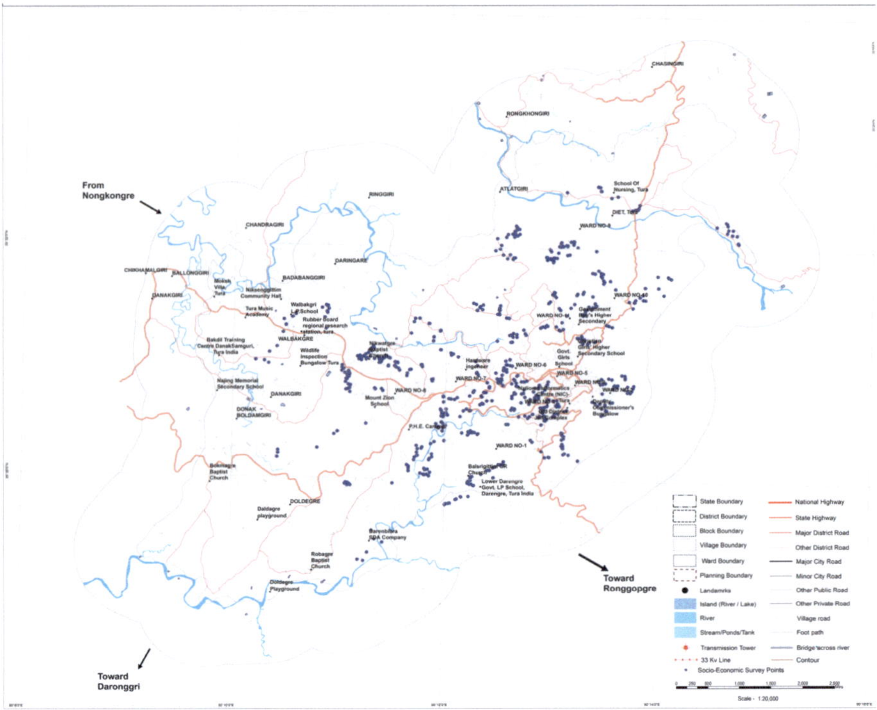

Fig. 13.11 Socio-economic household survey points for vulnerability assessment in Tura town, Meghalaya, India

constructed on steep slopes or unstable soils are more susceptible to landslides. The land's topography is crucial in determining the likelihood of landslide occurrence (Fig. 13.12).

13.5.2 Flood Building Vulnerability

A robust framework for collecting socioeconomic data has been developed, encompassing critical areas such as building information, hazard information, identifying details, and general environmental information. These categories are systematically divided into detailed parameters. For instance, data related to buildings typically includes the structure type, construction age, and the materials used for components such as the floor, walls, and roof (Jaman et al. 2025). Hazard information decisively addresses key factors like the distance from rivers and potential flood levels. Building vulnerability to floods is a critical concern, especially in areas prone to heavy rainfall, storm surges, or river overflow. Various factors, including location, design, construction materials, and mitigation measures, influence the vulnerability of structures to

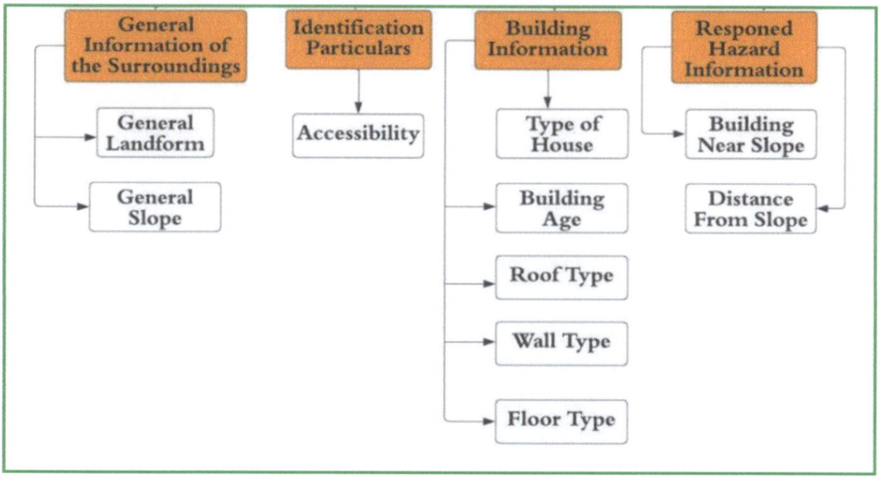

Fig. 13.12 Parameters considered for vulnerability of landslide susceptibility

flood damage. The elevation of a building relative to the base flood level is a key factor in determining vulnerability. Buildings elevated above the expected flood level are less susceptible to inundation. Additionally, the design and construction of foundations play a critical role in ensuring a structure's stability during and after a flood event. The choice of construction materials and techniques significantly impacts a building's resilience to flooding. Water-resistant and flood-resistant materials, as well as elevated construction techniques, can reduce the extent of damage caused by floodwaters (Figs. 13.13 and 13.14).

13.5.3 Building Vulnerability for Lightning

Lightning primarily revolves around the risk of lightning strikes causing damage to structures and the potential for subsequent fires or other safety hazards. Buildings in areas with a high frequency of thunderstorms are more susceptible to lightning strikes. Additionally, increased exposure may make structures on elevated terrain or open areas more vulnerable. The vulnerability of buildings, along with other hazard factors, has been evaluated and assigned ratings from 1 to 9 based on their potential impact on danger characteristics. Furthermore, the weights of various factors have been determined according to their relative importance, as outlined in Table 13.2. An assessment of building vulnerability has been performed, incorporating several socioeconomic data within developed areas. The region has been categorised into four hazard levels: high, moderate, low, and very low, based on the assessed vulnerability (Fig. 13.15).

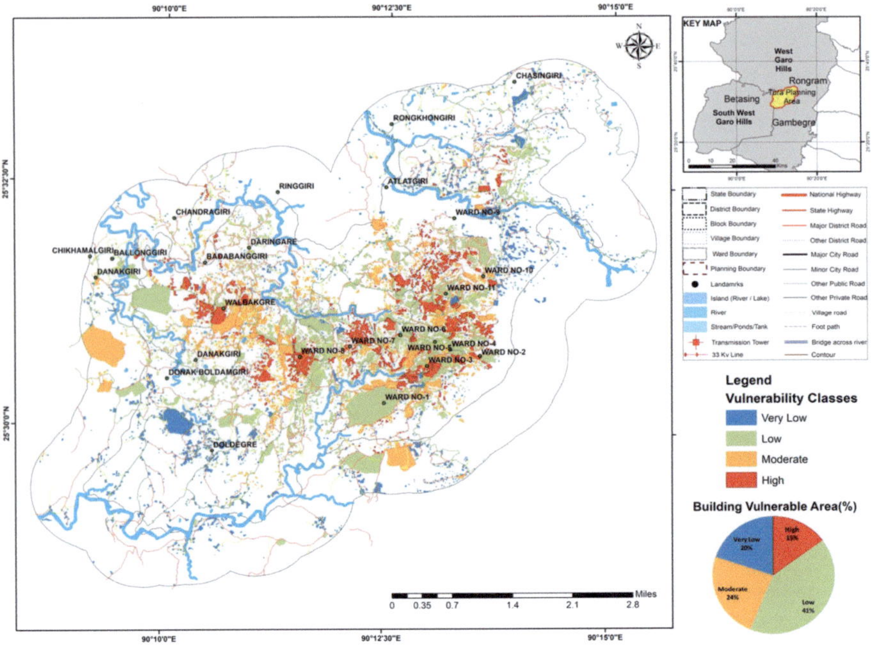

Fig. 13.13 Building vulnerability zones for landslide of Tura town, Meghalaya, India

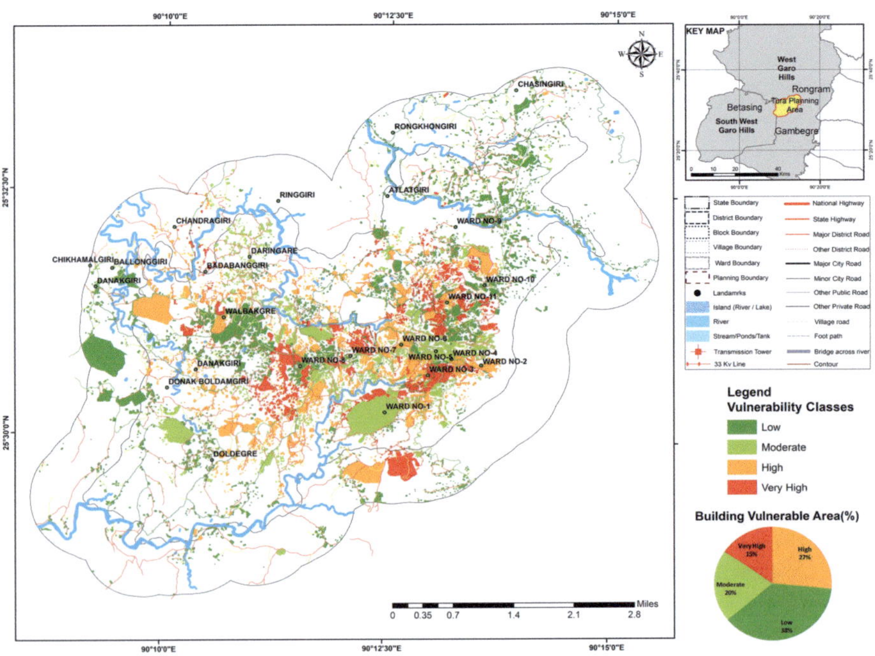

Fig. 13.14 Building vulnerability zones for flood hazard of Tura town, Meghalaya, India

Table 13.2 Vulnerability criteria for deriving building vulnerability for lightning hazard

Sl. No.	Building vulnerability parameters	Criteria	Rating	Score
1	Accessibility	Very good	1	8
		Good	3	
		Medium	6	
		Bad	9	
2	Type of house	RCC	1	9
		Masonry	6	
		Timber building	9	
3	Building age	Less than 5 years	1	6
		5–10 years	3	
		10–15 years	6	
		More than 15 years	9	
4	Roof type	Tin sheet	9	8
		CGI	6	
		Mixed	3	
		Cement	1	
5	Wall type	CGI	9	7
		Bamboo	6	
		Wooden	4	
		Brick	3	
		Stone	2	
		Mixed	1	
6	Floor type	Mud	9	5
		Bamboo	7	
		Wooden	5	
		Mixed	3	
		Cement	1	
7	Thunderstorm incident	Yes	9	14
		No	1	
8	Frequency of thunderstorm	After 5 years	9	12
		Once in 2 years	5	
		Every year	1	
9	Thunder incident damage	Less than Rs. 10,000	1	10
		Rs. 10,000–50,000	5	
		More than Rs. 50,000	9	
10	Lightning incident	Yes	9	11
		No	1	
11	Tower near house	Yes	9	10
		No	1	

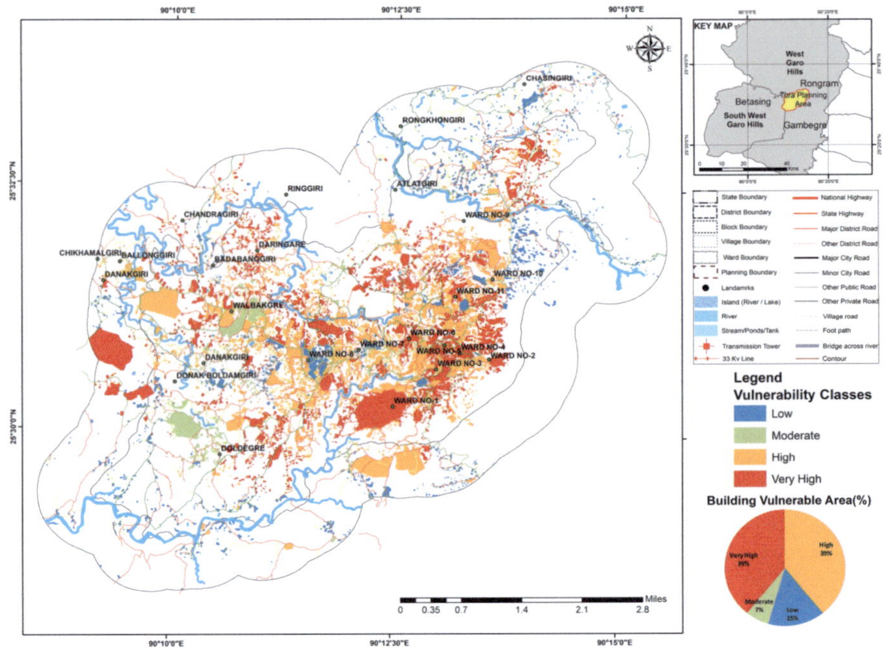

Fig. 13.15 Building vulnerability for lightning hazard Tura town, Meghalaya, India

13.6 Multi-hazard Risk Analysis

For Multi-Hazard risk assessment, in this case, the hazards considered are landslides, floods, and lightning in a given area (Fig. 13.16). "Risk is shaped by the dynamic relationship between hazard, vulnerability, and exposure".

$$\text{Risk} = \text{Hazard} \times \text{Element at Risk} \times \text{Vulnerability}$$

The following steps are integral to the risk assessment process.

13.6.1 Multi-hazard Risk Indexing

A multi-hazard risk index can be developed by integrating hazard maps and vulnerability assessments. This composite index will help rank areas based on their risk from earthquakes, landslides, floods, and extreme weather events (Marzocchi et al. 2012). Each hazard should be weighted based on its historical impact and probability of occurrence.

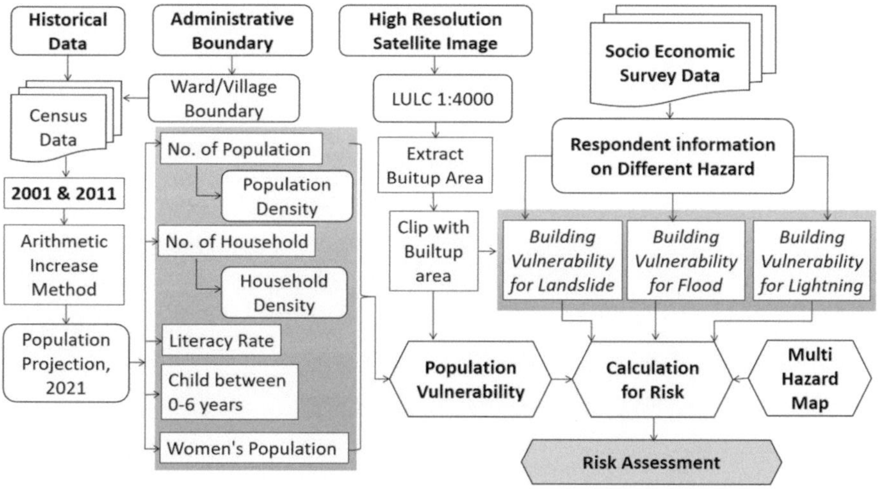

Fig. 13.16 Vulnerability and risk assessment for multi-hazard risk assessment

Several natural and human-induced disasters worldwide result in the loss of property, human life, and the environment. Risk is that potential loss expected to be sustained due to the occurrence of a hazard within a given space over an exposed time period; it relies on both the likelihood of a hazard and the potential damage it could cause. The scholars call for sustainable risk management of disaster, which emphasises spatial planning in disaster risk mitigation strategies owing to the potential effectiveness in handling long-term risks. The term "at risk" refers to a community or locality that is exposed to risks and is likely to experience negative effects from these occurrences. Scientific literature has widely reported that the severity of hazardous events depends on the vulnerability of the affected area and its inhabitants. The probability of a multi-hazard event occurring in that area and the vulnerability of the relevant community are common in most parts of NER. Usually, parts of risk assessment characterise a geographic investigation within GIS and are used as a basis for decision support tasks to optimise disaster preparedness and response.

13.6.2 Risk Evaluation for Landslide, Flood and Lightning Hazard

Geographical representations of vulnerability and hazard zones have been used to estimate risk. Details about the objects that are vulnerable, such as buildings, the economy, and the population, are required to assess an area's risk in relation to a specific hazard. Specific construction data, such as place of residence, building category, number of floors, construction type, roof type, and so on, are required for hazard risk assessment. This data was acquired from field data gathered in various

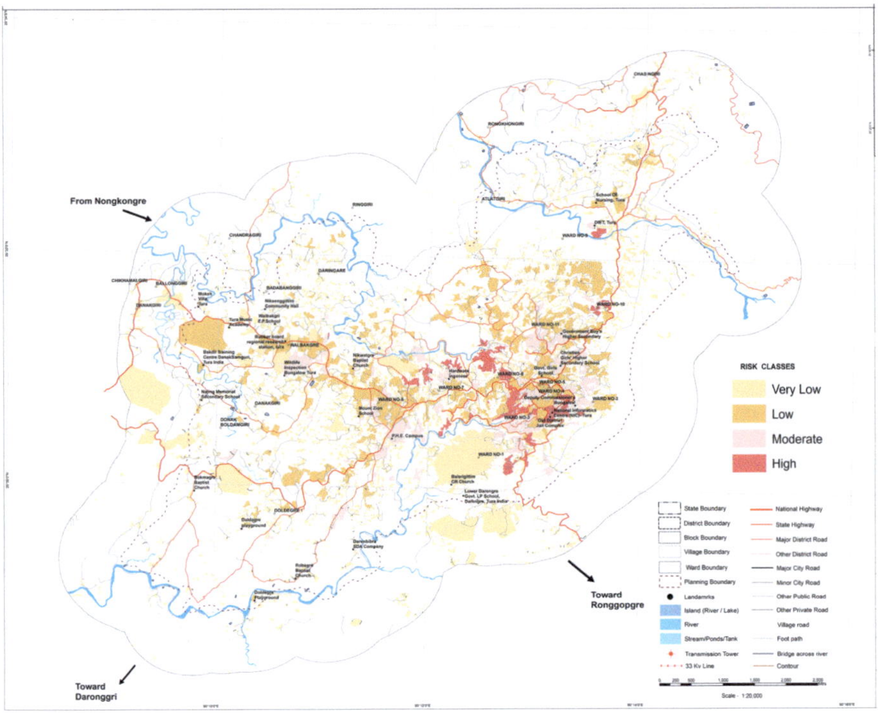

Fig. 13.17 Landslide risk zones of Tura town, Meghalaya, India

locations, and it was based on the type and density of the building structure/footprint, established using high-resolution imagery from satellites, supplemented with ground truth and census data. The MCDA technique has been used for multi-hazard risk assessment of Tura town by integrating hazard and vulnerability layers. Multi-hazard risk evaluation and mapping integrate many types of hazards that a region may face, providing a holistic framework for proactive planning. Regarding the integration of population vulnerability, building vulnerability, and other hazard layer risk factors in landslide (Fig. 13.17), flood (Fig. 13.18), and lightning (Fig. 13.19) risk assessments in the Tura town area, the thresholds for each hazard risk are synthesised using Weighted Overlay Analysis as a tool.

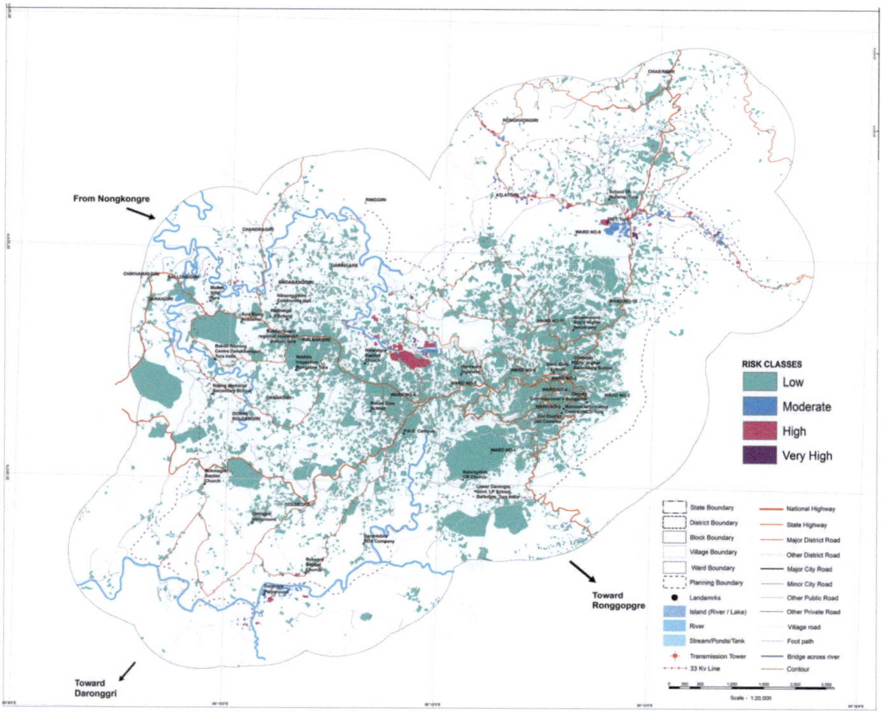

Fig. 13.18 Flood risk zones of Tura town, Meghalaya, India

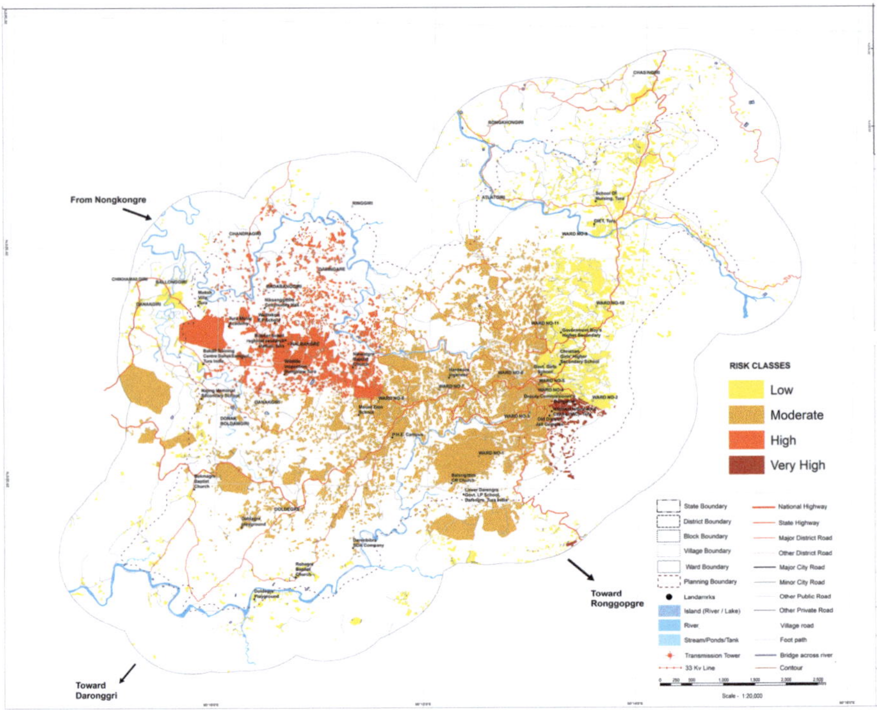

Fig. 13.19 Lightning risk zones of Tura town, Meghalaya, India

13.7 Conclusion

The North Eastern Region (NER) of India is particularly vulnerable to a several hazards, including floods, landslides, and lightning strikes. Due to its complex topography, growing urban population, and fragile ecosystems, the region requires a comprehensive and integrated approach to assess and mitigate disaster risks. This chapter has shown that a multi-hazard risk assessment, enhanced by advanced geospatial technologies such as remote sensing, GIS, and high-resolution satellite imagery, creates a solid framework for understanding the spatial patterns of hazards, vulnerabilities, and risks. By integrating various geospatial datasets, it is possible to accurately map areas prone to hazards, identify vulnerable communities, and evaluate both physical and socio-economic exposures. Utilising digital elevation models (DEMs), vulnerability surveys, and socio-economic data through mobile-based tools enriches this analytical process. Additionally, multi-criteria decision analysis (MCDA) allows for the synthesis of various parameters into vulnerability and risk indices, enabling evidence-based prioritisation of disaster mitigation efforts. Ultimately, conducting a multi-hazard risk assessment is crucial for informed urban planning and disaster preparedness in the NER. It empowers policymakers, planners, and local authorities to develop targeted interventions and promote disaster-resilient communities. The successful implementation of such assessments, particularly with the support institutional support, can significantly contribute to the goals of the Sendai Framework for Disaster Risk Reduction and the broader objectives of sustainable regional development. Finally, geospatial tools employed during multi-hazard risk assessments would strengthen disaster resilience in NER and contribute to sustainable development through improved preparedness, response capabilities, and long-term recovery strategies in the region. An active, evidence-based strategy to disaster risk reduction and climate change adaptation in NER will be made possible by implementation through government interagency and private sector partnerships with local stakeholders. An essential component of disaster risk reduction is multi-hazard risk assessment. Moreover, with the integration of geospatial techniques on a very high-resolution scale, implementation at the local level becomes effective for spatial planning. These techniques mentioned in this chapter can further be utilised for disaster mitigation. Detailed planning, which is more suitable to the physical and socio-economic characteristics of the region.

Acknowledgements The authors gratefully acknowledge the financial support provided by the North Eastern Council (NEC), Government of India, and the Meghalaya State Disaster Management Authority for facilitating the successful execution of the projects. We also extend our sincere thanks to the Tripura Space Applications Centre and the various line departments of Meghalaya and Tripura for their valuable inputs and consistent cooperation throughout the project duration.

References

D.E. Alexander, *Confronting Catastrophe: New Perspectives on Natural Disasters* (Terra and Oxford University Press, 2000)

J. Birkmann, Risk and vulnerability indicators at different scales: applicability, usefulness and policy implications. Environ. Hazards **7**, 20–31 (2007)

K.T. Chau, Y.L. Sze, M.K. Fung, W.Y. Wong, E.L. Fong, L.C.P. Chan, Landslide hazard analysis for Hong Kong using landslide inventory and GIS. Comput. Geosci. **30**(4), 429–443 (2004)

R. Chauhan, A.K. Saha, R.P. Singh, Geospatial analysis of deforestation and landslide risks in the Indian Himalayas. Nat. Hazards **86**(2), 649–673 (2017)

S. Chouhan, M. Mukherjee, Design and application of a multi-hazard risk rapid assessment questionnaire for hill communities in the Indian Himalayan region, in *Estimating and Predicting Natural Hazards and Vulnerabilities in the Himalayan Region*, vol. 23, ed. by A. Agarwal, K. Cook, U. Öztürk, W. Schwanghart, R. Shukla, S. Fuchs. Natural Hazards and Earth System Sciences (2023), pp. 1267–1286. https://doi.org/10.5194/nhess-23-1267-2023

P.J. Das, Floods in Assam: magnitude, frequency, and magnitude-frequency analysis. J. Geol. Soc. India **80**(5), 685–692 (2012). https://doi.org/10.1007/s12594-012-0156-7

L. Dasallas, T.W. Hsu, M.H. Hsu, Modelling typhoon-induced coastal flooding and erosion under climate change scenarios—a case study in Taiwan. Water **11**(1), 82 (2019). https://doi.org/10.3390/w11010082

J.R. Elliott, Earth observation for the assessment of earthquake hazard, risk and disaster management. Surv. Geophys. **41**, 1323–1354 (2020). https://doi.org/10.1007/s10712-020-09606-4

P. Gamba, F. Dell'Acqua, G. Lisini, G. Trianni, Exploiting spectral and spatial analysis for urban area data interpretation, in *Proceedings of the 27th EARSeL Symposium* (Bolzano, Italy, 2007), pp. 117–124

S. Ghosh, C.J. Van Westen, E.J.M. Carranza, T.B. Ghoshal, N.K. Sarkar, M. Surendranath, A quantitative approach for improving the BIS (Indian) method of medium-scale landslide susceptibility. J. Geol. Soc. India **74**(5), 625–638 (2009)

M. Habib, A. Habib, M. Abboud, Multi-aspect critical assessment of applying digital elevation models in environmental hazard mapping. Revue Internationale de Géomatique, **33**, 247–271 (2024). https://doi.org/10.32604/rig.2024.053857

J. Hackl, B.T. Adey, M. Heitzler, An overarching risk assessment process to evaluate the risks associated with infrastructure networks due to natural hazards. Int. J. Perform. Eng. **11**, 153–168 (2015). https://hal.science/hal-01575239/

Hindustan Times, *Four Bodies Trapped in Landslide in Meghalaya recovered* (2024, November 1). https://www.hindustantimes.com/cities/others/four-bodies-trapped-in-landslide-in-meghalaya-recovered-101717233486644.html

M. Ibrahim, A. Al-Mashaqbah, B. Koch, P. Datta, An evaluation of available digital elevation models (DEMs) for geomorphological feature analysis. Environ. Earth Sci. **79**(13), 336 (2020)

T. Jaman, J. Nongkynrih, B. Sumanth, Influence of multi-hazard on socio-economic vulnerability of Tura Town area of Meghalaya, India: an urban modeling in respect of disaster risk reduction using geospatial technologies. J. Indian Soc. Remote Sens. (2025). https://doi.org/10.1007/s12524-025-02125-y

S. Manfreda, M. Di Leo, A. Sole, Detection of flood-prone areas using digital elevation models. J. Hydrol. Eng. **16**(10), 781–790 (2011). https://doi.org/10.1061/(ASCE)HE.1943-5584.0000367

W. Marzocchi, A. Garcia-Aristizabal, P. Gasparini, M.L. Mastellone, A. Di Ruocco, Basic principles of multi-risk assessment: a case study in Italy. Nat. Hazards **62**(2), 551–573 (2012). https://doi.org/10.1007/s11069-012-0093-4

G.A. Mendoza, H. Martins, Multi-criteria decision analysis in natural resource management: a critical review of methods. For. Ecol. Manag. **120**(1–3), 17–23 (1999). https://doi.org/10.1016/S0378-1127(99)00048-X

A. Mihu-Pintilie, I. Nistor, The influence of climate change on the flood risk of small river basins: a case study of Ozana River in Romania. Water **11**(4), 813 (2019). https://doi.org/10.3390/w11040813

T. Nakata, Active faults of the Himalaya of India and Nepal. Spec. Pap. Geol. Soc. Am. **232**, 243–264 (1989). https://doi.org/10.1130/SPE232-p243

R.D. Oldham, Report on the great earthquake of 12th June, 1897. Mem. Geol. Surv. India **29**, 379 (1899)

H.R. Pourghasemi, A. Gayen, M. Edalat, M. Zarafshar, J.P. Tiefenbacher, Is multi-hazard mapping effective in assessing natural hazards and integrated watershed management? Geosci. Front. **11**(4), 1203–1217 (2020)

J.A. Ramirez, N. Lacey, B. Rajagopalan, Land-use changes and their impact on frequency and intensity of flooding: a modelling approach. J. Hydrol. **540**, 143–156 (2016). https://doi.org/10.1016/j.jhydrol.2016.05.014

S. Sekhri, P. Kumar, C. Fürst, R. Pandey, Mountain specific multi-hazard risk management framework (MSMRMF): assessment and mitigation of multi-hazard and climate change risk in the Indian Himalayan Region. Ecol. Ind. **118**, 106700 (2020). https://doi.org/10.1016/j.ecolind.2020.106700

S.S.V. Sharma, P.S. Roy, V. Chakravarthi, S.R. Goru, Flood risk assessment using multi-criteria analysis: a case study from Kopili River basin, Assam, India. Geomat., Nat. Hazards Risk (2017). https://doi.org/10.1080/19475705.2017.1408705

J. Teng, A.J. Jakeman, J. Vaze, B.F.W. Croke, D. Dutta, S. Kim, Flood inundation modelling: a review of methods, recent advances and uncertainty analysis. Environ. Model. Softw. **90**, 201–216 (2017). https://doi.org/10.1016/j.envsoft.2017.01.006

The Assam Tribune, *24 Hrs On, Deluge Continues in Guwahati; Schools, Colleges Shut* (2024, November 1). https://assamtribune.com/guwahati/24-hrs-on-deluge-continues-in-guwahati-schools-colleges-shut-1546888

The Hindu, *South Lhonak Lake Flood: Teesta Dam, Climate and Disaster* (2024, October 5). https://www.thehindu.com/sci-tech/energy-and-environment/south-lhonak-lake-flood-teesta-dam-climate/article67503729.ece

UN/ISDR, *Annual Report 2007* (United Nations International Strategy for Disaster Reduction, 2007). https://www.preventionweb.net/files/7801_ISDRAnnualReport2007LR1.pdf

UNDRR, *Sendai Framework for Disaster Risk Reduction 2015–2030* (2015). https://www.preventionweb.net/go/sfdrr

UNDRR, *Developing Multi-Hazard Early Warning Systems: A Checklist* (United Nations Office for Disaster Risk Reduction, 2019a)

United Nations Office for Disaster Risk Reduction (UNDRR), *Global Assessment Report on Disaster Risk Reduction (GAR)* (2019b). https://gar.undrr.org

C.J. Van Westen, D. Alkema, M.C.J. Damen, N. Kerle, N.C. Kingma, *Multi-Hazard Risk Assessment Distance Education Course—Guide book* (ITC University, Netherlands, 2011)

D.J. Varnes, W. Savage (eds.), *The Slumgullion Earth Flow: A Large-Scale Natural Laboratory* (U.S. Geological Survey Bulletin 2130, 1996)

J.C. Villagran de Leon, An improved understanding of natural hazards and disasters through the use of satellite technologies: contributions from UN-SPIDER. Sustainability **15**(13), 10624 (2023). https://doi.org/10.3390/su151310624

Jenita M. Nongkynrih is currently Head Urban and Regional Planning Division at North Eastern Space Applications Centre under Department of Space, Government of India. She has more than 21 years of experience in space technology applications for urban and regional planning, natural resource management and disaster management. She has been leading mainly user specific projects relating to Risk-informed Master plan preparations, city disaster management plans and hazard risk vulnerability assessment for urban centers in North Eastern India. She has published research papers in national and international journals. She has guided a number of M.Sc./M.Tech. and Ph.D. students. She has been awarded ISRO Young Scientist Merit Award for 2011.

B. C. Sumanth is a Scientist/Engineer-SD, at North Eastern Space Applications Centre (NESAC), joined in March 2021. Holds a Bachelor's Degree in Civil Engineering, A Master's in Urban & Regional Planning and a PG Diploma in Urban & Regional Studies. At NESAC, he has contributed to key projects such as the Master Plan for Shillong, Village Level Mapping for Meghalaya, Border Village Planning in Arunachal Pradesh and Hazard Risk Vulnerability Assessments. He is also involved in National initiatives like AMRUT 1.0 and AMRUT 2.0, focusing on sustainable urban & Regional Development.

Santanu Das, a Scientist/Engineer-SE at NESAC, has 13 years of Experience across government administration and the private sector. He specializes in leveraging space technology to drive sustainable urban development. His expertise lies in applying geospatial intelligence to urban and regional planning, enabling data-driven decision-making for large-scale operations in planning, scheduling, and implementation. With a strong engineering foundation and a Master's degree in City Planning from the Indian Institute of Technology, Kharagpur, He brings an in-depth understanding of how satellite data and spatial analytics can transform urban systems, support infrastructure development, and enhance governance efficiency.

K. K. Sarma, Ph.D., leads the Remote Sensing Applications Group (RSAG) at the North Eastern Space Applications Centre (NESAC), where he has been a key figure since 2002. Specializing in remote sensing for forestry and biodiversity, his work leverages space technology for critical applications in Biodiversity, Ecosystems, Forestry, and Natural Resource Management. Dr. Sarma has spearheaded over 60 diverse geospatial projects, contributing significantly to natural resource mapping and monitoring across the North Eastern Region (NER). Currently, he coordinates 110 "Plan of Action" projects, focused on using space technology to improve livelihoods in the NER. He also served as the Principal Investigator and NER Coordinator for all ISRO natural resource management projects, including comprehensive Land Use Land Cover Mapping and National Wasteland Updating, Monitoring, and Change Analysis. With over 75 research publications, Dr. Sarma has successfully supervised six Ph.D. students and actively mentors many others across NER Universities.

Chapter 14
Geospatial Technology Support for Water Resources Management in NE India

Ranjit Das, Shanbor Kurbah, B. M. Arjun, and Diganta Barman

Abstract Water resource management in Northeast India presents complex challenges due to its diverse topography, intense monsoonal rainfall, frequent floods, and seasonal water scarcity. This chapter highlights the application of geospatial technologies, such as remote sensing (RS), geographic information systems (GIS), and digital elevation models (DEM) in managing water resources across the region. It provides a comprehensive overview of geospatial approaches for mapping and assessment of water resources. Case studies from Assam, Meghalaya, and other northeastern states demonstrate the successful use of satellite data (e.g., Landsat, Sentinel, IRS, Cartosat) for large-scale inventory, monitoring, and planning of water conservation structures like Jalkunds and multipurpose reservoirs. Advanced methods such as multi-criteria decision analysis (MCDA), normalized difference indices (NDWI, NDSI), and hydrological modeling are discussed for identifying suitable sites for interventions. The chapter also covers real-time flood monitoring using SAR data, riverbank erosion tracking through NDWI and DSAS tools, and impact assessments under IWMP. The integration of space-based and field-level data facilitates evidence-based decision-making and sustainable development of water resources. This geospatial framework serves as a model for climate-resilient water management in hilly and flood-prone regions.

Keywords Geospatial technology · Water resources management · Remote sensing · Flood monitoring · Watershed prioritization · Northeast India

---

R. Das (✉) · S. Kurbah · B. M. Arjun · D. Barman
North Eastern Space Applications Centre (NESAC), Umiam, Shillong, Meghalaya, India
e-mail: ranjit.das@nesac.gov.in

S. Kurbah
e-mail: s.kurbah@nesac.gov.in

B. M. Arjun
e-mail: arjun_bm@nrsc.gov.in

D. Barman
e-mail: diganta.barman@nesac.gov.in

S. P. Aggarwal et al. (eds.), *Remote Sensing of the North Eastern Himalayan Ecosystem*,
https://doi.org/10.1007/978-981-95-3221-6_14

14.1 Introduction

Water is precious for life and more so for humans. The need of water is felt more and more for better living with modern services. The water demand for irrigation and industrial purpose also increased correspondently to meet the demand of growing population. The over exploitation of water resources has resulted in a condition of unsustainability and environmental degradation. Water is one of the key drivers of socio-economic development and basic resource for environmental sustainability. Since water is a renewable resource, it should be managed in a sustainable manner. Therefore, assessment of water resources is a great concern for managers for management and development.

Many of the water resources issues are not isolated; they are linked with different human activities like human well-being, economic development and preservation of ecosystems. Water pollution, soil erosion, siltation, drying up of surface water bodies, and ground water depletion are the major issues that are intensifying along with increase in human activities. Frequent occurrence of extreme events is the evidence for some fundamental changes which are affecting the water resources scenario worldwide including India. Therefore, proper management of these precious resources are extremely important for its sustainability.

Distribution of water resources is not uniform in spatial and temporal domain. Many places experience flood whereas many places in the world are affected by water scarcity and drought. The flood and drought of some places persist for certain period of the year. The flood and drought scenario are also different in different places and season. Due to climate change scenario, the occurrence of hydrological extreme events is increasing. NE India is a region characterized by its mountainous terrain, abundant rainfall, steep slopes and delicate soil composition (Choudhury et al. 2022). The region receives very high annual average rainfall induced by the South West monsoon. Places like, Mawsynram and Sohra (erstwhile Cherrapunji) in Meghalaya receive the world's highest rainfall (11,872 mm and 11,777 mm of annual average rainfall respectively). In spite of a very high monsoon rainfall, availability of water is highly uneven in both space and time. The Brahmaputra valley has around 2000 mm average rainfall with local variations. Guwahati, being in the rain shadow of the Meghalaya plateau, receives only 1,717 mm of rain (Dikshit and Dikshit 2014). In this regard, water resource management is the most crucial issue and needed in many countries worldwide (Bouman 2007) and hence NE India also.

During monsoon, the river Brahmaputra, Barak and their tributaries experience high flood to create havoc in the region. Along with the flood, siltation and bank erosion are also a severe problem in the region. Bank erosion is a significant environmental and socio-economic issue along the rivers of Assam (Sarma and Phukan 2006). It is the main factor influencing the devastating loss of valuable land, destruction of infrastructure and environmental degradation (Konosoer et al. 2016).

In a contradictory situation, the hilly regions of Assam and other northeast Indian states faces scarcity of water for irrigation, drinking etc. during lean period. The heavy rainfall drains out of the region rapidly via the dense drainage network and create water scarcity for drinking and irrigation in uphill areas. It also instigates soil erosion from the slope lands of the region. In India, about 45% of the total land area is prone to erosion (Mythili and Goedecke 2016) which may be higher in NE India due to its hilly terrain and high rainfall. On average, a total of 36.64% of land degradation has been estimated in the NER of India, which is almost double the national average of 20.17% (Saha et al. 2012). Soil erosion results in the loss of important soil nutrients required for cultivation. Further, it also causes siltation in reservoirs and natural streams (Kothyari 1996; Biswas et al. 1999). The consequences of siltation in reservoirs, adversely affect water availability for agriculture, drinking water, and industrial uses, ultimately impacting crop yields and energy costs. To cope up with such issues a proper management plan for development of water resources is required. Implementation of advance geospatial technology has made water resource management programmes easier. It provides valuable support especially in regions like Northeast India, which face unique challenges due to its diverse topography, heavy rainfall, and varied climate. Geospatial technology has been extensively used in this region to enhance implementation of water resources management programmes.

14.2 Assessment of Water Resources

As of recent times, space technology advancements have provided satellite data which are highly accurate and capable of retrieving information on a large scale. Satellite products like National Aeronautics and Space Agency's (NASA) Landsat series, ESA's Copernicus Sentinel series and Indian Space Research Organization (ISRO) Resourcesat-2A LISS IV provide valuable Earth observation data which are available as high-resolution images and multispectral data. These satellite data are highly helpful for mapping out water resources which serves as an important pre-requisite in planning for management of water resources.

14.2.1 Surface Water Mapping

Surface water mapping using satellite imagery leverages remote sensing data, primarily from multispectral and Synthetic Aperture Radar (SAR) sensors, to detect and delineate water bodies. High-resolution optical imagery from satellites like Landsat, Sentinel-2 and Resourcesat-2A enables the identification of surface water based on spectral indices such as the Normalized Difference Water Index (NDWI), which utilizes near-infrared and green bands to distinguish water from land. SAR data is particularly effective for mapping surface water during cloudy or rainy conditions, as it can penetrate through clouds and detect changes in water surface elevation or flood extents. These satellite products allow for temporal analysis of dynamic behaviour of water in rivers and lakes, providing insights into seasonal fluctuations, land use changes, and the impacts of hydrological events like flood, drought, and sedimentation. Techniques such as time-series analysis and change detection further enhance the ability to track and quantify surface water variability over large geographical areas. With the existing satellite data, the surface water bodies can be mapped at different scale. IRS Resourcesat-2 LISS III and Landsat data can be utilized for regional level mapping and for large scale mapping Sentinel 2, Resourcesat-2A and Cartosat data can be used. Under the project "National Wetland Inventory and Assessment (NWIA)", the wetlands in the NE India were mapped twice on 2006–2007 and 2017–2018 at the scale of 1:50,000 using multi-date IRS-LISS-III data to assess the status of the wetlands in the region and their changes. The wetland map prepared for Lakhimpur district of Assam for the year 2017–2018 is shown in Fig. 14.1. The study showed that 27.44% of the geographical area in NE region is covered by wetlands which is 2.23% in Arunachal Pradesh, 8.66% in Assam, 7.12% in Manipur, 0.99 in Meghalaya, 2.8% in Mizoram, 0.93% in Nagaland, 1.38% in Sikkim and 3.75% in Tripura. Under this project the different types of wetlands like rivers/streams, lakes, marsh, reservoirs, ponds, etc. were mapped out. From the results of the wetland assessment, it was found that Assam has the highest percentage (12.3%) of rivers/streams area in the country.

Rivers/streams are also considered as surface water body. Since the state of Assam has highest percentage of rivers/streams area in the country, it is required to have a large scale mapping of it for proper management of water resources in the state. The major and minor rivers at a large scale along with riverbank, embankments, hydro-meteorological observatories, sluice gates, bunds, major locations, roads, railway

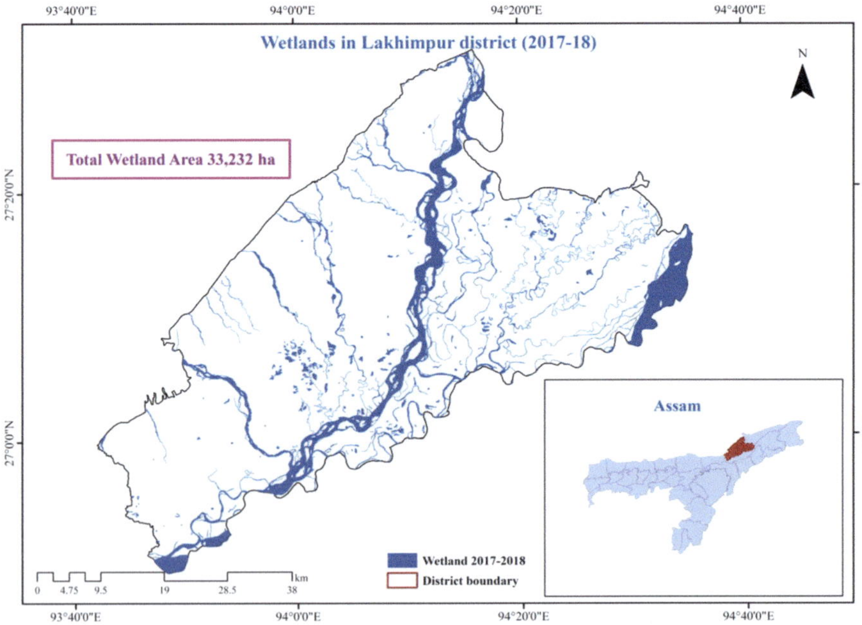

Fig. 14.1 Wetland map of Lakhimpur district of Assam (2017–2018)

lines etc. can be mapped successfully using high resolution satellite data and sufficient ground information. For mapping of all the river networks in Assam, high resolution satellite imagery like KOMPSAT imagery from Bhuvan web services and Bing hybrid maps (basemap) were used. These satellite imageries have spatial resolutions of 1 m and 2.39 m respectively. For generation of river networks outside Assam and delineation of all the river catchments inside Assam, CartoDEM v3 with a spatial resolution of 10 m and ASTER global Digital Elevation Model (DEM) with a spatial resolution of 30 m were used. The LULC maps showing details of LULC classes on both sides of the river networks were mapped out using LISS III satellite imageries for all the districts of Assam at a scale of 1:10,000. A user-friendly web portal was developed which incorporates all the layers generated from this project.

The state of Meghalaya receives a very high amount of rainfall during the monsoon periods but water shortage is prevalent during the lean season creating hardships for the people of the state. Geospatial technology can be utilized at a great extent for planning of soil and water conservation which will help for land and water resources development to combat the water scarcity. Different government departments worldwide have been implementing different such conservation measures. Different states of India have also been implementing various water conservation measures like construction of ponds, jalkunds, and small multipurpose reservoirs, etc. under "Integrated Watershed Management Programme" (IWMP). The documentation of such water conservation measures is crucial and will serve as a strong foundation for implementation of future projects.

Considering this a project was proposed by the government of Meghalaya to map out the existing waterbodies and to identify potential sites for construction of additional water harvesting structures. High resolution satellite imagery was used to map out small multipurpose reservoirs, ponds, jalkunds etc. at a large scale using visual interpretation. The state has a total of 15,572 water bodies, out of which 14,688 (94.32%) are in sizes less than 0.5 ha whereas the remaining 884 are in sizes greater than 0.5 ha. A total of 1828 suitable sites for construction of rainwater harvesting structures were also identified. For this purpose, geospatial layers such as LULC, soil texture, slope, stream order, euclidean distance, runoff potential map and lithology maps were prepared and utilized as important inputs in the multi-criteria decision analysis.

Multi-Criteria Decision Analysis (MCDA) refers to making a decision involving several criteria (Wątróbski et al. 2018). One of the effective methods of MCDA is the Analytical Hierarchy Process (AHP) (Baruah et al. 2022). The AHP method is commonly employed for selection of suitable sites for construction of water harvesting structures (Shao et al. 2020; Yasser et al. 2013). Seven criteria were considered for generation of a suitability map for construction of SMRs in Meghalaya. The thematic raster layers required as inputs was reclassified using the reclassification tool in the GIS platform into 4 classes (Shao et al. 2020) except for runoff potential, classified into 3 classes (Ramakrishnan et al. 2009) as per the expert view and opinion. The methodology followed in AHP analysis is as per flowchart given in Fig. 14.2.

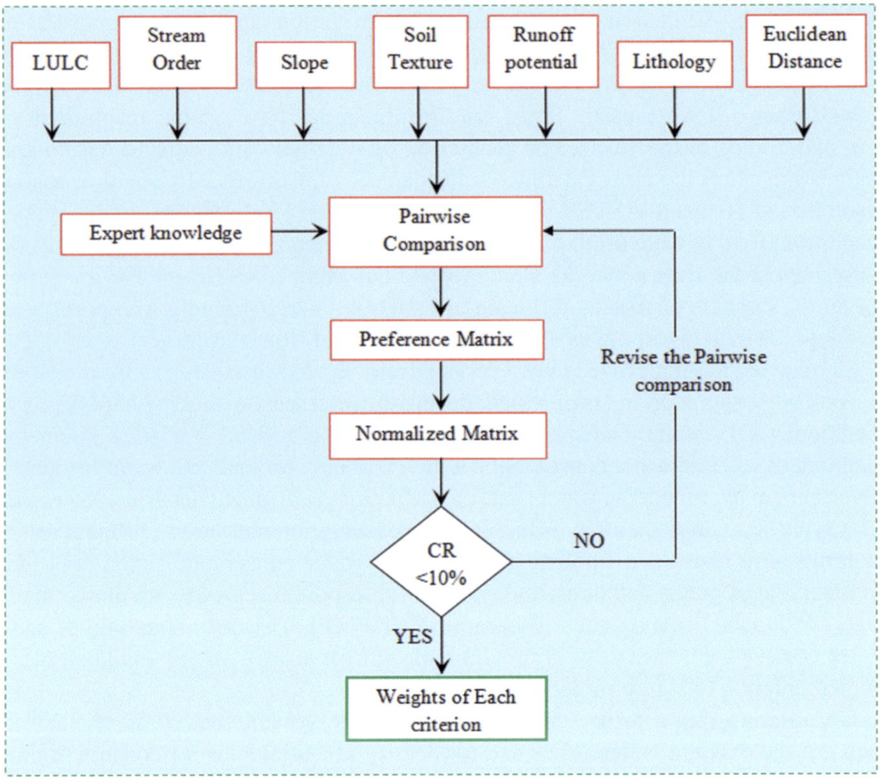

Fig. 14.2 AHP flowchart for identification of water harvesting potential sites

The reclassified layers are kept at a grid size of 10 m × 10 m. The pair wise comparison between the seven criteria against each other was constructed using Saaty's scale which ranges from 1 to 9 (Saaty 1980). When designing the weights, the relative importance of each parameter is taken into consideration. The next step is to calculate the relative weights of each criterion after the weight of each criterion has been assigned. Once the relative weights of each criterion were calculated, the normalized weights (b_{ij}; Eq. (14.1)), relative weights (Eq. (14.2)) and the sum of normalized weights (Eq. (14.3)) are also calculated simultaneously (Hammami et al. 2019). The normalized comparison matrix, weights of each criterion and the Consistency Ratio (CR) are presented in Tables 14.1 and 14.2.

$$b_{ij} = \frac{a_{ij}}{\sum_{i=1}^{n} a_{ij}}, \quad a_{ij} \text{ is the element of the comparision matrix} \tag{14.1}$$

$$w_i = \frac{\sum_{j=1}^{n} b_{ij}}{\sum_{i=1}^{n} \sum_{j=1}^{n} b_{ij}}, \quad i = 1, 2, 3 \ldots n \tag{14.2}$$

$$\sum_{i=1}^{n} w \cdot w_i(x, y) = 1 \quad (14.3)$$

The relative weights are calculated by using the highest eigen value ($\lambda_{\max}$) (Eq. (14.4)) also called as the normalized right eigen vector (w)

$$MnW = \lambda_{\max} x_W \quad (14.4)$$

To understand the information about ambiguity in assessment of AHP analysis, a consistency ratio was employed (Saaty 1977). Then the consistency ratio (CR) is calculated using Eq. (14.5). The inconsistency of less than or equal to 10% is acceptable, however, if it is larger than 10%, the pair wise comparison matrix should be revised (Saaty 1977).

$$\mathrm{CR} = \frac{\mathrm{CI}}{\mathrm{RI}} \quad (14.5)$$

where, RI is the Random Index which is the standard value depending on the number of criteria used as shown in Table 14.3 and CI is the consistency index given by Eq. (14.6).

$$\mathrm{CI} = \frac{\lambda_{\max} - n}{n - 1} \quad (14.6)$$

Table 14.1 Standard consistency index for randomly generated matrix

No. of criteria	2	3	4	5	6	7	8
RI	0	0.52	0.9	1.12	1.24	1.32	1.41

Table 14.2 Normalized matrix for SMRs site suitability

Criteria	Stream order	LULC	Slope	Soil texture	Lithology	Euclidean distance	Runoff potential
Stream order	0.321	0.2	0.308	0.267	0.324	0.5	0.273
LULC	0.107	0.067	0.077	0.067	0.027	0.05	0.045
Slope	0.16	0.133	0.154	0.133	0.081	0.2	0.182
Soil texture	0.16	0.133	0.154	0.133	0.162	0.05	0.182
Lithology	0.08	0.2	0.154	0.067	0.081	0.05	0.045
Euclidean distance	0.064	0.133	0.077	0.267	0.162	0.1	0.182
Runoff potential	0.107	0.133	0.077	0.067	0.162	0.05	0.091

Table 14.3 Respective weights of normalized matrix for SMRs site suitability

Criteria	Stream order	Land use	Slope	Soil texture	Lithology	Euclidean distance	Runoff potential
Weights	31.314	6.28	14.91	13.926	9.674	14.072	9.813

Consistency Ratio (CR) = 0.077

Here, λ_{max} is the maximum Eigen value of matrix A and n is the number of elements within a branch being compared.

14.2.2 Groundwater Prospect Mapping

Groundwater prospect maps can be prepared using space technology with support of field data. Different thematic maps like lithology, geomorphology, structures, and hydrology may be generated using suitable satellite data through visual image interpretation techniques supported by field/existing data. These thematic maps need to be analyzed in GIS environment to derive the groundwater prospect map. For preparation of the lithology layer, all the rock formations occurring in the study area need to be mapped in a single layer. Different lithological parameters may be represented by polygons and assigned with corresponding numeric codes. While preparing the lithology map by interpreting satellite imagery, the existing collateral data may be referred to understand the general geological formation and rock types of the area. So a preliminary investigation is useful for differentiation of rock types without which mapping of lithology is very difficult.

For preparation of structural maps, the geological structures present in the area can be classified into two major categories like (i) faults, shear zones, thrusts, fractures, dykes, veins, etc. which behaves as channel and barriers for the groundwater movement and (ii) structural elements such as bedding, schistocity/foliation, folds etc. which are examples of either litho unit or landform. Both these major features need to be mapped separately. For the geomorphology layer, all landforms/geomorphic units occurring in the study area are to be mapped as a single layer (NRSC 2008). These features are represented as polygons and assigned with alphabetic codes. Toposheets are also useful in mapping of the geomorphic units as they help in understanding the relief variations and topographic features. Drainage, water bodies, canals, rainfall data, irrigated areas, springs and wells need to be considered for mapping of hydrological features. In GIS, the drainage is represented as line as well as polygon features, the water spread area of water bodies is represented as polygon features whereas the canals are represented as lines, the rainfall data is marked as point feature with amount of rainfall, the irrigated area as polygon feature and the springs and wells are represented as point features.

14.2.3 Mapping Snow and Glacier

Snow and glacier covers are found in the lower Himalayan regions of north east India particularly in the states of Sikkim and Arunachal Pradesh. Runoff from glaciers provide a stable source of water for cultivation and for potable drinking water in semi-arid climates. However, a large built-up of snow cover and sudden melting of snow can cause severe flooding, leading to loss of life and destruction of properties in the downstream area. Therefore it is required to properly map out and identify the geographic extent of snow cover and water equivalent in critical glacial areas on a seasonal basis. When cloud cover is minimal, identification of snow cover distribution on a visible and/or near infrared image is a straightforward task as snow appears brighter than surrounding vegetation, soil and water. In case of confusion between snow cover and cloud cover, multiple images of the study area over a small time period can be obtained using AVHRR or GOES. Snow cover will be the one which doesn't change in a small time period.

Snow cover is found in high altitude terrain, so mountain shadow on snow makes it difficult to accurately map it out. In this situation a normalized difference snow index (NDSI) method can be utilized to identify the extent of snow cover. Snow exhibits a higher reflectance in the optical region in comparison to other features such as vegetation, rock and water but has lower reflectance in the SWIR region than rock and vegetation. Due to this, snow appears bright in visible and dark in SWIR region in satellite images. This characteristic forms the basis for the NDSI for snow cover mapping and proves to be a useful technique in the Himalayan region where mountain shadows appear more (Kulkarni et al. 2002; Hall et al. 1995). The NDSI is the ratio between the difference and sum of visible and SWIR band as given in Eq. (14.7).

$$\mathrm{NDSI} = \frac{\text{Green Band (B2)} - \text{SWIR Band (B5)}}{\text{Green Band(B2)} + \text{SWIR Band (B5)}} \tag{14.7}$$

A study was carried out for assessment of snow fed and rain fed catchment area in the Lohit basin using satellite data and the glacial lakes in the basin were mapped. Snow indices were used to identify and map out the snow extent using co-registered temporal satellite data (winter and summer), and glacial lakes were identified using optical satellite data acquired in summer season. From this study, 276 glacial lakes of varying sizes were identified covering a total area of 26.3 km^2 area. The snow indices analysis estimates 7997.9 km^2 snow cover area in winter and 7595.2 km^2 in summer. This shows that the snow cover area in Lohit basin reduces by 402.6 km^2 in summer than winter season. The snow cover map of Lohit catchment area during the summer is shown in Fig. 14.3.

A study of the snow cover was also carried out for the Tawang, Subansiri and Dibang sub-basins which are part of the Brahmaputra basin (SAC 2013). The analysis of snow cover data from the hydrological years 2004–05 to 2013–14 reveals that the snow cover extent in 2008–09, 2009–10, 2010–11, and 2011–12 was higher than

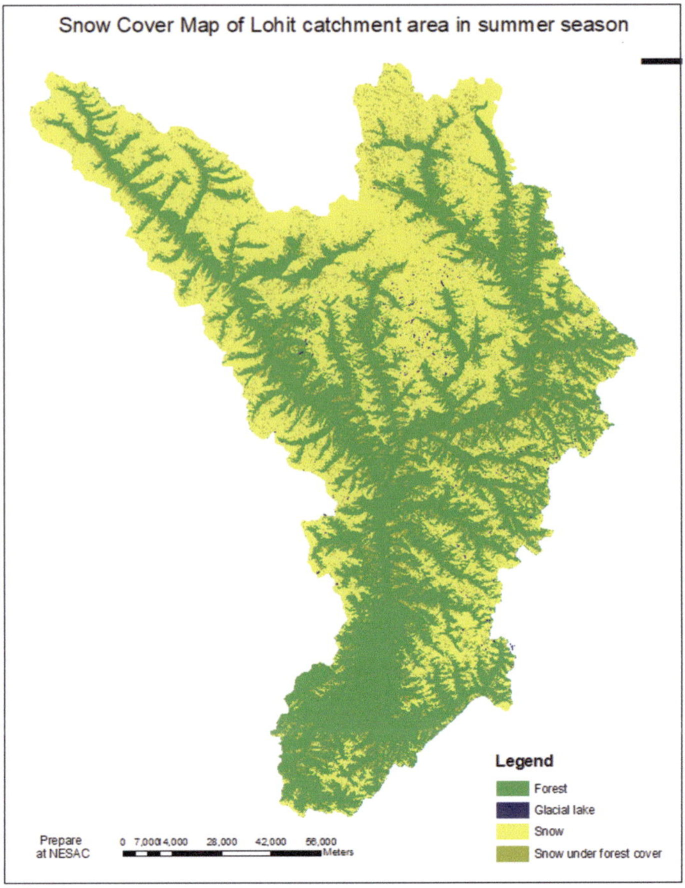

Fig. 14.3 Snow cover map of Lohit

the average across the years for the sub-basins. While the minimum snow cover showed minimal variation across the years and sub-basins, the maximum snow cover varies significantly. The mean maximum snow cover differed among the three sub-basins. The standard deviation values indicate that the minimum snow cover exhibits low temporal variability. However, there is a clear contrast between minimum and maximum snow cover across the sub-basins, suggesting that different areas within the Brahmaputra basin receive varying amounts of precipitation in different years. The analysis also indicates that the Tawang sub-basin experiences the highest variation in snow cover.

14.2.4 Assessment of Rainfall

Rainfall is the primary source of fresh water on Earth, essential for agriculture, ecosystems and human needs. It governs the hydrological planning of a region. Other hydrological measurements depend on precise measurement of rainfall. The amount and distribution of rainfall vary globally. While rainfall in moderation sustains life, excessive rainfall can lead to floods and insufficient rainfall can cause droughts. Rainfall assessment is the measurement and evaluation of rainfall data to understand its impact on the environment and water resources. It aids in forecasting water level in rivers and assessing the likelihood of natural disasters like floods and landslides. The traditional method of rainfall data collection involves measuring the amount of rainfall that collects on rain-gauges over a period of time or for a particular rainfall event. The measurement of rainfall using rain-gauges can be done manually or by automatic measurement through the use of Automatic Weather Stations (AWS). Although measurement of rainfall using rain-gauges gives accurate data, it does not represent precisely the rainfall characteristics of a region due to its insufficient distribution. To overcome this limitation, space-based techniques have emerged as vital tools for monitoring rainfall. Satellite-based rainfall estimation allows for wide-area coverage, frequent updates, and better insight into rainfall distribution over land and oceans—particularly useful for data-scarce regions.

Rainfall estimation from satellites is based on a variety of sensors, each suited for measuring specific atmospheric and surface conditions. These sensors fall into broad categories such as passive microwave sensors, active microwave sensors, visible and infrared imagers, and atmospheric sounders. Passive microwave sensors observe natural radiation emitted by Earth and clouds to detect precipitation, while active microwave sensors like radars, transmit signals and analyze their reflection from raindrops. Visible and infrared sensors measure cloud properties and temperature, indirectly inferring rainfall intensity and spread. Atmospheric sounding instruments, on the other hand, help derive vertical temperature and humidity profiles, useful in rainfall modeling. Each technique has its strengths and limitations depending on spatial resolution, temporal frequency, and sensitivity to different types of precipitation.

One of the most significant advancements in rainfall monitoring came with the launch of the Tropical Rainfall Measuring Mission (TRMM) in 1997, a joint initiative of NASA and JAXA. TRMM was the first satellite mission to combine both active and passive microwave instruments, enabling improved detection of tropical and subtropical rainfall. It introduced innovations like the Precipitation Radar (PR) and the TRMM Microwave Imager (TMI), which provided reliable measurements of rainfall rates and structure. The success of TRMM laid the groundwork for the Global Precipitation Measurement (GPM) mission, launched in 2014, which significantly improved global coverage and sensitivity—particularly for light rain and snowfall. These missions have contributed large-scale, high-resolution datasets that are used extensively in climate studies, hydrological modeling, and disaster forecasting. Refer to Chap. 16 for a more comprehensive discussion on TRMM, GPM, and related rainfall data products.

In India, the Indian National Satellite System (INSAT), developed by ISRO, has played a critical role in meteorological observations since the early 1980s. The latest INSAT series, including INSAT-3D, INSAT-3DR, and INSAT-3DS, are equipped with advanced meteorological payloads like Very High Resolution Radiometer (VHRR) and Atmospheric Sounders. These instruments operate across visible, infrared, and water vapor bands, enabling rainfall estimation across the Indian subcontinent and adjoining oceanic regions. The data collected are processed into quantitative precipitation estimates (QPEs) and other key meteorological products such as wind vectors, outgoing long-wave radiation, and temperature profiles. These inputs are extensively used by the India Meteorological Department (IMD) for real-time weather prediction and early warning systems. The key advantage of INSAT satellites lies in their high temporal resolution, with the capability to provide data every 30 min and even at 5 min intervals during cyclonic events.

Together, these space-based systems have revolutionized the way rainfall is measured and analyzed. They provide critical datasets for national weather services, research institutions, and disaster management agencies, enabling informed decisions and timely responses to hydrological extremes. The coupling of satellite observations together with ground-based measurements and hydrological measurements have improved rainfall assessment for both local and global scales.

14.2.5 Estimation of Reservoir Capacity

Reservoirs are hydrological structures constructed across rivers to store water and divert water for multiple uses like generation of hydropower, flood control, irrigation, drinking water supply, etc. Reservoirs are one of the crucial structures in water resource management, acting as storage facilities of large amounts of water, regulating river discharge and creating a reliable water supply. One of the prominent problems affecting reservoirs is siltation. Due to soil erosion in the upstream area of the river, sediments are transported and deposited into the reservoirs, leading to reduction of live storage capacity and functionality. Estimation of reservoir capacity can provide a clear picture of the amount of sediments entering the reservoir and help in formulation of soil erosion mitigation measures in the upstream areas.

A study to estimate reservoir capacity carried out in the Dumboor reservoir in Tripura, a multipurpose reservoir used for hydro-electric power generation elaborates the effective use of geospatial technology. The hilly terrain of Tripura is highly susceptible to soil erosion, causing a high silt load into the reservoirs reducing its life span. For estimation of reservoir capacity, the reservoir extent can be mapped using multi-temporal satellite data at different known depths of the reservoir. Microwave satellite imagery, Sentinel 1 series was used to map the water spread area of Dumboor reservoir with some known depth with which the water surface area and corresponding depth (collected from field) relationship was established (Table 14.4). To generate the water spread area from Sentinel 1 data (SAR) density slicing technique was used. Water spread area at different water levels for Dumboor reservoir are illustrated in Fig. 14.4.

By plotting the values of the reservoir's water levels and its water spread extent as, shown in Table 14.4, a linear relationship, Eq. (14.8) is obtained with a coefficient of determination (R^2) of 0.97 (Fig. 14.5) where y is the water spread at water level x.

As seen from Fig. 14.5, the linear relationship between the water level x and water spread area y is

$$y = 324.37x - 26370 \tag{14.8}$$

Table 14.4 Reservoir water levels and its water spread area

Reservoir water level (m)	85.2	86.5	87.55	88.35	89.35	89.9	91.5	93.55
Water spread area (Ha)	1305.9	1498.9	2014.8	2365.2	2612.6	3030.9	3228.0	3908.1

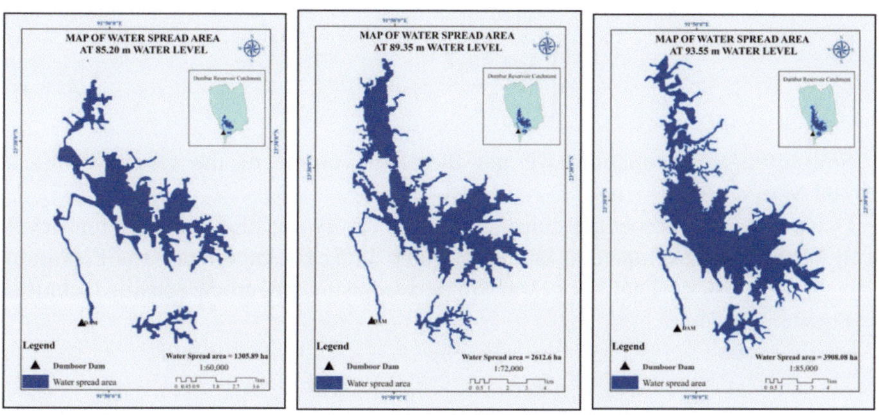

Fig. 14.4 Map of water spread area of Dumboor reservoir at different water levels

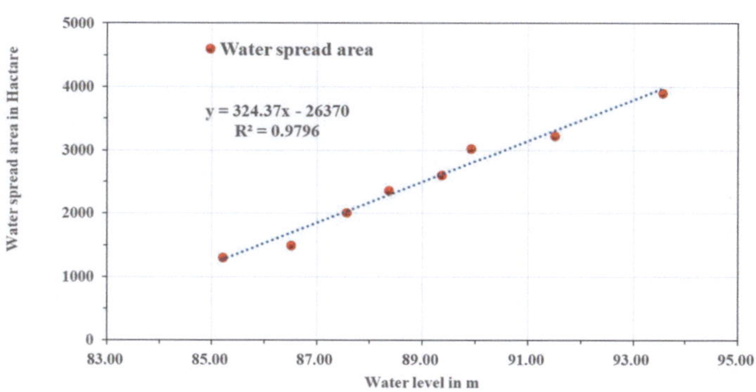

Fig. 14.5 Water spread area versus water level

The elemental change in area due to elemental change in water level (∂x) is

$$\frac{\partial y}{\partial x} = \frac{\partial}{\partial x}(324.37x - 26370) \tag{14.9}$$

The elemental change in volume due to elemental change in water level (∂x) is

$$\frac{\partial y}{\partial x}\partial x = \frac{\partial}{\partial x}(324.37x - 26370)\partial x \tag{14.10}$$

Since the water extent of the reservoir at the Minimum Drawdown Level (MDDL), is 83.60 m and the maximum at Full Reservoir Level (FRL), is 93.55 m. Therefore, the total volume of the reservoir (V) is

$$\int_{83.60}^{93.55} \frac{\partial y}{\partial x}\partial x = \int_{83.60}^{93.55} \frac{\partial}{\partial x}(324.37x - 26370)\partial x \tag{14.11}$$

$$V = \left[\frac{324.37x^2}{2} - 26370x\right]_{83.60}^{93.55} \tag{14.12}$$

Simplifying the equation, we get the total volume of the reservoir (V) as 234.92 M m^3.

Using the same procedure and method, the reservoir volume at different levels (1 m interval) are estimated as shown in Table 14.5. The modified Area-Elevation-Capacity curve of the Dumboor reservoir derived using the remote sensing technique is shown in Fig. 14.6.

14.3 Flood Management

The Brahmaputra and Barak basin are some of the most flood prone basins in India. Flooding occurs every year in the monsoon season in the plain areas of Assam due to overflowing of the Brahmaputra River causing loss of lives and severe damage to properties. Flood management is very important in this region to mitigate floods. Flood management can be classified into two measures, Structural and Non-structural measures and both these measures need to be implemented conjunctively for effective flood mitigation. Satellite data can be effectively used for site selection studies for construction of structural flood mitigation measures like embankments, flood walls, dams, flood diversion, etc. Space technology can also be utilized for non-structural measures of flood mitigation like flood forecasting, flood early warning and flood hazard zonation.

Table 14.5 Estimated reservoir capacity at different water levels

Reservoir elevation in (m)	Water spread area from trend line (Ha)	Live capacity b/w contour elevation using RS in MCM	Cumulative live capacity using RS in MCM
83.6 (MDDL)	740	0.00	0.00
84	860	3.25	3.25
85	1200	10.39	13.64
86	1540	13.64	27.28
87	1850	16.88	44.16
88	2210	20.12	64.28
89	2520	23.37	87.65
90	2840	26.61	114.26
91	3180	29.85	144.12
92	3495	33.10	177.21
93	3800	36.34	213.56
93.55 (FRL)	3908	21.37	234.93

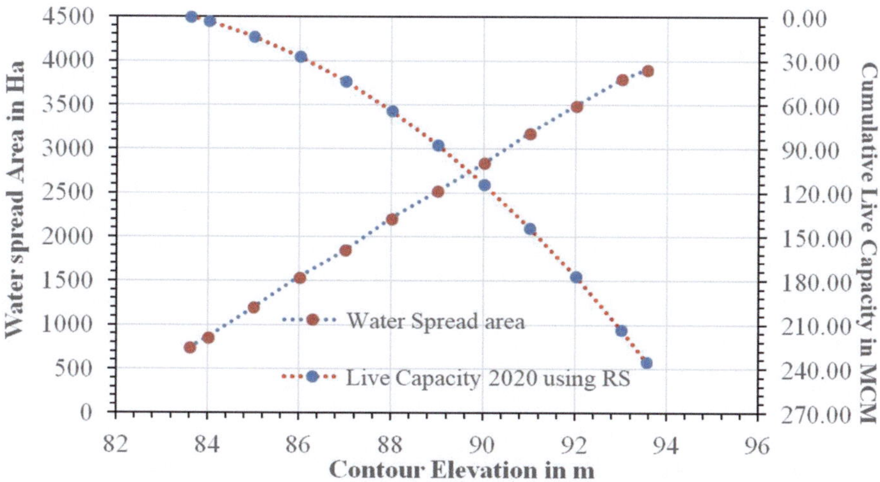

Fig. 14.6 Revised area-elevation-capacity curve of the Dumboor reservoir

14.3.1 Flood Hazard Zonation

Flood hazard zonation is the process of classifying an area into zones based on the frequency and severity of potential flooding. It involves evaluating flood risks across different regions and categorizing them according to the probability and potential impact of flooding. The resulting information supports planning for land use, disaster preparedness, and the development of effective alleviation strategies. It facilitates development of floodplains and is one of the most important non-structural measures to reduce flood impacts. Accurately mapping flood extents for predicted recurrence intervals requires integrating historical peak flow data, detailed topographic surveys (e.g., cross sections, slopes, contour maps), and various thematic maps, including soil types, land use, vegetation, and infrastructure. Hydraulic parameters must also be considered. Since collecting this data through ground surveys is time-consuming and difficult, remote sensing plays a crucial role in providing efficient, large-scale data for flood hazard mapping.

Systematic analysis of satellite imagery and spatial datasets to identify and delineate areas susceptible to flooding are the key components of flood hazard zonation. In this process, historical flood patterns and their recurrence over time are analyzed to estimate the frequency and extent of inundation across a given region. Effective flood zonation begins with the acquisition of satellite images covering a sufficiently long time frame—preferably over two decades—to capture temporal variations in flood events. In the Indian subcontinent, where the monsoon season and large river systems like the Brahmaputra contribute to frequent flooding, satellite data acquired from Indian Remote Sensing (IRS) missions and microwave sensors such as those onboard Radarsat are particularly valuable. Microwave imagery is especially advantageous in cloud-covered conditions, which are common during peak flooding periods.

Once the satellite data is collected, it must be processed to ensure spatial consistency. This involves geometric rectification of all imagery to a common projection system, which facilitates accurate overlay and integration with other geographic information system (GIS) layers. Timing is crucial in flood monitoring, and image acquisition should coincide with peak water levels as recorded at hydrological gauge stations. Additional imagery captured during the rising and receding stages of flood events provides a more comprehensive understanding of the flood's dynamics, including its spatial progression and temporal duration.

The next step involves the application of image classification techniques to extract water features from the satellite images. These classified water bodies represent the flood extent at a given point in time and are referred to as flood inundation layers. For each year within the observation period—such as 1998 to 2023—multiple inundation layers may be generated, depending on the number of flood events. These are subsequently integrated to produce an annual maximum flood layer that reflects the most extensive inundation observed during that year. Repeating this process annually results in a time series of flood extents that form the basis for hazard assessment.

By overlaying and analyzing the annual flood layers, a composite flood hazard layer can be generated. This layer categorizes regions based on the observed frequency of flooding, allowing for the classification of areas into high, moderate, or low flood hazard zones. Areas that experience recurrent flooding over the years are marked as high-risk zones, whereas areas with occasional or rare flooding fall into lower risk categories. This frequency-based stratification provides a quantitative foundation for understanding long-term flood risk.

To evaluate the broader implications of flooding, the flood hazard layer is integrated with various thematic GIS datasets like administrative boundaries; LULC classifications, transportation networks such as roads and railways, and settlement data including village and urban extents. This integration enables detailed exposure assessments, allowing planners and disaster management agencies to identify vulnerable infrastructure and population centers. The spatial analysis supports both emergency response planning and long-term risk mitigation strategies.

The final output of the flood zonation process is a series of hazard maps produced at different spatial scales, including state, district, and local levels. These maps use intuitive visual representations—typically color gradients—to indicate zones of differing flood frequencies. Critical geographic features such as rivers, water bodies, roadways, railways, and settlements are prominently included to aid interpretation and application. Flood hazard maps derived from this methodology serve as essential tools for policymakers, engineers, urban planners, and emergency responders, providing them with reliable, data-driven information to prepare a flood management plan.

Flood hazard zonation mapping and preparation of a flood hazard atlas have been carried out in the state of Assam in 2010 to aid in flood management. In this study, 10 years (1998–2007) of IRS and Radarsat microwave satellite data was acquired during the monsoon season. The flood hazard zonation map was prepared based on the flood frequency and categorized into 5 classes ranging from very high to very low. As per this study results, 28.31% (22.21 lakh hectares) of land in Assam was affected by flood during 1998–2007. Out of the total flood affected area, 3.53 lakh hectares of land fall under high to very high flood hazard categories. The flood hazard zonation map prepared for Dibrugarh district is shown in Fig. 14.7.

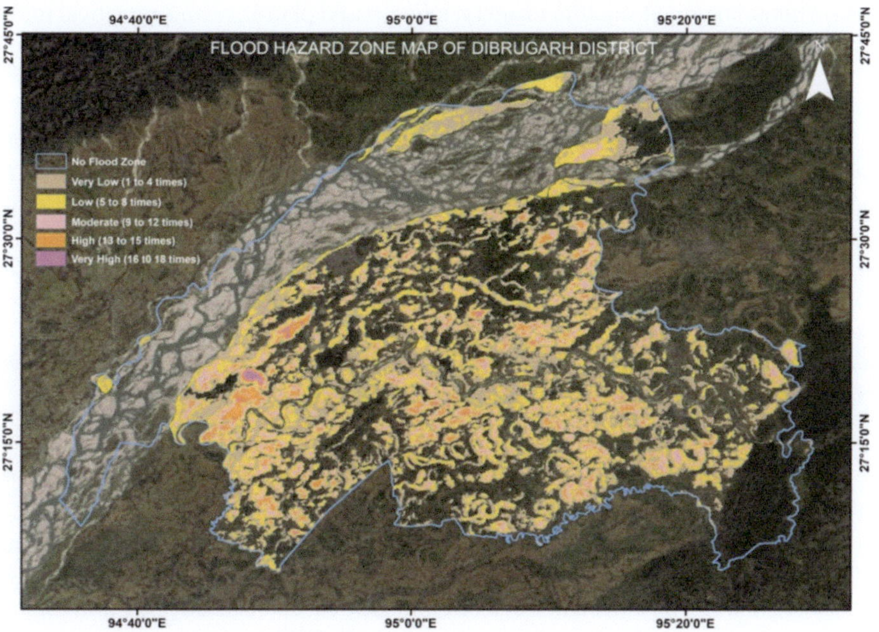

Fig. 14.7 Flood hazard zonation map

14.3.2 Monitoring of River Bank Erosion

Riverbank erosion is a natural geo-morphological process in which the flowing water of a river gradually wears away and removes soil and sediment from its banks, resulting in the loss of land. This phenomenon is intensified by factors such as high river flow velocity, excessive rainfall, deforestation, and anthropogenic activities near riverbanks. The consequences include loss of agricultural land, displacement of communities, and damage to infrastructures. River bank erosion is more pronounced in regions with highly meandering rivers or during floods. In the state of Assam, the Brahmaputra River has a history of severe and persistent river bank erosion. This massive river, with its dynamic flow and braided channels, erodes vast tracts of fertile land every year. Figure 14.8 shows the river bank erosion and deposition of Brahmaputra and selected tributaries during the period of 2016–2022. The problem is pronounced during the monsoon season when heavy rainfall causes the river to swell and it's current to strengthen. As a result, thousands of hectares of land and many villages are lost annually, displacing communities and affecting livelihoods. Areas like Dibrugarh, Majuli (the world's largest riverine island), and Dhubri are particularly affected. The erosion is not only a socio-economic challenge but also an ecological concern, impacting the fragile ecosystems of the Brahmaputra basin. Monitoring river bank erosion is essential in understanding fluvial processes, planning river management and mitigating damage to land and infrastructure. With advanced geospatial technology, tracking river morphology changes over time become possible using a combination of satellite imagery, cloud-based platforms and analytical tools. An effective approach for monitoring bankline erosion combines multiple-temporal remote sensing analysis with geospatial methods, as demonstrated in recent studies on large dynamic rivers such as Brahmaputra.

The process of monitoring river bank erosion starts with the acquisition of satellite imagery over a span of multiple years. High resolution multispectral images (such as those from Sentinel-2A and Sentinel-2B) are selected due to their fine spatial resolution and availability across various spectral bands. The satellite imagery should be acquired during the post-monsoon periods to ensure consistent seasonal conditions and water levels. The Normalized Difference Water Index (NDWI) is a proven method for distinguishing water bodies from surrounding land surfaces and is useful in extracting the bank lines of the river. In this method, the green and near-infrared (NIR) bands are used for calculation of NDWI as shown in Eq. (14.13):

$$\mathrm{NDWI} = \frac{\mathrm{Green} - \mathrm{NIR}}{\mathrm{Green} + \mathrm{NIR}} \tag{14.13}$$

The index takes advantage of the strong absorption of NIR radiation by water and reflectance of the green band to accurately delineate water bodies. Satellite image processing and NDWI calculation can be efficiently executed in Google Earth Engine (GEE), a cloud-based platform. After calculation of NDWI for each target year or sub-period, a change detection analysis is conducted to identify areas affected by erosion

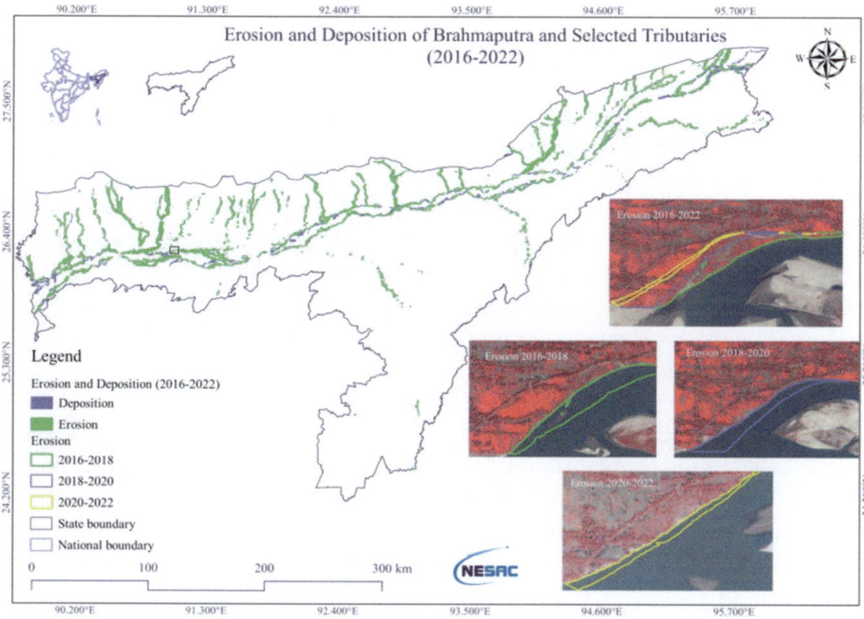

Fig. 14.8 Erosion and deposition status in Assam

and deposition. This involves comparing NDWI values across two time periods—for example, a pixel classified as land in one year and water in the subsequent year is identified as eroded, whereas a transition from water to land indicates deposition. The result of this analysis provides spatial maps showing changes in the river banks over the specified study period.

Following the identification of erosion and deposition zones, the next critical step is the extraction of river banklines for each study year. These banklines are derived by thresholding NDWI images to delineate the water–land interface. Once extracted, they are visually cross-checked against satellite based images to ensure their accuracy. For quantifying the rate and direction of river bank migration, the Digital Shoreline Analysis System (DSAS) can be employed. This tool enables the statistical analysis of bankline changes by computing the End Point Rate (EPR) of migration. Using the DSAS tool, a baseline is created inland from the banklines and runs parallel to the river's general course. From this baseline, transects are automatically generated at fixed intervals (usually 30 m) intersecting the banklines at multiple points. DSAS calculates the EPR at each transect using the Eq. (14.14)

$$\text{EPR} = \frac{\text{Distance of bankline movement}}{\text{Time elapsed between the oldest and most recent banklines}} \tag{14.14}$$

Positive EPR values indicate accretion (bankline growth) while negative EPR values indicate erosion (bankline retreat). By averaging the EPR values across all transects, the overall migration rate for each river bankline segment is determined. The results of the analysis may be cross-checked with high resolution imagery and a confusion matrix. The reliability of the bankline migration estimates can be assessed using bootstrap resampling, which is a statistical method that generates thousands of re-samples from the original data to compute confidence intervals. This technique does not assume a specific data distribution and provides robust estimates of mean migration rates with associated uncertainty bounds. To interpret and apply the findings, the study area is divided into zones or administrative units such as blocks or districts enabling localized analysis of erosion severity and migration trends. The resulting data can be visualized through thematic maps showing bankline positions, erosion and deposition areas, and migration rates over time. These maps are vital for identifying high-risk zones and can guide river training works or bank protection measures.

A study "Utilizing geospatial tools for the assessment of river bank erosion and migration patterns in complex braided and meandering river systems" (*Earth Surface Processes and Landforms*, Vol. 50 Issue 1, 2025) was carried out by a team of researchers at NESAC to investigate the bank erosion of the Brahmaputra during the years 2016–2022 using high resolution Sentinel 2A/2B imagery. The study found that the left bank of Brahmaputra experienced a higher erosion and migration rate than the right bank. Year wise analysis revealed that the sub-period of 2018–2020 showed severe bank erosion and migration rate due to intense flood events in 2020. Among all the 3 zones, Zone 1 has a higher net erosion (8440 ha) and migration (−28.76 m/year) which may be attributed to scouring caused by impact of flood waters as the river enters Assam with a high erosive velocity. Zone 3 of the right bank experiences higher net erosion (7640 ha) and migration (−24.38 m/year). The effects of bank erosion in Brahmaputra River are prominent in 21 districts of Assam. During the study period, it was found that Barpeta district and Tinsukia district suffers highest and lowest bank erosion respectively.

Apart from the Brahmaputra, many of its major tributaries and Barak river is also prone to bank erosion and migration. Among the tributaries, Subansiri river experiences highest erosion and deposition followed by Beki, Manas, Lohit and Jia Bharali. River Subansiri, Beki and Manas also show highest shifting rate.

Another aspect of space technology utilization for bankline monitoring is site selection for river training works. River training works are structural or non-structural interventions constructed to guide, stabilize or protect river channels from erosion, flooding and uncontrolled migration. In erosion-prone regions of Northeast India, especially along dynamic rivers like the Brahmaputra, such measures are essential to protect adjoining agricultural lands, infrastructure, and human settlements. Traditional methods of planning river training works involve field surveys which are time-consuming and spatially limited. The utilization of geospatial technology in planning river training works enables a more comprehensive, data-driven approach to identify, assess and prioritize vulnerable sites for intervention. The integration of satellite imagery, DEM and GIS-based spatial analysis enables the monitoring of river dynamics over time, detection of erosion hotspots and simulate future scenarios. This

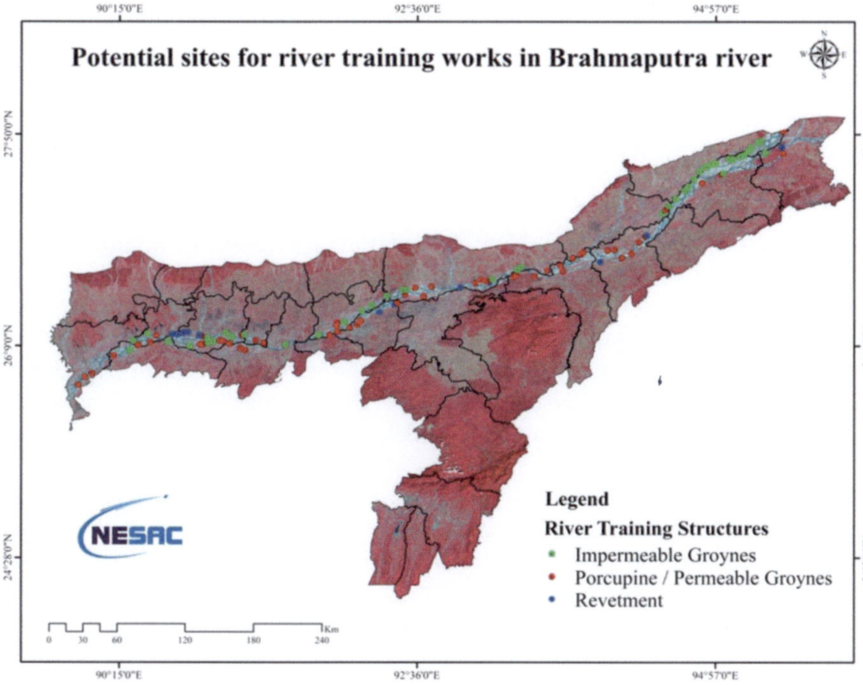

Fig. 14.9 Selected sites for different river training work

not only enhances the precision of site selection for training works such as embankments, revetments, spurs, or groynes, but also supports the sustainable design of river management strategies aligned with hydrological and geomorphologic conditions. For improving the morphometric situation and channel configuration, different river training works are suggested using geospatial technology for selected sites in the Brahmaputra river as shown in Fig. 14.9.

14.3.3 Monitoring of Embankments

Embankments are among the most widely used structural measures for flood control in many riverine and coastal regions. Typically constructed using locally available materials such as soil, sand and rocks, embankments serve to confine the river flow within a defined channel and prevent floodwaters from overflowing into the surrounding populated areas and agricultural lands. In cases where enhanced strength and durability are required, concrete and reinforced steel may also be incorporated into the design. Positioned strategically along river banks, floodplains or coastal margins, embankments are a primary defense structure against seasonal flooding and its associated risks.

Geospatial technology has become an invaluable tool for monitoring the status of embankments and identifying breaches. The use of satellite imagery for pre and post flood assessments of embankments enables a wide-scale, efficient and timely way to monitor embankment conditions. Satellite images captured after flood events are compared with those from before the floods to identify changes in water spread and locate breach points. Visual interpretation techniques are used to detect anomalies, such as sudden water entry into protected areas indicating a possible breach. Standard visual interpretation keys such as shape, tone, texture, size, association, collateral data and ground truth to pinpoint embankment breach locations and damaged segments. In cases where multispectral optical imagery is unavailable due to cloud cover, Synthetic Aperture Radar (SAR) data from Sentinel-1 (VH and VV polarizations) are used. SAR images have the advantage of cloud penetration and can capture surface changes even during unfavourable weather conditions. These radar datasets are pre-processed using the Sentinel-1 algorithm to enhance detection accuracy and map the breach locations effectively.

A prominent example of using geospatial technology for embankment monitoring is in Assam, where NESAC has been working in collaboration with the state government to map and monitor embankment breaches in near-real time. NESAC conducts operational mapping using satellite imagery each flood season, supporting the disaster management authorities in flood response and mitigation planning. Visual evidence of embankment breaches, captured through comparative analysis of pre- and post-flood satellite images, serves as crucial input for updating flood risk zones. In many cases, breach locations identified in the previous flood season remain unrepaired by the time the next monsoon arrives. Hence, before the onset of the successive flood season, these previously mapped breach points are re-verified using updated satellite imagery. This ensures that State Disaster Management Authority can anticipate flood behavior in vulnerable areas and remain prepared for floods resulting from unrepaired or recurring breaches. Figure 14.10 shows the comparison of pre and post flood scenario due to embankment breaching observed from satellite imagery.

Another study was carried out for mapping and monitoring of embankments and identification of its vulnerable reaches due to river bank erosion using multi-date satellite data. High resolution Sentinel-2A satellite images for the time period

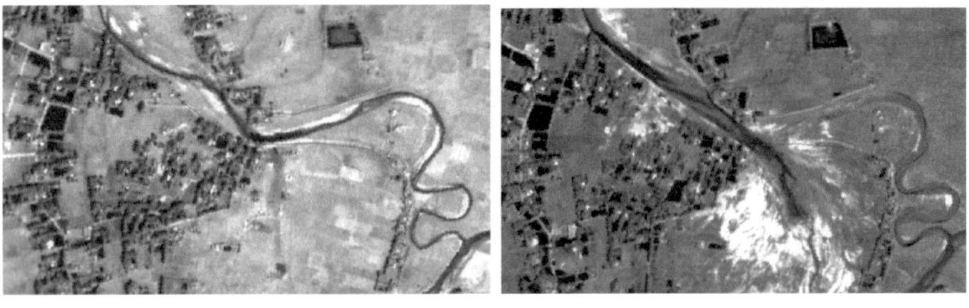

Fig. 14.10 Pre and post flood satellite image for a breach location for Borchola area as on February 2011 (left) and Borchola area as on October 2011 (right)

of 2016–2022 were used in NDWI to map the bank line for Brahmaputra, Barak and selected tributaries. Analysis was done to determine continuous erosion and deposition patches along the Brahmaputra, Barak and selected tributaries. Based on the proximity to the river banks with continuous erosion throughout the study period, the vulnerable embankment patches were identified. In total, 22 embankment patches (50.87%) along the Brahmaputra River and 37 patches (29.33 km) along the major/ sub-tributaries were identified as vulnerable to breaching and 3 embankment patches in Brahmaputra and 4 in the major tributaries have already breached. The districts of Lakhimpur, Dibrugarh and Dhemaji were identified to have the highest number of vulnerable patches in the state.

14.3.4 Near-Real Time Flood Monitoring

Flood disasters occur frequently in throughout India, causing widespread devastation to settlements, infrastructure, croplands, and natural ecosystems. In such situations, rescue and relief operations become extremely difficult. To support timely disaster response, space-based technologies are increasingly used to monitor and assess flood conditions. Near real-time flood mapping offers critical information for damage assessment, evacuation planning, and minimizing loss of life and property.

Near real-time flood monitoring and mapping involves the use of both optical and microwave satellite data to detect inundation. Optical imagery enables spectral analysis, where water bodies are identified based on their unique reflectance properties—typically through indices like the NDWI. However, optical sensors are often limited during monsoon seasons due to cloud cover. To overcome this, SAR imagery is employed, as it can penetrate clouds and provide consistent data regardless of weather or lighting conditions. In SAR data, flooded areas are identified using changes in backscatter coefficients, with inundated zones typically exhibiting low backscatter and appearing darker in the imagery.

Flood inundation layers are generated by comparing pre-flood and post-flood imagery to detect changes in surface water extent. These layers are then integrated with GIS-based administrative boundaries and LULC layer to estimate affected areas at detailed spatial levels, such as villages or blocks. The resulting maps provide actionable insights for emergency response and post-disaster recovery.

In India, the National Remote Sensing Centre (NRSC) under ISRO has been conducting near real-time flood monitoring since 1995. NESAC utilizes NRSC's flood layers to generate detailed flood impact maps for the state of Assam. These include overlays with village boundaries and land cover data, enabling authorities to estimate flood-affected settlements and agricultural zones. The Fig. 14.11 shows flood inundation map of Dibrugarh district of Assam which was derived from EOS-04 on 03-07-2024. This operational approach has become an essential tool for flood disaster management in the region.

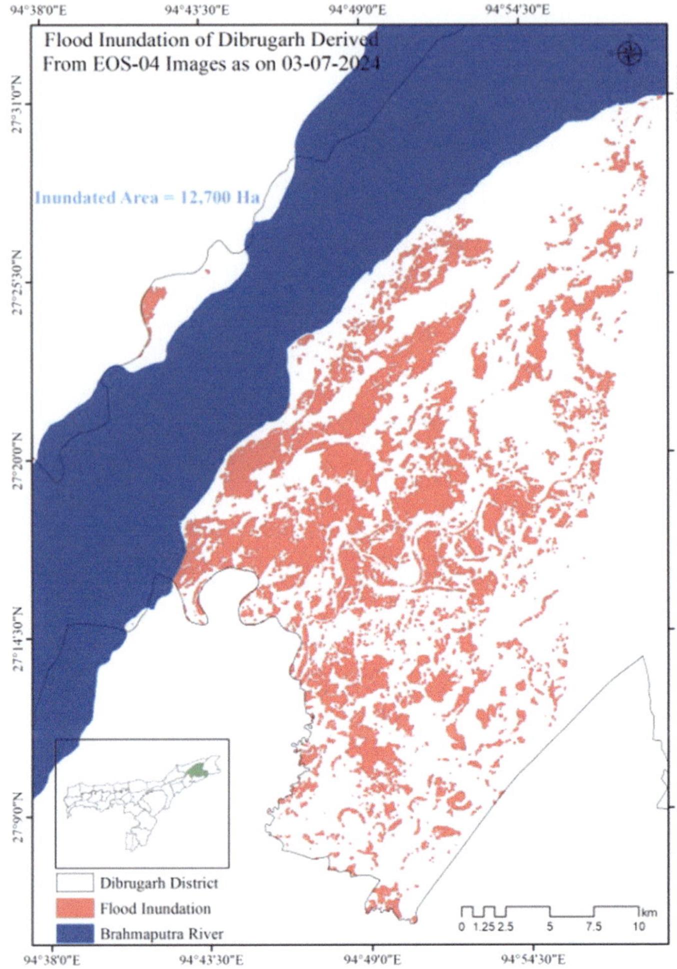

Fig. 14.11 Flood inundation map of Dibrugarh derived from EOS-04

14.4 Watershed Management

A watershed is a distinct geographical area bordered by ridgelines, where all surface runoff drains into a common outlet. It serves as a natural unit for managing and developing essential natural resources. Watershed management is an integrated approach that combines both natural resources management and social factors within the watershed. By implementing local resource conservation planning, the adverse effects of climate change can be mitigated, while contributing to improvements in the region's microclimate. When applied at a national scale, such efforts will help significantly in ensuring sustainable food security. The identification, delineation, and mapping of catchments are essential tasks for various activities like drainage basin planning and management, soil and water conservation planning. The objective of Watershed management program is to keep balance between nature and human needs by implementing strategies that address internal diversities. It covers critical aspects such as natural disasters, LULC, land stability, soil erosion, and agricultural productivity. Additionally, it supports a wide extent of allied economic activities including fisheries, dairy, horticulture, and agro-forestry, which contribute to the region's economy. Integrated watershed management focuses on developing a comprehensive approach to sustain and enhance watershed functions, ensuring the delivery of products, services, and values that meet the needs of the community within the watershed boundary (Rawat 2014).

Geospatial technology plays an important function in watershed management by making high-resolution satellite imagery available for LULC mapping and land use change analysis. It helps identify critical zones prone to soil erosion, deforestation, or land degradation. These insights guide interventions such as afforestation, soil conservation, and improved agricultural practices. The different aspects of watershed management are as follows.

14.4.1 Delineation of Watershed

Delineation of a watershed is the process of identifying the boundary of a watershed. This enables the understanding of water movement through the landscape and makes implementation of water resource management and conservation plans easier. Therefore, Department of Land Resources designed a methodology for categorization and coding of the catchment areas of India through Soil and Land Use Survey of India (SLUSI 2013). In this methodology, the catchment area of the country is divided into 6 regions, 35 basins, 112+ catchments, 500+ sub-catchments and 3237+ watershed based on area and developed an eight-digit alphanumeric codification scheme. This system facilitates efficient data management, analysis, and comparison, particularly in water resource management, monitoring, and planning. By assigning each watershed a unique code based on its position within a larger river basin hierarchy, the process ensures better organization and accessibility of hydrological information.

Delineation of watersheds can be done by using DEM in GIS software. In absence of high-resolution DEM, the watershed delineation can be carried out using Survey of India (SOI) toposheets in GIS environment. The watershed delineation process breaks down an area into smaller drainage networks that contribute all the flow into a single outlet.

As part of a detailed assessment of water resources, high-resolution elevation data has been utilized to delineate watershed hierarchies in Dhemaji district, Assam—a region prone to frequent flooding and soil erosion. The analysis employed the Airborne Laser Terrain Mapper (ALTM) DEM with a spatial resolution of 1 m, along with a Digital Terrain Model (DTM) of 5 m resolution. These datasets provided the necessary topographic precision to accurately delineate drainage patterns and watershed boundaries at a micro scale.

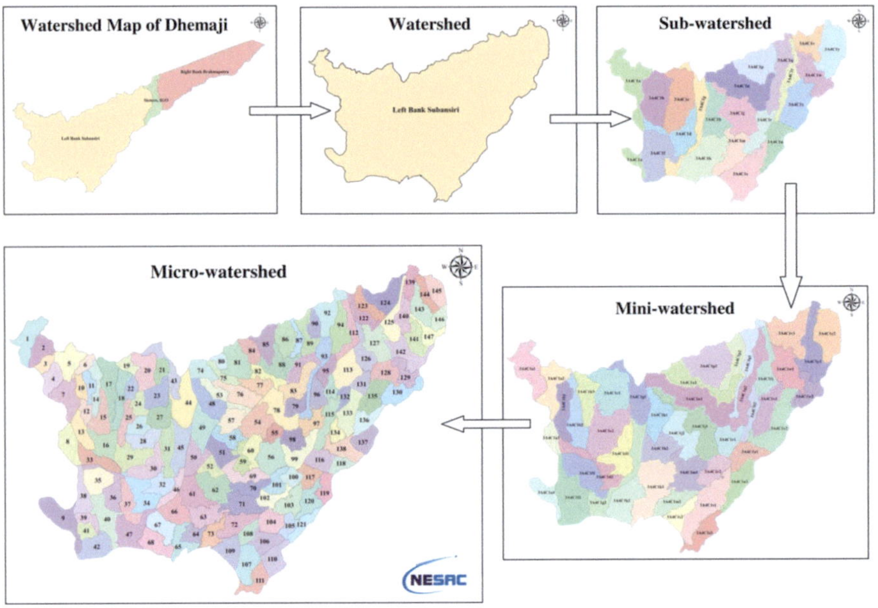

Fig. 14.12 Micro-watershed for left bank Subansiri watershed in Dhemaji district

Using these elevation models, a systematic delineation of watershed units was carried out. The process followed the standard classification system prescribed by the SLUSI, which enables consistent codification and hierarchy of watershed units across the country. Within the Dhemaji district, the study successfully identified and codified 29 sub-watersheds, 63 mini-watersheds, and 201 micro-watersheds.

This hierarchical delineation supports localized water resource planning, soil conservation strategies, and watershed management interventions. By breaking down the landscape into smaller, manageable hydrological units, the approach facilitates targeted planning—particularly valuable in regions like Dhemaji that are environmentally sensitive and hydrologically complex. The results, visualized in Fig. 14.12, provide a foundational layer for further geospatial analysis, including runoff estimation, erosion risk assessment, and the planning of water harvesting structures.

14.4.2 Watershed Characterization

Watershed characterization is the process of assessing a watershed's physical, hydrological, ecological, and socio-economic attributes to support sustainable management and conservation efforts. It involves analyzing topography, climate, drainage patterns, water quality, land use and biodiversity to understand the watershed's condition and challenges. This information is useful for effective water resource planning, land management, environmental monitoring, and disaster mitigation. By identifying critical areas for intervention, watershed characterization helps policymakers, environmentalists, and communities implement sustainable development strategies for long-term ecological and economic benefits.

Space technology inputs were utilized for preparation of land resources inventory maps for planning of WDC-PMKSY projects in both the states of Assam and Meghalaya for selected watersheds. The maps prepared represented the characteristics of the watersheds like LULC, sheet and gully erosion, detailed drainage network, soil (texture, pH, depth, organic carbon etc.), geology, topography, slope, rainfall and soil erosion. It also included identification of water bodies, agriculture and build up areas in proximity to the sheet erosion areas. The LULC maps have been prepared using LISS III and LISS IV along with PAN merged satellite data. Figure 14.13 shows the soil properties of selected Micro-watershed in Charaideo district of Assam.

Soil erosion is a severe environmental problem that reduces fertility and the agricultural yield of an area (Koirala et al. 2019) (Parveen and Kumar 2012). In Assam, water erosion affects 1,929 thousand ha (about 24% of TGA) of agricultural land and 437 thousand ha (about 5.5% of TGA) of open-forest area; thus affecting 29.5% of the total area of the state (Reddy et al. 2010). Soil erosion is the process of detachment and transport of soil particles due to geomorphologic factors (Saroha 2017). The rate of erosion is affected by variations in precipitation, run-off excess, soil texture, topography, and land cover conditions. Loss of fertile soil in agricultural land and decline

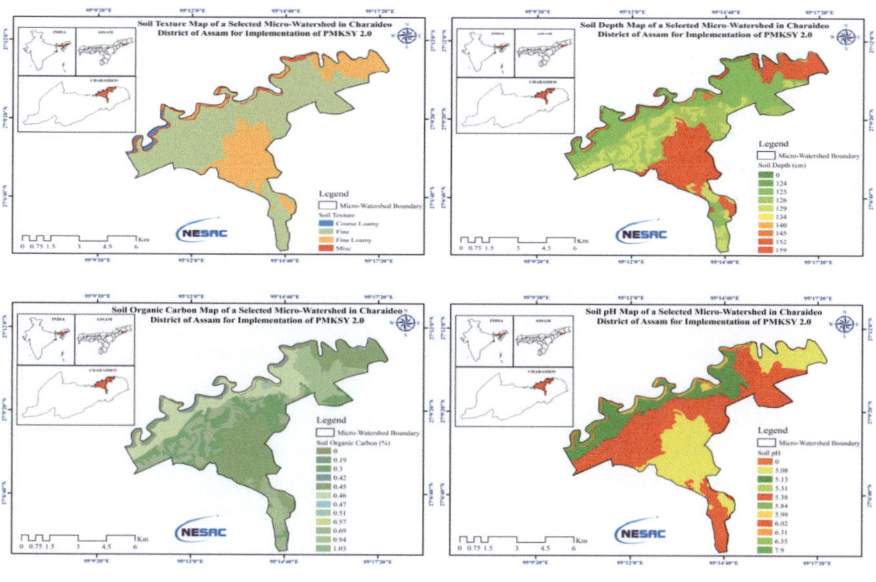

Fig. 14.13 Map of soil texture (top left), soil depth (top right), soil organic carbon (bottom left), and soil ph (bottom right) of selected micro-watersheds in Charaideo district of Assam

in the quality of soil resources are the main consequences of on-site soil erosion (Morgan 2005). Siltation of water bodies is a serious effect of off-site soil erosion (Navas et al. 2004). Complete destruction of farmland was also observed in certain extreme cases, such as gully erosion (Kirkby and Bracken 2009). Understanding the current situation of soil erosion will contribute to controlling soil erosion and restoring ecology. Therefore, soil erosion from watershed is also a serious problem that needs to be addressed. The basic objective of different watershed management programs in India is land and water conservation. The most advanced soil erosion models currently available are the Universal Soil Loss Equation (USLE) (Wischmeier and Smith 1978), Revised Universal Soil Loss Equation (RUSLE) (Renard et al. 1997) and the revised Morgan-Morgan-Finney model (MMF) (Morgan 2001) which incorporates geospatial data to estimate the soil erosion from an area. The geospatial inputs required for soil erosion estimation using these models are rainfall, soil data, slope, LULC, etc. which can be directly interpreted or derived from different satellite data using GIS techniques. The soil loss from the catchment areas of the Dumboor reservoir in Tripura and Loktak Lake (a RAMSAR site) in Manipur was estimated using the revised MMF model for prioritization of micro-watersheds and implementation of conservation measures.

14.4.3 Prioritization of Watershed

Prioritization of watershed is a process of ranking or categorizing watersheds based on various factors to identify for urgent conservation, management or restoration efforts. It can help in efficient resource allocation for water management, soil conservation, flood control and sustainable land use planning. Since in India watershed management program aims at soil and water conservation, watersheds are prioritized based on vulnerability to soil erosion and socio-economic conditions. The vulnerability to soil erosion can be estimated using various physical models with the help of remote sensing and GIS. The states in the North east region have a hilly to undulating terrain with intense rainfall. Due to modern advancements, vegetation covers are diminishing in hilly areas leading to increased soil erosion. Due to constraint of time and resources, all the degraded watersheds cannot be planned for management simultaneously. Therefore, prioritization of watersheds is a primary requirement for watershed management and development. A study carried out shows that out of 2776 micro-watersheds in Meghalaya, 6, 50 and 764 micro-watersheds are classified as essential, high and moderate priority respectively (Fig. 14.14). Another study shows out of 516 micro-watersheds in the catchment area of Loktak Lake, 11, 66, and 439 micro-watersheds are high, moderate and low priorities respectively. Similarly, out of 66 micro-watersheds in the catchment area of Dumboor reservoir, 1, 26 and 39 micro-watersheds are prioritized as high, moderate and low priority respectively.

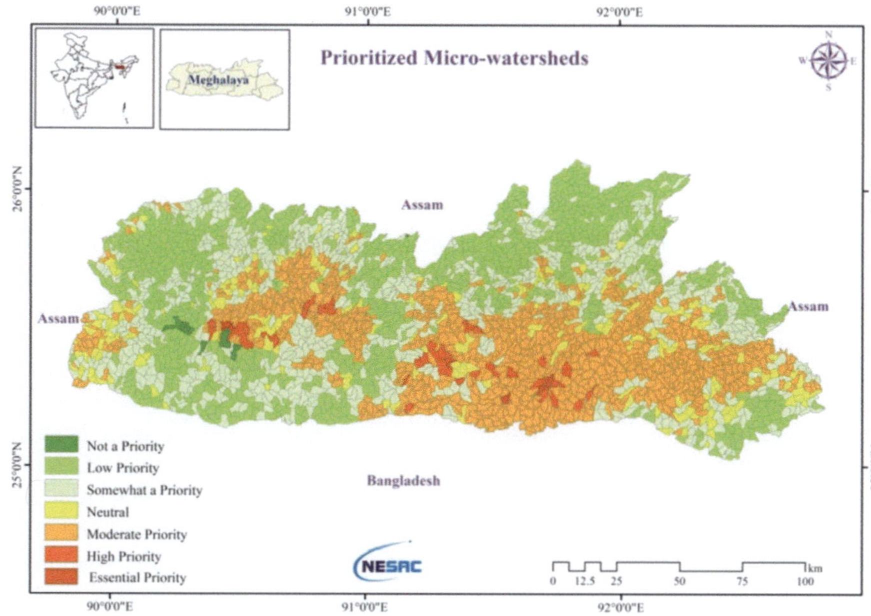

Fig. 14.14 Prioritized micro-watersheds map of Meghalaya

14.4.4 Monitoring and Evaluation of IWMP Watersheds

IWMP is a flagship initiative of the Government of India aimed at restoring ecological balance in degraded and rainfed areas through effective watershed development. Launched in 2009–10 under the Department of Land Resources (DoLR), IWMP consolidates and scales up various earlier watershed development efforts to promote land and water resource management sustainably, enhancing crop productivity, and supporting rural livelihoods. The selection and planning of IWMP projects are grounded in the principles of watershed prioritization, where geospatial data is used to identify critical watersheds that require immediate intervention based on rate of land degradation, water scarcity, and socio-economic vulnerability. Following the prioritization and implementation of watershed projects, systematic monitoring and evaluation (M&E) becomes crucial to ensure that interventions are effective, outcomes are measurable, and strategies remain adaptive. In this context, geospatial technology plays a central role by providing objective, scalable, and repeatable methods to assess changes over time.

A national-level initiative for geospatial monitoring of IWMP projects is being implemented by the National Remote Sensing Centre (NRSC) using tools such as Srishti—a web-based GIS platform on Bhuvan—and Drishti, an Android-based mobile application. NESAC is responsible for implementing this monitoring framework in the northeastern states of India. The scope includes the monitoring of IWMP projects sanctioned between 2009–10 and 2014–15, with each project observed over a five-year period from its start of implementation. High-resolution satellite datasets, including LISS-IV and Cartosat, are processed to generate time-series data for each watershed. These datasets support the refinement of watershed boundaries using the SIS-DP database, and the generation of thematic layers such as LULC and Normalized Difference Vegetation Index (NDVI). Change detection analysis is conducted using multi-date satellite imagery to track landscape transformations attributable to watershed interventions. Field-level validations are supported through geo-tagged photographs collected via the Drishti mobile app, which enables near real-time documentation of soil and water conservation works. This integration of remote sensing data with limited but representative ground truth ensures a cost-effective and scientifically sound approach to performance assessment. The results are a set of spatial outputs and statistical indicators that inform both mid-course corrections and post-project evaluations.

This approach to monitoring is not just a compliance measure, but a continuation of the watershed prioritization framework, now focused on evaluating how effectively prioritized interventions are performing on the ground. By linking planning and monitoring through geospatial technologies, the IWMP demonstrates a model for data-driven watershed management tailored to the diverse agro-ecological conditions of India. The Fig. 14.15 shows the monitoring of IWMP implemented watershed using multi-date satellite imagery.

Fig. 14.15 Monitoring of watershed using multi-date satellite imagery

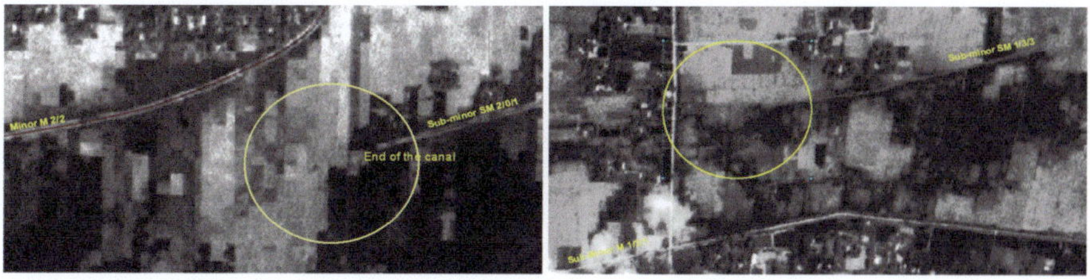

Fig. 14.16 Monitoring of irrigation infrastructure using Cartosat-1 imagery

14.5 Infrastructure Planning

14.5.1 Monitoring of Irrigation Potential Creation

Major National programs in India such as the Accelerated Irrigation Benefit Programme (AIBP) are advancing the creation of new irrigation infrastructures. High resolution pan chromatic image of Cartosat-1 is being utilized to supervise the physical progress of new irrigation potentials created. Figure 14.16 shows the capability of Cartosat-1 images for monitoring the creation of irrigation infrastructures. A study was carried out for assessing irrigation potential created in Dhansiri (Major) and Bordikorai (Minor) irrigation project in Assam funded by AIBP through the mapping of irrigation infrastructure using cartosat satellite (2.5 m resolution) data. The irrigation infrastructure in terms of canal network up to minor/sub minor along with irrigation and cross drainage structures and status was derived from the satellite data. Based on such inventory in terms of number of canal, their lengths and off-take chainages along with hydraulic conductivity depending on the status, irrigation potential created was assessed under AIBP component in the project. The overall irrigation potential created under AIBP components of Dhansiri project was estimated using cartosat-1 satellite images. The potential created was found to be 14,210 ha which was proposed to be 26,365 ha.

Acknowledgements The authors express gratitude to the Director, NESAC for his vision and insightful guidance in completing this work. The authors gratefully acknowledge the NRSC and ISRO for providing multi-temporal satellite data for water resource management initiatives in North-east India, flood layer for near real time flood monitoring and flood hazard zonation for flood management. Special appreciation is also due to the Soil and Land Use Survey of India (SLUSI) for their watershed delineation protocols and codification framework, which served as a foundation for various planning activities. The authors acknowledge the National Bureau of Soil Survey and Land Use Planning (NBSS&LUP) for their contributions in providing soil resource data essential for land capability assessment and erosion modelling. We also thank the Departments of Water Resources, Soil and Water Conservation, and Agriculture of the NE states for their valuable support in field data collection, validation, and coordination.

References

A. Baruah, D. Barman, Arjun B M, B.L. Chyne, S. Kurbah, S.P. Aggarwal, Habitat response of adult fish species under the influence of ecological flow and hydrodynamic regime in perennial river system. Ecohydrology **16** (2022). https://doi.org/10.1002/eco.2481

S. Biswas, S. Sudhakar, V.R. Desai, Prioritisation of sub-watersheds based on morphometric analysis of drainage basin: a remote sensing and GIS approach. J. Indian Soc. Remote Sens. **27**(3), 155–166 (1999)

B.A.M. Bouman, A conceptual framework for the improvement of crop water productivity at different spatial scales. Agric. Syst. **93**, 43–60 (2007)

B.U. Choudhury, G. Nengzouzam, M.A. Ansari, A. Islam, Causes and consequences of soil erosion in northeastern Himalaya, India. Curr. Sci. **122**(7), 772–789 (2022). https://doi.org/10.18520/cs/v122/i7/772-789

B.L. Chyne, R. Das, R. Sarmah, A. Saha, K.K. Sarma, S.P. Aggarwal, Utilizing geospatial tools for the assessment of river bank erosion and migration patterns in complex braided and meandering river systems, in *Earth Surface Processes and Landforms*

K.R. Dikshit, J.K. Dikshit, Natural vegetation: forests and grasslands of North East India, in *North-East India: Land, People and Economy, Advances in Asian Human-Environmental Research*, ed. by K.R. Dikshit J.K. Dikshit (Springer, Dordrecht, 2014), pp. 213–255. https://doi.org/10.1007/978-94-007-7055-3_9

D.K. Hall, G.A. Riggs, V.V. Salomonson, Development of methods for mapping global snow cover using moderate resolution imaging spectroradiometer data. Remote Sens. Environ. **54**, 127–140 (1995)

M.J. Kirkby, L.J. Bracken, Gully processes and gully dynamics. Earth Surf. Proc. Land. **34**, 1841–1851 (2009)

P. Koirala, S. Thakuri, S. Joshi, R. Chauhan, Estimation of soil erosion in Nepal using a RUSLE modelling and geospatial tool. Geoscience **9**, 147 (2019)

K.M. Konsoer, B.L. Rhoads, E.J. Langendoen, J.L. Best, M.E. Ursic, J.D. Abad, M.H. Garcia, Spatial variability in bank resistance to erosion on a large mixed meandering, bedrock–alluvial river. Geomorphology **252**, 80–97 (2016). https://doi.org/10.1016/j.geomorph.2015.08.002

U.C. Kothyari, Erosion and sedimentation problems in India, in *IAHS Publications—Series of Proceedings and Reports*, vol. 236 (International Association of Hydrological Sciences, 1996), pp. 531–540

A.V. Kulkarni, J. Srinivasulu, S.S. Manjul, P. Mathur, Field-based spectral reflectance to develop NDSI method for snow cover monitoring. J. Indian Soc. Remote Sens. **30**, 221–228 (2002)

S.K. Lahiri, A. Sarma, R.J. Wasson, Bed-bank relationship and flood characterisation in the upper reach of the Brahmaputra Valley, Assam, in *Water Security and Sustainability: Proceedings of Down to Earth 2019* (Singapore, Springer, 2021), pp. 191–206

R.P. Morgan, A simple approach to soil loss prediction: a revised Morgan–Morgan–Finney model. CATENA **44**, 305–322 (2001)

R.P. Morgan, *Soil Erosion and Conservation*, 3rd edn. (Blackwell Publishing, Cornwall, UK, 2005)

G. Mythili, J. Goedecke, Economics of land degradation in India, in *Economics of Land Degradation and Improvement—A Global Assessment for Sustainable Development*, ed. by E. Nkonya, A. Mirzabaev, J. von Braun (Springer, Switzerland, 2016), pp. 431–468. https://doi.org/10.1007/978-3-319-19168-3

National Remote Sensing Centre (NRSC), Ground water prospect mapping using remote sensing and geographic information system for Rajiv Gandhi National Drinking Water Mission (2008)

A. Navas, B.L. Valero-Garcés, J. Machín, An approach to integrated assessment of reservoir siltation: the Joaquín Costa reservoir as case study. Hydrol. Earth Syst. Sci. **8**, 1193–1209 (2004)

K. Pareta, Brahmaputra River embankment failures and bank failures in Assam: a geotechnical study. Am. J. Inf. Sci. Comput. Eng. **7**(2), 16–30 (2021)

R. Parveen, U. Kumar, Integrated approach of universal soil loss equation (USLE) and geographical information system (GIS) for soil loss risk assessment in upper South Koel Basin, Jharkhand. J. Geogr. Inf. Syst. **4**, 588–596 (2012)

D. Ramakrishnan, A. Bandyopadhyay, K.N. Kusuma, SCS-CN and GIS-based approach for identifying potential water harvesting sites in the Kali Watershed, Mahi River Basin, India. J. Earth Syst. Sci. **118**(4), 355–368 (2009). https://doi.org/10.1007/s12040-009-0034-5

M. Rawat, Prospects of water resources management for sustainable development in Nagaland, North East India. Int. J. Dev. Stud. Res. **3**(2), 123–145 (2014)

L.I.P. Ray, P. Bora, A. Singh, N.J. Singh, R. Singh, F. Sheikh, Rainfall characteristics, pattern and distribution of central Meghalaya. J. Indian Water Works Assoc. **34**, 9–16 (2014)

O.G. Reddy, D. Sarkar, A.K. Maji, *Degraded and Wastelands of India: Status and Spatial Distribution* (Indian Council of Agricultural Research, New Delhi, 2010)

K.G. Renard, G. Foster, G. Weesies, D. McCool, D. Yoder, *Predicting Soil Erosion by Water: A Guide to Conservation Planning with the Revised Universal Soil Loss Equation (RUSLE)* (U.S. Department of Agriculture, Washington, DC, 1997)

R. Saha, R.S. Chaudhary, J. Somasundaram, Soil health management under hill agroecosystem of North East India. Appl. Environ. Soil Sci. **2012**, Article ID 696174 (2012)

S. Hammami, L. Zouhri, D. Souissi, A. Souei, A. Zghibi, A. Marzougui, M. Dlala, Application of the GIS based multi-criteria decision analysis and analytical hierarchy process (AHP) in the flood susceptibility mapping (Tunisia). Arab. J. Geosci. **12**(21), 653 (2019). https://doi.org/10.1007/s12517-019-4754-9

J.N. Sarma, M.K. Phukan, Bank erosion and bank line migration of the Brahmaputra River in Assam during the twentieth century. J. Geol. Soc. India **68**(6), 1023–1028 (2006). (Exact end page assumed)

D.J. Saroha, Soil erosion: causes, extent and management in India. Int. J. Creat. Res. Thoughts (IJCRT) **5**(4), 1321–1330 (2017)

T.L. Saaty, A scaling method for priorities in hierarchical structures. J. Math. Psychol. **15**(3), 23–281 (1977). https://doi.org/10.1016/0022-2496(77)90033-5

T.L. Saaty, *The analytical hierarchy process: planning, priority setting, resources allocation* (MacGraw-Hill Co., New York, NY, USA, 1980)

Z. Shao, Z. Jahangir, Q. Muhammad Yasir, S. Mahmood, Identification of potential sites for a multi-purpose dam using a dam suitability stream model. Water **12**(11), 3249 (2020)

SLUSI, *Micro-Watershed Atlas of India* (SLUSI, 2013)

Space Applications Centre (ISRO), Ahmedabad, *Snow Cover Atlas of the Brahmaputra Basin, Sub-Basins: Dibang, Subansiri and Tawang* (ISRO, 2013)

Space Applications Centre (ISRO), Ahmedabad, *Monitoring Snow and Glaciers of the Himalayan Region* (ISRO, 2016)

E.R. Thieler, E.A. Himmelstoss, J.L. Zichichi, A. Ergul, *The Digital Shoreline Analysis System (DSAS) Version 4.0—An ArcGIS Extension for Calculating Shoreline Change (No. 2008-1278)* (U.S. Geological Survey, 2009

J. Wątróbski, J. Jankowski, P. Ziemba, A. Karczmarczyk, M. Zioło, Generalised framework for multi-criteria method selection: rule set database and exemplary decision support system implementation blueprints. Data Brief **22**, 639–642 (2019). https://doi.org/10.1016/j.dib.2018.12.015

W. Wischmeier, D. Smith, *Predicting Rainfall Erosion Losses—A Guide to Conservation Planning* (USDA Agricultural Handbook, Washington, DC, 1978)

M. Yasser, K. Jahangir, A. Mohmmad, Earth dam site selection using the analytic hierarchy process (AHP): a case study in the west of Iran. Arab. J. Geosci. **6**, 3417–3426 (2013). https://doi.org/10.1007/s12517-012-0602-x

Ranjit Das holds a Master of Engineering in Watershed Management and Flood Control from Assam Engineering College, Guwahati. With over 20 years of experience at the North Eastern Space Applications Centre (NESAC), he specializes in applying geospatial technology to water resources, watersheds, and flood disaster management. He received the ISRO Team Excellence Award in 2009 for his work on assessing irrigation potential in AIBP-funded projects using Cartosat data. Currently, he leads several major regional and national projects at NESAC and has published research in leading national and international journals and conferences.

Shanbor Kurbah is currently serving as a Scientist in the Water Resources Division at the North Eastern Space Applications Centre (NESAC), Department of Space, Umiam, Meghalaya. He holds an M.Tech. in Hydrology (1st Class with Distinction) from the Indian Institute of Technology (IIT) Roorkee, Uttarakhand. With over eight years of experience in the field of hydrology and water resources, his core expertise lies in hydrological data analysis, flood forecasting and modeling, hydrodynamic and snowmelt runoff modeling, Glacial Lake Outburst Flood (GLOF) simulation, and assessment of downstream releases from reservoirs. He is well-versed in the application of integrated spatial technologies, GIS, and advanced modeling software for effective water resources management. He has played a key role in the implementation of several user-funded projects and research initiatives across North East India, delivering actionable insights for regional water governance. He has also guided and mentored numerous M.Tech. dissertations, and has authored/co-authored publications in peer-reviewed national and international journals and conferences.

B. M. Arjun is currently working as a Scientist at the National Remote Sensing Centre (NRSC), ISRO, Hyderabad, focusing on strategic applications in the water resources domain. Prior to joining NRSC, he served as a Scientist at NESAC, Shillong, for seven years, where he contributed significantly to key water resources projects. He holds a Ph.D. from IIT Kanpur and an M.Tech. from NIT Surathkal, specializing in Water Resources Engineering. His areas of expertise include hydrologic and hydrodynamic modelling, flood inundation modeling, river system analysis, hydro-meteorological modelling for reservoir optimization, and hydrography. He was a core team member of the first ever GIS based operational Flood Early Warning System (FLEWS) established in India for the Brahmaputra and Barak river systems in North East India.

Diganta Barman, a senior scientist with over 22 years of experience, heads the Water Resources Division at NESAC. He holds a Ph.D. in Civil Engineering (Water Resources and Hydrology) from IIT and specializes in hydro-modelling, watershed management, flood forecasting, and reservoir operations. He led the development of India's first operational Hydro-met Flood Early Warning System (FLEWS) for the Brahmaputra and Barak basins, a system he has steered since 2009. Dr. Barman has also led major projects like the River Atlas (Assam and NER), SWaRA, NEEPCO's Hydro-met monitoring, and flood hazard zonation for Assam. He currently mentors the HIU Team under the World Bank-funded AIRBMP project. A member of various expert committees on water and flood management in NER, he has authored 20+ peer-reviewed papers and supervised multiple Ph.D. and postgraduate theses. He collaborates with national and international agencies including CWC, IMD, IRADe, Columbia Water Center, and ICIMOD.

Chapter 15
GIS Based Hydrological Flood Early Warning System (FLEWS) in North Eastern Region

Diganta Barman, Shanbor Kurbah, Manbhalang Dakermi Shylla, B. M. Arjun, and Ranjit Das

Abstract Floods are among the most common and deadly natural disasters, especially in Northeast India, where heavy monsoon rains and complicated river systems such as the Brahmaputra and Barak frequently create widespread flooding. Due to the region's difficult geography and the paucity of hydro-meteorological data, traditional forecasting techniques have had difficulty. On the request from Government of Assam, The Flood Early Warning System (FLEWS) has been developed since 2009 by the North Eastern Space Applications Centre (NESAC) integrating semi-distributed hydrological models, remote sensing, GIS, and atmospheric numerical rainfall forecasting. This system leverages geospatial technologies to forecast flood in a real time with a parallel decision support input from the synoptic weather monitoring system. Initially focused on the highly dynamic and sediment-laden valley of Assam, FLEWS have proven effective even in remote and high-altitude areas. With support from the North Eastern Council, NESAC is now expanding FLEWS to other North Eastern Region. Despite infrastructure and data limitations, the ongoing refinement of FLEWS-driven by technological advancement and improved resource allocation-marks a major leap in flood preparedness. Integrating GIS and remote sensing has enhanced predictive capabilities, making FLEWS a cornerstone

D. Barman (✉) · S. Kurbah · M. D. Shylla · R. Das
North Eastern Space Applications Centre (NESAC), Umiam, Shillong, Meghalaya, India
e-mail: diganta.barman@nesac.gov.in

S. Kurbah
e-mail: s.kurbah@nesac.gov.in

M. D. Shylla
e-mail: manbhalangdakermishylla@gmail.com

R. Das
e-mail: ranjit.das@nesac.gov.in

B. M. Arjun
National Remote Sensing Centre (NRSC), Balanagar, Hyderabad, Telangana, India
e-mail: arjun_bm@nrsc.gov.in

S. P. Aggarwal et al. (eds.), *Remote Sensing of the North Eastern Himalayan Ecosystem*,
https://doi.org/10.1007/978-981-95-3221-6_15

of modern flood resilience, protecting lives, infrastructure, and ecosystems across Northeast India.

Keywords Flood early warning systems · Hydrologic and hydraulic model · Remote sensing and GIS · Disaster preparation · HEC-HMS · HEC-RAS · MIKE

15.1 Introduction

Floods are among the most frequent and deadly natural disasters, having killed countless people and seriously damaged infrastructure all over the world. Over 90% of people in Asia are vulnerable to natural disasters, and two-thirds of those affected by them are victims of flooding (Pilon 2002). Every year, various rivers in the Indus–Ganga–Brahmaputra plains of India reported 5–11 floods on average that lasted anywhere between two and six days (Kale 2003). Being one of the biggest rivers in the world, the Brahmaputra River flowing through Assam is renowned for its unpredictability and frequent alteration of its course, which causes floods in the nearby communities. Nearly every year, three to four waves of flooding wreak havoc in Assam's flood-prone areas (Water Resource, Govt. of Assam). It is therefore self-evident that flood early warnings have become an essential component of disaster risk reduction strategies to take timely action to protect lives and property (Rogers and Tsirkunov 2010).

Reed (1984) defined flood forecasting as "*the prediction of stage, discharge, time of occurrence, and duration of a food-especially of peak discharge at a specific point on a stream-resulting from precipitation and/or snowmelt*" and defined flood warning as "*advance notice that a flood may occur in the near future at a certain station or in a certain river basin*". While the urban flood warning systems focus on underground pipeline flow, channelized river flow, and urban surface flow based on the inundation area and the inundation index, which includes the water flow velocity and inundation water depth (Chen et al. 2015). River flood warnings come into effect when it is anticipated that rivers or streams will rise above their banks and endanger nearby property, infrastructure, and people.

The frequency and intensity of extreme weather events are expected to rise due to climate change, increasing the likelihood of floods worldwide and the concentration of flood-prone areas among urban populations as they develop. As a result of melting glaciers and more rainfall brought on by rising global temperatures, rivers and drainage systems are overloaded. This increases the unpredictability and hazard of flood episodes. The establishment of efficient FLEWS is essential in view of these increased hazards. Vulnerable populations can receive timely notifications from these systems, allowing them to take precautionary measures, lessen damage, and save lives. Therefore, by integrating state-of-the-art technologies with climate data, FLEWS can mitigate the impacts of floods in a changing climate.

Historically, basic gauge readings and local observations that mostly relied on river gauge data and in-situ meteorological data were used to predict floods (Bates et al. 1997). But as technology advances, attention has turned to more complex, real-time flood predicting techniques using satellite-based remote sensing, Geographic Information Systems (GIS), Numerical Weather Predictions (NWP), hydrologic and hydraulic models. Remote sensing data, such as satellite imagery and aerial photographs that swiftly covers large geographical areas; offer precise information on the scope and severity of floods. As a result, the shift from conventional to new approaches has greatly increased the predictive accuracy and timeliness of FLEWS, which use a variety of monitoring technologies, including satellite imagery, weather forecasts, and hydrological data, to forecast floods.

Many researches on satellite meteorology, climatology, weather forecasting and analysis have been conducted since the beginning of the twenty-first century highlighting the improvements in meteorological forecasting skills that are thought to offer more accurate estimates of the quantity, timing, and spatial distribution of precipitation data (Smith et al. 1986; Kidd et al. 2009; Thies and Bendix 2011; Nandy and Kumar 2024). These refined data helps in accurately determining rainfall intensity, duration, and distribution, which are key factors in predicting flood events and managing flood control systems. Furthermore, integrating geographic information systems (GIS), high-resolution digital elevation models (DEMs), and distributed hydrologic models, it is possible to improve flood simulation and provide a reliable application to ungauged basins by accurately representing the efficient runoff generation and translation processes functioning at a certain grid or catchment size (Freeze and Harlan 1969; Beven 1989; Guo et al. 2021; Horton et al. 2022).

Around the world, numerous flood early warning systems and methods have been developed such as ALERT (Automated Local Evaluation in Real Time) in Australia, FFG (Flash Flood Guidance) or GFFG (Gridded FFG) in Northern America, and EFFS (European Flood Forecasting System) in Europe (de Roo et al. 2003). India also has implemented several successful flood early warning systems in the past few years. One prime example is the FLEWS in Assam, India developed by North Eastern Space Applications Centre (NESAC). This system forecasts flood events using probabilistic Numerical Weather Prediction (NWP), an automated network of hydro-meteorological stations, real-time hydro-meteorological readings from weather radar satellites and a GIS based distributed hydrological model that give the river discharge to decides the flood early warning with a parallel decision support input from the synoptic weather monitoring system.

There are two basic approach flood early warning systems namely "Top-down" and "Bottom-up" (Baudoin et al. 2016). A top-down flood early warning approach is a system in which decisions and actions regarding flood early warnings, risk management, monitoring, and response are initiated at higher levels of government or authoritative bodies (such as national or regional levels) and then communicated down to local communities and stakeholders. The Bottom-up approach is a community centric warning system such as local people wisdom that is passed down through generations to suggest whether prolonged heavy rain increases the likelihood of flooding; or the bio-indicators, such as plant phenology, animal and insect behaviour to predict climate-related hazards (Zommers 2014). But due to the inherent uncertainty of river-catchment dynamics and the issues with model over-parameterization, bottom-up approaches may not always be suitable for real-time forecasting (Young 2002). As a result, two-way communication between the top-down and bottom-up approaches of flood early warning systems must be combined, as community involvement can be critical to the system's overall effectiveness; otherwise, warnings will only be reacted to by a tiny percentage of the recipients (Rana et al. 2021).

15.1.1 Flood Meteorology in North East India

The southwest monsoon (June to September) is the primary source of rainfall in North East India. Large volumes of rainfall with notable variations make the area a breeding ground for both localised and large-scale flood (Varikoden and Revadekar 2020; Zahan et al. 2021; Bora et al. 2022). Heavy rainfall of 250–510 cm in the Brahmaputra floodplain and 510–640 cm in the Abor and Mishmi hills of Arunachal Pradesh provides the majority of the Brahmaputra River's runoff (Murthy 1981; Dhar and Nandargi 2000).

Orographic lifting caused by Northeast India's steep terrain results in heavy rainfall on the windward side of mountains due to the Madden–Julian oscillation, which is critical in the development of intense rainfall events during years with exceptionally strong monsoons, particularly over the Meghalaya Hills which is renowned for having the largest rainfall in the world, measuring 26,461 mm between August 1860 and July 1861 (Jennings 1950; Murata et al. 2007; Prokop and Walanus 2015). For example, two deadly flash floods occurred in the Goalpara and Sonitpur districts of Assam, which border Meghalaya and Arunachal Pradesh, respectively, in 2004 as a result of two extremely powerful cloud bursts of unprecedented intensity, one in the western Meghalaya hills and the other in Western Arunachal Pradesh (Das et al. 2009).

The spatial distribution of land-based rainfall observation networks is uneven in many developing regions of the world (Nhi et al. 2019a, b). These shortcomings are compensated for by satellite observations, which use sophisticated infrared (IR) and microwave (MW) detectors to provide more timely and spatially uniform coverage across large regions of the world (Sun et al. 2018). Since the launch of the first meteorological satellite, the Television and Infrared Observation Satellite (TIROS-1) by NASA, in April 1960, satellite observations have been able to make quasi-global observations, which is perhaps the most significant contribution to climatological rainfall estimates (Kidd 2001). Some other satellite-derived data sets available are the Tropical Rainfall Measuring Mission (TRMM) (Huffman et al. 2007), Precipitation Estimation from Remotely Sensed Information using Artificial Neural Networks (PERSIANN) (Ashouri et al. 2015), and the Climate Prediction Centre (CPC) morphing technique (CMORPH) (Joyce et al. 2004) products. Furthermore, our own ISRO domestic satellites INSAT series, provide continuous, high-resolution multispectral imagery for real-time monitoring of cloud formation, precipitation patterns, and surface water levels, have made flood meteorology in

India easier (Kelkar 1994). The data assists flood predictions by providing precise hydro-meteorological modelling and early warning systems. A Geographic Information System (GIS) is a tool that stored geographical information in a database that can be queried and graphically displayed for analysis (Khanna et al. 2018). Additionally, GIS is used to analyse observed and real-time spatial distribution of rainfall, temperature, wind patterns, and humidity over specific regions. It integrates remote sensing data with meteorological station data from Doppler weather radar (DWR) to create precise models of weather systems and their potential to cause flooding.

15.1.2 Flood Hydrology in North East India

Flooding in Northeast India is primarily caused by heavy monsoon rainfall and the region's unique topography, which leads to the overflow of rivers such as the Brahmaputra and Barak. The Yarlung Tsangpo River, known as the Brahmaputra river in Assam, the Siang or Dihang in Arunachal Pradesh and the Jamuna in Bangladesh, is one of the largest rivers of the world with a catchment area of 5,80,00 km^2 that traverse through China (Tibet), India and Bangladesh for a total distance of 2880 km. In Assam, it travels for a distance of 640 km from Sadiya to Dhubri. On entering the Indian part, the river system shows a sharp drop in elevation from approximately 3000 m in Chinese province to 660 m near Pasighat, India. The flood has become a chronic problem in Brahmaputra valley for two main reasons—primarily due to sudden fall in profile elevation in the border region of China and India triggering bed scouring effect resulting in sediment deposition in the downstream flat profile of the riverbed. Secondly, due to the massive earthquake of 1950 in Assam, the riverbed itself has raised an average of approximately 5 m in major part of the river in Assam resulting in paramount decrease in the channel capacity which further results in inundation of its floodplains during heavy flow season of peak monsoon. Moreover, due to high spate of Brahmaputra main channel, its tributaries both in the north and the south bank of the river also inundates due to their individual high discharges as well as due to not being able to effectively discharge their flow into the Brahmaputra main channel. Since flood as a natural phenomenon, cannot be prevented, can only be attempted to mitigate the disastrous effect due to its occurrence. Flood Forecasting is an important component of overall non-structural mitigation measures of flood. The River network map of North East India is shown in the Fig. 15.1.

The Barak basin covers parts of India, Bangladesh and Myanmar. In India it spreads over states of Meghalaya, Manipur, Mizoram, Assam, Tripura and Nagaland having an area of 41,723 km^2. The Barak River rises from the Manipur hills, south of Mao in Senapati district of Manipur at an elevation of 2,331 m. It flows then along Nagaland-Manipur border through hilly terrains and enters Assam. It further enters Bangladesh where it is known by the name of the Surma and the Kushiyara and later called the Meghna before receiving the combined flow of the Ganga and the Brahmaputra. The length of the Barak River from its origin upto the border of Assam along the Kushiyara is 564 km. The principal tributaries of Barak joining from north

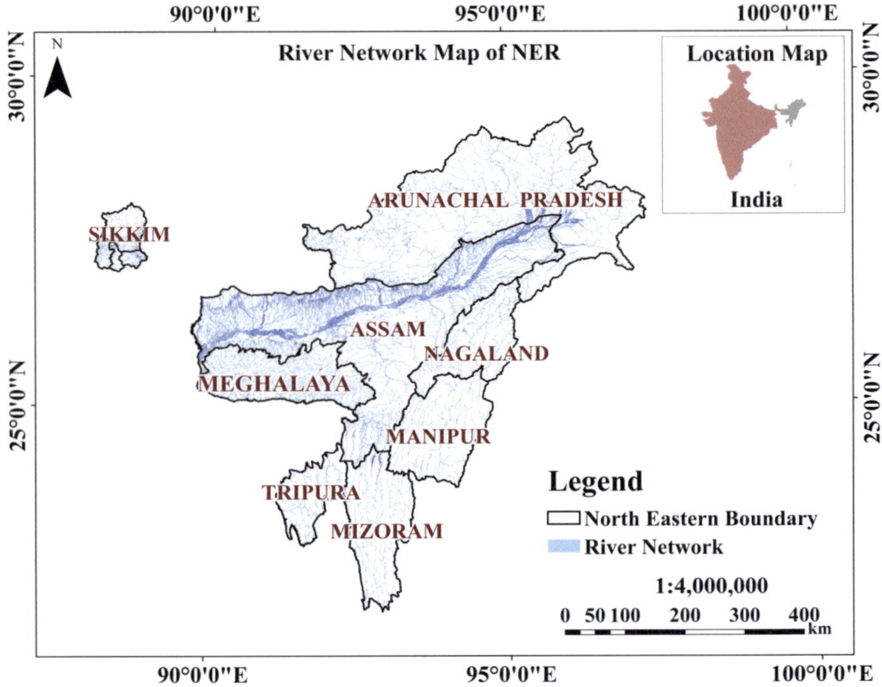

Fig. 15.1 River network for north east India

bank are the Jiri, the Chiri, the Modhura, the Jatinga, the Harang, the Kalain and the Gumra whereas the Dhaleswari, the Singla, the Longai, the Sonai and the Katakhal joins from south bank.

15.1.3 Flood Plain Topography in North East India

Because of its varied geography, which includes both high mountain ranges and lowland plains, Northeast India has a wide range of elevations. The elevations of Assam's lowlands and the neighboring regions range from 20 to 150 m above sea level. States like Arunachal Pradesh, Sikkim and Nagaland have higher Himalayan ranges that reach elevations of 2,000–7,000 m (6,561–22,965 feet). Some of the highest points in the area are found in Sikkim, where peaks like Kangchenjunga (8,586 m) are located. The Digital Elevation Model for NER and the LISS 4 Imagery showing braided River floodplain in the Rivers are shown in Fig. 15.2A, B respectively. Northeast India, known for its diverse geography and rich natural resources, features extensive floodplains formed by several major river systems, including the Brahmaputra, Barak, and numerous tributaries, have created vast floodplains in North

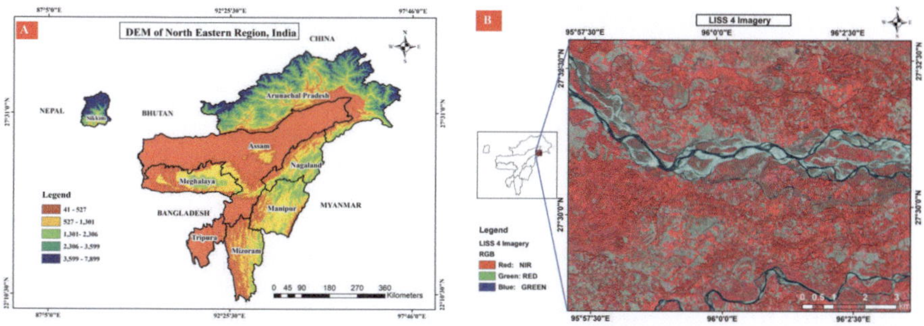

Fig. 15.2 **A** Digital elevation model for NER. **B** LISS 4 imagery showing braided river floodplain

Eastern India (Gogoi et al. 2023). The floodplain topography in this region is dynamic and plays a crucial role in shaping the ecology, economy, and culture of the area.

15.1.4 The Characteristics of the Floodplains of the Major River

The Brahmaputra River floodplain: The Brahmaputra River is one of the largest rivers and dominates the floodplain landscape in Northeast India. The Brahmaputra flows for 2880 km out of which 670 km through Assam receives 103 tributaries—65 on the right (north) bank and 38 on the left (south) bank along the valley (Gupta 2003; Goodbred et al. 2003). Its floodplain, known as the Brahmaputra Valley, stretches across Assam and is characterized by alluvial deposits, sandbars, and braided river channels (Lahiri and Sinha 2012). The Meghalaya plateau to the south and the Eastern Himalayas to the north enclose the valley (Goswami 1985).

The Barak River floodplain: The Barak River flows through the Cachar Plains in Assam, it is bounded by the Barail range separating it from the Brahmaputra basin on the north, by the Naga and Lushai hills on the east and by Mizo hills and territory of Bangladesh on the south and west (Nath and Ghosh 2022; Sil et al. 2022). The floodplains of the Barak are narrower than those of the Brahmaputra but are equally significant in terms of agricultural productivity and biodiversity. Floods in Barak Valley are mostly caused by drainage constraints (Gupta 2003).

Floodplains of the other Tributaries: Tributaries such as the Subansiri, Manas, Dibang, Lohit, Teesta and Dhansiri also contribute to the intricate floodplain networks, creating a mosaic of wetlands, oxbow lakes and fertile lands.

15.1.5 Topographical Characteristics of Rivers in NER

Braided Channels: The Rivers in Northeast India, especially the Brahmaputra, exhibit braided channels due to high sediment load and seasonal variability in water flow. This creates shifting sandbars, or "chars," which are often used for temporary settlements and agriculture (Sarma and Acharjee 2018; Lahiri and Sinha 2012). Most of the tributaries of Brahmaputra are braided in nature.

Alluvial Plains and fans: The floodplains are composed of alluvial soil deposited by the rivers, making them highly fertile and suitable for crops like rice, jute, and mustard. These plains also host wetlands and marshy areas, which are crucial for biodiversity (Sinha and Tandon 2014).

Wetlands and Oxbow Lakes: The region is dotted with numerous wetlands and oxbow lakes formed due to the meandering nature of the rivers (Devrani et al. 2022). These water bodies serve as habitats for migratory birds and support fishing communities. Some notable lakes are the Dumbur, Rudrasagar lake and Loktak lake.

Natural Levees and Flood Basins: Along the banks of major rivers, natural levees are formed due to the deposition of coarser sediments. Behind these levees, flood basins or backswamps are created, which remain inundated for longer periods and support aquatic vegetation.

15.2 Institutional Framework for Flood Management

The successful implementation of FLEWS requires coordination between several agencies, including the Central Water Commission (CWC), National Disaster Management Authority (NDMA), and local state authorities like Assam State Disaster Management Authority (ASDMA). These agencies collaborate to ensure that the data is collected, analyzed, and disseminated to the public in a timely manner.

The Central Water Commission (CWC) was established by the Government of India in 1945 with the aim of promoting and advancing efforts in flood control, water conservation, and the utilization of water resources across the country. This includes areas like irrigation, hydropower generation, flood management, and river conservation. As a national apex engineering organization in the field of water resources development, the CWC has developed significant expertise in the planning, investigation, management, and design of water resources development schemes. In addition to sharing its knowledge with developing countries worldwide, the CWC has made significant contributions to the nation's remarkable progress in this area.

The Brahmaputra Board was established by the Government of India through the Brahmaputra Board Act of 1980, originally under the Ministry of Irrigation (now the Ministry of Water Resources), to oversee the Brahmaputra and Barak Basins across all northeastern states. Its primary functions include conducting surveys and investigations in these valleys, developing master plans for flood control, bank erosion, and drainage improvements, and preparing Detailed Project Reports (DPRs) for dams and other infrastructure projects. The Board also sets standards for the construction, operation, and maintenance of dams, handles the construction and upkeep of multipurpose dams, and performs other tasks related to the implementation of the Brahmaputra Board Act of 1980. It has developed flood management master plans for the Brahmaputra and Barak rivers and has carried out studies to address issues such as flood control, erosion, and drainage congestion, alongside DPR preparation for multipurpose projects.

The National Disaster Management Authority (NDMA) was established in 2005 under the leadership of the Prime Minister of India to prevent and mitigate the effects of disasters, including floods, and to ensure a coordinated and efficient response to any disaster. Its responsibilities include setting disaster management policies, approving the national and departmental plans, guiding state authorities in creating state plans, integrating disaster prevention into development projects, coordinating the implementation of these plans, recommending funding for mitigation, offering support to other countries affected by disasters, and taking measures for disaster preparedness and capacity building. The NDMA also provides broad guidelines for the National Institute of Disaster Management. Guidelines for flood management were released by the NDMA in January 2008, outlining the responsibilities of several Central and state entities for creating strategies to mitigate floods and providing aid after flood catastrophes.

The Assam State Disaster Management Authority (ASDMA) was established in 2007 after the adoption of the Disaster Management Act of 2006, with its Secretariat becoming fully operational in 2010. ASDMA oversees disaster management at the state level and has set up District Disaster Management Authorities across all 27 districts, with officers dedicated to disaster-related activities. The authority's key focus areas include research and development, the establishment of Disaster Response and Information Centre, preparation of Disaster Management Plans at various levels, strengthening the techno-legal framework, capacity building, and integrating disaster risk reduction into development initiatives. It also emphasizes forming the State Disaster Response Force (SDRF), creating District-level Quick Response Teams, raising awareness, ensuring safer schools and hospitals, and building partnerships with national and international institutions. ASDMA works to maintain dialogue with stakeholders, organize regular meetings, provide relevant data and maps, and review embankment statuses pre- and post-flood seasons. Furthermore, it disseminates flood advisories, monitors flood inundation through field data, and validates flood advisories issued by NESAC, ensuring a coordinated and effective disaster response across the state.

The Water Resources Department has been actively involved in flood control and riverbank erosion management in Assam, particularly following the announcement of the National Water Policy. In response, a detailed flood control plan was developed, highlighting priority areas that required urgent attention. The department has focused on the overall development of rural areas and the protection of major townships in both the Brahmaputra and Barak valleys. Additionally, efforts have been made to alleviate drainage congestion in cities such as Guwahati and other important towns. Key initiatives include the construction of embankments and flood walls, river training and bank protection works, anti-erosion measures, town protection projects, river channelization with pro-siltation devices, drainage improvements, sluices, raised platforms, and flood forecasting and warning systems. The department also undertakes flood zoning and implements measures based on the recommendations of the Rastriya Barh Ayog (RBA).

15.2.1 FLEWS Mechanisms and Policy Framework

India has developed a robust policy framework for disaster management, including flood forecasting. The government has established systems to ensure effective data sharing among various agencies, promoting a standardized approach to disaster management. The NESAC (North Eastern Space Applications Centre) plays an important role in providing space-based data and technical support for FLEWS implementation in the North East Region. It also plays a vital role in the collection and analysis of hydro-meteorological data, using satellite imagery and ground-based observations to monitor weather patterns, river levels, and flood risks. In addition, NESAC supports the monitoring of embankment breaches both before and after flood events, enabling timely interventions. It also took on the responsibility for developing a flood forecasting model and issuing early warnings. By utilizing advanced forecasting techniques, it provides timely flood alerts to local authorities and communities during the flood season, ensuring that necessary precautions can be taken to reduce flood-related damage and casualties. Through its contributions, NESAC enhances the ability of India's disaster management framework to respond effectively to flood threats, ultimately improving community preparedness and safety.

15.3 Technology Development in Flood Prediction and Forecasting

New developments in remote sensing, information technology, social media and artificial intelligence (AI) are changing the mechanisms of flood early warning systems and giving them the potential to become more capable (Perera et al. 2020). Few notable scientific developments in flood prediction and forecasting are listed below.

15.3.1 Remote Sensing

Remote sensing technology has revolutionized flood forecasting by providing real-time, high-resolution data on rainfall, river levels, and soil moisture. Increased resolution of precipitation input greatly enhanced the accuracy and dynamics of the simulated stream flow. It is noteworthy, however, that the impact of regional rainfall data on flow modeling differs according to the catchment, scale, and event characteristics (Lobligeois et al. 2014). Adding remotely sensed soil moisture to discharge measurements improves flood awareness by reducing false positive alert and increasing dependability (Wanders et al. 2014).

15.3.2 Advance Weather Radar Systems

The addition of wind data is expected to enhance forecasting by incorporating radial winds, profiling in the rain volume, and producing multi-parameter radar products that supplement Doppler and polarization measurements to provide more accurate rain-rate estimates and to nowcast severe weather events (Meischner et al. 1997).

15.3.3 Numerical Weather Prediction (NWP)

The Numerical weather prediction (NWP) or "*Weather Prediction by Numerical Process*" uses atmospheric dynamical equations to predict weather (Richardson 1922). The expansion of our Earth intelligence through the use of a global constellation of low-Earth-orbit and geosynchronous-equatorial-orbit satellites is anticipated to further progress data assimilation (DA) by utilizing these observations to enhance NWP capabilities which further help in flood forecasting (Brotzge et al. 2023).

15.3.4 AI, ML, IoT, Cloud Computing, and Robotics Techniques

It is anticipated that prediction performance will improve with the use of cutting-edge technologies for flash flood early warnings and susceptibility predictions, including artificial intelligence (AI), machine learning (ML), Internet of Things (IoT), cloud computing and robotics technology (Mosavi et al. 2018; Al-Rawas et al. 2024).

15.3.5 Big Data Analytics

The use of big data-high performance computing (HPC) convergence and social collaborative Industrial Internet of Things (IIoT) have led to more vibrant and efficient flood monitoring and forecasting (Sood et al. 2018; Nagendra et al. 2022).

15.3.6 Improved Climate Models

With growing computational capacity, dynamical climate model simulations can generate high resolutions data which improves the representation of orographic precipitation (Chan et al. 2013). This is expected to be beneficial for flood forecasting in the mountainous regions of the North East India.

15.3.7 Hydrological and Hydraulic Models

The advancement of computer-aided technology has led to the creation of numerous detailed numerical models for the accurate simulation of flood through catchments and river systems, allowing for more accurate flood prediction and scenario analysis. It is anticipated that the performance of the distributed hydrologic model will significantly increase with the availability of finer input forcing resolution for designated Catchment area (Shrestha et al. 2006). Similarly, in areas with complex topography, it is also anticipated that the higher-resolution grids for hydraulic models would yield the greatest results (Shustikova et al. 2019). Therefore, flood maps, which are produced using GIS and LiDAR data, improved emergency preparedness by visualizing possible flood depths and extent.

15.4 FLEWS-Assam and North Eastern Region

This section describes the overall methodological framework of FLEWS in the State of Assam and the other States in NER.

15.4.1 FLEWS for Assam

A novel flood forecasting method utilizing geospatial data/products, numerical rainfall prediction, and a grid-based distributed hydrological model has been developed and operationalized since 2009 at the request of the Government of Assam, India due to the traditional river gauge correlation-based flood forecasting system's inability to accurately predict impending flood waves.

The overall methodology of this exercise comprises of two major components of Meteorology and Hydrology. In the meteorology component along with a geographic grid based numerical rainfall prediction model known as the Weather Research Forecast (WRF) model, an innovatively developed synoptic weather monitoring system has also been incorporated. Subsequently, in the hydrology component, along with a distributed hydrologic model known as the Hydrologic Engineering Centre's—Hydrologic Modeling System (HEC-HMS), a lumped hydrological model in the form of rational formula, has been incorporated. Rainfall predicted by WRF being the dynamic input to both the distributed and the lumped hydrologic models give the river discharge that decides the flood early warning with a parallel decision support input from the synoptic weather monitoring system.

15.4.2 FLEWS for Other North Eastern States (Other Than Assam)

After the successful operationalization of FLEWS for Assam, NESAC has taken up the expansion of FLEWS services to the flood prone districts of North Eastern States with funding from the North Eastern Council (NEC). As a part of this study, the flood prone rivers have been identified and 36 semi-distributed hydrological models are developed for the flood prone rivers of NER. For generating the flood discharge hydrograph using the 24 h rainfall from the Weather Research Forecast (WRF), one of the commonly established Numerical Weather Prediction model, HEC-HMS model is used. The mandate of the study is to generate the flood alerts at the block/revenue circles level with a sufficient lead time. Henceforth, with the support from different stake holders, both state and central, the system will be beneficial for the effective effort on disaster preparedness in all the states. The framework for the Flood Early Warning Systems is show on the Fig. 15.3.

15.4.3 About Study Area and Challenges of Individual Basins

Northeast India frequently experiences flooding, which is made worse by the region's distinct topography, abundant rainfall, and multitude of rivers, such as the Brahmaputra and its tributaries. Every year, the region is hit by monsoon floods that severely harm houses, farms, and infrastructure. Deforestation, soil erosion, and climate change exacerbate the problem by causing more erratic rainfall and erosion of riverbanks. Flood control is a major issue for the area since these floods causing thousands of people to be displaced annually, disrupt livelihoods, and result in fatalities. The overall challenges of individual basins and states are mentioned below.

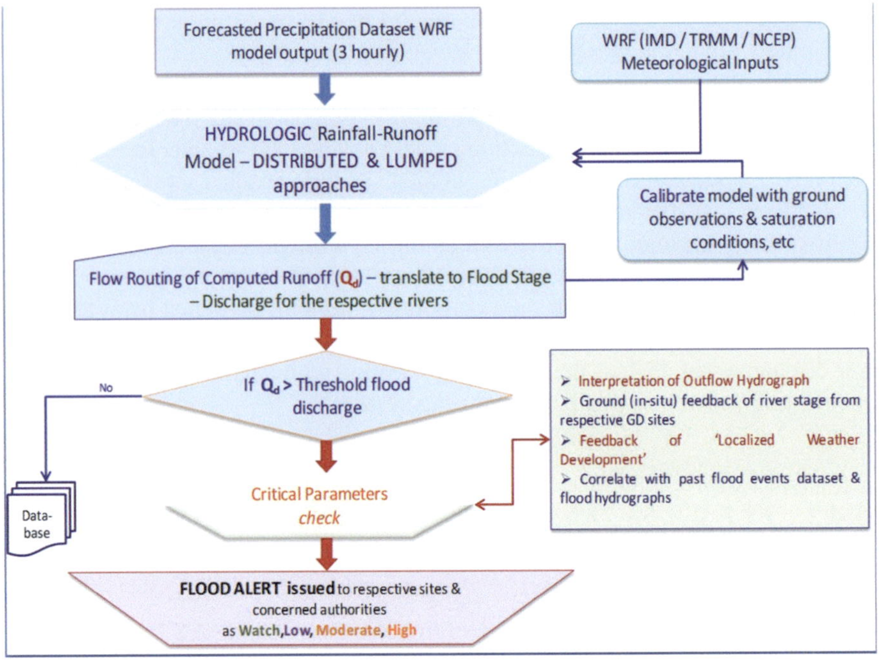

Fig. 15.3 Methodological framework for flood early warning systems

15.4.3.1 Assam: The Brahmaputra and the Barak Basin

Being one of the biggest rivers in the world, the Brahmaputra River flowing through Assam is renowned for its unpredictability and frequent alteration of its course, which causes floods in the nearby communities. Nearly every year, three to four waves of flooding wreak havoc in Assam's flood-prone areas (Water Resource, Govt. of Assam). The main problem of flooding in Assam is inundation caused by the spilling of the rivers Brahmaputra and Barak as well as their tributaries (Mohapatra and Singh 2003). Flooding in the state is also a cause of heavy rainfall during monsoons and rising water levels from the Brahmaputra and its tributaries. The other causes of flooding are the failure of embankments in various locations in the state resulting in fatalities, evictions, the destruction of crops and animals, damage to infrastructure, and health risks, to name just a few of the serious issues that the affected populations in Assam have to deal with.

While dams can be useful for producing electricity and irrigation, they can also alter river flow and increase the risk of flooding in areas downstream. Numerous dams have been laid on rivers that flow into the state of Assam. Water that is released by the hydroelectric facilities constructed on these rivers has a significant impact on the amount of water released, which can result in flooding in the districts downstream. Some of the dams erected on tributaries of the Brahmaputra River are the Subansiri Lower Hydroelectric Project constructed on the Subansiri River, the Rangandi

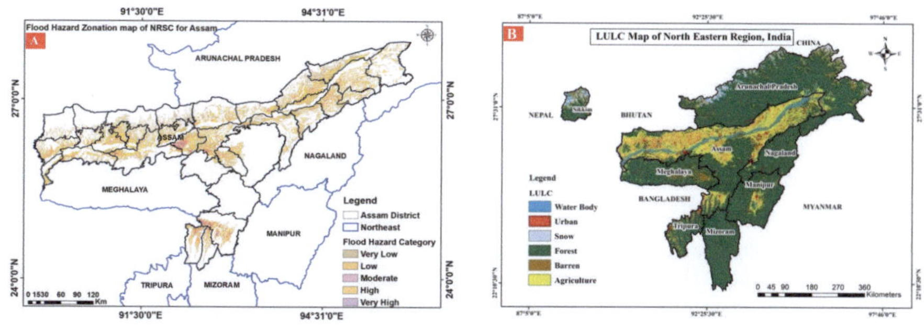

Fig. 15.4 **A** Flood hazard zone map in Assam (NRSC). **B** LULC map of north eastern region

Hydroelectric Project at Ranganadi River, the Kurichhu Hydroelectric Project on the Kurichhu River which is a tributary of the Manas River; the Tala Hydroelectric Project at Wangchu River, the Punatsangchhu Hydroelectric Project at Punatsangchhu River, the Doyang Hydroelectric Project in the Dhansiri River, Kopili Hydroelectric Project in Kopili River and the KarbiLangpi Hydroelectric Project in the Langpi River which is also a tributary of Kopili River. The sudden release of large amounts of water from these dams can have significant impacts on the downstream which includes adverse effects on the livelihoods of those living downstream of the dam, with long-term societal repercussions including an increase in poverty and migration. Rapid and severe flooding downstream can have an adverse effect on the ecosystem of the river by causing erosion, sedimentation, and other changes to the river channel and surrounding environment. The increased water velocity has the potential to undermine river banks, rip out vegetation, and harm fish and other wildlife habitat. The quality of the water may also be harmed and aquatic life endangered by the dam's release of sediment and other pollutants. Therefore, implementing comprehensive strategies for managing flood disasters is of great importance. The Fig. 15.4A, B shows the flood hazard zonation map in Assam and the LULC map of NER respectively, which highlight the vulnerable area to flooding and the distribution of natural resources in the regions.

15.4.3.2 North East: The Barak and Other Basin

The section below describes the challenges in Meghalaya, Arunachal Pradesh, Tripura, Mizoram, Nagaland, Sikkim, and their districts that are prone to flooding.

Meghalaya: The Kynshi, the Umkhrah, the Umngot, the Umiam, the Krishnai, the Umngi, the Ganol and the Simsang are Meghalaya's principal rivers. The elevation of this hilly state ranges from 200 m to more than 2000 m above sea level. The state is separated into the Garo Hills, Khasi Hills, and Jaintia Hills physiographic zones. The state know for having the highest recorded point rainfall in the world has had major catastrophic floods in 2014, 2017, 2018, and 2020 resulted in the loss of life, eviction of communities and significant infrastructural damage specially the western part of the state (Jennings 1950; Barman et al. 2024).

Arunachal Pradesh: Numerous Rivers, Riverine systems, and streams, including Subansiri, Kameng, Siang, Dibang, Lohit, and NoaDihing, traverse the state of Arunachal Pradesh. The primary flood prone Rivers are Subansiri, Lohit, and Kameng (Bhadra et al. 2011, 2015). The state is made up of several plains and valleys, and the physiography of Arunachal Pradesh is defined by the Patkai Hills in the south and the Himalayas in the north. Arunachal Pradesh has a monsoon-influenced humid subtropical climate with subtropical humidity having varying amounts of rainfall (Kumar 2013). The eastern half of the state receives a lot of rain because of its proximity to the Bay of Bengal, as compared to the western part of the state. With an average of 2500–3000 mm of rainfall per year, the State has seen catastrophic floods and flash floods in 1961, 1994, 2001, 2004, 2008, and 2014, and they now happen every year (SDMA, Govt. of Arunachal Pradesh).

Tripura: Bangladesh borders the state of Tripura on its north, west, and south; Mizoram borders it on its east; and Assam borders it on its northeast. The Gumti (Gomti), the Khowai, the Manu, the Haora, the Dhalai and the Muhuri are a few of the prominent rivers in Tripura. The Manu River frequently floods. With an average elevation of 183 m above sea level, Tripura is a hilly state. West Tripura and Dhalai are the areas which are mostly affected by flood. In 1993 and 2003 West Tripura severly affected by flood whereas Dhalai had experienced continuous flood since 1962 and the last major flood occurred in 2003 as per the report of Tripura Disaster Management Authority.

Mizoram: The state is covered with up to 21 major hill ranges or peaks, each of varying heights, with a few plains thrown in here and there. The Tlawng (also known as Dhaleswari or Katakhal), Tut (Gutur), Tuirial (Sonai), and Tuivawl are the most significant rivers, and they all flow through the northern region before eventually joining the river Barak in Cachar. The Koldoyne (Chhimtuipui), which flows through southern Mizoram and has its source in Myanmar, is a significant river. The river is patchy and has four tributaries. One of the main rivers in Eastern India, the Mahi, is frequently prone to significant floods, which has a detrimental effect on the local infrastructure and economy. It frequently experienced flooding. Heavy rains in June 2012 caused damage to roads, bridges, and livelihoods. Similar to how it did in 2015, it experienced disastrous floods as a result of constant rain, but in 2018 the Tlawng River and its tributaries spilled, causing flooding.

Nagaland: The Doyang and Dikhu in the north, the Barak in the southwest, and the tributaries of the Chindwin River (in Myanmar) in the southeast, significantly covered in Nagaland. The huge rivers in Nagaland are flourishing because of their proximity to the foothills and the heaving monsoon rains. The Barak River frequently floods. The weather in Nagaland is monsoonal (wet-dry). The annual rainfall averages between 2000 and 2500 mm. Rugged hills and small valleys make to Nagaland's physiography. All of Nagaland's low-lying districts that border Assam are affected by flooding. The Township of Tuli and the surrounding areas were severely impacted by flooding in the year 2005. The various colonies of Dimapur Town were underwater from September 11 to September 24, 2008. Dobhinalla, Supermarket, Nagarjan, Burma Camp, Walford, Sachu Colony, Nagagaon, Khermahal, Netaji Colony, Naharbari, and Airport Areas are included in this [District Disaster Management Plan for Dimapur District, 2018–19].

Sikkim: Sikkim shares borders with the Indian state of West Bengal to the south, Bhutan to the east, and Nepal to the west and the Republic of China in the North. Teesta, Rangeet, and Rangpo chu are Sikkim's three main rivers. Sikkim receives about 3,500 mm of rain on average each year, and the Teesta River is prone to floods. While floods also happened in 2017 as a result of strong rainfall, Sikkim had heavy rainfall that caused a flash flood in September 2018 as the Teesta river and its tributaries rose up. In 2022, Gangtok, Sikkim, saw a flash flood that claimed one life and also the subsequent catastrophic flood event in 2023 along the Teesta River caused by glacier lake burst and heavy rainfall in the Northern upstream, claiming multiple lives and damage of infrastructures.

15.4.4 The Weather and Climate Component Used for FLEWS

Near real time meteorological parameters over the North East (NE) sector are available in different sources and are listed in Table 15.1. In this section the description for dissemination of station rainfall, synoptic rainfall, forecasted rainfall, and extraction of gridded rainfall datasets for using as an input into the hydrologic model stream flow simulation are also reported.

15.4.4.1 Observed Station Rainfall

The observed station rainfall occurring during the past 24 h is obtained from the state authorities daily; providing a crucial input to analyze the overall idea of rainfall intensity and distribution across a catchment area; and potential flow into rivers and streams. Table 15.2 shows an example of observed rainfall obtained for a specific event in Lakhimpur district.

15.4.4.2 Automatic Weather Stations

An automated variant of the conventional weather system that allows for readings from far-off places is called an Automated Weather Station (AWS). The data logger, solar panel, rechargeable battery, telecommunication system, and meteorological sensors are all part of the weatherproof system. Using a satellite-based telecommunications system, the system transmits data in almost real-time, and it is stored on a computer at the Space Applications Centre in Ahmadabad called MOSDAC (Meteorological and Oceanographic Satellite Data Archival Centre). Rainfall, humidity,

Table 15.1 Meteorological parameters

Sl. No.	Data/products name		Source	Time resolution	Available
1	Observed rainfall		State Dept	24 h	Email/SMS
2	AWS rainfall		AWS-ISRO	1 h	www.mosdac.gov.in
			AWS-IMD	1 h	www.imdaws.com
3	Satellite images (IR&WV)-INSAT-3D		ISRO	Half hourly	www.mosdac.gov.in
4	850 mb relative vorticity		University of Wisconsin-Madison	3 h	http://tropic.ssec.wisc.edu/tropic.php
5	Cloud motion vector and atmospheric motion vector		University of Wisconsin-Madison and IMD	3 h	http://tropic.ssec.wisc.edu/tropic.php www.imd.gov.in
6	Qualitative precipitation estimation (QPE)		IMD	3 h/daily/ monthly	www.imd.gov.in
7	DWR products	Precipitation accumulation (PAC)	IMD	Daily	
		Surface rainfall intensity (SRI)	IMD	Near real time	

Table 15.2 Station rainfall obtained for district authority

Date	Rainfall (mm) at different station in Lakhimpur district		
	WRD rest house	Lilabari airport	Bihpuria WR Sub division
01/July/2024	31	103.4	70

temperature, pressure, wind speed, and wind direction sensors are commonly found in AWSs. Hourly, near-real-time rainfall data collected from AWSs is used for FLEWS. Two distinct networks—ISRO-AWS (www.mosdac.gov.in) and IMD-AWS (www.imdaws.com)—provide the data. In North East India, ISRO and IMD currently operate 120 and 64 AWSs, respectively as shown in the Fig. 15.5.

15.4.4.3 Weather Observation Satellite Images

ISRO's meteorological satellite INSAT 3D designed for enhanced meteorological observations and monitoring of land and ocean surfaces, weather forecasting and disaster warning that provided products of 6 channel multi-spectral Imager viz.

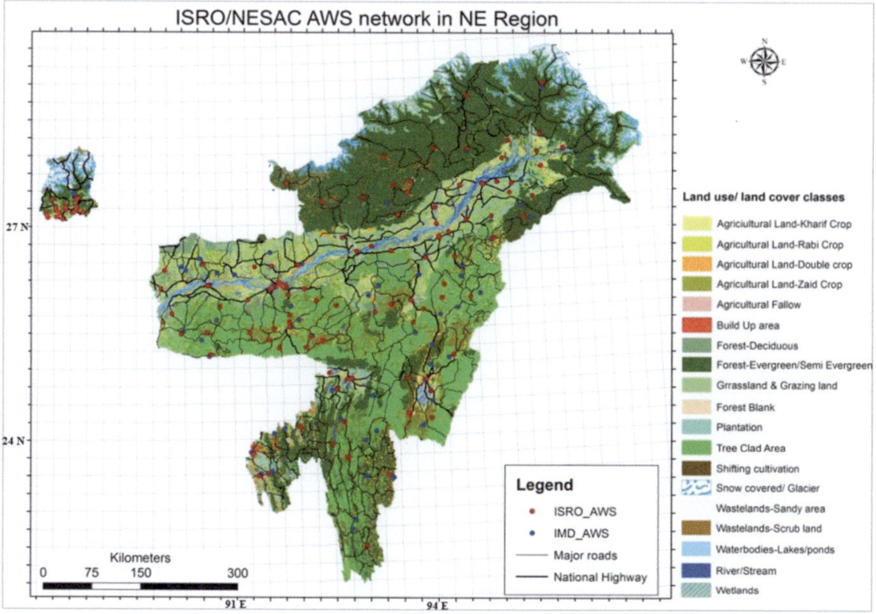

Fig. 15.5 Spatial distribution of all rainfall measuring stations over NE sector by IMD, ISRO and other agencies

VIS (0.55–0.75 μm), SWIR (1.55–1.70 μm), MIR (3.80–4.00 μm), WVP (6.50–7.10 μm), TIR 1 (10.2–11.3 μm), TIR 2 (11.5–12.5 μm) (which are uploaded at every half an hour) is used to understand cloud dynamics over the FLEWS covered areas. These imagery of different channel help us in identifying clouds types and their intensity. A typical example of TIR2 and WV images on 14th Aug 2014 at 10:30 h IST is shown in Fig. 15.6.

15.4.4.4 Atmospheric Vorticity

The curvature of atmospheric flow is measured by vorticity. A positive value indicates clockwise motion in the northern hemisphere and counter clockwise motion in the southern hemisphere. Storm formation is aided by an atmosphere with positive vorticity, which indicates upward air flow; whereas negative vorticity, which indicates downward air flow, indicates air subsidence over the area, so reducing convection. It is therefore feasible to predict the general synoptic weather over a certain area for a minimum of 12 h using this vorticity product.

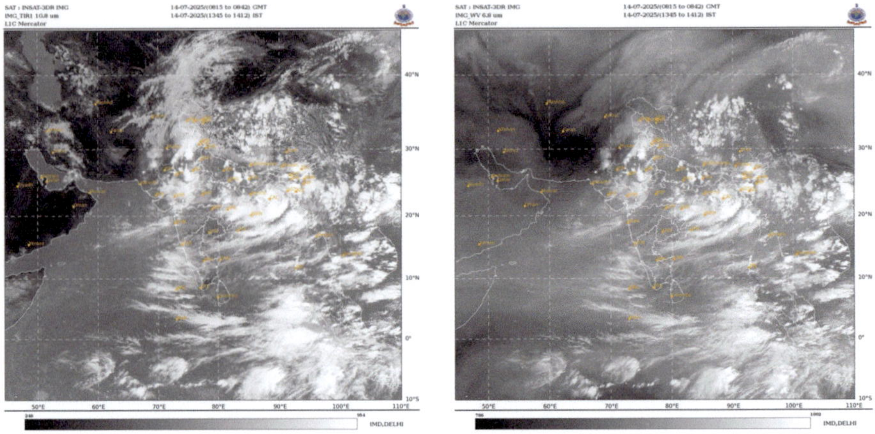

Fig. 15.6 Satellite image in TIR 2 (left) and WV (right) channel over NE sector on 14th Aug 2014 at 10:30 h IST reveals cloud growth

15.4.4.5 Cloud Motion Vector

The cloud motion vector (CMV) is available for both the lower and upper level atmospheres on various sources such as the IMD, MOSDAC, CIMSS (USA), and others as shown in Fig. 15.7. The CMV provides speed, direction of cloud and wind at different pressure level. This information is used to determine if a specific cloud system will pass over the study area or disappear within a given time frame. When real-time observation is not possible, this diagnosis is used to forecast synoptic weather scenarios at night.

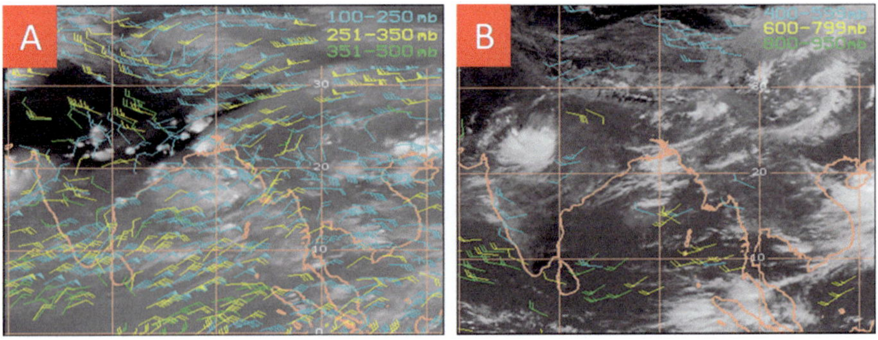

Fig. 15.7 CMV for upper atmosphere (**A**) and lower atmosphere (**B**)

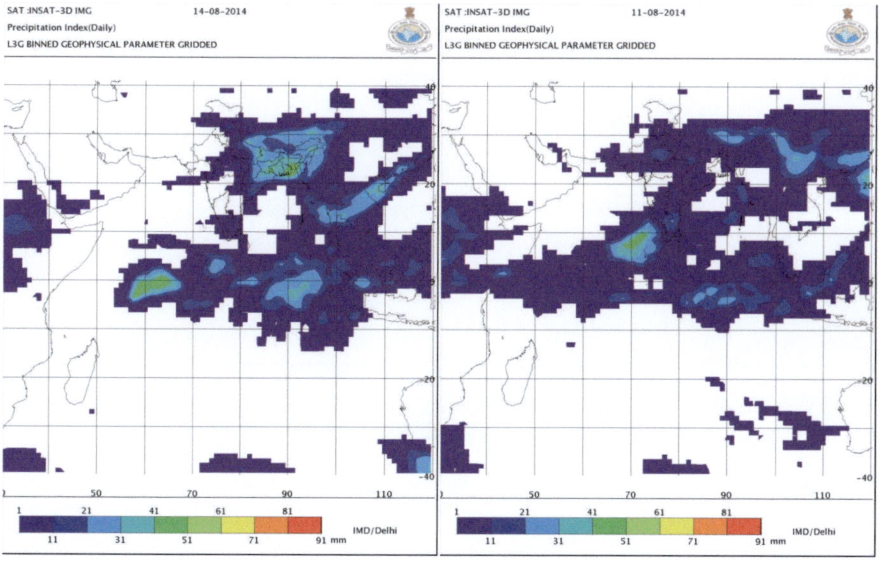

Fig. 15.8 QPE product shows approx. low rain (left) and moderate rain (right) over NE sector during last 24 h ending at 08:30 h IST

15.4.4.6 Quantitative Precipitation Estimate (QPE)

QPE is an essential indicator that is obtained from satellite imagery and is accessible on the IMD website on a three-hourly, daily, monthly, and seasonal basis. It provides a quantitative assessment of the precipitation's spatial distribution. The daily QPE obtained from the INSAT 3D imager is utilized for FLEWS to determine the daily cumulative precipitation ending at 08:30 h (IST). A typical example of the product is shown in Fig. 15.8.

15.4.4.7 Nowcasting Using RADAR Observations (Doppler Weather Radar DWR)

Nowcast systems employ RADAR to generate quantitative precipitation forecasts, which may be very beneficial for short-term forecasting and flood warning in metropolitan areas associated with intense storm occurrences. RADAR uses electromagnetic waves in microwave region to detect location (range and direction), height (altitude), intensity (in case of weather systems) and movement of moving and non-moving targets. On the other hand; DWR works in the Doppler principle where the ranges, direction and intensity of an object can be accessed with the changing frequency of receiving signal from the target with respect to its relative motions. On the IMD website, there are six primary products: Plan Position Indicator PPI(Z), PPI(V), Max(Z), Volume Velocity Processing (VVP2), Surface Rainfall Intensity

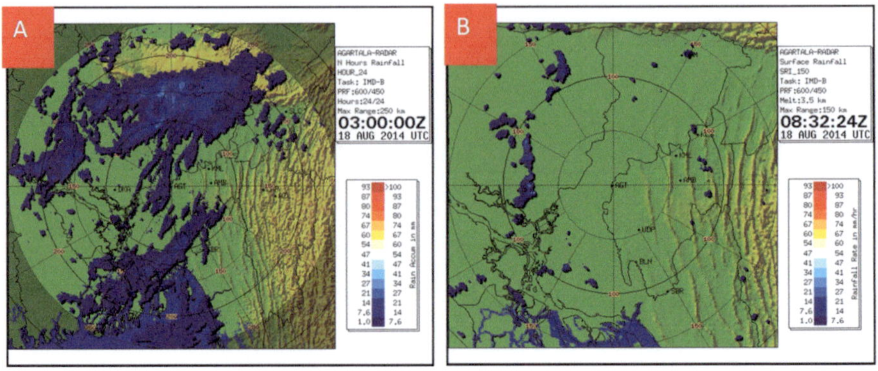

Fig. 15.9 **A** DWR precipitation accumulation and **B** surface rainfall intensity product. *Source* https://mausam.imd.gov.in/agartala/index_radar.php?id=Agartala

(SRI), and Precipitation Accumulation (PAC). For FLEWS dissemination the PAC (Precipitation Accumulation) available daily at 08:30 IST that gives accumulated rain for last 24 h and Surface Rainfall Intensity (SRI) at every 10 min is mostly used as shown in the Fig. 15.9.

15.4.5 Extraction of Rainfall Products for Hydrologic Models

The overall framework for rainfall extraction is shown in the Fig. 15.10. AWS Rainfall data hosted on cloud source can be accessed via an FTP server such as MobsXterm or WinSCP using provided credentials. Once connected, the compressed data files were downloaded usually in ZIP format. After downloading, the files are unzipped and extracted to NetCDF (.nc) files containing the rainfall data. These NetCDF files were then be processed and analyzed in MATLAB using the 'ncread' function into an Excel file format using the 'writetable' or 'xlswrite' function, depending on the data format. In python, the files were then converted into a DSSVue format which is primarily used for hydrologic and hydraulic data visualization and analysis (often in conjunction with models like HEC-RAS or HEC-HMS).

15.5 The Hydrology, Hydraulics and Hydro-dynamic Component

Two key elements make up the development of the Flood Early Warning System: the Weather Research Forecast (WRF), a numerical rainfall prediction model backed by a qualitative synoptic weather monitoring exercise, and the HEC-HMS, a distributed hydrological model based on Geographic Information Systems. In contrast, the first

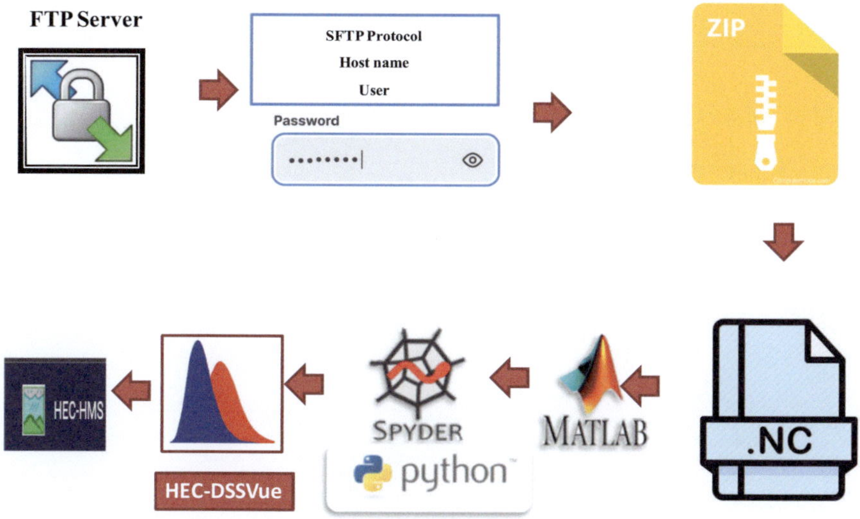

Fig. 15.10 Rainfall extraction and conversion for FLEWS

component forecasts the rainfall values for the next 24 h at hourly intervals with a grid resolution of 9 km. Predicted rainfall values are fed into a spatially dispersed mesh of interconnected watershed runoff models in the second component. The simulated discharge value of the models for each basin was critically analyzed and compared with the observed peak discharge and hydrograph at different outlet location. The hydrologic model inputs were then feed as an input into the hydraulic models i.e. HEC-RAS to simulate the river flow and propagation. The below section describes the individual components in detail.

15.5.1 Hydrologic Component

The main purposes of hydrologic models are to predict system behavior and understand various hydrological processes. A wide range of physical parameters, such as rainfall, air temperature, soil characteristics, terrain, vegetation, and hydrogeology, are used as inputs in a number of models that have recently been developed worldwide (Devi et al. 2015). Depending on the degree of physical principles used and their governing equations as a function of space and time, these hydrological models can be broadly categorized as either empirical or conceptual, determinist or stochastic, lump or semi-distributed, and kinematic or dynamic models (Clarke 1973; Jajarmizadeh et al. 2012). The US Army Corps of Engineers' hydrologic modelling program, the Hydrologic Engineering Centre-Hydrologic Modelling System (HEC-HMS), was used for the FLEWS. The HMS model has been used in a number of studies, showing that it can be used to simulate the rainfall-runoff processes of dendritic watershed

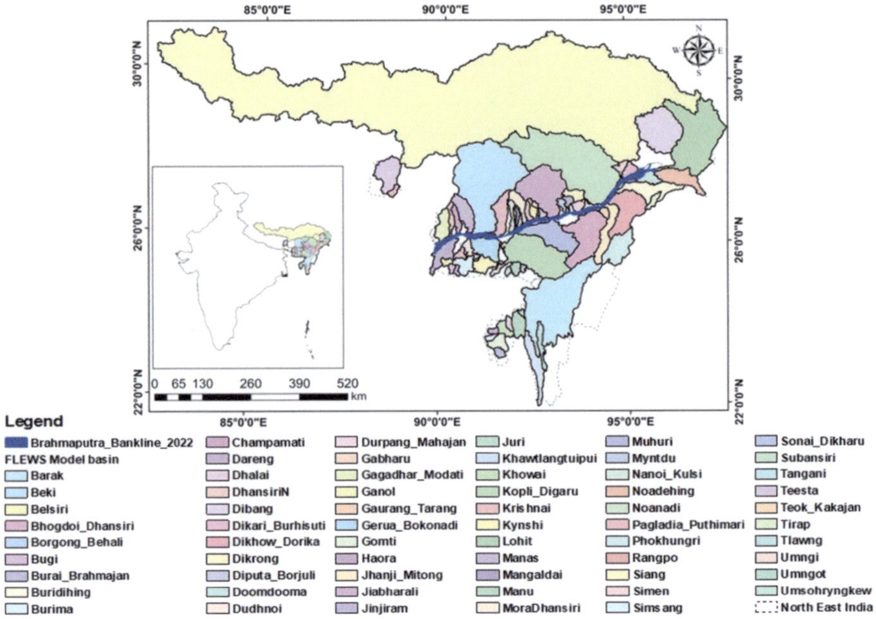

Fig. 15.11 FLEWS watershed

systems in event-based rainfall-runoff simulations and peak flood discharge estimations (Natarajan and Radhakrishnan 2019); evaluate the effects of rainfall products and satellite-based rainfall products on streamflow (Nhi et al. 2019a, b); and assess the hydrologic impacts of climate change (Meenu et al. 2013). The Fig. 15.11 depicts the watershed that has been incorporated into the FLEWS system.

15.5.1.1 HEC-HMS Model

The HEC-HMS (https://www.hec.usace.army.mil/software/hec-hms) is a semi-distributed hydrological model for precipitation-runoff-routing simulation of a dendritic watershed system developed by the United States Army Corps of Engineers (USACE) (Feldman 2000). In order to replicate catchment hydrological processes, HEC-HMS combines hydrological components such as a loss component, base flow component, transforming component, and channel routing component. The HEC-HMS model setup consists of input data (time series data), a basin model, a meteorological model, and a control specification (HEC 2006). The sub-basin, reach, junction, reservoir, diversion, source, and sink are the physical components of a basin. When working with watersheds in cold climates, the meteorological model takes into account precipitation, evapotranspiration, shortwave radiation, and melting (HEC 2006). Control requirements are one of the most important components of a HEC-HMS project, despite the fact that they don't contain a lot of parameter data.

This is because they specify the time interval that is used in the simulation as well as when it starts and stops (HEC 2006). ArcGIS and HEC-GeoHMS were used to generate the basin model and all of its hydrologic components, which were then put into the HEC-HMS model as a backdrop map file. The Table 15.3 shows the types of parameter set used for development of HEC-HMS used for FLEWS.

Loss Method

Typically, the loss models in HEC-HMS determine the runoff volume by deducting the amount of precipitation from the volume of water that is transpired, evaporated, intercepted, infiltrated, and stored (Tassew et al. 2019). For the study; the Soil Conservation Service (SCS) curve number (CN) loss model that uses cumulative rainfall, soil cover, land use, and antecedent moisture to determine precipitation excess was employed. Soils and CNs are closely related where each soil evaluated is grouped into one of four categories (A, B, C, or D), where A is the most porous, deepest, and least prone to runoff, while D is the shallowest, finest-textured, and most prone to runoff (Hawkins et al. 2008). The values of the dimensionless runoff index CN vary from 1 to 100. As per USDA (1986), increased CN values correspond to increased runoff. Potential retention S is the maximum depth of storm rainfall that could potentially be abstracted by a given site and initial abstraction I_a refers to interception, infiltration, and surface storage before runoff occurs. The SCS method uses the following formula to calculate the direct runoff depth:

$$Q = \frac{(P - I_a)^2}{(P - I_a) + S} \tag{15.1}$$

where

Q = direct surface runoff from an area or catchment [L];

P = accumulated rainfall depth at time t [L];

I_a = the initial abstraction [dimensionless];

S = the maximum potential retention after runoff begins [L].

Table 15.3 Parameter set selected for HEC HMS model

Sl. No.	Model	Parameter	Method
1	Loss	Curve number (CN), initial abstraction (Ia), imperviousness	SCS curve number
2	Transform	Lag time (min)	SCS unit hydrograph
3	Routing	Channel length (m), slope (m/m), manning's n, channel width (m)	Muskingum-Cunge

The Soil Conservation Service (SCS 1985) used a standard value of 0.2 to approach the value in Eq. (15.1). Substituting the given value in Eq. (15.1), the expression of direct surface runoff is given as (Ponce and Hawkins 1996):

$$Q = \frac{(P - 0.2S)^2}{P + 0.8S} \tag{15.2}$$

Subject to $P \geq 0.2S$ and $P_e = 0$ otherwise.

S (inch) is associated with the soil and cover conditions of the watershed via the CN as

$$S = \frac{1000}{CN} - 10. \tag{15.3}$$

Transform Method

Surface runoff is calculated using a transform approach within the sub-basin. In this study, the Soil Conservation Service *Unit Hydrograph* method that averages the rainfall from the watersheds was chosen as a transform method to calculate the actual surface runoff. The SCS unit hydrograph method makes use of a dimensionless, curvilinear unit hydrograph to route excess precipitation to the sub-basin outlet (Cronshey 1986). This transformation approach is based on the utilization of the time of concentration (T_c) and the lag time (minutes). Time of concentration refers to the time it takes for runoff to travel from the farthest point of a watershed to a specific place of interest (outlet). While time of concentration is the time between the centroid of rainfall excess and the inflection point of the recession of direct runoff; the time lag is the time from the centroid of rainfall excess to peak flow of a direct runoff hydrograph (Sultan et al. 2022). The relationship between the time lag and time of concentration as presented by Mockus (1957) is calculated as

$$L = 0.6T_c \tag{15.4}$$

where,

L = Time lag in minutes

T_c = Time of concentration in minutes.

The SCS method for estimating time lag is as follows:

$$L = \frac{l^{0.8}(S + 1)^{0.7}}{1900Y^{0.5}} \tag{15.5}$$

where,

L = time lag, h

l = flow length (the longest path along which water flows from the watershed divide to the outlet), ft

Y = the average watershed land slope, %

S = the maximum potential retention after runoff begins, inch.

Channel Routing Method

Channel storage effects attenuate flood runoff as it flows through the channel reach. This attenuation is accounted for by the HEC-HMS routing models. In this study the Muskingum-Cunge method of flood routing that combines the continuity equation with a simplified version of the momentum equation is used. This method is also referred to as a "variable coefficient" method, because the routing parameters are recalculated every time step based on channel features and flow depth. The use of this method in HEC-HMS requires the initial condition, the reach length [ft or m], the friction slope [ft/ft or m/m], Manning's n roughness coefficient, a space–time interval method and value, an index method and value, and a cross-section shape and parameters/dimensions (HEC 2022).

According to the Muskingum-Cunge method the outflow at the downstream end is calculated as follows (Bedient and Huber 1988):

$$Q_{j+1}^{n+1} = C_1 Q_j^n + C_2 Q_j^{n+1} + C_3 Q_{j+1}^n + C_4 \tag{15.6}$$

where

$$C_1 = \frac{Kx + \Delta t/2}{D} \tag{15.7}$$

$$C_2 = \frac{\Delta t/2 - Kx}{D} \tag{15.8}$$

$$C_3 = \frac{K(1-x) - \Delta t/2}{D} \tag{15.9}$$

$$C_4 = \frac{q\Delta t\Delta x}{D} \tag{15.10}$$

$$D = K(1-x) + \frac{\Delta t}{2} \tag{15.11}$$

Numerical stability requires 0, x, 1 > 2 (Cunge 1969). The Muskingum-Cunge method is advantage over Muskingum method as that it can be applied to assist in parameter selection (K, x, Δ, Δx) based on stream characteristics.

15.5.2 Hydraulic Component

After peak flood discharges and flow hydrographs have been derived from hydrologic models, the next step is the hydraulic analysis, which simulates river flow and flood water propagation in the floodplain or urban catchment. For the precise spatial distribution of the flow and velocity dynamics, flood hydraulic modelling is typically done using a 1-D or 2-D or coupling 1D–2D flood routing method that is based on different modified versions and solutions of St. Venant's hydrodynamic equations. For FLEWS the HEC-RAS hydraulic model was used.

15.5.2.1 HEC-RAS Model

The HEC RAS (Hydrologic Engineering Center—River Analysis System) model has the ability to execute one dimensional steady flow hydraulic calculation; one-dimensional unsteady flow hydraulic calculations and two-dimensional unsteady flow hydraulic calculations; as well as a combined 1D and 2D unsteady flow routing (HEC 2021). The HEC RAS model can solve either 2D Diffusion Wave equations or the 2D Shallow Water equations, with the option to incorporate momentum for turbulence, wind forces, mud and debris flows, and Coriolis effects (HEC 2021). For flood simulation, the HEC-RAS flood model employs the 2D Saint–Venant and 2D diffusive wave equations through an implicit finite-volume solution (HEC 2021). We chose the 2D diffusion wave solver to make use of more stable numerical solutions and shorten the computation time for the current situation. Due to the fact that 2D flow area can begin entirely dry and can withstand a sudden influx of water, the wetting and drying of 2D cells is particularly robust (HEC 2021). HEC-RAS have wide range boundary conditions that are made up of internal boundary conditions, global boundary conditions (meteorological data), and external boundary conditions along the 2D area's perimeter (i.e., precipitation, wind, etc.). The boundary of 2D Flow Areas can be directly connected to four different kinds of external boundary conditions viz. flow hydrograph, stage hydrograph, normal depth and rating curve (HEC 2021). The resulted HEC-RAS inundation map can be viewed and extracted from RAS-Mapper. The methodological framework for hydraulic flood forecast modeling is shown in the Fig. 15.12.

1D Governing Equation in HEC RAS

The principles of conservation of mass and momentum are the physical laws that control the movement of water in a stream. Both of these rules are theoretically described as partial differential equations, which are typically referred to individually as momentum equations and continuity equations and collectively as Saint–Venant equations. The Eq. (15.12) is the based on principle of conservation of mass and the Eq. (15.13) is based on the principle of conservation of momentum.

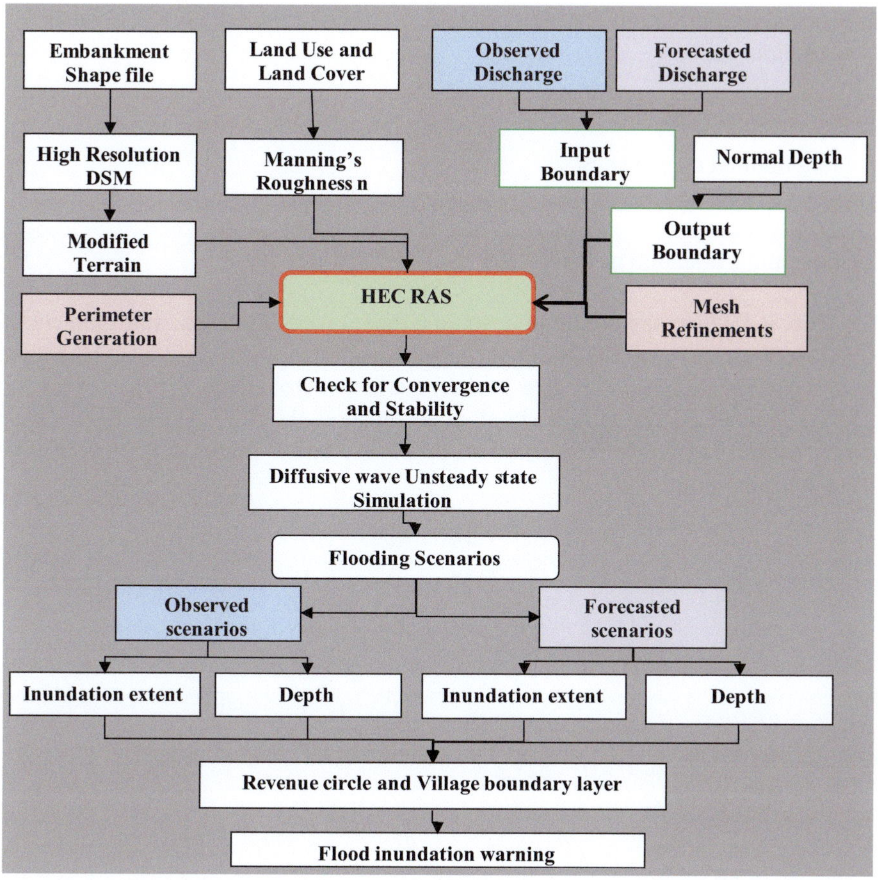

Fig. 15.12 Methodological framework for hydraulic flood forecast modeling

$$\frac{\partial A}{\partial t} + \frac{\partial Q}{\partial x} = 0 \tag{15.12}$$

$$\frac{\partial Q}{\partial t} + \frac{\partial \frac{Q^2}{A}}{\partial x} + gA\frac{\partial H}{\partial x} + gA\left(S_o - S_f\right) = 0 \tag{15.13}$$

where,

A = cross-sectional area normal to the flow;

q = lateral inflow/outflow of the channel;

Q = discharge;

g = acceleration due to gravity;

H = water surface elevation, also called stage;

S_o = bed slope;
S_f = friction slope;

t = temporal coordinates;

x = longitudinal coordinates.

2D Governing Equation in HEC RAS

The HEC RAS have the option to develop a model using either the full momentum equation, often known as the 2D Saint Venant equation, or the simplified 2D diffusive wave approximation, as indicated in the equations below (Eqs. (15.3), (15.4) and (15.5)) (Brunner 2016; Quirogaa et al. 2016; Patel et al. 2017). The diffusive wave approximation is used to reduce model instability and shortens computation time (Betsholtz and Nordlöf 2017).

$$\frac{\partial \zeta}{\partial t} + \frac{\partial p}{\partial x} + \frac{\partial q}{\partial x} = 0 \tag{15.14}$$

$$\frac{\partial p}{\partial t} + \frac{\partial}{\partial x}\left(\frac{p^2}{h}\right) + \frac{\partial}{\partial y}\left(\frac{pq}{h}\right) = -\frac{n^2 pq\sqrt{p^2+q^2}}{h^2} - gh\frac{\partial \xi}{\partial x} + pf + \frac{\partial}{\rho \partial x}(h\tau_{xx}) + \frac{\partial}{\rho \partial y}(h\tau_{xy}) \tag{15.15}$$

$$\frac{\partial q}{\partial t} + \frac{\partial}{\partial y}\left(\frac{q^2}{h}\right) + \frac{\partial}{\partial y}\left(\frac{pq}{h}\right) = -\frac{n^2 qg\sqrt{p^2+q^2}}{h^2} - gh\frac{\partial \xi}{\partial y} + qf + \frac{\partial}{\rho \partial y}(h\tau_{yy}) + \frac{\partial}{\rho \partial y}(h\tau_{xy}) \tag{15.16}$$

where,

h = water depth (m);

p and q = specific flow in the x and y direction ($m^2\ s^{-1}$);

ξ = surface elevation (m);

g = acceleration due to gravity ($m^2\ s^{-2}$);

n = Manning's resistance;

ρ = density of water ($kg\ m^{-3}$);

τ_{xx}; τ_{yy} and τ_{xy} = the components of the effective shear stress;

f = the Coriolis (s^{-1}).

15.6 The DSS for Flood Early Warning Alert Generation

A flood early warning alert generation Decision Support System (DSS) is a platform or tool that helps anticipate and lessen the effects of floods by giving fast, accurate, and useful information. To send out alerts and warnings for possible flooding events, this system combines a number of data sources, analytical models, and decision-making procedures. Key components of a DSS for flood early warning alert generation are shown in the Fig. 15.13.

Validation of rainfall forecast: Flood early warning Decision Support Systems (DSS) rely on the 3-hourly rainfall estimates produced by the model that forecasts' high-resolution, time-sensitive rainfall projections. Using measurements from the ground, the rainfall forecasts are verified for accuracy and dependability during peak and off-peak season. This validation is a crucial step in enhancing the DSS since it guarantees the reliability of the forecast data used in the system and aids in gradually increasing the accuracy of the hydrologic model.

Flood forecasting hydrologic models: The validated 3-hourly rainfall estimate from the model is fed into the hydrological model, which used a distributed approach to drive runoff simulations and predict flow and flood potential in the watershed. The hydrological models combine the physical features of the watershed and river system, such as slope, drainage patterns, land use, and soil type. This enables the model to

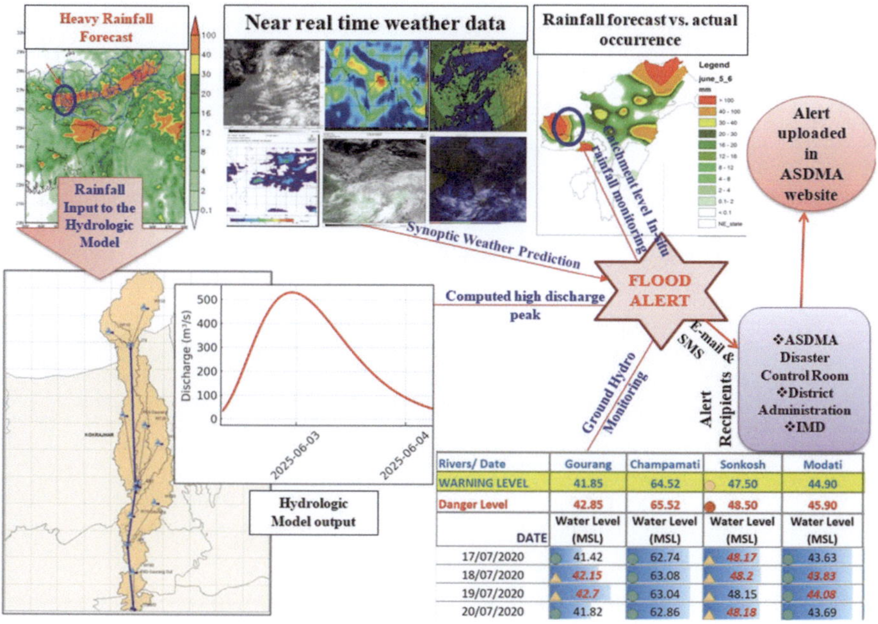

Fig. 15.13 Key components of a DSS for flood early warning alert generation

consider both the hydro-meteorological input and the physical characteristics of the watershed.

Validation of the forecast hydrograph: By comparing the model-based discharge projections with the observed river gauge readings in real time, the accuracy of the model is evaluated and any necessary short-term forecast adjustments is made to ensure that the expected flood risk matched the actual situation. Validation is typically done through statistical metrics like the Nash–Sutcliffe Efficiency (NSE) and RMSE, and discrepancies are addressed comparing the model obtained hydrograph peak discharge with the historical flood peak threshold as shown in the Fig. 15.14. Flooding is possible if the water level is above the warning level or if the model simulates a discharge that is higher than the specified threshold.

Synoptic Weather data: Near real-time synoptic weather analysis plays a crucial role in enhancing the confidence of the DSS, as it provides up-to-date meteorological data (e.g., precipitation, temperature, wind patterns, atmospheric pressure) that directly influences flood forecasting.

In-situ catchment level rainfall: Rainfall monitoring interpolated from IMD or MOSDAC at the in-situ catchment level yields vital local data that improves the precision and promptness of FLEWS. It enables the early detection of isolated storms that may be missed by more comprehensive satellite or meteorological data.

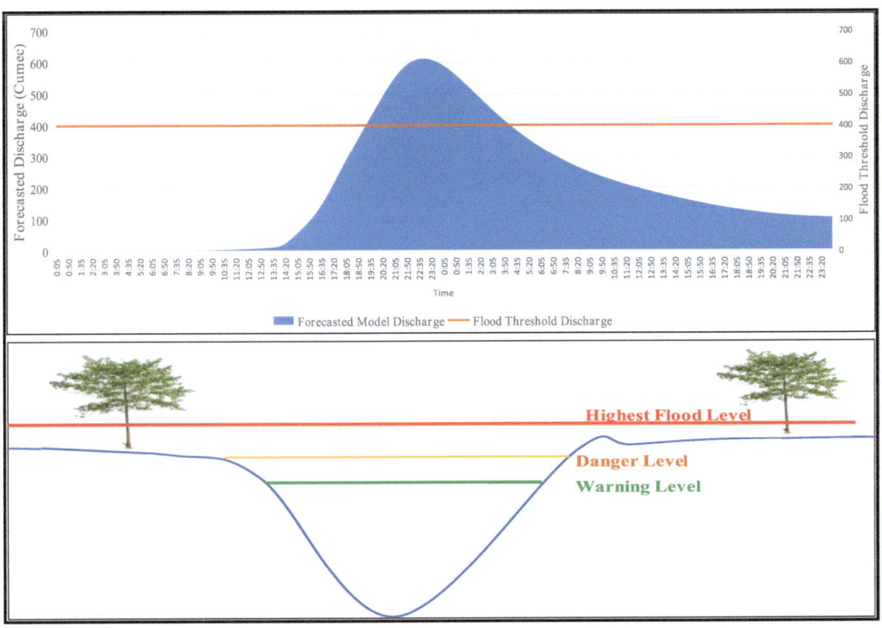

Fig. 15.14 Validation with flood threshold

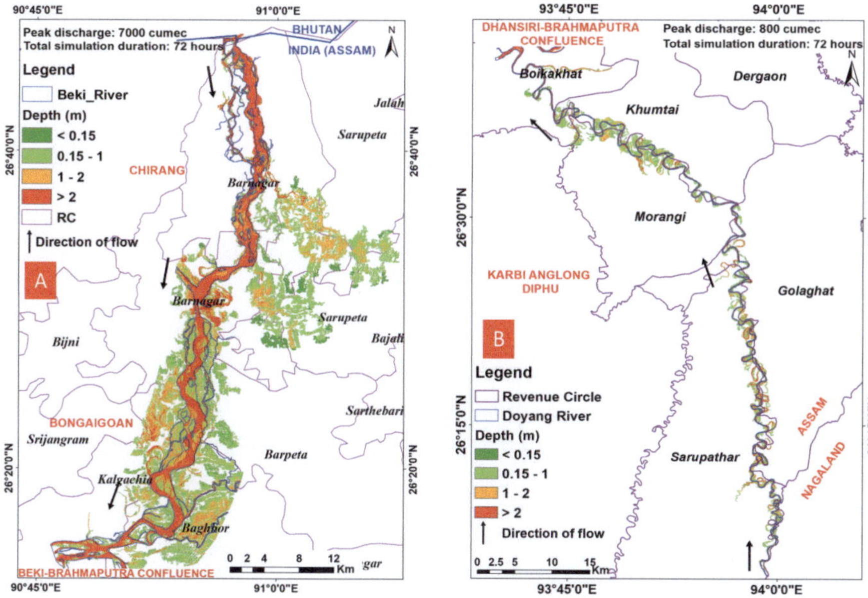

Fig. 15.15 Hydraulic simulation for **[A]** Beki River and **[B]** Dhansiri River

Hydraulic simulation for the flood-concern River: Using a hydraulic model that takes boundary conditions, river geometry, and channel roughness into consideration the flood simulation for the flood-concerned Rivers is examined to check the floodplain impacts, water surface elevations, and flow conditions as shown in the Fig. 15.15. Accurate flood forecasts are ensured by calibration with past data.

Flood alert generation: FLEWS teams issue flood alerts when river level or rainfall intensity thresholds are exceeded; these alerts are distributed to stakeholders such as State Disaster authority, Deputy Commissioners of concern districts and Nodal agencies via pre-established channels of communication, guaranteeing prompt warnings for impacted areas. Ongoing updates are integrated into the system to enhance forecasts and provide extra notifications as conditions change.

Alert posted on AASDMA websites: A flood alert issued by the FLEWS team is immediately posted on the AASDMA website. Information about the impacted areas, the expected intensity of the flood, and safety precautions are all included in the advisory. To guarantee that the public and pertinent authorities receive accurate and timely information for action, the website is updated in real time.

15.7 FLEWS Alert Accuracy and Success

Reliable data, predictive modeling, real-time monitoring, and efficient communication must all be integrated for FLEWS to be accurate and successful in order to guarantee prompt and appropriate action prior to flooding events. The system's effectiveness is enhanced by longer lead times and rapid responses, though challenges like the dynamic nature of climate change, data shortages etc. makes forecasting of extreme weather events problematic, particularly in remote places.

15.7.1 Precision and Accuracy Assessment

FLEWS is most effective during peak monsoon seasons, when heavy rainfall and flooding patterns are more predictable, allowing for accurate forecasts and prompt alerts. However, during the lean season, when rainfall is less frequent or more erratic, these systems can struggle due to less reliable met forecast and the difficulty in predicting sporadic or localized flooding events due to local moisture condition.

FLEWS typically have the greatest success at the District or Revenue circle level due to data availability and communication systems. At the municipal or village level it is challenging to forecast localized flooding in small, isolated areas due to the coarse spatial resolution of input data, such as DEM, rainfall and river flow measurements. The accuracy of flood forecasts at the village level is also hampered by the absence of comprehensive data for microclimates and regional topography differences. Poor infrastructure, limited data collection, and monitoring gaps further hinder accurate flood forecasts. Communication barriers, such as inadequate cell coverage, lack of local readiness, public awareness, and resource crunch further impede the success of FLEWS.

15.7.2 Success Rate and User Feedback

Following the alert's transmission to the relevant authority, the success rate is assessed. ***Absolute success***: When a flood occurs within the warning's specified validity duration, the event is deemed an absolute success. ***Partial success***: An event is deemed partially successful if, within the warning's validity, there is no flooding but increase in the water level above a certain threshold. ***Failure***: If a flood warning is provided and then deemed inaccurate or delayed, it is considered a failure.

The average success score of this project during 2012–2024 period has been around 92% (absolute and partial combined) with an achieved alert lead time ranging between 12 and 36 h, and occasionally going up to 48 h. The FLEWS performance for the state of Assam is shown in the Fig. 15.16 and the number of FLEWS alert sent to the other states of North East India is shown in Fig. 15.17.

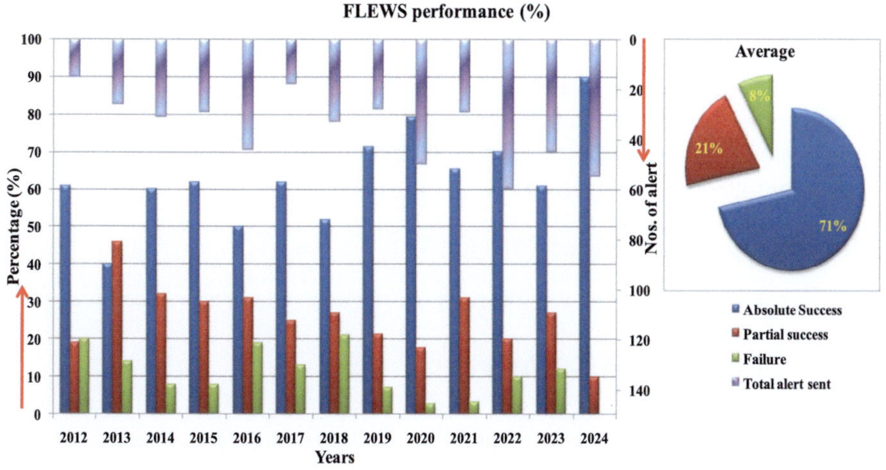

Fig. 15.16 FLEWS performance from 2012 to 2024 for the state of Assam

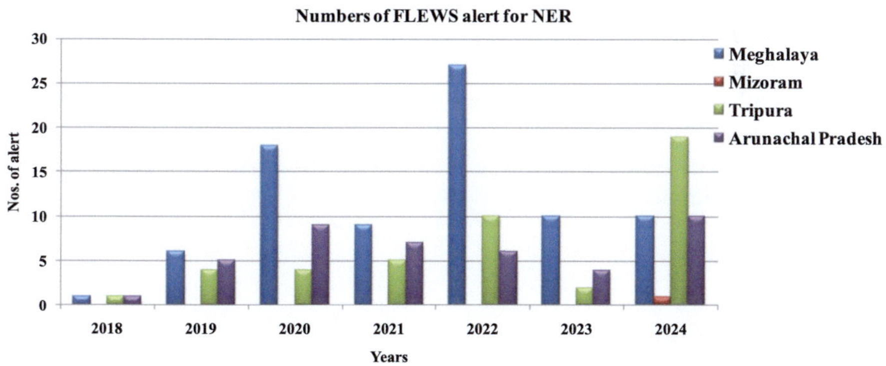

Fig. 15.17 Numbers of FLEWS alert sent to other states of NE

15.7.2.1 User Feedback

The FLEWS activity was taken up at the backdrop of the 14th June, 2008 flood event in Lakhimpur district of Assam due a catastrophically high discharge that came down from the Ranganadi hydro-electric power project of NEEPCO. At the request of NEC and Govt. of Assam, its scope was extended to the whole of Lakhimpur district as a pilot study in 2009 and subsequently to fifteen other districts of Assam by 2013 in phase manner. FLEWS is currently in 5th operational phase, covering the entire state of Assam. After the successful operationalization of FLEWS for Assam, NESAC has taken up the expansion of FLEWS services to the flood prone districts of North Eastern States at the request and funding from the North Eastern Council (NEC) under plan of action (POA) programme.

15.7.3 Challenges in FLEWS

The lack of a sufficient number of hydro-met observatories, as well as the difficulty in maintaining these hydro-met observatories at most times, poses significant challenges. Obtaining the discharge/stage data, high resolution DEM is also a challenging task. Flood mitigation structures such as embankments are not available in digital format. For most rivers, the designated DL, WL, and HFL are not updated regularly with fluvial changes in bathymetry.

15.7.4 Awards and Recognitions

- Winner of the "E-north east award" for E-Governance and Citizen services delivery, 2013.
- Declaration as "Best governance practice and funding for professional documentation" by the Department of Administrative Reforms under Ministry of Public Grievances and Pensions, Govt. of India in the year 2012.
- Shortlisted as a finalist for the prestigious Prime Minister's Award for Innovations in Public administration (2012).
- Good number of Parliament questions answered on FLEWS for implementation in rest of the country.
- Formal Letter of Appreciation from the CEO of ASDMA (2024) requesting for the 5th Phase implementation.
- Nominated from ASDMA to SKOCH Summit, 2018.

15.8 Conclusion

FLEWS represents a vital advancement in Flood disaster risk reduction, leveraging cutting-edge hydro-meteorological data, real-time monitoring, and predictive modeling to significantly enhance flood forecasting accuracy and response time. By integrating diverse data sources, including river stage, rainfall, and topographical information, FLEWS can provide timely alerts that enable communities, authorities, and decision-makers to take proactive measures to mitigate flood risks. The system's ability to forecast flood events, even in remote or high-altitude regions, and its capacity to adapt to evolving climate patterns, marks a transformative step forward in flood management. Despite challenges such as infrastructure limitations and data gaps, the continuous refinement of FLEWS, through technological advancements and improved resource allocation, holds immense potential to save lives, protect property, and safeguard ecosystems, making it a cornerstone of modern flood resilience strategies.

Acknowledgements The authors of this book chapter are thankful to all the past and present scientists and researchers who have contributed directly and indirectly to this GIS based Flood Early Warning system in various stages of its development during the last more than a decade period. Special acknowledgement is due for the meteorology team of NESAC being an integral part of the overall activity. Team would like to place in record, its sincere appreciation for both administrative and financial support from Assam State Disaster Management Authority (ASDMA) and North Eastern Council (NEC) in implementation of this activity in NER of India. Lastly, the authors are grateful to all the other authors from whom references have been drawn while preparing the manuscript of this book chapter.

References

G. Al-Rawas, M.R. Nikoo, M. Al-Wardy, T. Etri, A critical review of emerging technologies for flash flood prediction: examining artificial intelligence, machine learning, internet of things, cloud computing, and robotics techniques. Water **16**(14), 2069 (2024). https://doi.org/10.3390/w16142069

H. Ashouri, K.L. Hsu, S. Sorooshian, D.K. Braithwaite, K.R. Knapp, L.D. Cecil et al., PERSIANN-CDR: daily precipitation climate data record from multisatellite observations for hydrological and climate studies. Bull. Am. Meteor. Soc. **96**(1), 69–83 (2015). https://doi.org/10.1175/BAMS-D-13-00068.1

J. Barman, B.S. Marak, K.S. Rao, B. Biswas, GIS-based frequency ratio model for flood susceptibility zonation in the state of Meghalaya, Northeast India. Proc. Indian Natl. Sci. Acad. 1–14 (2024). https://doi.org/10.1007/s43538-024-00290-1

P.D. Bates, M.S. Horritt, C.N. Smith, D. Mason, Integrating remote sensing observations of flood hydrology and hydraulic modelling. Hydrol. Process. **11**(14), 1777–1795 (1997). https://doi.org/10.1002/(SICI)1099-1085(199711)11:14%3c3C1777::AID-HYP543%3e3E3.0.CO;2-E

M.A. Baudoin, S. Henly-Shepard, N. Fernando, A. Sitati, Z. Zommers, From top-down to "community-centric" approaches to early warning systems: exploring pathways to improve disaster risk reduction through community participation. Int. J. Disaster Risk Sci. **7**, 163–174 (2016). https://doi.org/10.1007/s13753-016-0085-6

P.B. Bedient, W.C. Huber, *Hydrology and Floodplain Analysis* (Addison-Wesley, Reading, Massachusetts, USA, 1988)

A. Betsholtz, B. Nordlöf, Potentials and limitations of 1D, 2D and coupled 1D-2D flood modelling in HEC-RAS. TVVR17/5003 (2017)

K. Beven, Changing ideas in hydrology—the case of physically-based models. J. Hydrol. **105**(1–2), 157–172 (1989). https://doi.org/10.1016/0022-1694(89)90101-7

A. Bhadra, A. Bandyopadhyay, S. Hodam, C.Y. Yimchungru, R. Debbarma, Assessment of vulnerability of Arunachal Pradesh (India) to floods, in *Congress Proceedings of International Water Resource Association* (2015), pp. 1–16

A. Bhadra, S. Choudhury, D. Kar, Flood hazard mapping in Dikrong basin of Arunachal Pradesh (India). World Acad. Sci., Eng. Technol. **60**, 1614–1619 (2011)

S.L. Bora, K. Bhuyan, P.J. Hazarika, J. Gogoi, K. Goswami, Analysis of rainfall trend using non-parametric methods and innovative trend analysis during 1901–2020 in seven states of North East India. Curr. Sci. **122**(7), 801–811 (2022). https://doi.org/10.18520/cs/v122/i7/801-811

J.A. Brotzge, D. Berchoff, D.L. Carlis, F.H. Carr, R.H. Carr, J.J. Gerth et al., Challenges and opportunities in numerical weather prediction. Bull. Am. Meteor. Soc. **104**(3), E698–E705 (2023). https://doi.org/10.1175/BAMS-D-22-0172.1

G.W. Brunner, *HEC-RAS River Analysis System: Hydraulic Reference Manual, Version 5.0* (US Army Corps of Engineers—Hydrologic Engineering Center, 2016), p. 547

S.C. Chan, E.J. Kendon, H.J. Fowler, S. Blenkinsop, C.A. Ferro, D.B. Stephenson, Does increasing the spatial resolution of a regional climate model improve the simulated daily precipitation? Clim. Dyn. **41**, 1475–1495 (2013). https://doi.org/10.1007/s00382-012-1568-9

Y. Chen, H. Zhou, H. Zhang, G. Du, J. Zhou, Urban flood risk warning under rapid urbanization. Environ. Res. **139**, 3–10 (2015). https://doi.org/10.1016/j.envres.2015.02.028

R.T. Clarke, A review of some mathematical models used in hydrology, with observations on their calibration and use. J. Hydrol. **19**(1), 1–20 (1973)

R. Cronshey, *Urban Hydrology for Small Watersheds*, No. 55 (US Department of Agriculture, Soil Conservation Service, Engineering Division, 1986)

J.A. Cunge, On the subject of a flood propagation computation method (Musklngum method). J. Hydraul. Res. **7**(2), 205–230 (1969). https://doi.org/10.1080/00221686909500264

A. Das, P.K. Ghosh, B.U. Choudhury, D.P. Patel, G.C. Munda, S.V. Ngachan, P. Chowdhury, Climate change in North East India: recent facts and events–worry for agricultural management, in *Proceedings of the Workshop on Impact of Climate Change on Agriculture*, vol. 2009 (2009, December), pp. 32–37

A.P.J. de Roo, B. Gouweleeuw, J. Thielen, J. Bartholmes, P. Bongioannini-Cerlini, E. Todini et al., Development of a European flood forecasting system. Int. J. River Basin Manag. **1**(1), 49–59 (2003). https://doi.org/10.1080/15715124.2003.9635192

G.K. Devi, B.P. Ganasri, G.S. Dwarakish, A review on hydrological models. Aquat. Procedia **4**, 1001–1007 (2015)

R. Devrani, P. Srivastava, R. Kumar, P. Kasana, Characterization and assessment of flood inundated areas of lower Brahmaputra river basin using multitemporal synthetic aperture radar data: a case study from NE India. Geol. J. **57**(2), 622–646 (2022)

O.N. Dhar, S. Nandargi, A study of floods in the Brahmaputra basin in India. Int. J. Climatol. **20**(7), 771–781 (2000). https://doi.org/10.1002/1097-0088(20000615)20:73.0.CO;2-Z

A.D. Feldman, *Hydrologic Modeling System HEC-HMS: Technical Reference Manual* (US Army Corps of Engineers, Hydrologic Engineering Center, 2000)

R.A. Freeze, R.L. Harlan, Blueprint for a physically-based digitally-simulated hydrologic response model. J. Hydrol. **9**(3), 237–258 (1969). https://doi.org/10.1016/0022-1694(69)90020-1

B. Gogoi, R. Khargharia, M. Uzir, J.J.K. Basumatary, S. Sarma, K. Pathak et al., *Agricultural Sustainability and Resource Management in the Silt-Deposited Floodplains of Assam in North-East Region of India* (2023)

S.L. Goodbred Jr., S.A. Kuehl, M.S. Steckler, M.H. Sarker, Controls on facies distribution and stratigraphic preservation in the Ganges-Brahmaputra delta sequence. Sed. Geol. **155**(3–4), 301–316 (2003)

D.C. Goswami, Brahmaputra River, Assam, India: physiography, basin denudation, and channel aggradation. Water Resour. Res. **21**(7), 959–978 (1985)

Y. Guo, Y. Zhang, L. Zhang, Z. Wang, Regionalization of hydrological modeling for predicting streamflow in ungauged catchments: a comprehensive review. Wiley Interdiscip. Rev. Water **8**(1), e1487 (2021). https://doi.org/10.1002/wat2.1487

A. Gupta, Flood and floodplain management in North East India: an ecological perspective, in *Proceedings of the 1st International Conference on Hydrology and Water Resources in Asia Pacific Region*, vol. 1 (2003, January), pp. 231–236

R.H. Hawkins, T.J. Ward, D.E. Woodward, J.A. Van Mullem (eds.), *Curve Number Hydrology: State of the Practice* (American Society of Civil Engineers, 2008, November)

HEC, *Hydrological Modelling (HEC-HMS) User's Manual* (US Army Corps of Engineers, 2006)

HEC, *HEC-RAS User's Manual* (U.S. Army Corps of Engineers, Davis CA, 2021, April)

HEC, U.S. Army Corps of Engineers, Hydrologic Engineering Center. 2012. *HEC-HMS Hydrologic Modeling System, User's Manual, Version 4.0, CPD-74A* (Hydrologic Engineering Center, Davis, CA, 2022)

P. Horton, B. Schaefli, M. Kauzlaric, Why do we have so many different hydrological models? A review based on the case of Switzerland. Wiley Interdiscip. Rev. Water **9**(1), e1574 (2022). https://doi.org/10.1002/wat2.1574

G.J. Huffman, D.T. Bolvin, E.J. Nelkin, D.B. Wolff, R.F. Adler, G. Gu et al., The TRMM multi-satellite precipitation analysis (TMPA): quasi-global, multiyear, combined-sensor precipitation estimates at fine scales. J. Hydrometeorol. **8**(1), 38–55 (2007). https://doi.org/10.1175/JHM560.1

M. Jajarmizadeh, S. Harun, M. Salarpour, A review on theoretical consideration and types of models in hydrology. J. Environ. Sci. Technol. **5**(5), 249–261 (2012)

A.H. Jennings, World's greatest observed point rainfall. Mon. Weather Rev. **78**, 4–5 (1950). https://doi.org/10.1175/1520-0493(1950)078%3c3C0004:WGOPR%3e3E2.0.CO;2

R.J. Joyce, J.E. Janowiak, P.A. Arkin, P. Xie, CMORPH: a method that produces global precipitation estimates from passive microwave and infrared data at high spatial and temporal resolution. J. Hydrometeorol. **5**(3), 487–503 (2004). https://doi.org/10.1175/1525-7541(2004)005%3c3C0487:CAMTPG%3e3E2.0.CO;2

V.S. Kale, Geomorphic effects of monsoon floods on Indian rivers, in *Flood Problem and Management in South Asia* (Springer, Dordrecht, 2003), pp. 65–84

R.R. Kelkar, *Satellite Meteorology in India—An Overview* (1994). http://nopr.niscpr.res.in/handle/123456789/35877

R.K. Khanna, C.K. Agrawal, P. Kumar, C.E. Director, *Remote Sensing and GIS Applications in Flood Management* (Central Water Commission, New Delhi, India, 2018)

C. Kidd, Satellite rainfall climatology: a review. Int. J. Climatol.: J. R. Meteorol. Soc. **21**(9), 1041–1066 (2001). https://doi.org/10.1002/joc.635

C. Kidd, V. Levizzani, P. Bauer, A review of satellite meteorology and climatology at the start of the twenty-first century. Prog. Phys. Geogr. **33**(4), 474–489 (2009).https://doi.org/10.1177/0309133309346647

G. Kumar, *Geology of Arunachal Pradesh* (GSI, 2013)

S.K. Lahiri, R. Sinha, Tectonic controls on the morphodynamics of the Brahmaputra River system in the upper Assam valley, India. Geomorphology **169**, 74–85 (2012)

F. Lobligeois, V. Andréassian, C. Perrin, P. Tabary, C. Loumagne, When does higher spatial resolution rainfall information improve streamflow simulation? An evaluation using 3620 flood events. Hydrol. Earth Syst. Sci. **18**(2), 575–594 (2014). https://doi.org/10.5194/hess-18-575-2014

R. Meenu, S. Rehana, P.P. Mujumdar, Assessment of hydrologic impacts of climate change in Tunga-Bhadra river basin, India with HEC-HMS and SDSM. Hydrol. Process. **27**(11), 1572–1589 (2013)

P. Meischner, C. Collier, A. Illingworth, J. Joss, W. Randeu, Advanced weather radar systems in Europe: the COST 75 action. Bull. Am. Meteor. Soc. **78**(7), 1411–1430 (1997). https://doi.org/10.1175/1520-0477(1997)078%3c3C1411:AWRSIE%3e3E2.0.CO;2

V. Mockus, *Use of Storm and Watershed Characteristics in Synthetic Hydrograph Analysis and Application* (US Department of Agriculture, Soil Conservation Service, 1957)

P.K. Mohapatra, R.D. Singh, Flood management in India, in *Flood Problem and Management in South Asia* (Springer, Dordrecht, 2003), pp. 131–143

A. Mosavi, P. Ozturk, K.W. Chau, Flood prediction using machine learning models: literature review. Water **10**(11), 1536 (2018). https://doi.org/10.3390/w10111536

F. Murata, T. Hayashi, J. Matsumoto, H. Asada, Rainfall on the Meghalaya plateau in Northeastern India—one of the rainiest places in the world. Nat. Hazards **42**, 391–399 (2007). https://doi.org/10.1007/s11069-006-9084-z

Y.K. Murthy, *Water Resource Potentials of the Himalaya. Himalaya, Aspects of Change* (1981)

N.P., Nagendra, G. Narayanamurthy, R. Moser, Management of humanitarian relief operations using satellite big data analytics: the case of Kerala floods. Ann. Oper. Res. **319**(1), 885–910 (2022). https://doi.org/10.1007/s10479-020-03593-w

A. Nandy, A. Kumar, Application of geographic information system and remote sensing on flood hazard assessment and its impact on agriculture: a review, in *Flood Risk Management, Springer Natural Hazards*, ed. by B. Biswas, B.B. Ghute (Springer, Singapore, 2024). https://doi.org/10.1007/978-981-97-2688-2_13

S. Natarajan, N. Radhakrishnan, Simulation of extreme event-based rainfall–runoff process of an urban catchment area using HEC-HMS. Model. Earth Syst. Environ. **5**, 1867–1881 (2019)
A. Nath, S. Ghosh, The influence of urbanization on the morphology of the Barak River floodplain in Cachar District, Assam. Water Policy **24**(12), 1876–1894 (2022)
P.T.T. Nhi, D.N. Khoi, N.X. Hoan, Evaluation of five gridded rainfall datasets in simulating streamflow in the upper Dong Nai river basin, Vietnam. Int. J. Digit. Earth **12**(3), 311–327 (2019a). https://doi.org/10.1080/17538947.2018.1426647
P.T.T. Nhi, D.N. Khoi, N.X. Hoan, Evaluation of five gridded rainfall datasets in simulating streamflow in the upper Dong Nai river basin, Vietnam. Int. J. Digit. Earth **12**(3), 311–327 (2019b)
D.P. Patel, J.A. Ramirez, P.K. Srivastava, M. Bray, D. Han, Assessment of flood inundation mapping of Surat city by coupled 1D/2D hydrodynamic modeling: a case application of the new HEC-RAS 5. Nat. Hazards **89**(1), 93–130 (2017)
D. Perera, O. Seidou, J. Agnihotri, H. Mehmood, M. Rasmy, Challenges and technical advances in flood early warning systems (FEWSs), in *Flood Impact Mitigation and Resilience Enhancement* (2020)
P.J. Pilon, *Guidelines for Reducing Flood Losses* (United Nations International Strategy for Disaster Reduction (UNISDR), 2002)
V.M. Ponce, R.H. Hawkins, Runoff curve number: has it reached maturity? J. Hydrol. Eng. **1**(1), 11–19 (1996)
P. Prokop, A. Walanus, Variation in the orographic extreme rain events over the Meghalaya Hills in northeast India in the two halves of the twentieth century. Theor. Appl. Climatol. **121**, 389–399 (2015). https://doi.org/10.1007/s00704-014-1224-x
V.M. Quirogaa, S. Kurea, K. Udoa, A. Manoa, Application of 2D numerical simulation for the analysis of the February 2014 Bolivian Amazonia flood: application of the new HEC-RAS version 5. Ribagua **3**(1), 25–33 (2016)
I.A. Rana, S.S., Bhatti, A. Jamshed, Effectiveness of flood early warning system from the perspective of experts and three affected communities in urban areas of Pakistan. Environ. Hazards **20**(3), 209–228 (2021). https://doi.org/10.1080/17477891.2020.1751031
D.W. Reed, *A Review of British Flood Forecasting Practice* (1984). https://nora.nerc.ac.uk/id/eprint/5955/1/IH_090.pdf
L.F. Richardson, *Weather Prediction by Numerical Process* (University Press, 1922)
D. Rogers, V. Tsirkunov, *Cost and Benefits of Early Warning Systems* (United Nations International Strategy for Disaster Reduction UNISDR; The World Bank, 2010) (Contributing Paper to the Global Assessment Report on Disaster Risk Reduction). https://documents1.worldbank.org/curated/en/609951468330279598/pdf/693580ESW0P1230aster0Risk0Reduction.pdf
J.N. Sarma, S. Acharjee, A study on variation in channel width and braiding intensity of the Brahmaputra River in Assam, India. Geosciences **8**(9), 343 (2018)
SCS, *National Engineering Handbook, Section 4, Hydrology* (Soil Conservation Service (SCS), US Department of Agriculture, Washington, DC, 1985)
R. Shrestha, Y. Tachikawa, K. Takara, Input data resolution analysis for distributed hydrological modeling. J. Hydrol. **319**(1–4), 36–50 (2006). https://doi.org/10.1016/j.jhydrol.2005.04.025
I. Shustikova, A. Domeneghetti, J.C. Neal, P. Bates, A. Castellarin, Comparing 2D capabilities of HEC-RAS and LISFLOOD-FP on complex topography. Hydrol. Sci. J. **64**(14), 1769–1782 (2019). https://doi.org/10.1080/02626667.2019.1671982
B.S. Sil, K. Ashwini, W. Annayat, J. Debnath, M. Farooq, G. Meraj, Observing spatiotemporal inconsistency of erosion and accretion in the Barak river using remote sensing and GIS techniques. Conservation **3**(1), 14–31 (2022)
R. Sinha, S.K. Tandon, Indus-Ganga-Brahmaputra plains: the alluvial landscape, in *Landscapes and landforms of India* (2014), 53–63
W.L. Smith, W.P. Bishop, V.F. Dvorak, C.M. Hayden, J.H. McElroy, F.R. Mosher et al., The meteorological satellite: overview of 25 years of operation. Science **231**(4737), 455–462 (1986). https://doi.org/10.1126/science.231.4737.455

S.K. Sood, R. Sandhu, K. Singla, V. Chang, IoT, big data and HPC based smart flood management framework. Sustain. Comput.: Inform. Syst. **20**, 102–117 (2018). https://doi.org/10.1016/j.suscom.2017.12.001

State Disaster Management Authority, Govt of Arunachal Pradesh. https://sdma-arunachal.in/history-of-vulnerability-of-the-state-to-the-disasters-of-different-types. Access 28 Jan 2025

D. Sultan, A. Tsunekawa, M. Tsubo, N. Haregeweyn, E. Adgo, D.T. Meshesha et al., Evaluation of lag time and time of concentration estimation methods in small tropical watersheds in Ethiopia. J. Hydrol.: Reg. Stud. **40**, 101025 (2022). https://doi.org/10.1016/j.ejrh.2022.101025

Q. Sun, C. Miao, Q. Duan, H. Ashouri, S. Sorooshian, K.L. Hsu, A review of global precipitation data sets: data sources, estimation, and intercomparisons. Rev. Geophys. **56**(1), 79–107 (2018). https://doi.org/10.1002/2017RG000574

B.G. Tassew, M.A. Belete, K. Miegel, Application of HEC-HMS model for flow simulation in the Lake Tana basin: the case of Gilgel Abay catchment, upper Blue Nile basin, Ethiopia. Hydrology **6**(1), 21 (2019)

B. Thies, J. Bendix, Satellite based remote sensing of weather and climate: recent achievements and future perspectives. Meteorol. Appl. **18**(3), 262–295 (2011). https://doi.org/10.1002/met.288

USDA, *Soil Conservation Service National Engineering Handbook* (USA, 1986)

H. Varikoden, J.V. Revadekar, On the extreme rainfall events during the southwest monsoon season in northeast regions of the Indian subcontinent. Meteorol. Appl. **27**(1), e1822 (2020). https://doi.org/10.1002/met.1822

N. Wanders, D. Karssenberg, A. De Roo, S.M. De Jong, M.F.P. Bierkens, The suitability of remotely sensed soil moisture for improving operational flood forecasting. Hydrol. Earth Syst. Sci. **18**(6), 2343–2357 (2014). https://doi.org/10.5194/hess-18-2343-2014

P.C. Young, Advances in real-time flood forecasting. Philos. Trans. R. Soc. London Ser. A: Math., Phys. Eng. Sci. **360**(1796), 1433–1450 (2002). https://doi.org/10.1098/rsta.2002.1008

Y. Zahan, R. Mahanta, P.V. Rajesh, B.N. Goswami, Impact of climate change on North-East India (NEI) summer monsoon rainfall. Clim. Chang. **164**, 1–19 (2021). https://doi.org/10.1007/s10584-021-02994-5

Z. Zommers, "Follow the Spiders": ecosystems as early warnings, in *Reducing Disaster: Early Warning Systems for Climate Change* (2014), pp. 339–353. https://doi.org/10.1007/978-94-017-8598-3_18

Diganta Barman a senior scientist with over 22 years of experience, heads the Water Resources Division at NESAC. He holds a Ph.D. in Civil Engineering (Water Resources and Hydrology) from IIT and specializes in hydro-modelling, watershed management, flood forecasting, and reservoir operations. He led the development of India's first operational Hydro-met Flood Early Warning System (FLEWS) for the Brahmaputra and Barak basins, a system he has steered since 2009. Dr. Barman has also led major projects like the River Atlas (Assam and NER), SWaRA, NEEPCO's Hydro-met monitoring, and flood hazard zonation for Assam. He currently mentors the HIU Team under the World Bank-funded AIRBMP project. A member of various expert committees on water and flood management in NER, he has authored 20+ peer-reviewed papers and supervised multiple Ph.D. and postgraduate theses. He collaborates with national and international agencies including CWC, IMD, IRADe, Columbia Water Center, and ICIMOD.

Shanbor Kurbah is currently serving as a Scientist in the Water Resources Division at the North Eastern Space Applications Centre (NESAC), Department of Space, Umiam, Meghalaya. He holds an M.Tech. in Hydrology (1st Class with Distinction) from the Indian Institute of Technology (IIT) Roorkee, Uttarakhand. With over eight years of experience in the field of hydrology and water resources, his core expertise lies in hydrological data analysis, flood forecasting and modeling, hydrodynamic and snowmelt runoff modeling, Glacial Lake Outburst Flood (GLOF) simulation, and assessment of downstream releases from reservoirs. He is well-versed in the application of integrated spatial technologies, GIS, and advanced modeling software for effective water resources management. He has played a key role in the implementation of several user-funded projects and research initiatives across North East India, delivering actionable insights for regional water governance. He has also guided and mentored numerous M.Tech. dissertations, and has authored/co-authored publications in peer-reviewed national and international journals and conferences.

Manbhalang Dakermi Shylla is a Research Scientist at NESAC's Water Resources Division, working on the Flood Early Warning Systems-Assam project. He holds an M.Tech. from IIT Kharagpur and a B.Tech. from Central Agricultural University. He specializes in hydrologic and hydraulic modeling, remote sensing, and GIS. He helped automate FLEWS models by integrating hydro-meteorological data and programming. Over four years, he contributed to projects including FLEWS-NER, Assam's Integrated River Basin Management flood forecasting, flood impact studies for hydropower projects (Ranganadi, Doyang, Kurichhu), and reservoir surveys.

B. M. Arjun is currently working as a Scientist at the National Remote Sensing Centre (NRSC), ISRO, Hyderabad, focusing on strategic applications in the water resources domain. Prior to joining NRSC, he served as a Scientist at NESAC, Shillong, for seven years, where he contributed significantly to key water resources projects. He holds a Ph.D. from IIT Kanpur and an M.Tech. from NIT Surathkal, specializing in Water Resources Engineering. His areas of expertise include hydrologic and hydrodynamic modelling, flood inundation modeling, river system analysis, hydro-meteorological modelling for reservoir optimization, and hydrography. He was a core team member of the first ever GIS based operational Flood Early Warning System (FLEWS) established in India for the Brahmaputra and Barak river systems in North East India.

Ranjit Das holds a Master of Engineering in Watershed Management and Flood Control from Assam Engineering College, Guwahati. With over 20 years of experience at the North Eastern Space Applications Centre (NESAC), he specializes in applying geospatial technology to water resources, watersheds, and flood disaster management. He received the ISRO Team Excellence Award in 2009 for his work on assessing irrigation potential in AIBP-funded projects using Cartosat data. Currently, he leads several major regional and national projects at NESAC and has published research in leading national and international journals and conferences.

Chapter 16
Integrated Spatio-Hydrological Modeling for Hydro-Power Management in North East India

Shanbor Kurbah, Diganta Barman, B. M. Arjun, and Ranjit Das

Abstract North East India, a part of Eastern Himalaya, despite holding an estimated 40% of India's total hydro-power potential (62,000 MW), struggles with electricity shortages and complex water management. The region faces significant climatic shifts, including decreasing summer monsoon rainfall, rising temperatures, and Himalayan glacier retreat, which affect river flows and hydro-power generation. Anthropogenic activities like deforestation and road construction exacerbate issues such as increased peak discharge and reservoir siltation, further complicating management. This chapter advocates for a holistic, integrated spatio-hydrological modeling approach to optimize hydro-power management. A case study at the Panyor Lower Hydro-Electric Project (HEP) demonstrates this by combining in-situ measurements from Automatic Weather Stations (AWS) and Doppler Water Level Recorders (DWLR) with advanced geospatial data from satellites (e.g., CartoDEM, KOMPSAT-3, GPM IMERG, MODIS, SMAP). The HEC-HMS hydrological model was used to simulate rainfall-runoff processes, rigorously calibrated and validated. This integration establishes a decision support system that provides crucial insights for optimal dam gate operations, mitigating sudden flood releases, and enhancing reservoir management, including water balance assessment periodically for the upstream reservoir catchment. This approach is vital for addressing data scarcity, improving flood mitigation, and ensuring sustainable hydro-power generation in the Eastern Himalayan region.

S. Kurbah (✉) · D. Barman · R. Das
North Eastern Space Applications Centre (NESAC), Umiam, Shillong, Meghalaya, India
e-mail: s.kurbah@nesac.gov.in

D. Barman
e-mail: diganta.barman@nesac.gov.in

R. Das
e-mail: ranjit.das@nesac.gov.in

B. M. Arjun
National Remote Sensing Centre (NRSC), Balanagar, Hyderabad, Telangana, India
e-mail: arjun_bm@nrsc.gov.in

S. P. Aggarwal et al. (eds.), *Remote Sensing of the North Eastern Himalayan Ecosystem*,
https://doi.org/10.1007/978-981-95-3221-6_16

Keywords Hydro-power potential · Spatio-temporal data · Hydrological modeling · Holistic approach · Water resource management

16.1 Introduction

North East India, a part of Eastern Himalaya is remarkably rich and gifted with a variety of natural resources. This region, often faces issues with electricity shortages in both generation and distribution, resulting in inadequate power supply in certain areas and hindering economic development (Handique and Dutta 2014). North East India, despite accounting for only 8% of the country's total land area, is exceptionally rich in surface water resources, holding an estimated 652.3 billion cubic meters, which constitutes approximately 34% of India's entire water capital (Anonymous 2010). NE India has the country's highest water availability per person and per hectare. For instance, the Sabarmati basin in western India has 240 m^3 of water available per person, but the Brahmaputra River basin has 17,000 m^3 (Amarasinghe et al. 2005; Manish and Rao 2018). The Brahmaputra-Barak river systems, which dominate this landmass, have the potential to generate a significant amount of hydro-power, benefiting not just this part of the country but the entire nation (Pranab 2013). North eastern states has been endowed with enormous hydro-power potential estimated to be of the order of 62,000 MW, which is about 40% of the hydro-power potential of the country, i.e. 145,320 MW from large hydro projects above 25 MW capacity (Jaideep et al. 2022). The state wise hydro-power potential and development status in North east India is shown in Table 16.1.

Hydro-power development and management comes with a lot of challenges. Reports indicate that summer monsoon rainfall in this region has been significantly

Table 16.1 Hydro-electric capacity status in NE India (those with capacity greater than 25 MW as on 2023) *source* Anonymous 2023

State	Expected capacity (MW)	Capacity developed (MW)	Capacity in development (MW)	Balance capacity	
				(MW)	(%)
Arunachal Pradesh	50,064	1115	4880	44,069	88.00
Assam	650	350	120	180	27.67
Manipur	1761	105	0	1,656	94.00
Meghalaya	2298	322	0	1,976	85.98
Mizoram	2131	60	0	2,071	97.18
Nagaland	1452	75	0	1,377	94.83
Tripura	0	0	0	0	0
Sikkim	4248	2282	1037	929	21.86
Total	62,604	4309	6037	52,258	83.47

falling at a rate of about 11 mm every ten years throughout the past century (Das 2004; Mirza et al. 1998). Furthermore, the average annual temperature is increasing by 0.04 °C per decade (Das 2004; Pant and Kumar 1997). This trend may reflect the regional effects of global warming and climate change. Over the past decade, 67% of the glaciers in the Himalayan region have retreated, e.g. a receding rate of 28 m/year was observed in the Gangotri glacier (Das 2009). With the warming climate, these glaciers are expected to melt at a faster rate, resulting in higher summer flows and flooding in the short term. However, over the coming decades, as the glaciers that feed the rivers continue to shrink and eventually vanish, there will be a gradual decrease in river flow (Das 2009). Additionally, in glacier-contributing rivers, glacial melt flow helps to increase seasonal discharge during the lean period. As the contribution from glaciers decreases over time, future lean/dry season flows will decrease, leading to greater water stress in the Brahmaputra basin and negatively impacting hydro-power generation and management. The retreat of Himalayan glaciers may lead to substantial alterations in the hydrological regime of the Brahmaputra river system, as the main river and several of its tributaries, such as the Subansiri and Jia-Bharali are partially contributed by snowmelt runoff (Das 2009). However, estimating and understanding the complex and diverse nature of the numerous components of the hydrological and meteorological systems within the hydro project catchment area is not ideally possible. Therefore, a holistic approach with the available limited spatial information from space technology and observed field measurements can be extrapolated through hydrological modeling to provide optimum solution for efficient hydro-power management.

According to the IPCC's Sixth Assessment Report (AR6) (IPCC 2023), if global temperatures rise by 1.5 °C, an increase in heavy rainfall and flooding events will be expected. These events are expected to increase in both intensity and frequency across various regions of the world, including Asia, Africa, Europe, and North America. The report is very confident about this happening in Africa and Asia, somewhat confident about it in North America, and moderately confident about it in Europe. This means that as the planet warms, areas that are already prone to heavy rain and floods might face even more extreme weather conditions, leading to greater risks of damage and disruption.

16.2 Hydrological Processes in North East India

NE India is distinguished by its complex topography, diverse land use and land cover, and largely monsoon-driven climate, all of which combine to influence runoff generating processes (Bhattacharya et al. 2020; Saha et al. 2017; Singh et al. 2018). This region is dominated with one of the largest river system in India, the Brahmaputra river, followed by the Barak river system, rivers like Haora, Gumti, Muhuri, Khowai, Dhalai, Manu, Juri and Burima in Tripura, the Imphal/manipur river system which drains the Manipur valley, the Kolodyne river system draining the southern part of Mizoram, and the Tizu river system in Nagaland. The powerful Brahmaputra

Table 16.2 Water resources potential in NE India (Adapted from Jain et al. 2007)

	Mean yearly surface water potential (km^3)	Ground water potential (km^3)	Total potential (km^3)
Brahmaputra Basin	537.20	27.90	565.10
Barak Basin	48.40	1.80	50.20
Manipur, Mizoram and Tripura basins	31.00	1.50	32.50
Total	616.60	31.20	647.80

River begins in the vast glacier region north of the Kailash Range, at an elevation of 5,300 m, situated south of Konggyu Tsho lake (Jain et al. 2007). The Brahmaputra river extends a basin area of 580,000 km^2 up to its confluence within Bangladesh, of which 194,413 km^2 (33.5%) is within India. The Brahmaputra River has an average annual discharge of 19,820 m^3/s, transports approximately 735 million metric tonnes of sediment load per year, and exhibits a specific flood discharge of 0.149 m^3/s per km^2 (Jain et al. 2007). The overall basin wise water resources potential in NER is shown in Table 16.2.

The hydrological processes in this region are governed by the intricate interaction of topography, rainfall, temperature, evapotranspiration, and surface runoff, all of which are critical to comprehending water availability and distribution.

16.2.1 Topography of North East India

NE India, with a total geographical area of 262,230 km^2, comprises approximately 8% of the total land area of the country. Additionally, it shares 90% of its boundary with international borders, including those of Bhutan, China, Bangladesh, and Myanmar (Deka and Sit 2023). Arunachal Pradesh covers the largest area of 83,743 km^2 in the region and consists of mountainous ranges sloping to the plains of Assam with an elevation ranges of 100 m to 7000 m approximately. Due to its high altitude nature of topography, the state of Arunachal Pradesh is home to some of the important glaciers, such as Kangto, Mazgol, etc., glacier lakes and all year snow covered area. Sikkim spans an area of 7,098 km^2, with elevations ranging from 300 m to 8,500 m, and is known for its prominent glaciers like Zemu, Rathong, and Lhonak. Meghalaya, covering 22,429 km^2, shares borders with Bangladesh to the southwest and Assam to the north and northeast, with elevation levels varying from 60 m to 1,961 m. Mizoram, covers a total area of 21,087 km^2 sharing the international boundaries along Bangladesh in the West and Myanmar in the East and South, with an average altitude ranges from 500 to 800 m, where Phawngpui is recorded as the highest peak (2157 m). Manipur, in the other hand, possess the Imphal valley in the middle while being surrounded by hills on the state boundary, which covers the total area of 22,327 km^2 with an elevation range from 790 to 2020 m. Nagaland, is another state which is hilly in nature and covers the area of 16,579 km^2 with an altitude range from 200 m in the plains to 3840 m in the hills. Tripura, the third smallest state of India, contains a geographical area of 10,491 km^2, with the highest elevation of 975.36 m above sea level at Bethliangchhip. Assam spans a geographical area of 78,438 km^2 and is broadly divided into three major physiographic regions: the Brahmaputra Valley in the north, the central hilly terrains of Karbi-Anglong and North Cachar, and the Barak Valley in the south. Assam has the highest elevation of 1713 m near Haflong. These topographic variations contribute to both intense monsoonal flooding and groundwater recharge dynamics. Studies have shown that the complex terrain of this region dictates spatial differences in water flow and seasonal variations in water availability (Chakravarty and Saikia 2015; Mahanta and Sharma 2017). Understanding these interactions is crucial for effective water resource management and flood mitigation strategies (Bhuyan and Das 2018; Sharma and Vaidya 2020).

16.2.2 Climate and Rainfall Distribution in North East India

The region is defined by substantial rainfall, moderate to high temperatures, and delightful seasonal change. Its climate is primarily influenced by the southwest monsoon, geographical factors, and elevation, resulting in significant variations in rainfall distribution (Kumar and Borah 2019; Sarma and Singh 2023). Northeast India has three climate zones: tropical, subtropical, and temperate. The tropical wet climate dominates the plains of Assam, Tripura, and Manipur, where temperatures frequently exceed 35 °C during summer and moderate winters (Saha and Srivastava 2020). On the other hand, the mountainous parts of Meghalaya and Arunachal Pradesh have a subtropical highland climate, with colder temperatures and copious rainfall due to orographic lifting. Higher elevations, such as Sikkim and areas of northern Arunachal Pradesh, have a temperate climate with cooler temperatures and snowfall in the winter. (Ramesh and Mukherjee 2021).

The southwest monsoon, which arrives in early June and retreats by September, is the predominant source of rainfall in the region. Meghalaya, notably in locations like Mawsynram and Sohra (Cherrapunjee), receives some of the highest rainfall in the world, with annual precipitation reaching over 11,000 mm due to the orographic effect (Sharma and De 2018). The monsoon winds are forced upward by the Khasi Hills, causing moisture to condense and fall as rain. In contrast, the lower rainfall areas, like Assam's Brahmaputra valley, receive comparatively low intense but still substantial rainfall, typically ranging between 2000 to 3000 mm annually (Das and Roy 2019). After the monsoon season, north east India enters the post-monsoon period from October to November, characterized by reduced rainfall and moderate temperatures. Winter, from December to February, experiences a dry spell across the region, although some higher altitudes receive snow due to western disturbances, particularly in Sikkim and Arunachal Pradesh.

Although this region has an abundance of water resources that offer great potential for hydro-power, the climate and rainfall patterns of the region call for sophisticated management strategies to handle seasonal variations, flood hazards, and siltation issues in order to ensure effective and dependable power generation (Choudhury et al. 2015; Sharma and Deka 2018; Roy et al. 2021). These issues are aggravated by irregular monsoon seasons, which generate both surpluses and deficits in water flow, affecting the efficiency of hydroelectric plants (Bhattacharya et al. 2017). Furthermore, siltation in reservoirs, a significant problem in the region, exacerbates operating challenges (Mishra and Pandey 2019). The IPCC sixth assessment report, working group I (2021), noted that in Southeast Asia, the patterns of mean precipitation are not consistent or uniform across different datasets and seasons. While much of the region has seen an increase in rainfall intensity, the number of wet days has decreased.

16.2.3 River Systems and Hydrological Characteristics in North East India

The region is blessed with intricate and complicated river networks in the country. Notably, the Brahmaputra river and Barak river, are two of the major rivers in the region. The Brahmaputra river originates from the Kailash ranges of Himalayas. The Brahmaputra River, known as the "Tsangpo" in Tibet, begins its course on the vast Tibetan Plateau, flowing roughly 1,625 km before reaching the Indo-China border. In India, the river traverses another 916 km across the country before draining into Bangladesh. In Arunachal Pradesh, it is called the Siang River in its upper reaches and the Dihang River as it flows downstream. From the Indo-China border to the town of Pasighat, the river flows southward for about 226 km. At Kobo, the Dihang River leaves the hilly terrain and enters the plains of Assam, where it merges with two major Himalayan tributaries; the Dibang from the northeast and the Lohit from the east. From this confluence onward, the river is known as the Brahmaputra, flowing westward across Assam. Near Dhubri, it takes an abrupt shift southward into Bangladesh, eventually emptying into the Bay of Bengal. The gradient of the Brahmaputra River is steepest in the gorge section upstream of Pasighat, with a slope ranging from 4.3 to 16.8 m/km. However, near Guwahati, the gradient becomes much flatter, at just 0.1 m/km (Anonymous 2012). The longitudinal profile of the Brahmaputra River adapted from the Water and Power Consultancy Services (India) Limited, 1993 (WAPCOS) is shown in Fig. 16.1. The Brahmaputra River, together with its vast network of tributaries, plays a crucial role in shaping the geomorphic features of the region, with a particularly strong influence on the Brahmaputra Valley. In the North bank, the major tributaries are river Subansiri, Jiadhal, Ranganadi, Jiabharali, Dhansiri (North), Puthimari, Pagladiya, Beki, Manas-Aei, Saralbhanga, Sonkosh, and Champamati. Whereas, river Buridehing, Dikhow, Kopili, Dhansiri, Krishnai, Disang, Bhogdoi, Jinari forms the major tributaries in the South Bank of the Brahmaputra. The important tributaries from both banks of the Brahmaputra River are listed in the Table 16.3.

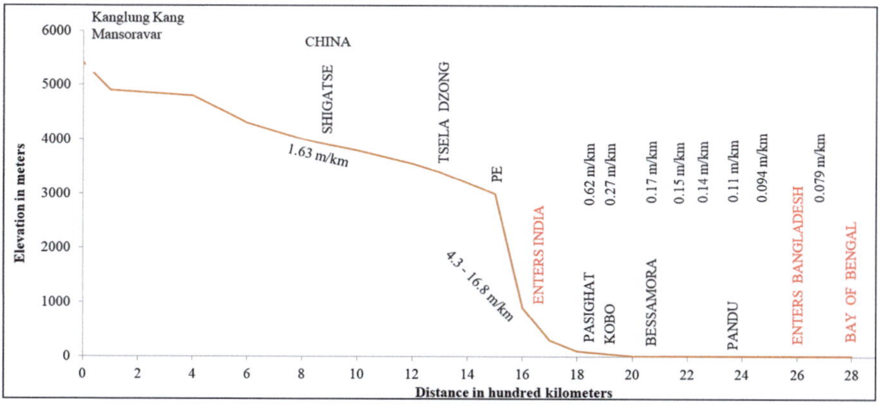

Fig. 16.1 Adapted from WAPCOS, 1993, L-profile of the Brahmaputra *Source* Anonymous (2012)

The Barak River originates from the hill near Mao at the border of Nagaland and Manipur, southern of japvo peak. The Barak flows through state of Assam, Manipur and Mizoram. It's important contributing tributaries in the right bank include Makru and Jiri in Manipur, Dalu, Labak, Madhura, Jatinga and Larang in the Barak plains. Whereas, the Irang, Tuivai in Manipur, and Sonai, Rukni, Kathakal, Dhaleswari, Singla, Langai in Cachar plains form the important tributaries in the left bank. Also, notable rivers in Meghalaya such as Kynshi, Simsang, Kangra, Dareng, Bugi, Umngi, Myntdu, Umngot, Umsohryngkew, forms the major sources of river system in the Meghna basin. Likewise, glacier-fed Teesta river, which originates from Pauhungri glacier in the Northern Sikkim, Rangeet river and Rangpo river forms the major rivers in the state of Sikkim which ultimately joins the Brahmaputra river downstream.

In terms of the average total annual discharge, Subansiri River ranks first among all the principal major tributaries with a discharge of 755,771 m^3/s. From the Table 16.3, it is observed that 9 tributaries in the Northern river bank and 3 tributaries in the Southern river bank of the Brahmaputra are attaining sediment yield > 500 tons/km^2/year, with Jiabharali basin producing the maximum sediment yield of 4721 tons/km^2/year. This implies acceleration of sedimentation process in reservoirs, which may lead to low reservoir's ability to regulate water flow, impacting energy generation during the dry periods and reducing the economic lifespan of hydro-power plants (Palmieri et al. 2001). However, with proper and efficient mitigation strategies, these can be controlled. For instance, implementing erosion control measures in upstream catchments, such as, controlled land use, terracing, afforestation, can reduce sediment yield at the source (Pimentel and Burgess 2013).

Table 16.3 Hydrological characteristics of some major tributaries of Brahmaputra *Source* Anonymous (2012) and Anonymous (2020)

Tributaries	Basin area (km^2)	Sources of origin	Average total annual discharge (m^3/s)	Sediment yield (tons/km^2/year)
North Bank tributaries of Brahmaputra River				
Subansiri	28,000	Comai, China	755,771	959
Ranganadi	2941	Dafla hills, Arunachal Pradesh	74,309	1598
Jiabharali	11,716	Bisai, Arunachal Pradesh	349,487	4721
Gabharu	577	Doimara, Arunachal Pradesh	8450	520
Belsiri	751	Doimara, Arunachal Pradesh	9300	477
Dhansiri (N)	1657	Trashigang, Bhutan	26,577	463
Nanoi	860	Samdrup Jongkhar district, Bhutan	10,281	228
Barnadi	739	Near Sakteng Wildlife sanctuary, Bhutan	5756	323
Puthimari	1787	Kangpar, Bhutan	26,324	2887
Pagladiya	1674	Rading, Bhutan	15,201	1883
Manas-Aei-Beki	41,350	Kula Kangri glaciers, China	307,947	1581
Champamati	1038	Bhurgaon, Bhutan	32,548	386
Gaurang	1379	Southern slopes of Bhutan hills	22,263	506
Tipkai	1364	Ripu, Bhutan	61,786	598
South Bank tributaries of Brahmaputra River				
Buri dehing	8730	Patkai hills, Arunachal Pradesh	141,539	1129
Disang	3950	Patkai hills, Arunachal Pradesh	55,101	622
Dikhow	3610	Tuensang, Nagaland	41,892	252
Jhanji	1130	Tuensang, Nagaland	8797	366
Bhogdoi	920	Mokokchung district, Nagaland	6072	639

(continued)

Table 16.3 (continued)

Tributaries	Basin area (km^2)	Sources of origin	Average total annual discharge (m^3/s)	Sediment yield (tons/km^2/year)
Dhansiri (S)	10,242	Nsenlwa, Peren, Nagaland	68,746	379
Kopili	13,556	Near Ladmynkseh, Meghalaya	90,046	230
Kulsi	400	Nongbudum, Meghalaya	11,643	135
Krishnai	1615	Near Samin Rongal, Meghalaya	22,452	131
Jinari	594	Near Baljek Songitcham, Meghalaya	7783	96

16.2.4 Impact of Anthropogenic Activities and Human-Induced Climate Change on Hydrological Regimes

North East India, characterized by its unique hydrological systems and abundant water resources, is experiencing significant changes due to anthropogenic activities and climate change. Deforestation, a prevalent issue in the region, alters the hydrological processes by reducing canopy interception and evapotranspiration, is a major concern, with a reduction in forest cover of approximately 1020 km^2 during 2019 to 2021 (Anonymous 2021). Figure 16.2 shows state wise forest cover reduction in this region. Also, construction of roads in different parts of North East India, has lead to a widespread deforestation. Around 2500 hectares of forest land were diverted for road construction projects between 2015 to 2020 as per the Ministry of Environment, Forest and Climate Change (MoEFCC), 2021. This contributes to a rise in peak discharge flow rates in rivers by 20% to 40% during heavy rainfall events (Das et al. 2019).

Similarly, road cutting and excavation in hilly states such as Sikkim and Arunachal Pradesh aggravates soil erosion with increase in sediment yield by 40–60 tons/ha/year in road construction zones (Pandit et al. 2012). Tripathi et al. (2012), found that 86% of rural households in the region, practiced shifting cultivation extensively, which has lead to severe soil erosion rates of 30–50 tons/ha/year, reducing infiltration rates and increasing sediment loads in rivers. This sedimentation not only degrades the aquatic environment but also diminishes the reservoir capacity and hydro-power plants, which are very crucial for the region's energy needs and development (Bhattacharjee et al. 2020). Rapid urbanization in cities such as Guwahati has increased impervious surfaces, leading to poor ground water recharge. The ground water levels

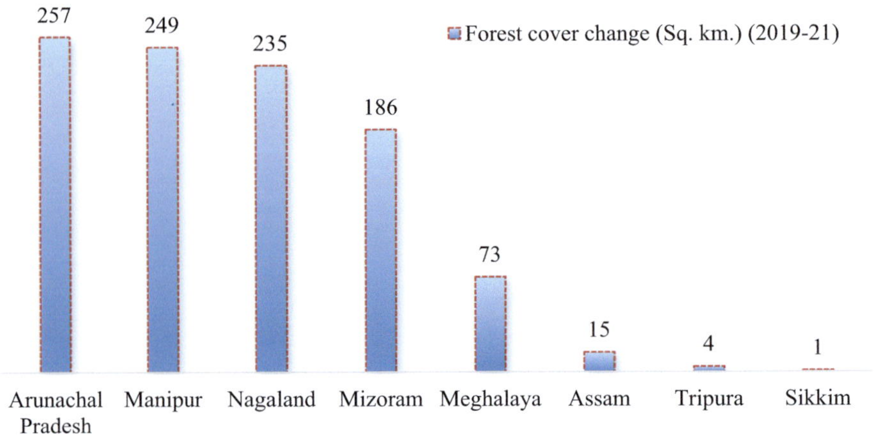

Fig. 16.2 Forest cover change in NE India (2019–21) *Source* Values adapted from Anonymous (2021)

have reduced by 2–3 m in Guwahati alone due to over exploitation and insufficient infiltration process (Sharma et al. 2014).

As reported by the IPCC's AR6 Working Group I (IPCC 2021), most land regions with dependable observational records have experienced a significant rise in both the frequency and intensity of heavy rainfall events. Many areas have also seen a reduction in available water during dry seasons, a trend linked to human-induced climate change and changes in evapotranspiration patterns. Global hydrological models suggest that, in the future, more regions will experience an increase in river floods rather than a decrease. Additionally, extreme rainfall and pluvial flooding (flooding caused by intense rainfall) are expected to rise in many parts of the world across almost all continents. However, predicting changes in river floods is more uncertain compared to pluvial floods, as river flooding is influenced by complex factors such as land use changes and human water management practices.

16.3 Spatio-Hydrological Modeling Concepts

16.3.1 Basic Principles of Hydrological Modeling and Types of Hydrological Models

Hydrological models have become indispensable for effectively managing water resources. These models provide a holistic understanding of the hydrological cycle, allowing for accurate and precise predictions of surface runoff, streamflow and other critical parameters (Alipour and Hosseini 2018). The foundation of hydrological modeling is built upon the fundamental water balance equation, which is expressed as follows:

$$\text{Precipitation (P)} = \text{Runoff (Q)} + \text{Evapotranspiration (ET)} + \text{Change in storage (S)}$$

This equation reflects the dynamic interactions between the different elements of the hydrological cycle, providing the foundation for the creation of more advanced intricate models. Hydrological models can be classified into different types based on factors such as their structure, spatial representation, stochastic characteristics, and spatio-temporal dynamics. Structurally, they are commonly grouped into four categories: empirical, conceptual, physically based, and hybrid models. The empirical, also known as metric or black-box models, does not have the capacity to describe the physical processes in detail but instead use the abundant observed data to predict the output based on the statistical relationships. Examples of such models are unit hydrograph theory built for catchment-scale limited to event-based simulations of ungauged catchments (Sherman 1932), regression models, Artificial Neural Networks (ANNs), and Long Short-Term Memory (LSTM). ANNs and LSTM, work well for streamflow prediction, especially in areas with scarce historical data. ANNs, and LSTM can be trained and simulate on historical streamflow data to predict the future flow based on important variables such as rainfall, temperature, and past streamflow measurements (Dawson and Wilby 2001; Nguyen et al. 2025).

In contrast to empirical models, conceptual models are constructed mainly on an understanding of many relevant natural processes, representing precipitation, evaporation, infiltration, runoff and groundwater flow that affect the input and generate the corresponding outcome. In other words, these models rely on the parameters (such as coefficients) that are refined through calibration to match observed data. The prominent examples for the conceptual models are the HBV Model (Hydrological Bureau of Vatten), Soil Conservation Service–Curve Number (SCS-CN) model, Stanford Watershed Model (Lindström 1997; USDA Soil Conservation Service 1986; Crawford and Linsley 1960). Physics based also known as processed based model, use the principles of physics and its governing equations like the laws of mass and energy conservation to better represent the underlying physical processes in order to better simulate the movement and the transformation of flow within the catchment. Some of the commonly used physics-based models are the MIKE SHE model, Soil and Water Assessment Tool (SWAT), TOPMODEL (Topography-based hydrological model) (Abbott 1986; Arnold and Fohrer 2005; Beven and Kirkby 1979).

Hydrological models can also be categorized into lumped, semi-distributed, and fully distributed models, depending space discretization criteria. Lumped models represents the entire watershed as a single homogeneous unit and hence spatial variability of all the state variables within the unit are averaged or simplified (Beven and Binley 1992; McDonnell and Woods 2004). On the other hand, distributed hydrological models simulate the land–atmosphere interactions and hydrological processes in the catchment at sub-divided units or elements, which accounts for spatial variability, and mathematical equations that represents the processes are applied and solved to determine the underlying state variables linked with each unit or element (Beven and Kirkby 1979). On the other hand, semi-distributed models are those in which the model divides the catchment into sub-units based on land use, soil types or hydrological characteristics. Within each sub-unit, the hydrological processes such as precipitation, infiltration, evaporation, channel routing, are represented by lumped parameters. Semi-distributed models combine features from both lumped and fully distributed models, which are commonly known as hybrid approach (Doe 2022). This approach is more computationally efficient than fully distributed but still captures spatial variability within the catchment. A prime example of this type of model is the Variable Infiltration Capacity (VIC) model, as well as the Soil and Water Assessment Tool (SWAT).

Moreover, hydrological models can be defined as event-based models, which simulates a single storm phenomena ranging from few hours to few days. The best and commonly used event-based model is the Hydrologic Engineering Center–Hydrological Modeling System (HEC-HMS). Whereas, models that represents the study of hydrological processes for both wet and dry period of the water year is known as continuous model. Further, the input data requirement, the approach, the output reliability and accuracy mostly vary from one model to another. Though, it is well known that distributed hydrological model is the most reliable and accurate model but representation of such model with intensive data is highly impractical and impossible to collect field data to carry out study on water resources related phenomenon. However, in such cases, Geo- spatial tools and techniques plays a vital role in fulfilling the requirement for preparation of the spatio-temporal data at formidable accuracy. The spatial and temporal data collected via remote sensing approach can be easily fed into the selected hydrological models and hence derive the important targeted hydrological outputs. Table 16.4 shows the most widely used hydrological models ans their capabilities.

16.3.2 Spatio-Temporal Data for Hydrological Modeling

Hydrological modeling plays a vital role in the effective planning and management of water resources, and the availability of accurate and comprehensive spatio-temporal data is essential for the development and refinement of these models. Hydrological models, such as the SWAT and the Sacramento Soil Moisture Accounting model, rely on a wide range of datasets to simulate the complex processes within a catchment, including precipitation, temperature, soil characteristics, land use, and other variables (Gharbia et al. 2020; Brouziyne et al. 2020). The integration of these diverse datasets into a decision support system can facilitate the understanding of the water cycle and other dynamics, while also enabling the assessment of the impacts of human activities and climate change on both the quantity and quality of water resources. However, the availability of these datasets can be a significant challenge, particularly in regions with limited monitoring infrastructure (Gharbia et al. 2020). Moreover, the assumption of time-invariant parameters in many hydrological models is often inappropriate,

Table 16.4 Examples of various hydrological models emphasizing their strength and applicability

Models	Advantage	Disadvantage	Streamflow generation	Baseflow/ Groundwater	Climate framework	Snow/glacial melt	Main findings
HEC-HMS (small to large scale watershed)	Emphasize on direct runoff, channel routing and water control structures	Mainly event-based simulations	Empirically based	Empirically based	Precipitation	Present	H, PF, YY, WB, ET, RO, SM, IF, GF, SF, OF, LF, SW
SWAT (small to large scale watershed)	Emphasize water quantity and quality, along with the groundwater representation	Snow processes requires further refinement for greater accuracy	Empirically based	Physically based	Precipitation	Present	H, SE, NF, YY, WT, SW, IF, GF, RO, SM, LF, SF, OF, ET, PF, WQ, WB
MIKE-SHE (small to large scale watershed)	Simulates the full land phase of the water cycle	High data and computational demands; high complexity	Physically based	Physically based	Precipitation	Present	H, SM, LF, YY, SW, PF, IF, GF, OF, RO, WB, SF, WT, WQ, ET
VIC (medium to large scale watershed)	Capturing land–atmosphere interactions and sub-grid variability	Simplified representation of groundwater processes	Physically based	Physically based	Precipitation	Present	H, OF, YY, RO, WB, PF, SW, SF, GF, ET, IF, SM, LF
TOPMODEL (small to large scale watershed)	Focus on water balance, topography based approach and simple structure	Limited in dry and arid climate, assumption of homogeneous soil properties	Conceptual	Conceptual	Rain dominated	Absence	H,WB, OF, SF, GF, SM, RO, ET, IF, PF

H hydrograph, PF peak flow, ET evapotranspiration, WT water table, SF shallow subsurface flow, GF groundwater flow, WQ water quality, YY yearly yield, WB water balance, IF infiltration, OF overland flow, RO basin total runoff, LF low flow, NF nutrient fluxes, SE sediment soil erosion, SW snow water equivalent, SM soil moisture as the dynamic nature of catchment processes is not accurately represented by a single optimized parameter set (Lan et al. 2020). To address this issue, researchers have proposed the development of integrated modeling frameworks that combine conceptual, empirical, and physical approaches to better capture the spatio-temporal variability of hydrological processes (Gharbia et al. 2020). The use of spatio-temporal data in hydrological modeling can also be complicated by the presence of missing or erroneous data, as well as the challenges of scaling up from local to regional or global scales. One approach to addressing these challenges is the use of remote sensing and geospatial technologies, which can provide valuable spatio-temporal data on precipitation, soil moisture, snow cover, and other parameters (Pellicciotti et al. 2012). Another approach is the integration of in-situ measurements, remote sensing data, and model outputs through data assimilation techniques (Kucera and Jovanovic 2006; Wang and Liang 2009).

By incorporating these advanced techniques and leveraging the resources of spatio-temporal data available, hydrological models can be improved to better support water resource management decision-making and address the complex challenges facing the watersheds.

16.3.2.1 Precipitation/Rainfall

The precipitation is any type of water that descends from the atmosphere to the Earth's surface which consists of different forms such as rainfall, snowfall, hail, sleet, frost, dew, etc. Precipitation is a fundamental element of the hydrological system, with far-reaching implications for various domains, including water resource management, agricultural practices, and climate analysis (Sharannya et al. 2020). Rainfall and snowfall are the two main precipitation that transforms into streamflow in majority of rivers. The magnitude of precipitation varies both in time and space based on various predominant factors such as topography, physiography and climatology of the region. The diversity in regional variation in precipitation leads to hydrological problems including floods or droughts. Rainfall and snowfall are usually measured in terms of depth and equivalent of depth of water (in mm). Measuring precipitation accurately is a challenging task, especially in regions with complex topography and limited ground-based observation networks. Indian Meteorological Department, though have put a lot of effort in representing the precipitation depth with in-situ observatories across the country as shown in Fig. 16.3.

In the context of NE India, the density of the rainfall stations is limited by the region's diverse terrain and difficult conditions, as compared to the other parts of India. In case of the water resources related studies, proper coverage of the rainfall stations across the catchments is highly favoured. However, due to the limited data distribution over the NE India, researchers have been prompted to explore the potential of remote sensing or satellite-based precipitation products as an alternative or supplementary data source. The incorporation of in-situ based rainfall data and satellite-derived precipitation data is an intricate process that requires cautious consideration of significant factors. The spatial density of rain gauges and the quality

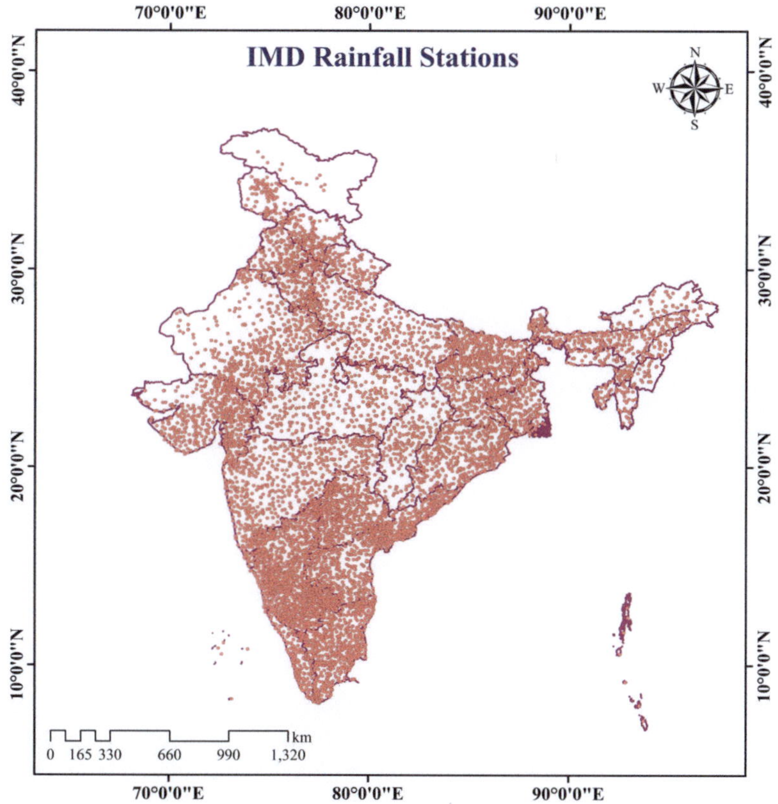

Fig. 16.3 Location of 9785 rainfall stations in India as in 2024 *Sources* IMD Data Service Portal; https://dsp.imdpune.gov.in/home_station_list.php#

of the satellite precipitation product can significantly impact the predictive performance of the integrated approach. Previous studies have highlighted the challenges associated with the integration of coarse-resolution satellite data and rain gauge observations, particularly in regions with complex terrain and sparse station networks (Sharannya et al. 2020; Prakash et al. 2018).

The Indian National Satellite System (INSAT) is a series of multifunctional geostationary satellites serving various operational needs, which includes the launch of INSAT 1 in 1983 followed by INSAT 2, INSAT 3, INSAT 3D, INSAT 3DR and the latest launch of INSAT 3DS in 2024, operated by the Indian Space Research organization (ISRO). These satellites have become a crucial tool in rainfall estimation, a critical component of weather forecasting and disaster management (Prakash et al. 2018; Sharannya et al. 2020). INSAT satellites carry a meteorological payload called the Very High Resolution Radiometer (VHRR), which features sensors operating across visible, infrared, and water vapor channels. These sensors observe in specific spectral bands, including visible (0.55–0.75 μm), thermal infrared (10.5–12.5 μm),

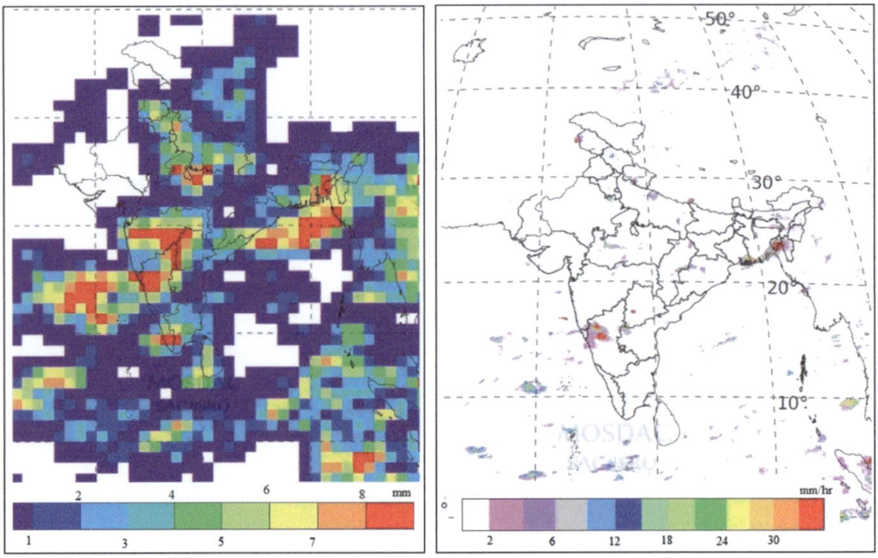

Fig. 16.4 Quantitative Precipitation Index (left) and Hydro Estimator Precipitation for India (right) during the August 20, 2024 (https://mosdac.gov.in/gallery/index.html)

and water vapor (6.3 μm). A key benefit of INSAT data is its ability to cover the entire India and surrounding areas, as well as provide valuable information for data-scarce oceanic regions. The satellite is capable of collecting data every half hour and can also focus on specific areas at 5-min intervals during cyclonic events (Manikiam 2003). The quantitative products derived from INSAT data include numerical outputs such Quantitative Precipitation Index, Hydro Estimator Precipitation, and more that can be utilized even for near real time hydro-power monitoring and management as shown in Fig. 16.4.

The Tropical Rainfall Measuring Mission (TRMM), launched in 1997, is a joint initiative between NASA and the Japan Aerospace Exploration Agency (JAXA). It was the first satellite to integrate both active and passive microwave instruments aboard a low-inclination (35°) orbit to study precipitation, storms, and climate processes in the tropics. TRMM's core instruments include the Precipitation Radar (PR), the TRMM Microwave Imager (TMI), and a multi-channel passive microwave radiometer that complements the PR by measuring the total hydrometeor content (both liquid and ice) within storm systems. Additionally, the Visible and Infrared Scanner (VIRS) provides cloud imagery to contextualize precipitation structures. A detailed overview of TRMM's rain-sensing instruments is provided in Table 16.5.

These instruments feed into an algorithm that produces the TRMM Combined Instrument (TCI) calibration dataset (TRMM 2B31), which supports the TRMM Multi-satellite Precipitation Analysis (TMPA). Among the most widely used outputs for climate research are the monthly precipitation averages (TMPA 3B43) and the daily to sub-daily (3-hourly) estimates provided by TMPA 3B42. These data set

Table 16.5 TRMM rain package sensor summary (Adapted from Scott 2011)

Instrument	Frequencies	Spatial resolution	Scanning technique	Swath width (Km)
Microwave Imager (TMI)	10.7, 19.3, 21.3, 37.0, 85.5 GHz (both Vertical & Horizontal except 21.3 GHz which is vertical)	11 km × 8 km at 37 GHz	Conical scanning (53° inclination)	880
Precipitation Radar (PR)	13.8 GHz	5 km footprint & 250 m vertical	Cross-track scanning	250
Visible and Infrared Scanner (VIRS)	0.63, 1.61, 3.75, 10.8, 12 μm	2.5 km	Cross-track scanning	830

are available in 0.25° spatial resolution, covering 50° N and 50° S. Further, TRMM 3B42 is a gridded record of precipitation with high resolution in both space and time, and are available in binary format. It provides a non-instantaneous rain rate in mm/hour and contributes to the monthly rainfall estimates available in TRMM 3B43. For instance, Fig. 16.5 shows the TRMM 3B42 daily rain rate (mm/day) for May 09, 2013 over NE India.

On the onset of highly successful rain sensing package of the TRMM, Global Precipitation Measurement (GPM) mission is an extension with the GPM Core Observatory designed primarily to focus on heavy to moderate rain over tropical and sub-tropical oceans. The GPM Core Observatory carries two key instruments: the first-ever advanced Dual-Frequency Precipitation Radar (DPR), which operates in both the Ku- and Ka-bands, and a highly sensitive multi-channel GPM Microwave Imager (GMI). The DPR includes two radar components: KuPR and KaPR, which operate at 13.6 GHz and 35.5 GHz, respectively, allowing for more detailed and accurate measurements of precipitation from space. The DPR, high sensitive to light rain (< 0.5 mm hr-1) and snowfall is better performed relative to the TRMM Precipitation Radar. The DPR's capability allows for three-dimensional measurements of precipitation structure and characteristics, along with the micro-physical measurements from the instrument to complement cloud and aerosols observations. Additionally, the Integrated Multi-satellite Retrievals for Global Precipitation Measurement (IMERG), a comprehensive algorithm developed by NASA, is used to combine data from a constellation of rain-sensing satellites in Earth's orbit at a given time, enabling the estimation of precipitation across the Earth's surface. The GPM IMERG daily rainfall data (June 27, 2024) for NE India is shown in Fig. 16.6. With the advancement of precipitation retrieval methods over a wide range of high resolution satellite-based precipitation measurement in time and space, this leads to improved weather forecast and climate models, authentic hydrological or streamflow simulation models over the catchments, and estimates of various extreme events (floods, droughts, cyclone, etc.).

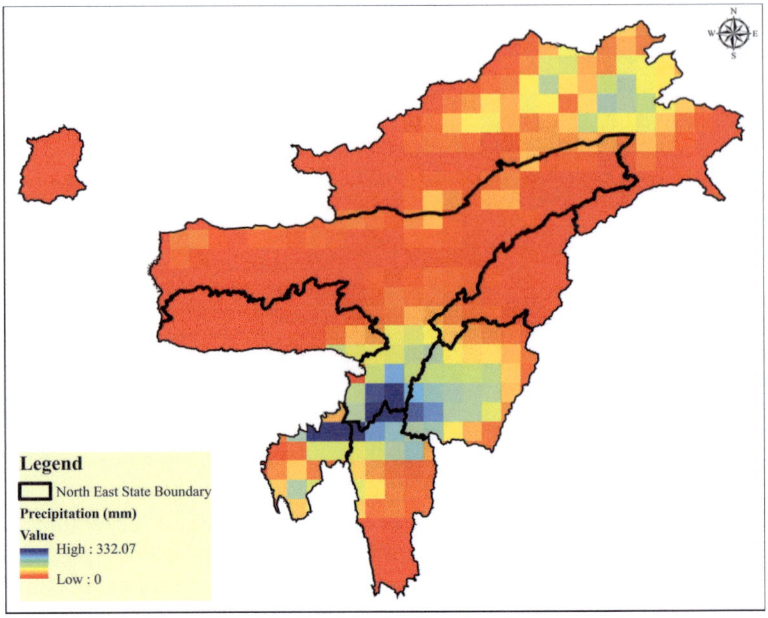

Fig. 16.5 Daily rain rate (mm/day) from TRMM 3B42 for May 9, 2013 *Source* https://disc.gsfc.nasa.gov/datasets/TRMM_3B42_Daily_7/summary

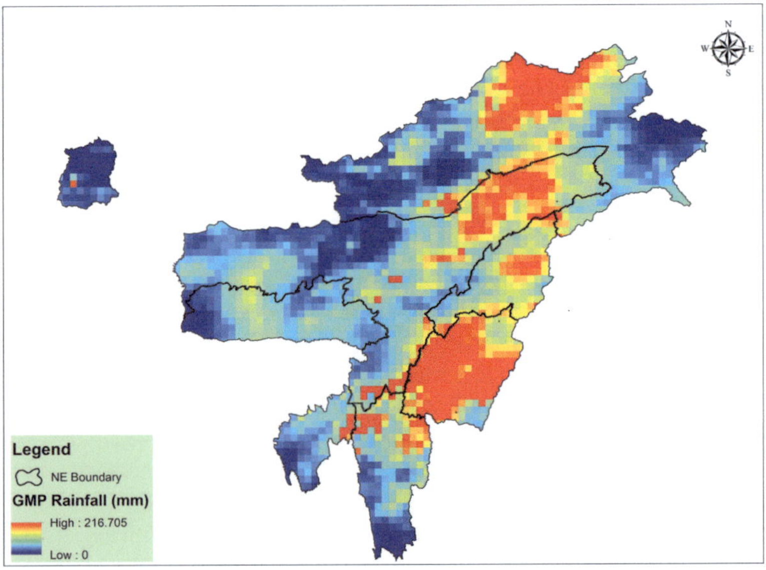

Fig. 16.6 Daily rainfall from GPM IMERG V07B for NE India (June 27, 2024)

16.3.2.2 Soil Moisture

Soil Moisture, a vital component among other hydrological processes, plays a crucial role in providing meaning information for various applications (agricultural, ecological and environmental). Conventional methods of soil moisture measurement, such as gravimetric sampling and Time-Domain Reflectometry (single probe or multiple probe TDR), have been widely used, but they are usually time consuming, labour intensive and have limited spatial coverage. In the context of agricultural conditions, particularly east and north east India, which are characterized by the the extreme seasonal variability leading to prominent wet to dry conditions, the spatio-temporal analysis of soil moisture and the development of an optimal sampling design for regional-scale soil moisture monitoring are essential (Singh et al. 2019). However, recent advancements in space technologies have opened up new opportunities to exploit monitoring and measurement of soil moisture at larger scales, enabling a more comprehensive understanding of its spatio-temporal variability.

Now, surface soil moisture can be effectively estimated using a combination of optical and thermal sensors. Using the optical electromagnetic radiation band, the soil moisture can be indirectly estimated by analysing the abrupt changes in the spectral signatures of the soil surface. On the other hand, thermal sensors also plays an important role in estimating soil moisture by determining the change in thermal inertia of the wet and dry soil soils over a particular land surface, but these are constrained with the limited continuous data, low spatial resolution and poor vegetation penetration. As such, soil moisture estimation requires the knowledge of the subsurface information, which can only be done using both active and passive remote sensing data. Wagner et al. (2007), Henderson and Lewis (2008), Wang and Qu (2011), Tan and Wang (2013) also reported that L-band, C- band and X-band of the microwave spectrum are highly sensitive towards soil moisture. One best example is the Advanced Microwave Scanning Radiometer for EOS (AMSR-E) sensor, a 6 different frequency band, microwave radiometer embedded with dual polarization (horizontal and vertical), making 12 measurement channels that flew on-board NASA's Aqua satellite is used for deriving products such as snow depth, snow water equivalent, water vapor, soil moisture, sea ice concentration, sea surface temperature, and other parameters. The spatial resolution of each measurements varied from 5.4 km at 89.0 GHz to 74 $\times$ 43 km at 6.9 GHz.

In 2009, in order to have a better coverage of the data throughout the entire globe, the Soil Moisture and Ocean Salinity (SMOS) was launched at the low earth orbit by the European Space Agency. This was the first ever microwave L-band (1.413 GHz) measurements, pertaining to observations in soil moisture and sea surface salinity to capture the global variability exposing to continuous interactions between the 3 components of hydrological cycle viz., land, ocean and atmosphere. The SMOS mission utilizes the Microwave Imaging Radiometer using Aperture Synthesis (MIRAS) to deliver global soil moisture data with a spatial resolution of 35–50 km, updated every 2–3 days, and an accuracy of 0.04 m^3/m^3. To improve spatial detail, NASA launched the Soil Moisture Active Passive (SMAP) mission in 2015, which integrates an L-band radar and L-band radiometer to measure soil

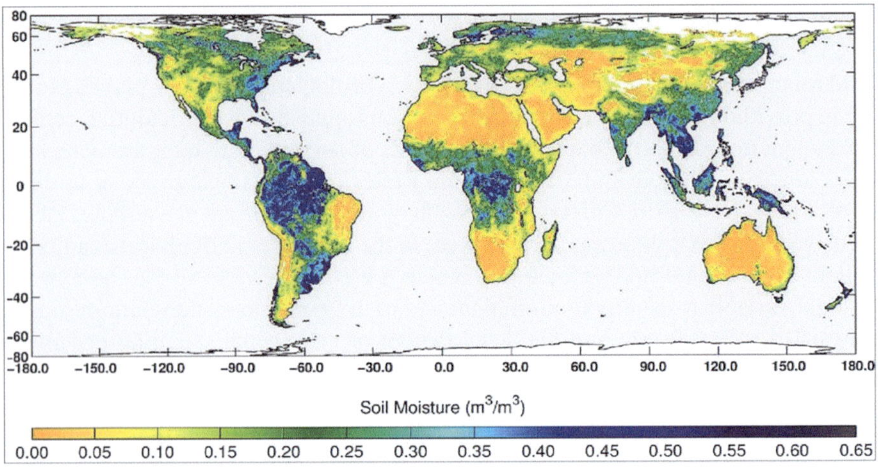

Fig. 16.7 SMAP radiometer only global soil moisture between October 3 to 5, 2015 *Source* NASA/JPL-Caltech/GSFC

moisture globally at finer resolutions of 1 km and 3 km. This mission leverages both radar and radiometer observations using an active–passive approach, aiming to benefit from the higher resolution provided by the Synthetic Aperture Radar (SAR) and the greater sensitivity to soil moisture offered by the radiometer. Unfortunately, the L-band radar on the SMAP satellite ceased functioning just two and a half months after launch. To overcome the limitations such as low spatial resolution, vegetation interference, and data gaps caused by cloud cover, etc., several techniques have been developed to enhance the spatial detail of SMOS and SMAP observations. These approaches leverage the integration of multi-sensor data and apply various physical assumptions to improve accuracy and coverage. Figure 16.7 shows the SMAP's radiometer capture view of global soil moisture.

Active satellites such as Seasat, European Remote Sensing (ERS) satellite, RADARSAT, Advanced Land Observation Satellite (ALOS), ISRO's follow-on Radar Imaging Satellite EOS-04 (C-band) operates through the mechanism of SAR and Scatterometers, are commonly utilized using a number of different algorithms to estimate soil moisture. Singh et al. (2021) explored the feasibility of integrating RISAT-1 C-band SAR data into the SMAP active–passive algorithm to estimate soil moisture at spatial resolutions of 1 km, 3 km, and 9 km. Their study focused on an agricultural region dominated by paddy fields, which experience seasonal flooding. Further, to cater to the requirement for agriculture conditions in India, a high resolution (~ 500 m) soil moisture map is derived by using the active (EOS-04 C-band SAR) and passive (SMAP radiometer) measurements as Level-4 value added products. The spatio-temporal dynamics of soil moisture over India developed by Space Applications Centre, ISRO, as analysis ready data is shown in Fig. 16.8.

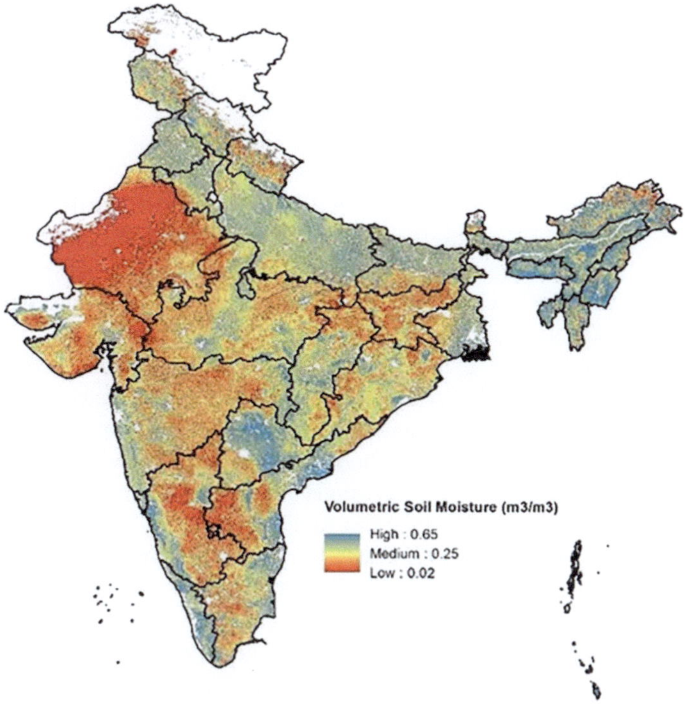

Fig. 16.8 EOS-04 C-band SAR surface soil moisture over India (January 04–21, 2023) *Source* ISRO/High-Resolution Operational Soil Moisture Product

Interestingly, the microwave based remote sensing soil moisture products will eventually provide the information on the surface level or few centimeters in subsurface level which may vary in space and time due to impact of varying patterns in rainfall, evaporation, etc., but yet complete and thorough information regarding the soil moisture profile in the unsaturated zone is highly crucial to carry out the study related to the hydrological response, climatic and agriculture. However, Haas and Williams (2017), Nikam et al. (2015), estimated the soil moisture at multi-depth using physically-based semi-distributed VIC Land Surface Model (LSM).

16.3.2.3 Evapotranspiration (ET)

ET is a key component of the water cycle, playing a significant role in regulating the exchange of mass and energy among the soil, water, vegetation, and the atmosphere (Singh and Singh 2023; Senay et al. 2012). Consumptive use and ET is synonymous in nature in terms of practicality. ET is usually expressed in terms of depth (mm or cm) for a particular duration. The important variables affecting the rate of change in ET with the most varying weather conditions are temperature, solar radiation, relative humidity, vapor-pressure gradient and wind speed. Moreover, Land use land cover classes and soil types with varying soil moisture also influence the actual ET (ET_a). ET is an important component in hydrological studies and estimation of ET is highly crucial for various applications, such as catchment modeling, determination of water balance, assessment of hydrological responses with changing land use land cover, agricultural water management, etc. Despite the concept of Evapotranspiration (ET) is widely applied, considerable confusion has persisted around the definitions of key terms such as potential ET and reference crop ET. To clarify these distinctions, the Food and Agriculture Organization (FAO) of the United Nations released the FAO-56 report (Allen et al. 1998), which established standardized definitions and provided clarity on terms such as potential ET and reference crop ET.

ET can be measured on field using weighing Lysimeter method, eddy correlation (also known as eddy covariance system), Soil depletion method, etc. However, these approach is time consuming, expensive and restricted with poor spatial coverage. But to counter this problem, empirical methods are developed based on meteorological input for the estimation of ET. These methods are FAO-56 Penman–Monteith (FAO-56 PM) (Allen et al. 1998), Bowen Ratio Energy Balance (BREB) (Bowen 1926), temperature based equations such as the Hargreaves-Samani method (Hargreaves and Samani 1985), the Thornthwaite Method (Thornthwaite 1948) and the Blaney-Criddle method. Among the mentioned methods, the application of FAO-56 Penman–Monteith method for estimation of ET is the most common and widely used method which depend on the spatio-temporal variables, namely, solar radiation, wind speed, air temperature, vapor pressure, and humidity. However, these important variables are not easily available for a particular location. Therefore, remote sensing technology play an important role in providing information of all the related variables on surface, radiation and climate at different time and spatial scale. Energy balance model is one of the methods for estimation of ET by dividing the energy into three fluxes viz., sensible heat, latent heat and ground heat. Surface Energy Balance Algorithm for Land (SEBAL) is one of the most widely used energy balance model. The algorithm utilizes the thermal infrared band to obtain surface temperature and hence derive ET based on the energy fluxes. SEBAL has been applied successfully in various studies at regional and global scale (Bastiaanssen et al. 1998).

An advancement over the SEBAL approach, the Mapping Evapotranspiration at High Resolution with Internalized Calibration (METRIC) model offers improved spatial resolution and greater accuracy in estimating ET (Allen et al. 2007). METRIC was validated in cultivated areas and has been used to monitor crop water use across various climatic conditions (Allen et al. 2007). Other notable remote sensing-based methods for estimating ET include the Vegetation Index approach (Glenn et al. 2011) and the Two-Source Energy Balance (TSEB) model (Colaizzi et al. 2012). The Aqua Moderate Resolution Imaging Spectroradiometer (MODIS) MYD16A2 Version 6.1 provides product (L4) such as total ET, Potential ET, Latent heat flux, etc., globally. The ET layer is a 8-day composite product derived based on the improved algorithm of Penman-Montieth equation, with a product of 500 m resolution (Running et al. 2021). Figure 16.9 shows the net ET in $kg/m^2/8$ day from MODIS (MYD16A2) for North East India and the value 32,766 represents NoData value. This data when

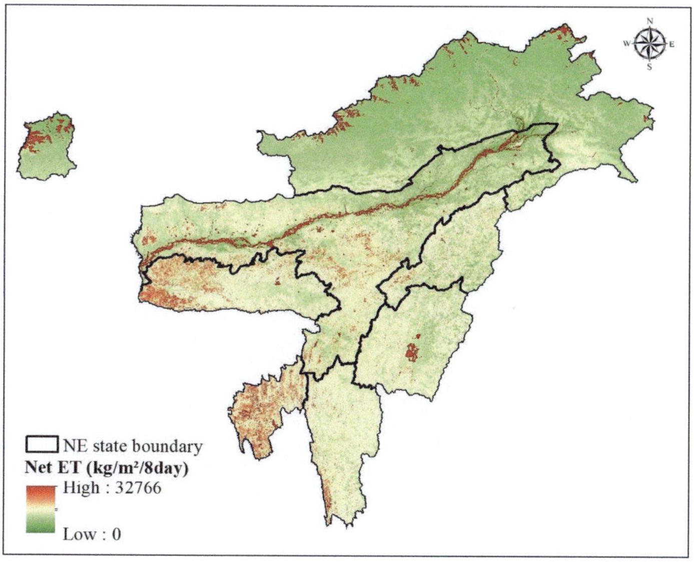

Fig. 16.9 Net ET from Aqua MODIS (MYD16A2) for NE India (28 January, 2024)

divided with the density of water (1000 kg/m^3), the ET will be in the measurement unit of meter (m). Hence, estimating Evapotranspiration will eventually help to understand the impact of its changing rate with respect to irrigation and agriculture water use, accurate water balance calculation, river flow and reservoir management and supporting long term water resource planning.

16.3.3 Calibration and Validation of Hydrological Model

Hydrological Models are essentially the sets of mathematical equations that represent the land phase of the hydrologic processes (Devi et al. 2015). These models are widely used to understand the hydrologic processes of the basin for efficient water resources management, flood control, drought management, reservoir management, etc. (Beckers et al. 2009; Bihon et al. 2024; Kumar et al. 2023). Since these models are mathematical models, they contain parameters that represent the characteristics of the system. Therefore, before the implementation of these hydrological models in any basins, it is important to calibrate and validate these models so that they satisfactorily represent the hydrologic processes of the basin.

Model calibration involves adjusting the model parameters-either manually or through automated techniques so that the simulated outputs closely align with observed field data, ensuring the model accurately represents real-world conditions (Shen et al. 2022). The calibrated model is tested using a set of inputs that were not part of the calibration process before being applied in water resources management.

In this context, validation refers to the process of confirming that a site-specific model can produce sufficiently accurate simulations (Henriken et al. 2003).

16.3.3.1 Manual Calibration

Manual calibration is the trial-and-error method in which the parameters are adjusted so that the simulated outputs matched the field observations satisfactorily. This process requires complex, knowledge-driven analysis to align the observed hydrological behavior of the watershed with the corresponding conceptual and mathematically linked representations embedded within the model's structure. To successfully calibrate the model using manual approach, it is important to understand the physical characteristics and hydrological behaviour of the watershed as well as the physical basis of the different mathematical components and parameters of the models (Yilmaz et al. 2010). Major limitations of the manual calibration approach are tedious and labour intensive process involving subjective judgement of the modeller.

16.3.3.2 Automatic Calibration

Automatic calibration method is the use of a search algorithm to determine the best-fit values of the parameter (Sharma et al. 2006). This method relies on objective functions for performance evaluation, as opposed to visual and subjective ones, and utilizes the computers capacity to efficiently and systematically explore the compatible parameter space of the model (Yilmaz et al. 2010). Algorithms such as the Shuffled Complex Evolution (SCE) method (Duan et al. 1992) and Genetic Algorithm are commonly employed in automatic calibration techniques. SCE, a global optimization method, integrates elements of the simplex algorithm, controlled random search, and competitive evolution, along with the innovative concept of complex shuffling to enhance calibration efficiency and robustness (Yilmaz et al. 2010). However, automatic calibration might lead to the estimation of model parameters that are not hydrologically realistic, resulting in unrealistic simulations.

The SCE-University of Arizona algorithm is a global optimization technique that combines parts of the simplex approach with evolutionary techniques. This technique was widely implemented in different hydrological model for automatic calibration such as in MIKE 11 NAM (DHI, 2017), Lumped Hydrologiska Byrans Vattenavdelning (HBV) model, SWAT, VIC model and etc. (Rahnamay et al. 2019).

The SCE process begins with initializing parameters and selecting the number of complexes (p) and points per complex (m). The overall sample size is calculated as $s = p.m$. A set of s points is uniformly sampled from the feasible search space, and their objective function values are calculated. These points are ordered in ascending order based on their function values and divided into p complexes to ensure search diversity. Each complex is then evolved individually, with solutions refined using the Competitive Complex Evolution (CCE) approach. The evolved complexes are mixed, fused, and rearranged, allocating points to new complexes. This process iterates, with convergence criteria checked after each cycle. The algorithm stops when the convergence criteria are met, providing an effective and efficient optimization strategy for complex, high-dimensional problems. Regarding the discussion of SCE and CCE can be referred to Duan et al. (1992) for further detail understanding.

16.3.3.3 Validation of Hydrological Model

To assess the effectiveness of the calibrated model parameters, it is important to test their performance using a set of observed data that were not involved in the calibration phase. This process, known as model validation, is a vital step that confirms whether the calibrated hydrologic or hydraulic model can reliably replicate the actual processes occurring within the catchment. Validation involves comparing the simulated outputs with the corresponding observed responses, employing both qualitative evaluations and quantitative statistical techniques. Commonly used quantitative performance indicators include the Nash–Sutcliffe Efficiency (NSE), Root Mean Square Error (RMSE), Correlation Coefficient (R), and Percent Bias (PBIAS), among others (Moriasi et al. 2007).

16.4 Case Study on Catchment Level Hydro-Met Monitoring and Decision Support System for Panyor Lower HEP, Arunachal Pradesh

NE India is blessed with unique and diverse topography along with enormous hydro-power potential. Currently, there are a number of operational hydro-power projects established in this part of the country by various state and central agencies, such as, Panyor Lower HEP (erstwhile Ranganadi HEP), Bichom HEP, Tenga HEP, Subansiri Lower HEP, Doyang HEP, Dumbur HEP, Tuirial HEP, Umiam HEP, Leshka HEP, Loktak HEP, Teesta III HEP, etc. These projects across NE India have benefited the region to a greater extend helping to expand both social and economical development. However, with the changing and varying climatic conditions, the anthropological development and exploitation over the respective catchments, implications is viable in case of power supply assurance in long term due to its diminish storage life with time, or high dam inflow due to extreme rainfall events leading to sudden release of high flood discharge downstream causing devastation. Therefore, to counter this problem, an effective monitoring and management mechanism can be adopted for proper decision support system.

The Panyor Lower HEP (erstwhile Ranganadi HEP) is the project constructed by the North Eastern Electric Power Corporation Limited (NEEPCO) in Arunachal Pradesh with an installed capacity of 405 MW. The scheme as such in this hydel project is a runoff river scheme having minimal pondage with limited provision for flood cushioning. The sudden gate openings due to high intensity rainfall in its catchment areas result in heavy discharge entering its pondage is a regular phenomenon. This not only creates trouble for people residing in downstream areas of the North Lakhimpur district in Assam but also creates enormous difficulty for the dam operators in optimizing both the penstock and spillway releases, especially during peak monsoon season. Another major challenge is due to the very steep gradient of the upper catchment areas and a relatively smaller catchment size of 1883 km^2 resulting in a very fast catchment response, making it difficult for the dam managers to optimize and pre-program the gate operations. The impact of water releases from the Panyor Lower Dam on downstream areas in North Lakhimpur has been a subject of continued engagement among stakeholders, including NEEPCO, the state government, and civil society. To address this issue, an integrated approach using in-situ hydro-met measurements along with spatio-hydrological modeling techniques and the dashboard for visualization of the in-situ data was established. The subsequent discussion deals with the collaborative efforts from the North Eastern Space Applications Centre, and NEEPCO to establish a state of the art technology for effective reservoir monitoring and management. Figure 16.10 shows the study area for this particular study.

In this study, a total of 17 numbers of Automatic Weather Stations (AWSs) were installed, sufficiently distributed over the catchment. The rainfall stations are equipped with tipping bucket rain gauges, each having a resolution of 0.2 mm and recording data at a one-hour time interval. The temperature sensor has a range from –40 °C to + 50 °C and the relative humidity sensor measures the values from 10 to 100% RH. The data recorded in the data logger is transmitted through the INSAT series of satellites via the existing national ISRO AWS network, with data reception facilities located at the Space Applications Centre (SAC) in Ahmedabad. The sample rainfall data at hourly for the AWS station 11 and 14 is shown in Fig. 16.11. Along

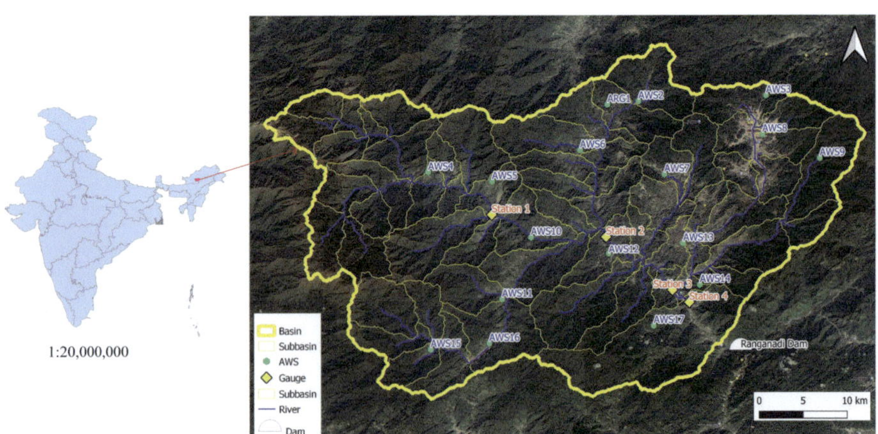

Fig. 16.10 Study area of the Panyor Lower HEP in Arunachal Pradesh

the main river channel in the Panyor Lower HEP catchment, 4 Doppler Water Level Recorders (DWLR) were installed to keep track of the river level rise at a time interval of one hour. The installed DWLRs operate using impulse radar technology, wherein the transmitting antenna emits short radar pulses in the 24 GHz frequency band. These pulses reflect off the water surface and are captured by the receiving antenna to determine the distance between the sensor faceplate and the water surface. This measurement allows for accurate tracking of river water level fluctuations, which are automatically recorded by a data logger at predefined time intervals and transmitted to the local station via satellite communication (ISRO AWS network of MOSDAC). The sample time series of water level is shown in Fig. 16.12 for the station 1, i.e., Mangio village located at 27.4938° N and 93.5322° E at an altitude of 957 m. Figure 16.12 illustrates the seasonal trend of river water level fluctuations throughout the year, with a dynamic rise and fall observed during the monsoon months. The dam authority relies on real-time hourly water level data from the upstream Mangio station and other subsequent downstream 3 DWLR stations to assess the incoming flow. This information plays a vital role in guiding operational decisions, particularly in determining the optimal timing and extent of dam gate adjustments to manage downstream discharge effectively. Both the DWLR and AWS data can be visualized through the dashboard developed as shown in Fig. 16.13, using the Apache and Django framework with the interface of Web Server Gateway Interface (WSGI) to fetch the data from MOSDAC sftp, then processes the data cleaning process, stores the PostgreSQL and response the user request by http request/response.

To simulate the near real time rainfall-runoff process in the catchment, the HEC-HMS model has been used. For development of the hydrological model, CartoDEM version 3 having a spatial resolution of 10 m was used for determining the topographic features and catchment characteristics. Delineation of river networks, identification of suitable sites for installation of 17 AWSs and 4 DWLRs, and preparation of LULC map, was done using a very high resolution KOMPSAT-3 satellite data, having 0.5 m panchromatic resolution and 2.8 m multispectral resolution, and CartoSat 2 data with spatial resolution less than 2 m. On the other hand, the soil texture and its corresponding/representative hydrologic soil groups have been derived from the

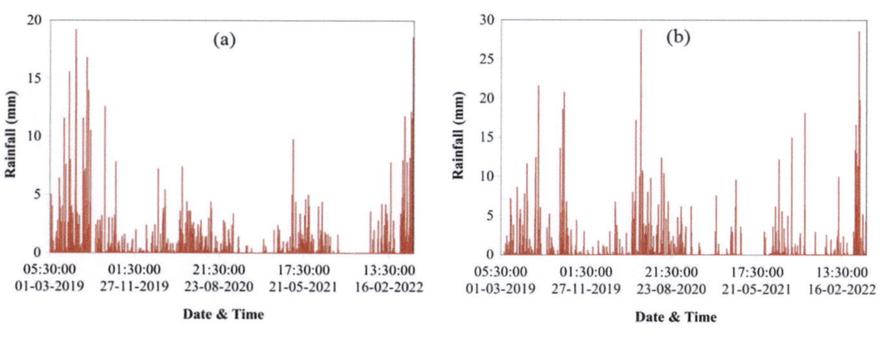

Fig. 16.11 Time series plot of rainfall for **a** AWS 11 and **b** AWS 14

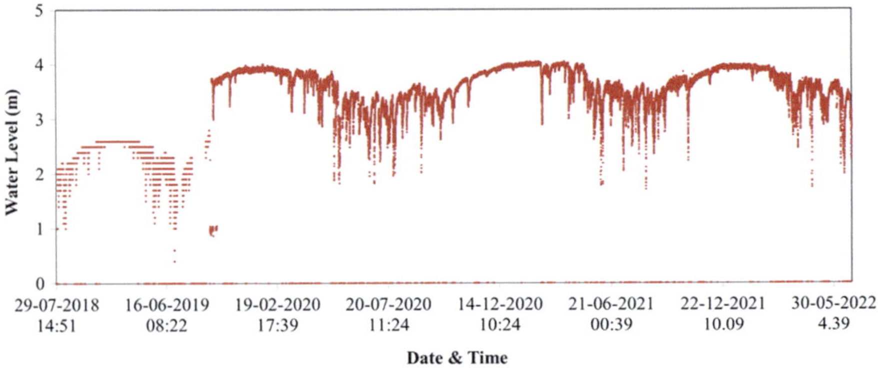

Fig. 16.12 Time series plot of water level for station 1 (Mangio)

Fig. 16.13 In-situ data visualization in dashboard

National Bureau of Soil Survey and Land Utilization Planning (NBSS&LUP) which was analyzed and prepared in ArcGIS platform. The LULC and the soil data are two critical layers for the generation of the curve number grid, a dimensionless parameter representing the degree of a catchment in respect to runoff potential. In addition, NEEPCO records discharge data at the dam inflow five times daily, specifically at 07:00 h, 10:00 h, 12:30 h, 15:00 h, and 17:00 h. These discharge measurements, along with the stage data obtained from the DWLRs, are utilized for the calibration and validation of the hydrological model. The overall methodology of the study incorporating the in-situ measurements and spatio-hydrological model running to cater to the gate opening/closing operations based on the decision support system developed is shown Fig. 16.14. Figure 16.15 shows the spatial data used in the study.

The DWLR installed, explicitly provide the river level rise or fall for a particular time. Using these data and the cross-section data generated by boat mounted acoustic

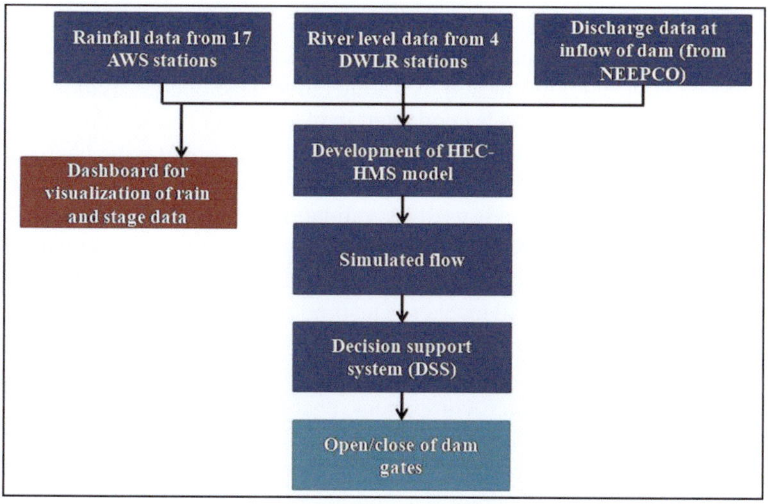

Fig. 16.14 The overall methodology for hydro-met monitoring for optimized gate operations for Panyor Lower HEP

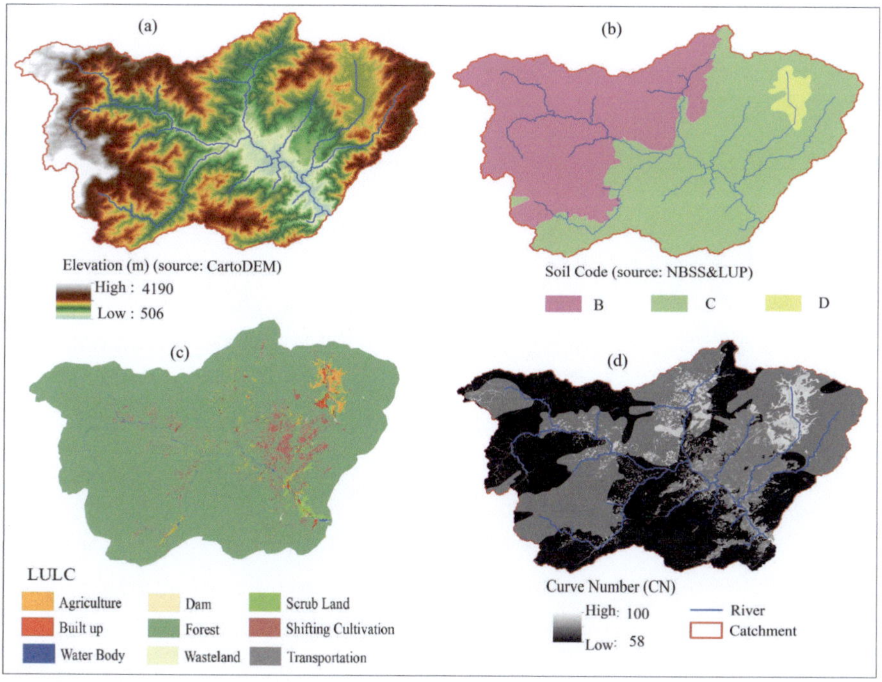

Fig. 16.15 **a** DEM map, **b** Soil map, **c** LULC map and **d** CN map for the study area

doppler current profilers, a stage-discharge relationship was developed for all the 4 DWLR stations. The model calibration and validation are performed using observed discharge data at the dam inflow and the rating curve developed at Station 4, located near the dam inlet. The process is conducted in two distinct phases: Phase I involves calibrating the model with 2020 monsoon data and validating it with 2021 monsoon data, and Phase II expands the calibration period to include 2020–2021 monsoon data, while 2022 monsoon data is used for validation. The critical parameters which are calibrated are curve number, initial abstraction, lag time, and Muskingum-Cunge parameter. A total of 229 parameters have been calibrated, which comprises parameters from 63 sub-basins and 40 reaches. The time series plot for both the calibration and validation of the model is shown in Fig. 16.16 representing a satisfactory model performance with R and NSE of 0.87 and 0.748 for the calibration, respectively.

The hydrological modeling component enables the accurate simulation of discharge at the dam inflow. Moreover, the magnitude and duration of the inflow are critical factors in assessing the rate and extent of water level rise within the reservoir. Accordingly, dam authorities can act on the operation of gates. Hence, the existing reservoir Area-Elevation-Capacity curve of 2019 from NEEPCO has been analyzed to understand the level rise in the reservoir. The area capacity curve was useful for correlating between modeled inflow, duration of flow, volume of the

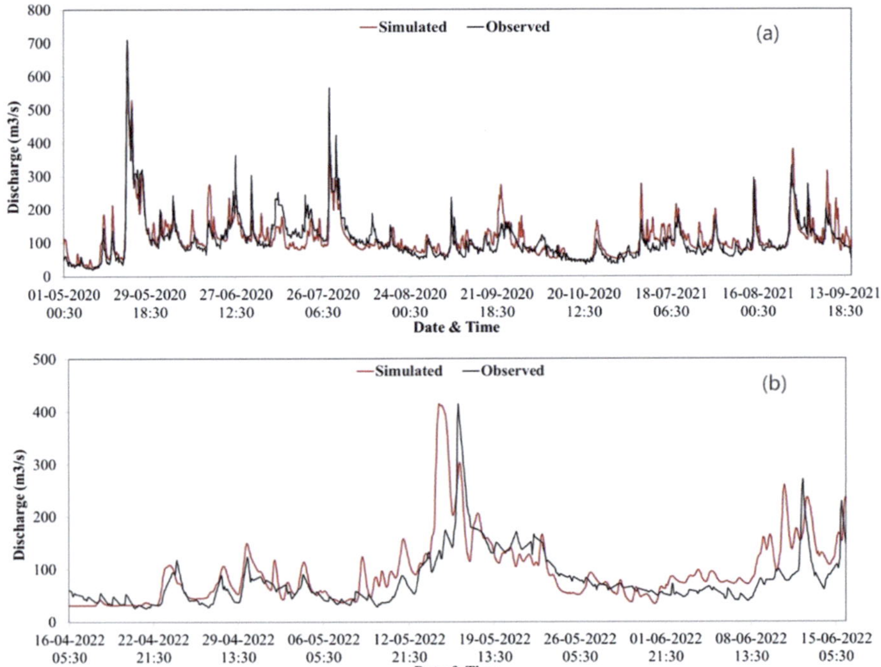

Fig. 16.16 Time series plot between simulated and observed discharge during calibration (left) and validation (right) of the hydrological model

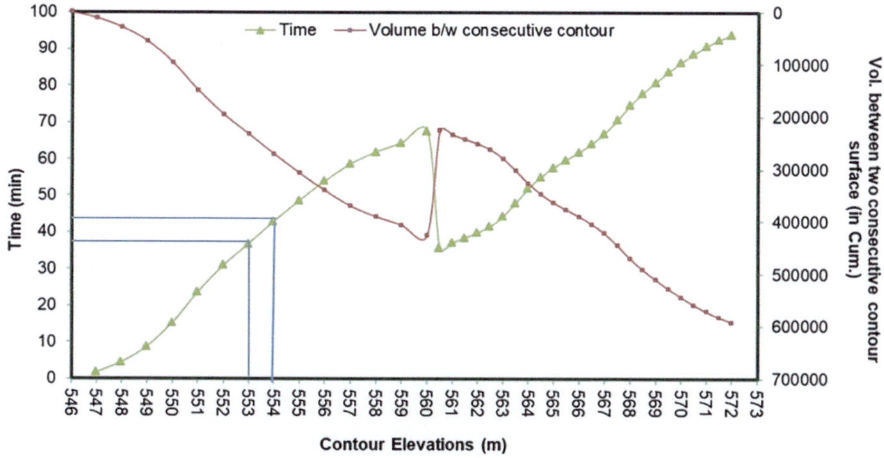

Fig. 16.17 Relationship between duration, level and volume for 103 m^3/s inflow

reservoir and level rise in the reservoir. Figure 16.17 shows the relationship between modeled inflow, duration, reservoir level rise, and volume. From the figure, it is noted that for an inflow of 103 m^3/s to the reservoir, the time taken for reservoir level rise from 553 to 554 m is nearly 15 min. This information helps the dam authorities for optimal gate operation. Similarly, the scenarios for other discharge values can be generated to aid in quick decision process, specially during intense monsoon season. Therefore, with the integrated approach using both in-situ measurements, spatio-hydrological modeling and the reservoir level monitoring have extensively helped the dam operators in Panyor lower HEP in Arunachal Pradesh in optimizing and managing the gate operations especially during the South-West monsoon season, thus avoiding the sudden unprecedented flow release in the downstream of the HEP.

To enable a comprehensive and overall understanding of water availability in support of hydropower operations, a water balance assessment was performed for the Ranganadi catchment, with the outlet delineated at the DWLR station, located just upstream of the Panyor Lower Hydro-Electric Project (HEP). This analysis employed satellite-derived datasets for rainfall, ET, and soil moisture-each aligned spatially and temporally to estimate monthly runoff, a key indicator of inflow availability for hydropower generation. Rainfall input was derived from the GPM IMERG V07B dataset, characterized by a spatial resolution of approximately 0.1° (~9 km). Daily rainfall values in millimeters (mm) were then aggregated to obtain the total monthly rainfall for July 2024 (Fig. 16.18). Net ET estimates were taken from the MODIS MYD16A2 product (8-day composites). These were converted to monthly totals and resampled from a resolution of 500 m to 9 km using bilinear interpolation to match the rainfall grid. Given that ET was expressed in kg/m^2, it was directly interpreted in mm, assuming the density of water as 1000 kg/m^3. The resultant monthly net ET distribution is shown in Fig. 16.19. Further, soil moisture data, representing the change in storage (ΔS), was extracted from the SMAP enhanced daily product,

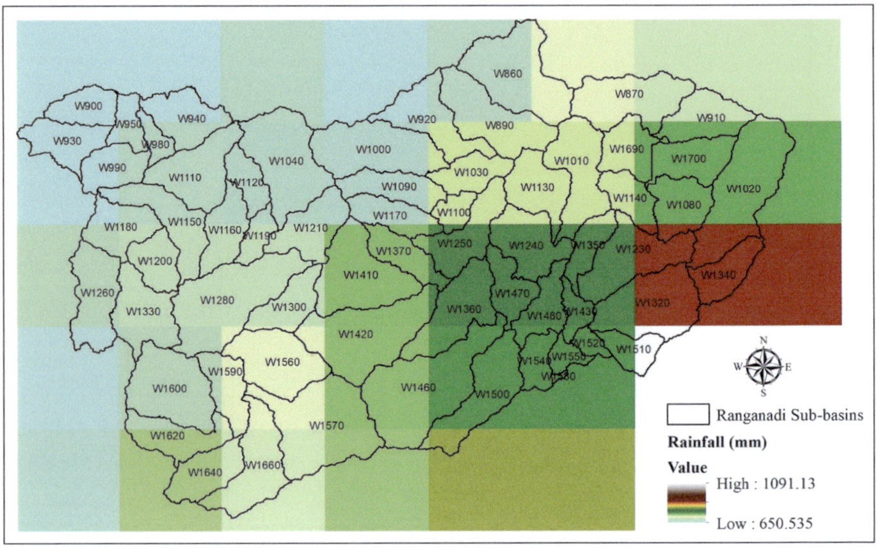

Fig. 16.18 Monthly total rainfall (mm) from GPM IMERG V07B (July 2024) for the Ranganadi catchment

also resampled to a 9 km × 9 km grid. The unit of m^3/m^3 was converted to mm by multiplying with the assumed soil depth of 50 mm (in the case of SMAP dataset). The aggregated monthly soil water storage is shown in Fig. 16.20. Using these datasets, the monthly runoff (mm) (Fig. 16.21) was estimated through the water balance equation as described in Sect. 16.3.1.

This integrated satellite-based hydrological balance provides important insight into the available surface runoff, which is a primary factor for hydroelectric generation. The spatially explicit runoff estimates allow reservoir managers at the Panyor Lower HEP to assess inflow variability, optimize turbine operations, and improve water release scheduling, especially during monsoonal peaks. Furthermore, by monitoring the balance between rainfall, ET, and soil moisture, potential deficits or surpluses can be identified early, supporting drought preparedness or flood mitigation strategies. The estimated monthly runoff observed in Fig. 16.21 reveals a predominantly surplus surface water condition across the basin, suggesting favorable hydrological availability that can contribute positively to sustained hydropower generation. However, while satellite-based hydrological datasets offer valuable, scalable tools for basin-wide monitoring, their limitations; particularly in spatial resolution, accuracy over complex terrain, and latency, must be carefully addressed through data assimilation, ground validation, and hybrid modeling approaches for robust and adaptive hydropower management.

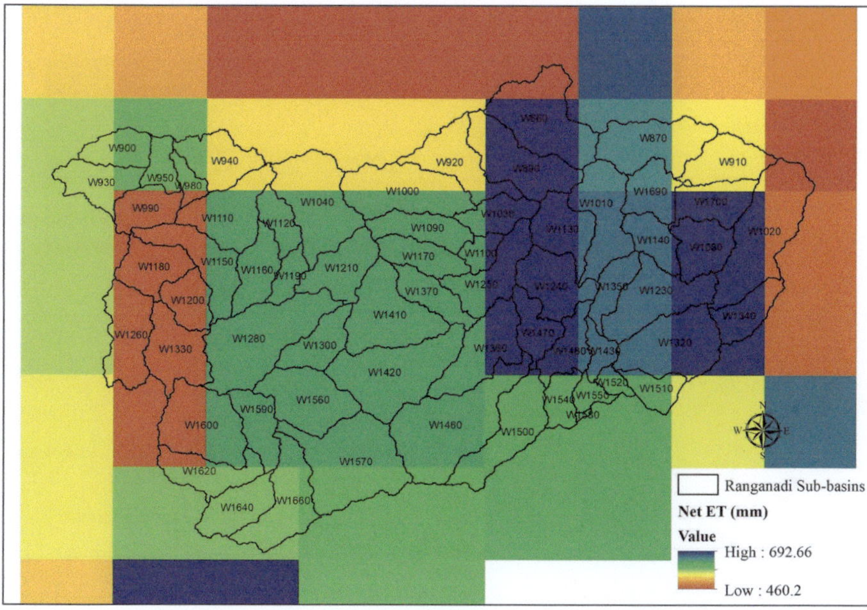

Fig. 16.19 Monthly total net ET (mm) from MODIS (July 2024) for the Ranganadi catchment

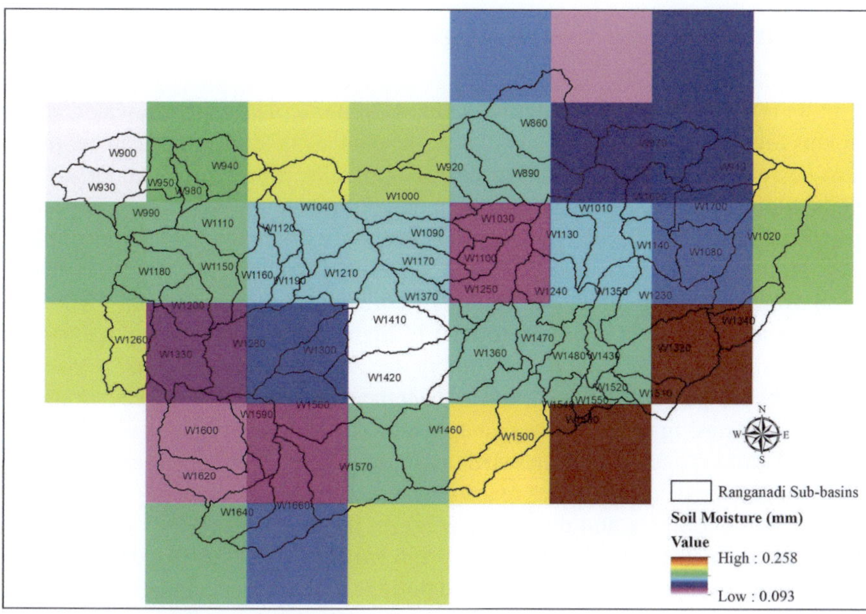

Fig. 16.20 Monthly total soil moisture (mm) from SMAP (July 2024) for the Ranganadi catchment

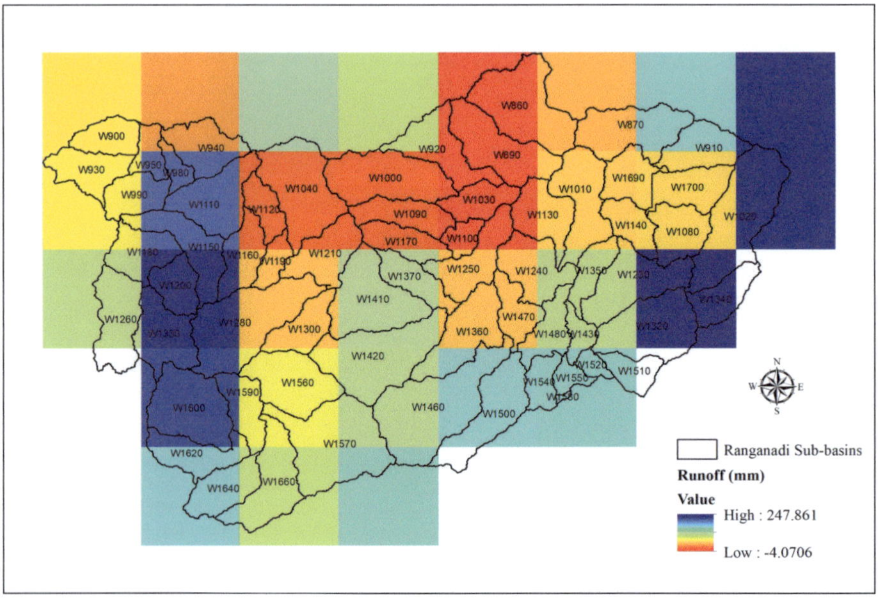

Fig. 16.21 Monthly runoff (mm) for July 2024 over the Ranganadi catchment

16.5 Summary and Conclusions

NE India is immensely rich in natural resources, particularly surface water, which accounts for about 34% of India's total water capital. This region, holds an enormous hydro-power potential of approximately 62,000 MW, representing about 40% of India's total. However, realizing this potential and managing existing hydro-power projects is characterized with challenges. The region experiences significant climatic shifts, including a historical decrease in summer monsoon rainfall and a rise in annual average temperatures. The retreat of Himalayan glaciers, which feed rivers like the Brahmaputra, is expected to lead to higher monsoon flows initially but a gradual decrease in river flow in the coming decades, thereby impacting dry season flows and hydro-power generation and management. Additionally, issues like sudden dam releases causing downstream damage, high siltation rates in reservoirs, and the impacts of anthropogenic activities such as deforestation, road construction, shifting cultivation, etc., exacerbate water management complexities.

Given these complex and diverse challenges, the chapter strongly advocates for a holistic approach with available spatial information from space technology and observed field measurements, extrapolated through hydrological modeling to provide an optimum solution for efficient hydro-power management. The case study of the Panyor Lower Hydro-Electric Project (HEP) in Arunachal Pradesh serves as a powerful demonstration of this integrated spatio-hydrological modeling approach.

16.5.1 Integrated Data Collection: The Study Emphasizes the Critical Role of Spatio-Temporal Data. It Combines

- In-situ measurements: Installation of 17 AWSs and 4 DWLRs distributed across the Panyor lower HEP catchment. These stations provide hourly rainfall, temperature, humidity, and river level data, transmitted via INSAT series satellites to a central data reception facility and visualized through a dedicated dashboard.
- Geospatial data: Utilization of advanced satellite data like CartoDEM (10 m spatial resolution) for topography, KOMPSAT-3 (0.5 m panchromatic, 2.8 m multi-spectral) and CartoSat 2 (less than 2 m spatial resolution) for river network delineation, AWS/DWLR site identification, and LULC mapping. Soil texture and hydrologic soil groups were derived from the NBSS&LUP. These critical layers were used to generate the curve number (CN) grid, a parameter for runoff potential.
- Hydrological modeling for prediction: The HEC-HMS model, an event-based hydrological model, was employed to simulate the near real-time rainfall-runoff process in the catchment. This model was rigorously calibrated and validated using observed discharge data from the dam inflow and stage data from DWLRs, achieving satisfactory performance with a correlation coefficient of 0.87 and Nash–Sutcliffe Efficiency of 0.74 during calibration. Key parameters such as curve number, initial abstraction, lag time, and Muskingum-Cunge parameters were calibrated.
- Decision Support System: The integration of real-time in-situ data, spatio-hydrological modeling, and pre-existing reservoir Area-Elevation-Capacity curves allowed for a comprehensive understanding of reservoir level rise in relation to modeled inflows and duration. This provides crucial information for dam authorities to make optimal gate operation decisions, particularly during intense monsoon seasons, thereby preventing sudden, unprecedented flood releases downstream that have historically caused significant damage.

16.5.2 Utility in the Eastern Himalayan Region

The successful implementation and demonstrated benefits of this integrated approach at Panyor Lower HEP hold significant utility for the broader Eastern Himalayan region:

- Addressing Data Scarcity: The complex topography and difficult conditions in North East India often limit the density of ground-based observation networks for rainfall and other hydrological parameters. The reliance on remote sensing and satellite-based precipitation products (like INSAT, TRMM, GPM IMERG), along with other geospatial technologies for soil moisture and evapotranspiration estimation, provides a viable alternative or supplementary data source, overcoming the limitations of traditional monitoring. This is crucial for accurate hydrological modeling across vast and inaccessible terrains. As elaborated in Sect. 16.4, water balance estimation using satellite-derived datasets offers vital insights into hydrological availability, supporting informed decision-making for hydropower generation and management, particularly in data-deprived regions. Despite inherent limitations, this approach also enhances our understanding of catchment-scale water dynamics and serves as an important tool for monitoring and planning.

- Mitigating Flood Hazards: The Eastern Himalayan region is highly susceptible to intense monsoonal flooding The capability to simulate accurate discharge at dam inflows and predict reservoir level changes in near real-time, as demonstrated, can enable proactive flood mitigation strategies and improve flood management for downstream communities, avoiding devastating impacts from sudden water releases.
- Sustainable Hydropower Management: The integrated system allows for a better understanding of the hydrological regime, including the impact of climate change on river flows (e.g., glacier melt contribution) and the effects of anthropogenic activities (e.g., deforestation-induced peak discharge increases, siltation). By providing insights into water availability and distribution, it supports efficient water resource management and ensures dependable power generation despite seasonal variations and increasing siltation issues. This directly contributes to the long-term viability and economic lifespan of crucial hydro-power plants.
- Enhancing Economic Development: By optimizing power generation and mitigating flood risks, such integrated monitoring and management systems can significantly contribute to expanding social and economic development in a region that frequently faces electricity shortages.

Overall, the chapter not only details the available space technology data, intricate hydrological processes and challenges of North East India but also provides a tangible solution through the demonstration of an integrated spatio-hydrological modeling system. This powerful combination of advanced geospatial technologies, in-situ monitoring, and robust hydrological models offers a blueprint for enhanced hydro-power management, flood control, and sustainable water resource planning across the geographically challenging and climatically sensitive Eastern Himalayan region. It highlights how cutting-edge technology can be leveraged to safeguard both vital energy infrastructure and vulnerable downstream populations in the face of a changing climate and increasing human pressures.

Acknowledgements The authors gratefully acknowledge the North Eastern Electric Power Corporation Limited (NEEPCO), a Government of India Enterprise for funding the study. We sincerely thank the NEEPCO officials for their invaluable and constant support in various aspect of field work, surveys, sensor installations, hydrological data for model calibration and validation, and many more. We also extend our appreciation to the Space Applications Centre (SAC), ISRO, Ahmedabad, for hosting the sensor data into the MOSDAC platform.

References

M.B. Abbott, J.C. Bathurst, J.A. Cunge, P.E. O'Connell, J. Rasmussen, An introduction to the European hydrological system–systeme Hydrologique Europeen, "SHE" 1: history and philosophy of the system. J. Hydrol. **87**(1–2), 45–59 (1986)

M. Alipour, M. Hosseini, Simulation of surface runoff in Karaj dam basin, Iran. Appl. Water Sci. **8** (2018). https://doi.org/10.1007/s13201-018-0782-y

R.G. Allen, L.S. Pereira, D. Raes, M. Smith, Crop evapotranspiration: guidelines for computing crop requirements. Irrigation and Drainage Paper No. 56, FAO, Rome, Italy (1998)

R.G. Allen, M. Tasumi, R. Trezza, Satellite-based energy balance for mapping evapotranspiration with internalized calibration (METRIC)–model. J. Irrig. Drain. Eng. **133**(4), 380 (2007). https://doi.org/10.1061/(ASCE)0733-9437

U.A. Amarasinghe, B. R. Sharma, N. Aloysius, C. Scott, V. Smakhtin, C. de Fraiture, Spatial variation in water supply and demand across river basins of India. Rep. 83. Colombo, Sri Lanka: International Water Management Institute (2005)

Anonymous, Central electricity authority Annual Report 2009–10 (2010)

Anonymous, Report on study of Brahmaputra River Erosion and its control. National Disaster Management Authority (NDMA) of India (2012)

Anonymous, Report on forest cover change in NE India (2019–21). Forest Survey of India (2021)

Anonymous, Report on Assam River Atlas. North Eastern Space Applications Centre, Department of Space, Govt. of India (2020)

Anonymous, Report of review of performance of hydro power stations 2022–23. Hydro Project Planning and Investigation Division. Central Electricity Authority, Ministry of Power, Govt. of India (2023)

J.G. Arnold, N. Fohrer, SWAT 2000: current capabilities and research opportunities in watershed modeling. Hydrol. Process. **19**(3), 263–265 (2005)

W.G.M. Bastiaanssen, H. Pelgrum, J. Wang, Y. Ma, J. Moreno, G. Roerink, T. van der Wal, A remote sensing surface energy balance algorithm for land (SEBAL). J Hydrology J Hydrol. **212**, 213–229 (1998)

J. Beckers, B. Smerdon, M. Wilson, Review of hydrologic models for forest management and climate change applications in British Columbia and Alberta. Forrex Forum for Research and Extension (2009). Accessed. http://epe.lac-bac.gc.ca/100/200/300/forrex/forrex_series/FS25.pdf.

K.J. Beven, A. Binley, The future of distributed models: Model calibration and uncertainty analysis. Hydrol. Process. **6**(3), 283–298 (1992)

K.J. Beven, M.J. Kirkby, A physically-based variable contributing area model of basin hydrology. Hydrol. Sci. Bull. **24**(1), 43–69 (1979)

S. Bhattacharjee, A.K. Bora, P. Das, Impacts of sedimentation on hydropower projects in North East India. Environ. Monit. Assess. **192**(5), 316 (2020)

S. Bhattacharya, A. Verma, R. Singh, Effects of seasonal rainfall variability on hydroelectric power plants in North East India. Journal of Renewable Energy **41**(5), 467–481 (2017)

P. Bhattacharya, S. Dutta, S. Paul, Influence of forest cover on runoff in the Brahmaputra basin. J. Hydrol. **582** (2020)

P. Bhuyan, S. Das, Land use and topography: Implications for flood vulnerability and water management in Northeast India. Environ Earth Sci **77**(6), 1–15 (2018)

Y.T. Bihon, T.K. Lohani, A.T. Ayalew, B.G. Neka, A.K. Mohammed, G.B. Geremew, E.G. Ayele, Performance evaluation of various hydrological models with respect to hydrological responses under climate change scenario: a review. Cogent Engineering **11**(1), 2360007 (2024)

I.S. Bowen, The ratio of heat losses by conduction and by evaporation from any water surface. Phys. Rev. **27**, 779–787 (1926)

Y. Brouziyne, L. Benaabidate, A. Abouabdillah, R. Bouabid, A. Chehbouni, Modeling hydrologic processes and potential responses to climate change in an agro-silvo-pastoral watershed in the Mediterranean area, Proc. IAHS, 383, 151–158

P. Chakravarty, R. Saikia, Hydrological dynamics in the Brahmaputra Basin: Role of topography and seasonal rainfall. Hydrol. Process. **29**(10), 2257–2269 (2015)
S. Choudhury, A. Das, S. Borthakur, Water resource management for hydroelectric power generation in North East India: challenges and opportunities. J. Hydroelectr. Power **25**(3), 178–192 (2015)
P. Colaizzi, W. Kustas, M. Anderson, J. Tolk, S. Evett, T. Howell, P. Gowda, S. O'Shaughnessy, Two-source energy balance model estimates of evapotranspiration using component and composite surface temperatures. Adv. Water Resour. **50**, 134–151 (2012)
N.H. Crawford, R.K. Linsley, Computation of a synthetic streamflow record on a digital computer. Int. Assoc. Sci. Hydrol. Publication **51**, 526–538 (1960)
P.J. Das, Rainfall regime of Northeast India: a hydrometeorological study with special emphasis on the Brahmaputra Basin. Unpublished Ph.D. Thesis. Gauhati University (2004)
P.J. Das, Water and climate induced vulnerability in northeast India: Concerns for environmental security and sustainability. WATCH Research Report 1. AARANYAK. Guwahati, Assam, India (2009). www.aaranyak.org
S. Das, B. Gupta, B.U. Choudhury, Impact of deforestation on hydrological regimes in Northeast India. Hydrol. Process. **33**(4), 542–556 (2019)
S. Das, A. Roy, Climate variability and change in Northeast India: current trends and future projections. Environ. Res. Lett. **14**(3)
C.W. Dawson, R.L. Wilby, An artificial neural network approach to rainfall-runoff modeling. Hydrol. Sci. J. **46**(5), 709–720 (2001)
P. Deka, N. Sit, A review of the traditional cereal based foods and beverages of north-east India. Int. J. Gastron. Food Sci. **31**,100632 (2023)
G. Devi, G. Bigganahalli Puttaswamigowda, G.S. Dwarakish, A review on hydrological models. Aquatic Procedia **4**, 1001–1007 (2015)
DHI (Danish Hydraulic Institute). MIKE 11: a modeling system for rivers and channels–Reference Manual. DHI, Denmark (2017)
J. Doe, Semi-distributed hydrological models: a hybrid approach for watershed modeling. Hydrol. Sci. J. **67**(3), 234–245 (2022)
Q. Duan, S. Sorooshian, V. Gupta, Effective and efficient global optimization for conceptual rainfall-runoff models. Water Resour. Res. **28**(4), 1051–1061 (1992)
S.S. Gharbia, L. Gill, P. Johnston, F. Pilla, GEO-CWB: GIS-based algorithms for parametrising the responses of catchment dynamic water balance regarding climate and land use changes. Hydrology **7**(3), 39 (2020)
E. Glenn, C. Neale, D. Hunsaker, P. Nagler, Vegetation index-based crop coefficients to estimate evapotranspiration by remote sensing in agricultural and natural ecosystems. Hydrol. Process. **25**, 4050–4062 (2011)
E. Haas, A.P. Williams, Multi-depth soil moisture simulation using the VIC land surface model Evaluating the representation of root-zone moisture dynamics. J. Hydrol. **544**, 99–112 (2017)
K. Handique, A. Dutta, Power and North East: the hydro power scenario of North East. Int. J. Sci. Res. **3**(12), 2319–7064. ISSN (Online) (2014)
G. Hargreaves, Z. Samani, Reference crop evapotranspiration from temperature. Appl. Eng. Agric. **1** (1985). https://doi.org/10.13031/2013.26773
F.M. Henderson, A.J. Lewis, Principles and applications of microwave remote sensing for soil moisture estimation. J. Hydrol. **350**(1–2), 145–159 (2008)
H.J. Henriksen, L. Troldborg, P. Nyegaard, T. Sonnenborg, J. Refsgaard, B. Madsen, Methodology for construction, calibration, and validation of a national hydrological model for Denmark. J. Hydrol. **280**(1–4), 52–71 (2003)
IPCC, Summary for policymakers. In: Climate Change 2021: The physical science basis. contribution of working group i to the sixth assessment report of the intergovernmental panel on climate change, V.P. Masson-Delmotte Zhai, A. Pirani, S.L. Connors, C. Péan, S. Berger, N. Caud, Y. Chen, L. Goldfarb, M.I. Gomis, M. Huang, K. Leitzell, E. Lonnoy, J.B.R. Matthews, T.K. Maycock, T. Waterfield, O. Yelekçi, R. Yu, B. Zhou (eds.). Cambridge University Press, Cambridge, United Kingdom and New York, NY, USA, pp. 3–32 (2021)

IPCC, Summary for policymakers. In: climate change 2023: synthesis report. Contribution of working groups i, ii and iii to the sixth assessment report of the intergovernmental panel on climate change, Core Writing Team, H. Lee, J. Romero (eds.). IPCC, Geneva, Switzerland, pp. 1–34 (2023)

S.B. Jaideep, A.R. Singal, V. Awasthi, Identify and harness hydropower potential in North East Region, in *International conference on Hydropower and dam development for water and energy security–under changing climate*, 7–9, Rishikesh, India (2022)

S.K. Jain, P.K. Agarwal, V.P. Singh, Brahmaputra and Barak Basin, *in Hydrology and Water Resources of India. Water Science and Technology Library*, vol 57. Springer, Dordrecht (2007)

P. Kucera, M. Jovanovic, Data assimilation in hydrological modeling: a review. Hydrol. Earth Syst. Sci. **10**(6), 745–757 (2006)

P. Kumar, S. Borah, Impact of Southwest monsoon and elevation on rainfall distribution in the north eastern region of India. J. Geophys. Res: Atmos. **124**(14), 7521–7536 (2019)

V. Kumar, K. Sharma, T. Caloiero, D. Mehta, K. Singh, Comprehensive overview of flood modeling approaches: a review of recent advances. Hydrology **10**(7), 141 (2023)

T. Lan, K. Lin, C.-Y. Xu, X. Tan, X. Chen, Dynamics of hydrological-model parameters: mechanisms, problems and solutions. Hydrol. Earth Syst. Sci. **24**, 1347–1366 (2020)

G. Lindström, B. Johansson, M. Persson, M. Gardelin, Development and testing of the HBV-96 hydrological model. J. Hydrol. **201**(1–4), 293–315 (1997)

C. Mahanta, J. Sharma, Influence of topographic and climatic factors on the hydrology of Northeast India. J. Hydrol. **551**, 470–485 (2017)

B. Manikiam, Evolution of Indian satellite meteorological programme. Mausam **54**(1), 1–12 (2003)

K.G. Manish, Y.S. Rao, Impact of Climate Change on Water Resources in India. J. Environ. Eng. **144**(7), 04018054 (2018). https://doi.org/10.1061/(ASCE)EE.1943-7870.0001394

J.J. McDonnell, R.A. Woods, On the need for catchment classification. J. Hydrol. **299**(1–2), 21–39 (2004)

Ministry of Environment, forest and climate change report, MoEFCC. Forest land diversion for road projects (2021)

M. Mishra, V. Pandey, Siltation control techniques in hydro-power reservoirs of North East India: a review. Int. J. Environ. Eng. **14**(2), 85–98 (2019)

M.M.Q. Mirza, R.A. Warrick, N.J. Ericksen, G.J. Kenny, Trends and persistence in precipitation in the Ganges, Brahmaputra and Meghna river basins. Hydrol. Sci.-J.-Des Hydrol. **43**(6), 845–858 (1998)

D.N. Moriasi, J.G. Arnold, M.W. Van Liew, R.L. Bingner, R.D. Harmel, T.L. Veith, Model evaluation guidelines for systematic quantification of accuracy in watershed simulations. Trans. ASABE **50**(3), 885–900 (2007)

V. Nikam, B.R. Nikam, V. Garg, P.K. Thakur, S.P. Aggarwal, S, Assimilation of remote sensing derived soil moisture in macroscale hydrological model, in Proceedings of HYDRO-2015, IIT Roorkee, Uttarakhand, India during (2015)

N.Y. Nguyen, D.D. Kha, L.V. Ninhb, V.T. Anhc, T.N. Anha, Streamflow prediction using Long Short-Term Memory networks: a case study at the Kratie Hydrological Station, Mekong River Basin. J. Hydroinf. **27**(2), 275 (2025)

M.K. Pandit, R. Kumar, S. Dutta, Soil erosion in road construction zones in Arunachal Pradesh. J. Mt. Sci. **9**(3), 350–362 (2012)

G.B. Pant, K. Rupa Kumar, *Climates of South Asia* (John Wiley and Sons, New York, 1997), p.320p

A. Palmieri, F. Shah, A. Dinar, Economics of reservoir sedimentation and sustainable management of dams. J. Environ. Manage. **61**(2), 149–163 (2001)

F. Pellicciotti, C. Buergi, W.W. Immerzeel, M. Konz, A.B. Shrestha, Challenges and Uncertainties in Hydrological Modeling of Remote Hindu Kush–Karakoram–Himalayan (HKH) Basins: Suggestions for Calibration Strategies,". Mt. Res. Dev. **32**(1), 39–50 (2012)

D. Pimentel, M. Burgess, Soil erosion threatens food production. Agriculture **3**(3), 443–463 (2013)

S. Prakash, A.K. Mitra, A. AghaKouchak, Z. Liu, H. Norouzi, D.S. Pai, A preliminary assessment of GPM-based multi-satellite precipitation estimates over a monsoon dominated region. J. Hydrol. **556**, 865–876 (2018)

D. Pranab Kumar, North –East, 'The Power House of India': prospects and problems. IOSR J Hum.Ities Soc. Sci. **18**(3), 36–48. e-ISSN: 2279–0837. (2013)

M. Rahnamay Naeini, B. Analui, H. Gupta, Q. Duan, S. Sorooshian, Three decades of the Shuffled Complex Evolution (SCE-UA) optimization algorithm: Review and applications', Scientia Iranica, 26(Special Issue Dedicated to Professor Abolhassan Vafai), pp. 2015–2031 (2019)

V. Ramesh, A. Mukherjee, Spatial distribution of rainfall and its impact on water resources in northeast India. J. Hydrol, **595** (2021)

S. Roy, P. Kumar, M. Pandit, Flood risks and seasonal fluctuations in hydro-power generation in North East India. J. Water Resour. Manag. **37**(6), 1053–1067 (2021)

S. Running, Q. Mu, M. Zhao, MODIS/Aqua Net Evapotranspiration 8-Day L4 Global 500m SIN Grid V061. NASA EOSDIS Land Processes Distributed Active Archive Center (2021)

S.K. Saha, A.K. Srivastava, Rainfall and temperature variability in northeast India: implications for agriculture. Int. J. Climatol. **40**(3), 1570–1582 (2020)

D. Saha, S. Dey, P. Mukherjee, Monsoon-driven flood patterns and runoff generation in the Brahmaputra basin. J. Flood Risk Manag. **10**(1), 85–94 (2017)

A. Sarma, R.K. Singh, Climatic patterns and seasonal rainfall distribution in the north eastern region of India: influence of the southwest Monsoon. J. Clim. Atmos. **47**(9), 765–778 (2023)

A.B. Scott, TRMM Senior Review Proposal (2011)

G.B. Senay, N.M. Velpuri, S. Bohms, M. Budde, C. Young, J. Rowland, J.P. Verdin, Drought monitoring and assessment: remote sensing and modeling approaches for the famine early warning systems network, in *Hydro-Meteorological Hazards, Risks, and Disasters*, eds. Paron, P, Baldassarre, G.D., pp. 233–262, Elsevier (2015)

T.M. Sharannya, N. Al-Ansari, S. Deb Barma, A. Mahesha, Evaluation of satellite precipitation products in simulating streamflow in a humid tropical catchment of india using a semi-distributed hydrological model. Water **12**(9), 2400 (2020)

V. Sharma, D. Swayne, D. Lam, W. Schertzer, Auto-calibration of hydrological models using high-performance computing. Int. Congr. Environ. Model. Softw. **299** (2006). https://scholarsarchive.byu.edu/iemssconference/2006/all/299

D. Sharma, M. Jain, S. Sharma, Urbanization and groundwater depletion in Guwahati. Environ Earth Sci **73**(2), 777–785 (2014)

P. Sharma, S. Deka, Impact of climate change on hydroelectric power potential in North East India. Environ. Sci. Policy **45**(4), 295–308 (2018)

R. Sharma, U.S. De, Monsoon dynamics and rainfall patterns in the Northeastern Region of India. Theoret. Appl. Climatol. **131**(3–4), 1209–1219 (2018)

S. Sharma, A. Vaidya, The role of topography in shaping hydrological extremes in Northeast India under changing climate scenarios. Hydrol. Res. **51**(3), 451–467 (2020)

L.K. Sherman, Streamflow from rainfall by the unit hydrograph method. Eng. News-Record **108**, 501–505 (1932)

H. Shen, B.A. Tolson, J. Mai, Time to update the split-sample approach in hydrological model calibration. Water Resour. Res. **58**, e2021WR031523 (2022)

A. Singh, A. Kumar, M. Jha, Topography and its influence on soil erosion and runoff in the hilly regions of Meghalaya. Geogr J India **43**(2), 215–229 (2018)

G. Singh, R.K. Panda, B.P. Mohanty, Spatiotemporal analysis of soil moisture and optimal sampling design for regional-scale soil moisture estimation in a tropical watershed of India. Water Resour. Res. **55**(3), 2057–2078 (2019)

G. Singh, N.N. Das, R.K. Panda, B.P. Mohanty, D. Entekhabi, B.K. Bhattacharya, Soil moisture retrieval using SMAP L-band radiometer and RISAT-1 C-band SAR data in the paddy dominated tropical region of India. IEEE JSTAR **14**, 10644–10664 (2021)

G. Singh, S.K. Singh, Evapotranspiration over the Indian region: implications of climate change and land use/land cover change. Nat. Environ. Pollut. Technol. **22**, 211–219 (2023)

Y. Tan, J. Wang, Application of microwave remote sensing for soil moisture monitoring at L-band, C-band, and X-band frequencies. Int. J. Remote Sens. **34**(4), 1147–1161 (2013)

C.W. Thornthwaite, An Approach toward a rational classification of climate. Geogr. Rev. **38**, 55–94 (1948). https://doi.org/10.2307/210739

N. Tripathi, R.K. Panda, S. Poonia, Soil erosion in shifting cultivation areas. For. Ecol. Manage. **278**, 123–129 (2012)

USDA Soil Conservation Service, Urban Hydrology for Small Watersheds. Technical Release No. 55 (1986)

W. Wagner, G. Bloschl, P. Pampaloni, J.C. Calvet, B. Bizzarri, J.P. Wigneron, Y. Kerr, Operational readiness of microwave remote sensing of soil moisture for hydrologic applications. Nord. Hydrol. **38**(1), 1–20 (2007)

L. Wang, X. Liang, A hybrid approach for data assimilation of remote sensing and in-situ measurements into hydrological models. J. Hydrol. **367**(1–2), 138–151 (2009)

L. Wang, Y. Qu, Soil moisture estimation using L-band, C-band, and X-band radar data. Remote Sens. Environ. **115**(4), 1052–1062 (2011)

K. Yilmaz, J. Vrugt, H. Gupta, S. Sorooshian, Model calibration in watershed hydrology, in *Advances in Data-Based Approaches for Hydrologic Modeling and Forecasting*, pp. 43–74 (2010)

Shanbor Kurbah is currently working as scientist in the Water Resources Division of North Eastern Space Applications Centre (NESAC), Department of Space, Umiam, Meghalaya. He holds an M.Tech degree in Hydrology from the Indian Institute of Technology (IIT) Roorkee, Uttarakhand. His research expertise includes hydrological data analysis, flood modeling and forecasting, hydrodynamic modeling, snowmelt runoff processing and GLOF modeling, and reservoir downstream release assessment. With over 8 years of experience in hydrology, flood monitoring and modeling, and water resources management using the integrated spatial technology, modeling software and GIS, he has ccontributed to successfully implement user-funded projects and researches in North East India. He has also guided and mentored over several M.Tech project thesis and authored publications in some of the peer-reviewed national and international journals and conferences.

Diganta Barman a senior scientist with over 22 years of experience, heads the Water Resources Division at NESAC. He holds a PhD in Civil Engineering (Water Resources & Hydrology) from IIT and specializes in hydro-modelling, watershed management, flood forecasting, and reservoir operations. He led the development of India's first operational Hydro-met Flood Early Warning System (FLEWS) for the Brahmaputra and Barak basins, a system he has steered since 2009. Dr. Barman has also led major projects like the River Atlas (Assam & NER), SWaRA, NEEPCO's Hydro-met monitoring, and flood hazard zonation for Assam. He currently mentors the HIU Team under the World Bank-funded AIRBMP project. A member of various expert committees on water and flood management in NER, he has authored 20+ peer-reviewed papers and supervised multiple PhD and postgraduate theses. He collaborates with national and international agencies including CWC, IMD, IRADe, Columbia Water Center, and ICIMOD.

B. M. Arjun is currently working as a Scientist at the National Remote Sensing Centre (NRSC), ISRO, Hyderabad, focusing on strategic applications in the water resources domain. Prior to joining NRSC, he served as a Scientist at NESAC, Shillong, for seven years, where he contributed significantly to key water resources projects. He holds a Ph.D. from IIT Kanpur and an M.Tech from NIT Surathkal, specializing in Water Resources Engineering. His areas of expertise include hydrologic and hydrodynamic modelling, flood inundation modeling, river system analysis, hydro-meteorological modelling for reservoir optimization, and hydrography. He was a core team member of the first ever GIS based operational Flood Early Warning System (FLEWS) established in India for the Brahmaputra and Barak river systems in North East India.

Ranjit Das holds a Master of Engineering in Watershed Management and Flood Control from Assam Engineering College, Guwahati. With over 20 years of experience at the North Eastern Space Applications Centre (NESAC), he specializes in applying geospatial technology to water resources, watersheds, and flood disaster management. He received the ISRO Team Excellence Award in 2009 for his work on assessing irrigation potential in AIBP-funded projects using Cartosat data. Currently, he leads several major regional and national projects at NESAC and has published research in leading national and international journals and conferences.

Chapter 17
Emerging Geospatial Technology for Empowering Governance Activities in NER

Dibyajyoti Chutia, Nilay Nishant, Puyam S. Singh, Victor Saikhom, Siddhartha Bhuyan, Avinash Chouhan, Ritu Anilkumar, Md Abu Sahin, and K. K. Sarma

Abstract The North Eastern Region (NER) of India, with its distinct geographical, and socio-cultural dynamics, poses considerable governance challenges due to its rugged terrain, remote location, and infrastructural limitations. Geospatial technologies including geographic information systems (GIS), remote sensing, Global Navigation Satellite Systems (GNSS), and AI-driven spatial analytics hold transformative potential for enhancing governance in the region. These technologies facilitate efficient land-use planning, disaster management, infrastructure development, and resource allocation while improving transparency and decision-making processes.

D. Chutia (✉) · N. Nishant · P. S. Singh · V. Saikhom · S. Bhuyan · A. Chouhan · M. A. Sahin · K. K. Sarma
North Eastern Space Applications Centre (NESAC), Umiam, Shillong, Meghalaya, India
e-mail: d.chutia@nesac.gov.in

N. Nishant
e-mail: nilay.nishant@nesac.gov.in

P. S. Singh
e-mail: ss.puyam@nesac.gov.in

V. Saikhom
e-mail: v.saikhom@nesac.gov.in

S. Bhuyan
e-mail: siddhartha.bhuyan1@nesac.gov.in

A. Chouhan
e-mail: avinash.chouhan@nesac.gov.in

M. A. Sahin
e-mail: abu.sahin@nesac.gov.in

K. K. Sarma
e-mail: kk.sarma@nesac.gov.in

R. Anilkumar
University of Bristol, Bristol, United Kingdom
e-mail: ritu.anilkumar@bristol.ac.uk

S. P. Aggarwal et al. (eds.), *Remote Sensing of the North Eastern Himalayan Ecosystem*,
https://doi.org/10.1007/978-981-95-3221-6_17

This chapter examines the complete geospatial data ecosystem, spanning open satellite constellations, IoT edge devices, and advanced cloud-based analytics. Key technological advancements include GeoAI, big data frameworks, scalable cloud GIS, IoT edge architectures, and Digital Twins, which collectively enable near-real-time decision-making and governance. Case studies illustrate practical applications, such as the e-ATLAS system optimizing election logistics through spatial planning, the project monitoring portal improving monitoring capability, and GeoTourism platforms promoting sustainable travel by mapping cultural and ecological assets. Additionally, SMART AXOM delivers geofenced disaster alerts for floods and lightning, while IoT-enabled edge devices monitor road conditions for real-time pothole detection. Public health initiatives like Fever Tracker and Geo-Health DSS enable targeted interventions, and the Transboundary Animal Disease Dashboard enhances rapid outbreak containment. These innovations demonstrate the critical role of geospatial technologies in advancing governance and development in the NER.

Keywords Geospatial governance · North Eastern India · GeoAI · Cloud GIS · IoT Edge monitoring · Digital twins

17.1 Introduction

Geospatial applications are increasingly utilising advanced technologies to address critical global challenges, including predictive analytics for natural disasters, sustainable urban planning with minimal ecological impact, and precision-driven natural resource management (Koldasbayeva et al. 2024). With robust spatial data acquisition, analytical, and visualization capabilities, geospatial technologies have become integral to governance, ensuring informed decision-making and optimized policy execution. Its ability to provide a comprehensive and integrated perspective of geographic areas makes it invaluable for addressing complex challenges and informing decision-making across diverse sectors (Hugo and Viertel 2024). From managing natural resources and mitigating disasters to planning infrastructure and delivering public services, geospatial technologies offer powerful solutions for enhancing efficiency, transparency, and accountability in governance (Smith and Mennis 2020). Core components of geospatial technology include geographic information systems (GIS), remote sensing, and Global Positioning Systems (GPS), further augmented by emerging technologies such as drones, the Internet of Things (IoT), and artificial intelligence (AI) (Dietmar Müller et al. 2016). Initially limited to mapping and exploration, geospatial technology has evolved into an indispensable tool across diverse sectors, including governance, environmental management, infrastructure development, and public services.

The driving forces behind the technological evolution in the public sector are multifaceted viz. (i) Cost reduction, (ii) Performance, (iii) Transparency & accountability and (iv) Crises. Governments increasingly face pressure to enhance operational efficiency and reduce costs; leveraging technology for task automation, workflow optimization, and process streamlining ensures sustained service delivery while minimizing expenditures. Secondly, there is an increasing demand for improved performance in public services. Citizens expect services to be faster, more reliable, accurate, and user-friendly. Digital platforms enhance government performance by reducing response times, increasing data accuracy, and providing continuous access to services, ultimately restoring public trust in government institutions. Thirdly, technology plays a crucial role in promoting transparency and accountability. Advances in secure data management, sharing, and visualization allow governments to make data accessible to the public, empowering citizens to monitor resource allocation, evaluate program outcomes, and hold decision-makers accountable. Lastly, crises often accelerate technological innovation. For instance, the COVID-19 pandemic forced governments to rapidly adopt digital solutions, enabling remote work, virtual services, and data-driven decision-making. Similarly, geospatial technologies provide critical real-time data for emergency response and recovery efforts, making them indispensable in disaster management. Figure 17.1 depicts a framework for advancing governance through space Technology and data analysis.

The North Eastern Region (NER) of India, characterised by its unique geographical features, rich biodiversity, and diverse socio-cultural landscape, presents significant opportunities and distinct challenges for governance (Basumatary and Panda

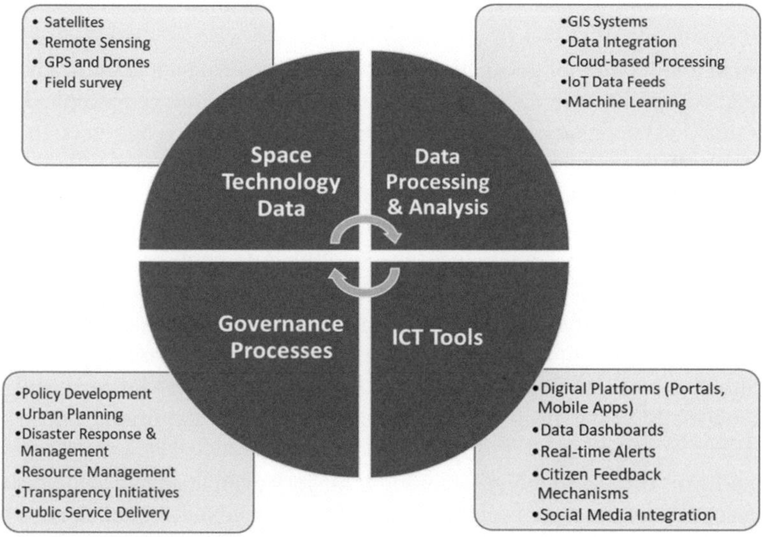

Fig. 17.1 Framework for advancing governance through space technology, data analysis, and information and communication technology (ICT) integration

2020). The region's complex terrain, susceptibility to natural hazards, and developmental disparities necessitate innovative approaches to governance that can effectively address these issues. Geospatial technology, with its capacity to provide spatially explicit information and insights, holds immense potential for empowering governance activities in the NER (Dangermond and Goodchild 2019). It can facilitate better planning and decision-making by providing accurate and up-to-date information on land use, infrastructure, demographics, and environmental conditions.

This chapter explores the transformative role of emerging geospatial technologies in empowering governance activities in NER. It explores how these technologies can be strategically utilized to tackle the region's distinct challenges while unlocking its unique opportunities. This chapter starts by outlining an overview of the various geospatial data sources and management platforms relevant to the NER context, emphasising the importance of open-source data and tools. Subsequently, the chapter explores the practical applications of geospatial technologies in governance, spanning critical areas like disaster response, land management, infrastructure planning, environmental oversight, agriculture, and public health. Real-world case studies of successful geospatial projects in the NER are showcased to highlight best practices and key takeaways. The chapter concludes by exploring the challenges and future prospects of geospatial technology in the region, providing policy recommendations to drive wider adoption and optimize its impact on governance.

17.2 Geospatial Data Sources and Management Platforms for NER

The effective utilization of geospatial technology in governance activities requires a robust and comprehensive data infrastructure. With its unique geographical characteristics and diverse data needs, the NER necessitates a tailored approach to geospatial data collection, management, and dissemination (Aggarwal et al. 2024). Remote sensing satellite data from a wide array of platforms, including Landsat, Sentinel-2, MODIS, and indigenous satellites like the Resourcesat series and Microwave EOS-04 (Wang et al. 2020), has augmented the satellite data pool, enabling comprehensive studies across multiple domains. These datasets provide valuable information on land cover, vegetation, water bodies, and infrastructure, enabling planners and decision-makers to monitor and analyze the region's dynamic landscape effectively. The Sentinel-1 satellite, with its Synthetic Aperture Radar (SAR) capabilities, and the Microwave EOS-04, designed to operate in all weather conditions, play a critical role in monitoring the region during cloud cover and monsoon seasons. These radar-based platforms can capture high-resolution imagery even in adverse weather conditions, ensuring uninterrupted monitoring of floods, landslides, and other dynamic phenomena (Jaman et al. 2025).

In addition to satellite imagery, the NER benefits from an extensive network of ground-based data collection systems, including weather stations, hydrological monitoring stations, and agricultural monitoring systems, along with Unmanned Aerial Vehicles (UAV)-derived data. These in-situ data sources complement the remote sensing data, providing a more comprehensive and high-accuracy understanding of the region's environmental and socioeconomic conditions. By integrating data from satellite platforms, UAV surveys and ground-based systems, governance agencies can create a holistic framework for effective planning, monitoring, and decision-making in the region.

To effectively manage and leverage this wealth of geospatial data, the NER has seen the emergence of various open-source, indigenous data management platforms. Platforms such as the North Eastern Spatial Data Repository (NeSDR), developed by the North Eastern Space Applications Centre (NESAC) under Department of Space serve as central repositories for geospatial data, providing access to a wide range of datasets and enabling users to discover, access, and analyze relevant information easily. Additionally, platforms like Bhuvan, developed by Indian Space Research Organization (ISRO), offer detailed satellite imagery and thematic data layers, supporting diverse applications in governance, planning, and disaster management. Similarly, tools such as Bhoondhi facilitate crowd-sourced geospatial data collection, enhancing citizen engagement in decision-making processes (Yordanov et al. 2021).

Other platforms, including MOSDAC (Meteorological & Oceanographic Satellite Data Archival Centre) and NDEM (National Database for Emergency Management), provide specialized data for meteorological studies and disaster management, respectively, aiding in risk mitigation and response planning. VEDAS (Visualization of Earth Observation Data and Archival System) offers advanced analytical tools for monitoring natural resources and environmental conditions.

UAVs have emerged as a critical tool in augmenting geospatial data collection in the NER. UAVs provide high-resolution, real-time data for areas that are difficult to access due to challenging terrain and weather conditions. These platforms are particularly effective for capturing detailed information on infrastructure development, agricultural monitoring, and disaster management. Equipped with advanced sensors, including multispectral, thermal, and LiDAR, UAVs offer flexible, low-cost alternatives to traditional aerial surveys and ground-based monitoring.

For instance, in Arunachal Pradesh, UAVs have been deployed to map road networks in remote, mountainous regions, improving the planning of infrastructure projects. Similarly, during the 2022 floods in Assam, UAVs were utilized to assess flood damage in real-time, allowing emergency response teams to prioritize relief efforts. Another notable case study involved the use of UAVs to map the damage caused due to landslides in Dima Hasao district of Assam, providing critical information to North East Frontier Railways.

By integrating UAV data with satellite imagery and ground-based systems, governance agencies in the NER can enhance situational awareness and improve the accuracy of decision-making processes. This integration has proven particularly valuable in high-risk scenarios, such as landslide-prone regions, where near real-time data is critical for early warnings and disaster mitigation.

The adoption of open-source geospatial software, such as QGIS, Google Earth Engine, and PostGIS, has further empowered governance agencies in the NER to efficiently manage, analyze, and visualize geospatial data (Wang et al. 2020). These tools, along with platforms like Google Earth Engine that enable planetary-scale geospatial analysis, offer cost-effective and flexible alternatives to proprietary software, allowing customization and integration with other systems to meet the region's specific needs. By leveraging these diverse platforms and tools, governance in the NER is becoming increasingly data-driven, enabling effective planning, monitoring, and decision-making.

17.3 Key Geospatial Technologies

While the core geospatial technologies like Remote Sensing, GIS, and GNSS remain fundamental, several emerging technologies like artificial intelligence, big data analytics, cloud computing, Internet of things, unmanned aerial vehicles or drones are rapidly transforming the field and hold immense potential for governance (Boutayeb et al. 2024). These advancements offer enhanced capabilities for data acquisition, processing, analysis, and visualization, enabling more sophisticated and impactful geospatial applications. The key geospatial technologies that are being incorporated in transforming governance are being depicted in Fig. 17.2.

17.3.1 Geospatial AI

AI with geospatial data, commonly known as GeoAI, brings powerful capabilities for processing and interpreting complex geospatial information. AI and machine learning (ML) algorithms, particularly advances such as deep learning, are increasingly being integrated with geospatial tools to automate data analysis, extract insights from complex datasets, and generate predictive models. These technologies can be applied to tasks such as image classification, object detection, change detection, and spatial pattern recognition, significantly enhancing the efficiency and accuracy of geospatial analysis (Zhu et al. 2017). The Fig. 17.3 depicts utilisation of deep learning based models to detect regions of deforestation in Manipur.

Recent advances in AI, such as deep learning-based convolutional neural networks (CNNs), have achieved exceptional accuracy in geospatial tasks, including land cover classification and object detection from satellite imagery (Prakash et al. 2020). These models excel in extracting fine-grained details, enabling precise mapping of land use changes and supporting better land management decisions. Super-resolution techniques have further advanced geospatial analysis by enhancing the resolution of satellite imagery, enabling the downscaling of coarse imagery to provide more

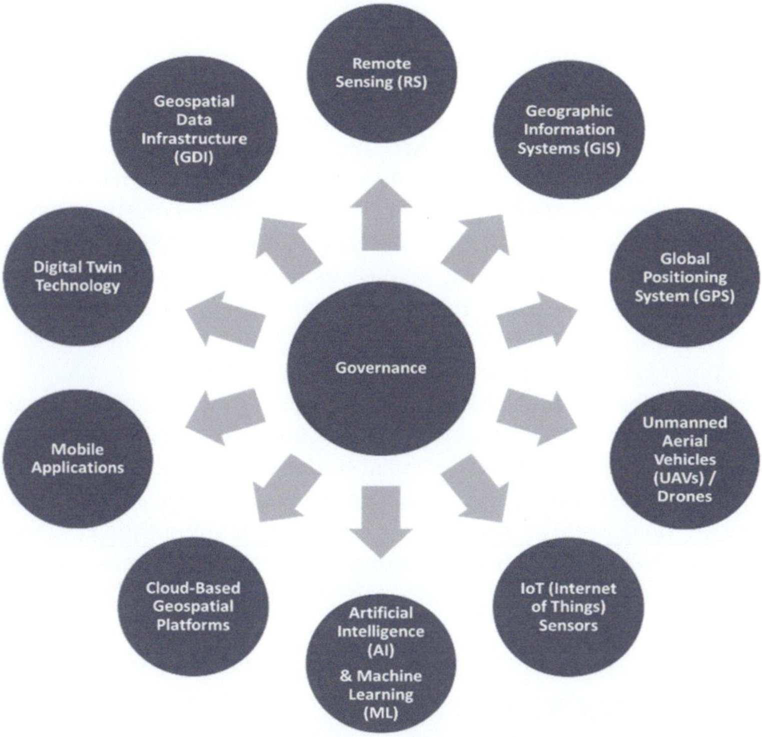

Fig. 17.2 A conceptual schematic showing key geospatial technologies supporting governance

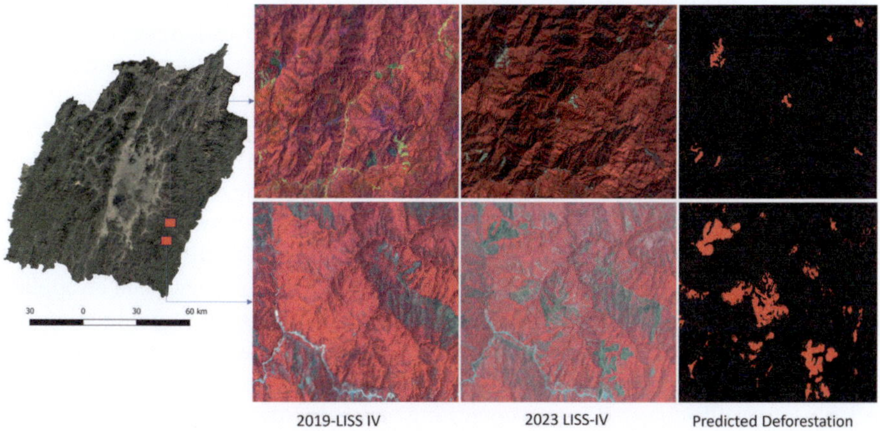

Fig. 17.3 CNN based deforestation detection on LISS-IV data for Tengpoupal district of Manipur

detailed and actionable information. Moreover, Vision Transformers (ViTs) are being employed for zero-shot learning in geospatial applications, where models can classify or interpret new types of data without extensive retraining (Kashyap et al. 2025). This capability significantly reduces the time and resources required for geospatial data analysis. Large Language Models (LLMs), such as those designed for geospatial data queries, allow users to interact with spatial databases and analytical tools using natural language, streamlining complex geospatial workflows and making them more accessible to non-experts (William 2025). In the NER of India, GeoAI has transformative applications in areas like predictive land use modelling, urban growth forecasting, and disaster risk assessment. For instance, AI-powered systems can analyze historical data and terrain using deep learning models to predict flood patterns, facilitating early warning systems and preparedness strategies. These models can incorporate data from super-resolution-enhanced satellite imagery for higher precision and ViTs for adaptive learning of new disaster patterns. Additionally, LLMs can simplify the querying of geospatial databases for rapid retrieval of actionable insights during emergencies (Adriano et al. 2021). Such predictive insights and advancements in AI are especially valuable in the disaster-prone and ecologically sensitive regions of the NER, where proactive environmental resource management and accurate, real-time analysis are essential for sustainable development and disaster resilience.

17.3.2 Geospatial Big Data Analytics

The increasing volume and velocity of geospatial data, generated from various sources like satellites, sensors, and social media, necessitate the use of big data analytics techniques (Maganathan et al. 2020). These techniques enable the processing and analysis of massive datasets to identify trends, patterns, and anomalies that would be difficult to discern using traditional methods. Big data analytics can be applied to various governance challenges in the NER, such as real-time traffic monitoring, urban planning, and disaster response. Tools such as Rasdaman, PySpark, GeoTrellis, and Hadoop are enabling the efficient processing, analysis, and visualization of large-scale geospatial data. These technologies facilitate the identification of trends, patterns, and anomalies that would be difficult to discern using traditional methods. These big data analytics techniques and technologies can be applied to various governance challenges in the NER, such as real-time traffic monitoring, urban planning, and disaster response. For instance, Hadoop and PySpark can process real-time traffic data collected from GPS-enabled devices and IoT sensors, providing insights into traffic congestion and suggesting optimal routes. GeoTrellis can be used in urban planning to handle large-scale raster datasets, enabling planners to simulate and evaluate infrastructure scenarios. Rasdaman is particularly effective for disaster response, allowing rapid analysis of satellite imagery to identify affected areas and allocate resources efficiently.

17.3.3 Cloud Computing for Scalable Geospatial Data Processing

Cloud computing platforms provide scalable and cost-effective solutions for storing, processing, and disseminating large geospatial datasets. Cloud-based GIS platforms and geospatial processing tools enable users to access and analyze geospatial data from anywhere with an internet connection, fostering collaboration and data sharing among different stakeholders (Li et al. 2020). Google Earth Engine, Sentinel Hub, and the Microsoft Planetary Computer are notable examples of advanced cloud computing tools that have revolutionized geospatial data processing and analysis (Yang et al. 2016).

In NER of India, platforms like NeSDR and North-Eastern Region Disaster-Risk-Reduction portal (NER-DRR), developed by the NESAC, complement these global tools by providing government agencies and stakeholders with unified geospatial data search, visualization, and access systems tailored to regional needs. These tools collectively streamline spatial data infrastructure, enabling stakeholders to make timely, data-driven decisions.

By combining global platforms like Google Earth Engine, Sentinel Hub, and Microsoft Planetary Computer with localized systems like NeSDR, cloud-based solutions minimize redundancies, maximize data usage, and increase efficiency in developmental planning. The integration of such technologies fosters real-time data sharing, centralized geospatial resource management, and enhanced collaboration across agencies, driving better outcomes in governance and developmental activities.

17.3.4 IoT and Edge Computing for Real Time Data Collection and Analytics

Integrating the Internet of Things with geospatial platforms has revolutionised data collection and decision-making processes (Chaterji et al. 2020). In the Northeast Region, where the challenging terrain and environmental vulnerabilities necessitate robust monitoring systems, IoT plays a pivotal role in tracking environmental parameters, managing infrastructure, and enhancing disaster preparedness. Sensors deployed in rivers continuously collect real-time data on water levels, temperature, and flow rates. When this data is integrated with geospatial models, it enables the prediction of flood inundation areas and the development of early warning systems to protect vulnerable communities (Shi et al. 2021).

A key technological enabler in this ecosystem is edge computing, which allows IoT devices to process data locally, reducing latency and minimizing the reliance on centralized cloud infrastructure (Kimovski et al. 2018). Instead of transmitting raw data to remote servers, edge computing enables analytics on the fly, ensuring immediate insights and faster decision-making. This is particularly beneficial in disaster response, smart agriculture, and asset tracking, where real-time processing can optimize resource deployment and mitigate risks.

Modern AI-powered edge devices, such as Google Coral Dev Board and NVIDIA Jetson Nano, bring sophisticated ML models to IoT networks. These compact yet powerful devices can run AI models for image recognition, anomaly detection, and predictive analytics directly on-site (Boroumand et al. 2021). For instance, in precision agriculture, UAVs and IoT sensors equipped with AI-driven edge devices can identify crop stress, detect pests, and analyze soil health in real time, allowing farmers to take immediate corrective actions. Similarly, in disaster management, AI models deployed on edge devices can analyze satellite and drone imagery to detect landslides, monitor deforestation, and assess damage in real time, providing actionable intelligence for response teams.

17.3.5 UAV-Based Geospatial Services for Governance in NER

UAV equipped with advanced sensors, such as Light Detection and Ranging and multispectral instruments, offer enhanced capabilities for geospatial data acquisition (Greenwood and Lynch 2019). The integration of these technologies on UAVs significantly expands their functionality and utility for governance and development planning in the NER of India. LiDAR sensors enable the generation of highly accurate three-dimensional point clouds, facilitating detailed topographic mapping, forest canopy analysis, and infrastructure monitoring (Indirabai et al. 2020). This is particularly valuable for large-scale terrain mapping in the hilly and forested landscapes of the NER. Multispectral sensors, on the other hand, capture data across multiple wavelengths, enabling applications like vegetation health monitoring, crop stress analysis, and land use classification. These capabilities are essential for precision agriculture and environmental management in the region (Singh et al. 2020).

Complementing the advanced sensor suites, high-resolution imagery from UAVs provides detailed visuals for tasks such as disaster assessment, urban planning, and infrastructure inspections. The ability of Vertical Take-Off and Landing (VTOL) UAVs to operate efficiently over large areas and their capacity to capture both high-resolution imagery and advanced sensor data make them crucial tools for geospatial data acquisition at a large scale (Tmušić et al. 2020). These UAVs can access remote and difficult-to-reach areas in the NER, capturing comprehensive datasets that support three-dimensional modeling of terrain and infrastructure. Users can generate precise 3D point clouds and analyze multispectral data at unprecedented scales. This combination of technologies facilitates applications such as floodplain modeling, forest resource mapping, and monitoring infrastructure in disaster-prone

Fig. 17.4 UAV-derived 3-D point cloud depicting a landslide along a railway line (Dima Hasao, Assam)

and ecologically sensitive regions (Mano et al. 2020). These advancements are transforming geospatial data acquisition and analysis, empowering governance and developmental planning in the NER. Figure 17.4 depicts a 3D point cloud of landslide along the railway line in Dima Hasao district of Assam.

17.4 Geospatial Applications for Governance in the NER

The NER of India faces a wide array of complex governance challenges that can be effectively addressed through the strategic application of geospatial technologies (Basumatary 2020). These challenges span a diverse range of domains, including disaster management, land administration, infrastructure development, environmental monitoring, agriculture, and healthcare, among others. By leveraging the power of geospatial data, analysis, and visualization, governance agencies in the NER can enhance their decision-making capabilities, improve service delivery, and promote sustainable development across the region (Wang et al. 2020). Some major areas where these emerging technologies can play a vital role are as follows:

The NER is highly susceptible to natural disasters, including floods, landslides, earthquakes, and cyclones. Geospatial techniques can play a crucial role in disaster management by providing real-time monitoring, early warning systems, and comprehensive risk assessment (Jaman et al. 2025). Remote sensing data can be used to identify flood-prone areas, monitor disaster impacts, and assess the vulnerability of local communities (2021). GIS-based modeling and analysis can help in the development of robust disaster response plans, optimizing resource allocation, and enhancing coordination among various stakeholders (Abbas et al. 2009).

Land-related issues, including land disputes, encroachment, and inefficient land records management, are significant challenges in the NER. Geospatial technologies offer effective solutions for modernizing land administration and promoting sustainable land use planning. Establishing a comprehensive land information system based on geospatial data can provide accurate and up-to-date land records, reducing land disputes and facilitating transparent land transactions. Geospatial tools can also be used for land use planning, zoning, and the identification of suitable areas for development, ensuring optimal utilization of land resources (Alsabhan et al. 2012).

Effective infrastructure development is essential for economic growth and social progress in the NER. Geospatial technologies can significantly enhance the planning, monitoring, and management of infrastructure projects. By providing detailed information on terrain, topography, and existing infrastructure, geospatial tools enable optimized route planning for roads, railways, and pipelines. During project implementation, geospatial monitoring systems can track progress, identify potential delays, and ensure adherence to quality standards. Furthermore, geospatial platforms can facilitate asset management and maintenance of infrastructure, optimizing resource allocation and extending the lifespan of infrastructure assets.

The NER is a biodiversity hotspot, and its rich natural resources are vital for the region's ecological balance and economic well-being. Geospatial technologies provide powerful tools for monitoring environmental changes, assessing biodiversity, and managing natural resources sustainably (Sergeant et al. 2012). Satellite imagery and remote sensing techniques can be used to monitor forest cover change, detect deforestation and illegal logging activities, and assess the impact of climate change on ecosystems. Geospatial tools can also be employed for wildlife monitoring, habitat mapping, and the identification of critical biodiversity areas, supporting conservation efforts and promoting sustainable resource management.

Geospatial technologies offer significant potential for enhancing agricultural productivity and promoting sustainable farming practices. Precision agriculture techniques, utilizing geospatial data on soil properties, weather patterns, and crop health, can optimize input application, reduce waste, and improve crop yields (Vali et al. 2020). Geospatial tools can also be used for crop monitoring, yield forecasting, and market analysis, empowering farmers with valuable information for informed decision-making.

Access to quality healthcare is a critical challenge in the NER, particularly in remote and underserved areas. Geospatial technologies can play a vital role in improving healthcare delivery and addressing health disparities. Mapping health facilities, identifying areas with limited access to healthcare, and tracking disease outbreaks are some of the key applications of geospatial technology in the health sector (Smith and Mennis 2020). Geospatial tools can also be used for health resource planning, optimizing the distribution of medical supplies, and improving the efficiency of healthcare delivery systems.

Geospatial technology has transformed various governance projects across Northeast India, empowering local governments and communities to tackle unique regional challenges. We explore some of the key success stories and case studies where platforms like NeSDR, SMART AXOM, e-ATLAS and other geospatial tools have enhanced governance outcomes in disaster management, land records modernisation, and tourism development. These examples illustrate the practical benefits of geospatial technology for NER's diverse and dynamic needs.

17.4.1 Empowering Electoral Process Through Space Technology

Efficient election management in NER poses significant challenges due to the vast geographical diversity, scattered polling stations, and the need for real-time communication and monitoring. Ensuring seamless coordination across administrative levels, addressing mobile connectivity gaps, and providing timely support to election officials during the polling process are critical requirements. Additionally, the safe transit of election officials and sensitive equipment, as well as the need for rapid response to incidents, demands robust technological solutions.

To address these challenges, the e-Advanced Tracking and Location-based Alerting System (e-ATLAS) e-ATLAS was developed by NESAC as a comprehensive geo-enabled platform that provides geo-located data on all polling stations within the state. The dashboard integrates information on polling station facilities and essential infrastructure such as police stations, health centers, and road networks. It also includes multiple GIS layers such as administrative boundaries and satellite image from Bhuvan application programming interfaces (APIs) as base maps, to streamline election management processes. The platform has successfully supported key elections, including the Meghalaya and Mizoram Assembly Elections (2018), Lok Sabha Elections across states except Arunachal and Assam (2019), Manipur Bye Elections (2020), Manipur Assembly Elections (2022), Tripura and Mizoram Assembly Elections (2023), and is prepared for the upcoming Lok Sabha Elections for Mizoram, Manipur, and Sikkim (2024).

Using multi-level filters, election officials can locate and assess detailed polling station information. Heat maps highlight mobile connectivity gaps, enabling effective communication planning. The Live Dashboard provides real-time updates from Presiding Officers, reporting on safe arrivals, polling status, queue lengths, and any Electronic Voting Machine (EVM) and Voter-Verifiable Paper Audit Trail (VVPAT) issues through a Mobile App. This live, geo-tagged data allows Chief and District Electoral Officers to take prompt action on incidents.

Additionally, the e-Tracker feature, leveraging open-source GPS technology, monitors the safe transit of election officials and equipment. It enables Sector Officers to respond swiftly to issues during the transit process. With continuous real-time monitoring, the system empowers election officials to make quick, informed

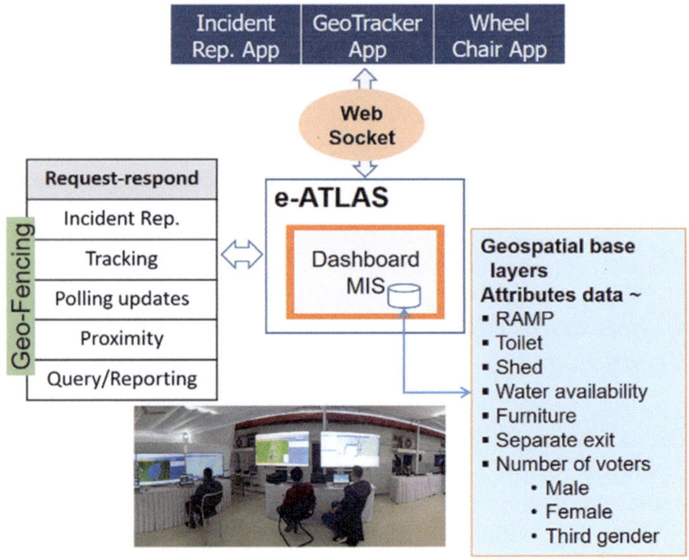

Fig. 17.5 e-ATLAS framework for real-time electoral monitoring with incident reporting, tracking, polling updates, and geospatial data

decisions, ensuring smooth election operations. The architectural framework for implementing the e-ATLAS platform is depicted in Fig. 17.5.

17.4.2 Transformative Project Monitoring with Space Technology Solutions

Monitoring large-scale development projects in the Northeast region of India has traditionally faced several challenges. The region's diverse geography, vast distances, and limited connectivity can impede the timely collection of accurate project data, making it difficult to assess progress, ensure accountability, and manage resources effectively.

In response to these challenges, the Project Monitoring Portal was developed with an objective to geotag and monitor NEC/MDoNER funded projects using geospatial technology and tools. This unique application, developed by the NESAC with support from the North Eastern Council (NEC) and Ministry of DoNER (MDoNER), leverages IT-enabled geospatial technology to streamline governance processes.

Currently, 588 projects spanning over 1,664 locations are mapped and monitored through this platform, including 37 major MDoNER projects distributed across 90 sites in the Northeast region. State Remote Sensing Applications Centres (SRSAC) actively contribute to geotagging project sites, utilizing satellite and drone imagery to capture progress for large infrastructure developments like roads and stadiums.

The portal uses a multi-pronged approach of collecting data through mobile apps, UAVs, and satellite imagery and integrates with a geo-enabled Management Information System (MIS) to track project status. By providing an end-to-end workflow, ranging from site geotagging by implementing agencies to report approval, tracking, and monitoring by NEC/MDoNER officials it ensures continuous oversight and accountability. Additionally, the platform facilitates Beneficiary Analysis (pinpointing the villages and populations benefiting from projects) and Gap Analysis (identifying infrastructure requirements like road connectivity). It also aids in preparing Detailed Project Reports (DPRs) by supplying critical base information on land use, slope, and connectivity, while linking with PM GatiShakti for integrated planning. Some of the outputs of the platform are depicted in Fig. 17.6a and b.

The Project Monitoring Portal was officially launched by Shri Ram Nath Kovind, the Hon'ble President of India, on May 4, 2022, during the valedictory ceremony of the North East Festival under the Azadi Ka Amrit Mahotsav at Sankaradeva Kalakshetra, Guwahati. The key benefits of the portal include:

- Enhanced Transparency and Accountability: Real-time, geotagged data ensures that stakeholders can readily verify project locations, statuses, and fund utilization.
- Streamlined Decision-Making: Integrated workflows and comprehensive data analytics enable quicker approvals, modifications, and interventions by NEC/MDoNER officials.

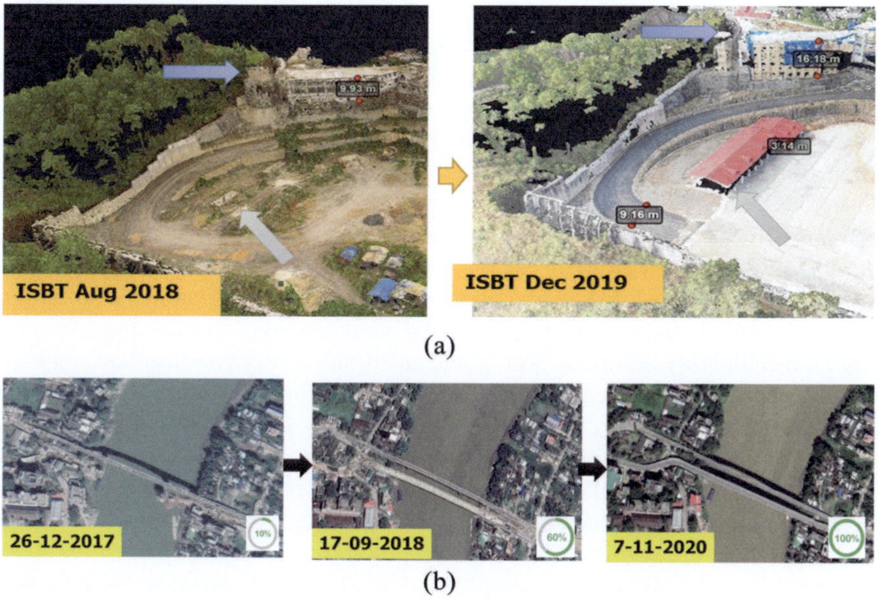

Fig. 17.6 **a** Monitoring of infrastructure projects using time series drone data. **b** Monitoring of NEC funded projects for using high resolution time series satellite images

- Efficient Resource Allocation: Beneficiary and Gap Analysis tools help identify priority areas and direct resources effectively, ensuring equitable development across the region.
- Improved Coordination: The portal's end-to-end coverage—from data collection to project completion—reduces overlaps, ensuring smoother collaboration between multiple agencies.
- Robust Planning Support: Integration with PM GatiShakti and availability of crucial spatial data (such as land use and slope information) aid in formulating more informed DPRs and long-term infrastructure strategies.

17.4.3 Tourism Development: Identifying and Promoting Tourism Potential

Northeast India's diverse cultures, breathtaking landscapes, and rich heritage offer significant potential for tourism growth. However, limited infrastructure and visibility have historically constrained the sector. While the region's scenic trails, heritage sites, and cultural festivals attract interest, both domestic and international tourists often lack sufficient information to fully plan and enjoy these offerings. This creates a disconnect between the region's tourism assets and the expectations of prospective travelers seeking unique, immersive experiences. To address this gap, the North Eastern Space Applications Centre has developed a platform called GeoTourism, which maps 1,115 tourist locations and over 20,000 utility points, enhancing awareness about the region domestically and internationally. Through GeoTourism applications, key sites such as monasteries, scenic trails, historical landmarks, and ecotourism areas are now accessible on a WebGIS platform, enabling potential tourists to explore the offerings of Arunachal Pradesh and neighbouring states. The platform integrates 38 tourist circuits, including 8 circuits identified from the MDoNER, and 20 trekking trails, creating a comprehensive database with reference to tourism infrastructure of the region. A sample tourist circuit for Meghalaya is represented in Fig. 17.7. Key functionalities include tourist location visualisation, navigation and routing, nearest utility point identification, weather forecasts, AI based travel itinerary generator, 360° photo and drone survey visualisations, tourism circuit mapping, festival details, and analyses of the Tourism Potential Index (TPI) and Travel Comfort Index (TCI). These features allow for an enriched user experience and informed trip planning. GIS analysis also identifies infrastructure gaps, such as the need for roads, accommodations, and utilities in popular areas, enabling data-driven decisions about tourism infrastructure investment. Additionally, the initiative has the potential of raising awareness about preserving ecologically sensitive sites, thus minimising the environmental impact of tourism. Through geospatially informed planning and management, Northeast India's tourism sector is poised for sustainable growth, benefiting visitors and local communities.

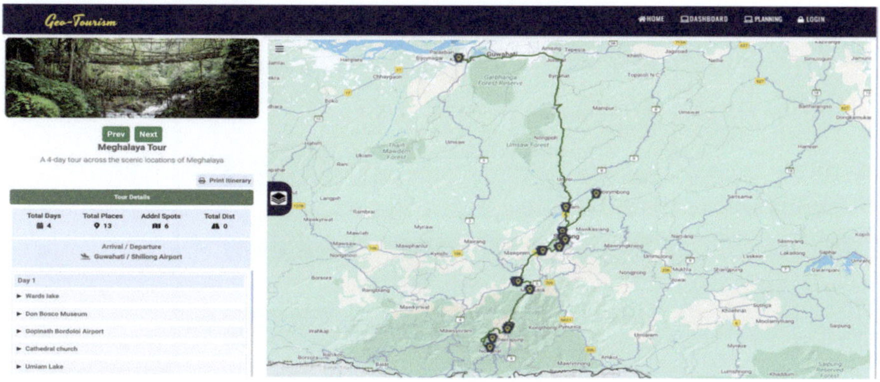

Fig. 17.7 WebGIS-based visualization of the tourism circuit identified by MDoNER, showcasing a 4-day Meghalaya tour with mapped routes and itinerary details

17.4.4 ICT Tools for Alert Dissemination: Real-Time Disaster Alerts and Citizen Engagement

The state of Assam frequently faces natural disasters such as floods, landslides, and other extreme events. Managing these hazards is a significant challenge due to the vast geography of the region, limited real-time situational awareness, and the need for rapid and accurate alert dissemination. Timely warning systems, accurate vulnerability assessments, and effective post-disaster analytics are critical for saving lives and minimizing damage. However, the absence of a centralized, integrated platform has often hindered the efficiency of emergency response and recovery efforts.

In response to these challenges, the NESAC, in collaboration with the Assam State Disaster Management Authority (ASDMA), developed SMART AXOM—an advanced, integrated disaster management platform. Designed to enhance alert dissemination, support post-disaster analytics, and strengthen overall preparedness, SMART AXOM funnels alerts from key agencies such as the India Meteorological Department (IMD), NESAC, and the Central Water Commission (CWC) through the SACHET platform. The solution encompasses three major modules:

- Geo-Fencing Based Alert Dissemination System: Delivers highly targeted alerts to specific geographical areas for more efficient and focused communication.
- Event Visualization Platform: Displays ongoing events—like earthquakes, forest fires, and floods—providing real-time situational awareness.
- Crowd-Sourcing Module: Gathers field data directly from users, enhancing the accuracy and timeliness of information.

Available on both Android and iOS, the SMART AXOM mobile application provides location-specific disaster alerts (e.g., for floods and lightning) and guides users to the nearest relief facilities or hospitals through optimal routes. A comprehensive dashboard management system enables the sending of alerts—including SMS

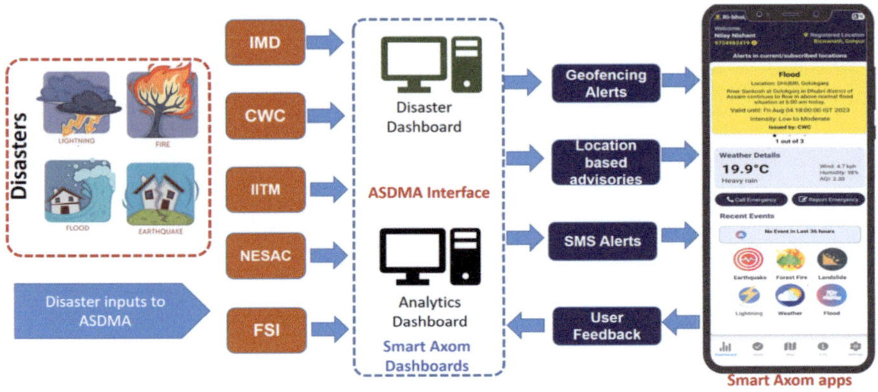

Fig. 17.8 High level workflow of geofencing based alert dissemination using SMART AXOM mobile application

notifications—to mobile devices, and integrates geospatial big data analytical tools for damage and vulnerability assessment. The platform further supports disaster preparedness through regularly updated information on relief facilities, hospitals, police stations, fire stations, disaster volunteers (Aapada Mitra), and regions prone to floods, forest fires, lightning, and other hazards. Major benefits of platform include:

- Timely and targeted warnings through geo-fencing based alerts to reach communities at-risk quickly and accurately.
- Near real-time event visualization and crowd-sourced data collection allow emergency responders and authorities to make well-informed decisions.
- Improved Disaster Preparedness with help of up-to-date information on infrastructure, volunteers, and vulnerable areas enables proactive planning and risk reduction.
- The integration of satellite imagery, big data analytics, and a central dashboard helps identify the most critical areas, optimizing the deployment of relief efforts and resources.
- The intuitive Android/iOS application empowers the public with vital alerts and navigation features, reducing panic and improving safety outcomes.

A schematic framework of the alert dissemination process is depicted in Fig. 17.8, illustrating how SMART AXOM delivers an end-to-end solution for disaster risk reduction and management in Assam.

17.4.5 Edge Computing for Pothole Detection on Roads

The road infrastructure in Northeast India faces significant challenges due to its unique terrain and climatic conditions. Frequent landslides, heavy rainfall, and poor road maintenance often result in the rapid formation of potholes, which pose serious safety risks to commuters. These issues are further compounded by limited connectivity in remote areas, making centralized data collection and analysis ineffective for timely intervention. Conventional road monitoring systems relying on manual surveys or cloud-based solutions are not feasible due to high latency, limited internet access, and high operational costs.

To address these challenges, there was a need for a cost-effective, scalable, and real-time solution to detect and report road conditions. Edge computing emerged as a viable solution by providing localized data processing, thereby reducing the dependency on stable network connectivity and centralized cloud infrastructure.

To develop a robust pothole detection system, we employed edge computing technology using the Google Coral Dev Board, a high-performance platform optimized for ML applications at the edge. The project involved deploying a pre-trained MobileNet model, fine-tuned for pothole detection, onto the Coral Dev Board. This model was selected for its balance between accuracy and efficiency, making it ideal for resource-constrained environments.

The system was integrated with video feeds from dashcams mounted on vehicles traveling across the road network. As vehicles moved along their routes, the dashcams captured real-time video footage. The edge device processed this footage on-site, using the MobileNet model to detect potholes in real-time. Detected pothole locations were immediately tagged with GPS coordinates and stored locally. In areas with adequate connectivity, the data was periodically uploaded to a central monitoring system for further analysis and visualization. Figure 17.9 depicts, utilisation of AI models deployed on google coral dev board for pothole detection in near real time.

This edge-based setup ensured real-time detection and immediate response capabilities without relying on constant internet access. It also minimized data transfer costs by filtering and processing relevant data at the source. The use of edge computing for pothole detection offered several key benefits:

1. The system enabled instantaneous detection and reporting of potholes, allowing road maintenance teams to respond promptly to critical issues.

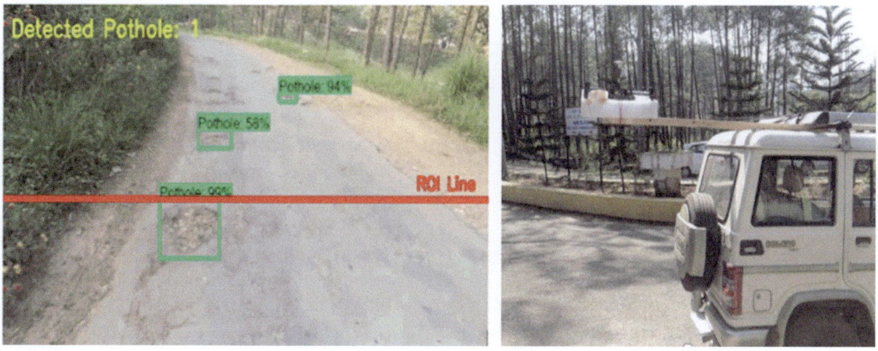

Fig. 17.9 Road pothole detection using IoT sensor mounted on vehicle camera

2. Since data processing occurred locally on the Google Coral Dev Board, the system operated efficiently even in remote areas with limited or no network access.
3. By processing data at the edge, the system reduced the volume of data that needed to be transmitted to central servers, thereby lowering data transfer and cloud storage costs.
4. The solution's lightweight design made it easily deployable across multiple vehicles, enabling wide-scale monitoring of road networks.

17.4.6 Health Sector Intervention

Monitoring and mapping for vector-borne diseases

Malaria remains a critical health issue in Tripura's rural and border areas, where access to healthcare is limited. The disease, caused by Plasmodium parasites and spread by Anopheles mosquitoes, continues challenging state-led control efforts. Despite government initiatives like distributing mosquito nets and indoor spraying, the lack of near real-time data has hindered effective targeting of high-risk areas. Traditional paper-based surveillance methods are slow and inefficient, particularly in remote regions.

To address this, the Fever Tracker app and dashboard (Fig. 17.10) were developed to streamline malaria case tracking and enable data-driven response efforts. Using a three-tier service-oriented architecture, the mobile app facilitates real-time data collection on fever cases and vector control, while the dashboard displays this data spatially on a map of Tripura. Deployed across 19 villages in Dhalai district, the app empowers health workers and volunteers to report cases and vector control data seamlessly, enhancing timely intervention efforts. The app's design prioritizes ease of use, supporting multilingual data and offline mode to meet the needs of users in remote, low-connectivity areas.

The Fever Tracker system's impact has been promising, with its user-friendly interface receiving positive feedback from health workers, who find it effective for identifying high-risk zones. Real-time GIS visualisation on the dashboard gives health officials actionable insights to manage disease outbreaks more effectively. Looking ahead, incorporating predictive analytics based on weather, breeding sites, and case trends could enhance forecasting abilities, enabling preemptive measures against malaria. Scaling the app to additional high-risk areas and other states could significantly boost malaria control efforts in India and beyond, with the potential global impact on public health.

Geospatial technology for health resource planning and accessibility analysis

Limited accessibility to health-care services happens because of its geographic isolations, poor economic status, and traditional beliefs. Further, there are sizable vulnerable populations located in remote villages with no accessibility of health care. There is therefore a need to properly assess, analyze, understand and plan better health care

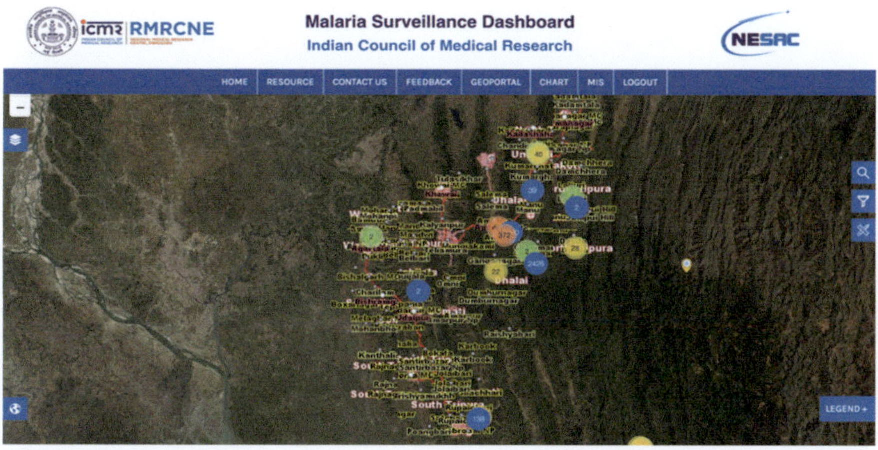

Fig. 17.10 Geoportal based visualization of malaria incidence captured via fever tracker dashboard

resources across the state to give wider accessibility to health care and its resources. To overcome healthcare infrastructure challenges in hilly terrains, it's vital to identify gaps in coverage. This involves pinpointing areas with limited transportation access, guiding new healthcare facility planning or enhancements. Strategic allocation in densely or scattered populated areas ensures optimal resource utilization, especially crucial for serving marginalized communities effectively. Recognizing the need for an integrated platform is essential for cohesive health resource planning and monitoring across the state, promoting efficiency in the healthcare system. Achieving comprehensive and precise health facility mapping is essential for effective healthcare planning, involving meticulous documentation of all infrastructure details. Mapping the road network and transportation infrastructure plays a pivotal role in assessing the accessibility of health facilities. In the context of hilly terrains, it becomes crucial to analyze population distribution meticulously, especially focusing on identifying vulnerable populations residing in remote or hard-to-reach areas. An essential step in this process is the performance of Accessibility/Gap Analysis, which considers factors such as population density, topography, and accessibility to pinpoint optimal locations for healthcare facilities.

Meghalaya Geospatial Health Resource Information System has been developed to integrate all these insights into a Unified Geo-Health Information System across the state that connects maps, applications, data, and people together for holistic decision making. Its end-to-end workflow—data ingestion, gap analysis and decision dashboards—is summarised schematically in Fig. 17.11 This interconnected system should enhance overall healthcare planning and ensures a more effective response to the healthcare needs of diverse populations, fostering a more resilient and accessible healthcare infrastructure.

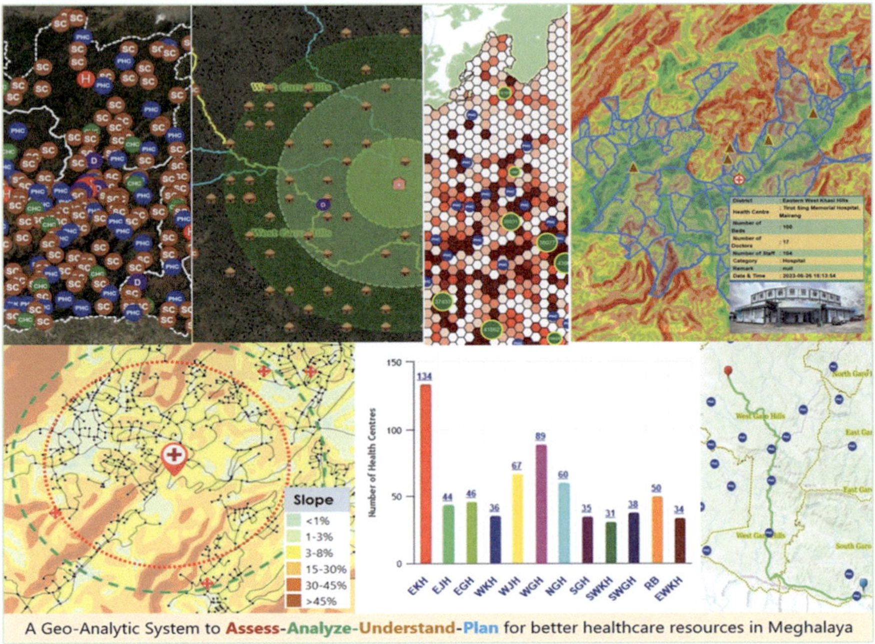

Fig. 17.11 Unified framework for health resource planning and accessibility analysis

Geospatial solutions for managing the threat of transboundary animal diseases

Transboundary animal diseases (TADs) are a huge threat for Assam and the NER of India due to the close proximity of foreign countries with neighbouring states and the situation is made more serious due to the free movement of animals into India through unfenced sections of the international border and riverine routes leading to rapid spread of diseases. Traditionally mitigating the effects of deadly diseases like African Swine Fever, Lumpy Skin Disease etc. has proved to be a long drawn out, challenging process. A snapshot of the prototype decision-dashboard is presented in Fig. 17.12, giving field officers an at-a-glance view of current outbreaks. This was evident during March–April 2020, when Assam reported an outbreak of African swine fever—a severe viral transboundary disease affecting domestic and wild pigs. It caused the death of 15,000 + pigs within the State in a matter of weeks and casualties continued even months after the initial outbreak, spreading to 22 districts over time. In the aftermath it was understood that there was absence of a means of quickly obtaining live outbreak data from the field with minimal transmission risk; there were no means of visualizing information like risk zones, outbreak areas/ hotspots, alerts, locations of veterinary facilities and centres etc. on a live map with safe navigation/geofencing facilities; there was absence of a decision support system (DSS) with latest relevant information like livestock census data, water body map, forest cover map, land use map, administrative divisions like village and district boundaries etc. in the geospatial domain along with enhanced analytical tools and

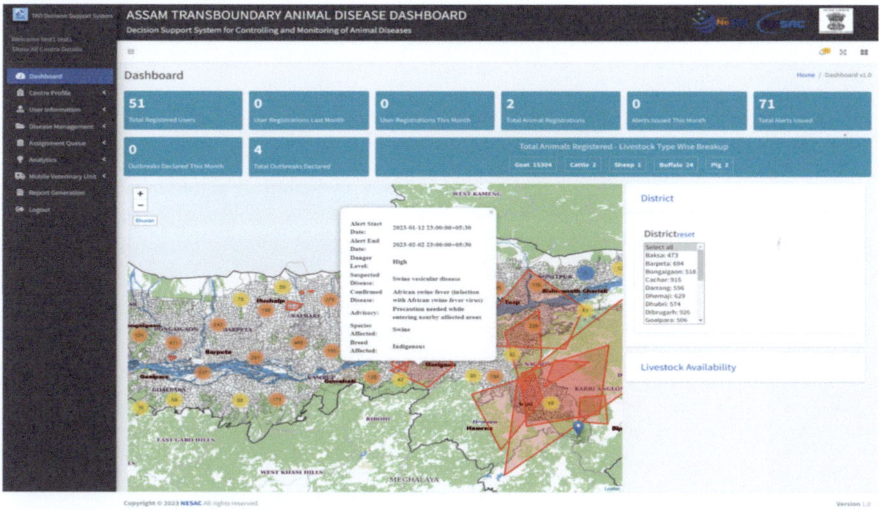

Fig. 17.12 Dashboard to visualize and track the outbreak of TADs in Assam

historical disease/outbreak data; there was no means of quickly obtaining an accurate estimate of the actual number of animal owners and animals owned, their distribution in the different villages and districts of Assam as well as historical information on animal patients; and there was no mechanism by which alerts and advisories could be quickly generated, validated and disseminated to the public as well as veterinary staff in the field.

This is in essence what the Assam Transboundary Animal Disease Dashboard application does through a mobile and web enabled interface. The application uses a geospatial technology enabled mobile app for collecting data from public users and the field using crowdsourcing as well as a web-based dashboard platform to perform analysis and decision making using this data. The presence of different roles for veterinary, district level and administrative users makes it possible to implement a hierarchy-based workflow for approving decisions made through the application. The entire veterinary infrastructure data has also been converted into digital maps and by use of interactive, on-the-fly chart based geospatial analytics and query tools, comprehensive statistics can be obtained regarding villages affected by outbreaks, livestock that needs to be culled, geofencing based risk zones and advisories, safe routing and navigation, geotagged information and data obtained from the field and strength of immunization against specific diseases etc. With several such advanced and useful features, the TAD DSS application is expected to be a path breaking intervention to manage and control TAD outbreaks in the region (Fig. 17.13).

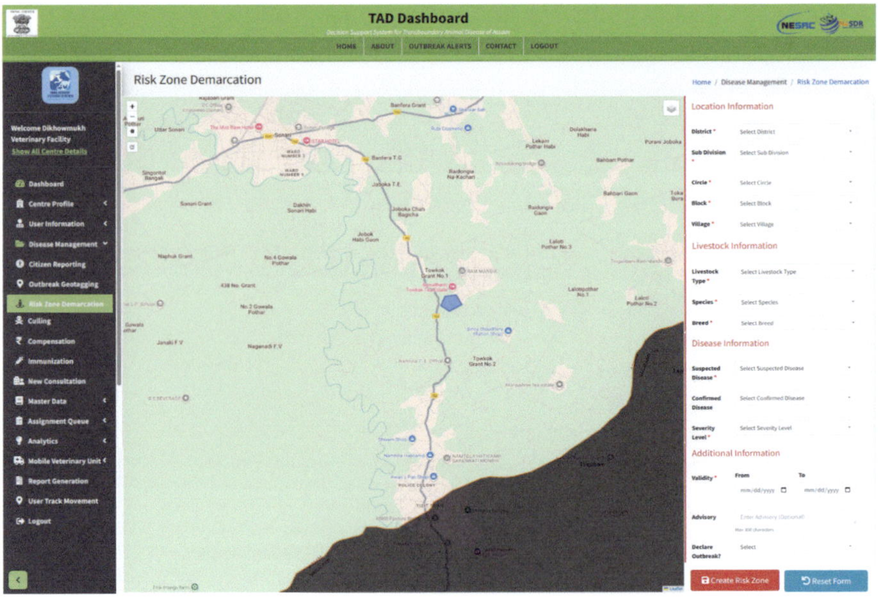

Fig. 17.13 Risk zone demarcation map generated from multi attribute analysis of TAD outbreaks

17.4.7 Connectivity and Gap Area Assessment

Many communities across the region struggle with limited access to essential healthcare and education services. Despite the presence of numerous healthcare facilities and educational institutions, connectivity gaps and inadequate infrastructure leave underserved populations without timely or convenient care. Traditional, non-data-driven approaches to infrastructure planning have been insufficient to accurately pinpoint critical service shortfalls or effectively allocate resources.

In response, a targeted geospatial initiative has mapped 9,244 healthcare facilities and 103,432 educational institutions, analyzing accessibility for 36,070 census villages. This effort is driven by a spatial decision support system (SDSS) that integrates real-time geospatial data, providing on-the-fly analytics to inform policymakers. By highlighting where infrastructure deficits are most acute such as the need for new healthcare facilities in 6,940 villages and improved roads in 828 decision-makers can prioritize interventions. The DSS further identifies 151 villages requiring new schools and 282 where better road connectivity is crucial. By leveraging these data-driven insights, stakeholders can implement targeted solutions that enhance healthcare and education outcomes, ultimately fostering broader socio-economic development in underserved areas. The spatial distribution of these gaps—and villages earmarked for new facilities or road upgrades—is mapped in Fig. 17.14. A critical component of this initiative is the development of an advanced Spatial DSS, designed to incorporate geospatial data and provide on-the-fly analytics. This

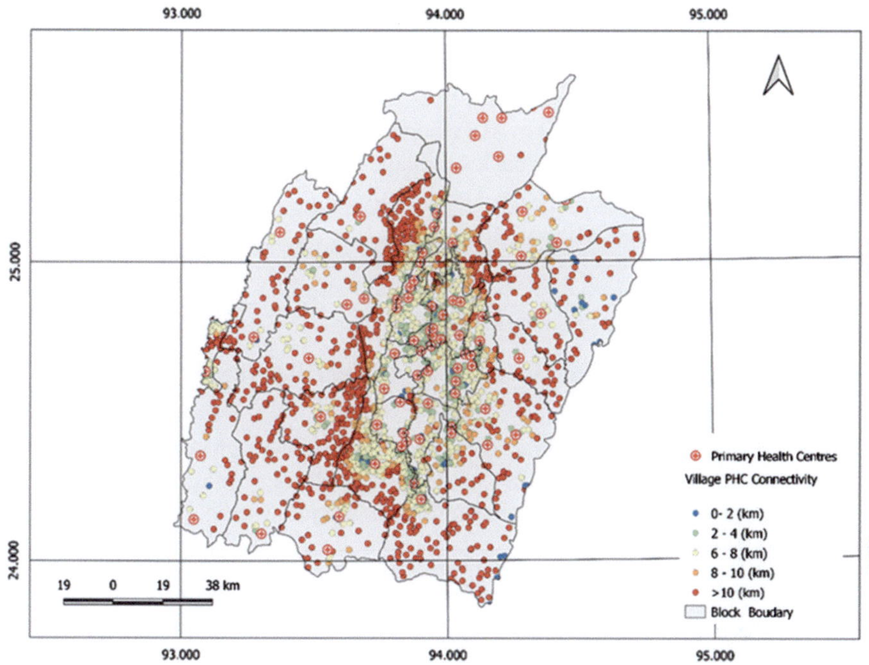

Fig. 17.14 Connectivity and gap area assessment of Health facilities in Manipur

DSS equips policymakers and stakeholders with robust tools for resource allocation and data-driven decision-making, fostering a responsive governance framework. The integration of real-time data within this system bridges the gap between spatial analysis and actionable insights, empowering authorities to address accessibility challenges and optimize infrastructure planning quickly.

The findings from this analysis highlight the need for targeted infrastructure improvements. Specifically, an estimated 6,940 villages could benefit from the establishment of new healthcare facilities, addressing vital gaps in health accessibility. Additionally, improved road construction is recommended for 828 villages to facilitate better access to healthcare services. In the education sector, 151 villages have been identified as having significant needs for new schools, while road construction for enhanced connectivity to educational facilities is suggested for 282 villages. By addressing these critical accessibility challenges through a data-driven approach, the recommendations supported by the DSS have the potential to create substantial improvements in health and education outcomes, contributing to the region's broader socio-economic development.

17.5 Future Directions

As geospatial technology evolves, new possibilities emerge for improving governance in NER. Technologies like AI, cloud-based GIS, IoT, and Digital Twins can transform how government agencies manage resources, monitor environmental conditions, and engage with citizens. These future directions point to a more interconnected, data-driven governance framework that is adaptive to NER's unique geographical, environmental, and socio-economic needs.

17.5.1 Integrating AI with Geospatial Technologies

AI has already begun to revolutionize geospatial analysis through automation, pattern recognition, and predictive capabilities. AI, especially deep learning-based workflows, has tremendous potential to enhance spatial data processing, providing insights that support proactive governance.

- AI can process historical and real-time geospatial data to predict trends such as urban growth, deforestation, and the potential impact of natural disasters. For instance, in flood-prone areas of Assam, ML models can predict flood events by analysing weather patterns, topography, and past flood data, enabling more accurate early warnings and preparedness measures.
- **Automated feature detection**: Deep learning models can be trained to recognize features like buildings, roads, and water bodies in satellite imagery, which aids in rapid infrastructure assessments, land use monitoring, and environmental conservation. This capability is particularly useful in NER, where dense forests and varied topographies make manual mapping challenging.
- **Enhanced land use and environmental monitoring**: AI can detect subtle changes in land use, forest cover, and water bodies, supporting the early identification of illegal logging, encroachments, or environmental degradation. By leveraging these insights, government agencies can implement timely interventions that prevent large-scale environmental impacts.

The integration of AI with geospatial technology represents a powerful tool for data-driven governance, helping NER address environmental challenges, infrastructure planning, and disaster resilience with improved efficiency.

17.5.2 Cloud-Based GIS Platforms for Data Sharing and Collaboration

Cloud-based GIS platforms enable real-time access to geospatial data, making it easier for multiple agencies and stakeholders to collaborate on governance projects.

This shared infrastructure approach allows for scalable data storage, centralized processing, and efficient data sharing across departments.

- **On-Demand access to spatial data**: Cloud-based GIS platforms provide government officials and stakeholders with on-demand access to data, regardless of their location. This is crucial for remote areas in NER, where local officials and community members may otherwise face limited access to spatial data resources.
- **Enhanced collaboration across agencies**: With cloud-based GIS, data sharing becomes seamless, promoting collaboration between departments such as disaster management, urban planning, health, and environmental conservation. NeSDR's cloud capabilities, for instance, can facilitate coordinated responses during emergencies by allowing all agencies to access the same real-time data.
- **Scalability and cost efficiency**: Cloud-based GIS reduces the need for costly on-site infrastructure, making it a scalable and cost-effective option for government agencies. As data needs grow, the platform can easily expand storage and processing capabilities, accommodating larger datasets and complex analyses.

By embracing cloud technology, NER's governance systems can operate more flexibly and cost-effectively, benefiting from a centralized, easy-to-access repository of geospatial data that enhances decision-making and resource allocation.

17.5.3 Role of IoT in Geospatial Data Collection

The IoT refers to networks of interconnected devices that collect and transmit data in real-time. When combined with geospatial technology, IoT provides invaluable, continuous streams of environmental, infrastructure, and public service data that can be analyzed spatially.

- **Infrastructure and environmental monitoring**: IoT sensors placed on bridges, roads, and dams can monitor structural health and environmental parameters like humidity, temperature, and vibration. This information, overlaid on GIS maps, supports proactive maintenance, especially in regions prone to landslides and floods.
- **Smart agriculture and resource management**: In agriculture, IoT devices monitor soil moisture, temperature, and crop health, supporting precision farming techniques. This is particularly beneficial in NER, where diverse climates and terrains require customized agricultural approaches. IoT data, combined with geospatial analysis, helps farmers optimize irrigation, fertilization, and harvesting schedules.
- **Enhanced public health surveillance**: IoT devices can support health surveillance by tracking disease vectors and environmental conditions that impact public health. Integrated with GIS, this data can help health agencies predict disease outbreaks, as seen with the FeverTracker app used to monitor vector-borne diseases in NER.

IoT-enabled geospatial solutions hold great potential for monitoring infrastructure, optimizing agriculture, and enhancing public health surveillance, creating a foundation for smart governance across NER's varied terrains and climates.

17.5.4 Expanding Use of Digital Twins for Simulated Governance

Digital Twin technology creates a virtual replica of physical assets, infrastructure, or entire cities. By simulating real-world dynamics, Digital Twins provide insights into how changes in one area—such as traffic, infrastructure wear, or environmental conditions—may affect others.

- **Simulating urban growth and environmental changes**: Digital Twins allow urban planners to simulate scenarios for urban growth, traffic management, and environmental changes. In NER, where cities are expanding into ecologically sensitive zones, Digital Twins help predict potential impacts, allowing planners to design eco-friendly and sustainable urban spaces.
- **Infrastructure Monitoring and maintenance**: Virtual models of infrastructure assets (e.g., bridges, dams, and roads) allow officials to assess structural health in real-time. By simulating stressors like floods or earthquakes, Digital Twins can predict how infrastructure will respond, enabling proactive maintenance and preventing catastrophic failures.
- **Disaster preparedness and response simulation**: Digital Twins are highly effective for disaster management planning. By simulating disaster scenarios—such as floods, landslides, or fires—governments can develop more robust response strategies, ensure efficient resource allocation, and minimize potential damages. These virtual simulations also enable risk-free training for emergency responders.

The expansion of Digital Twins across NER would allow local governments to make data-driven decisions that optimize infrastructure longevity, improve urban resilience, and protect natural resources, supporting sustainable growth and disaster readiness.

17.5.5 Integration of Citizen-Centric Geospatial Applications

As geospatial technology becomes more accessible, citizen-centric applications that involve communities in governance are gaining momentum. These applications empower individuals to contribute to data collection, provide feedback, and participate in decision-making processes.

- **Community mapping and participatory GIS**: Platforms that allow citizens to contribute spatial data—such as reporting infrastructure damage, illegal logging,

or environmental hazards—improve data accuracy and foster community engagement. Participatory GIS tools can be particularly effective in NER, where local communities often have valuable knowledge of their environments.
- **Mobile applications for citizen feedback**: Apps that enable citizens to report issues (e.g., potholes, garbage disposal, or water contamination) enhance governance responsiveness. By integrating this feedback with GIS, governments can track problem areas, prioritize resource allocation, and ensure that public concerns are addressed swiftly.
- **Crowdsourced data for environmental monitoring**: Citizens can play a critical role in monitoring natural resources, such as tracking changes in forest cover or water quality. By using geospatial mobile applications, communities in NER can document environmental changes, helping authorities address degradation early.

17.6 Conclusion

Geospatial technology has emerged as a transformative tool for governance in NER, enabling data-driven decision-making, fostering transparency, and enhancing citizen engagement. Through platforms like the NeSDR, governments in NER are addressing unique regional challenges, from disaster preparedness and natural resource management to infrastructure planning and public service delivery. As illustrated in this chapter, geospatial solutions—such as GIS, remote sensing, GeoAI, and Digital Twins—have brought significant advancements, allowing for more resilient, efficient, and responsive governance. The integration of geospatial technology in NER has led to several benefits that are reshaping the governance landscape:

1. **Enhanced Disaster Resilience**: Real-time monitoring, predictive analytics, and early warning systems have enabled proactive disaster management, saving lives, minimizing property damage, and building community resilience.
2. **Improved Resource Management**: GIS and remote sensing allow authorities to monitor and manage forests, water bodies, and agricultural land, promoting sustainable use of natural resources and environmental conservation.
3. **Informed Urban and Infrastructure Planning**: Spatial data and analytical tools are helping government agencies plan and manage urban growth, infrastructure projects, and connectivity in ways that meet local needs and respect ecological limitations.
4. **Streamlined Land Administration**: Digitized land records, geotagging, and cadastral mapping have brought transparency and efficiency to land administration, reducing disputes and empowering local communities with recognized land ownership.
5. **Increased Citizen Engagement and Transparency**: By providing public access to geospatial data and enabling citizen participation in data collection, governments are fostering a collaborative environment where transparency and accountability are the norm.

Acknowledgements The authors gratefully acknowledge the financial and technical support received from the NEC and the MDoNER, Government of India for funding NeSDR, Project-Monitoring Portal and Geo-Tourism; ASDMA for sponsoring SMART AXOM; and Chief electoral officers of the states for e-ATLAS and user departments of NER other initiatives presented in this chapter.

References

S.H. Abbas, R.K. Srivastava, R.P. Tiwari, B.P. Ramudu, GIS-based disaster management: a case study for Allahabad Sadar sub-district (India). Manag. Environ. Qual. **20**(1), 33–51 (2009). https://doi.org/10.1108/14777830910922433

B. Adriano, N. Yokoya, J. Xia, H. Miura, W. Liu, M. Matsuoka, S. Koshimura, Learning from multimodal and multitemporal earth observation data for building damage mapping. ISPRS J. Photogramm. Remote Sens. **175**, 132–143 (2021). https://doi.org/10.1016/j.isprsjprs.2021.02.016

S.P. Aggarwal, S.S. Kundu, K.K. Sarma, Geospatial technology for effective disaster risk reduction: best practices in capacity building. Int. Arch. Photogramm. Remote Sens. Spatial Inf. Sci. **XLVIII-5-2024**(5), 147–153 (2024). https://doi.org/10.5194/isprs-archives-XLVIII-5-2024-147-2024

W. Alsabhan, O. Ben Ali, Mobile land information system (MLIS): a GIS-based e-government application. Int. J. Mob. Learn. Organ. **6**(3–4), 260–284 (2012). https://doi.org/10.1504/IJMLO.2012.050037

N. Basumatary, B. Panda, A review of institutional and developmental issues in North-East India. Indian J. Public Adm. **66**(2), 206–218 (2020). https://doi.org/10.1177/0019556120916884

A. Boroumand, S. Ghose, B. Akin, R. Narayanaswami, G.F. Oliveira, X. Ma, E. Shiu, O. Mutlu, Google neural network models for edge devices: analyzing and mitigating machine learning inference bottlenecks, in *Proceedings of 30th International Conference on Parallel Architectures and Compilation Techniques (PACT)* (2021), pp. 159–172. https://doi.org/10.1109/PACT52795.2021.00019

A. Boutayeb, I. Lahsen-cherif, A. El Khadimi, A comprehensive GeoAI review: progress, challenges and outlooks (2024), arXiv:2412.11643. https://arxiv.org/abs/2412.11643

S. Chaterji, N. DeLay, J. Evans, N. Mosier, B. Engel, D. Buckmaster, R. Chandra, Artificial intelligence for digital agriculture at scale: techniques, policies, and challenges (2020), arXiv:2001.09786. https://arxiv.org/abs/2001.09786

Y. Choi, GeoAI: integration of artificial intelligence, machine learning, and deep learning with GIS. Appl. Sci. **13**(6), 3895 (2023). https://doi.org/10.3390/app13063895

J. Dangermond, M.F. Goodchild, Building geospatial infrastructure. Geo-Spatial. Inf. Sci. **23**(1), 1–9 (2019). https://doi.org/10.1080/10095020.2019.1698274

R. Dietmar Müller, M. Seton, S. Zahirovic, S.E. Williams, K.J. Matthews, N.M. Wright, G.E. Shephard, K.T. Maloney, N. Barnett-Moore, M. Hosseinpour, D.J. Bower, J. Cannon, Ocean basin evolution and global-scale plate reorganization events since Pangea breakup. Annu. Rev. Earth Planet. Sci. **44**(1) 107–138 (2016). https://doi.org/10.1146/earth.2016.44.issue-1; https://doi.org/10.1146/annurev-earth-060115-012211

W.W. Greenwood, J.P. Lynch, D. Zekkos, Applications of UAVs in civil infrastructure. J. Infrastruct. Syst. **25**(2), 04019002 (2019). https://doi.org/10.1061/(ASCE)IS.1943-555X.0000464

N. Hugo, D. Viertel, The use of GIS for land use planning: recommendations for PPGIS in the United States. Int. J. Eng. Sci. Technol. **6**(3), 329–337 (2024). https://doi.org/10.46328/ijonest.220

I. Indirabai, M.V.H. Nair, J.R. Nair, R.R. Nidamanuri, Direct estimation of leaf area index of tropical forests using LiDAR point cloud. Remote Sens. Appl. Soc. Environ. **18**, 100295 (2020). https://doi.org/10.1016/j.rsase.2020.100295

T. Jaman, S. Bhaskar, S. Swain, V. Saikhom, R. Gogoi, K.K. Sarma, S. Aggarwal, Integrating hydrospatial variables with machine learning for groundwater potential zonation in Dhalai District, Tripura. Water Conserv. Sci. Eng. **10** (2025). https://doi.org/10.1007/s41101-025-00401-z

G.S. Kashyap, M. Kulahara, N. Joshi, U. Naseem, How can multimodal remote sensing datasets transform classification via SpatialNet-ViT? (2025), arXiv:2506.22501. https://arxiv.org/abs/2506.22501

D. Kimovski, H. Ijaz, N. Surabh, R. Prodan, An adaptive nature-inspired fog architecture. arXiv preprint arXiv:1803.03444 (2018)

D. Koldasbayeva, P. Tregubova, M. Gasanov, A. Zaytsev, A. Petrovskaia, E. Burnaev, Challenges in data-driven geospatial modeling for environmental research and practice. Nat. Commun. **15**, 10700 (2024). https://doi.org/10.1038/s41467-024-55240-8

Z. Li, W. Tang, Q. Huang, E. Shook, Q. Guan, Introduction to big data computing for geospatial applications. ISPRS Int. J. Geo-Inform. **9**(8), 487 (2020). https://doi.org/10.3390/ijgi9080487

T. Maganathan, S. Senthilkumar, V. Balakrishnan, Machine learning and data analytics for environmental science: a review prospects and challenges. Abstract IOP Conf. Ser.: Mater. Sci. Eng. **955**(1), 012107 (2020). https://doi.org/10.1088/1757-899X/955/1/012107

K. Mano, K. Sakai, K. Tachibana, K. Sakita, S. Nishiyama, The measurement accuracy and measurement characteristics of green LiDAR drone. Int. Arch. Photogramm. Remote Sens. Spatial Inf. Sci. **XLIII-B1-2020**, 479–483. ISPRS Congress (2020)

J. Mennis, D. Guo, Spatial data mining and geographic knowledge discovery—an introduction. Comput. Environ. Urban Syst. **33**(6), 403–408 (2009). https://doi.org/10.1016/j.compenvurbsys.2009.11.001

N. Prakash, A. Manconi, S. Loew, Mapping landslides on EO data: performance of deep learning models vs. traditional machine learning models. Remote Sens. **12**(3), 346 (2020). https://doi.org/10.3390/rs12030346

C.J. Sergeant, B.J. Moynahan, W.F. Johnson, Practical advice for implementing long-term ecosystem monitoring. J. Appl. Ecol. **49**, 969–973 (2012). https://doi.org/10.1111/j.1365-2664.2012.02149.x

B. Shi, S. Catsamas, P. Kolotelo, M. Wang, A. Lintern, D. Jovanovic, P.M. Bach, A. Deletic, D.T. McCarthy, A low-cost water depth and electrical conductivity sensor for detecting inputs into urban stormwater networks. Sensors **21**(9), 3056 (2021). https://doi.org/10.3390/s21093056

P. Singh, P.C. Pandey, G.P. Petropoulos, A. Pavlides, P.K. Srivastava, N. Koutsias, K.A.K. Deng, Y. Bao, Hyperspectral remote sensing in precision agriculture: present status, challenges, and future trends, in *Earth Observation, Hyperspectral Remote Sensing* (Elsevier, 2020), pp. 121–146. https://doi.org/10.1016/B978-0-08-102894-0.00009-7

C.D. Smith, J. Mennis, Incorporating geographic information science and technology in response to the COVID-19 pandemic. Prev. Chronic Dis. **17**, 200246 (2020). https://doi.org/10.5888/pcd17.200246

G. Tmušić, S. Manfreda, H. Aasen, M.R. James, G. Gonçalves, E. Ben-Dor, A. Brook, M. Polinova, J.J. Arranz, J. Mészáros, R. Zhuang, K. Johansen, Y. Malbeteau, I.P. de Lima, C. Davids, S. Herban, M.F. McCabe, Current practices in UAS-based environmental monitoring. Remote Sens **12**(6), 1001 (2020). https://doi.org/10.3390/rs12061001

A. Vali, S. Comai, M. Matteucci, Deep learning for land use and land cover classification based on hyperspectral and multispectral Earth observation data: a review. Remote Sens. **12**, 2495 (2020). https://doi.org/10.3390/rs12152495

T. Wang, G. Giuliani, A. Lehmann, Y. Jiang, X. Shao, L. Li, H. Zhao, Supporting SDG 15, life on land: identifying the main drivers of land degradation in Honghe Prefecture, China, between 2005 and 2015. ISPRS Int. J. Geo-Inf. **9**(12), 710 (2020). https://doi.org/10.3390/ijgi9120710

I.O. William, Geo-spatial analysis with large language model. Int. J. Adv. Nat. Sci. Eng. Res. **9**(1), 1–9 (2025), https://as-proceeding.com/index.php/ijanser/article/view/2409

C. Yang, Q. Huang, Z. Li, K. Liu, F. Hu, Big data and cloud computing: innovation opportunities and challenges. Int. J. Digit. Earth **10**(1), 13–53 (2016). https://doi.org/10.1080/17538947.2016.1239771

V. Yordanov, L. Biagi, X.Q. Truong, V.A. Tran, M.A. Brovelli, An overview of geoinformatics state-of-the-art techniques for landslide monitoring and mapping. Int. Arch. Photogramm. Remote Sens. Spatial. Inf. Sci. **XLVI-4/W2-2021**, 205–212 (2021). https://doi.org/10.5194/isprs-archives-XLVI-4-W2-2021-205-2021

X.X. Zhu, D. Tuia, L. Mou, G.S. Xia, L. Zhang, F. Xu, F. Fraundorfer, Deep learning in remote sensing: a comprehensive review and list of resources. IEEE Geosci. Remote Sens. Mag. **5**(4), 8–36 (2017). https://doi.org/10.1109/MGRS.2017.2762307

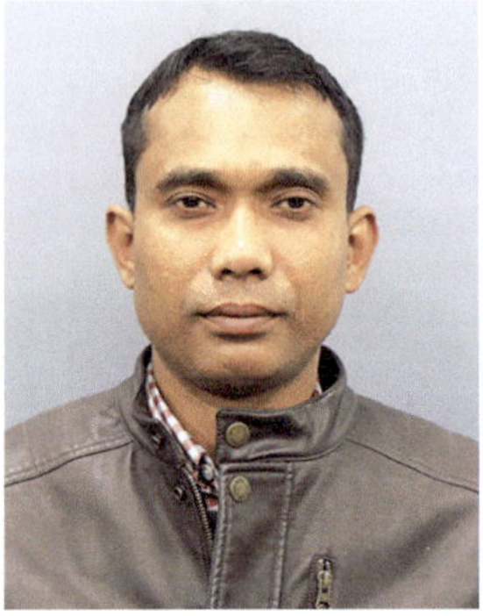

Dibyajyoti Chutia received the M. Tech. degree in Information Technology and Ph.D. degree in computer science & engineering with specialization in soft computing technique for classification of geospatial data from Tezpur Central University, Tezpur, India, in 2000 and 2015, respectively. He is currently serving as Head of Geoinformatics and IT division at North Eastern Space Applications Centre (NESAC), Department of Space, Government of India. His research interests include soft computing, machine learning, and deep learning applications for remotely sensed data. He has authored or co-authored more than 65 research articles in peer-reviewed journals, conferences, book chapters, etc. He has received the National e-Governance Award (2018, 2022), National Geospatial award (2020) and the eNorth-East Award (2020, 2023) for his and his team's contributions to geospatial innovations in governance and regional development.

Nilay Nishant is currently serving as Scientist/Engineer-SE at the North Eastern Space Applications Centre (NESAC), Department of Space, Government of India. He holds a Master's degree in Remote Sensing from Birla Institute of Technology, Mesra, and Bachelor's degree in Electronics and Communication Engineering from Pondicherry University. His research interests include geospatial big data analytics, e-Governance, and the application of Artificial Intelligence and deep learning techniques to earth observation data. He has played a key role in the development of more than 25 governance-focused geospatial applications for various departments across the North Eastern Region of India. He has authored 30 research publications in reputed national and international journals and conference proceedings. In recognition of his significant contributions to the field of geospatial technology for governance, he has been honored with the National e-Governance Award (2022) and the e-North East Award (2023, 2020).

Puyam S. Singh currently working as Scientist/Engineer–SF at the North Eastern Space Applications Centre (NESAC), specializing in Geo-spatial Intelligence and Computer Vision. He holds an M.E. in Computer Science from PEC University of Technology, Chandigarh. Previously, he worked as a research fellow at DRDO's Snow and Avalanche Study Establishment (SASE), contributing to avalanche forecasting models. At NESAC, he works on the development of space-based geospatial applications for societal benefit and e-Governance, integrating satellite and aerial data. His research focuses on computer vision algorithms for automated feature extraction from remote sensing imagery. He has mentored over 54 postgraduate and undergraduate students and has published 43+ papers in reputed journals and conferences. His innovative contributions have earned him National and regional e-Governance Awards in 2015, 2018, and 2023. Puyam continues to advance AI-driven remote sensing solutions for operational and research applications in governance and development sectors.

Victor Saikhom is currently serving as Scientist/Engineer-SF at the Geoinformatics & IT Division, NESAC, Department of Space, with a Ph.D. in Remote Sensing, M.Tech in Remote Sensing & GIS, and B.E. in Electrical & Electronics Engineering. With over two decades of combined experience in industry and research, he specializes in Photogrammetry, 3D GIS, UAV Remote Sensing, and Geoinformatics. He has led key geospatial projects in land use mapping, infrastructure planning, terrain modeling, and precision agriculture using satellite, UAV, DGPS, and TLS data. His expertise includes DEM generation, orthomosaics, point cloud modeling, and multispectral analysis for disaster response and environmental monitoring. Dr. Saikhom also mentors students and professionals and serves as Course Director for UAV Remote Sensing at NESAC. He has published widely and received the National e-Governance Award (2018) and the 8th eNorthEast Award (2023) for his and his team's contributions to geospatial solutions for governance and regional development.

Siddhartha Bhuyan is presently serving as Scientist/Engineer-SD at the North Eastern Space Applications Centre (NESAC). He holds an M.Tech from VIT, Vellore, and a B.Tech from ICFAI University, Dehradun both in Computer Science and Engineering. With over 12 years of experience in the IT industry and academia, his expertise spans AI, Algorithms, Machine/Deep Learning, System and Database Design, and Natural Language Understanding. At NESAC, he has worked on mobile app development, cyber security, automation, geo web services, and decision support systems for e-governance. He has authored 5 journal publications, 2 conference papers, and holds a patent for a swarm UAV-based data acquisition algorithm developed with KJ Somaiya College of Engineering. He has also contributed several innovative and impactful ideas in national hackathons / brainstorming events and also pioneered the use of IoT for disaster management at NESAC. He is currently focusing on developing novel algorithms and applications in the field of Health GIS.

Avinash Chouhan is a Scientist/Engineer-SE at the North Eastern Space Applications Centre (NESAC), Department of Space, Shillong, India. He began his research career at NESAC in 2014. He earned his Ph.D. in Computer Science and Engineering from the Indian Institute of Technology Guwahati in 2024 and holds a B.E. in Computer Science and Engineering from the Government Engineering College, Jabalpur. His research interests include GeoAI, Generative AI, and Deep Learning, with a particular focus on developing domain-adaptive models for remote sensing applications. Dr. Chouhan actively works on bridging foundational AI with earth observation data to address real-world geospatial challenges in the North Eastern Region and beyond, contributing to the advancement of space-based remote sensing solutions.

Ritu Anilkumar is currently a Senior Research Associate at the University of Bristol, working in the domain of glaciology and Earth system science. She holds a Bachelor's degree from the Indian Institute of Space Science and Technology (IIST) and has recently submitted her PhD thesis at the Indian Institute of Technology Guwahati, where her research focused on Glacier Mass Balance Modelling using machine learning techniques. Prior to this, she served as Scientist/Engineer at the North Eastern Space Applications Centre (NESAC), Department of Space, Government of India, from 2017 to 2025. During her tenure, she was actively involved in developing geospatial solutions for environmental monitoring and natural resource management, with a focus on applying data-driven and AI-based approaches to remote sensing. Her research interests include geospatial data science, machine learning for Earth observation, and environmental modeling. She has contributed to multiple interdisciplinary projects aimed at advancing the application of AI in geosciences and has authored several publications in reputed journals and conferences.

Md Abu Sahin has been a Scientific Assistant at the North Eastern Space Applications Centre (NESAC) since December 2021. He holds an M.Sc. in Physics from Presidency University, Kolkata. His work focuses on Remote Sensing, GIS, and Geoinformatics, with expertise in geospatial data analysis, satellite image interpretation, and deep learning for change detection. He contributes to deforestation monitoring, landslide studies, and disaster risk assessment, as well as network analysis, thematic atlas preparation, and site suitability modeling. His efforts support regional projects in tourism, infrastructure planning, and sustainable development in Northeast India. He also handles hyperspectral instruments for field data collection and calibration.

K. K. Sarma Ph. D, leads the Remote Sensing Applications Group (RSAG) at the North Eastern Space Applications Centre (NESAC), where he has been a key figure since 2002. Specializing in remote sensing for forestry and biodiversity, his work leverages space technology for critical applications in Biodiversity, Ecosystems, Forestry, and Natural Resource Management. Dr. Sarma has spearheaded over 60 diverse geospatial projects, contributing significantly to natural resource mapping and monitoring across the North Eastern Region (NER). Currently, he coordinates 110 "Plan of Action" projects, focused on using space technology to improve livelihoods in the NER. He also served as the Principal Investigator and NER Coordinator for all ISRO natural resource management projects, including comprehensive Land Use Land Cover Mapping and National Wasteland Updating, Monitoring, and Change Analysis. With over 75 research publications, Dr. Sarma has successfully supervised six Ph.D. students and actively mentors many others across NER Universities.

Chapter 18
GeoAI: Exploring Trends and Potential Applications

Avinash Chouhan, Puyam S. Singh, Ritu Anilkumar, Siddhartha Bhuyan, Nilay Nishant, and Dibyajyoti Chutia

Abstract Geospatial Artificial Intelligence (GeoAI) is an emerging interdisciplinary field that synergizes geospatial data science with advanced artificial intelligence (AI) and machine learning techniques. This chapter provides a comprehensive overview of the foundational concepts, core methodologies, and impactful real-world applications of GeoAI, with a particular focus on its transformative role in remote sensing and geospatial analytics. The integration of deep learning has substantially enhanced automation, accuracy, and scalability in tasks such as land cover classification, object detection, semantic segmentation, and change detection. A comparative analysis of benchmark geospatial datasets is presented, illustrating their significance and usage in recent AI-driven research across various domains. The chapter also examines the key challenges facing the field, along with ethical considerations related to privacy, fairness, and transparency. Finally, it explores future research directions, emphasizing the need for explainable, domain-adaptive, and sustainable GeoAI systems to support responsible decision-making in the era of big Earth observation data.

Keywords Geospatial artificial intelligence · Deep learning · Machine learning · Automated geospatial image analysis · Computer vision · Natural language processing

A. Chouhan (✉) · P. S. Singh · S. Bhuyan · N. Nishant · D. Chutia
North Eastern Space Applications Centre (NESAC), Umiam, Shillong, Meghalaya, India
e-mail: avinash.chouhan@nesac.gov.in

P. S. Singh
e-mail: ss.puyam@nesac.gov.in

S. Bhuyan
e-mail: siddhartha.bhuyan1@nesac.gov.in

N. Nishant
e-mail: nilay.nishant@nesac.gov.in

D. Chutia
e-mail: d.chutia@nesac.gov.in

R. Anilkumar
University of Bristol, Bristol, UK

S. P. Aggarwal et al. (eds.), *Remote Sensing of the North Eastern Himalayan Ecosystem*,
https://doi.org/10.1007/978-981-95-3221-6_18

18.1 Introduction

As of February 2025, the Indian Space Research Organization has launched 130 Indian satellites, 433 foreign satellites, and 18 private satellites over 100 launches (as per records on ISRO website last accessed on 3 February 2025). These satellites generate terabytes of data daily, supporting applications in natural resource management, disaster risk reduction, governance, and planning. Unmanned aerial vehicle (UAV) data address critical challenges in the northeastern region of India by improving disaster management capabilities, supporting agriculture and forestry, strengthening security measures, facilitating infrastructure development, revolutionizing healthcare delivery, aiding environmental monitoring, and driving economic growth (Aggarwal et al. 2023). Leveraging this wealth of data, Artificial Intelligence (AI) has emerged as a transformative tool used to identify large-scale patterns, model complex natural systems, and address a wide range of challenges. AI applications in geospatial domains include predicting and monitoring natural disasters, optimizing agricultural practices through crop health analysis, and supporting urban planning by modeling infrastructure growth. It is also instrumental in tracking environmental changes, such as glacier retreat and deforestation, optimizing transportation networks, managing water resources, and conserving wildlife habitats. Furthermore, AI aids in public health by mapping disease outbreaks and improving resource allocation during epidemics. This new area where AI and geospatial technology intersect is called Geospatial Artificial Intelligence (GeoAI) (Zhu et al. 2017). The details of various components of GeoAI are presented in Fig. 18.1. The term "GeoAI" was initially used during the 2017 ACM SIGSPATIAL conference to describe a combination of AI and Geographic Information Systems (GIS) techniques, including spatial data processing, big data analysis, Geo-DBMS (Database Management System), and algorithm analysis and synthesis (Mao et al. 2018).

Computer science-based AI research aims to create computer systems and algorithms with machine intelligence that can think and act like humans do to imitate human perception, reasoning, and social interactions (Russell and Norvig 2016). These algorithms when extended to geospatial datasets permitted improved techniques to analyze and derive insight from an extensive database including, but not limited to, location-tagged posts on social media, sensor readings, administrative and public information records and satellite images. Analyzing this massive amount of data, sometimes called "big data", using traditional methods can be a complex process. AI and Machine Learning (ML) techniques can identify and recognize patterns that would otherwise be very difficult to do manually on geospatial data. The first few studies of GeoAI at the turn of the century were using simplistic ML algorithms for satellite image classification. By the 2010s, the integration of geospatial data processing with big data technologies and real-time data analysis began. This enabled the processing of a vast amount of geospatial data and dynamic mapping and analysis. Also, using AI and ML techniques in geospatial analysis enabled more advanced pattern recognition and predictive modeling. In the 2020s, advances in AI such as the development of complex transformer architectures, foundation models

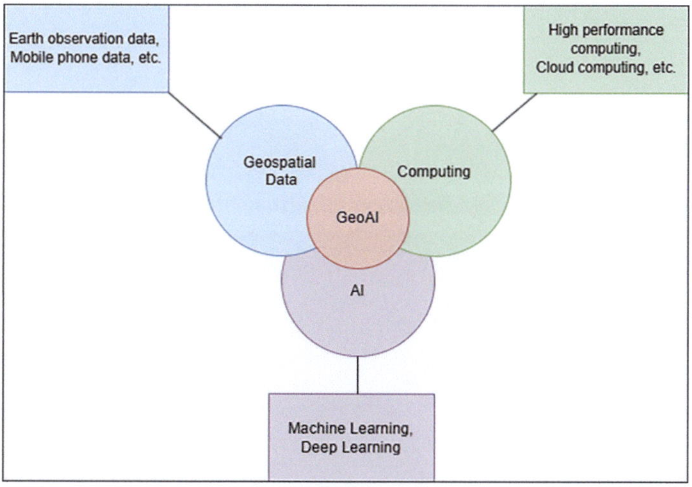

Fig. 18.1 Fundamental components of GeoAI, comprising geospatial data, computational infrastructure, and artificial intelligence techniques

and generative AI models have resulted in big leaps in geospatial data analysis. This evolution of geospatial data processing over the last two decades has resulted in a remarkable record of technological advancement and enhanced data analytics skills.

Recent trends in GeoAI include integrating multispectral and hyperspectral data for vegetation monitoring and land cover classification, and using Synthetic Aperture Radar (SAR) data for terrain mapping, crop yield prediction, and disaster management. ML models enhance geospatial analysis by predicting future trends such as floods, traffic congestion, or crop yields through the evaluation of historical data and external factors like weather patterns. AI also automates satellite image analysis, streamlining tasks like deforestation monitoring and urban expansion mapping, which were traditionally labor-intensive. By extracting features such as roads, buildings, and vegetation cover with high accuracy, AI significantly reduces processing time while improving efficiency in geospatial data applications. One such example is presented in Fig. 18.2, where a GeoAI technique (Chouhan et al. 2023) has been employed for the automatic generation of Level-1 Land Use/Land Cover (LULC) maps for the North Eastern Region (NER) of India using Sentinel-2 imagery from 2023–2024.

18.2 Core Concepts and Enabling Technologies in GeoAI

Geospatial data refers to information that describes the location, characteristics, and spatial relationships of objects on Earth. It integrates geographic coordinates with attribute data, enabling the analysis of spatial patterns and trends across

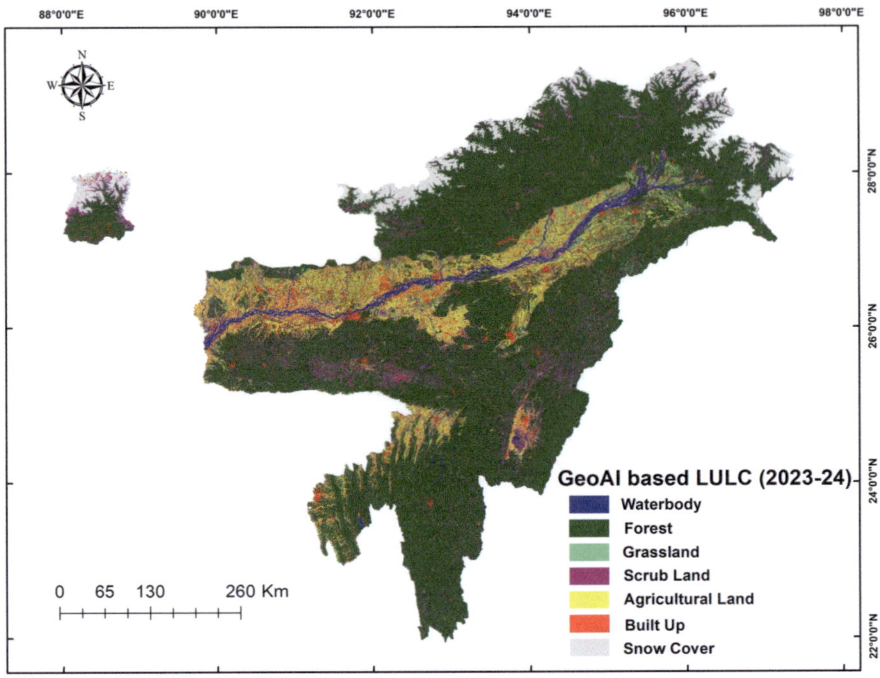

Fig. 18.2 Level-1 land use/land cover (LULC) map of the North Eastern Region (NER) of India, automatically generated using a GeoAI technique

various domains. Geospatial data can be derived from multiple sources, including satellite imagery, aerial surveys, GPS sensors, and geotagged social media posts. With advancements in remote sensing and Geographic Information Systems (GIS), geospatial data has become a fundamental resource for decision-making in diverse fields such as environmental monitoring, disaster management, urban planning, and agriculture. The ability to capture and analyze spatial information allows for better resource allocation, risk assessment, and policy development. In recent years, the rise of AI has transformed how geospatial data is processed, enabling automated feature extraction, predictive modeling, and real-time analysis. By leveraging vast amounts of spatial data, AI-driven geospatial applications are improving accuracy, efficiency, and scalability across multiple disciplines. Figure 18.3 illustrates the evolution of artificial intelligence, highlighting its progressive development over the years and its growing potential to perform a wide range of tasks across diverse domains.

AI is a technology that allows computers to learn and perform tasks like human intelligence using a set of algorithms. The idea is to make machines function like the human brain and reduce manual processes. AI systems can be classified based on capabilities and functionalities (Matter and Gado 2024) as shown in Fig. 18.4.

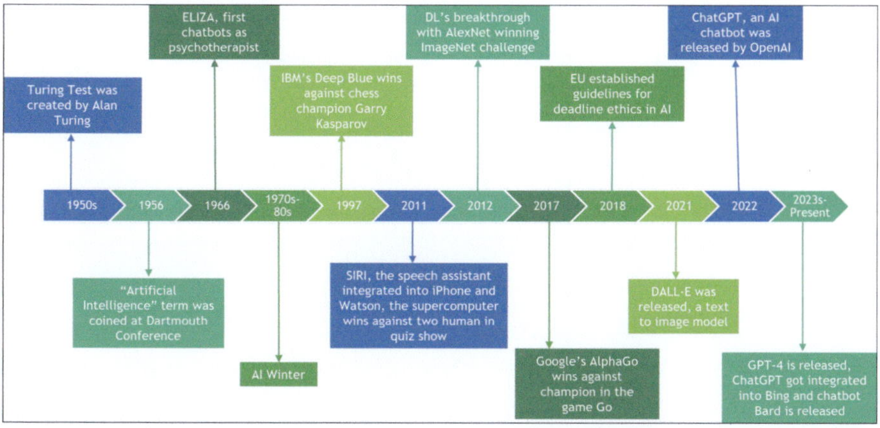

Fig. 18.3 Evolution of artificial intelligence, highlighting key milestones from early theoretical concepts to modern-day developments

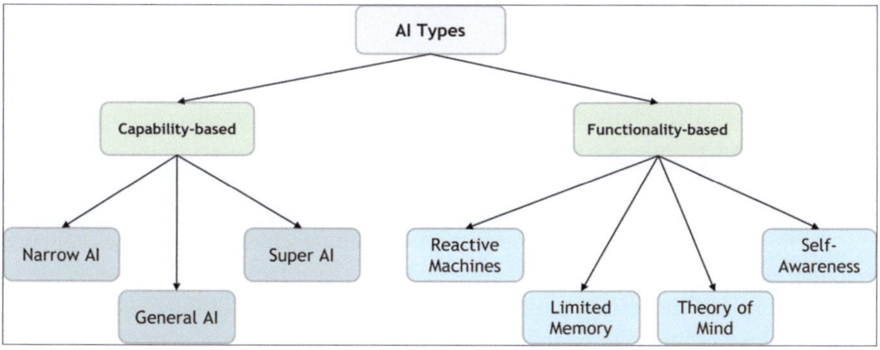

Fig. 18.4 Types of artificial intelligence categorized by their capabilities and functional characteristics in performing tasks

18.2.1 Machine Learning and Deep Learning

ML is a subset of AI that enables machines to learn patterns from data and improve their performance over time without being explicitly programmed for every scenario. Within ML, Deep Learning (DL) represents a more advanced and specialized subset that uses multi-layered artificial neural networks to model complex patterns in large-scale data. Figure 18.5 presents the visual depiction of this relation. While AI encompasses the overall goal of intelligent behavior, ML provides the tools to achieve it through data-driven learning, and DL further enhances this capability by enabling highly accurate learning from vast and unstructured datasets, such as images, speech, and natural language.

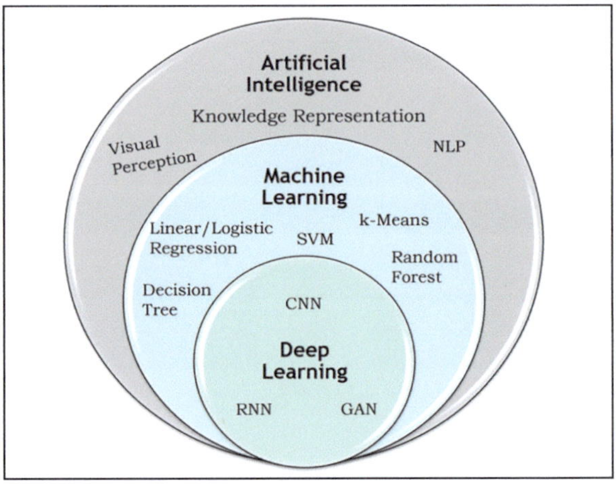

Fig. 18.5 Hierarchical relationship between AI, ML, and DL, illustrating DL as a subset of ML, which in turn is a subset of AI

18.2.2 Geospatial Data and AI

Geospatial data, comprising spatially referenced information such as satellite imagery, sensor readings, and vector layers, serves as the foundational input for AI-driven geospatial applications. The vast volume, heterogeneity, and complexity of such data necessitate the adoption of advanced AI techniques to uncover meaningful patterns, trends, and predictive insights. By integrating geospatial data with AI and GIS, it becomes feasible to analyze large-scale spatial datasets more effectively and intelligently. AI augments traditional GIS workflows by enhancing the detection of spatial patterns, improving predictive accuracy, and facilitating the selection of appropriate spatial modeling approaches for diverse datasets. Ultimately, this convergence enables a deeper understanding of geographic processes and supports informed decision-making across domains such as urban planning, disaster management, environmental monitoring, and infrastructure development.

18.3 Current Trends in GeoAI for Computer Vision and Natural Language Processing

The integration of geospatial data with AI which formed GeoAI has completely changed how geographical data is processed, examined, and understood. Through the use of technologies like Natural Language Processing (NLP), Computer Vision (CV), DL, and ML, GeoAI enhances accuracy and efficiency in a variety of performing tasks by enabling autonomous geospatial intelligence. With the usage of AI models in the field of UAV and remote sensing, it enables faster and more accurate assessments of classifying, detecting and tracking objects presented in them. GeoAI can track urbanization and deforestation by automating change detection, offering vital information for sustainable development and climate change mitigation. GeoAI can also detect objects and perform segmentation techniques in various fields by utilizing data such as multispectral, SAR, thermal images, etc. Future intelligent decision-making is being shaped by GeoAI, which is utilizing geospatial data to its utmost potential and encouraging innovation in a variety of fields.

18.3.1 AI Based CV for RGB Inputs

Analyzing and interpreting visual imagery, such as aerial and satellite images, is made possible by CV. Urban planning, disaster relief, and land use mapping are just a few of the uses for CV algorithms' ability to recognize and categorize structures, infrastructure, and changes over time from images.

18.3.1.1 Image-Level Classification

Image classification is a process of categorizing and labeling pixels in an image or remotely sensed satellite data based on their spectral values to obtain distinct land cover classes or themes, such as forests, water bodies, urban areas, and agricultural fields. This process enabled the analysis and interpretation of complex patterns and information in the images, transforming raw data into more meaningful insights that can be used for various applications. This technique is also referred to as multi-label classification and is capable of predicting both natural and artificial features as well as the categories of land use or land cover inside remotely sensed images. The combination of convolutional neural networks (CNNs) with other ML models, including graph neural networks (GNNs) (Khan et al. 2019), is used in remote sensing image analysis to support multi-label classification. The GNN is a promising model that can increase DL's capacity to handle non-Euclidean data (Li et al. 2020a, b). The GNN is widely used in the domains of knowledge graphs (Nathani et al. 2019), recommender systems (Wu et al. 2019), social networks (Wang et al. 2019), etc. Table 18.1 shows different remote sensing public benchmark datasets for image classification along with their details such as image size, the total number of images, images in each class, and year they were created. Figure 18.6 shows the visual sample result of multi-labels classification using aerial imagery from part of Shillong as input image data. The result is shown with Black predicted labels for correct predictions according to ground truth from the data. Red and Blue predicted labels for false positive and false negative respectively.

Table 18.1 Summary of datasets used for image classification (Mehmood et al. 2022)

Dataset	Classes	No. of images	Image size	Year
UCM (Shi et al. 2022)	21	2100	256 × 256	2010
RSSCN7 (Zou et al. 2015)	7	2800	400 × 400	2015
Brazilian coffee scene (Penatti et al. 2015)	2	2876	64 × 64	2015
AID (Xia et al. 2017)	30	10,000	600 × 600	2017

Input Images			
Ground Truth	plane, vehicle, buildings, roads, grass	vehicle, buildings, roads, trees	playgrounds, vehicle, buildings, roads, trees, grass
Predicted Labels	plane, pavement, buildings, roads, grass	vehicle, buildings, roads, trees, pavement	playgrounds, court, vehicle, buildings, roads, trees, grass

Fig. 18.6 Sample results of multi-label classification using image data from parts of Shillong, Meghalaya. Predicted labels shown in red indicate false positives, while those in blue represent false negatives

18.3.1.2 Object Detection

Object detection is a process of identifying objects by their classes and bounding box location in an image. Object detection comprises two kinds of detection methods: single-stage and two-stage. Single-stage detectors consist of bounding boxes and object categorization are directly provided by a single feed-forward fully convolutional network. The single-stage object detector obtains object categorization and location data straight from the source, bypassing the application area generating step whereas the two-stage object detection method separates the detection process into two stages: the classification step and the region proposal stage. The model first chooses the object candidate region known as region of interest (ROI) using algorithms like Selective Search or Edge Boxes and then the proposals are classified and further refine the object's location. Some popular DL methods such as OverFeat (Sermanet et al. 2013), R-CNN (Ren et al. 2016), YOLO (Redmon and Farhadi 2017), SSD (Liu et al. 2016) and M2Det (Zhao et al. 2019) are used in object detection. Object detection can be used to identify both natural and manmade features from remote sensing imagery for urban planning (Kamusoko 2017), and search and rescue operations (Bejiga et al. 2017), Table 18.2 shows the details of some publicly available dataset for object detection (Zhang et al. 2023). Result of the applied object detection technique is shown in Fig. 18.7, where cars are detected using UAV imagery of Kohima, Nagaland. Vehicles are identified and enclosed within yellow bounding boxes, each labeled with a confidence score indicating detection accuracy. The UAV-based detection provides an aerial perspective, useful for traffic monitoring, urban planning, and autonomous navigation applications.

Table 18.2 Summary of publicly available datasets for object detection in remote sensing imagery, highlighting key characteristics

Datasets	Classes	No. of images	Width	Instances
FAIR1M (Sun et al. 2022)	37	42,796	600–10,000	1,020,579
DIOR (Li et al. 2020a, b)	20	23,463	800	192,518
DOTA (Xia et al. 2018)	15	2806	800–4000	188,282
DOSR (Han et al. 2021)	20	1066	600–1300	6172

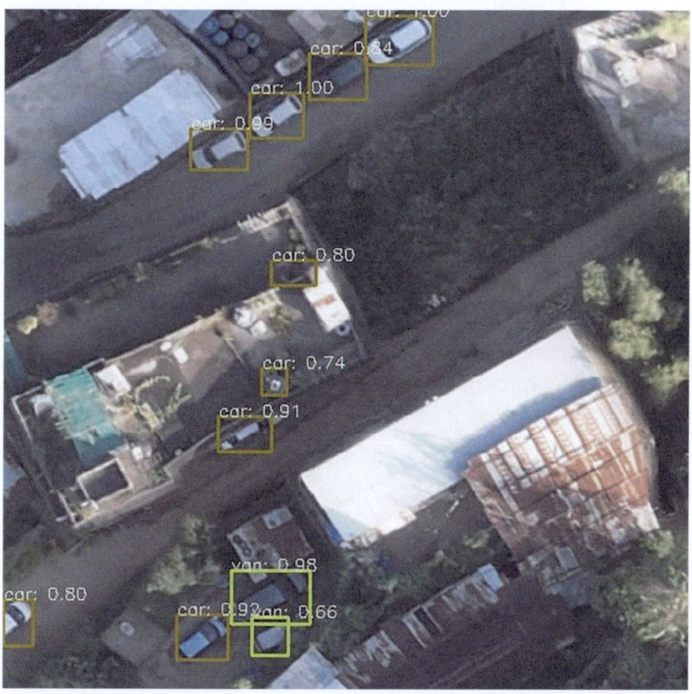

Fig. 18.7 Visualization of car detection results using UAV imagery of Kohima, Nagaland. Detected cars are highlighted with yellow bounding boxes, each annotated with the corresponding confidence score

Table 18.3 Overview of remote sensing datasets for semantic segmentation (Dimitrovski et al. 2024)

Datasets	No. of labels	No. of images	Image size
Potsdam (Hanyu et al. 2024)	6	5,472	6000 × 6000
LoveDA (Wang et al. 2021)	7	5,987	1024 × 1024
INRIA (Maggiori et al. 2017)	2	18,180	5000 × 5000

18.3.1.3 Semantic Segmentation

Semantic Segmentation is a process of classifying each pixel on an image as belonging to a specific class, which divides or categorizes the whole image into semantically distinct sections that represent various objects or classes. Class imbalance is one of the challenges found by existing works in aerial data semantic segmentation. To address the issue, a proposed network (Chouhan et al. 2022a, b) used a negative mining based training technique in their work. Semantic segmentation has a wide range of applications in remote sensing areas which include precision agriculture, disaster management, urban planning, and land cover mapping (Li et al. 2022). The framework for semantic segmentation includes general CNN-based techniques, attention mechanism-based techniques, transformer-based techniques, multi-scale strategy-based techniques, generative adversarial network (GAN)-based techniques, and fusion-based techniques.

Table 18.3 shows details of some publicly available semantic segmentation datasets (Dimitrovski et al. 2024) such as number of labels, number of images and the image size which are present in each dataset. Figure 18.8 shows the semantic segmentation example using NESAC_UAV_Seg (Chouhan et al. 2023) datasets where image showcases distinct classes, including buildings, trees, roads, water bodies, cars, low vegetation, and other features, segmented using DL-based techniques. Each category is represented with unique colors, enabling clear differentiation of land cover types. This segmentation aids in precise urban planning, environmental monitoring, and resource management.

18.3.1.4 Change Detection

Change detection involves the process of detecting changes in specific areas that have seen changes by simultaneously determining whether the changes between two or more registered images is the result of urbanization or natural catastrophes. Change detection has two types of algorithm: homogeneous and heterogeneous. Homogeneous approaches utilize data obtained from sensors that are the same whereas heterogeneous approaches utilize data from sensors that are different. DRMNet (Chouhan et al. 2022a, b) used a homogeneous approach for changes analysis. Change detection plays a significant role in damage assessment, deforestation assessment, and land use land cover (LULC) analysis. The major types of change detection methods can be

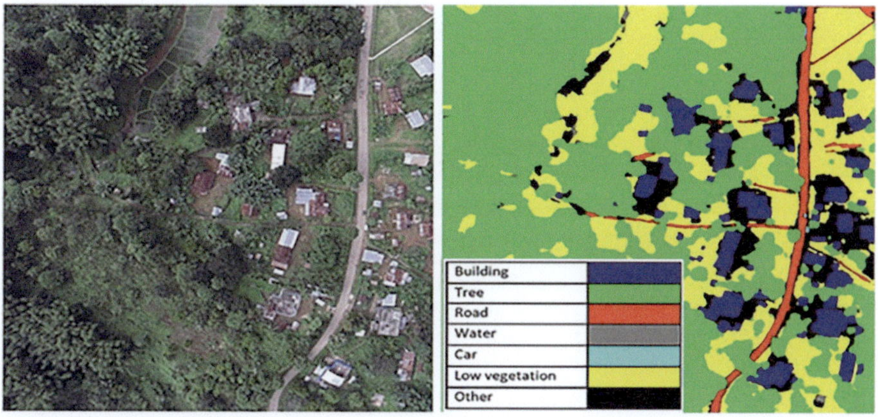

Fig. 18.8 Visualization example of semantic segmentation using the NESAC_UAV_Seg dataset (Chouhan et al. 2023), along with the corresponding classification results obtained using a deep learning method. Distinct land cover classes are represented with unique colors

Table 18.4 Overview of benchmark datasets for remote sensing change detection (Wang et al. 2024)

Datasets	Spatial resolution	Image size (pixels)	No. of image
LEVIR-CD (Chen and Shi 2020)	0.5 m	1024 × 1024	637
WHU-CD (Ji et al. 2018)	0.3 m	32,507 × 15,354	7620
SVCD (Lebedev et al. 2018)	3–100 cm	256 × 256	16,000

categorized as Transformation based, Statistics-based, Algebra-based and DL based (Parelius et al. 2023). Some recent work has used multitask learning based technique (Chouhan et al. 2024) for multi objects change detection.

Table 18.4 shows the description of some change detection's datasets. In Fig. 18.9, the detected changes are highlighted in the output image (Change map), showing differences between two temporal images of different timestamps consisting of size 256 × 256 pixels in each patch. The white colors labels in the output depicts the changes between the two images of different timestamps. Areas experiencing changes, such as urban expansion, vegetation loss, or infrastructure development, are distinctly marked. This analysis helps in monitoring land-use transformations and environmental changes over time.

18.3.1.5 Object Tracking

Object tracking is about determining an object's size and location as it travels across a scene which is a challenging and complex task. It was even more challenging in the context of remote sensing due to object movement and background occlusions,

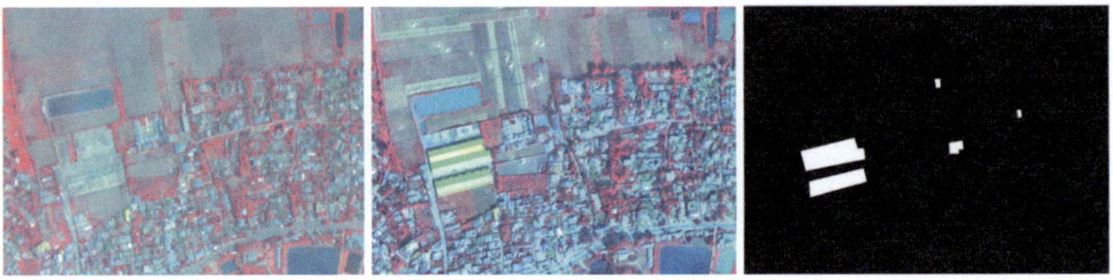

Fig. 18.9 Visualization of change detection results using 256 × 256 pixel Cartosat-3 image patches from parts of Imphal, Manipur. The white regions in the output represent detected changes, specifically building modifications, between two input images captured at different timestamps

Table 18.5 Summary of publicly available datasets for object tracking in remote sensing and aerial imagery (Sun et al. 2024)

Dataset	No. of frames	No. of sequences
UAVDT (Du et al. 2018)	37,084	50
DTB70 (Li and Yeung 2017)	15,777	70
UAV123 (Mueller et al. 2016)	112,578	123

and low-resolution objects in the target region. Object tracking is used in a variety of applications, including video indexing, automated surveillance, human–computer interaction and vehicle tracking. DL approaches can be used to solve the issues regarding object tracking. Augmentation on negative data was performed in CNN model during training. GANs method can be used to generate negative data such that the network will be more discriminate. Table 18.5 shows the description between some object tracking's datasets (Sun et al. 2024). An illustration of multi-target tracking is shown in Fig. 18.10, where moving targets are indicated by blue boxes with azure lines representing their trajectories, and stationary targets are encircled by green boxes. The figure depicts an aerial object tracking where two cars, labeled "car1" and "car2," are detected and enclosed in yellow bounding boxes. Red lines indicate their movement trajectories. This tracking approach is useful for traffic monitoring, surveillance, and urban planning by analyzing vehicle movement patterns.

18.3.1.6 3D Scene Reconstruction

3D scene reconstruction (Zhou et al. 2024) is a technique in CV applications that is applied in various fields including geographic mapping, environmental monitoring, and urban planning. Drones, equipped with high resolution imaging capabilities, have emerged as invaluable tools for acquiring vast quantities of 2D images over large geographical areas. However, the transformation of these 2D images into accurate and comprehensive 3D representations poses significant challenges. The initial step in 3D scene reconstruction involves the extraction of features (Chen et al. 2020)

Fig. 18.10 Visual example of multi-target object tracking using UAV imagery from a region in Kohima, Nagaland. The red lines represent the movement trajectories of the tracked objects

from the overlapping 2D images acquired by drones. This process demands precision and accuracy to ensure the precision of the resulting 3D model. Key correspondences extracted from these features serve as the foundation for generating an initial sparse 3D point cloud, laying the 3D dataset for subsequent dense reconstruction efforts. However, there are challenges in seamless extraction and matching of features (Ingale and Udayan 2021), particularly in environments characterized by mixed natural landscapes and rugged terrains, such as the NER of India. The complexity of such scenes introduces difficulties in separating distinct features amidst varying textures and topographies. Moreover, aerial drone imagery is susceptible to a multitude of environmental factors, including changes in scale, orientation, rotation, and illumination. These variations further aggravate the challenge of feature detection and matching, necessitating the adaptation and refinement of CV algorithms to accommodate diverse imaging conditions. Addressing these challenges requires a finer understanding of the intricacies involved in scene analysis and feature extraction. It requires judicious selection and optimization of CV algorithms tailored to the specific characteristics of the scenes captured by drones. This necessitates an adaptive choice of feature detectors, descriptors, and matching techniques capable of robust performance across a spectrum of environmental conditions (Wang et al. 2022). Furthermore, intelligent processing strategies must be devised to enhance the efficiency and accuracy of 3D reconstruction pipelines. This may involve the integration of ML techniques to refine feature extraction algorithms or the implementation of parallel processing frameworks to expedite computational tasks. The advancement of 3D scene reconstruction techniques along with development of sophisticated methodologies that use cutting-edge CV algorithms and intelligent processing techniques can help in better 3D reconstruction allowing accurate 3D spatial analysis and environmental modeling (Flores-Fuentes et al. 2023). Figure 18.11 illustrates the process of 3D scene reconstruction, which starts with the adaptive selection of suitable feature detectors and descriptors, followed by feature matching. An incremental Structure from Motion (SfM) approach is then applied, involving the simultaneous estimation of camera rotation, translation, and scene structure. The initial model is further refined or bundle adjusted using additional overlapped image pairs. Finally, the sparse 3D point cloud is densified using patch and cluster-based multi-view stereo methods.

18.3.2 AI Based CV for SAR Data

SAR is a technology that is utilized in remote sensing and also one of the most effective methods for creating high-resolution images. SAR is not affected by flying attitude or weather conditions because it can choose frequencies to prevent weather-induced signal attenuation. It is usually attached to a moving platform, such a spacecraft or an aircraft. SAR images are widely used for mapping the surfaces of Earth and other planets and for remote sensing purposes. The benefit of SAR is that, because it is less impacted by weather, lighting, and other factors, it can artificially create higher quality images in any weather situation. SAR images can be used in ship tracking technology (Yasir et al. 2024), military vehicle detection using MSTAR (Coman 2018) dataset, environmental monitoring, disaster management such as floods, earthquakes and landslides, InSAR for changes detection in deformations and elevations, crop monitoring, estimate yields, and evaluate the condition of agricultural land. Figure 18.12 shows the visual output example of military vehicles classification using the MSTAR SAR ATR images dataset.

18.3.3 AI Based CV for Hyperspectral Data

A more sophisticated method called hyper-spectral imaging (HSI) (Manolakis et al. 2016) uses hundreds of narrow bands (Sun and Du 2019) for collecting data from ground objects across the electromagnetic spectrum with very high spectral resolution. A data cube or hypercube with spectral and spatial information emerges after the image is captured. The number of bands in the image and its resolution make up the dimensions of the data cube. Hyperspectral imaging approach has been used in many remote sensing areas like change detection and classification (Ma et al. 2019). Using hyperspectral images, detection of vegetation, natural elements, satellites or soil properties can be made (Kwon et al. 2020). Table 18.6 shows the widely used

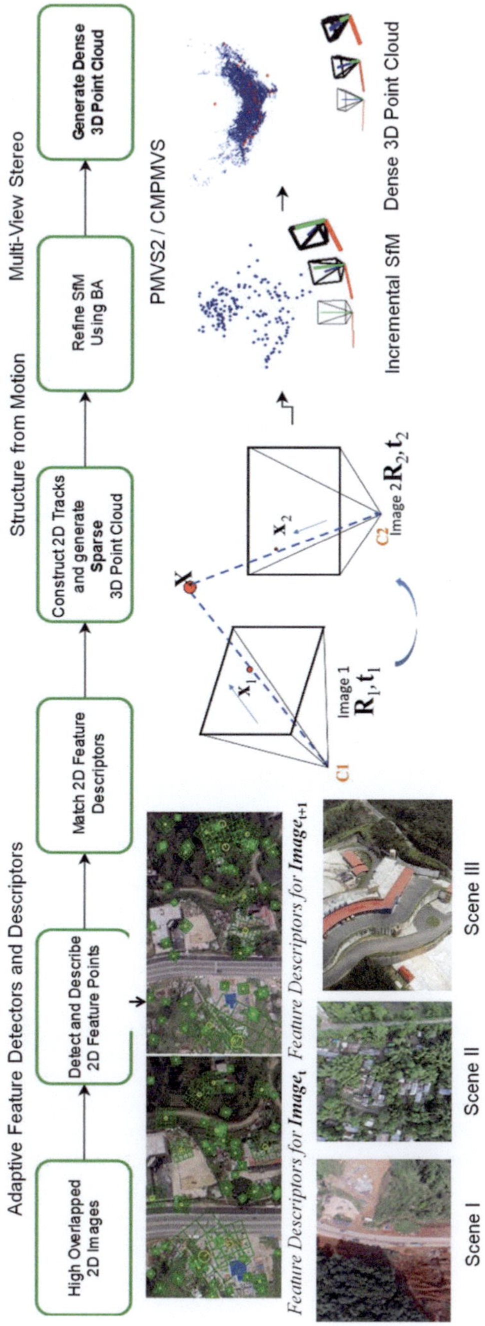

Fig. 18.11 Process pipeline illustrating the selection of suitable feature detectors for 3D sparse and dense reconstruction

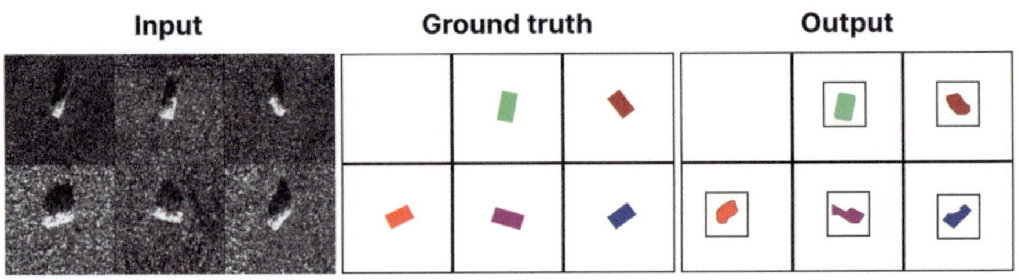

Fig. 18.12 Segmentation results on SAR images using the MSTAR dataset (Coman 2018)

datasets of hyperspectral images (Ozdemir and Polat 2020). Figure 18.13 shows the visual example for semantic segmentation where hyperspectral image is being classified into 16 different ground truth classes in unique colors using Indian Pines dataset (Gao et al. 2018) which consists of one-third forest or other naturally occurring perennial vegetation and two-thirds agriculture scene.

Table 18.6 Datasets for Hyperspectral Images (Ozdemir and Polat 2020)

Datasets	No. of bands	Image size
Pavia Centre (Samat et al. 2014)	102	1096 × 1096
Pavia University (Luo et al. 2020)	115	610 × 640
Indian Pines (Gao et al. 2018)	220	145 × 145
Salinas (Chen et al. 2018)	224	517 × 217

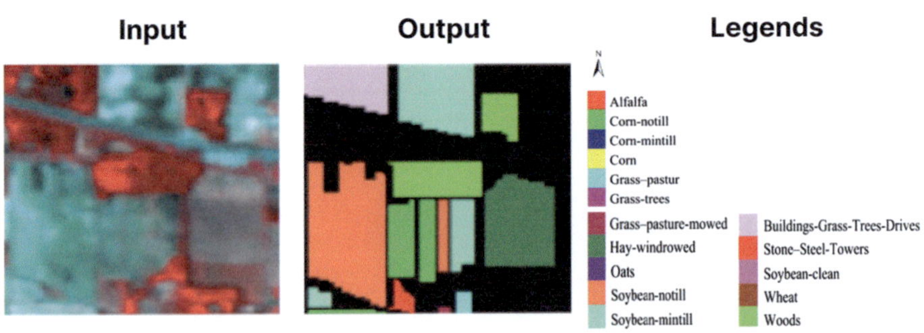

Fig. 18.13 Visualization example of hyperspectral image classification using data from the Indian Pines dataset (Gao et al. 2018). The input is shown as a false color composite (FCC) image, with the corresponding classification output displayed using uniquely colored legends for each class

18.3.4 *AI Based Virtual Reality for Intelligent 3D Point Cloud Visualisation and Analytics*

There is a growing trend in AI-driven immersivity and the metaverse (Soliman et al. 2024), which aim to deliver highly interactive and immersive experiences for users to explore complex, multidimensional data and its real-world applications. AI-powered CV techniques, such as structure from motion, enable the generation of high-resolution 3D point clouds from drone-captured images (Zhou et al. 2024). Integrating AI with VR enhances visualization, interaction, and automated analysis of 3D environments, providing a more intuitive and intelligent understanding of complex data and structures (Cao and Jian 2024). AI-driven VR environments improve training, collaboration, and design evaluation by enabling dynamic, real-time interaction with 3D geospatial data (Pavelka et al. 2024). ML algorithms can optimize rendering, automate object recognition, and enhance spatial analytics, increasing engagement and accessibility across various applications, including education, tourism, infrastructure monitoring, and simulated exploration. However, challenges remain in effectively rendering and seamlessly visualizing massive 3D point cloud datasets within VR environments (Casado-Coscolla et al. 2023), particularly on standard hardware.

To address these challenges, AI-based optimization techniques are crucial for accelerating data processing, reducing computational overhead, and enabling efficient 3D point cloud rendering, interaction, and analysis. AI-augmented virtual reality platforms are essential to import, render, manipulate, and analyze 3D mesh and point cloud data using advanced AI-driven rendering techniques. Further, intelligent optimization algorithms, such as AI-based denoising, adaptive level-of-detail (LOD) rendering, and data structuring techniques like octrees, must be integrated to enhance performance, reduce loading times, and ensure a seamless user experience (Forsberg et al. 1997). Figure 18.14 shows the immersive navigation, interaction, teleportation and analysis of drone imagery derived high density 3D mesh models.

Fig. 18.14 3D point cloud interaction and guided navigation in VR

18.3.5 Natural Language Processing (NLP)

Natural Language Processing (NLP) is a subset of AI which deals with natural language interaction between humans and computers. NLP techniques in GeoAI enable applications like geospatial trend analysis, opinion-based urban planning and monitoring disaster by facilitating the unstructured geospatial text data analysis, sentiment analysis and object recognition from text data. The technique of grouping data points called Geospatial clustering can be used for finding clusters of data points which are related objects. We can apply both NLP and group clustering to study geotags with the help of social media (Mehta et al. 2023). Retrieval-Augmentation-Generation (RAG) is an advanced NLP framework that combines information retrieval, contextual augmentation, and generative modeling to enhance query generation and response accuracy (Lewis et al. 2020). When applied to text-to-SQL tasks, particularly with large language models (LLMs) like LLAMA, RAG improves the process of generating dynamic SQL queries, including PostgreSQL queries, by effectively utilizing relevant database schema and metadata. In this approach, the model first retrieves relevant database structure information, such as table names, column attributes, and relationships, from the database schema. This retrieved information is then used to augment the input query context, allowing the model to better understand and structure the SQL output. Using LLAMA's generative capabilities, RAG models can dynamically create optimized SQL queries that cater to complex user requirements, even when the input is a natural language question. By integrating retrieval modules with schema-aware contextual prompts, this method supports efficient and scalable query generation, enabling real-time data extraction and analysis in applications such as geospatial databases, where structured data needs to be queried based on unstructured user inputs. Figure 18.15 presents a visual interface of RAG based map interaction.

18.4 GeoAI Applications

18.4.1 Environmental Monitoring and Conservation

Environmental monitoring and conservation is a must in today's world in order to save the planet for the future generation to come from monitoring deforestation to conserving endangered species. Satellite imagery can be analyzed using ML approach for land cover change detection and identify land use changes, for example like deforestation and look for conservation if required. Sensor networks also can be used to monitor water or air quality with deployment of AI sensor technology (Sahu et al. 2024). For wildlife tracking, use of AI technology enables conservationists to map migration routes and identify habitats of endangered species by monitoring the movement pattern of the species (Hauenstein et al. 2022). It also helps to prevent wildlife trafficking and poaching as it can detect activities on a real-time basis.

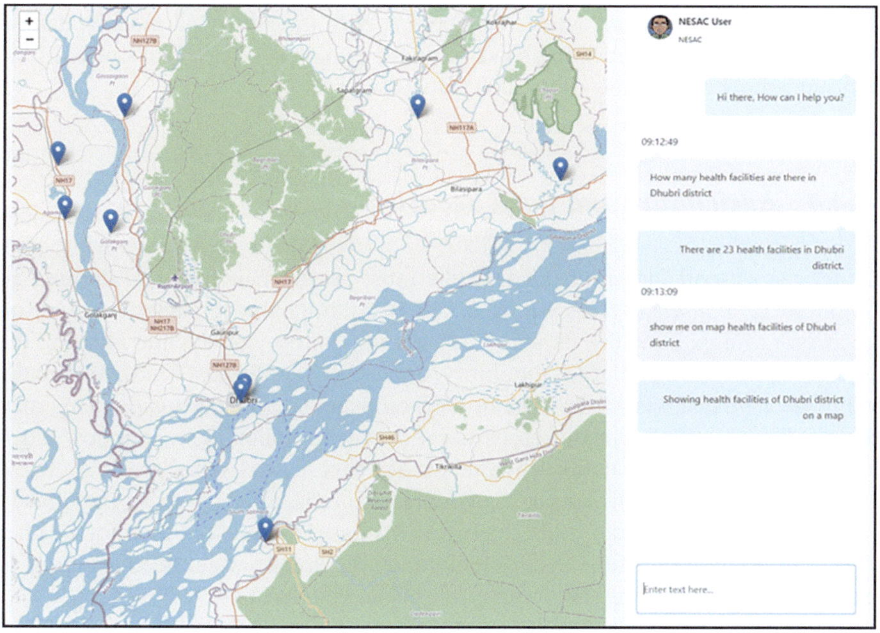

Fig. 18.15 User interface showcasing retrieval augmentation generation based data extraction from PostgreSQL database

18.4.2 Urban Planning and Smart Cities

Using GeoAI in urban planning and smart cities (Mortaheb and Jankowski 2023), city planners can better understand the patterns of urban growth as time goes on each and every year. Consequently, this can support sustainable growth, enhance infrastructure development, and better allocate resources. It is also helpful in the development of smarter cities with the optimization of land use and network transportation. A few AI use cases that can improve a smart city's attributes include smart homes, service delivery, smart parks and crowd management. Smart cities (NITI Aayog 2018) developed features based on IT solutions, in an effort to address the problems caused by urbanization.

18.4.3 Disaster Response and Management

GeoAI is essential to disaster response because it provides real-time data analysis, which enables fast and well-informed decision-making during emergencies. It assists in estimating the damage, forecasting the effects of a disaster, and recommending the most effective emergency response plans. It also facilitates automatic mapping, which is helpful in regions affected by crises. With the use of AI models, natural calamities can be predicted beforehand through the analysis of past data and real-time environmental observations (Sufi 2022).

18.4.4 Agriculture and Food Security

GeoAI can support farmers in making the most use of their land and keeping an eye out for crop diseases (Hosseini et al. 2023). Additionally, it may offer advice on precision farming, assisting farmers in maximizing crop yields while utilizing the fewest resources possible. It has the ability to assess geographic data and give farmers comprehensive insights on the state of their crops and soil. Additionally, they may obtain advice on what kinds of crops to grow, when to plant them, and how much water to give them. A data-driven strategy like this lessens environmental effects while simultaneously increasing sustainability and productivity. AI-driven automated irrigation systems, integrated with Internet of Things (IoT) sensors, monitor weather conditions and soil moisture levels to determine the precise water needs of crops in real time, conserving water and promoting sustainable farming.

18.4.5 Public Health and Epidemiology

GeoAI has a lot of potential in the field of healthcare and can aid in precision medicine as well as public health (Sahana et al. 2022). Finding patterns in infectious diseases and environmental health, forecasting the impacts and spatial distribution of diseases, and mapping and analyzing spatial data for disease tracking, predictive modeling, and resource allocation in epidemiological research are all possible with it. Enhancing the knowledge of the spatial distribution of diseases also allows for taking more proactive steps in the healthcare system. In epidemiology, GeoAI enables researchers to examine the impact of location-based determinants on disease outcomes as well as to characterize and analyze the geographical distribution of diseases (Boulos et al. 2019). ML methods were employed for determining the spatiotemporal patterns of gestational age (Byrnes et al. 2015) and to extract mobility and connectivity information from mobile phone data that was georeferenced (Brdar et al. 2016).

18.5 Challenges in GeoAI

18.5.1 Data Quality and Availability

Data quality and availability are fundamental to AI's efficiency. AI applications' ability to acquire accurate and thorough geospatial data is essential to their success (Chisom et al. 2024). However, this basic criterion poses serious challenges in many parts of the world. The inconsistent quality of data makes it more difficult to create and implement strong AI models, which in turn limits the insights and predictions that can be obtained from them. In order to address this issue, data collection processes should be done properly.

18.5.2 Computational Complexity and Scalability

Processing large amounts of spatial data can be computationally challenging in GeoAI as we need to go for high end computing technologies which are very costly and also not affordable to some small or remote organizations. Increase in the amount and complexity of spatial data makes a challenging issue in scalability (Ossai and Oliha 2024). So algorithms need to be developed efficiently to deal with the large volume datasets. To make AI more accessible, techniques like utilizing edge or cloud computing and developing efficient algorithms are very essential. Optimization of infrastructure and algorithms for scalability should be the main focus.

18.5.3 Ethical and Privacy Considerations

For AI to be used ethically, it is imperative that data collecting procedures respect individuals' right to privacy, follow ethical norms, and maintain away from harming biases (Rabbani et al. 2022). Training programs also will be helpful to get the knowledge about ethical guidelines. Concerns over widespread monitoring are raised by the ongoing tracking of individuals as it may expose secret information about their daily routines and activities. Information that is sensitive like the personal address, office, connections, and so on can be mistreated if not properly secured. Misuse of GeoAI tools such as stalking or checking someone's location without their idea or knowledge leads to privacy and security issues.

18.6 Future Directions in GeoAI

The field of GeoAI (Hu et al. 2019) has experienced remarkable development and innovation over the years. Using spatiotemporally explicit models is essential for fully understanding complex geospatial settings and processes, requiring assessments that integrate human and machine intelligence, improve model generalizability by considering sensors and geography, accommodate uncertainty in datasets and spatial challenges, and combine geospatial data from multiple sources to enhance knowledge discovery. Development of GeoAI tools and applications for agriculture, carbon capture, scalable GeoAI processes to support disaster relief and management, enhancing monitoring systems, methods for climate science, reliable and robust models.

18.6.1 Integration of Multi-Source Data

Future AI powered geospatial will progressively draw on a wider range of data sources, including as social media feeds, Internet of Things (IoT) devices, satellite imaging, aerial surveys, and ground-based sensors. Stakeholders may gain a thorough knowledge of complicated systems and dynamics by integrating multi-sources of data to allow comprehensive and real-time study of spatial phenomena.

18.6.2 Edge Computing and Real-Time Analytics

GeoAI will increasingly leverage edge computing for real-time, decentralized decision-making to the forefront, especially for remote or resource-constrained environments. By processing the data closer to the source will reduce the latency and requirement of cloud-based computation. Also, developing lightweight, efficient models will enable real-time analytics.

18.6.3 Explainable AI for Geospatial Applications

The wide adoption of GeoAI models, particularly in high-stakes situations like disaster management or urban planning, will depend on how interpretable their predictions and decisions are as they get more complicated. The integration of transparency and interpretability elements into AI models may improve trust, accountability, and usability. This will allow stakeholders to confidently comprehend, validate, and interpret choices made by AI.

18.7 Conclusion

In this chapter, we have discussed the foundational concepts, enabling technologies, and diverse applications of GeoAI. We examined the integration of geospatial data with artificial intelligence techniques, which collectively constitute the field of GeoAI, and explored its current trends in computer vision and natural language processing. The growing adoption of GeoAI has shown significant potential to support researchers, government agencies, and both public and private stakeholders by enhancing decision-making, optimizing resource allocation, and improving operational efficiency. Nevertheless, challenges related to data quality, model interpretability, and privacy must be carefully addressed to fully realize the potential of GeoAI and to ensure its ethical and responsible use. NESAC has undertaken several strategic initiatives to remain aligned with the evolving GeoAI landscape by developing state-of-the-art models, implementing advanced techniques, and integrating them into user-centric applications.

Acknowledgements We gratefully acknowledge the institutional support provided by the North Eastern Space Applications Centre (NESAC), Department of Space, Government of India, during the preparation of this chapter. This work was carried out as part of our official responsibilities at NESAC and did not receive any external funding.

References

S. Aggarwal, P. Gupta, N. Mahajan, S. Balaji, K.J. Singh, B. Bhargava, S. Panda, Implementation of drone based delivery of medical supplies in North-East India: experiences, challenges and adopted strategies. Front. Public Health **11**, 1128886 (2023). https://doi.org/10.3389/fpubh.2023.1128886

M.B. Bejiga, A. Zeggada, A. Nouffidj, F. Melgani, A convolutional neural network approach for assisting avalanche search and rescue operations with UAV imagery. Remote Sens. **9**(2), 100 (2017). https://doi.org/10.3390/rs9020100

M.N.K. Boulos, G. Peng, T. VoPham, An overview of GeoAI applications in health and healthcare. Int. J. Health Geogr. **18**(1), 7 (2019). https://doi.org/10.1186/s12942-019-0171-2

S. Brdar, K. Gavrić, D. Ćulibrk, V. Crnojević, Unveiling spatial epidemiology of HIV with mobile phone data. Sci. Rep. **6**(1), 19342 (2016). https://doi.org/10.1038/srep19342

J. Byrnes, R. Mahoney, C. Quaintance, J.B. Gould, S. Carmichael, G.M. Shaw, A. Showen, C. Phibbs, D.K. Stevenson, P.H. Wise, Spatial and temporal patterns in preterm birth in the United States. Pediatr. Res. **77**(6), 836–844 (2015). https://doi.org/10.1038/pr.2015.55

F.F. Cao, Y. Jian, The role of integrating AI and VR in fostering environmental awareness and enhancing activism among college students. Sci. Total. Environ. **908**, 168200 (2024). ISSN 0048-9697. https://doi.org/10.1016/j.scitotenv.2023.168200

A. Casado-Coscolla, C. Sanchez-Belenguer, E. Wolfart et al., Rendering massive indoor point clouds in virtual reality. Virtual R.ity **27**, 1859–1874 (2023). https://doi.org/10.1007/s10055-023-00766-3

H. Chen, Z. Shi, A spatial-temporal attention-based method and a new dataset for remote sensing image change detection. Remote Sens. **12**(10), 1662 (2020). https://doi.org/10.3390/rs12101662

L. Chen, F. Rottensteiner, C. Heipke, Feature detection and description for image matching: from hand-crafted design to deep learning. Geo-Spat. Inf. Sci. **24**(1), 58–74 (2020). https://doi.org/10.1080/10095020.2020.1843376

Z. Chen, J. Jiang, X. Jiang, X. Fang, Z. Cai, Spectral-spatial feature extraction of hyperspectral images based on propagation filter. Sensors **18**(6), 1978 (2018). https://doi.org/10.3390/s18061978

O.N. Chisom, P.W. Biu, A.A. Umoh, B.O. Obaedo, A.O. Adegbite, A. Abatan, Reviewing the role of AI in environmental monitoring and conservation: a data-driven revolution for our planet. World J. Adv. Res. Rev. **21**(1), 161–171 (2024). https://doi.org/10.30574/wjarr.2024.21.1.2720

A. Chouhan, D. Chutia, S.P. Aggarwal, Deep learning approach for multi-class semantic segmentation of UAV images. Int. J. Artif. Intell. Tools **32**(07), 2350033 (2023). https://doi.org/10.1142/S0218213023500331

A. Chouhan, A. Sur, D. Chutia, Aggregated context network for semantic segmentation of aerial images, in *2022 IEEE International Conference on Image Processing (ICIP)* (2022), pp. 1526–1530. IEEE. https://doi.org/10.1109/ICIP46576.2022.9898016

A. Chouhan, A. Sur, D. Chutia, DRMNet: Difference image reconstruction enhanced multiresolution network for optical change detection. IEEE J. Sel. Top. Appl. Earth Obs. Remote Sens. **15**, 4014–4026 (2022b). https://doi.org/10.1109/JSTARS.2022.3174780

A. Chouhan, A. Sur, D. Chutia, S.P. Aggarwal, TMLNet: triad multitask learning network for multi-objective based change detection. Neurocomputing 128080 (2024). https://doi.org/10.1016/j.neucom.2024.128080

C. Coman, A deep learning SAR target classification experiment on MSTAR dataset, in *2018 19th International Radar Symposium (IRS)* (2018), pp. 1–6. IEEE. https://doi.org/10.23919/IRS.2018.8448048

I. Dimitrovski, V. Spasev, S. Loshkovska, I. Kitanovski, U-net ensemble for enhanced semantic segmentation in remote sensing imagery. Remote Sens. **16**(12), 2077 (2024). https://doi.org/10.3390/rs16122077

D. Du, Y. Qi, H. Yu, Y. Yang, K. Duan, G. Li, W. Zhang, Q. Huang, Q. Tian, The unmanned aerial vehicle benchmark: object detection and tracking, in *Proceedings of the European Conference on Computer Vision (ECCV)* (2018), pp. 370–386. https://doi.org/10.1007/978-3-030-01249-6_23

W. Flores-Fuentes, G. Trujillo-Hernández, I.Y. Alba-Corpus, J.C. Rodríguez-Quiñonez, J.E. Mirada-Vega, D. Hernández-Balbuena, F.N. Murrieta-Rico, O. Sergiyenko, 3D spatial measurement for model reconstruction: a review. Measurement **207**, 112321 (2023). https://doi.org/10.1016/j.measurement.2022.112321

A.S. Forsberg, J.J. LaViola, L. Markosian, R.C. Zeleznik, Seamless interaction in virtual reality. IEEE Comput. Graph. Appl. **17**(6), 6–9 (1997). https://doi.org/10.1109/38.626956

Q. Gao, S. Lim, X. Jia, Hyperspectral image classification using convolutional neural networks and multiple feature learning. Remote Sens. **10**(2), 299 (2018). https://doi.org/10.3390/rs10020299

Y. Han, X. Yang, T. Pu, Z. Peng, Fine-grained recognition for oriented ship against complex scenes in optical remote sensing images. IEEE Trans. Geosci. Remote Sens. **60**, 1–18 (2021). https://doi.org/10.1109/TGRS.2021.3123666

T. Hanyu, K. Yamazaki, M. Tran, R.A. McCann, H. Liao, C. Rainwater, M. Adkins, J. Cothren, N. Le, Aerialformer: multi-resolution transformer for aerial image segmentation. Remote Sens. **16**(16), 2930 (2024). https://doi.org/10.3390/rs16162930

S. Hauenstein, N. Jassoy, A.C. Mupepele, T. Carroll, M. Kshatriya, C.M. Beale, C.F. Dormann, A systematic map of demographic data from elephant populations throughout Africa: implications for poaching and population analyses. Mammal Rev. **52**(3), 438–453 (2022). https://doi.org/10.1111/mam.12291

F.S. Hosseini, M.B. Seo, S.V. Razavi-Termeh, A. Sadeghi-Niaraki, M. Jamshidi, S.M. Choi, Geospatial artificial intelligence (GeoAI) and satellite imagery fusion for soil physical property predicting. Sustainability **15**(19), 14125 (2023). https://doi.org/10.3390/su151914125

Y. Hu, S. Gao, D. Lunga, W. Li, S. Newsam, B. Bhaduri, GeoAI at ACM SIGSPATIAL: progress, challenges, and future directions. Sigspatial Spec. **11**(2), 5–15 (2019). https://doi.org/10.1145/3377000.3377002

A.K. Ingale, J. Udayan, Real-time 3D reconstruction techniques applied in dynamic scenes: a systematic literature review. Comput. Sci. Rev. **39**, 100338 (2021). ISSN 1574-0137. https://doi.org/10.1016/j.cosrev.2020.100338

S. Ji, S. Wei, M. Lu, Fully convolutional networks for multisource building extraction from an open aerial and satellite imagery data set. IEEE Trans. Geosci. Remote Sens. **57**(1), 574–586 (2018). https://doi.org/10.1109/TGRS.2018.2858817

C. Kamusoko, Importance of remote sensing and land change modeling for urbanization studies, in *Urban Development in Asia and Africa*, ed. by Y. Murayama, C. Kamusoko, A. Yamashita, R. Estoque. The Urban Book Series (Springer, Singapore, 2017). https://doi.org/10.1007/978-981-10-3241-7_1

N. Khan, U. Chaudhuri, B. Banerjee, S. Chaudhuri, Graph convolutional network for multi-label VHR remote sensing scene recognition. Neurocomputing **357**, 36–46 (2019). https://doi.org/10.1016/j.neucom.2019.05.024

Y.S. Kwon, J. Pyo, Y.H. Kwon, H. Duan, K.H. Cho, Y. Park, Drone-based hyperspectral remote sensing of cyanobacteria using vertical cumulative pigment concentration in a deep reservoir. Remote Sens. Environ. **236**, 111517 (2020). https://doi.org/10.1016/j.rse.2019.111517

M.A. Lebedev, Y.V. Vizilter, O.V. Vygolov, V.A. Knyaz, A.Y. Rubis, Change detection in remote sensing images using conditional adversarial networks. Int. Arch. Photogramm. Remote. Sens. Spat. Inf. Sci. **42**, 565–571 (2018). https://doi.org/10.5194/isprs-archives-XLII-2-565-2018

P. Lewis, E. Perez, A. Piktus, F. Petroni, V. Karpukhin, N. Goyal, H. Küttler, M. Lewis, W.T. Yih, T. Rocktäschel, S. Riedel, D. Kiela, Retrieval-augmented generation for knowledge-intensive NLP tasks, in *Proceedings of the 34th International Conference on Neural Information Processing Systems (NIPS 20)*. Curran Associates Inc., Red Hook, NY, USA, Article 793, pp. 9459–9474 (2020). https://doi.org/10.5555/3495724.3496517

K. Li, G. Wan, G. Cheng, L. Meng, J. Han, Object detection in optical remote sensing images: a survey and a new benchmark. ISPRS J. Photogramm. Remote Sens. **159**, 296–307 (2020a). https://doi.org/10.1016/j.isprsjprs.2019.11.023

S. Li, D.Y. Yeung, Visual object tracking for unmanned aerial vehicles: a benchmark and new motion models, in *Proceedings of the AAAI Conference on Artificial Intelligence*, vol. 31, no. 1. https://doi.org/10.1609/aaai.v31i1.11205

W. Li, C.Y. Hsu, GeoAI for large-scale image analysis and machine vision: recent progress of artificial intelligence in geography. ISPRS Int. J. Geo Inf. **11**(7), 385 (2022). https://doi.org/10.3390/ijgi11070385

Y. Li, R. Chen, Y. Zhang, M. Zhang, L. Chen, Multi-label remote sensing image scene classification by combining a convolutional neural network and a graph neural network. Remote Sens. **12**(23), 4003 (2020b). https://doi.org/10.3390/rs12234003

W. Liu, D. Anguelov, D. Erhan, C. Szegedy, S. Reed, C.Y. Fu, A.C. Berg, Ssd: single shot multibox detector, in *Computer Vision–ECCV 2016: 14th European Conference*, Amsterdam, The Netherlands, October 11–14, 2016, Proceedings, Part I 14 (Springer International Publishing, 2016), pp. 21–37. https://doi.org/10.1007/978-3-319-46448-0_2

F. Luo, Y. Huang, W. Tu, J. Liu, Local manifold sparse model for image classification. Neurocomputing **382**, 162–173 (2020). https://doi.org/10.1016/j.neucom.2019.11.084

L. Ma, Y. Liu, X. Zhang, Y. Ye, G. Yin, B.A. Johnson, Deep learning in remote sensing applications: a meta-analysis and review. ISPRS J. Photogramm. Remote Sens. **152**, 166–177 (2019). https://doi.org/10.1016/j.isprsjprs.2019.04.015

E. Maggiori, Y. Tarabalka, G. Charpiat, P. Alliez, Can semantic labeling methods generalize to any city? The Inria aerial image labeling benchmark, in *2017 IEEE International Geoscience and Remote Sensing Symposium (IGARSS)* (2017), pp. 3226–3229. IEEE. https://doi.org/10.1109/IGARSS.2017.8127684

D.G. Manolakis, R.B. Lockwood, T.W. Cooley, Hyperspectral imaging remote sensing: physics, sensors, and algorithms. Cambridge University Press (2016). https://doi.org/10.1017/CBO9781316017876

H. Mao, Y. Hu, B. Kar, S. Gao, G. McKenzie, Geoai 2017 workshop report: the 1st acm sigspatial international workshop on geoai:@ ai and deep learning for geographic knowledge discovery: Redondo beach, ca, usa-november 7, 2016. Sigspatial Spec. **9**(3), 25–25 (2018). https://doi.org/10.1145/3178392.3178408

N.M. Matter, N. Gado, Artificial intelligence in architecture: integration into architectural design process. Eng. Res. J. **181**, 1–16 (2024). https://doi.org/10.21608/erj.2024.344313

M. Mehmood, A. Shahzad, B. Zafar, A. Shabbir, N. Ali, Remote sensing image classification: a comprehensive review and applications. Math. Probl. Eng. **2022**(1), 5880959 (2022). https://doi.org/10.1155/2022/5880959

S. Mehta, G. Jain, S. Mala, Natural Language processing approach and geospatial clustering to explore the unexplored geotags using media, in *2023 13th International Conference on Cloud Computing, Data Science & Engineering (Confluence)* (2023), pp. 672–675. IEEE. https://doi.org/10.1109/Confluence56041.2023.10048848

R. Mortaheb, P. Jankowski, Smart city re-imagined: city planning and GeoAI in the age of big data. J. Urban Manag. **12**(1), 4–15 (2023). https://doi.org/10.1016/j.jum.2022.08.001

M. Mueller, N. Smith, B. Ghanem, A benchmark and simulator for UAV tracking, in *Computer Vision–ECCV 2016: 14th European Conference*, Amsterdam, The Netherlands, October 11–14, 2016, Proceedings, Part I 14 (Springer International Publishing, 2016), pp. 445–461. https://doi.org/10.1007/978-3-319-46448-0_27

D. Nathani, J. Chauhan, C. Sharma, M. Kaul, Learning attention-based embeddings for relation prediction in knowledge graphs, in *Proceedings of the 57th Annual Meeting of the Association for Computational Linguistics* (2019, Florence, Italy), pp. 4710–4723. Association for Computational Linguistics. https://doi.org/10.18653/v1/P19-1466

NITI Aayog, National strategy for artificial intelligence. Paper, pp. 2019–01 (2018), https://www.niti.gov.in/sites/default/files/2023-03/National-Strategy-for-Artificial-Intelligence.pdf

E.N. Ossai, A. Oliha, Integration of geoinformatics and artificial intelligence: enhancing surveying applications through advanced data analysis and decision-making. Int. J. Adv. Multidiscip. Res. Stud. **4**(1) (2024). https://doi.org/10.62225/2583049X

A. Ozdemir, K. Polat, Deep learning applications for hyperspectral imaging: a systematic review. J. Inst. Electron. Comput. **2**(1), 39–56 (2020). https://doi.org/10.33969/JIEC.2020.21004

E.J. Parelius, A review of deep-learning methods for change detection in multispectral remote sensing images. Remote Sens. **15**(8), 2092 (2023). https://doi.org/10.3390/rs15082092

K. Pavelka Jr., M. Landa, Using virtual and augmented reality with GIS data. ISPRS Int. J. Geo-Inf. **13**, 241 (2024). https://doi.org/10.3390/ijgi13070241

O.A. Penatti, K. Nogueira, J.A. Dos Santos, Do deep features generalize from everyday objects to remote sensing and aerial scenes domains? in *Proceedings of the IEEE Conference on Computer Vision and Pattern Recognition Workshops* (2015), pp. 44–51. https://doi.org/10.1109/CVPRW.2015.7301382

M.R. Rabbani, A. Sarea, S. Khan, Y. Abdullah, Ethical concerns in artificial intelligence (AI): the role of RegTech and Islamic finance, in *Artificial Intelligence for Sustainable Finance and Sustainable Technology: Proceedings of ICGER 2021 1* (Springer International Publishing, 2022), pp. 381–390. https://doi.org/10.1007/978-3-030-93464-4_38

J. Redmon, A. Farhadi, YOLO9000: better, faster, stronger, in *Proceedings of the IEEE Conference on Computer Vision and Pattern Recognition* (2017), pp. 7263–7271. https://doi.org/10.48550/arXiv.1612.08242

S. Ren, K. He, R. Girshick, J. Sun, Faster R-CNN: towards real-time object detection with region proposal networks. IEEE Trans. Pattern Anal. Mach. Intell. **39**(6), 1137–1149 (2016). https://doi.org/10.1109/TPAMI.2016.2577031

S.J. Russell, P. Norvig, Artificial intelligence: a modern approach. Pearson (2016)

K.S. Sahana, C.B. Karthik, M.M. Sree, M. Chaithra, S.K. Abhinandithe, B. Snehalatha, R. Sridhar, S. Balasubramanian, B. Madhu, Geospatial artificial intelligence (GeoAI): applications in health care. Int. J. Health Allied Sci. **11**(4), 4 (2022). https://doi.org/10.55691/2278-344X.1044

S. Sahu, N. Mallick, S. Patnaik, Mitigating climate change: AI and the revolution of environmental regulatory frameworks, in *Maintaining a Sustainable World in the Nexus of Environmental Science and AI* (2024), pp. 161–200. IGI Global. https://doi.org/10.4018/979-8-3693-6336-2.ch007

A. Samat, P. Du, S. Liu, J. Li, L. Cheng, ${{\rm E}^{2}}{\rm LMs} $: ensemble extreme learning machines for hyperspectral image classification. IEEE J. Sel. Top. Appl. Earth Obs. Remote. Sens. **7**(4), 1060–1069 (2014). https://doi.org/10.1109/JSTARS.2014.2301775

P. Sermanet, D. Eigen, X. Zhang, M. Mathieu, R. Fergus, Y. LeCun, OverFeat: integrated recognition, localization and detection using convolutional networks. CoRR, abs/1312.6229 (2013)

C. Shi, X. Zhang, J. Sun, L. Wang, Remote sensing scene image classification based on self-compensating convolution neural network. Remote Sens. **14**(3), 545 (2022). https://doi.org/10.3390/rs14030545

M.M. Soliman, E. Ahmed, A. Darwish et al., Artificial intelligence powered metaverse: analysis, challenges and future perspectives. Artif. Intell. Rev. **57**, 36 (2024). https://doi.org/10.1007/s10462-023-10641-x

F.K. Sufi, AI-SocialDisaster: An AI-based software for identifying and analyzing natural disasters from social media. Softw. Impacts **13**, 100319 (2022). https://doi.org/10.1016/j.simpa.2022.100319

L. Sun, X. Li, Z. Yang, D. Gao, Visual object tracking based on the motion prediction and block search in UAV videos. Drones **8**(6), 252 (2024). https://doi.org/10.3390/drones8060252

W. Sun, Q. Du, Hyperspectral band selection: a review. IEEE Geosci. Remote. Sens. Mag. **7**(2), 118–139 (2019). https://doi.org/10.1109/MGRS.2019.2911100

X. Sun, P. Wang, Z. Yan, F. Xu, R. Wang, W. Diao, J. Chen, J. Li, Y. Feng, T. Xu, M. Weinmann et al., FAIR1M: a benchmark dataset for fine-grained object recognition in high-resolution remote sensing imagery. ISPRS J. Photogramm. Remote Sens. **184**, 116–130 (2022). https://doi.org/10.1016/j.isprsjprs.2021.12.004

H. Wang, T. Xu, Q. Liu, D. Lian, E. Chen, D. Du, H. Wu, W. Su, MCNE: an end-to-end framework for learning multiple conditional network representations of social network, in *Proceedings of the 25th ACM SIGKDD International Conference on Knowledge Discovery & Data Mining* (2019), pp. 1064–1072. https://doi.org/10.1145/3292500.3330931

J. Wang, Z. Zheng, A. Ma, X. Lu, Y. Zhong, Loveda: a remote sensing land-cover dataset for domain adaptive semantic segmentation, in *Proceedings of the Neural Information Processing Systems Track on Datasets and Benchmarks*, ed. by J. Vanschoren, S. Yeung, (2021) vol. 1

J. Wang, Z. Gong, X. Liu, H. Guo, D. Yu, L. Ding, Object detection based on adaptive feature-aware method in optical remote sensing images. Remote Sens. **14**, 3616 (2022). https://doi.org/10.3390/rs14153616

X. Wang, Z. Wang, J. Jiang, L. Chen, X. Wang, B. Jiang, Treat stillness with movement: remote sensing change detection via coarse-grained temporal foregrounds mining. ArXiv, abs/2408.08078 (2024)

S. Wu, Y. Tang, Y. Zhu, L. Wang, X. Xie, T. Tan, Session-based recommendation with graph neural networks. Proc. AAAI Conf. Artif. Intell. **33**(01), 346–353 (2019). https://doi.org/10.1609/aaai.v33i01.3301346

G.S. Xia, X. Bai, J. Ding, Z. Zhu, S. Belongie, J. Luo, M. Datcu, M. Pelillo, L. Zhang, DOTA: a large-scale dataset for object detection in aerial images, in *Proceedings of the IEEE Conference on Computer Vision and Pattern Recognition* (2018), pp. 3974–3983. https://doi.org/10.1109/CVPR.2018.00418

G.S. Xia, J. Hu, F. Hu, B. Shi, X. Bai, Y. Zhong, L. Zhang, X. Lu, AID: a benchmark data set for performance evaluation of aerial scene classification. IEEE Trans. Geosci. Remote Sens. **55**(7), 3965–3981 (2017). https://doi.org/10.1109/TGRS.2017.2685945

M. Yasir, L. Shanwei, X. Mingming, W. Jianhua, S. Hui, S. Nazir, X. Zhang, A.T.I. Colak, YOLOv8-BYTE: ship tracking algorithm using short-time sequence SAR images for disaster response leveraging GeoAI. Int. J. Appl. Earth Obs. Geoinf. **128**, 103771 (2024). https://doi.org/10.1016/j.jag.2024.103771

X. Zhang, T. Zhang, G. Wang, P. Zhu, X. Tang, X. Jia, L. Jiao, Remote sensing object detection meets deep learning: a metareview of challenges and advances. IEEE Geosci. Remote. Sens. Mag. (2023). https://doi.org/10.1109/MGRS.2023.3312347

Q. Zhao, T. Sheng, Y. Wang, Z. Tang, Y. Chen, L. Cai, H. Ling, M2det: a single-shot object detector based on multi-level feature pyramid network. Proc. AAAI Conf. Artif. Intell. **33**(01), 9259–9266 (2019). https://doi.org/10.1609/aaai.v33i01.33019259

L. Zhou, G. Wu, Y. Zuo, X. Chen, H. Hu, A comprehensive review of vision-based 3D reconstruction methods. Sensors **24**, 2314 (2024). https://doi.org/10.3390/s24072314

X.X. Zhu, D. Tuia, L. Mou, G.S. Xia, L. Zhang, F. Xu, F. Fraundorfer, Deep learning in remote sensing: a comprehensive review and list of resources. IEEE Geosci. Remote. Sens. Mag. **5**(4), 8–36 (2017). https://doi.org/10.1109/MGRS.2017.2762307

Q. Zou, L. Ni, T. Zhang, Q. Wang, Deep learning based feature selection for remote sensing scene classification. IEEE Geosci. Remote Sens. Lett. **12**(11), 2321–2325 (2015). https://doi.org/10.1109/LGRS.2015.2475299

Avinash Chouhan is a Scientist-SE at the North Eastern Space Applications Centre (NESAC), Department of Space, Shillong, India. He began his research career at NESAC in 2014. He earned his Ph.D. in Computer Science and Engineering from the Indian Institute of Technology Guwahati in 2024 and holds a B.E. in Computer Science and Engineering from the Government Engineering College, Jabalpur. His research interests include GeoAI, Generative AI, and Deep Learning, with a particular focus on developing domain-adaptive models for remote sensing applications. Dr. Chouhan actively works on bridging foundational AI with earth observation data to address real-world geospatial challenges in the North Eastern Region and beyond, contributing to the advancement of space-based remote sensing solutions.

Puyam S. Singh currently works as Scientist/Engineer–SF at the North Eastern Space Applications Centre (NESAC), specializing in Geo-spatial Intelligence and Computer Vision. He holds an M.E. in Computer Science from PEC University of Technology, Chandigarh. Previously, he worked as a research fellow at DRDO's Snow and Avalanche Study Establishment (SASE), contributing to avalanche forecasting models. At NESAC, he works on the development of space-based geospatial applications for societal benefit and e-Governance, integrating satellite and aerial data. His research focuses on computer vision algorithms for automated feature extraction from remote sensing imagery. He has mentored over 54 postgraduate and undergraduate students and has published 43+ papers in reputed journals and conferences. His innovative contributions have earned National and regional e-Governance Awards in 2015, 2018, and 2023. Puyam continues to advance AI-driven remote sensing solutions for operational and research applications in governance and development sectors.

Ritu Anilkumar is currently a Senior Research Associate at the University of Bristol, working in the domain of glaciology and Earth system science. She holds a Bachelor's degree from the Indian Institute of Space Science and Technology (IIST) and has recently submitted her PhD thesis at the Indian Institute of Technology Guwahati, where her research focused on Glacier Mass Balance Modelling using machine learning techniques. Prior to this, she served as Scientist/Engineer at the North Eastern Space Applications Centre (NESAC), Department of Space, Government of India, from 2017 to 2025. During her tenure, she was actively involved in developing geospatial solutions for environmental monitoring and natural resource management, with a focus on applying data-driven and AI-based approaches to remote sensing. Her research interests include geospatial data science, machine learning for Earth observation, and environmental modeling. She has contributed to multiple interdisciplinary projects aimed at advancing the application of AI in geosciences and has authored several publications in reputed journals and conferences.

Siddhartha Bhuyan is presently serving as Scientist/Engineer-SD at the North Eastern Space Applications Centre (NESAC). He holds an M.Tech from VIT, Vellore, and a B.Tech from ICFAI University, Dehradun both in Computer Science and Engineering. With over 12 years of experience in the IT industry and academia, his expertise spans AI, Algorithms, Machine/Deep Learning, System and Database Design, and Natural Language Understanding. At NESAC, he has worked on mobile app development, cyber security, automation, geo web services, and decision support systems for e-governance. He has authored 5 journal publications, 2 conference papers, and holds a patent for a swarm UAV-based data acquisition algorithm developed with KJ Somaiya College of Engineering. He has also contributed several innovative and impactful ideas in national hackathons / brainstorming events and also pioneered the use of IoT for disaster management at NESAC. He is currently focusing on developing novel algorithms and applications in the field of Health GIS.

Nilay Nishant is currently serving as Scientist/Engineer-SE at the North Eastern Space Applications Centre (NESAC), Department of Space, Government of India. He holds a Master's degree in Remote Sensing from Birla Institute of Technology, Mesra, and a Bachelor's degree in Electronics and Communication Engineering from Pondicherry University. His research interests include geospatial big data analytics, e-Governance, and the application of Artificial Intelligence and deep learning techniques to earth observation data. He has played a key role in the development of more than 25 governance-focused geospatial applications for various departments across the North Eastern Region of India. He has authored 30 research publications in reputed national and international journals and conference proceedings. In recognition of his significant contributions to the field of geospatial technology for governance, he has been honored with the National e-Governance Award (2022) and the e-North East Award (2023, 2020).

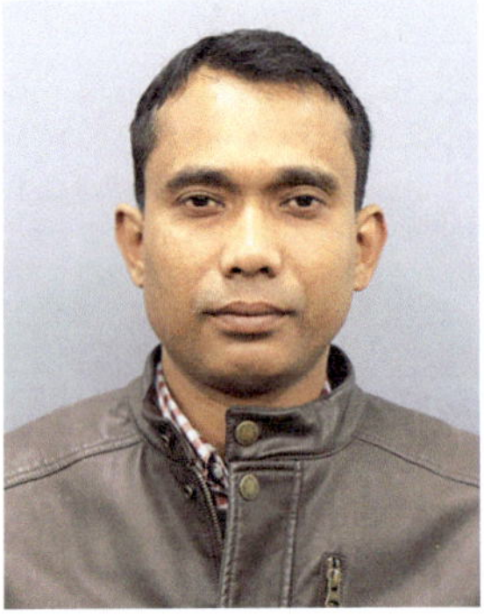

Dibyajyoti Chutia received the M. Tech. degree in information technology and the Ph.D. degree in computer science & engineering with specialization in soft computing technique for classification of geospatial data from Tezpur Central University, Tezpur, India, in 2000 and 2015, respectively. He is currently serving as head of Geoinformatics and IT division at North Eastern Space Applications Centre (NESAC), Department of Space, Government of India. His research interests include soft computing, machine learning, and deep learning applications for remotely sensed data. He has authored or coauthored more than 65 research articles in peer-reviewed journals, conferences, book chapters, etc. He has received the National e-Governance Award (2018, 2022), National Geospatial award (2020) and the eNorthEast Award (2020, 2023) for his and his team's contributions to geospatial innovations in governance and regional development.

Chapter 19
Advancement in Drone Data Processing Analytics and Applications

Victor Saikhom, Puyam S. Singh, Nilay Nishant, and Dibyajyoti Chutia

Abstract The advent of Unmanned Aerial Vehicles (UAVs) has revolutionized geospatial data acquisition, processing, and analysis, particularly in challenging terrains like the North Eastern Region (NER) of India. This chapter presents a comprehensive overview of UAV remote sensing, encompassing data acquisition workflows, advanced image processing techniques, and analytics powered by Geographic Information Systems (GIS) and Artificial Intelligence (AI). Emphasis is placed on preprocessing, orthomosaic generation, 3D modeling (DSM and point clouds), and data management. Applications across diverse sectors such as agriculture, environmental monitoring, infrastructure planning, disaster management, and urban development are discussed through real-world case studies from NER. Notable examples include precision farming using UAV-derived NDVI, flood impact assessment on Umiam Reservoir, and volumetric analysis for earthwork estimation. Emerging trends such as edge AI, IoT integration, swarm intelligence, and regulatory advancements are also explored, indicating a shift toward more autonomous, scalable, and sustainable UAV operations. The chapter highlights the growing potential of UAV technology in delivering high-resolution, timely, and cost-effective geospatial intelligence for informed decision-making and sustainable development.

The original version of the chapter has been revised. A correction to this chapter can be found at https://doi.org/10.1007/978-981-95-3221-6_26

V. Saikhom (✉) · P. S. Singh · N. Nishant · D. Chutia
North Eastern Space Applications Centre (NESAC), Shillong, Meghalaya, India
e-mail: victor.saikhom@nesac.gov.in

P. S. Singh
e-mail: ss.puyam@nesac.gov.in

N. Nishant
e-mail: nilay.nishant@nesac.gov.in

D. Chutia
e-mail: d.chutia@nesac.gov.in

S. P. Aggarwal et al. (eds.), *Remote Sensing of the North Eastern Himalayan Ecosystem*,
https://doi.org/10.1007/978-981-95-3221-6_19

Keywords UAV remote sensing · Photogrammetry · Orthomosaic · 3D point cloud · Digital surface model · Geospatial analytics · Artificial intelligence · IoT integration · Edge computing

19.1 Introduction

The North Eastern Region (NER) of India, which includes Arunachal Pradesh, Assam, Manipur, Meghalaya, Mizoram, Nagaland, Tripura, and Sikkim, is characterized by complex terrain, dense forests, high-altitude areas, and has frequent natural disasters such as landslides, floods or forest-fires. These geographical and environmental challenges make infrastructure development, resource management, and disaster preparedness difficult and complex. To address these issues effectively, large-scale survey and mapping is essential to provide accurate and detailed geospatial information (Aggarwal et al. 2024).

Unmanned Aerial Vehicles (UAVs), commonly referred to as drones, have revolutionized the way geospatial data is collected, processed, and analyzed across multiple domains. Originally developed for military applications, UAV technology has since expanded into diverse civilian and industrial sectors, including agriculture, environmental monitoring, infrastructure assessment, and disaster response. This transition has been facilitated by advancements in miniaturization, sensor technology, artificial intelligence (AI), and data analytics, enabling UAVs to perform highly specialized and automated tasks with precision (Saikhom et al. 2024; Khosiawan et al. 2018).

The evolution of UAV technology has been significantly influenced by breakthroughs in key areas such as GPS-enabled navigation, high-resolution imaging, and real-time data processing. Early UAVs were limited by short flight durations, minimal payload capacity, and rudimentary data acquisition capabilities. However, recent improvements in battery efficiency, aerodynamics, and sensor integration have enabled UAVs to operate over longer durations, capture high-fidelity data, and perform autonomous missions with minimal human intervention. These enhancements have expanded UAV applications, making them indispensable tools in domains requiring high-resolution, near real-time data acquisition (Zhang et al. 2023; Khosiawan et al. 2018).

One of the most transformative aspects of UAV development has been the integration of sophisticated data processing techniques. Traditional remote sensing methods relied heavily on satellite and manned aerial platforms, often constrained by high operational costs, restricted revisit times, and weather-dependent imaging conditions. UAVs, on the other hand, provide on-demand, cost-effective, and high-precision data capture, making them an ideal solution for time-sensitive and localized studies (Saikhom et al. 2024; Jafarbiglu et al. 2022; Phang et al. 2023). Advanced techniques such as LiDAR (Light Detection and Ranging) and UAV-based photogrammetry can overcome the limitations of satellite imaging by capturing high-resolution terrain data, even in heavily forested and cloudy areas. Large-scale mapping using UAVs, aerial surveys, and LiDAR technology can provide resolutions as fine as 4–5 cm/

pixel, making it highly suitable for infrastructure planning, disaster management, and environmental conservation.

The increasing role of UAVs in geospatial applications underscores the importance of advancements in UAV data processing and analytics (Quamar et al. 2023). Emerging technologies such as machine learning, cloud computing, and edge AI have further enhanced UAV capabilities, enabling automated feature extraction, object detection, and predictive analytics (Pal et al. 2024). The integration of UAV data with Geographic Information Systems (GIS) has also allowed for more accurate mapping, 3D modeling, and spatiotemporal analysis.

Accurate and detailed large-scale mapping has numerous applications in NER. It plays a vital role in infrastructure and urban planning by assisting in the alignment of roads and railways, supporting smart city development in areas like Guwahati and Shillong, and aiding in the planning of power transmission lines and hydropower projects. In disaster management, high-resolution terrain data helps in landslide susceptibility mapping, flood modeling, and post-disaster damage assessment, ensuring better preparedness and response strategies. Agriculture and forestry also benefit significantly from large-scale mapping, as it aids in identifying shifting cultivation areas for sustainable land-use planning, implementing precision agriculture for crops such as tea and bamboo, and monitoring forest degradation and illegal logging activities. Additionally, biodiversity conservation and ecotourism efforts are strengthened through detailed wildlife habitat mapping, the development of ecotourism circuits, and monitoring the impact of climate change on fragile ecosystems. High-resolution geospatial data enhances surveillance, facilitates the mapping of remote border villages, and supports the development of security infrastructure. The rapid adoption of UAV technology has been driven by the need for better geospatial intelligence, which supports informed decision-making in various sectors. In agriculture, UAVs equipped with multispectral and hyperspectral sensors facilitate precision farming by monitoring crop health, soil conditions, and irrigation needs. In environmental monitoring, UAVs assist in deforestation tracking, biodiversity assessments, and pollution detection. Infrastructure projects benefit from UAV-based surveys for route alignment, volumetric analysis, and construction progress monitoring. Additionally, in disaster management, UAVs are invaluable for damage assessment, search-and-rescue operations, and early warning system support (Makam et al. 2024; Zhang et al. 2023).

This chapter delves into the comprehensive aspects of UAV remote sensing, focusing on data acquisition, processing methodologies, analytical techniques, and practical applications across different industries. By examining the latest developments in UAV technology, this discussion aims to highlight how UAVs are shaping the future of geospatial intelligence and transforming industries that rely on precise, high-resolution aerial data.

19.2 UAV Remote Sensing

19.2.1 Principles of UAV Remote Sensing

UAV remote sensing involves the acquisition of information about objects or areas from a distance, typically using sensors mounted on UAVs. The principles of UAV remote sensing are based on detecting and measuring the reflected or emitted electromagnetic radiation from the Earth's surface and atmosphere (Quamar et al. 2023; Garg 2021). The sensors capture data in various spectra, including visible, infrared, and thermal. This capability is enhanced by the diverse range of UAV platforms and sensors available. Fixed-wing UAVs are ideal for extended flight times and large-area surveys due to their aerodynamics and efficiency. Rotary-wing UAVs, on the other hand, excel in maneuverability and are suited for detailed inspections in confined or complex environments. Hybrid UAVs combine the strengths of both designs, offering vertical takeoff and landing (VTOL) capabilities alongside efficient forward flight for broader coverage (Ke et al. 2018).

The sensors mounted on UAVs, including optical cameras, multispectral and hyperspectral cameras, thermal cameras, and LiDAR systems, allow for detailed and accurate data collection tailored to specific applications. For instance, optical cameras capture high-resolution images for general monitoring, multispectral and hyperspectral cameras support precision agriculture, vegetation analysis, and environmental monitoring, thermal cameras enable heat detection and energy audits, and LiDAR provides precise 3D mapping for applications like urban planning, forestry management, and flood modelling (Zhang et al. 2023).

19.2.2 Data Acquisition Workflow

Data acquisition through UAVs involves several critical steps to ensure that the collected data is accurate and meets the required standards. The process begins with pre-flight planning, where the type of UAV platform and sensor is selected based on the specific application and mission requirements (Liu et al. 2020). For instance, fixed-wing UAVs are ideal for large-area surveys due to their extended flight times and speed, while rotary-wing UAVs are suited for detailed inspections in confined spaces. The choice of sensor also depends on the data needed, such as optical cameras for general imagery, multispectral or hyperspectral cameras for vegetation analysis, thermal cameras for heat detection, and LiDAR for detailed mapping and surface imaging. During this phase, mission objectives are clearly defined, flight paths are planned using specialized software, and regulatory compliance is ensured (Collins 2024; Cao et al. 2023).

Furthermore, the establishment of Ground Control Points (GCPs) are crucial for accurately georeferencing the collected data. Suitable locations for GCPs are selected and marked with visible targets, and their coordinates are precisely recorded using

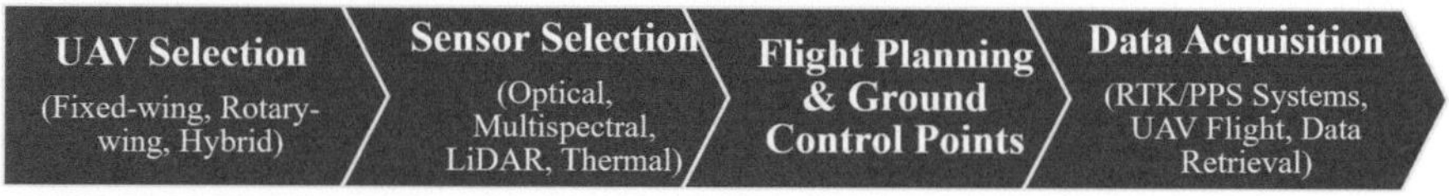

Fig. 19.1 UAV data acquisition workflow

DGPS methods or total stations (JOUAV 2024). The number and placement of GCPs depend on the accuracy required by the application, with more GCPs needed for high-precision tasks.

Flight execution involves conducting the UAV flight according to the pre-planned path. Before the flight, thorough pre-flight checks are performed to ensure that the UAV and its sensors are functioning correctly. Weather conditions are also monitored to ensure they are suitable for flying. During the flight, the UAVs performance and data collection are closely monitored to address any issues that arise. After the flight, post-flight data retrieval takes place. The data collected during the flight is downloaded from the UAVs onboard storage system, and backup copies are made to secure storage to prevent data loss.

The use of RTK (Real-Time Kinematic) and PPS (Post-Processed Kinematic) systems enhances the accuracy of the collected data. RTK systems provide real-time corrections to GPS data by using ground-based reference stations, offering high positional accuracy during the flight. This is particularly useful for applications like surveying and mapping. PPS systems involve collecting raw GPS data during the flight and applying corrections after the flight using data from reference stations. This post-processing improves positional accuracy and is often used in conjunction with RTK to achieve even higher precision (Xia et al. 2021).

Overall, these steps—pre-flight planning, establishment of GCPs, flight execution, post-flight data retrieval, and the use of RTK and PPS systems—are essential for effective data acquisition through UAVs, ensuring high-quality and reliable data collection for various applications as shown in Fig. 19.1.

19.3 Data Processing for Product Generation

Data preprocessing is a critical phase in UAV data processing, aimed at transforming raw data into a usable format for analysis (Garg 2021; Pargieła et al. 2023; Ram´ırez et al. 2021; Aierken 2024; Saikhom et al. 2024). The overall UAV data processing workflow is shown in the Fig. 19.2 given below:

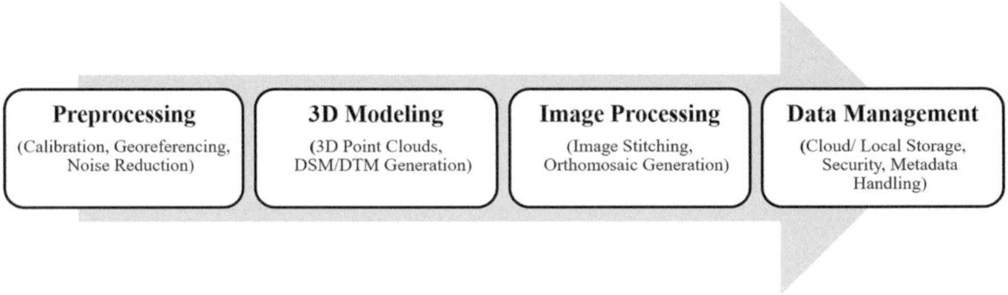

Fig. 19.2 UAV data processing workflow

19.3.1 Preprocessing

Data preprocessing is a crucial phase in UAV data processing, aimed at transforming raw data into a usable format for analysis. The process begins with *sensor calibration*, which corrects any distortions or inaccuracies introduced by the UAV's sensors. This includes adjusting for lens distortions, sensor drift, and other sources of error to ensure the data is as accurate as possible. Radiometric calibration follows, which standardizes the intensity measurements in imagery to reflect true values by adjusting for variations in sensor response and lighting conditions.

Subsequently, *data alignment* is performed to ensure the data is accurately mapped to geographical coordinates. This involves georeferencing the data using ground control points (GCPs) or GPS data, which aligns the imagery with real-world locations. Orthorectification is then applied to correct geometric distortions caused by the UAV's flight path or terrain variations, producing accurate and scale-corrected representations of the terrain.

Noise reduction is another key step, aimed at improving data quality by removing unwanted artifacts. Filtering techniques are used to clean up the data while smoothing algorithms help reduce variability and enhance clarity. This ensures that the data is less affected by random fluctuations or noise.

Data integration involves combining data from various sources and ensuring consistency in coordinate systems. This includes converting data into a common coordinate system or projection and resampling the data to align with the desired resolution. Coordinate system conversion ensures that all data is compatible while resampling adjusts the resolution to meet analysis requirements.

Data correction addresses additional adjustments needed for accurate analysis. Radiometric correction adjusts for variations in sensor response and atmospheric effects, while atmospheric correction removes interference caused by scattering and absorption in the atmosphere.

Finally, *data structuring* organizes the data into a format suitable for analysis. This involves converting data into standardized formats compatible with analysis tools and maintaining detailed metadata that includes information about data acquisition conditions and processing steps. Proper metadata management ensures that all relevant information about the data is recorded and available for future reference.

Overall, these pre-processing steps—sensor calibration, data alignment, noise reduction, data integration, data correction, and data structuring—are essential for preparing raw UAV data for accurate and reliable analysis. Each step plays a crucial role in enhancing the quality and usability of the data, ensuring that it provides meaningful insights for various applications.

19.3.2 Image Stitching and Orthomosaics

Image stitching is a critical step in UAV data processing, where multiple overlapping images are combined to create a single, seamless orthomosaic (Fig. 19.3). Orthomosaics provide a geospatially accurate representation of the surveyed area, corrected for distortions caused by camera angle and terrain (Garg 2021; Shammi et al. 2024). This process involves:

Feature Matching: It is the initial step in both image stitching and creating orthomosaics. It involves identifying and linking common features across overlapping images. The process includes:

- *Keypoint Detection*: Algorithms detect keypoints or distinct features in each image, such as edges, corners, or textures. These keypoints are the basis for matching images.
- *Feature Extraction*: Once keypoints are identified, descriptors are extracted to characterize these features. Descriptors are unique patterns or attributes that help in matching keypoints across images.

Fig. 19.3 UAV-derived RGB orthomosaic image (GSD: 3.5 cm/pix)

- *Feature Matching*: Using descriptors, the algorithm matches keypoints from one image to corresponding keypoints in another. This matching process is crucial for aligning and stitching the images accurately.

Alignment and Warping: After matching features, the next step is *alignment and warping*, which involves adjusting the images so that they fit together seamlessly. This process includes:

- *Image Alignment*: The matched keypoints are used to estimate a transformation matrix that aligns the images. This matrix adjusts for differences in perspective, rotation, and scale, ensuring that the images overlap correctly.
- *Homography Estimation*: For images taken from different angles or heights, a homography transformation is estimated. This mathematical transformation corrects for geometric distortions and aligns the images accurately.
- *Image Warping*: Using the estimated transformation matrix, each image is warped to fit the alignment. Warping adjusts the pixel positions according to the transformation, ensuring that the images overlap correctly and align with one another.

Blending and Seamless Integration: Once the images are aligned and warped, the next step is *blending and seamless integration*, which involves creating a smooth and coherent final image. This process includes:

- *Blending*: To eliminate visible seams and differences in color or exposure between overlapping images, blending techniques are applied. Methods such as feathering or gradient blending gradually transition between images, smoothing out any noticeable boundaries.
- *Exposure Correction*: Differences in exposure between images are corrected to ensure uniform brightness and contrast across the final stitched image. This may involve adjusting the intensity values of the pixels to match the overall lighting conditions.
- *Seamless Integration*: The final step involves combining all the warped and blended images into a single seamless mosaic. This process ensures that the edges of the images blend smoothly, creating a continuous and coherent view.

19.3.3 3D Point Clouds and Digital Surface Models (DSM)

UAVs generate 3D point clouds (Fig. 19.4) and DSMs (Fig. 19.5) using photogrammetry and LiDAR. These models provide detailed topographical information and surface representations, enabling various applications (PK Garg 2021; Bouziani et al. 2021; Žabota et al. 2023).

3D Point Clouds: These consist of numerous data points with coordinates (X, Y, Z) and sometimes additional attributes such as colour or intensity. They are generated by LiDAR, which uses laser pulses to measure distances, or by photogrammetry, which analyzes overlapping images to derive depth information. LiDAR provides

Fig. 19.4 UAV-derived 3D point cloud (average density: 23.91 points/m^3)

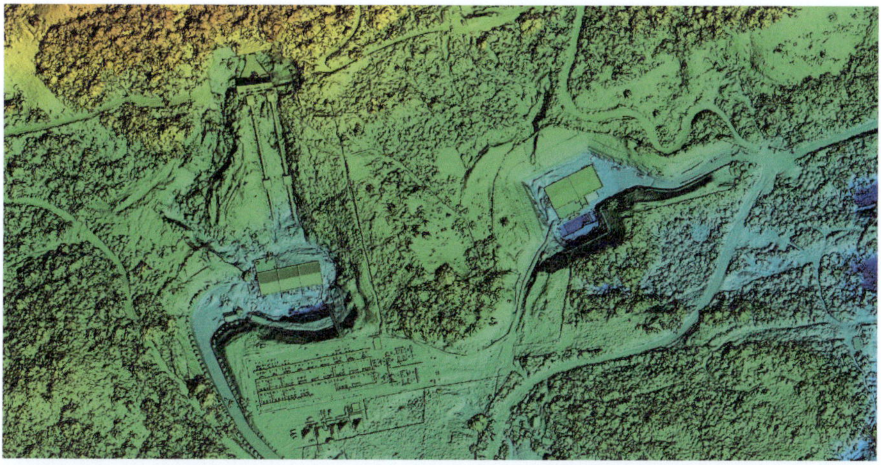

Fig. 19.5 UAV-derived digital surface model (DSM) with 3.5 cm posting

high accuracy in elevation, while photogrammetry offers detailed surface textures but can be less precise in height.

Digital Surface Models (DSM): *DSMs* are raster representations showing elevation data, including all objects on the surface, such as buildings and vegetation. DSMs are created from point clouds through processes like noise reduction, rasterization, and interpolation. They offer a detailed view of terrain and surface features, with resolution dependent on the point density of the input data.

In essence, 3D point clouds provide detailed spatial data through discrete points, while DSMs offer a comprehensive elevation view organized in a raster grid, both being crucial for mapping, planning, and analysis.

19.3.4 Data Management and Storage

Effective data management and storage solutions are essential for handling the large volumes of data collected by UAVs. Key considerations in this process include data integrity, accessibility, security, and the choice of storage solutions (Khan 2024; OPSWAT 2024; Marco et al. 2024; Emilien et al. 2021).

Data Integrity: Ensuring data integrity involves maintaining accuracy and consistency throughout the data lifecycle. Validation checks during data collection and transfer help ensure data accuracy, while error detection algorithms correct any inconsistencies. Implementing version control is vital for tracking changes and updates to datasets, and keeping backup versions allows for recovery from data corruption or accidental deletions. Metadata management is another important aspect, where descriptive metadata, such as date, time, location, sensor specifications, and flight parameters, provides essential context for data usability. Automated tools can capture and store metadata during data acquisition, streamlining the process.

Accessibility: Efficient data retrieval and easy accessibility are crucial for maximizing the use of UAV-collected data. Indexing strategies facilitate quick data retrieval, while user-friendly interfaces, such as web portals or GIS applications, make it easy for users to access data. Data sharing is enhanced through collaborative platforms that allow multiple users to access and work on data simultaneously. APIs can further facilitate data integration with other systems and applications. Scalability is another important consideration, with cloud storage solutions offering flexible and scalable options. A hybrid approach, combining local and cloud storage, can balance speed and scalability effectively.

Security: Data security involves implementing robust measures to protect data from unauthorized access and cyber threats. Access control mechanisms, such as strong authentication and role-based access control (RBAC), ensure that only authorized users can access the data. Encrypting data at rest and in transit adds an extra layer of protection against unauthorized access and interception. Compliance with regulatory standards, such as GDPR or HIPAA, is crucial for legal and ethical data management. Maintaining audit trails helps track access and modifications, ensuring accountability and compliance.

Storage Solutions: Choosing the right storage solutions is critical for managing large volumes of UAV data. Local storage offers quick access and control over data without relying on internet connectivity, but it has limited scalability and a higher risk of data loss without proper backup systems. Cloud storage provides high scalability, redundancy, and accessibility from any location with an internet connection, though it comes with ongoing subscription costs and potential data privacy concerns. A hybrid storage solution, combining local and cloud storage, balances speed, scalability, and redundancy. Frequently accessed data can be stored locally, while large datasets and backups are kept in the cloud. Implementing redundant storage systems, such as RAID (Redundant Array of Independent Disks), protects against hardware failures, and scheduling regular backups ensures data can be recovered in case of corruption or loss.

Best Practices for Data Management and Storage: Effective data lifecycle management is vital for efficient data handling. This includes data ingestion, preprocessing raw data for analysis, archiving older data in cost-effective storage solutions, and ensuring quick and efficient data retrieval through well-organized indexing and storage systems. Securely deleting data that is no longer needed, in compliance with relevant regulations and standards, is also important. Regular audits and reviews help maintain data integrity and identify any vulnerabilities. Data integrity checks ensure the data remains accurate and consistent, while security audits address potential security weaknesses. Training users on best practices for data management, including secure handling and access protocols, is essential for maintaining data security. Comprehensive documentation of data management procedures, including metadata standards, access protocols, and backup schedules, supports effective and consistent data management practices.

19.4 UAV Data Analytics

19.4.1 Geographic Information Systems (GIS)

Geographic Information Systems (GIS) provide a powerful platform for integrating and analyzing data captured by Unmanned Aerial Vehicles (UAVs), which can convert raw imagery and geospatial data into useful insights. The integration of UAV data into GIS allows for comprehensive GIS analysis and data visualization, which are crucial for a variety of applications, such as environmental monitoring, urban planning, agriculture, disaster management etc. (Quamar et al. 2023; Ozturk et al. 2024).

Spatial Analysis: Spatial analysis in GIS (Ding et al. 1992) is crucial for understanding the relationships between different features. After UAV data is integrated into GIS, it can be analyzed to identify spatial patterns, trends, and correlations that would have been difficult to detect otherwise with traditional methods. For example, spatial analysis can reveal how certain features, such as vegetation or buildings, are distributed across the given area. Techniques such as buffer analysis, proximity analysis, and overlay analysis can highlight relationships between features like roads, rivers, or land use. This allows decision-makers to detect trends such as urban sprawl, deforestation etc. Additionally, spatial statistics tools like hot spot analysis can help identify areas with statistically significant patterns.

Data Visualization: One of the key advantages of using GIS with UAV data is the ability to visualize the details of the map. Further, GIS tools can be used to generate detailed maps, charts, and 3D models that provide insights into complex high spatial UAV data. These visualizations help users to understand the data and make better decisions.

Integration with Other Data: A significant strength of GIS is its ability to integrate UAV data with other geospatial data sources, creating a multi-dimensional view of the area being studied. UAV data having high-resolution imagery, LiDAR, and thermal data, can be combined with satellite imagery along with the existing GIS layers (such as roads, zone maps, or land use data), and even real-time data from sensors or IoT devices. This multi-data fusion allows better analysis for more accurate and comprehensive assessments.

19.4.2 Image Analysis and AI

Advanced image analysis techniques are essential for extracting meaningful information from UAV imagery, utilizing the latest developments in AI, cloud computing, and high-performance geospatial processing. Deep learning models, such as U-Net, DeepLabV3+, and Transformer-based architectures, have greatly improved the accuracy of recognition, reorganisation and reconstruction, enabling precise identification of features such as land use types, vegetation health, or infrastructure details (Osco et al. 2021). Object detection algorithms like YOLO (Ma et al. 2023) and Faster R-CNN (R-Convolutional Neural Networks) further enhance UAV-based monitoring by automating the identification of vehicles, buildings, and anomalies in urban and natural environments. Moreover, hyperspectral and multispectral image analysis enables advanced feature extraction by utilizing spectral signatures for applications such as crop health assessment or soil nutrient identification.

Recent advancements in automated processing pipelines integrate cloud-based geospatial platforms and edge AI that allows real-time analytics and scalability of platforms (McEnroe et al. 2022). Platforms that support cloud-based deep learning environments support large-scale UAV data analysis, reducing computational overhead on local systems. Some advanced technologies such as Edge AI allow real-time processing onboard UAVs, making it possible to detect and classify objects instantly while the UAV is flying. More innovative approaches like multimodal fusion techniques that combine optical, LiDAR, and SAR (Synthetic Aperture Radar) data could enhance UAV image analysis capabilities for more robust insights towards applications such as disaster management, precision agriculture, and smart city development. All these image analysis techniques ensure faster, more accurate, and capable of handling huge amounts of geospatial data in more efficient ways.

Machine learning and artificial intelligence (AI) significantly enhance UAV data analytics (Khan et al. 2019) by automating complex data processing tasks, improving analytical accuracy, and enabling predictive modeling. AI-driven automation streamlines processes such as image classification and object detection, reducing manual effort and increasing efficiency. Advanced AI algorithms enhance the precision of data analysis, minimizing errors and improving the reliability of insights. Additionally, AI enables predictive modelling, allowing for the forecasting of trends and outcomes based on UAV data, which is particularly valuable for applications such as environmental monitoring, disaster management, and precision agriculture.

Object Detection: Object detection in high-resolution UAV data involves identifying and locating features representing different objects using various techniques of machine learning and deep learning. Traditional methods like edge detection and feature extraction are useful but limited in complex environments such as in the north-eastern region with varying topography. Deep learning approaches, such as CNNs, are more powerful. YOLO (You Only Look Once) provides real-time detection of objects by segmenting the images into grids and predicting individual object locations, while Faster R-CNN uses a two-stage process to propose regions and classify objects. UAV data captured by RGB sensors is commonly processed with CNN-based techniques, while multispectral sensor data allows for spectral band analysis, enhancing the detection of crop health, soil conditions, or feature differentiation. The object detection could also be applied to UAV-based 3D point clouds generated by image-based approach or LiDAR sensors, using techniques such as voxelization, segmentation, and machine learning algorithms like PointNet to accurately identify and classify objects based on their spatial characteristics. All these methods could then be applied in different applications such as agriculture for detecting weeds, pests, and plant health issues, in infrastructure for monitoring project progress or monitoring the different segments of construction sites, or in disaster management for assessing damaged structures and affected roads etc.

Classification: Segmentation and categorizing of land cover types such as forests, water bodies, and urban areas can be carried out through basic image interpretation techniques and advanced feature extraction algorithms. High-resolution UAV images could be used to extract finer visual characteristics such as colour, texture, shape, and spatial patterns which can be used to manually distinguish features like forests, water bodies, and urban areas from remote sensing imagery. To automate the classification of land features, advanced algorithms, such as machine learning techniques or deep learning approaches can be used. These methods can handle large datasets, improve accuracy, and detect fine differences in spectral signatures, making them highly efficient for diverse and complex landscapes.

Change Detection: Recent trends in change detection using high-resolution UAV imagery focus on utilization of advanced computational techniques and integrating multispectral and hyperspectral data for greater precision. One popular method is the use of deep learning algorithms, such as CNNs, to automatically detect and classify changes over time. These models are trained on large datasets to recognize fine changes in land cover, infrastructure, or environmental conditions, and work better than traditional pixel-based methods. This method enhances accuracy in detecting changes in complex environments, especially in areas with high spatial variability like urban or forested landscapes. The integration of multispectral and hyperspectral data from UAVs can help in detailed analysis of crop health, water quality, and other environmental conditions. This is coupled with time-series analysis, where multiple images taken over time are compared to detect gradual or abrupt changes. The combination of AI-driven techniques with high-resolution, multispectral UAV imagery is enabling more precise and efficient monitoring of land cover changes, infrastructure development, and environmental degradation, making it a powerful tool for applications in agriculture, urban planning, and environmental management.

19.5 Applications

UAVs have become a valuable asset in remote sensing, enabling high-resolution and cost-effective data acquisition for various applications and centres like NESAC have demonstrated their effectiveness across diverse landscapes in the Eastern Himalayan region, especially the North Eastern Region of India. These UAVs provide high-resolution, near real-time geospatial data, supporting various sectors such as disaster management, infrastructure planning, agriculture, forestry, and environmental monitoring. In disaster management, The UAV plays an important role in rapid damage assessment during floods, landslides, etc., aiding emergency response and relief efforts. For infrastructure development, high-resolution imagery supports road and power transmission line alignment, urban planning, and construction monitoring. In agriculture and forestry, UAV-based multispectral imaging facilitates crop health assessment, precision farming, and forest cover mapping. Additionally, UAVs contribute to environmental monitoring by assisting in wetland conservation, river morphology studies, and biodiversity assessments. The Centre's UAV-based initiatives significantly enhance decision-making by providing cost-effective, detailed, and up-to-date spatial information, driving sustainable development in the region. Some of the practical applications of UAV remote sensing are discussed in the subsequent section:

19.5.1 Precision Agriculture

19.5.1.1 Smart Farming with Agro-Advisory Services

Agriculture in Northeast India is marked by diverse topography, fragmented landholdings, variable climate, and frequent weather extremes, which collectively pose significant challenges to sustainable farm productivity. Conventional extension mechanisms and satellite-based remote sensing often fall short in addressing the information needs of farmers, particularly due to limited connectivity, data latency, and the unavailability of cloud-free satellite imagery during critical crop phenological phases. UAVs equipped with RGB and multispectral sensors were deployed to generate real-time, high-resolution data for precision agriculture. UAV flights were planned according to key crop growth stages, ensuring that critical phenological phases were effectively captured. Parallelly, field sampling of soil and plant specimens was conducted for laboratory analysis to obtain biochemical and environmental parameters that complements the data captured by the UAV. The collected UAV data after processing generates 2D and 3D models of the agricultural fields. These models help visualize crop health, field conditions, and terrain features. These outputs, combined with cadastral and topographic maps, enabled precise delineation of farm boundaries and identification of problem zones. Ground truthing was performed to validate UAV-derived indices with actual field conditions and sample results. Reflectance maps were used to compute custom vegetation and soil health indices, providing critical insights into crop vigor and stress patterns (Shammi et al. 2024).

These indices formed the basis for evidence-based agro-advisory services tailored to specific crops and locations. The advisories included recommendations on irrigation, nutrient application, pest and disease control, and other essential practices.

The generated agro-advisory services are further disseminated through a dedicated web platforms and the geoportal DHaBReT *(Drone Haka Buit Rep Thymai)*, a localized agro-advisory system tailored for Meghalaya State in the Northeastern region of India (Figs. 19.6 and 19.7). The platform delivers advisory reports and data to farmers, while the geoportal provides access to spatial data and maps. All these agro-advisory services are shared directly to the farmers and for their actions to optimize their crop management and improve productivity.

Fig. 19.6 DHaBReT showing RGB and NDVI indices of Cabbage Crop, Thynroit, Meghalaya

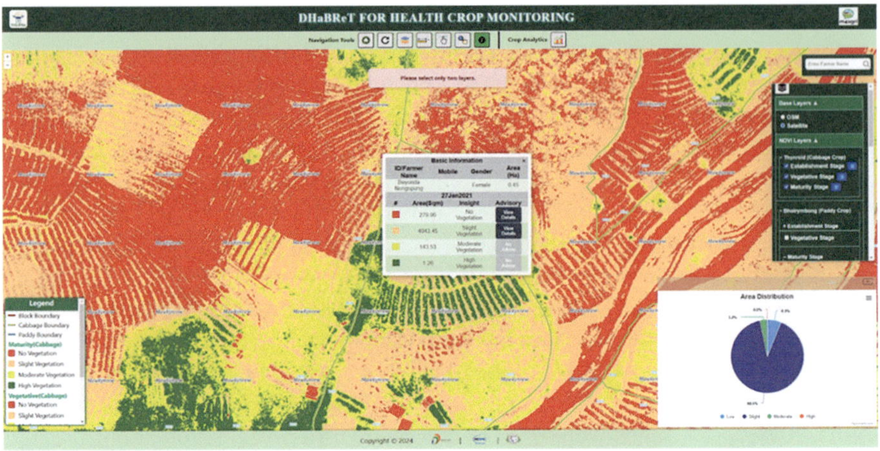

Fig. 19.7 DHaBReT showing zonal statistics on crop health and farm level advisories

19.5.1.2 Crop Damage Assessment Using UAV

Insect infestation poses a significant threat to agricultural productivity, leading to drastic crop yield losses. Traditional methods for assessing crop damage due to insect infestation can be time-consuming, labor intensive, and costly. The use of UAV Remote Sensing in Agriculture sector has transformed, enabling real-time and high-resolution data acquisition. UAVs equipped with multispectral and thermal sensors offer rapid, precise, and scalable solutions to assess crop damage due to insect infestations. UAVs allows to capture images at resolutions of a few centimetres, revealing subtle signs of infestation that are not detectable by satellite imagery. Early detection of infestation patterns enables targeted interventions, minimizing crop losses and reducing pesticide use.

In April 2016, farmers in Morigaon and Nagaon districts of Assam State faced severe pest infestations in their Boro paddy fields, particularly in Morigaon district. The District Agricultural Department and Regional Agricultural Research Station identified the pest as the Brown Plant Hopper (BPH). To assess the damage, UAV was used to survey four severely infested locations: Naramari, Mikirbheta, Bhurbandha, and Jaluguti. The UAV captured high-resolution images, covering areas between 24 to 54.94 ha. These images were processed to create seamless mosaics and digitize the rice fields. The survey identified infested areas based on tonal variations in the images. Fields were categorized as completely affected (over 50% infestation) or partially affected (less than 50% infestation). Completely affected fields showed no yield potential, while immediate intervention could save partially affected areas (Gupta et al. 2018).

19.5.2 Environmental Monitoring

19.5.2.1 Flood Impact Analysis of Reservoir

Flood impact analysis on reservoirs is critical for water resource management, flood mitigation, and disaster response. Reservoirs serve as a buffer to control water flow, but during extreme weather events, overflow or structural issues can arise which can impact and flood nearby communities and ecosystems. The use of UAVs with high-resolution cameras will ensure rapid, precise, and cost-effective flood impact analysis on reservoirs by capturing real-time data and generation of high resolution 2D and 3D data products for emergency response and management.

In a project focused on the Umiam Reservoir, UAV technology was deployed to conduct a detailed flood impact analysis. The high-resolution data products such as DSM were used to delineate the High Flood Level (HFL) and create crucial geospatial data products. The HFL was extracted from the DSM, and overlaid on orthomosaic images (Fig. 19.8) to visualize potential flood impact zones, enhancing flood level analysis and aiding in comprehensive planning and risk management.

Fig. 19.8 High flood level map, Umiam reservoir

Fig. 19.9 Tree crown delineation by automatic segmentation method (Chakraborty et al. 2019)

19.5.2.2 Canopy Delineation Using UAVs

Canopy delineation helps in identifying and mapping the extent, shape, and structure of individual tree canopies within a forest or agricultural area. This information is useful for applications in forestry, precision agriculture, ecological monitoring, and biodiversity conservation. UAVs equipped with high-resolution sensors, provide a cost-effective and accurate solution for capturing highly detailed canopy data, allowing for precise mapping of canopy boundaries and characteristics (Fig. 19.9). This way, detailed assessment on canopy structure and health can be carried out to provide insights on tree growth, canopy density, and probable signs of disease or stress. Further, we can monitor temporal changes in canopy size and density to track growth rates, deforestation, and reforestation efforts. This will also help in supporting ecosystem and biodiversity studies by providing data on tree species distribution and habitats etc.

At the garden of Botanical Survey of India in Umroi, Meghalaya, UAVs data were used to automatically map a 10-hectare area's canopy cover, generating accurate 3D models. UAVs are especially beneficial in Eastern Himalayan region, where they reduce the need for ground inventory in challenging terrains.

19.5.2.3 Enhancing Tea Garden Management with UAV Technology

The UAV project for tea garden management, initiated by the Tea Development Centre in Umsning, Meghalaya, focuses on creating a detailed digital database for a 10-hectare tea garden. High-resolution UAV images of the entire garden provide comprehensive information to the garden management authority (Fig. 19.10a

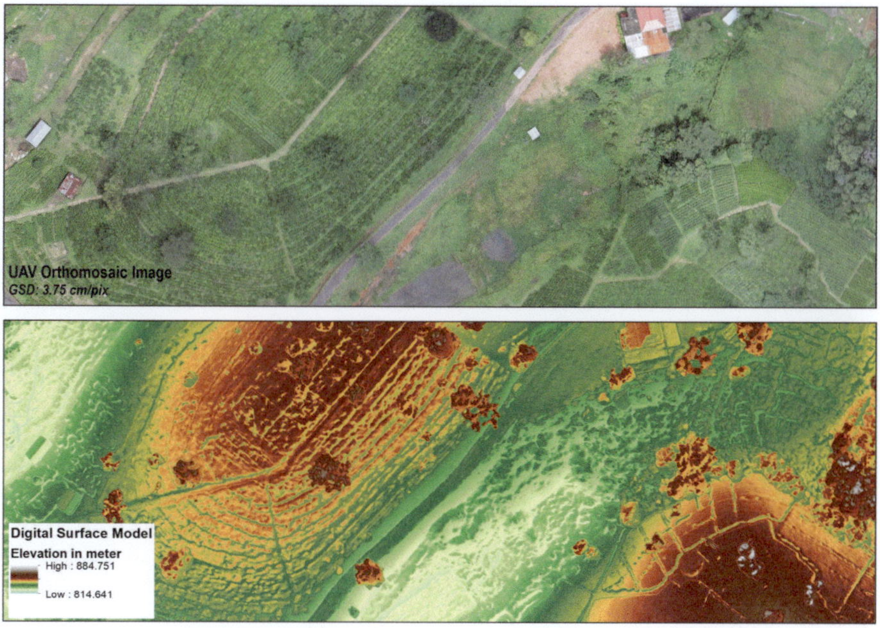

Fig. 19.10 UAV orthomosaic image (top) and DSM (bottom) for tea development centre in Umsning, Meghalaya

and b). These images help delineate management blocks, map the areas under different tea varieties, and document the garden's infrastructure, such as irrigation systems, pathways, and buildings. The data is integrated into a digital database, allowing for advanced analysis and informed decision-making. This UAV application enhances operational efficiency, optimizes resource allocation, and supports sustainable agricultural practices in the tea garden.

19.5.2.4 Survey and Mapping of Community Reserve Forests

UAVs with DGPS survey and mapping are essential tools for managing community reserve forests (CRF). They provide high-resolution data to accurately map forest boundaries, document biodiversity, and monitor forest health, which is crucial for conservation and sustainable resource management. This technology enables precise, up-to-date mapping of land use, vegetation cover, and natural resources, supporting effective community-based forest management and conservation planning. Figure 19.11 below shows an example of Chengni CRF map in South Garo Hills district of the state of Meghalaya. Such, UAV-based mapping improves decision-making by offering a visual and data-driven approach to monitor human activities, prevent encroachment, and promote sustainable practices, empowering communities to protect and maintain their forest resources.

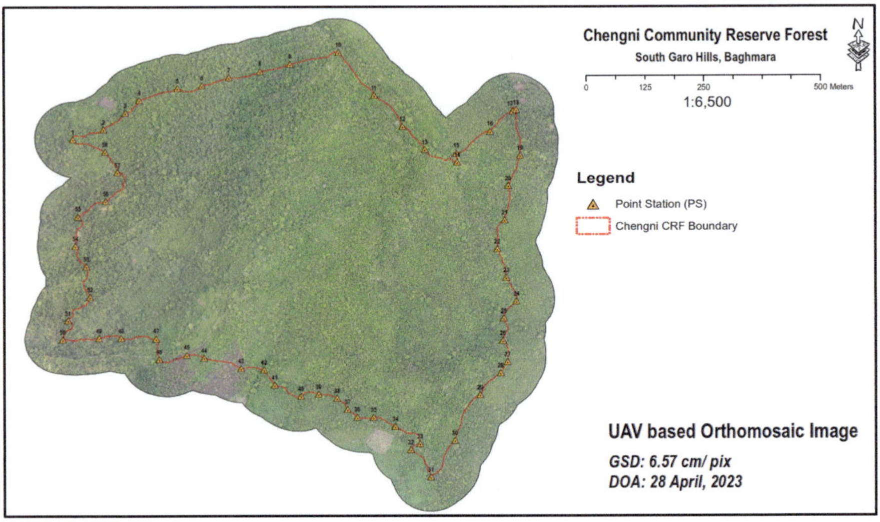

Fig. 19.11 Chengni community reserve forest map

19.5.3 *Infrastructure and Construction*

19.5.3.1 Estimation of Earth Work Using Volumetric Analysis

UAVs along with DGPS survey and mapping technology for earthwork estimation are highly beneficial, especially for infrastructure projects like airport extensions, road constructions, etc. This technology allows for accurate, high-resolution aerial surveys, producing detailed digital surface models (DSMs) and orthomosaics. These datasets facilitate precise calculations of existing ground levels and help estimate the volume of earth to be excavated or filled.

The estimation of earthwork for the Shillong Airport extension was carried out using UAV and DGPS survey and mapping technologies to achieve high precision and efficiency in project planning. The data collected from UAV surveys were processed to create Digital Elevation Models (DEMs) and contour maps, allowing for a comprehensive analysis of earthwork requirements, including cut and fill calculations (Fig. 19.12). The method not only streamlines the estimation process but also reduces the time and costs associated with traditional surveying techniques. Further, the integration of UAV and DGPS technologies in estimating earthwork for the Shillong Airport extension supports informed decision-making, effective resource allocation, and timely project execution.

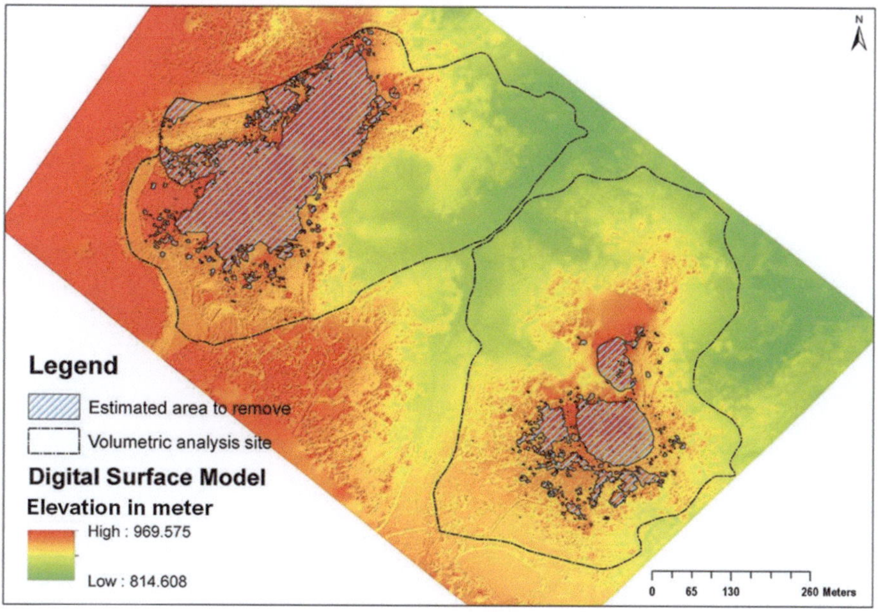

Fig. 19.12 2D map of cut and fill area for earthwork estimation—Shillong airport extension project

19.5.3.2 Resource Assessment in Mining Lease Areas

DGPS and UAV surveying have been effectively utilized in the mining lease areas, to enhance data accuracy and efficiency in resource management and planning. The combination of DGPS and UAVs allows for precise georeferencing and high-resolution aerial imagery, critical for assessing the geographical features and resources within the lease areas.

UAVs equipped with advanced sensors capture detailed aerial data, facilitated to create high-quality topographic maps and 3D models of the terrain for Dalmia Cement lease areas in Thangskai, East Jaintia Hills, Meghalaya (Fig. 19.13). With DGPS and UAV technologies, Dalmia Cement could achieve effective monitoring and management of its lease areas, ultimately contributing to informed decision-making, improved operational efficiency, and sustainable development within the lease areas following the Indian Bureau of Mines guidelines.

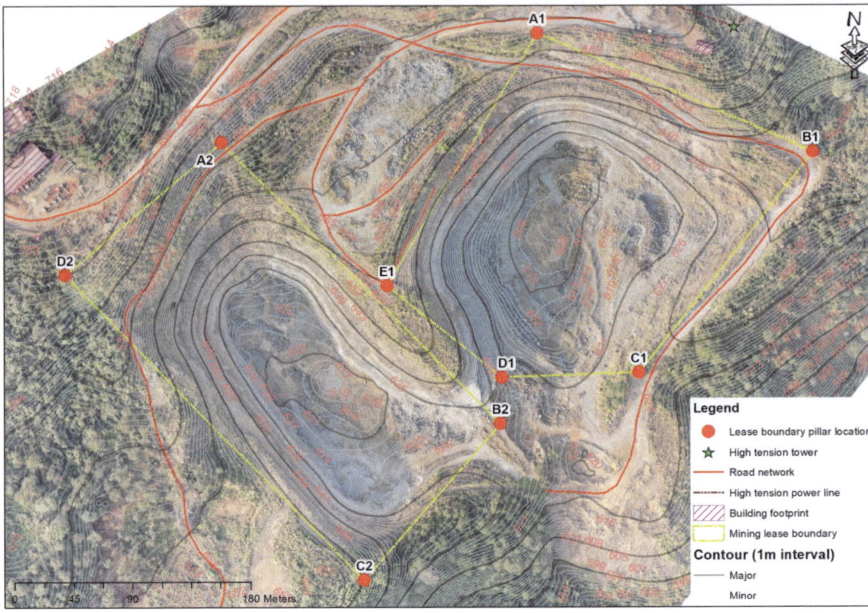

Fig. 19.13 Contour & mine lease boundary overlaid with UAV based orthomosaic image

19.5.4 Disaster Management

19.5.4.1 Mapping of Flood-Affected Areas

Using UAVs to map flood-affected areas provides quick, accurate data essential for disaster response. UAVs equipped with high-resolution cameras can efficiently survey flooded regions, creating detailed maps and digital models to assess flood extent, infrastructure damage, and locate isolated communities. This data also aids flood modelling and planning for future events, offering a safe, cost-effective alternative to traditional surveys, especially in remote areas.

During 2022, Slichar town in Assam was severely impacted by flooding (Fig. 19.14), UAVs captured high-resolution imagery (5 cm/pixel GSD) that documented extensive road and infrastructure damage. This precise data helped assess flood impacts, guide repairs, and improve disaster response for the affected communities.

19.5.4.2 Embankment Survey and Monitoring Using UAVs

Floods cause extensive damage globally, often resulting from rivers overflowing their banks. Traditional embankment monitoring methods are costly and time-intensive. UAVs provide a more affordable and efficient solution, delivering high-resolution

Fig. 19.14 UAV based ortho-image showing breaching location

data for flood assessment and embankment management. Using these data, the longitudinal and cross-sectional profiles of the embankments can be plotted to evaluate their condition and pinpoint vulnerable areas (Fig. 19.15). Such information is essential for assessing damage and planning repairs on the embankments before the monsoon season (Das et al., 2024).

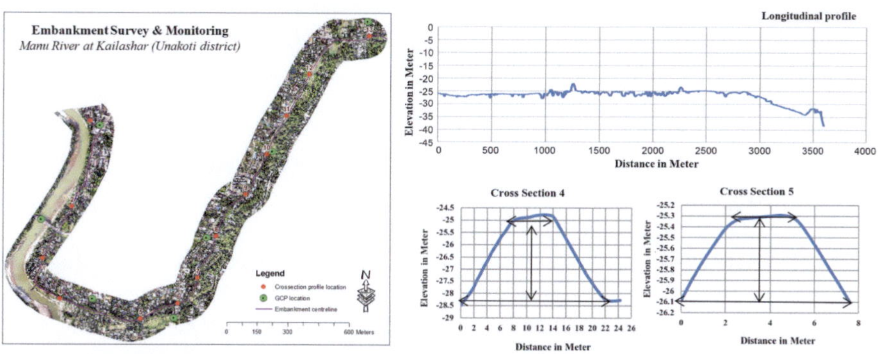

Fig. 19.15 Embankment longitudinal and cross section profile

Fig. 19.16 UAV image of part of Shillong city & landuse

19.5.5 *Urban Planning*

19.5.5.1 High Resolution Mapping for Land Use Planning Using UAV Data

Urban planning focuses on land development, environmental protection, public welfare, and urban design. Employing UAV technology, a project was initiated to analyze the existing urban land use and land cover, and to assess the suitability of locations for amenities in Shillong city, covering 0.75 sq.km. High-resolution UAV images were captured to prepare detailed maps of existing urban features at a 1:1000 scale (Fig. 19.16). This data allowed for precise identification of land use patterns, infrastructure, and open spaces, aiding in effective urban planning and decision-making. The work provided valuable insights for sustainable development, optimizing land use, and enhancing public welfare and environmental protection.

19.5.5.2 Village Resource Mapping Using UAV in Garo Hills, Meghalaya

In Garo Hills, Meghalaya, a pilot project was carried out to use UAV technology for mapping and demarcating village boundaries to support better development planning. The project involved aerial and DGPS surveys over a 15 sq.km area, producing high-resolution digital surface models (DSMs) and 5 cm/ pixel geo-referenced orthomosaics. Thematic maps were created to show roads, water bodies, forests, plantations, and agriculture (Fig. 19.17). A mobile app and a web-based decision support system were developed for asset geo-tagging and map analysis. Funded by the Meghalaya Basin Development Agency (MBDA) and GHADC, the project aims to enhance village boundary definitions and planning efficiency.

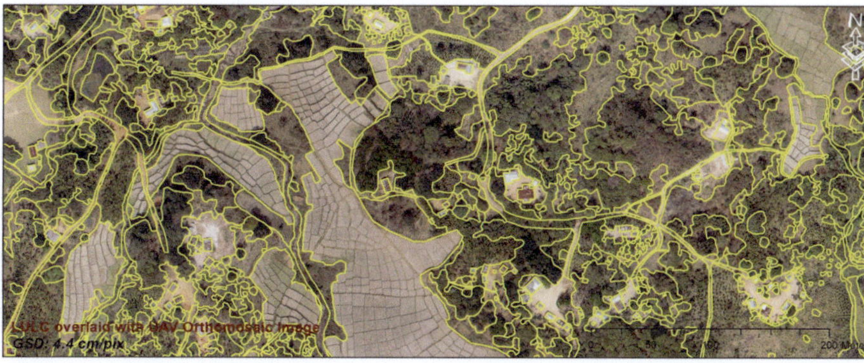

Fig. 19.17 Village resource map

19.6 Emerging Trends and Technological Advancements in UAV Applications

19.6.1 Technological Innovations

Recent technological innovations have significantly enhanced the capabilities of UAVs and their onboard sensors, making them more efficient and versatile across various applications. The miniaturization of sensors has led to the development of smaller, lighter, and more energy-efficient payloads, allowing UAVs to carry advanced instruments without compromising flight performance. This has enabled the integration of high-resolution LiDAR and microwave sensors equipped with precise GNSS receivers, facilitating the generation of detailed velocity maps through InSAR analysis. Such advancements improve the accuracy of terrain deformation studies, infrastructure monitoring, and disaster assessment. Additionally, improvements in UAV flight capabilities, including extended endurance, enhanced stability, and autonomous navigation, allow them to cover larger areas and operate effectively in challenging environments. These technological strides make UAVs more reliable for applications in geospatial mapping, environmental monitoring, and precision agriculture, among others (Mohsan et al. 2023; Laghari et al. 2024).

19.6.2 Integration with IoT

The integration of UAVs with IoT-based sensors significantly enhances real-time data collection, remote monitoring, and connectivity with smart systems, enabling more precise and efficient applications across various domains. In precision agriculture, IoT sensors deployed on the ground can collect in-situ data, such as soil moisture, temperature, and nutrient levels, which, when combined with UAV-based remote sensing, provide a comprehensive understanding of crop health and field conditions. This synergy allows for data-driven decision-making, optimizing irrigation, fertilization, and pest control strategies. Similarly, in bathymetry mapping, UAVs equipped with water level sensors enable accurate depth measurement of water bodies, facilitating the creation of high-resolution bathymetric models for hydrological studies, flood risk assessment, and aquatic ecosystem monitoring. The enhanced connectivity between UAVs and IoT devices ensures seamless data transmission, enabling real-time analysis and remote operation from centralized control centers (Telli et al. 2023; Rajak et al. 2023).

19.6.3 Regulatory Developments

Regulatory developments in the UAV industry are evolving to ensure safety, privacy, and responsible operations while enabling innovation. Safety regulations focus on airspace management, operational limits, and collision avoidance systems to prevent accidents and conflicts with manned aircraft. Privacy concerns are being addressed through data collection guidelines, consent requirements, and legal frameworks to prevent unauthorized surveillance. Certification and licensing requirements are being standardized, mandating training and competency tests for UAV operators and compliance with safety standards for manufacturers. Additionally, permits for specialized operations, such as beyond-visual-line-of-sight (BVLOS) flights, are being introduced. These evolving regulations aim to create a structured UAV ecosystem that balances security and privacy concerns with the growing demand for UAV applications across industries. By implementing clear guidelines, authorities are ensuring a safer and more accountable UAV industry while supporting technological advancements and commercial expansion (Lee et al. 2022; Bauranov et al. 2021; IAD 2020).

19.6.4 Future Trends

Recent advancements in UAV technology have introduced swarm intelligence, artificial intelligence (AI) integration, enhanced autonomy, and sustainability-focused innovations, revolutionizing their applications in various domains. Swarm technology enables the coordinated operation of multiple UAVs, allowing for large-scale and high-resolution data collection over vast areas. This is particularly beneficial in applications such as coordinated aerial mapping, disaster response, and environmental monitoring, where multiple UAVs can simultaneously cover different regions, reducing time and increasing efficiency. The integration of advanced AI further enhances UAV capabilities by enabling real-time data analysis, object detection, and tracking. AI-driven applications include identifying changes in urban landscapes, tracking wildlife movements, and monitoring agricultural fields for early detection of crop stress, pests, and diseases. Additionally, enhanced autonomy has led to the development of fully autonomous UAVs capable of executing complex missions without human intervention, improving efficiency in applications such as infrastructure inspection, precision agriculture, and security surveillance. Sustainability is also a key focus, with innovations in electric propulsion systems, energy-efficient designs, and environmentally friendly materials making UAV operations more eco-friendly. These advancements collectively enhance the reliability, efficiency, and versatility of UAVs, making them indispensable tools for geospatial analysis, resource management, and environmental conservation (Lt Col Narendra Tripathi (r) 2024; Pal et al. 2024; Caballero-Martin et al. 2024).

19.7 Conclusion

UAV remote sensing has emerged as a transformative technology across various sectors, driven by advancements in sensor technology, data processing, and analytics. The continuous evolution of UAV capabilities and applications promises to further enhance their impact, offering new opportunities for innovation and efficiency in diverse fields. As the technology advances, it will be essential to address challenges related to data management, regulatory compliance, and ethical considerations to fully harness the potential of UAV remote sensing. By staying abreast of technological innovations and regulatory developments, stakeholders can maximize the benefits of UAV remote sensing while mitigating associated risks and challenges.

Acknowledgements The authors sincerely acknowledge the contributions of the technical and non-technical staff of the North Eastern Space Applications Centre (NESAC), Department of Space, Government of India, whose dedicated efforts in UAV data acquisition, processing, and application development have been fundamental to the works presented in this chapter. This chapter, *"Advancement in UAV Data Processing Analytics and Applications"*, is entirely based on in-house studies, operational activities, and research initiatives carried out at NESAC under its institutional mandate. While no external funding was received specifically for the preparation of this chapter, the content reflects the cumulative experience and outcomes of NESAC's ongoing engagements across multiple geospatial domains in the North Eastern Region of India.

References

S.P. Aggarwal, S.S. Kundu, K.K. Sarma, Geospatial technology for effective disaster risk reduction: best practices in capacity building. Int. Arch. Photogramm. Remote Sens. Spat. Inf. Sci. **XLVIII-5-2024**, in *ISPRS TC V Mid-term Symposium "Insight to Foresight via Geospatial Technologies"* (Manila, Philippines, 2024). https://doi.org/10.5194/isprs-archives-XLVIII-5-2024-147-2024

N. Aierken, B. Yang, L. Yongke, P. Jiang, G. Pan, L. Shijian, A review of unmanned aerial vehicle based remote sensing and machine learning for cotton crop growth monitoring. Comput. Electron. Agric. **227**, Part 2, 109601 (2024). ISSN 0168-1699. https://doi.org/10.1016/j.compag.2024.109601.

A. Bauranov, J. Rakas, Designing airspace for urban air mobility: a review of concepts and approaches. Prog. Aerosp. Sci. **125**, 100726 (2021). ISSN 0376-0421. https://doi.org/10.1016/j.paerosci.2021.100726

M. Bouziani, M. Amraoui, S. Kellouch, Comparison assessment of digital 3D model obtained by drone based LiDAR and drone imagery. Int. Arch. Photogramm. Remote Sens. Spat. Inf. Sci. **XLVI-4/W5-2021**, 113–118 (2021). https://doi.org/10.5194/isprs-archives-XLVI-4-W5-2021-113-2021

D. Caballero-Martin, J.M. Lopez-Guede, J. Estevez, M. Graña, Artificial intelligence applied to drone control: a state of the art. Drones **8**(7), 296 (2024). https://doi.org/10.3390/drones8070296

Z. Cao, L. Kooistra, W. Wang, L. Guo, J. Valente, Real-time object detection based on UAV remote sensing: a systematic literature review. Drones **7**, 620 (2023). https://doi.org/10.3390/drones7100620

F. Carvajal-Ramírez, F. Agüera-Vega, P. Martínez-Carricondo, UAV photogrammetry and remote sensing (2021). ISBN: 978-3-0365-1453-6. https://doi.org/10.3390/books978-3-0365-1453-6

K. Chakraborty, V. Saikhom, S.B. Borah, M. Kalita, C. Gupta, L.R. Meitei, K.K. Sarma, P.L.N. Raju, Forest biometric parameter extraction using unmanned aerial vehicle to aid in forest inventory data collection. Curr. Sci. **117**(7) (2019). https://doi.org/10.18520/cs/v117/i7/1194-1199

A. Collins, Evaluating the Accuracy of UAV-Obtained Horizontal and Vertical Coordinates in District Six, Cape Town. *EasyChair-Preprint,* 15157 (2024), https://easychair.org

R. Das, V. Saikhom, S. Pandit, A. Saha, Monitoring of embankment structures in urban local bodies (ULB) of Tripura. NESAC-SR-341-2024 (2024)

Y. Ding, A.S. Fotheringham, The integration of spatial analysis and GIS. Comput. Environ. Urban Syst. **16**(1), 3–19 (1992). ISSN 0198-9715. https://doi.org/10.1016/0198-9715(92)90050-2

A.V. Emilien, T. Corpetti, T. Houet, UAV & satellite synergies for optical remote sensing applications: a literature review. Sci. Remote. Sens. **3**, 100019 (2021). ISSN 2666-0172. https://doi.org/10.1016/j.srs.2021.100019

P.K. Garg, Introduction to Unmanned Aerial Vehicles. New Age International Publishers (2021). ISBN: 978-93-89802-45-0

C. Gupta, B.K. Handique, D. Chutia, V. Saikhom, P.S. Singh, A. Chouhan, K.K. Sarma, P.L.N. Raju, Applications of unmanned aerial vehicle (UAV) based remote sensing in North Eastern Region of India. ISG Newsletter, vol. 23, no.2–4 & vol. 24, no.1 & 2 (2018)

IAD, Beyond Visual Line Of Sight (BVLOS), (2020), https://indianacademyofdrones.com. Accessed 04 Jan 2025

H. Jafarbiglu, A. Pourreza, A comprehensive review of remote sensing platforms, sensors, and applications in nut crops. Comput. Electron. Agric. **197**, 106844 (2022). ISSN 0168-1699. https://doi.org/10.1016/j.compag.2022.106844

JOUAV, A Comprehensive Guide to Using Ground Control Points for Drone Surveying (2024), https://www.jouav.com. Accessed 02 Nov 2024

Y. Ke, K. Wang, B.M. Chen, Design and implementation of a hybrid UAV with model-based flight capabilities. IEEE/ASME Trans. Mechatron. **23**(3), 1114–1125 (2018). https://doi.org/10.1109/TMECH.2018.2820222

A.I. Khan, Y. Al-Mulla, Unmanned aerial vehicle in the machine learning environment. Procedia Comput. Sci. **160**, 46–53 (2019). ISSN 1877-0509. https://doi.org/10.1016/j.procs.2019.09.442

R.A. Khan, A Complete Guide to Managing Data Access (2024), https://www.astera.com. Accessed 16 Nov 2024

Y. Khosiawan, Y. Park, I. Moon, J.M. Nilakantan, I. Nielsen, Task scheduling system for UAV operations in indoor environment. Neural Comput. Appl. **31**, 5431–5459 (2018)

A.A. Laghari, A.K. Jumani, R.A. Laghari, H. Li, S. Karim, A.A. Khan, Unmanned aerial vehicles advances in object detection and communication security review. Cogn. Robot. **4**, 128–141. ISSN 2667-2413. https://doi.org/10.1016/j.cogr.2024.07.002

Lee, D., Hess, D.J., Heldeweg, M.A., (2022). Safety and privacy regulations for unmanned aerial vehicles: A multiple comparative analysis. *Technology in Society*, Volume 71, 102079. ISSN 0160–791X. https://doi.org/10.1016/j.techsoc.2022.102079.

Y. Liu, H.N. Dai, H. Wang, M. Imran, X. Wang, M. Shoaib, UAV-enabled data acquisition scheme with directional wireless energy transfer for internet of things. Comput. Commun. **155**, 184–196 (2020). ISSN 0140-3664. https://doi.org/10.1016/j.comcom.2020.03.020

Lt Col Narendra Tripathi (r), Swarm Intelligence & Collaborative Unmanned Drone Systems For Maritime Surveillance (2024), https://www.iadb.in. Accessed 04 Jan 2025

C. Ma, Y. Fu, D. Wang, R. Guo, X. Zhao, J. Fang, YOLO-UAV: object detection method of unmanned aerial vehicle imagery based on efficient multi-scale feature fusion. IEEE Access **11**, 126857–126878 (2023). https://doi.org/10.1109/ACCESS.2023.3329713

S. Makam, B.K. Komatineni, S.S. Meena et al., Unmanned aerial vehicles (UAVs): an adoptable technology for precise and smart farming. Discov. Internet Things **4**, 12 (2024). https://doi.org/10.1007/s43926-024-00066-5

L.S. Marco, S. Nicotri, G. Donvito, A. Italiano, R. Colacicco, G. Miniello, I. Lapietra, R. Roseto, P. Dellino, D. Capolongo, A paradigm shift in processing large UAV image datasets for emergency management of natural hazards. Int. J. Appl. Earth Obs. Geoinformation **132**, 103996 (2024). ISSN 1569-8432. https://doi.org/10.1016/j.jag.2024.103996

P. McEnroe, S. Wang, M. Liyanage, A Survey on the convergence of edge computing and AI for UAVs: opportunities and challenges. IEEE Internet Things J. **9**(17), 15435–15459 (2022). https://doi.org/10.1109/JIOT.2022.3176400

S.A.H. Mohsan, N.Q.H. Othman, Y. Li et al., Unmanned aerial vehicles (UAVs): practical aspects, applications, open challenges, security issues, and future trends. Intel. Serv. Robotics **16**, 109–137 (2023). https://doi.org/10.1007/s11370-022-00452-4

OPSWAT, Data Storage Security- Challenges, Risks, and Best Practices (2024), https://www.opswat.com. Accessed 16 Nov 2024

L.P. Osco, J.M. Junior, A.P.M. Ramos, L.A.C. Jorge, S.N. Fatholahi, J.A. Silva, E.T. Matsubara, H. Pistori, W.N. Gonçalves, J. Li, A review on deep learning in UAV remote sensing. Int. J. Appl. Earth Obs. Geoinformation **102**, 102456 (2021). ISSN 1569-8432,https://doi.org/10.1016/j.jag.2021.102456

M.Y. Ozturk, I. Colkesen, A novel hybrid methodology integrating pixel- and object-based techniques for mapping land use and land cover from high-resolution satellite data. Int. J. Remote Sens. **45**(16), 5640–5678 (2024). https://doi.org/10.1080/23312041.2018.1498180

O.K. Pal, M.S.H. Shovon, M.F. Mridha et al., In-depth review of AI-enabled unmanned aerial vehicles: trends, vision, and challenges. Discov. Artif. Intell. **4**, 97 (2024). https://doi.org/10.1007/s44163-024-00209-1

K. Pargieła, Optimising UAV data acquisition and processing for photogrammetry: a review. GeomatS Environ Eng **17**(3), 29–59 (2023). https://doi.org/10.7494/geom.2023.17.3.29

S.K. Phang, T.H.A. Chiang, A. Happonen, M.M.L. Chang, From satellite to UAV-based remote sensing: a review on precision agriculture. IEEE Access **11**, 12705710 (2023). 1109/ACCESS.2023.3330886

M.M. Quamar, B. Al-Ramadan, K. Khan, M. Shafiullah, S. El Ferik, Advancements and applications of drone-integrated geographic information system technology—a review. Remote Sens. **15**, 5039 (2023). https://doi.org/10.3390/rs15205039

P. Rajak, A. Ganguly, A. Adhikary, S. Bhattacharya, Internet of Things and smart sensors in agriculture: scopes and challenges. J. Agric. Food Res. **14**, 100776 (2023). ISSN 2666-1543. https://doi.org/10.1016/j.jafr.2023.100776

V. Saikhom, M. Kalita, UAV for remote sensing applications: an analytical review, in *Emerging Technology for Sustainable Development, ed. by* J.K. Deka, P.S. Robi, B. Sharma, EGTET 2022, Lecture Notes in Electrical Engineering (Springer, Singapore, 2024), vol. 1061. https://doi.org/10.1007/978-981-99-4362-3_6

S.A. Shammi, Y. Huang, G. Feng, H. Tewolde, X. Zhang, J. Jenkins, M. Shankle, Application of UAV multispectral imaging to monitor soybean growth with yield prediction through machine learning. Agronomy **14**, 672 (2024). https://doi.org/10.3390/agronomy14040672

K. Telli, O. Kraa, Y. Himeur, A. Ouamane, M. Boumehraz, S. Atalla, W. Mansoor, A comprehensive review of recent research trends on unmanned aerial vehicles (UAVs). Systems **11**(8), 400 (2023). https://doi.org/10.3390/systems11080400

X. Xia, Q. Honglei, L. Hui, High-precision time synchronization of kinematic navigation system using GNSS RTK differential carrier phase time transfer. Measurement **176**, 109132 (2021). ISSN 0263-2241. https://doi.org/10.1016/j.measurement.2021.109132

B. Žabota, F. Berger, M. Kobal, The potential of UAV-acquired photogrammetric and LiDAR-point clouds for obtaining rock dimensions as input parameters for modeling rockfall runout zones. Drones **7**(2), 104 (2023). https://doi.org/10.3390/drones7020104

Z. Zhang, L. Zhu, A review on unmanned aerial vehicle remote sensing: platforms, sensors, data processing methods, and applications. Drones **7**, 398 (2023). https://doi.org/10.3390/drones7060398

Victor Saikhom is currently serving as Scientist/Engineer-SF at the Geoinformatics & IT Division, NESAC, Department of Space, with a Ph.D. in Remote Sensing, M.Tech in Remote Sensing & GIS, and B.E. in Electrical & Electronics Engineering. With over two decades of combined experience in industry and research, he specializes in Photogrammetry, 3D GIS, UAV Remote Sensing, and Geoinformatics. He has led key geospatial projects in land use mapping, infrastructure planning, terrain modeling, and precision agriculture using satellite, UAV, DGPS, and TLS data. His expertise includes DEM generation, orthomosaics, point cloud modeling, and multispectral analysis for disaster response and environmental monitoring. Dr. Saikhom also mentors students and professionals and serves as Course Director for UAV Remote Sensing at NESAC. He has published widely and received the National e-Governance Award (2018) and the 8th eNorthEast Award (2023) for his and his team's contributions to geospatial solutions for governance and regional development.

Puyam S. Singh currently works as Scientist/Engineer–SF at the North Eastern Space Applications Centre (NESAC), specializing in Geo-spatial Intelligence and Computer Vision. He holds an M.E. in Computer Science from PEC University of Technology, Chandigarh. Previously, he worked as a research fellow at DRDO's Snow and Avalanche Study Establishment (SASE), contributing to avalanche forecasting models. At NESAC, he works on the development of space-based geospatial applications for societal benefit and e-Governance, integrating satellite and aerial data. His research focuses on computer vision algorithms for automated feature extraction from remote sensing imagery. He has mentored over 54 postgraduate and undergraduate students and has published 43+ papers in reputed journals and conferences. His innovative contributions have earned National and regional e-Governance Awards in 2015, 2018, and 2023. Puyam continues to advance AI-driven remote sensing solutions for operational and research applications in governance and development sectors.

Nilay Nishant is currently serving as Scientist/Engineer-SE at the North Eastern Space Applications Centre (NESAC), Department of Space, Government of India. He holds a Master's degree in Remote Sensing from Birla Institute of Technology, Mesra, and a Bachelor's degree in Electronics and Communication Engineering from Pondicherry University. His research interests include geospatial big data analytics, e-Governance, and the application of Artificial Intelligence and deep learning techniques to earth observation data. He has played a key role in the development of more than 25 governance-focused geospatial applications for various departments across the North Eastern Region of India. He has authored 30 research publications in reputed national and international journals and conference proceedings. In recognition of his significant contributions to the field of geospatial technology for governance, he has been honored with the National e-Governance Award (2022) and the e-North East Award (2023, 2020).

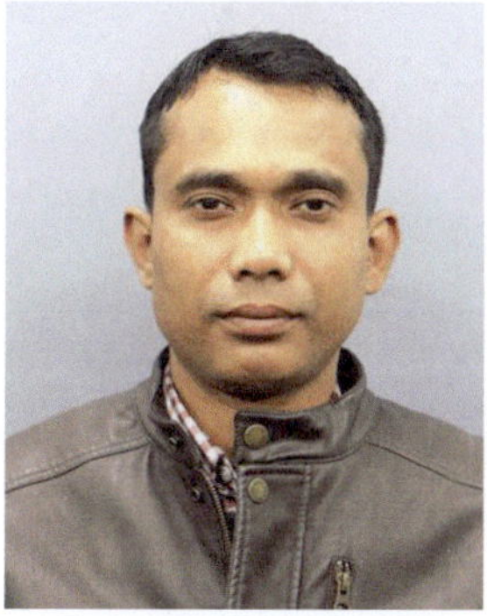

Dibyajyoti Chutia received the M. Tech. degree in information technology and the Ph.D. degree in computer science & engineering with specialization in soft computing technique for classification of geospatial data from Tezpur Central University, Tezpur, India, in 2000 and 2015, respectively. He is currently serving as head of Geoinformatics and IT division at North Eastern Space Applications Centre (NESAC), Department of Space, Government of India. His research interests include soft computing, machine learning, and deep learning applications for remotely sensed data. He has authored or coauthored more than 65 research articles in peer-reviewed journals, conferences, book chapters, etc. He has received the National e-Governance Award (2018, 2022), National Geospatial award (2020) and the eNorthEast Award (2020, 2023) for his and his team's contributions to geospatial innovations in governance and regional development.

Chapter 20
The Physics and Dynamics of Aerosols, Trace Gases and Atmospheric Boundary Layer Over the Eastern Himalayan Region

Arup Borgohain, Shyam S. Kundu, Arundhati Kundu, Manasi Gogoi, Mukunda M. Gogoi, Binita Pathak, and Nilamoni Barman

Abstract The Eastern Himalayan Region (EHR), including the North Eastern Region (NER) of India and adjoining areas, is a climatically sensitive zone influenced by both natural and anthropogenic emissions. This chapter investigates the complex interactions among aerosols, trace gases, and the atmospheric boundary layer (ABL) in the region. Ground-based measurements, satellite observations, and reanalysis products reveal marked seasonal and spatial variability in aerosol properties, with elevated levels of black carbon and aerosol optical depth during the pre-monsoon and winter seasons. Long-range transport from the Indo-Gangetic Plain, biomass burning, and forest fires significantly increase pollutant concentrations, affecting air quality, cloud microphysics, and regional climate. Trace gases such as CO, NO_x, and CH_4 show distinct seasonal trends driven by local emissions, meteorological

A. Borgohain (✉) · S. S. Kundu · A. Kundu
North Eastern Space Applications Centre (NESAC), Umiam, Shillong, Meghalaya, India
e-mail: arupborgohain@gmail.com

S. S. Kundu
e-mail: ss.kundu@nesac.gov.in

A. Kundu
e-mail: to_arundhati@outlook.com

M. Gogoi · B. Pathak
Department of Physics, Dibrugarh University, Dibrugarh, Assam, India
e-mail: manasi.gogoi93@gmail.com

B. Pathak
e-mail: binita@dibru.ac.in

M. M. Gogoi
Space Physics Laboratory (SPL), VSSC, Thiruvananthapuram, Kerala, India
e-mail: dr_mukunda@vssc.gov.in

N. Barman
Central Ground Water Board, Kolkata, West Bengal, India
e-mail: neelspinor@gmail.com

S. P. Aggarwal et al. (eds.), *Remote Sensing of the North Eastern Himalayan Ecosystem*,
https://doi.org/10.1007/978-981-95-3221-6_20

conditions, and transboundary influences. The chapter also highlights the crucial role of ABL dynamics, especially mountain-valley wind circulations in controlling pollutant dispersion. Observations from high-altitude sites like Umiam in Meghalaya and Yala Glacier in Sikkim show seasonal variations in ABL height, surface fluxes, and turbulence. Enhanced radiative forcing and reduced ABL height during winter contribute to pollutant accumulation and climatic feedbacks. These insights stress the importance of long-term monitoring, regional collaboration, and integrated modeling to address atmospheric and climatic challenges in the EHR.

Keywords Eastern Himalayan region · Aerosol optical depth · Black carbon · Atmospheric boundary layer · Mountain valley wind circulation · Radiative forcing · Trace gases

20.1 Introduction

The Eastern Himalayan Region (EHR) encompasses parts of North Eastern Region (NER) of India, Bhutan, Eastern Nepal, Northern Myanmar, and the Tibet Autonomous Region of China. With its distinctive heterogeneous topography and rich biodiversity, this region is situated at the confluence of various air masses, wind circulation patterns, and climatic systems. This region holds special significance for studying the Himalayan hydrosphere-cryosphere, ecosystems, hydrological cycles, biodiversity, public health, environment and climate. The rapid socio-economic developments in recent years have led to increased anthropogenic influences in the EHR, similar to many other parts of the world. Understanding the physics and dynamics of aerosols, trace gases, and the atmospheric boundary layer (ABL), as well as the interactions between them in this region, is crucial for predicting weather, assessing climate change, and addressing environmental challenges. The favorable topography of this region leads the exchange of aerosols and trace gases between the land surface, troposphere, and stratosphere. The southern slope of the high Himalayas acts like a atmospheric channel, capable of transporting pollutants up to 5 km above sea level, where the pristine atmospheric composition can be significantly affected, especially during the pre-monsoon season (Gogoi et al. 2014; Zhang et al. 2020).

The EHR experience significant climatic influence from both aerosols and trace gases. Aerosols released from natural sources like dust and volcanic activity as well as human activities such as biomass burning and industrial emissions, impact the region's climate directly by modifying sunlight through scattering and absorption, and indirectly by altering cloud properties and precipitation. Their deposition on snow surfaces, especially black carbon, reduces reflectivity and accelerates glacier melt, intensifying regional warming. These particles often originate far and are transported by prevailing winds and pressure differences. Trace gases though present in minute quantities affect atmospheric chemistry, radiation balance, and climate dynamics. In this mountainous terrain, the ABL plays a key role in controlling how aerosols and gases mix and disperse through Mountain Valley Wind Circulation (Barman et al.

2021; Kundu et al. 2025). The complex topography comprising valleys, forests, and urban areas further modifies these interactions, influencing local weather patterns and enhancing climate vulnerability in the Eastern Himalayas.

The study of atmospheric processes over the EHR faces numerous challenges. The rugged terrain makes it difficult to deploy and maintain monitoring networks and collect comprehensive data, limiting the availability of long-term observational data needed to draw robust conclusions about trends and impacts. Also, due to varying land-use land-cover, and terrain complexity, retrieving parameters from satellite measurements poses a challenge. Thus, there is a need for international cooperation to study and understand the trans-boundary nature of air pollution and atmospheric flows. Collaborating across borders can facilitate data sharing, coordinated research, and the implementation of policies aimed at mitigating the adverse effects of atmospheric changes in the Eastern Himalayan region.

20.2 Aerosol and Trace Gases Over the Eastern Himalayan Region

Over recent years, extensive research has been conducted to examine the constituents of the atmosphere, including aerosols and trace gases, over the EHR. The NER of India within the EHR is a unique location conducive to study the atmospheric chemistry due to the abundant solar insolation and reactive precursor gases, coupled with aerosols emitted from a variety of local as well as remote natural and anthropogenic sources. Studies have shown significant spatiotemporal variations in aerosol and trace gas properties over this region, with seasonal patterns influenced by both local and long-range transport mechanisms.

20.2.1 Characteristics of Aerosols

Aerosols over NER of India are diverse and have complex properties influencing air quality, human health, climate, and ecosystems. Understanding their characteristics helps scientists and policymakers develop strategies to mitigate their negative effects. Long-term observation of Aerosol Optical Depth (AOD) shows deeper insight into the aerosol variability over the region. Studies have revealed that aerosol loading over the region is significant within the South Asian region, with a regional average AOD of approximately 0.49, second only to the Indo-Gangetic Plains (IGP ~ 0.64) (Pathak et al. 2016). Four-dimensional mapping of aerosol characteristics reveals a west-to-east asymmetry/gradient in AOD (Dahutia et al. 2019) over the region. The highest regional columnar AOD is observed during the pre-monsoon, followed by winter. Aerosol burden in the NER, like most of the locations in South and Southeast Asia exhibits minimum values in the post-monsoon (October and November) season associated with preceding monsoon wet removal. In Dibrugarh, which is in the foothills of the Eastern Himalayas, the spectral AOD shows clear variation in wavelength, where it declines with wavelength from January to April and from October to December, but it is almost independent of the wavelength from May to September (Bhuyan et al. 2005). Wintertime AOD spectra are comparatively steeper than monsoon-season (Gogoi et al. 2008). At 500 nm, the average AOD ranges from 0.44 ± 0.07 to 0.56 ± 0.07 during the pre-monsoon season and from 0.19 ± 0.02 to 0.22 ± 0.02 during the retreating monsoon. The climatological mean values of the derived Ångström exponent (α) and turbidity coefficient (β) over the region are 1.08 ± 0.29 and 0.19 ± 0.11, respectively, indicating low columnar aerosol loading and a dominance of coarse-mode particles (Gogoi et al. 2009). The seasonal change in aerosol types is also observed at a remote hilly terrain in the Meghalaya plateau. The measurements of backscatter fraction from Integrating Nephelometer showed a decrease from winter to monsoon indicating the presence of bigger particles during the monsoon over the station. This was attributed to the hygroscopic growth of particles during the wet monsoon seasons (Barman et al. 2024).

Black Carbon (BC) also known as soot particles is another dominant aerosol over NER that have a significant impact on radiation budget, environment, health, and climate. Resulting from the incomplete combustion of fossil fuels, biofuels, and biomass, the surface mass concentration of BC over NER exhibited significant diurnal and seasonal variation. High levels of BC are found throughout the dry months of November through February, peaking in January at 6.7 ± 0.3 μgm^{-3}, and then gradually declining to 1.4 ± 0.17 μgm^{-3} in July (Pathak et al. 2016). Beginning in November, the concentration of BC begins to rise as a result of local production through biomass-burning activities. However, decreasing BC concentrations are seen over the region (Subba et al. 2023) as a result of the implementation of pollution control measures. In a study to examine seasonal variations in aerosol sources and concentrations in Agartala, Tripura, the findings indicated that during winter and post-monsoon seasons, fossil fuel combustion was the primary source of BC emissions, while biomass-burning aerosols dominated in the monsoon, influenced by humid atmospheric conditions (Barman et al. 2019b). Guwahati, the largest city of NER shows that traffic-derived BC emissions were dominant, as determined by a two-component mixing model using an Aethalometer (AE-33) and radiative transfer equations (Singh and Gokhale 2023).

At high-altitude observational sites in the eastern sub-Himalayan region, such as Tawang (~3000 m AMSL) and Geku (~452 m AMSL) in Arunachal Pradesh (AP), coarse-mode aerosols like dust and secondary organic aerosols dominate. In contrast, fine-mode aerosols, mainly from biomass burning, are more common at New Palin (~1080 m AMSL), AP. Notably, high-altitude sites recorded higher BC concentrations compared to the lower-altitude Geku site (Fig. 20.1). This discrepancy suggests that aerosol concentrations at higher altitudes are significantly influenced by transported aerosols, while the lower-altitude site, surrounded by mountains, exhibited lower concentrations, likely due to topographical barriers restricting

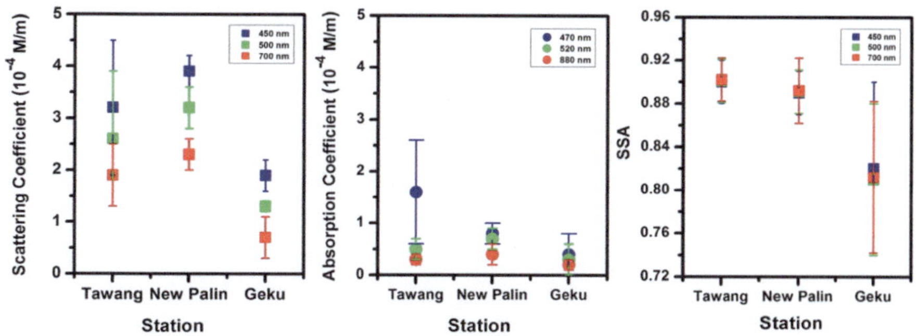

Fig. 20.1 Aerosol scattering coefficient, absorption coefficient and single scattering albedo (SSA) for the stations averaged over Tawang, New Palin and Geku during February 2018

aerosol transport from distant sources (Borgohain et al. 2020). Apart from BC, light-absorbing organic carbon namely brown carbon (BrC) aerosol, over a remote high-altitude Himalayan site in Lachung, were observed with the use of a high-volume sampler and spectrophotometer (Arun et al. 2024). This study showed high levels of BrC absorption coefficients in the spring season suggesting significant contributions by anthropogenic sources over the high-altitude region.

The above observations indicate that BC is not only a concern at plain locations only but at higher altitudes too. Figures 20.2 and 20.3 shows the presence of a very high amount of smoke and polluted dust over the NER as high as up to 4 km. Gogoi et al. (2023) have done extensive campaigns by deploying Micro-aethalometer, Optical Particle Counter and Dr. Pisharoty Radiosonde using tethered balloon up to 1 km altitude over several locations of NER in different seasons. They observed that vertical profiles of BC show multiple elevated layers which closely followed the vertical profiles of meteorological parameters. The Cloud Aerosol Interaction and Precipitation Enhancement Experiment (CAIPEEX) aerial campaigns observed an elevated BC layer over the Brahmaputra River Valley (BRV), contributed by both long-range transport and surface emissions. Surface BC over this region accumulates in the upper atmosphere by the ascending motions driven by large-scale Indian summer monsoon circulations (Rahul et al. 2014). Elevated aerosol layers (EALs) include clean marine dust, polluted continental smoke, clean continental aerosols, polluted dust, and elevated smoke. During the monsoon, these layers extend up to ~7.4 km.

The compositions of aerosols are usually characterized by a source apportionment study such as receptor modeling, chemical speciation etc. This helps in understanding the relative importance of various emission sources, such as vehicles, industries, biomass burning, and natural sources, in contributing to air pollution. Studies illustrate a regional heterogeneity in sources of aerosol emissions over NER. The western part of the region has a stronger influence from fossil fuel sources, possibly due to industrial activities, transportation, and urbanization. In contrast, the eastern part is more significantly impacted by biomass burning, which could be attributed to agricultural practices, forest fires, and rural cooking methods.

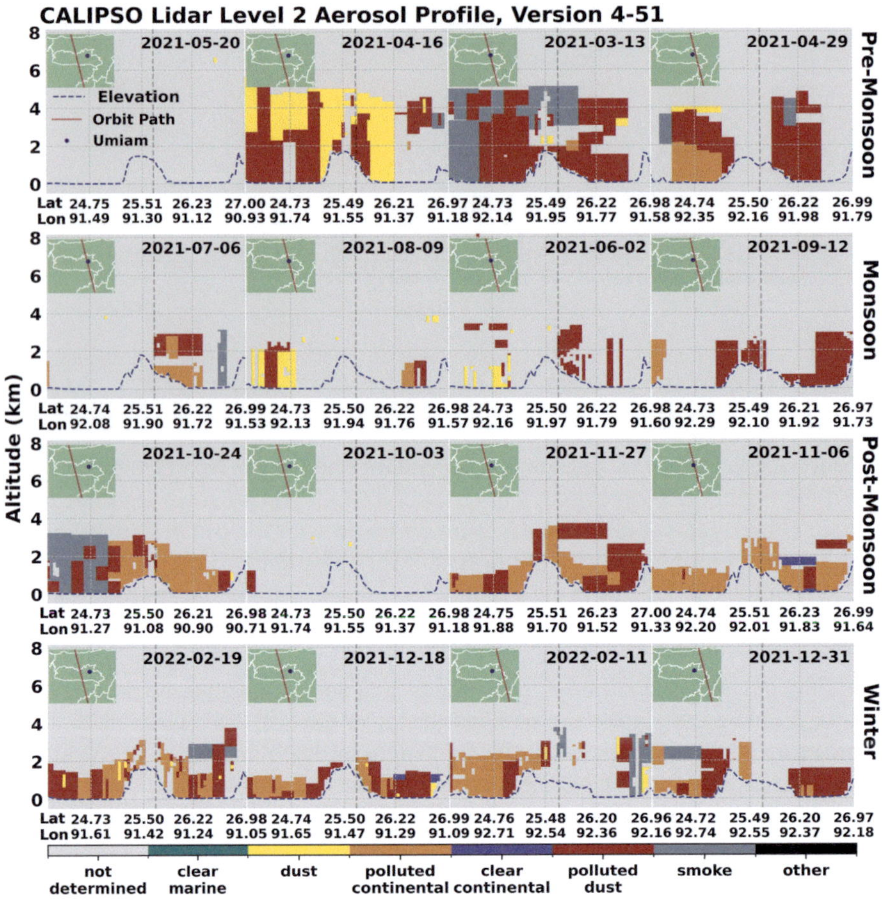

Fig. 20.2 Vertical profile of different aerosol species as observed by CALIPSO over NEI during different seasons

Figure 20.4 shows that Lachung which is a high-altitude station in the western part experiences lower biomass burning as seen from the low Absorption Ångstrom Exponent (AAE) compared to the Tawang which is situated at the eastern side of the NER. Considerable seasonal variation in the distribution of AAE can be observed over all the locations.

Eastern Himalayan site of Lachung, Sikkim shows that biomass burning, aqueous-phase processes, vehicular and industrial emissions, mineral dust (22.28%), and aged Secondary Organic Aerosols (SOA)s are the potential sources of aerosol in this region based on positive matrix factorization model (Arun et al. 2021a). Further, the winter and spring have the highest scattering and absorption values, driven by coarse mode aerosols and enhanced by organic carbon (OC) and sulfate aerosols. Also, elevated levels of OC and Elemental Carbon (EC) in winter and spring and high level of

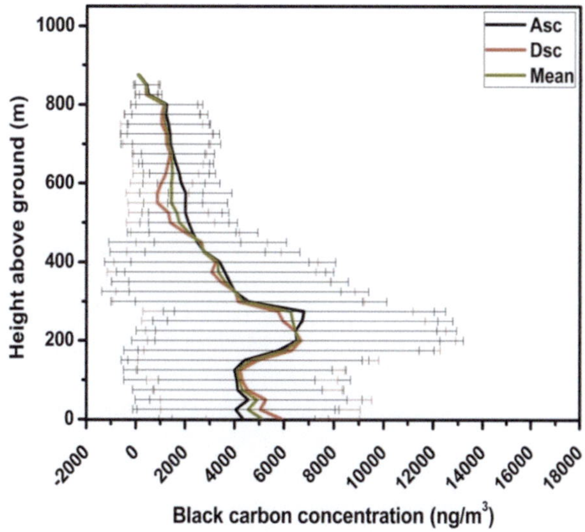

Fig. 20.3 Vertical profile of black carbon concentration at 880 nm observed over Umiam during the morning of Jan 2019 using a micro-aethalometer and tethered balloon. Asc indicates the profile when the balloon ascends, and dsc indicates descent

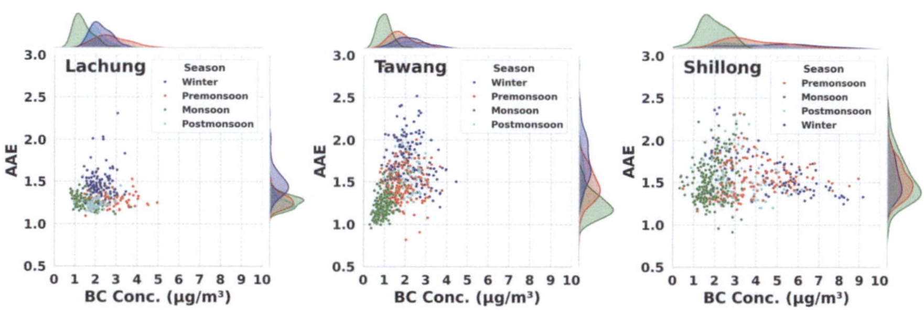

Fig. 20.4 Scatter plot of BC mass conc. against Absorption Ångstrom Exponent (AAE) along with their seasonal distributions as kernel density on the horizontal and vertical axes over Lachung (Sikkim), Umiam/Shillong (Meghalaya) and Tawang (Arunachal Pradesh)

sulphates during spring shows anthropogenic emissions. The dominance of SOA is indicated by an OC/EC ratio of 5.5 ± 2.5 and a WSOC/OC ratio of 0.74 ± 0.15.

Forest fires are considered one of the causes that significantly impact aerosol load in the atmosphere. When forests burn, they release a large amount of particulate matter, including BC, OC, trace gases, and other aerosols, that increases the AOD over any location. They also release gaseous pollutants such as carbon monoxide (CO), nitrogen oxides (NOx), sulfur dioxide (SO_2), and volatile organic compounds (VOCs). These pollutants can react in the atmosphere to form secondary aerosols, increasing the overall aerosol load. Large forest fires can even impact areas hundreds of kilometers downwind. Fire events are very common over NER during the dry seasons as evident from Fig. 20.5. Sometimes, natural causes like lightning or spontaneous combustion and friction between tree branches, dry leaves, and grasses initiate the fires. Interestingly, NER experiences strong winds during the dry seasons, which sustain the fires for prolonged periods that can be observed by satellites too. Other than natural causes, Jhum cultivations, slashes, and burn activities to clear out dry shrubs and secondary vegetation can create forest fires. As hills are common in NER, some native species like pine (Pinus Khasianas) are abundant which are prone to igniting quickly due to their physical and chemical properties. Fire events are more witnessed here in March than in any other month in a long-term study (Borgohain et al. 2023). Fire counts obtained from Moderate Resolution Imaging Spectroradiometer (MODIS) collection 6 active fire count products have been found to vary from 60 to 100 K in March and 5–30 K in January and February. Surface concentrations of BC over NER taken from Modern Era-Retrospective Analysis for Research and Applications (MERRA2) reanalysis data showed significant increases during forest fire events (Borgohain et al. 2023). The highest number of fire counts was recorded during the drought period of 2009–2011, which coincided with high concentrations of BC observed over the region. Drought conditions reduce moisture availability in vegetation, making it more flammable and increasing the likelihood of forest fires. High BC loading was particularly prominent during the pre-monsoon season in the EHR, especially over Mizoram and parts of Manipur, Tripura, and Assam, due to the increased occurrence of forest fires.

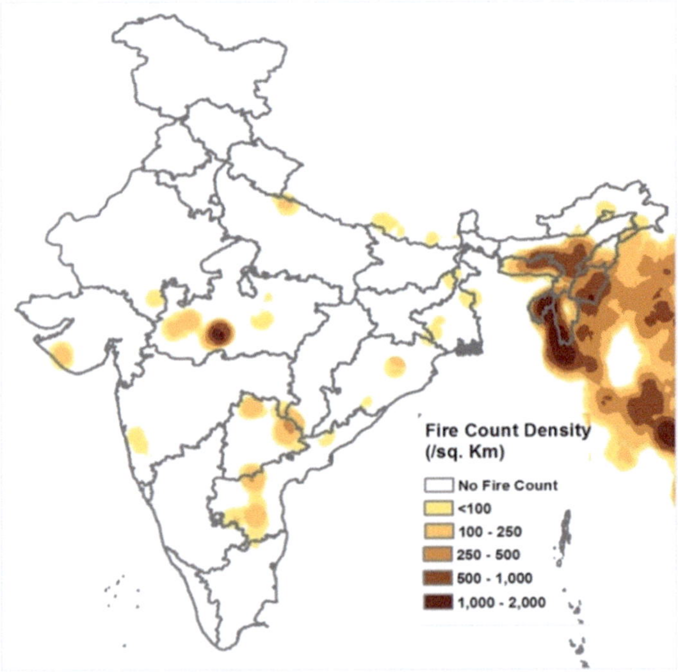

Fig. 20.5 Typical forest fire scenario observed over NEI during the dry season; yearly mean fire count density was computed using MODIS active fire count data for the years 2010-2020

However, it is not only local sources; many of contemporary studies have attributed the long-range transport as one of the main contributors behind aerosol spatiotemporal heterogeneity over NER. The usual approach to show the long-range transport characteristics of aerosols is back-trajectory analysis or concentrated weighted trajectory (CWT) analysis (Kundu et al. 2025) where air mass trajectories were computed using a trajectory model (Hybrid Single Particle Lagrangian Integrated Trajectory, HYSPLIT) and meteorological data.

It simulates the movement and spread of atmospheric pollutants across different spatial and temporal scales. Trajectories for five or seven days are computed keeping in mind the atmospheric lifetime of the aerosol at different arrival heights over a receptor location. Observed concentrations of aerosol at the receptor location are then assigned to each respective back-trajectory and CWT analysis is conducted to find the potential source regions. In recent literature, the atmospheric composition over NER is found altered by remote sources like dust advection from the Indian deserts, marine emissions from the Bay of Bengal (Pathak et al. 2012), and anthropogenic emissions from the IGP (Kundu et al. 2025) (Fig. 20.6). Pre-monsoon time high AOD observed over NER was mainly attributed to the transport of mineral dust from the arid regions of west Asia and northwest India across the IGP (Gogoi et al. 2014). A study based on CWT analyses showed that observed BC, in Tripura, in the winter is mainly of continental origin whereas it shifts to continental plus oceanic origin during pre and post-monsoon. Further, in the monsoon, the black carbon emissions are mainly of the local origin (Guha et al. 2015).

At the EHR site of Darjeeling (2115 m AMSL), fine-mode particles are primarily generated locally from anthropogenic activities such as fossil fuel burning during the pre-monsoon season and biomass burning in winter. In contrast, superfine aerosols (0.32–1.0 μm) are mainly transported over long distances. During winter, air masses reaching northern Arunachal Pradesh primarily originate from the IGP, Nepal, Bhutan, and the Brahmaputra Valley, carried by prevailing westerly winds. Even BC concentrations in remote areas in Gangtok (1650 m AMSL) during the spring season are similar to those in urban clusters, highlighting the strong atmospheric circulation influence on BC concentrations in the region (Sharma et al. 2022). In another study, air masses were found to predominantly originate from the northwest and southwest. These regions are highly populated and industrialized, potentially contributing to the presence of water-soluble organic aerosols and their precursors (Kumar et al. 2024). These aerosols likely originate from anthropogenic sources, including biomass burning. Several such studies assign the IGP region, which is one of the aerosol hotspot regions of the world, as the potential source region influencing NER. Interestingly, the prevailing synoptic wind is mostly westerly during the dry season that shows the potential to carry the pollutants from IGP to NER (Borgohain et al. 2023). Winter haze from IGP influences the entire Brahmaputra valley as established through a land campaign (Kundu et al. 2018). The maximum columnar glyoxal and its better correlation with formaldehyde over IGP and NER show the strong anthropogenic emissions from common sources, thus establishing the IGP influence over NER (Chutia et al. 2019).

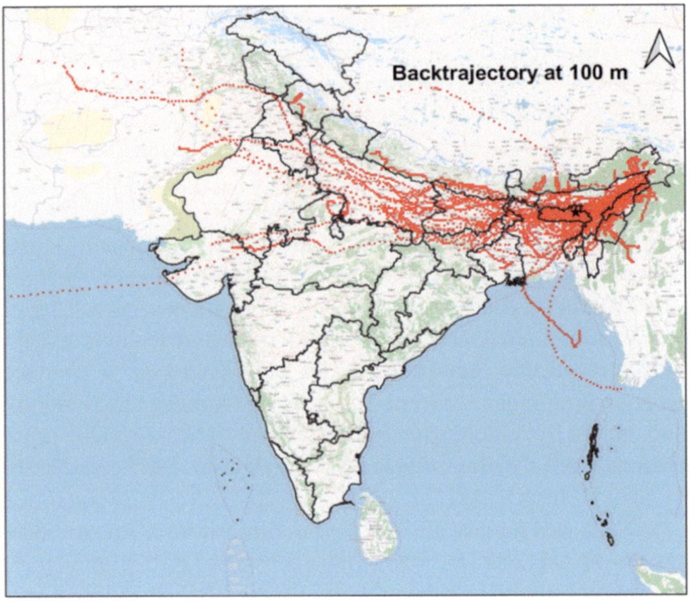

Fig. 20.6 Five days' backtreactories computed using HYSPLIT model using the receptor location as Umiam and global data assimilation system (GDAS) data as meteorological input at 100 m. Origin of the trajectories can clearly be seen over the IGP region

Strong aerosol-radiation interaction (ARI) and aerosol-cloud interaction (ACI) over NER have been highlighted in several studies. The region experiences significant surface forcing due to the presence of aerosols, with surface aerosol direct radiative forcing (DRF) causing surface cooling of around −56.5 Wm^{-2} and simultaneous atmospheric heating of up to 1.6 K day^{-1}, comparable to or exceeding many other locations in the Indian subcontinent. Synoptic-scale assessments of aerosol DRF reveal the highest surface dimming over West Asia, where dust storms are prevalent, followed by the IGP and western India (Subba et al. 2018). In NER, BC radiative forcing is highest in the western part of the region, particularly during winter and pre-monsoon seasons, due to the high BC mass concentration. This results in considerable heating in the lower atmosphere, reaching up to ~1.5 K day^{-1} during winter (Gogoi et al. 2017). In case of columnar loading, a phase reversal was noted during 2008 with a rapid increase in AOD after 2008, contributing to greater top-of-atmosphere (TOA) cooling over NER (Subba et al. 2023). Higher aerosol-induced surface forcing was observed during both winter and pre-monsoon seasons, with comparable levels of 32 W/m^2 and 33.45 W/m^2, respectively over a semi-continental site Tripura. During this period, the elevated atmospheric forcing was primarily attributed to increased columnar aerosol loadings (AOD ~0.71) and high BC concentrations, reflected by a single scattering albedo (SSA) of around 0.7. However, during the monsoon season, atmospheric forcing decreased due to lower BC magnitude (SSA ~0.94 in July) and columnar AOD, also, top of the atmosphere

(TOA) forcing is positive in pre-monsoon and monsoon months, with the highest positive value of 3.78 W/m^2 in June 2012 (Dhar et al. 2017). 10–15% of atmospheric BC content over Dibrugarh is found to contribute to ~3/4 part of atmospheric radiative absorption, causing potential warming with efficiency of +43.04 Wm^{-2} τ^{-1} and relative surface dimming by −26.14 Wm^{-2} τ^{-1} (τ is AOD) (Subba et al. 2018). The elevated aerosol layers over Dibrugarh are found to heat the atmosphere at 2–4 km altitude by ~0.3 K day^{-1} especially during March–April-May (MAM) and December-January–February (DJF) (Subba et al. 2018).

At the high-altitude site Lachung, summer monsoon observations reported the lowest scattering and absorption values, primarily influenced by fossil fuel-based aerosols, resulting notable surface cooling and atmospheric heating during spring, with a direct radiative forcing efficiency of 123.9 $Wm^{-2}\tau^{-1}$ due to absorbing aerosols (Arun et al. 2021b).

20.2.2 Characteristics of Trace Gases

The seasonal and diurnal variability in surface ozone (O_3), nitrogen oxides, and evapotranspiration are influenced by both local and regional factors. The local anthropogenic emissions are mainly from agricultural activities, biomass, and fossil fuel burnings, brick kilns, oil and gas fields, industries, etc., releasing various trace gases. Higher surface ozone levels were observed during pre-monsoon daytime and bimodal NO_x peaks linked to traffic emissions (Tyagi et al. 2020). Bali et al. (2021) observed that NER is under a VOC sensitive regime where O_3 formation is largely controlled by the VOC levels especially during the biomass burning. 80–90% of biomass burning generated CO over Dibrugarh during pre-monsoon mostly contributed locally (59%), as revealed by the Weather Research and Forecasting (WRF) model coupled with Sulfur Transport dEposition Model (STEM)model (Saikia et al. 2019) and 10–15% of its co-emitter BC contributes to ~3/4th of atmospheric warming (Subba et al. 2018). Lightning-generated NO_X is found to chemically form ~33% of measured O_3 over Dibrugarh (Bharali et al. 2015). This leads to higher levels of nighttime O_3 during lightning (~18 ppb) and thunderstorm (~12 ppb) days as compared to that on clear days (~3 ppb) in spring and summer seasons. The region is also conducive to stratospheric O_3 intrusion into the troposphere. Back trajectory and CWT analysis of NOx and O_3 concentrations show that in the post-monsoon and winter seasons, the trajectories ending over Dibrugarh are confined locally, whereas in the pre-monsoon and monsoon seasons, they originate at the Indo-Gangetic plains, Bangladesh, and Bay of Bengal (Bhuyan et al. 2014). Unlike the rest of India, where column O_3 peaks in post monsoon season, in NER, O_3 peaks in the pre-monsoon season, and the region sees the second-highest abundance of O_3 in South Asia next to IGP (Kalita and Bhuyan 2011).

All surface level pollutants, particulate matter (PM), BC, CO, and NOx, except O_3, possess similar diurnal variability, peak concentrations during dry months of November–March due to receding rainfall and the onset of the small to large scale biomass burning activities (Pathak et al. 2012; Bhuyan et al. 2014). Surface O_3 formation falls under a NOx-limited and VOC-sensitive regime (due to the nonlinearity of O_3 with NOx concentration with a negative correlation) like many parts of India (Chutia et al. 2019), and its concentration peaks in March due to the availability of precursors like VOCs, NOx, CO, etc., and higher solar insolation. Surface O_3 concentration exhibits a single peak at around 1500 h LT. Annual maximum O_3 occur in March due to the availability of biomass-burning generated precursor gases (NOx and CO) and higher solar insolation, the lowest being in July, mostly due to the washout of precursor gases. There is a significant contribution of CO from IGP in all the seasons along with the Southeast Asian region in monsoon and Central Asian region in pre and post-monsoon as revealed by the STEM WRF coupled with Chemistry model simulations (Saikia et al. 2019). During winter and post-monsoon, NER contributes significantly to CO levels. A notable spike in the concentrations of these pollutants is observed during the peak hours of fireworks activity during the Diwali festival in Dibrugarh. Over a period of 15 years, emissions of volatile organic compounds (VOC) and nitrogen oxides (NOx) were reported to be 431 mg/m^2/day and 69.5 mg/m^2/day, respectively (Bali et al. 2021). The seasonal and annual variations in NOx and CO mixing ratios closely follow those of BC.

The monthly mean concentrations of NOx and CO are lowest in July (3.6 $\pm$ 0.6 ppb and 368 $\pm$ 33 ppb, respectively) and highest in January (30.9 $\pm$ 3.16 ppb and 1029 $\pm$ 110 ppb). Although SO_2 shows no clear annual pattern, its levels are higher during the monsoon season, coinciding with increased transport from the IGP. During the COVID lockdown period, it has been observed that SO_2 Concentration has drastically reduced over the IGP as compared to the pre-lockdown period (Fig. 20.7). The reductions are more prominent over the cities, industrial areas and power plants. This indicates that SO_2 Concentration in the atmosphere is highly dependent on anthropogenic activities.

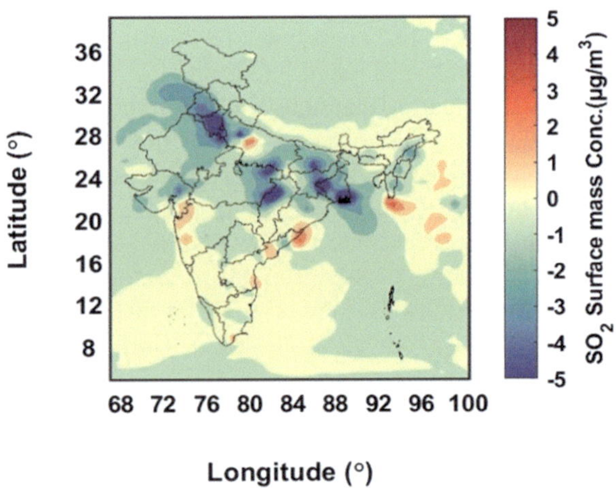

Fig. 20.7 Fifteen day's mean SO_2 difference before and after lockdown period obtained from MERRA2 reanalysis data

Gaseous pollutants released from forest fires exhibit similar behavior to aerosols during such events. Tropospheric Columnar Nitrogen dioxide (TCN) data from the Ozone Monitoring Instrument (OMI) indicated higher values during the months of March and April, when forest fires were most prevalent (Borgohain et al. 2023) (Fig. 20.8). These changes were more pronounced in the lower parts of Northeast India, including Meghalaya, Tripura, Mizoram, and parts of Assam and Nagaland. TCN peaked in March and April, while it reached its lowest in August and September, aligning with the monthly fire event patterns. Additionally, TCN was found to be higher during the drought period of 2009–2011. In a study examining Fire Radiative Power (FRP) and biomass burning emissions across Northeast India, the average FRP was 0.37 W/m^2, with values ranging from 0.07 to 0.7 W/m^2. Mizoram exhibited the largest contribution in FRP and fire emissions among the 7-states over NER. Emissions were higher in 2006 and 2010, but lower in 2017 (Bali et al. 2021).

Over Darjeeling, a high-altitude station in eastern India, extensive movement of pollutants from source areas has been identified as the cause of the enhancement observed in gases, such as Sulfur Hexafluoride (SF_6), methane (CH_4), and nitrous oxide(N_2O). It has been demonstrated that some SF_6 events are caused by transport from Southeast Asia, while many pollution events for the three gases during the summer can be attributed to transport from the IGP of northern India. The air histories of SF_6 wintertime pollution events are less consistent than those of CH_4 and N_2O since the sources of SF_6 are less.

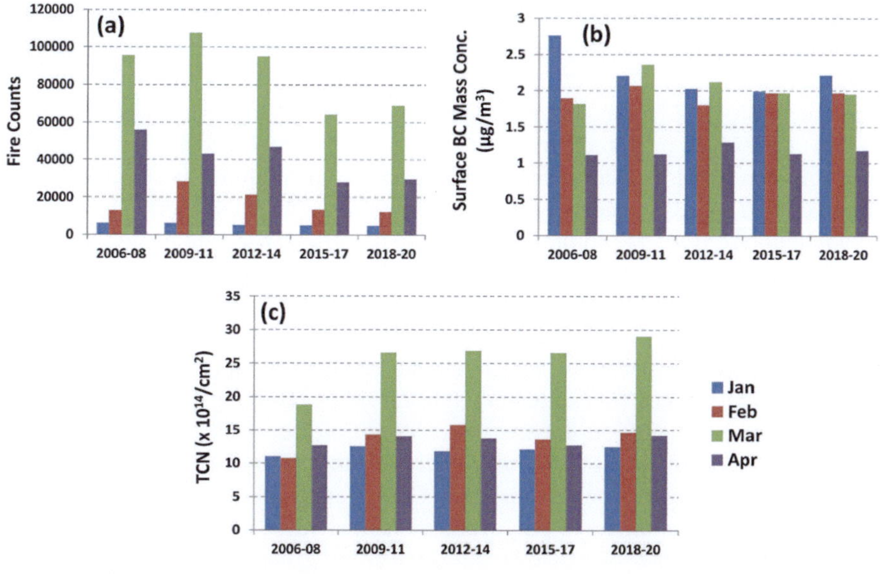

Fig. 20.8 Fire counts, total surface BC mass concentration and TCN over the whole NEI in Jan, Feb, Mar and Apr during 2006–2020

20.2.3 Impact of Aerosols and Trace Gases on Air Quality, Ecosystem, and Climate

The EHR domain, particularly the NER of India, is highly sensitive to the influence of aerosols, which play a crucial role in shaping the region's atmospheric and climatic dynamics. The anthropogenic contribution to the atmosphere over the NER and adjoining areas was observed to change significantly during the COVID-19 lockdown across India. There was a noticeable reduction in surface-level BC, PM, NO_x, and CO, accompanied by a simultaneous rise in O_3 levels. A study conducted in the hilly station of Darjeeling compared carbonaceous aerosol observations during the period of reduced anthropogenic activity. Measurements using a carbon aerosol analyzer indicated that, under normal conditions, the major sources of carbonaceous aerosols in Darjeeling were from Nepal and Northeast India. However, during the lockdown, the sources shifted to more regional and local origins. CH_4 emissions from both natural such as wetlands and paddy fields and anthropogenic including transportation and industrial activities were found to be increasing at a faster rate over the region, compared to the global rate as revealed by satellite and reanalysis datasets (Saikia et al. 2023). Trace gas studies in the eastern Himalayas further reveal complex interactions with ecosystems. It has been reported that 60% of CH_4 emissions are anthropogenic, contributing to a regional temperature increase nearly twice the global average. Additionally, coniferous forests in Darjeeling were identified as carbon sinks during spring, as CO_2 uptake from photosynthesis exceeded ecosystem respiration. These dynamics could pose serious threats to the fragile Himalayan ecosystem.

Although aerosols can influence meteorological conditions, the reverse is also true, with meteorological factors playing a significant role in shaping aerosol dynamics. Such was observed in a study where meteorological conditions significantly influenced elevated PM2.5 levels during winter, as high temperatures, low humidity, and minimal wind speeds create an environment conducive to aerosol formation and the accumulation of PM2.5 pollution. Exposure to harmful particulate matter, primarily from biomass burning, vehicular emissions, and industrial combustion poses significant health risks, including respiratory and cardiovascular diseases. Biomass burning from households near Tezpur University, Assam emitted 10–12 times higher PM2.5 and 6–7 times higher PM10 levels compared to those using liquefied petroleum gas (LPG), leading to three times greater exposure doses (Gogoi et al. 2021). Further, there is a seasonal variation in particulate deposition too as observed over Assam, with higher deposition in the pulmonary region during winter due to biomass burning, coal combustion, and vehicular emissions (Garaga et al. 2020). A study even identified the carcinogenic impact of polycyclic aromatic hydrocarbons (PAHs) in PM2.5 at Jorhat, attributing the sources to crop residue burning, coal, and wood combustion, with ambient temperature and planetary boundary layer height influencing pollutant concentrations (Vishwakarma et al. 2023).

Studies on aerosols over NER have highlighted their significant impacts on atmospheric processes and rainfall patterns. Aerosol properties and cloud condensation nuclei (CCN) activation illustrate their role in cloud processes and regional climate. One study has attributed BC to favoring the regional climate change experienced by the BRV region of the NER (Chakrabarty et al. 2012). Nighttime CCN activation in Darjeeling was found to be driven by biomass burning and aerosol hygroscopicity (Roy et al. 2017). Elevated aerosol layers and enhanced CCN during the pre-monsoon season further underscore the critical influence of aerosols on cloud dynamics and regional atmospheric conditions. A study pointed out that BC reduces short-wave radiation, surface evaporation, and convection, leading to suppressed rainfall during the pre-monsoon season (Barman and Gokhale 2022). Similarly, stronger aerosol-cloud interactions are observed during pre-monsoon, with elevated aerosol layers formed through convection and upper-air transport. Long-term trends indicate increasing anthropogenic aerosol contributions and their climatic effects over northeast India (Dahutia et al. 2019). Rising AOD and Ångström exponents from 2001 to 2014 are found leading to reduced shortwave radiation and rainfall. The role of sulfate and carbonaceous aerosols in enhancing radiative forcing and atmospheric heating was observed in Lachung, particularly during the spring season (Arun et al. 2021a) showing the potential to alter climate even over remote high-altitude areas. Evapotranspiration in Dibrugarh was found to closely follow net radiation, influenced by aerosols and soil moisture dynamics that also poses a concern for aerosol-induced climate change.

20.3 Atmospheric Boundary Layer in the Eastern Himalayan Region of NER

The ABL, also referred to as the planetary boundary layer (PBL), constitutes the lowest part of the troposphere, directly influenced by interactions with the Earth's surface (Stull 1988). The ABL plays a critical role in facilitating mass and energy exchange processes between the Earth's surface and the atmosphere, thereby governing the dynamics of energy and matter transfer within the Earth system. It exhibits temporal variability on scales ranging from seconds to hours and extends spatially and vertically from a few meters to several kilometres. Investigating ABL parameters in the EHR is particularly challenging due to the region's complex topography, which introduces pronounced spatial heterogeneity in ABL behaviour and complicates observational and modelling efforts. Mountainous terrains in the EHR foster localized weather systems and microclimates, where processes such as orographic lifting significantly influence ABL dynamics. These complexities emphasize the need for interdisciplinary approaches and advanced observational and computational techniques to achieve a comprehensive understanding of ABL processes in such intricate terrains.

20.3.1 *Instrumentations, Data and Techniques Used for the Study*

ABL studies require advanced instrumentation and techniques to measure various meteorological and turbulent parameters that define the dynamics and thermodynamics of the ABL. The key instruments include sonic anemometers, which measure three-dimensional wind components at a very high frequency and then use the eddy covariance technique to compute turbulence fluxes such as momentum, sensible heat, and latent heat flux. Sound Detection and Ranging (SODAR) may be employed to profile wind vector and turbulence at different heights, while a Light Detection and Ranging (LIDAR) ceilometer is useful to determine ABL height (ABLH) by analyzing variations in the backscattered profile of the atmosphere. Radiosonde is another equipment that is usually attached to a meteorological balloon that provides vertical profiles of temperature, relative humidity (RH), and pressure, crucial for understanding thermal stratification of the atmosphere. Additionally, radiometers measure radiative fluxes, while flux towers equipped with multiple sensors monitor near-surface fluxes and gradients over time. These tools, combined with data correction and quality control techniques, help researchers analyse diurnal and seasonal variations in ABL parameters over diverse terrains. The details on instrumentation are discussed below:

Three-dimensional sonic anemometer: A three-dimensional sonic anemometer is commonly used for flux measurements in eddy-covariance systems. It records three orthogonal wind components (u_x, u_{U+03B3}, u) and the speed of sound (c) at a maximum data rate of 60 Hz. This instrument captures turbulent fluctuations in both horizontal and vertical wind, which are essential for calculating surface layer parameters. In the complex terrains of the EHR, careful consideration is necessary when applying the eddy-covariance method. Since the sonic anemometer cannot always be perfectly levelled with gravity, its vertical w-axis may not align precisely perpendicular to the mean flow. Therefore, selecting an appropriate frame of reference is critical for accurately measuring vertical wind components and ensuring reliable flux calculations. This minimizes cross-contamination between velocity components (Lee et al. 2004), which can significantly contribute to errors in flux estimation. Therefore, before analyzing fluxes for meaningful interpretation, the measurements must be transformed into an appropriate terrain-following coordinate system (Yuan et al. 2011). Various methods can achieve this, including double rotation (DR), triple rotation (TR), and the general planar fit (GPF) coordinate system (Wilczak et al. 2001). The planar fit coordinate system is often influenced by terrain complexity, such as varying wind directions, meaning that GPF tilt correction will be affected by the prevailing wind sectors (Oldroyd et al. 2016). Wind flow is influenced by wind direction, and this issue can be partially addressed using the sector-wise planar fit (SPF) method (Barman et al. 2019a). The SPF technique helps minimize terrain-induced signals in time-series data and reduces the need for tilt correction between velocity components (Oldroyd et al. 2016).

Dr. Pisharoty GPS radiosonde: The radiosonde system is widely used to determine the atmospheric boundary layer height (ABLH) and vertical profile of meteorological parameters through in situ measurements of temperature, pressure, RH, wind speed, and wind direction. A typical radiosonde comprises a Global Positioning System (GPS) receiver, sensors, a microcontroller, and a transmitter. GPS-derived positional data is used to compute wind speed and direction. Ground-based receiving systems, equipped with antennas and data processing units, collect telemetry data at one-second resolution for real-time display and analysis. Radiosondes are launched using hydrogen-filled meteorological balloons and provide high-resolution vertical profiles of the atmosphere. The Dr. Pisharoty GPS radiosonde system, developed by the Indian Space Research Organization (ISRO), has played a crucial role in accurately characterizing ABL over several locations (e.g., Nair et al. 2011; Chandra et al. 2014). Radiosonde observations as shown in Fig. 20.9 enable the estimation of key boundary layer parameters such as potential temperature (θ), virtual potential temperature (θ_v), and specific humidity (q). The top of the mixing layer—often used as a proxy for ABLH—is typically identified by the maximum gradient in θ or θ_v (Wang and Wang 2014). Additionally, a minimum gradient in RH or q, combined with a decrease in atmospheric moisture, also serves as a reliable indicator of the ABLH.

Ceilometer: The ceilometer is an active remote sensing instrument equipped with a pulsed diode laser operating at a wavelength of approximately 910 nm. It is capable of measuring cloud base heights up to 7.5 km with a temporal resolution of 2 s (Kundu et al. 2023; Gogoi et al. 2023, 2024), and can determine the ABLH with a resolution of 16 s. To analyse and visualize local ABL dynamics, three commonly used algorithms are employed: the gradient method, the profile fit method, and the merged gradient and profile fit method. ABLH is typically identified by a sharp decline in the backscattered signal profile, where the intensity of the signal is influenced by atmospheric particles and affected by factors such as moisture content, horizontal advection, thermal convection, and chemical reactions (Emeis et al. 2004). The gradient

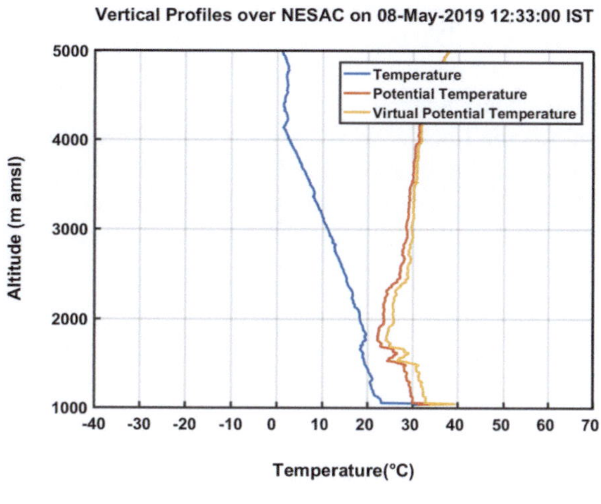

Fig. 20.9 Vertical profile of temperature, potential temperature and virtual potential temperature extracted and derived using Dr. Pisharoty radiosonde over Umiam

method is particularly effective for detecting shallow nocturnal boundary layers and provides associated uncertainty estimates. However, the merged gradient and profile fit method is generally preferred, as it combines the advantages of both techniques to improve the accuracy of ABLH detection.

Reanalysis products: Reanalysis products such as European Centre for Medium Range Weather Forecasts (ECMWF) Reanalyis Version-5 (ERA5), Indian Monsoon Data Assimilation and Analysis (IMDAA), and MERRA Version-2 (MERRA2) serve as essential tools for investigating long-term variations in the ABLH and provide a valuable comparison with observational data. ERA5, the latest reanalysis from ECMWF, offers a high-resolution dataset at 0.25° × 0.25° since 1940. It integrates radiosonde observations with advanced modelling and data assimilation systems, providing data at 37 pressure levels with a 1-h temporal resolution. ERA5 estimates ABLH using the bulk Richardson number (Ri) method, independent of turbulence parameterization (Seidel et al. 2012). The ABLH is identified as the first height where Richardson number(R_i) $\geq$ 0.25 (Seibert et al. 2000; Allabakash and Lim 2020; Guo et al. 2021). MERRA2, a National Aeronautics and Space Administration (NASA) atmospheric reanalysis dataset introduced in 1980 (Gelaro et al. 2017), utilizes the Goddard Earth Observing System Atmospheric Data Assimilation System (GEOS ADAS) and the Global Statistical Interpolation (GSI) analysis scheme. It assimilates advanced data from hyperspectral infrared radiance instruments, GPS radio occultation, and microwave observations. The dataset is originally provided at a 0.625° × 0.5° spatial resolution and later re-gridded to 0.25° × 0.25° using inverse distance-weighted interpolation for finer spatial resolution. MERRA2 uses the GEOS-5 model (Rienecker et al. 2011), with turbulence parameterization schemes from Lock et al. (2001) and Louis et al. (1982). The ABLH in MERRA2 is determined as the lowest level where heat diffusivity falls below 2 m^2/s (McGrath-Spangler and Molod 2014). IMDAA is a high-resolution (0.125° × 0.125°) reanalysis product specifically designed for the Indian subcontinent (1979–present), developed collaboratively by UK Met Office (UKMO), National Centre for Medium Range Weather Forecasts (NCMRWF), and India Meteorological Department (IMD) (Rani et al. 2021). IMDAA utilizes the Met office Unified Model (UM) with boundary conditions from ERA-Interim. It integrates a wide range of observations, including radiosondes, wind profilers, satellite radiances, and GPS data. ABLH estimation follows the Lock et al. (2001) scheme, with further improvements by Lock (2001). The method diagnoses unstable layers using buoyancy flux, while stable layers are identified through the gradient R_i with a threshold of 0.25. For unstable conditions, the parcel method and lifting condensation level are employed.

20.3.2 *Characteristics of Atmospheric Boundary Layer Height Over Eastern Himalayan Region*

ABLH exhibits significant spatial, temporal, and regional variability, typically ranging from a few hundred meters to a couple of kilometres over NER (Fig. 20.10). Factors such as soil characteristics, vegetation, topography, wind conditions, urban roughness, and agricultural practices influence ABLH properties over land. A study was carried out using a combination of Dr. Pisharoty radiosondes, a ceilometer, and global/regional reanalysis products (ERA5, MERRA-2 and IMDAA) to assess the diurnal and seasonal evolution of the ABLH over Umiam, Meghalaya (Gogoi et al. 2024).The ABLH exhibited clear seasonal variability, peaking during the pre-monsoon season, followed by winter, post-monsoon, and reaching its lowest during the monsoon season. The daytime seasonal average ABLH from radiosondes were ~1102 m in the pre-monsoon,

~966 m in winter, 822 m in post-monsoon, and ~494 m in the monsoon over Umiam. Lower ABLH during the monsoon is attributed to reduced wind speeds and the absence of diurnal evolution due to cloud cover. Additionally, monsoon precipitation increases soil moisture, raising its heat capacity. This leads to more latent and less sensible heat flux, resulting in a shallower ABL.

Ceilometer data over Umiam indicated a lower daytime ABLH on cloudy days compared to clear-sky conditions in all seasons. For clear-sky days, the seasonal

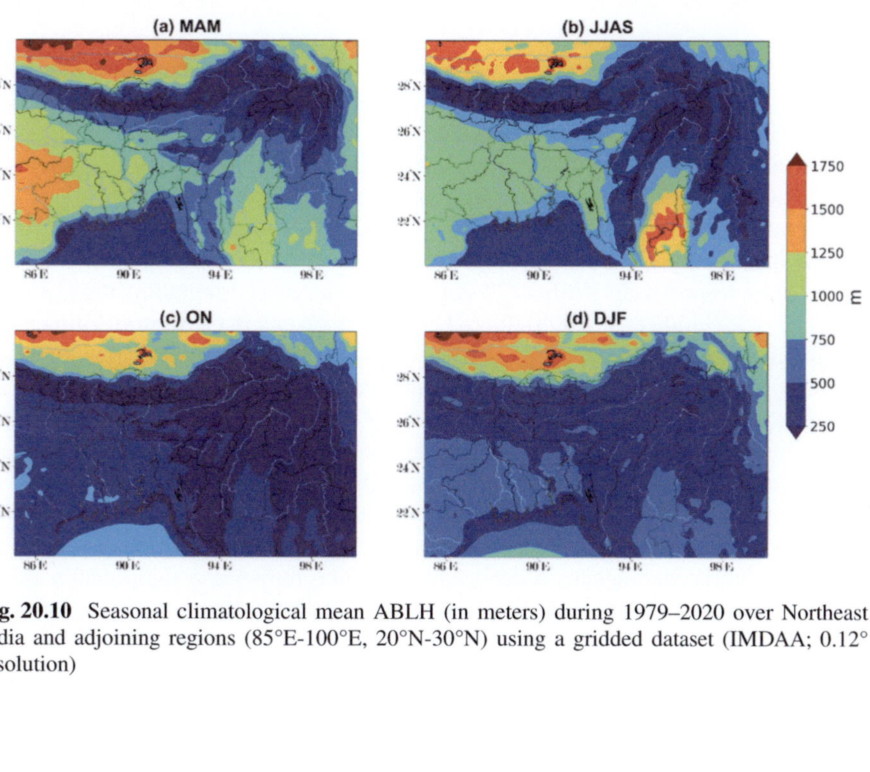

Fig. 20.10 Seasonal climatological mean ABLH (in meters) during 1979–2020 over Northeast India and adjoining regions (85°E-100°E, 20°N-30°N) using a gridded dataset (IMDAA; 0.12° resolution)

mean ABLH was ~1220 (±637) m in winter, ~1337 m in pre-monsoon, ~945 m in monsoon, and 1310 m in post-monsoon. Diurnal ABLH variations were consistent across the Ceilometer, ERA5, IMDAA, and MERRA2 datasets, with the ABL developing between 06:00 IST and 08:00 IST and peaking from 12:00 IST to 14:30 IST. A multi-decadal analysis of ABLH over the entire NER of India (1980–2020) also revealed significant spatial and temporal variability. Observations and reanalysis datasets showed higher ABLH during the pre-monsoon (~1000 m) and lower values during the monsoon, particularly over the western and valley regions of the NER of India. A decreasing trend of 2–4 m per year was observed in ABLH over the western and Brahmaputra valley regions, which is linked to declining ventilation coefficients and sensible heat flux (SHF). The seasonal analysis revealed a significant decreasing trend in ABLH during winter across the western parts of NER, extending over regions like West Bengal, Bangladesh, Meghalaya, and Assam. Similarly, during the pre-monsoon season, ERA5 indicated pronounced declines in ABLH over Sikkim, lower Assam, and Nagaland, while decreasing post-monsoon trends were more evident over West Bengal, Bangladesh, and Mizoram. ERA5 and IMDAA reanalysis data showed better alignment with observations compared to MERRA2, highlighting their reliability for ABLH studies in the region.

20.3.3 ABL Parameters Over the Mountainous Sites, Glaciers, and a Remote Hilly Station in the EHR

The ABL parameters were studied at an eastern Himalayan observational site, Yala Glacier (28.23°N, 85.62°E) and Umiam (25.65°N, 91.88°E), remote hilly stations using a 3D sonic anemometer (Barman et al. 2019a, 2019b, 2021, 2024). At an average elevation of 5330 m above sea level, Yala Glacier is a debris-free glacier in Nepal's Rasuwa district, which is part of the Langtang River watershed.

It has a northern international border with Tibet and is around 60 km north of Kathmandu, the country's capital. The highest and lowest points of the glacier are 5749 and 5128 m, respectively. With a 22° slope and a southwest orientation, the glacier is a plateau type (Fujita et al. 1998). It is a benchmark glacier in the Central Himalayas. The Indian summer monsoon, which brings strong solar radiation and high air moisture, has an effect on this summer accumulation of the glacier (Immerzeel et al. 2014). According to Immerzeel et al. (2014), 68–89% of the region's yearly precipitation fell between June and September when the monsoon circulation predominated. Temperature, humidity, and precipitation used to decrease from the end of August to October (Fujita et al. 1998). The up-valley breezes tend to carry moist air to higher elevations, which results in condensation and often overcast conditions in the afternoon. The debris coating causes surface temperatures to increase significantly above air temperatures during the day (Steiner and Pellicciotti 2016). An Integrated CO_2 and H_2O Open-Path Gas Analyzer and 3D Sonic Anemometer (IRGASON) is used to measure the wind speed components at a height of 3.42 m and 10 Hz. At a height

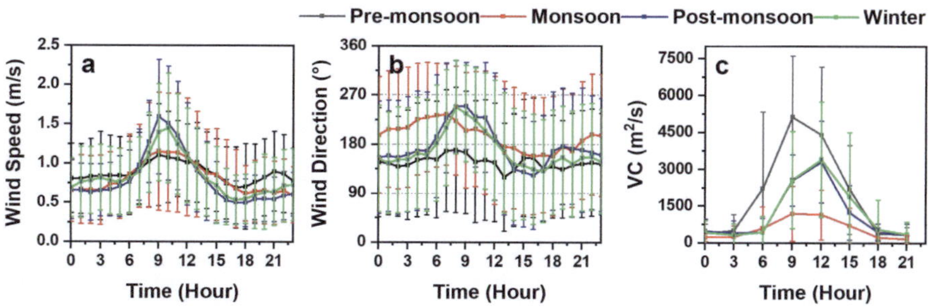

Fig. 20.11 Diurnal variation of a surface wind-speed (m/s) and b surface wind direction (°) measured over Umiam during different seasons, and c ventilation coefficient (m^2/s) derived over Umiam with the help of measured BLH by Ceilometer and vertical profile of horizontal wind speed from IMDAA reanalysis data

of 3.12 m an Automatic Weather Station(AWS) simultaneously monitored the air temperature and relative humidity (10-min interval). At a height of 1.6 m, incoming and outgoing shortwave and longwave radiation was also measured at 10-min interval (Barman et al. 2024).

Over Umiam, similar experiment was carried out using a 31-m meteorological flux tower with few results shown in Fig. 20.11. The setup is located on the NESAC campus above mean sea level in the Khasi Hills of NER. The hill slopes extend along the north–south axis, with the Khasi Hills' peak rising to approximately 1950 m above sea level, situated about 20 km southeast of the flux tower. The terrain slopes at an angle of 5°–6° toward the east and southeast within a horizontal distance of 1 km from the tower and is largely covered with pine trees. To the north, the slope is within 1 km, while to the southwest, the terrain slopes at about 2°, which primarily affects the upwind flow during the winter season. The average tree height in the southeast and southwest directions is around 5.5 m, with adjacent areas surrounding the tower covered by grasses and small bushes. January is the coldest month, with minimum temperatures ranging from 14.5 °C to 3 °C. RH varies between 45 and 89%, accompanied by moderate winds (1.5–4.0 m s^{-1}) predominantly blowing from the west and east during the season.

20.3.4 Factors Affecting ABL Dynamics Over Mountainous Sites

Solar angle: The solar angle is the angular measurement between the Sun's rays and the horizon. It plays a significant role in wind circulation and energy exchange, particularly in mountainous regions. In the Yala Glacier of EHR, the solar zenith angle (φ) and azimuth angle (z) started at approximately 3° and 100° in the morning, peaked at 53.5° and 179.5° around midday, and then decreased to 3° and 259° by evening. Although φ peaked at noon in both months, the zenith angle differed by about 10° in November compared to October. The azimuth angle (z) also showed a 9° difference between morning and evening, remaining consistent around 1130 LT. Due to the narrow φ, surface-reaching solar radiation intensity is moderate. Relative humidity in Yala declines in November as φ decreases compared to October, which reduces evaporation from the snow-covered surface (Barman et al. 2024).

Mountain-valley wind circulation: Over the Eastern Himalayas, mountain wind circulations play an important role in the development of the regional weather, climate, and ecosystem (Barman et al. 2019a, 2019b, 2021, 2024). Three distinct air mass exchange mechanisms between high terrain and adjoining valleys are mountain wind, wind along the valley axis, and larger-scale air circulation between the elevated plain and valley. When the larger-scale synoptic flow had no effect on the ground-level wind circulations, these thermally induced wind circulations appeared under weak synoptic wind circumstances. At the Yala Glacier site, mountain-valley wind circulation is evident due to the variation in wind direction observed during October and November. During the day, the predominant wind direction is from the northwest (270°–315°), while in the evening and night, winds are mainly from the east and southeast (60°–150°). November provides a favourable environment for surface-layer parameters to evolve, thanks to dry conditions, moderate wind speeds (3.8 ± 8 m/s), and clear skies.

At night, surface cooling within the valley creates a localized pressure gradient along the valley, which becomes the primary force for circulation. In the daytime, the temperature differential between the sun-facing and shadowed hillsides initiates a periodic rhythm of wind flows: down-slope winds (katabatic winds) dominate at night, while up-slope winds (anabatic winds) prevail during the day. The sun passes over the location at a smaller elevation angle (43° ± 4°), causing morning solar radiation to heat the east-facing slope, with the northwest wind predominating. In the afternoon, the solar radiation shifts to the west-facing slope, where the southeast winds become dominant.

Over a 24-h cycle, the wind direction shifts from southeast and east (60°–150°) in the morning to northwest (270°–315°) later in the day. During transitional periods, between 0930–1100 LT and 1700–1830 LT, wind speed decreases and the direction shifts. Ground heating in the morning reduces the katabatic flow and enhances the anabatic flow. As the wind speed increases, it travels along the main valley (approximately 5 km wide) before splitting into two branches, one moving north and the other toward the northeast station. The valley and sub-valley, situated between two mountains, have elevations of around 2500 and 3300 m above sea level, respectively. The wind flows toward the station via a sub-valley, which is about 20 km long and is situated between two mountains approximately 5765 m high.

In Umiam, mountain-valley wind circulation is presented by the wind direction variability in different seasons. Wind flows from the south-west (200°–330°) sector in the morning and the south-east (25°–170°) sector in the afternoon and for the rest of the day (Barman et al. 2019a). For the diurnal change of the surface layer characteristics at the mountainous site, winter clear skies are ideal. In winter, less moisture, moderate wind speeds, and strong solar radiation offer a favourable environment for the full development of the atmospheric boundary layer and surface-layer characteristics. Due to increased solar heating, there is a temperature differential between the valley air mass and the surrounding air mass during the day. At night, surface cooling within the valley generates a pressure gradient along the valley. During the pre-monsoon season, unlike in winter, wind speed increases at higher elevations (30 m), and wind direction is predominantly from the southeast direction (90°–180°) (Fig. 20.12). This indicates the influence of synoptic flow, which is more pronounced at higher altitudes. To further analyse the synoptic flow conditions, wind vectors at 10 m and 50 m above the surface were examined, with the 30 m level providing insight into the impact of synoptic flow. In the monsoon season, wind direction is confined to the south (180°–270°) during the day, and mountain-valley circulation is suppressed as synoptic flow dominates. However, in the post-monsoon period, mountain-valley wind circulation becomes evident again, with wind direction ranging from east (45°) to west (300°). In winter, the sun passes over the site at a lower elevation angle (43° ± 4°), causing morning solar radiation to face the east-facing slope and a south-west breeze to predominate.

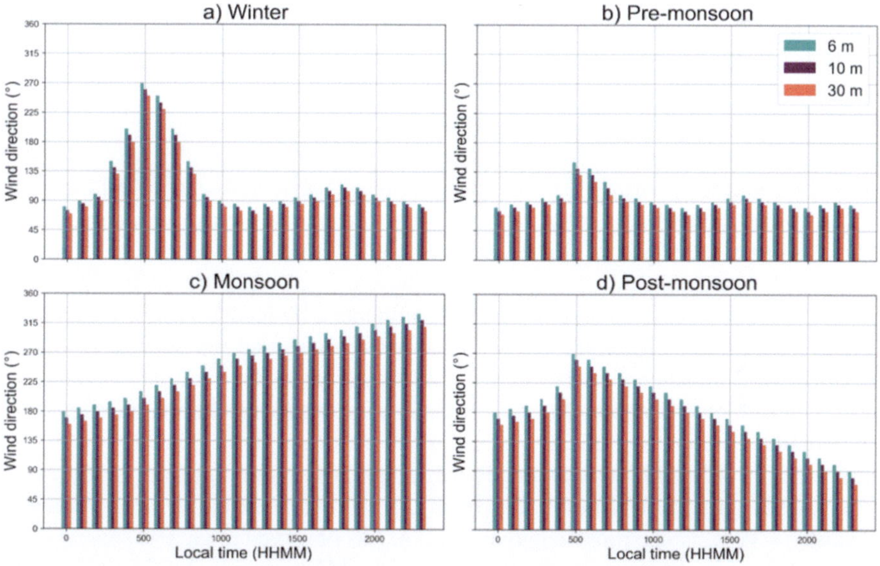

Fig. 20.12 Histograms showing seasonal mean diurnal variation of the mean wind direction (solid line) at heights of 6 m (turquoise), 10 m (purple), and 30 m (orange) during: **a** Winter, **b** Pre-monsoon, **c** Monsoon, **d** Post-monsoon seasons over Umiam

20.4 Surface Flux Variation

Sensible heat flux: Over the land surface, sensible heat flux (*H*) is an important factor in the microclimates, near the Earth's surface, influenced by topography, vegetation, soils, landforms, and structures (Bonan 2002). In this context, ρ and C represent the atmospheric density and the isobaric specific heat of air, respectively. The fluctuations of the vertical wind component and virtual potential temperature are denoted as w' and $\theta v'$. Water vapor exchange between the land surface and the atmosphere affects relative humidity, whereas sensible heat flux primarily regulates air temperature in the lower atmospheric layers. During the daytime, heat is transferred into the air, returning to the ground at night while maintaining a strong upward vapor flux.

Figure 20.13 shows the diurnal *H* over NER and nearby areas according to the MERRA2 reanalysis across different seasons. Over the Umiam station, the diurnal variation of *H* is most distinct during the winter (WN) and pre-monsoon (PM) seasons, with *H* rising from early morning (0730 LT), peaking between 1000–1100 LT, and declining by late afternoon (1530 LT). In the pre-monsoon (PM) season, H increases from 0630 LT, reaches a peak around 1000–1100 LT, and decreases by 1630 LT. During the monsoon (MN) season, H begins rising at 0600 LT, peaks between 1100–1200 LT, and reaches a minimum by 1700 LT. The peak in H typically occurs 1–2 h before the maximum solar elevation angle at the site. At 30-m height, peak H values were 219 ± 61 W m^{-2} (WN), 278 ± 120 W m^{-2} (PM), 152 ± 127 W m^{-2} (MN), and 166 ± 68 W m^{-2} (post monsoon; PMN). The highest value occurred in PM, attributed to stronger solar insolation from higher solar angles and enhanced convective activity. A drop in H around 1130 LT was observed when wind direction shifted from southwest to south, followed by a brief rise before declining again. This transition reflected the dominance of mountain-valley wind circulation over the westerly synoptic flow. As reported by Barman et al. (2019b), such wind shifts significantly affect surface layer dynamics, influenced by local thermal circulations.

The temperature difference between the atmosphere and the surface results in the downward transfer of thermal energy from the air to the land during the night time. The rate of this downward heat transfer increases with the magnitude of the temperature difference. Typically, nighttime *H* is small, ranging from −3 to −20 Wm^{-2}. In the WN and PM seasons, *H* values are larger, ranging from −10 to −20 Wm^{-2}, whereas in the MN and PMN seasons, *H* values are smaller, typically between − 3 and −7 Wm^{-2}. The most negative *H* value was observed in the PM season (−20 ± 9 Wm^{-2}), while the least negative was seen in the MN season (−3 ± 2 Wm^{-2}). Heat transfer from the atmosphere to the surface is higher during WN and PM, while it is lower during MN and PMN. Generally, the air layer above the surface remains warmer during the nighttime, leading to maximum downward heat transfer. However, in the MN season, the surface holds more moisture compared to other seasons, which results in higher evaporation rates due to significant solar insolation. The high vegetation growth in MN, driven by greater precipitation, further contributes to this process. Moist soil, having a higher specific heat capacity than dry soil, keeps the land surface warmer in MN compared to PM. The heat capacity of soil is influenced by both soil

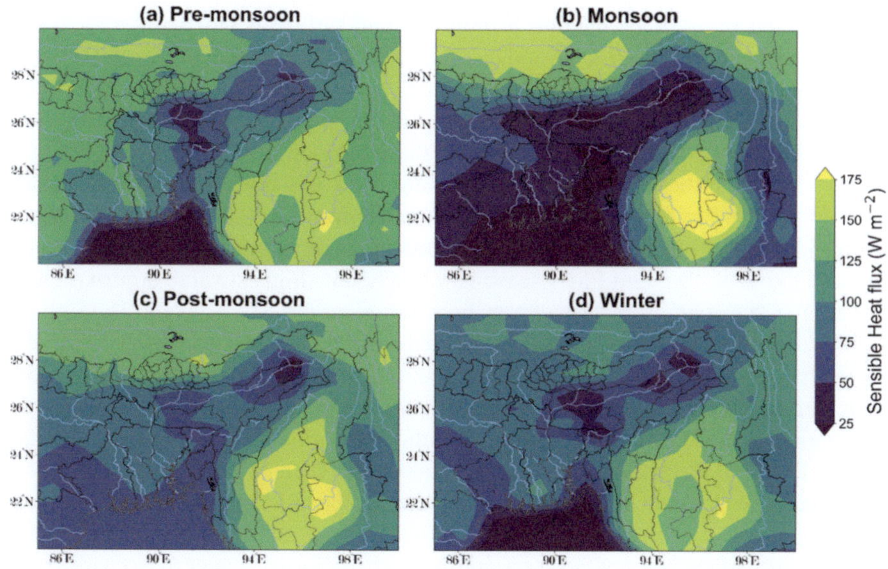

Fig. 20.13 Spatiotemporal distribution of sensible heat flux (from turbulence) over Eastern Himalayan Region during the daytime (0600–1800 LT) for the period 1980–2020 based on MERRA2

moisture content and soil density also noted that specific heat increases with the moisture content in the soil. Soil management practices that lead to compaction can increase soil density and reduce porosity, significantly affecting heat capacity. The net seasonal diurnal mean H from the land to the atmosphere was 42 Wm^{-2} in WN, 58 Wm^{-2} in PM, 35 Wm^{-2} in MN, and 34 Wm^{-2} in PMN.

In Yala Glacier of EHR, the mean H during the daytime (1100 LT-1700 LT) is 12.5 ± 4.9 Wm^{-2} in October and 17.1 ± 10.2 Wm^{-2} in November. During this period, heat is transferred from the snow surface to the atmosphere, while at other times of the day, heat moves from the atmosphere to the snow surface, as indicated by the negative H values (-13.9 ± 9 Wm^{-2} in October and -13 ± 5.8 Wm^{-2} in November). Similarly, the mean latent heat flux (LE) during the same daytime period is 59.6 ± 18.4 Wm^{-2} in October and 115.7 ± 28.9 Wm^{-2} in November. The higher H in November indicates a greater heat transfer from the snow surface to the atmosphere compared to October. However, due to the lower surface albedo in November, the solar radiation (S) is lower, and the snow surface absorbs more incoming shortwave radiation (S_{inc}), which leads to melting and evaporation, resulting in a higher LE value. The outgoing longwave radiation (L_{out}) is approximately the same in both October and November. Both snow and ice exhibit near-perfect black body behaviour in the infrared spectrum, with an emissivity of about 0.98–0.99 for snow and 0.97 for ice.

Momentum flux: The momentum flux (τ) exhibited a distinct diurnal variation across all seasons. Barman et al. (2019b) demonstrated that wind speed over the site increased by midday, leading to higher τ and turbulent kinetic energy (TKE) at the station during the WN season, with two distinct peaks of τ observed in both the morning and afternoon. The morning peak was higher than the afternoon peak at the 6-m level, while at the 10-m height, the morning and afternoon peaks were almost equal. At the 30-m level, however, the afternoon peak was higher than the morning peak. In the PM season, the morning peak was higher than the afternoon peak at the 6-m level, while the afternoon peak surpassed the morning peak at the 10-m and 30-m levels. During the MN season, the peak τ value occurred in the afternoon. In the WN, PM, and PMN seasons, τ reached its highest value approximately 2–3 h after the maximum H value at both the 6-m and 10-m levels. In contrast, during the MN season, the peak τ coincided with the peak H values. τ is closely related to the covariance of velocity perturbations, meaning that higher wind speeds at greater heights during the PM season contributed to elevated τ values at all levels. A variation in wind direction from the south-west to the southeast during the day in WN, PM, and PMN caused a drop in τ values at 1130–1230 LT. It was observed that a local thermal circulation dominated during the daytime, and wind direction fluctuated unpredictably. These wind direction changes played a crucial role in the evolution of surface layer parameters and ABL characteristics over the experimental site.

Turbulence kinetic energy: Turbulence is generated through thermal effects via buoyancy (daytime heat-driven mesoscale circulations) and mechanical effects through shear (nighttime low-level jets and mountain waves) in complex terrain (Serafin et al. 2018). Surface-level turbulence plays a crucial role in transporting energy and pollutants. This section explores the impact of mountain-valley wind circulation on turbulence features and turbulent fluxes at the station.

The seasonal mean daytime variation of TKE exhibited a clear pattern across WN, PM, MN, and PMN at the 6-m, 10-m, and 30-m levels. Barman et al. (2019b) demonstrated that the mountain-valley wind circulation affects surface-level turbulence or TKE. During the daytime, TKE values at higher altitudes (30-m level) were higher by 59%, 71%, 51%, and 41% than those at the lower 6-m level in WN, PM, MN, and PMN, respectively. The higher TKE in the PM season can be attributed to higher wind speeds at the 30-m level. In the PM season, synoptic flow also impacted the station, which is reflected in the TKE values. Daytime TKE values were consistently higher (99%, 56%, 72%, and 93%) than nighttime values in WN, PM, MN, and PMN, respectively. In PM, the smaller day-night difference in TKE suggests the influence of synoptic flow at higher levels. In the WN season, two peaks in TKE were observed during the morning and afternoon. The transition in TKE between 1200 and 1300 LT is associated with a rapid change in ground-level wind speed gradients and fluxes, dominated by westerly synoptic flow. The morning peak (MP) was higher than the afternoon peak (AP) at the 6-m level, while MP and AP were nearly equal at the 10-m level. At the 30-m level, the afternoon peak was higher than the morning peak. In the PM season, the TKE peak occurred in the afternoon across all levels. In MN, TKE was highest at the 30-m height, indicating stronger air mixing between the 10-m and 30-m layers. TKE and H peaked almost simultaneously in all seasons. The turbulence rate in a stable atmosphere increase from zero to the height of the convective boundary layer, a process known as the spin-up process (Stull 1988). Due to reduced wind speed and solar radiation, TKE values declined over time, particularly around noon. This decline in TKE, known as the spin-down process, occurs as turbulence generation dissipates (Stull 1988). Daily TKE variability showed an increase in turbulence intensity during the day and a decrease at night. TKE exhibited a linear increase in turbulence generation, corresponding to more intense land heating and faster wind conditions. Interestingly, nighttime TKE values increased from WN to PMN, implying higher turbulence intensity due to stronger wind speeds. The short-term transition in TKE between 1200 and 1300 LT was linked to dramatic changes in surface wind speed gradients and fluxes in the WN, PM, and PMN seasons.

20.5 Conclusion

The Eastern Himalayan Region (EHR) represents one of the most ecologically dynamic and climatically sensitive regions. In this chapter, an attempt has been made to understand the complex interactions between aerosols, trace gases, and atmospheric boundary layer (ABL) dynamics, with a particular emphasis on the North Eastern Region (NER) of India in the Eastern Himalayan Region (EHR). The findings are drawn from an integrated approach using ground-based observations, satellite remote sensing, and model simulations, with a focus on capturing the region's pronounced spatiotemporal variability and altitudinal gradients.

The seasonal characteristics of aerosols over the NER of India revealed a pronounced dominance of aerosol loading during the pre-monsoon and winter seasons, as evidenced by elevated aerosol optical depth (AOD), black carbon (BC) concentrations, and associated optical properties. Long-range transport from the Indo-Gangetic Plains (IGP), arid zones of western India, and marine sources significantly enhances the columnar aerosol burden during these periods. Concurrently, local emissions from biomass burning, vehicular traffic, and household combustion contribute substantially to near-surface concentrations. The radiative implications of such aerosol presence are considerable. Surface dimming and atmospheric heating observed over NER of India are comparable to those in major South Asian aerosol hotspots.

Trace gas dynamics in this region further amplify the complexity of the atmospheric system. Elevated levels of CO, NO_X, SO_2, and surface O_3, particularly during pre-monsoon and winter months, are strongly modulated by both anthropogenic and natural sources. The CH_4 trends indicate a rate of increase in the EHR that is significantly higher implicating this greenhouse gas as a major contributor to regional warming. In particular, the enhanced CH_4 and BC loading during the dry seasons points toward a feedback mechanism that affects both boundary layer processes and local climate regimes.

The atmospheric boundary layer (ABL) in the EHR, shaped by rugged terrain and localized circulation, displays marked diurnal and seasonal variability. Measurements from ceilometers, GPS radiosondes, and Eddy Covariance system, supported by ERA5 and IMDAA reanalysis products, show ABL heights peaking during the pre-monsoon season and declining sharply during monsoon months. Surface flux analyses from sites such as Umiam and Yala Glacier provide valuable insights into how terrain-induced wind systems, solar insolation, and land surface heterogeneity influence turbulence, radiation budgets, and moisture fluxes. The observed mountain-valley wind circulations, especially under weak synoptic forcing, govern the vertical exchange of aerosols and trace gases, modulating air quality and energy balance at various temporal scales. Another important aspect highlighted in this chapter is the impact of forest fires on aerosol and trace gas concentrations. Satellite-derived fire count data, elevated surface black carbon (BC) levels, and enhanced tropospheric NO_2 concentrations during peak fire months collectively point to a strong coupling between land-use practices; particularly biomass burning and atmospheric composition.

The cumulative impacts of aerosols and trace gases on regional climate and ecosystem health are profound. Beyond their radiative and chemical effects, these atmospheric constituents influence cloud microphysics, precipitation regimes, evapotranspiration, and broader biospheric processes. The suppression of convection, reduction in surface evaporation, and altered cloud condensation nuclei (CCN) characteristics observed over the region highlight the growing role of both anthropogenic and advected aerosols in modulating climate feedback. Furthermore, the decline in ABL height; particularly during the winter season may contribute to increased trapping of pollutants near the surface, thereby challenges in existing air quality.

In summary, this chapter highlights the need for continued and enhanced observational efforts in the EHR. In view of the region's complex terrain and climatic vulnerability, there is an urgent requirement for transboundary collaboration, data sharing, and long-term monitoring to better quantify the evolving aerosol-trace gas-ABL system. Such efforts will be pivotal for advancing our understanding of the regional climate dynamics.

Acknowledgements The authors sincerely acknowledge the Indian Space Research Organisation (ISRO) for supporting this work through the Atmospheric Forcing and Radiation Initiative (ARFI) and the Network for Observation of Boundary Layer Experiment (NOBLE) projects. The results and analyses presented in this report are derived from the data and research activities conducted under these ISRO-sponsored initiatives. The authors also extend their gratitude to all team members and collaborators for their valuable contributions to field observations, instrumentation, and data analysis.

References

S. Allabakash, S. Lim, Climatology of planetary boundary layer height-controlling meteorological parameters over the Korean Peninsula. Remote Sens. **12** (2020). https://doi.org/10.3390/RS12162571

B.S. Arun, M.M. Gogoi, A. Borgohain, P. Hegde, S.S. Kundu, S.S. Babu, Role of sulphate and carbonaceous aerosols on the radiative effects of aerosols over a remote high-altitude site Lachung in the Eastern Himalayas. Atmos. Res. **2021**(263), 105799 (2021a). https://doi.org/10.1016/j.atmosres.2021.105799

B.S. Arun, M.M. Gogoi, P. Hegde, A. Borgohain, S.K.R. Boreddy, S.S. Kundu, S.S. Babu, Carbonaceous aerosols over Lachung in the Eastern Himalayas: primary sources and secondary formation of organic aerosols in a remote high-altitude environment. ACS Earth Space Chem. **2021**(5), 2493–2506 (2021b). https://doi.org/10.1021/acsearthspacechem.1c00190

B.S. Arun, M.M. Gogoi, D.K. Deshmukh, P. Hegde, S.K.R. Boreddy, A. Borgohain, S.S. Babu, Enhanced light absorption by ambient brown carbon aerosols in the eastern Himalayas. Environ. Sci. Atmos. **4**(7), 782–801 (2024). https://doi.org/10.1039/d4ea00021h

K. Bali, A. Kumar, S. Chourasiya, Emission estimates of trace gases (VOCs and NOx) and their reactivity during biomass burning period (2003–2017) over Northeast India. J. Atmos. Chem. **78**, 17–34 (2021). https://doi.org/10.1007/s10874-020-09413-6

N. Barman, A. Borgohain, S.S. Kundu, B. Saha, R. Roy, R. Solanki, N.V.P. Kiran Kumar, P.L.N. Raju, Impact of atmospheric conditions in surface–air exchange of energy in a topographically complex terrain over Umiam. Meteorog. Atmos. Phys. (2019). https://doi.org/10.1007/s00703-019-00668-7

N. Barman, A. Borgohain, S.S. Kundu, B. Saha, R. Roy, R. Solanki, N.V.P. Kiran Kumar, P.L.N. Raju, Daytime temporal variation of surface-layer parameters and turbulence kinetic energy budget in topographically complex terrain around Umiam, India. Boundary Layer Meteorol. (2019). https://doi.org/10.1007/s10546-019-00443-6

N. Barman, A. Borgohain, S.S. Kundu, N.V.P. Kiran Kumar, Seasonal variation of mountain-valley wind circulation and surface layer parameters over the mountainous terrain of the northeastern region of India. Theor. Appl. Climatol. **143**, 1501–1512 (2021). https://doi.org/10.1007/s00704-020-03491-y

N. Barman, S. Gokhale, Aerosol influence on the pre-monsoon RF mechanisms over North-East India: a WRF-Chem study. Atmos. Res. **268**, 106002 (2022). https://doi.org/10.1016/j.atmosres.2021.106002

N. Barman, I. Roy, A. Gayen, Investigation of land–atmosphere interaction at a high altitude (5330 m a.s.l.) snow covered station of East Himalaya. Meteorol. Atmos. Phys. **136**, 49 (2024). https://doi.org/10.1007/s00703-024-01045-9

C. Bharali, B. Pathak, P.K. Bhuyan, Spring and sum mer night-time high ozone episodes in the upper Brahmaputra valley and their association with lightning. Atmos. Environ. **109**, 234–250 (2015). https://doi.org/10.1016/j.atmosenv.2015.03.035

P.K. Bhuyan, M.M. Gogoi, K. Krishnamurthy, Spectral and temporal characteristics of aerosol optical depth over a wet tropical location in Northeast India. Adv. Space Res. **35**, 1423–1429 (2005)

P.K. Bhuyan, C. Bharali, B. Pathak, G. Kalita, The role of precursor gases and meteorology on temporal evolution of O3 at a tropical location in north East India. Environ. Sci. Pollut. Res. 1–18 (2014). https://doi.org/10.1007/s11356-014-2587-3

G.B. Bonan, Ecological climatology. Cambridge University Press, UK (2002). https://doi.org/10.1017/CBO9781107339200

A. Borgohain, S.S. Kundu, N. Barman, P.L.N. Raju, R. Roy, B. Saha, P. Ajay, B. Pathak, P.K. Bhuyan, Investigation of physical and optical properties of aerosol over high altitude stations along the sub-Himalayan region of North-Eastern India. Atmos. Pollut. Res. **11**(2), 383–392 (2020). https://doi.org/10.1016/j.apr.2019.11.010

A. Borgohain, M. Gogoi, N. Barman, A. Kundu, T. Banik, S.S. Kundu, P.K. Bhuyan, S.P. Aggarwal, Impact of biomass burning on black carbon and NO2 over North Eastern region of India using multi-satellite observations. J. Indian Soc. Remote Sens. **51**, 1605–1617 (2023). https://doi.org/10.1007/s12524-023-01692-2

R.K. Chakrabarty, M.A. Garro, E.M. Wilcox, H. Moosmller, Strong radiative heating due to wintertime black carbon aerosols in the Brahmaputra River Valley. Geophys. Res. Lett. **39**(9), L09804 (2012). https://doi.org/10.1029/2012GL051148

S. Chandra, A. Dwivedi, M. Kumar, Characterization of the atmospheric boundary layer from radiosonde observations along eastern end of monsoon trough of India. J. Earth Syst. Sci. **123**, 1233–1240 (2014). https://doi.org/10.1007/s12040-014-0458-4

L. Chutia, N. Ojha, I.A. Girach, L.K. Sahu, L.M.A. Alvarado, J.P. Burrows, B. Pathak, P.K. Bhuyan, Distribution of volatile organic compounds over Indian subcontinent during winter: WRF-chem. Simul. Observations. Environ. Pollut. **252**, 256–269 (2019). https://doi.org/10.1016/j.envpol.2019.05.097

P. Dahutia, B. Pathak, P.K. Bhuyan, Vertical distribution of aerosol and clouds over North-East India and adjoining South-Asia: effect of aerosols on clouds. Atmos. Environ. (2019). https://doi.org/10.1016/j.atmosenv.2019.116882

P. Dhar, B.K. De, T. Banik, M.M. Gogoi, S. Suresh Babu, A. Guha, Atmospheric aerosol radiative forcing over a semi-continental location Tripura in north-East India: model results and ground observations. Sci. Total. Environ. **580**, 499–508 (2017). https://doi.org/10.1016/j.scitotenv.2016.11.200

S. Emeis, C. Münkel, S. Vogt, W.J. Müller, K. Schäfer, Atmospheric boundary-layer structure from simultaneous SODAR, RASS, and ceilometer measurements. Atmos. Environ. **38**, 273–286 (2004). https://doi.org/10.1016/j.atmosenv.2003.09.054

K. Fujita, N. Takeuchi, K. Seko, Glaciological observations of Yala Glacier in Langtang Valley, Nepal Himalayas, 1994 and 1996. Bull. Glacier Res. **16**, 75–81 (1998)

R. Garaga, S. Gokhale, S.H. Kota, Source apportionment of size-segregated atmospheric particles and the influence of particles deposition in the human respiratory tract in rural and urban locations of north-east India. Chemosphere **255**, 126980 (2020). https://doi.org/10.1016/j.chemosphere.2020.126980

R. Gelaro, W. McCarty, M.J. Suárez, R. Todling, A. Molod, L. Takacs, C.A. Randles, A. Darmenov, M.G. Bosilovich, R. Reichle, K. Wargan, L. Coy, R. Cullather, C. Draper, S. Akella, V. Buchard, A. Conaty, A.M. da Silva, W. Gu, G. Kim, R. Koster, R. Lucchesi, D. Merkova, J.E. Nielsen, G. Partyka, S. Pawson, W. Putman, M. Rienecker, S.D. Schubert, M. Sienkiewicz, B. Zhao, The modern-era retrospective analysis for research and applications, version 2 (MERRA-2). J. Clim. **30**, 5419–5454 (2017). https://doi.org/10.1175/JCLI-D-16-0758.1

M.M. Gogoi, P.K. Bhuyan, K. Krishna Moorthy, Estimation of the effect of long range transport on seasonal variation of aerosols over north eastern India. Ann. Geophys. **26**, 1365–1377 (2008)

M.M. Gogoi, K. Krishna Moorthy, S. Suresh Babu, P.K. Bhuyan, Climatology of columnar aerosol properties and the influence of synoptic conditions: first-time results from the northeastern region of India. J. Geophys. Res. Atmos. **114**(D8) (2009). https://doi.org/10.1029/2008JD010765

M.M. Gogoi, K.K. Moorthy, S.K. Kompalli, J.P. Chaubey, S.S. Babu, M.R. Manoj, V.S. Nair, T.P. Prabhu, Physical and optical properties of aerosols in a free tropospheric environment: results from long-term observations over western trans-Himalayas. Atmos. Environ. **84**, 262–274 (2014)

M.M. Gogoi, S.S. Babu, K.K. Moorthy, P.K. Bhuyan, B. Pathak, T. Subba, L. Chutia, S.S. Kundu, C. Bharali, A. Borgohain, A. Guha, B.K. De, B. Singh, M. Chin, Radiative effects of absorbing aerosols over northeastern India: observations and model simulations. J. Geophys. Res. **122**, 1132–1157 (2017). https://doi.org/10.1002/2016JD025592

D. Gogoi, A. Sazid, J. Bora, P. Deka, S. Balachandran, R.R. Hoque, Particulate matter exposure in biomass-burning homes of different communities of Brahmaputra Valley. Environ. Monit. Assess. **193**, 856 (2021). https://doi.org/10.1007/s10661-021-09624-8

M. Gogoi, A. Borgohain, S.S. Kundu, S. Sarma, A. Kundu, P.K. Bhuyan, N.V.P.K. Kumar, N. Barman, S.P. Aggarwal, Investigation of the atmospheric boundary layer characteristics over a high altitude station in North East India using measurements and reanalysis datasets. Theor. Appl. Climatol. **155**, 6265–6281 (2024). https://doi.org/10.1007/s00704-024-04997-5

M. Gogoi, A. Borgohain, A. Kundu, S.S. Kundu, P.K Bhuyan, B. Pathak, K. Bhuyan, N. Barman, T. Banik, A. Charri, A. Chakravorty, Atmospheric boundary layer and the vertical distribution of black carbon over plain and hilly terrain of North East India using in situ measurements, in *EGU General Assembly Conference Abstracts* (2023), pp. EGU-350

A. Guha, B.K. De, P. Dhar, T. Banik, M. Chakraborty, R. Roy, A. Choudhury, Seasonal characteristics of aerosol black carbon in relation to long range transport over Tripura in Northeast India. Aerosol Air Qual. Res. **15**(3), 786–798 (2015). https://doi.org/10.4209/aaqr.2014.02.0029

J. Guo, J. Zhang, K. Yang, H. Liao, S. Zhang, K. Huang, Y. Lv, J. Shao, T. Yu, B. Tong, J. Li, T. Su, S.H.L. Yim, A. Stoffelen, P. Zhai, X. Xu, Investigation of near-global daytime boundary layer height using high-resolution radiosondes: first results and comparison with ERA5, MERRA-2, JRA-55, and NCEP-2 reanalyses. Atmos. Chem. Phys. **21**, 17079–17097 (2021). https://doi.org/10.5194/acp-21-17079-2021

W.W. Immerzeel, L. Petersen, S. Ragettli, F. Pellicciotti, The importance of observed gradients of air temperature and precipitation for modeling runoff from a glacierized watershed in the Nepalese Himalayas. Water Resour. Res. **50**, 2212–2226 (2014). https://doi.org/10.1002/2013WR014506

I.S. Rani, T. Arulalan, J.P. George, E.N. Rajagopal, R. Renshaw, A. May cock, D.M. Barker, M. Rajeevan, IMDAA: high-resolution satellite-era reanalysis for the Indian monsoon region. J. Clim. **34**, 5109–5133 (2021). https://doi.org/10.1175/JCLI-D-20-0412.1

G. Kalita, P.K. Bhuyan, Spatial heterogeneity in tropospheric column ozone over the Indian subcontinent: long-term climatology and possible association with natural and anthropogenic activities. Adv. Meteorol. 924516 (2011). https://doi.org/10.1155/2011/924516

D. Kumar, P. Hegde, B.S. Arun, M.M. Gogoi, S.S. Babu, Anthropogenic sources and liquid water drive secondary organic aerosol formation over the eastern Himalaya. Sci. Total Environ. **949** (2024). https://doi.org/10.1016/j.scitotenv.2024.175072

S.S. Kundu, A. Borgohain, N. Barman, M. Devi, P.L.N. Raju, Spatial variability and radiative impact of aerosol along the Brahmaputra river valley in India: results from a campaign. J. Environ. Prot. **9**, 405–430 (2018). https://doi.org/10.4236/jep.2018.94026

A. Kundu, S.S. Kundu, S.K. Sharma, M. Gogoi, T. Banik, A. Borgohain, R. Mahanta, A. Debnath, The behavior of cloud base height over a hilly remote station of North-East India using ground based remote sensing technique. Atmos. Res. **282**, 106512 (2023). https://doi.org/10.1016/j.atmosres.2022.106512

A. Kundu, S.S. Kundu, A. Borgohain, M.M. Gogoi, S.S. Babu, S.K. Sharma, M. Gogoi, Role of atmospheric boundary layer dynamics in driving surface black carbon concentrations over a high-altitude station in north-east India. Atmos. Pol. Res. **16**(3), 102420 (2025). https://doi.org/10.1016/j.apr.2025.102420

X. Lee, W. Massman, B.E. Law, *Handbook of micrometeorology* (Kluwer Academic Press, Dordrecht, A guide for flux measurement and analysis, 2004). https://doi.org/10.1007/1-4020-2265-4

A.P. Lock, The numerical representation of entrainment in parameterizations of boundary layer turbulent mixing. Mon. Weather Rev. **129**, 1148–1163 (2001). https://doi.org/10.1175/1520-0493(2001)129%3C1148:TNROEI%3E2.0.CO;2

J.F. Louis, M. Tiedtke, J.F. Geleyn, A short history of the operational PBL—parameterization at ECMWF, in *Workshop on Planetary Boundary Layer Parameterization*, ECMWF, Reading, England, 5–27 November 1981, pp. 59–79 (1982)

E.L. McGrath-Spangler, A. Molod, Comparison of GEOS-5 AGCM planetary boundary layer depths computed with various definitions. Atmos. Chem. Phys. **14**, 6717–6727 (2014). https://doi.org/10.5194/acp-14-6717-2014

S.K. Nair, T.J. Anurose, B.D. Subrahamanyam, N.V.P.K. Kumar, M. Santosh, S. Sijikumar, M. Mohan, K.V.S. Namboodiri, Characterization of the vertical structure of coastal atmospheric boundary layer over Thumba (8.5oN, 76.9oE) during different seasons. Adv. Meteorol. **2011**, 1–9 (2011). https://doi.org/10.1155/2011/390826

H.J. Oldroyd, E.R. Pardyjak, H. Huwald, M.B. Parlange, Adapting tilt corrections and the governing flow equations for steep, fully three-dimensional, mountainous terrain. Bound.-Layer Meteorol. **159**, 539–565 (2016). https://doi.org/10.1007/s10546-015-0066-0

B. Pathak, P.K. Bhuyan, M.M. Gogoi, K. Bhuyan, Seasonal heterogeneity in aerosol types over Dibrugarh, north-eastern India. Atmos. Environ. **47**, 307–315 (2012). https://doi.org/10.1016/j.atmosenv.2011.10.061

B. Pathak, L. Chutia, C. Bharali, P.K. Bhuyan, Continental export efficiencies and delineation of sources for trace gases and black carbon in north-East India: seasonal variability. Atmos. Environ. **125B**, 474–485 (2016). https://doi.org/10.1016/j.atmosenv.2015.09.020

P. Rahul, R. Bhawar, D. Ayantika, A.S. Panicker, P.D. Safai, V. Tharaprabhakaran, B. Padmakumari, M.P. Raju, Double blanket effect caused by two layers of black carbon aerosols escalates warming in the Brahmaputra River Valley. Sci. Rep. **4**, 3670 (2014). https://doi.org/10.1038/srep03670

M.M. Rienecker, M.J. Suarez, R. Gelaro, R. Todling, J. Bacmeister, E. Liu, M.G. Bosilovich, S.D. Schubert, L. Takacs, G.-K. Kim, S. Bloom, J. Chen, D. Collins, A. Conaty, A. da Silva, W. Gu, J. Joiner, R.D. Koster, R. Lucchesi, A. Molod, T. Owens, S. Pawson, P. Pegion, C.R. Red der, R. Reichle, F.R., Robertson, A.G. Ruddick, M. Sienkiewicz, J. Woollen, MERRA: NASA's modern-era retrospective analysis for research and applications. J. Clim. **24**, 3624–3648 (2011). https://doi.org/10.1175/JCLI-D-11-00015.1

A. Roy, A. Chatterjee, C. Sarkar, S.K. Das, S.K. Ghosh, S. Raha, A study on aerosol-cloud condensation nuclei (CCN) activation over eastern Himalaya in India. Atmos. Res. **189**, 69–81 (2017). https://doi.org/10.1016/j.atmosres.2017.01.015

A. Saikia, B. Pathak, P. Singh, P.K. Bhuyan, B. Adhikary, Multi-model evaluation of meteorological drivers, air pollutants and quantification of emission sources over the Upper Brahmaputra basin. Atmosphere **10**(11), 703 (2019). https://doi.org/10.3390/atmos10110703

A. Saikia, B. Pathak, P.K. Singh, P.K. Bhuyan, B. Adhikary, Decadal changes in atmospheric methane emissions in the eastern Himalayan region: Source apportionment and impact assessment. Int. J. Environ. Res. **17**(1), 10 (2023). https://doi.org/10.1007/s41742-022-00501-x

P. Seibert, F. Beyrich, F. S.-E. Gryning, S. Joffre, A. Rasmussen, P. Tercier, Review and intercomparison of operational methods for the determination of the mixing height. Atmos. Environ. **34**(7), 1001–1027 (2000). https://doi.org/10.1016/S1352-2310(99)00349-0

D.J. Seidel, Y. Zhang, A. Beljaars, J.-C.Golaz, A.R. Jacobson, B. Medeiros, Climatology of the planetary boundary layer over the continental United States and Europe. J. Geophys. Research: Atmos. **117**, D17106 (2012). https://doi.org/10.1029/2012JD018143

S. Serafin, B. Adler, J. Cuxart, S.F.J. De Wekker, A. Gohm, B. Grisogono, N. Kalthoff, D.J. Kirshbaum, M.W. Rotach, J. Schmidli, I. Stiperski, Z. Vecenaj, D. Zardi, Exchange processes in the atmospheric boundary layer over mountainous terrain. Atmosphere **9**, 102 (2018). https://doi.org/10.3390/atmos9030102

R.K. Sharma, R. Kumar, P. Pradhan, A. Sharma, Climate-induced glacier retreats and associated hazards: need for robust glaciers and glacial lake management policy in Sikkim Himalaya, India, in Rani, S., Climate Change. Springer Climate ed. by R. Kumar (Springer, Cham, 2022). https://doi.org/10.1007/978-3-030-92782-0_8

S. Singh, S. Gokhale, Modeling the dispersion of traffic-derived black carbon emissions into hilly terrain. Environ. Monit. Assess. **1–13**(195), 958 (2023). https://doi.org/10.1007/s10661-023-11554-6

J. Steiner, F. Pellicciotti, On the variability of air temperature over a debris covered glacier, Nepalese Himalaya. Annals Glaciol. **57**, 1–13 (2016). https://doi.org/10.3189/2016AoG71A066

R.B. Stull, An introduction to boundary layer meteorology. Springer (1988). https://doi.org/10.1007/978-94-009-3027-8

T. Subba, M.M. Gogoi, B. Pathak, P. Ajay, F. Solmon, Assessment of 1D and 3D model simulated radiation flux based on surface measurements and estimation of aerosol forcing and their climatological aspects. Atmos. Res. **204**, 110–127 (2018). https://doi.org/10.1016/j.atmosres.2018.01.012

T. Subba, B. Pathak, M.M. Gogoi, P. Ajay, P. Dahutia, A. Chakraborty, P.K. Bhuyan, Observations on the decadal variability of aerosol in eastern Himalayan foothills: evidence of an anthropologically induced positive shift. Atmos. Environ. **299**, 119638 (2023). https://doi.org/10.1016/j.atmosenv.2023.119638

B. Tyagi, J. Singh, G. Beig, Seasonal progression of surface ozone and NOx concentrations over three tropical stations in North-East India. Environ. Pollut. **258**, 113662 (2020). https://doi.org/10.1016/j.envpol.2019.113662

P. Vishwakarma, P. Rajeev, S. Rabha, N. Islam, B.K. Saikia, T. Gupta, Wintertime trends of particulate-bound polycyclic aromatic hydrocarbons (PAHs) at north-east site of India: chemical characterization and source identification. J. Atmos. Chem. **80**, 251–269 (2023). https://doi.org/10.1007/s10874-023-09450-x

X.Y. Wang, K.C. Wang, Estimation of atmospheric mixing layer height from radiosonde data. Atmos. Meas. Tech. **7**, 1701–1709 (2014). https://doi.org/10.5194/amt-7-1701-2014

J.M. Wilczak, S.P. Oncley, S. Stage, Sonic anemometer tilts correction algorithms. Bound.-Layer Meteorol. **99**, 127–150 (2001). https://doi.org/10.1023/A:1018966204465

R. Yuan, M. Kang, S. Park, Expansion of the planar-fit method to estimate flux over complex Terrain. Meteorol. Atmos. Phys. **110**, 123–133 (2011). https://doi.org/10.1007/s00703-010-0113-9

Y. Zhang, S. Chen, S. Chen, H. Chen, P. Guo, A novel Lidar gradient cluster analysis method of nocturnal boundary layer detection during air pollution episodes. Atmos. Meas. Tech. **13**, 6675–6689 (2020). https://doi.org/10.5194/amt-13-6675-2020

Arup Borgohain is currently serving as a Senior Scientist in the Space and Atmospheric Science Division at the North Eastern Space Applications Centre, under the Department of Space, Government of India. He holds an M.Sc. in Physics, an M.Tech. in Space Science, and a Ph.D. in Physics with a specialization in upper atmospheric studies. His main areas of research include aerosol–trace gas dynamics, atmospheric boundary layer processes, mass–energy exchange mechanisms, and climate science. He has conducted several field campaign experiments across the North Eastern Region of India to investigate aerosol–trace gas interactions and boundary layer dynamics. Dr. Borgohain has authored more than 50 research papers in reputed international journals. In addition to his scientific contributions, he is deeply committed to science communication and regularly writes popular science articles in leading Assamese magazines and newspapers. He is also actively involved in guiding Ph.D. scholars from reputed institutions such as the Indian Institute of Space Science and Technology (IIST), Dibrugarh University, Gauhati University, and Assam Science and Technology University.

Shyam S Kundu did his M.Sc. in Physics from University of North Bengal and Ph.D. in Physics from Gauhati University. He has more than 20 years of professional experience at NESAC and currently working as the Head of the Space and Atmospheric Sciences Division at the centre. He was instrumental in setting up the Space and Atmospheric Science research facility at NESAC with several instruments, a S-band Doppler Weather Radar, largest network of weather stations in NER covering difficult terrains, three aerosol observatories including two in Himalayan region, network of lightning detectors, etc. His research interest include extreme weather forecasting in NER, lightning physics, aerosol-cloud-precipitation interaction studies and its climatic impact, climate change assessment, etc. He took lead role to develop an operational lightning early warning system for the NE region of India. He has more than 50 publications in peer reviewed journals and books. He is making significant contribution to professional bodies like ISRS as its Joint Secretary and ISPRS as the Secretary of the WG-4 under TC V. He also represents ISRO in the Committee on Earth Observation Satellites for Precipitation constellations.

Arundhati Kundu is currently working as a Research Scientist at NESAC under the North Eastern Regional node for Disaster Ridk Reduction (NER-DRR) program. She was born and brought up in Maldah district, West Bengal, India. She completed her schooling at native place and obtained her B.Sc. in Physics from the University of Calcutta, India. She has completed her masters in Atmospheric Science from the same university and currently persuing a PhD in Physics from Cotton University, India. Before joining as a Research Scientist, she served as a Research Fellow at NESAC for five years, making major contributions to Aerosol Radiative Forcing over India (ARFI) project under ISRO's Geosphere Biosphere Programme. Her research interests include aerosol cloud interaction and data assimilation for the prediction of extreme weather events. Outside her academic and research work, she loves reading psychological thrillers, watching anime, cooking, stitching and playing badminton.

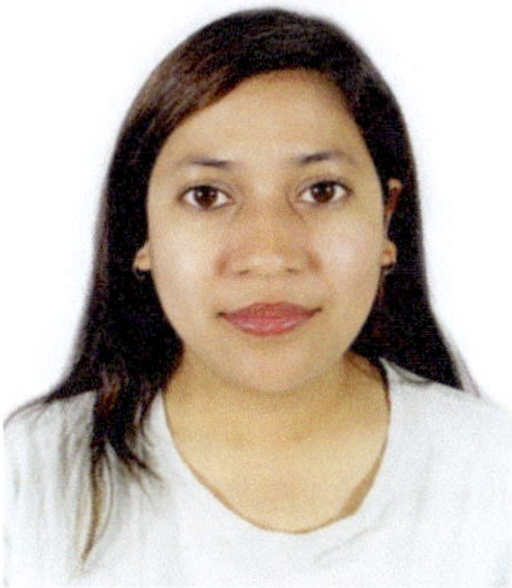

Manasi Gogoi is currently working as a Meteorologist at NeoGeoInfo Technologies Ltd. under the Assam State Disaster Management Authority. She is also a Ph.D. scholar at the Centre for Atmospheric Studies, Dibrugarh University, working on Atmospheric Boundary Layer Dynamics. She completed her B.Sc. and M.Sc. in Physics from Dibrugarh University. She previously worked for five years as a Research Fellow at NESAC, where she significantly contributed to ISRO's Network of Boundary Layer Experiments (NOBLE) project under the Geosphere Biosphere Programme. Her research interests include boundary layer processes, land-atmosphere interaction, aerosol science, and climate change. She is a passionate floraphile and orchid enthusiast, who finds joy in being close to nature.

Mukunda M. Gogoi is a Scientist at the Space Physics Laboratory, Vikram Sarabhai Space Centre, ISRO. His research focuses on developing and applying satellite- and ground-based techniques for aerosol characterization and climate impact assessment. He has been playing a key role in ISRO Geosphere-Biosphere Program, notably in operating ARFINET, a nationwide network of aerosol observatories. Dr. Gogoi has extensive experience in establishing and operating aerosol laboratories in the Arctic, Antarctic, and high-altitude Himalayas, studying aerosol effects on snow albedo. A seasoned field experimenter, he has led multi-platform campaigns using ships and aircraft to investigate aerosol–radiation and aerosol–cloud interactions. He has authored over 85 publications, book chapters, and technical reports, and has supervised numerous research scholars. His contributions have earned him several awards, including a guest associate professorship at the Atmospheric and Ocean Research Institute, University of Tokyo, and guest editorships of special issues in leading international journals.

Binita Pathak is an Assistant Professor in the Department of Physics at Dibrugarh University, Assam. She is an Associate of the Indian Academy of Sciences, Bangalore and a recipient of SERB Women Excellence Award in Earth and Atmospheric Sciences in 2022. She was a Junior Associate in ICTP, Italy during 2015-2021 and is currently a Regular Associate there. She holds a Ph.D. in Atmospheric Science, and her research interests span air quality, atmospheric aerosols, cloud microphysics, and climate change, with a special focus on the eastern Himalayan and northeastern Indian region. Dr. Pathak has been actively involved in several national and international research projects and has published extensively in peer-reviewed journals. Her work contributes significantly to understanding regional atmospheric processes and their implications for weather, environment, and health. In addition to her research pursuits, she is deeply committed to teaching and mentoring, guiding postgraduate and doctoral students in the field of atmospheric and environmental physics.

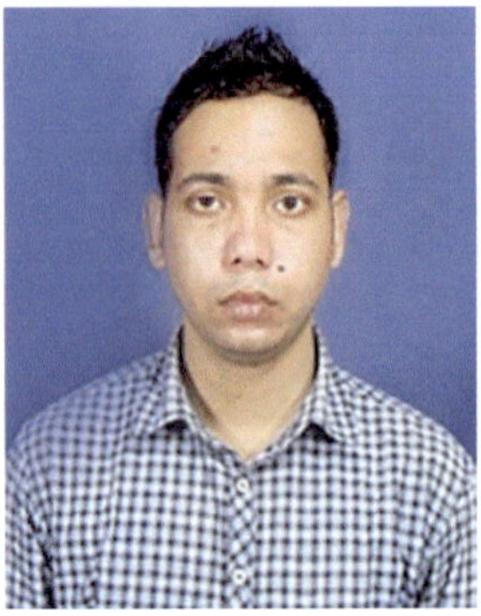

Nilamoni Barman is currently working as Scientist-'C' specializing in Hydrometeorology at the Central Ground Water Board (CGWB), Eastern Region, Kolkata, under the Ministry of Jal Shakti, Government of India. With over 11 years of experience in Atmospheric Physics, Meteorology, Aerosol Science, Climate Change, and Groundwater, he has made significant contributions to the field through both research and public service. He holds an M. Sc. and Ph. D. in Physics from the National Institute of Technology Agartala, Tripura, and has published 14 research articles in international peer-reviewed journals. Dr. Barman qualified the Union Public Service Commission (UPSC) examination in 2019 and began his tenure at CGWB as Scientist-B. His current work focuses on the intricate study about the land-atmosphere interaction, impact of climate change on groundwater resource and air pollution over East India. Originally, from Nalbari, Assam, Dr. Barman now resides in Kolkata, West Bengal. He enjoys serious thinking and occasionally the underrated activity of napping in his spare time.

Chapter 21
Space Technology to Improve Weather Prediction Over Mountainous Regions

Shyam S. Kundu, Rekha B. Gogoi, Aniket Chakravorty, Abhay Srivastava, Arundhati Kundu, Arup Borgohain, and Rosly B. Lyngdoh

Abstract Despite advances in numerical weather prediction (NWP), forecasting weather over Northeast India remains challenging due to several reasons including sparse ground-based observatories. The region's complex topography and expansive international borders complicates data collection, as atmospheric systems transcend human boundaries. In this context, space-based observations play a vital role by providing essential datasets such as cloud cover and dynamics, humidity, land use and land cover, radiance, atmospheric motion vectors, precipitation estimates, etc. These products support NWP model data assimilation, bias correction, and help improving disaster forecasting like floods, lightning, landslides, etc. Sensitivity tests with the WRF model revealed that ISRO's LULC datasets outperform USGS and MODIS inputs, particularly in Northeast India. The region-specific WRF setup showed high RMSE and positive bias along the windward slopes of the Shillong Plateau and Himalayan foothills. However, hybrid data assimilation helped reduce these errors. Additional experiments demonstrated that incorporating INSAT-3D radiance data significantly improved forecast skill over Meghalaya, central and

S. S. Kundu (✉) · R. B. Gogoi · A. Chakravorty · A. Srivastava · A. Kundu · A. Borgohain · R. B. Lyngdoh
North Eastern Space Applications Centre (NESAC), Umiam, Shillong, Meghalaya, India
e-mail: ss.kundu@nesac.gov.in

R. B. Gogoi
e-mail: rekha.bharali@nesac.gov.in

A. Chakravorty
e-mail: aniket.chakravorty@nesac.gov.in

A. Srivastava
e-mail: dr.abhaysrivastava@nesac.gov.in

A. Kundu
e-mail: arundhatikundu.ritu@gmail.com

A. Borgohain
e-mail: arup.borgohain@nesac.gov.in

R. B. Lyngdoh
e-mail: rosly.lyngdoh@nesac.gov.in

S. P. Aggarwal et al. (eds.), *Remote Sensing of the North Eastern Himalayan Ecosystem*,
https://doi.org/10.1007/978-981-95-3221-6_21

western Arunachal Pradesh and central Assam compared to GTS-only assimilation. The WRF-Elec model and a deep learning–based lightning nowcasting system demonstrated strong capability in nowcasting severe lightning events across the NE India. Synoptic scale weather advisories developed through the integrated use of satellite-derived meteorological products have remarkably reduced the limitations of NWP model over the complex terrain of Eastern Himalayan region.

Keywords NER · Severe weather · Lightning · WRF · Data assimilation · Satellite

21.1 Introduction

Weather prediction over mountainous regions, particularly in the eastern Himalayas, remains one of the most demanding challenges in contemporary meteorology. The region encompasses northeastern India, eastern Nepal, Bhutan, and adjoining parts of the Tibetan Plateau, a domain characterized by extreme topographic undulations, high to extreme precipitation variability, and complex atmospheric processes that interact over multiple scales (Barros et al. 2006; Thayyen and Gergan 2010). These geographical and atmospheric system related complexities have been the biggest challenge towards the development of robust weather forecasting systems over the region, despite significant advancements in numerical weather prediction (NWP) models, network of observatories, and space based observation systems at a global scale.

The eastern Himalayas stretch across an area of approximately 500,000 sq.km, with altitudes ranging from 10 m in some of the low lying river valleys of Assam to over 8,000 m above mean sea level in the high mountain peaks of Arunachal Pradesh, Bhutan, and eastern Nepal (Immerzeel et al. 2014). This steep altitudinal gradient over a relatively short horizontal distance gives rise to sharply varying microclimates, where temperature, humidity, wind vector, and precipitation can differ dramatically within a few kilometers. Orographic uplift and synoptic-scale systems frequently interact with the local terrain, intensifying deep convective activities and eventually leading to abrupt and localized weather phenomena, such as cloudbursts, intense thunderstorms, and heavy snowfall (Bhutiyani et al. 2010).

The region receives a substantial portion (as high as more than 80%) of its annual precipitation, ranging from 1,000 mm in some of the rain shadow areas like the central Assam region to more than 5,000 mm in the southern foothills of Meghalaya during the Southwest summer monsoon, which spans from June to September (Bookhagen and Burbank 2010). The region also hosts the place that receives world's highest rainfall, at Mawsynram, a village in Meghalaya, India (annual average rainfall of 11,872 mm). In addition to orographic rainfall, the eastern Himalayas are also prone to frequent mesoscale convective systems that manifest as severe thunderstorms, often accompanied by intense lightning, heavy rain, and damaging winds, sometimes exceeding 100 km/h (Das et al. 2019a; Dikhsit and Dikshit 2014). During the pre-monsoon months of March to May, the western parts of the regions become a global hotspot of lightning. These weather systems are further complicated by interactions with tropical low pressure system, western disturbances, and local diurnal heating processes, contributing to a complex and highly variable atmospheric environment.

The implications of this extreme weather variability are very deep over the region in terms of socio-economic conditions, ecology, etc. The region is home to millions of people and hosts sensitive ecological zones, critical biodiversity hotspots, and fragile glacial and permafrost systems. Natural hazards such as landslides, flash floods, cloudbursts, and glacial lake outburst floods (GLOFs) are common, driven by intense rainfall events and unstable geological conditions (Bajracharya and Mool 2009; Shrestha et al. 2025). More than 80% of natural disasters in the region are attributed to hydro-meteorological origin, with high social and economic costs, particularly in remote and under-privileged communities (Das et al. 2019b). In recent years, climate variability and anthropogenic influences, including deforestation and unplanned development, have further exacerbated the vulnerability of the region to extreme weather events.

In this context, accurate and timely weather forecasting over such complex region is not just a scientific pursuit but a critical necessity for disaster risk reduction, sustainable development, agricultural planning, protection of lives and property, and many other important causes. However, achieving this goal over mountainous terrain like the eastern Himalayas is far from straightforward. The sparse distribution of surface meteorological stations and the limited coverage of ground based weather radar networks implies that many high-altitude and remote areas remain virtually without any measurement (Barros and Lang 2003; Houze 2012). Ground based observations, while invaluable, are often logistically challenging to install and maintain in inaccessible terrains. Moreover, traditional numerical models, which are typically optimized for flat or gently undulating topographies, often struggle to resolve the fine-scale processes driving mountain weather (Weisman et al. 2008).

It is here that space-based observations comprising a range of Earth-observing satellites including meteorological satellites operating in different orbits and using various instruments offer an unparalleled advantage. Satellites provide repetitive and near-real-time data across multiple spatial and temporal scales, thus helping bridge the observational gaps in mountainous terrain (Kidder and Vonder Haar 1995; Kumar and Kamra 2012). In recent decades, significant progress has been made in deploying satellite instruments capable of monitoring key atmospheric parameters such as temperature, humidity, wind, cloud properties, and precipitation, with increasing spatial and temporal resolution.

Satellites like the INSAT 3D and Kalpana-1 from India, Polar-orbiting satellites such as Megha-Tropiques, NOAA (National Oceanic and Atmospheric Administration) series of satellites, MetOp, and Suomi-NPP series, as well as research satellites like NASA's Terra, Aqua, and TRMM (Tropical Rainfall Measuring Mission), have contributed invaluable datasets for hydrometeorological studies over mountainous regions. Currently, geostationary satellites such as INSAT-3DR/3DS and Himawari-8 offer high-frequency (collectively every 15 min or better) imagery for nowcasting and convective system tracking over the Eastern Himalayan region. The launch of GPM-core satellite (Global Precipitation Measurement) with dual frequency radar, Aeolus (for wind profiling), and EarthCARE (Earth Cloud, Aerosol, and Radiation Explorer) mission further enhances capabilities to monitor mountain weather with finer details, improved accuracy, and spatial coverage (Hou et al. 2014). This sector is seeing a lot of private participation as well both towards improving the ground based observation network as well as improving the space based observations with series of space based radar and microwave radiometers.

One of the most significant contributions of space technology to mountain weather prediction is in the area of precipitation estimation. Orographic enhancement of precipitation poses a challenge for both radar and ground stations due to partial beam blockage and undercatch issues. Satellite-based rainfall retrieval algorithms, particularly those using passive microwave and dual-frequency radar techniques (as in GPM), have shown marked improvement in detecting and quantifying rainfall over rugged terrain (Skofronick-Jackson et al. 2017). These satellite datasets serve as inputs for data assimilation in NWP models, bias correction of regional climate simulations, and validation of hydrological and landslide forecasting models (Yatagai et al. 2012). The upcoming network of space based radars and microwave radiometers is expected to improve the precipitation estimates further.

Satellite-based lightning detection systems, such as GLM (Geostationary Lightning Mapper) onboard GOES-R, Lightning Mapping Imager (LMI) onboard FY-4A and ISS (International Space Station, was available till November 2023), and lightning sensors under development by ISRO, are becoming increasingly important in nowcasting severe thunderstorms over regions like the eastern Himalayas, where total lightning activity often precedes severe weather onset. When integrated with Doppler weather radars, satellite images, and numerical models, these datasets offer a potential leap forward in early warning systems (Zhang et al. 2023; Hari Prasad et al. 2024). Further, the use of satellite derived high-resolution digital elevation models (DEMs), such as SRTM, TanDEM-X, and CartoDEM, enables better representation of terrain in numerical models and hydrological assessments. Remote sensing data also help identify vulnerable regions prone to landslides, soil erosion, and glacial lake development, thus contributing to multi-hazard risk mapping and mitigation planning (Frey et al. 2012).

Despite these advancements, challenges remain. The spatial resolution of satellite sensors, especially passive microwave instruments, is often too coarse to resolve small-scale convective cells in narrow valleys. The absence of space based radar with high revisit limits the complete understanding of cloud system over the hilly region. The geostationary orbit based satellites with imager and sounder is unable to resolve the clouds over the orography dominated regions. Additionally, high-frequency revisit times are essential for rapidly evolving weather phenomena, but no single satellite system can meet all these demands alone. Hence, there is a growing need for integrated observation systems combining data from satellites, ground-based instruments, airborne platforms, and model outputs, supported by machine learning techniques and advanced assimilation frameworks (Ravela et al. 2007; Karpatne et al. 2017).

In summary, the eastern Himalayas present an environment of high meteorological variability, low ground-based observation density, and heightened disaster risk. Space technology, with its ability to observe large, inaccessible areas with consistent accuracy, offers a powerful tool for enhancing weather prediction over these mountainous terrains. However, realizing its full potential will require sustained investments in sensor development, constellation design tailored for orographic regions, data assimilation systems, and regional capacity building. The subsequent sections of this chapter will explore the role of space-borne platforms in detail, highlight case studies over the north eastern region (NER) of India using NWP models, and suggest future directions to improve operational weather forecasting and climate resilience in mountain regions.

21.2 Challenges of Weather Prediction in Mountainous Regions

21.2.1 Complex Topography

Topography plays one of the most important role in weather forecasting by influencing wind patterns, generating turbulence, modifying the cloud micro-physical properties, abruptly changing the temperature and pressure variations, and modifying the storm movement. Mountains force moist air to rise, causing orographic rainfall on windward slopes and creating rain shadows on the leeward side. This is most important for the eastern Himalayan region. Figure 21.1 shows the elevation profile of the entire region that varies from 5 m to more than 8000 m and the annual rainfall over the NE region of India that is part of the eastern Himalayan region.

The Bay-of-Bengal acts as a vast resource of moisture for the region which is further augmented by the evaporation from the large water bodies and high moist soils in this region and evapotranspiration from the dense forest. The complex terrain in Eastern Himalaya can channel or block storms, intensify local winds, and trigger

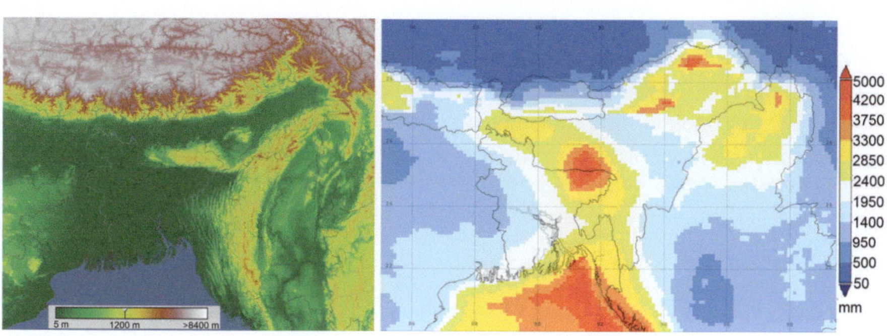

Fig. 21.1 Digital elevation model (DEM) of the Eastern Himalayas (in left) and climatology of annual rainfall over the NE region of India and its surroundings

extreme weather events, while valleys and slopes affect local wind and temperature variations due to differential heating. Such type of terrain can create a microclimatic effect that can lead to highly localized weather patterns (e.g., abrupt changes in temperature and precipitation over short distances) that are difficult to capture in broad-scale atmospheric models (Barros et al. 2018). The annual rainfall over the NER of India, that is a part of the eastern Himalayan region, closely follows the elevation profile with heavy rainfall seen along the foothills regions of the windward side. The primary wind during the monsoon seasons comes from south and south-west direction giving heavy rainfall over the southern Meghalaya, Himalayan foothills regions in Assam and sub-Himalayan West Bengal.

21.2.2 Sparse Data Networks

Observations are fundamental to atmospheric science and weather forecasting, serving as the backbone for understanding atmospheric processes and improving predictive capabilities. Accurate and timely observations from ground stations like automatic weather stations, radiosondes, radars, aircraft, and satellites are essential for monitoring weather systems, initializing NWP models, and validating forecasts (Kalnay 2003). The assimilation of observational data significantly enhances model accuracy, particularly in data-sparse regions such as mountainous terrains and oceans (Bauer et al. 2015). However, establishing and maintaining meteorological stations in remote, high-altitude regions is an extremely challenging task. The eastern Himalayan region, particularly the NER of India, suffers from data scarcity, particularly the ground based data, reducing the accuracy of conventional meteorological methods that rely on ground-based observations (Immerzeel et al. 2014). This is where satellite-based observations come into play as the major source of data. The density of observatories has always been sparse over the Eastern Himalayan region due to various factors like, inaccessible terrain posing serious difficulty in installing and maintaining observatories, few scientific studies taken up over this region for a long period, etc. The scenario changed over last one decade with significant improvement in instrumentation. The Monsoon Mission by the Ministry of Earth Sciences envisages enhancing this network further with installation of weather stations, weather radars, wind profilers, etc. ISRO and India Meteorological Department installed and operates a series of ground based instruments over NER of India. Figure 21.2 provides a comprehensive spatial distribution and coverage of surface based weather observatories over the Indian part of the Eastern Himalayan region as of 31 December, 2024.

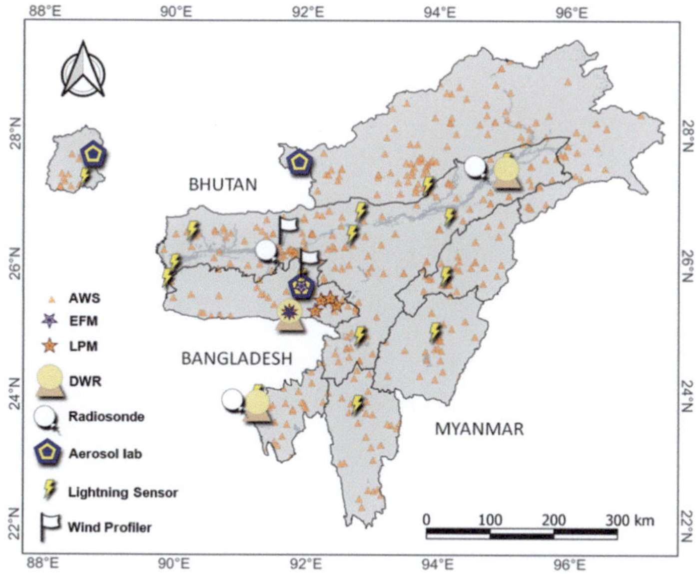

Fig. 21.2 Spatial distribution of weather observatories across the eastern Himalayan region within India

21.2.3 Rapidly Changing Weather Patterns

The eastern Himalayan region, particularly the southern part of the region, being in the Tropical belt, often experiences rapid shifts in weather, which are challenging to monitor and predict due to the influence of regional circulations and the interaction with the Indian monsoon system. This variability contributes to the frequent occurrence of extreme events that affect local populations severely (Thayyen and Gergan 2010). Cloud burst, severe storms and lightning, hails storms, small scale tornadoes, etc. are very common over this region that has caused some of the most severe disasters over this region. The complex topography and data scarcity adds to the problem. Given these challenges, space technology has emerged as a powerful tool to complement and enhance traditional meteorological observations. Satellite remote sensing, GPS-based measurements, and other space-based techniques provide a broader range of data that has the potential to improve the accuracy and timeliness of weather prediction.

21.2.4 Impact of Climate Change

Climate change is adding another major challenge to accurate weather forecasting over the eastern Himalayan region by increasing the frequency of extreme and localized events, such as intense rainfall and cloudbursts, which are difficult to predict due

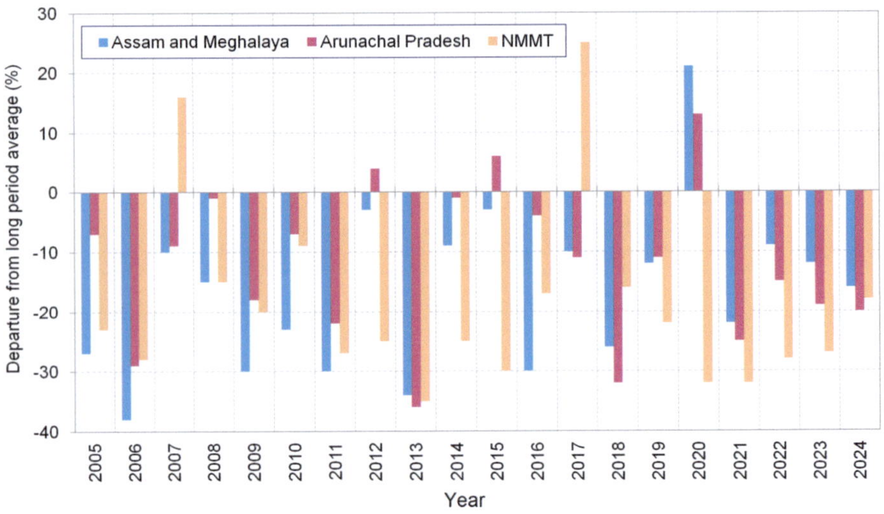

Fig. 21.3 Departure of monsoon rainfall over the three meteorological sub-divisions of the NER of India from long period average

to complex terrain and limited observations (Dimri et al. 2016). Warming temperatures alter snow cover, glacier dynamics, and higher moisture availability because of global warming, affecting local circulation and eventually model accuracy. Shifting monsoon patterns and climatic baselines further reduce the reliability of historical data-driven models. Figure 21.3 shows the departure of monsoon rainfall over the three meteorological sub-divisions (Assam & Meghalaya, Arunachal Pradesh, and Nagaland-Manipur-Mizoram-Tripura) of the NER of India that has been consistently less than normal, often lass by 30% or more. This is happening while the extremely heavy rainfall events are increasing in number (please refer to Chapter 22) over this region. Such dynamic and changing pattern will continue to pose additional challenge in forecasting and call for frequent adjustment in the models to capture the changing pattern.

21.3 The Physics of Mountain Meteorology

Mountain meteorology is fundamentally distinct from meteorology over flat terrain due to the strong interactions between complex topography and atmospheric processes. The presence of rugged terrain, steep slopes, deep valleys, and high mountain peaks significantly modifies the structure and evolution of weather systems. In the Eastern Himalayas, these interactions are particularly pronounced due to the region's extreme elevation gradient, proximity to the Bay of Bengal, and influence of both tropical and extratropical systems.

Mountains can have a multi-faceted effect on both weather and climate. Their effects depend on the physical dimensions and location of the mountains. The mountains influence the mean flow of large scale weather patterns by affecting their physics as well as their dynamics. Studies by Banta et al. (1990), Bruintjes et al. (1994), and Smith (1979) have all emphasized on the importance of mountains in influencing the vertical motions of air which are strongly related to cloud formation and precipitation. The height to which a body of air must be raised before condensation begins depends upon both the temperature and humidity of the air at sea level. Hence, in the regions of high humidity comparatively low mountains also become important agents in bringing about rainfall, whereas in regions of low humidity even very high mountains may have little influence. The job of the mountains is to force the warm moist winds into higher altitudes whereby expanding and cooling it and eventually bringing about condensation and precipitation. Mountains can only assist in the cloud formation and rainfall generation processes, their location and height are not the sole parameters affecting these processes. This section explores the major physical processes that define mountain meteorology and discusses their implications for the Eastern Himalayan region.

21.3.1 Orographic Forcing and Precipitation Enhancement

One of the primary drivers of mountain weather is orographic lifting, where air masses are forced to ascend as they encounter elevated terrain. The impact of altitude in modulating the local weather and climate can be easily quantified using the Hydrostatic equation (Eq. 21.1) that expresses the balance between the vertical pressure gradient force and the gravitational force in the atmosphere and gives an idea about how atmospheric pressure decreases with altitude under equilibrium conditions.

$$\frac{dP}{dz} = -\rho g \tag{21.1}$$

where, P is atmospheric pressure, z is altitude, ρ is air density, and g gravitational acceleration.

When moist air from lowlands (e.g., Bangladesh plains with respect to the NER of India) is forced to rise over foothills or mountain slopes, the vertical motion increases, disturbing the hydrostatic balance locally. As air rises, pressure decreases with height, and the air expands and cools. This adiabatic cooling leads to condensation and cloud formation once the dew point is reached. The continued uplift and moisture supply cause intense precipitation, especially if the atmosphere is already conditionally unstable. This process often results in orographic precipitation, particularly on the southern flanks of the Eastern Himalayas where moist monsoonal winds from the Bay of Bengal are intercepted (Barros and Lettenmaier 1994). However, this precipitation occurs on the windward (upwind) side of the mountain. After crossing the mountain, the air loses moisture and is forced to move down the slope of the mountain on the opposite side. While descending, the air becomes warm due to adiabatic compression. So it further dries up and loses more moisture resulting in less precipitation in the leeward (downwind) side of the mountain. This region is called a rain shadow area.

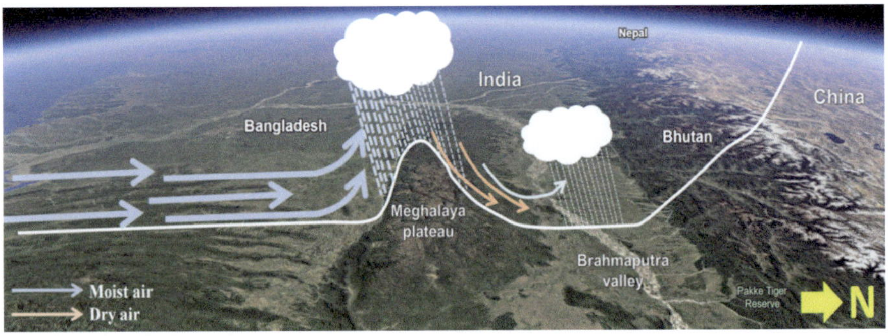

Fig. 21.4 Orographic lifting over the southern Meghalaya leading to heavy rainfall over the region and relatively less rainfall in the leeward regions. The image shows the Meghalaya plateau and parts of the Eastern Himalayan region (*Source* Google maps)

Figure 21.3 depicts the process of orographic lifting along the southern Meghalaya causing heavy rainfall over the foothill regions and significantly reduced rainfall in the leeward side of the wind.

The impact is so severe that many places along the Southern part of Meghalaya receive more than ten thousand mm of rain annually including having the world's highest rainfall place. Regions like southern Bhutan, Sikkim, Arunachal Pradesh, sub-Himalayan West Bengal and parts of eastern Nepal also receive excessive rainfall during the monsoon due to persistent orographic uplift. Studies have shown that localized rain intensification due to terrain effects can lead to daily rainfall totals exceeding 500 mm, contributing to flash floods and widespread landslides (Bookhagen and Burbank 2010). Additionally, barrier jets, low-level winds that flow parallel to mountain ridges can enhance moisture convergence and further amplify precipitation, occasionally leading to cloud burst like events (Fig. 21.4).

21.3.2 Diurnal Circulations: Mountain-Valley Winds

Diurnal heating and cooling of mountain slopes generate thermally driven circulations, commonly known as anabatic (upslope) and katabatic (downslope) winds. During the day, sunlit slopes heat up faster than adjacent valleys, causing air to rise upslope, drawing cooler valley air upward. At night, the process reverses, with cooler, denser air flowing downslope under gravity. In the Eastern Himalayas, these mountain-valley circulations play a crucial role in transporting moisture and pollutants and can initiate late afternoon convective storms due to localized convergence over ridges and summits. These circulations can also interact with large-scale flows (like monsoonal flow and westerlies), enhancing or suppressing convection depending on the alignment of local and regional winds (Whiteman 2000). Figure 21.5 depicts the impact of anabatic and katabatic flow in local scale weather

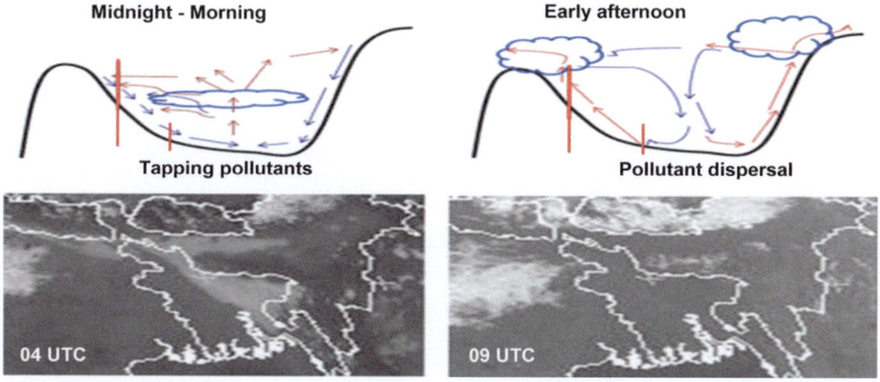

Fig. 21.5 Impact of anabatic and katabatic flow in local scale weather and pollution dispersal as seen over the hills and valley region that was confirmed by the Meteosat satellite data

and pollution dispersal, that could be observed during a intense campaign wherein series of GPS radiosonde was launched from Umiam, Meghalaya and Guwahati, Assam (marked as the vertical red lines). The Meteosat satellite data confirmed the presence of clouds over hills and valley regions during the times, as was predicted by the process.

21.3.3 Atmospheric Stability and Vertical Motion

Mountain slopes modify atmospheric stability through differential heating and surface fluxes. Convective available potential energy (CAPE) can be significantly higher on sunlit slopes, triggering deep convection when local uplift surpasses inhibition. Conversely, stable layers (such as temperature inversions) can become trapped in narrow valleys, leading to fog formation, cold air pooling, and pollutant accumulation. The Eastern Himalayan valleys, such as the Siang, Tawang, and Teesta valleys, often experience inversion layers during winter, particularly under clear-sky, low-wind conditions. These inversions are critical for forecasting radiation fog, frost, and poor air quality, especially in urban centers like Gangtok or Itanagar. The inversions, during the early pre-monsoon season over parts of the region, can tap huge amount of energy and cause formation of rapid severe weather systems once the inversion is broken.

21.3.4 Gravity Waves and Flow Blocking

When airflow encounters a mountain barrier, the stability of the atmosphere determines whether the air flows over the mountain or is diverted around it. Stable air tends to be blocked, while unstable air ascends and forms gravity waves, which is a kind of oscillations in the vertical caused by buoyancy. These waves can influence upper-level flow and trigger turbulence, especially at crest level and in the lee of ridges. Gravity wave breaking can lead to severe turbulence affecting aviation safety and create lee-side precipitation enhancement through wave-induced uplift. Flow blocking also leads to asymmetric precipitation patterns, with suppressed rainfall on the leeward side and heavy rainfall on the windward slope, often referred to as rain shadow effect (Smolarkiewicz and Rotunno 1989).

21.3.5 Cold Air Damming and Valley Cold Pools

Cold air can become trapped in high mountain valleys, especially during winter and pre-monsoon months. This leads to cold air damming, where dense air pools in valleys and basins, resisting mixing with warmer air above. These cold pools can influence cloud microphysics, leading to mixed-phase clouds, freezing drizzle, and enhanced snowfall in some areas. Over eastern Himalayan valleys, such phenomena are linked to seasonal snow events in places like Tawang or Lachen and to foggy conditions over lower valley areas such as in the Brahmaputra valley. These processes are also important for modeling glacier mass balance, as they affect local energy and moisture fluxes.

21.3.6 Convective Initiation and Mesoscale Storms

The complex terrain of the Eastern Himalayas frequently acts as a trigger mechanism for convection. Surface heating over the region, convergence zones formed by colliding of the warm and moist airmass with the dry and cool airmass, and upslope moist flow create ideal conditions for convective initiation. Orography can lock convection in place, leading to quasi-stationary storms and cloudbursts. These local scale convective systems often go undetected by coarse-resolution models, distant radars, or even in satellite images that are taken from geostationary orbits. The result is severe thunderstorm outbreaks, hailstorms, and localized flash floods, especially during the pre-monsoon and early monsoon periods (Shrestha et al. 2015). Satellite-based lightning detection, high-resolution satellite data in thermal bands, dense network of weather radars to cover the entire region, etc. are essential tools to capture these rapid developments.

21.3.7 Interaction with Large-Scale Systems

The Eastern Himalayas lie at a meteorological crossroads, where tropical monsoon dynamics interact with mid-latitude westerlies and occasional tropical cyclones. These interactions can amplify rainfall dramatically. For example, remnants of Bay of Bengal cyclones often reach the foothills and produce prolonged rainfall under blocked flow conditions. Similarly, western disturbances, when deflected southward, can bring snowfall and rain during winter. The presence of topography modulates these large-scale systems, producing orographically anchored vortices, stalled fronts, or enhanced jet streaks, which lead to persistent precipitation. These interactions are among the most challenging to simulate in weather models and are a major reason for forecast uncertainty in the Eastern Himalayas (Dimri et al. 2015).

21.4 Space Technology for Meteorological Observation Over Mountains and Weather Prediction

Space technology, primarily through satellite-based observations and remote sensing, has become indispensable in addressing the challenges of weather prediction over mountainous regions, like the Eastern Himalaya. Satellites equipped with sensors that provide high spatial and spectral resolution data on various meteorological parameters, including temperature, humidity, cloud cover, and precipitation, which are otherwise difficult to obtain in these regions (Sorooshian et al. 2020). Space based observation is of importance and strategic in nature also because this platform provides data across the international border, which otherwise is difficult to achieve in absence of international collaboration. Advanced meteorological satellites such as INSAT-3D/3DR/3DS (India's geostationary satellites) and international satellites like NOAA series, GOES, FY-4, and Himawari are increasingly used to track and monitor weather systems across the Himalayas and adjacent regions. The countries across the globe has been increasing its fleet of meteorological observatories in geostationary platform as well as low earth orbits to ensure high resolution and high repetitive data availability to the researchers and operational meteorologists.

International collaboration like the Global Precipitation Measurement (GPM) mission led by NASA and JAXA to provide advanced observations of rain and snow worldwide, has improved the understanding of the mountain meteorology. Launched in 2014, GPM's core observatory carries a Dual-frequency Precipitation Radar (DPR) and a GPM Microwave Imager (GMI), enabling high-resolution, three-dimensional measurements of precipitation structure and intensity (Hou et al. 2014). The mission unifies data from a constellation of partner satellites, offering near-global coverage every 30 min. GPM plays a crucial role in weather forecasting, climate research, hydrological modeling, and disaster monitoring by improving understanding of precipitation processes and enhancing prediction of extreme weather events.

Megha-Tropiques was another joint Indo-French satellite mission launched in 2011 by ISRO and CNES to study the water cycle and energy exchanges in the tropical atmosphere. Equipped with instruments like SAPHIR (a humidity sounder), MADRAS (a microwave imager), and ScaRaB (a radiometer for radiation budget), the satellite provided valuable data on tropical humidity, cloud dynamics, precipitation, and radiation fluxes (Desbois et al. 2007). Operating in a low-inclination orbit, the satellite enabled frequent observations over the tropics, enhancing understanding of monsoon processes, convective systems, and intra-seasonal oscillations including those over the complex mountainous regions of the Eastern Himalaya. The space platform provides several advantages and allows several activities to be carried out, which are highlighted below:

Global coverage: Satellites provide continuous, near-global observations, including remote and oceanic regions where ground stations are sparse or absent. The global to regional scale studies on weather, climate, and environment heavily depends on data from satellites.

Cloud and precipitation monitoring: Instruments like radiometers and radars onboard satellites help study clouds in three dimensions, detect cloud micro-physical properties, precipitation intensity, and storm development in real time.

Atmospheric profiling: Satellites based sounders can measure vertical profiles of temperature, humidity, and wind that are essential for initializing weather models and assess the atmospheric conditions for several other studies.

Tracking storms and cyclones: Geostationary satellites enable continuous monitoring of tropical cyclones and severe convective weather systems allowing their early warnings and impact assessments. The advent of space based observations have drastically improved the landfall forecasting of severe cyclones across the globe including in India leading to thousands of lives saved from such high impact weather events. Satellites with radar and multi-spectral sensors provide crucial data on indicators of extreme weather, like heavy rainfall or impending cloudbursts. This capability provides critical lead time for early warnings, which are essential for disaster preparedness and risk reduction in the Himalayan region (Das et al. 2019b). This is particularly useful in mountainous regions, where traditional monitoring is limited. The satellite radar based reflectivity (Z) can be used to estimate the rainfall (R) once the region specific constants like "a" and "b" are estimated using the Marshal-Palmer relationship as shown in Eq. 21.2.

$$Z = aR^b \tag{21.2}$$

where, Z is radar reflectivity, R is rainfall rate, a and b are empirical constants that are characteristic to the region for which it has been developed.

Surface observations: Satellites data can be used to estimate the land surface temperature, soil moisture, snow cover, sea surface temperature, total perceptible water, wind at different levels, and many other key parameters for weather prediction.

With enhanced utilization of artificial intelligence in geophysical parameter retrieval algorithm, the accuracy of such estimates using satellite data has been continuously improving, leading to such data being increasingly used for climate studies.

Data assimilation for NWP: Satellite data are assimilated into Numerical Weather Prediction models have been the major cause of continuous improvement in forecast accuracy and reliability (Phulpin et al. 2023). Such data are assimilated in the global circulation models as well as in regional models.

Monitoring rapid changes: The ability of high-frequency measurement makes satellite platform one of the most sought after data source for monitoring rapid changes at a meso-scale to higher scale events. Satellites like Himawari can provide data over a large part of the globe at every 10 min in multiple channels that can capture rapidly evolving weather systems like thunderstorms, cloudbursts, and frontal systems.

Support for disaster management: In today's world, satellite data is indispensible for disaster risk reduction. Real-time satellite data help assess and respond to extreme weather events such as floods, lightning, landslides, etc.

All major countries including India have its own meteorological satellite system. There are constellations of satellites as well by multiple counties as well as by Industries. Table 21.1 provides a comprehensive list of public funded satellites across the globe that can be directly or indirectly used for meteorological studies over the Eastern Himalayan region.

Satellites have come a long way in observing clouds and precipitation processes which help in the understanding of precipitation processes over the mountainous regions. The near infrared channel of 1.6 μm (NIR1.6) and 3.9 μm (NIR3.9), infrared channels of 10.8 μm (TIR10.8) and 12.0 μm (TIR12.0) are the examples of some of the most popular and useful channels to study clouds. The near infrared and infrared channels are effectively used for differentiating the ice clouds and water clouds and also for estimating precipitation. Rosenfeld and Gutman (1994) showed that using the information from visible and the difference between 10.7 μm and 6.9 μm the cirrus clouds can be differentiated from the raining clouds and using a combination of 3.9 μm, 10.7 μm and 12.0 μm during the day time, the cloud particle sizes can also be estimated. These abilities of NIR and TIR channels have been used in many satellites to monitor clouds, atmospheric motion vectors and precipitation.

The INSAT-3DR imager is an improved design of VHRR-2 (Very High Resolution Radiometer) which was present onboard the heritage satellites Kalpana-1 and INSAT-3A. The INSAT-3DR imager provides imaging capabilities of the Earth disc from the geostationary altitude in one visible (0.52–0.72 μm) and five infrared channels: 1.55–1.70 μm (SWIR), 3.80–4.00 μm (MIR), 6.50–7.00 μm (Water Vapor), 10.2–11.2 μm (TIR-1), and 11.5–12.5 μm (TIR-2) bands. The ground resolution of the satellite is 1 km × 1 km for Visible and SWIR, 4 km × 4 km for MIR, TIR-1, and TIR-2 and 8 km × 8 km for the WV channel. Along with INSAT-3DR and the follow-up mission of INSAT-3DS with improved version of the same instruments, is able to monitor the Indian subcontinent at every 15 min.

Table 21.1 List of global satellites that provide data over the Eastern Himalayan region

Satellite name	Launch year	Country/ agency	Main payloads	Major applications in mountain weather
INSAT 3D/ INSAT-3DR	2013/ 2016	India/ISRO	6 channel Imager and 19 channel (18 IR and one VIS) Sounder	Cloud motion vectors (CMV), vertical profiles of T & RH, precipitation, storm and cyclone monitoring
INSAT-3DS	2024	India/ISRO	6 channel imager and 19 channel (18 IR and one VIS) Sounder with advanced configuration	CMV, vertical profiles of T & RH, precipitation, storm and cyclone monitoring
SCATSAT-1	2016	India/ISRO	OSCAT (Ku-band scatterometer)	Ocean surface wind vectors, important for NWP
EOS-06 (Oceansat-3)	2022	India/ISRO	OCM-3, Ku-band Scatterometer-3), and SSTM	Ocean wind, ocean color, sea surface temperature, useful for monsoon system analysis
Megha-Tropiques	2011	India/France (ISRO/CNES)	SAPHIR, ScaRaB, MADRAS	Energy balance, humidity, precipitation, radiation, monsoonal convection
TRMM (tropical rainfall measuring mission)	1997	NASA/JAXA	Precipitation radar, TMI, lightning imaging sensor	Rainfall intensity, vertical structure of storms (used until 2015), Lightning
EarthCARE	2024	JAXA/ESA	ATLID, CPR, MSI, BBR	Vertical distribution of aerosols and clouds, cloud micro-physical property, cloud and aerosol optical properties, Radiation, etc
GPM-core (global precipitation measurement)	2014	NASA/JAXA	Dual-frequency precipitation radar (DPR), GMI	Accurate precipitation estimates over complex terrain, essential for hydrological modeling
Suomi NPP	2011	USA (NOAA/ NASA)	VIIRS, CrIS, ATMS, OMPS, & CERES	Cloud cover, T & RHy profiles, snow cover, aerosol, fire detection
NOAA-20 (JPSS-1)	2017	USA (NOAA/ NASA)	VIIRS, CrIS, ATMS, OMPS-N, CERES-FM6	Weather forecasting inputs, especially for data assimilation in NWP models

(continued)

Table 21.1 (continued)

Satellite name	Launch year	Country/ agency	Main payloads	Major applications in mountain weather
MetOp Series	2012 (B), 2018 (C)	ESA/ EUMETSAT	IASI, ASCAT, GRAS	Wind vectors, atmospheric sounding, sea ice, NWP input for global models
Himawari-8	2014	Japan (JMA)	AHI (Advanced Himawari Imager) with 16 channels (500 m to 2 km resolution)	10-min interval high-res imaging, nowcasting, cloud analysis over eastern Himalayas
FY-4A/B	2016/ 2021	China (CMA)	AGRI, GIIRS, LMI, SEP	Lightning studies, vertical profiling of T and RH, Cloud monitoring and tracking, space weather
Aeolus	2018	ESA	Atmospheric laser Doppler instrument (ALADIN)	Wind profile measurements globally, especially valuable over data-sparse mountain areas
Sentinel-5P	2017	ESA	TROPOMI	Trace gases, aerosols, important for studying atmospheric dynamics over the Himalayas
Terra/Aqua	1999/ 2002	NASA	MODIS, CERES, AIRS	Snow cover, temperature, vegetation, aerosols—supports multi-parameter monitoring
Landsat-8/9	2013/ 2021	USA (USGS/ NASA)	OLI, TIRS	Snow and land cover monitoring, useful for hydrology and terrain-atmosphere interactions
CloudSat	2006	NASA	Cloud profiling radar (CPR)	Improve understanding of cloud properties, distribution, and their role in climate
CALIPSO	2006	NASA/CNES	CALIOP, IIR	Vertical distribution of clouds and aerosols, radiation budget

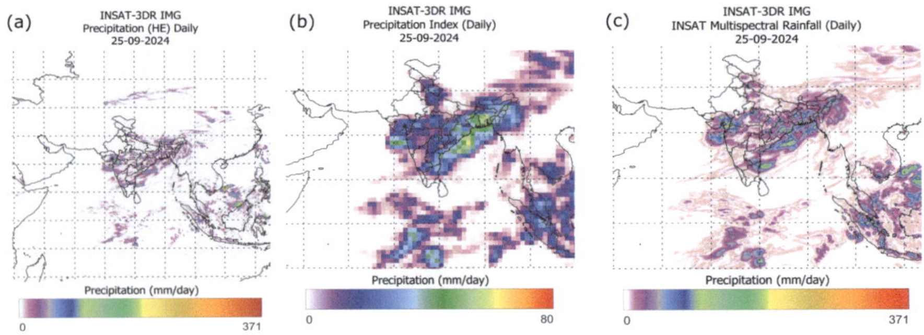

Fig. 21.6 Daily accumulated INSAT-3DR rainfall for 25–09-2024 from **a** HEM, **b** GPI and **c** IMR (*Source* MOSDAC)

Using the near infrared channels and the water vapor channel available in INSAT-3DR and –3DS, three quantitative precipitation estimates are generated for over the Indian sub-continent using three different retrieval schemes, namely, GOES Precipitation Index (GPI: Arkin 1979), INSAT Mulit-Spectral Rainfall Algorithm Technique (IMSRA: Ba and Gruber 2001) and Hydro-Estimator (HEM: Scofield et al. 1987). It was observed by Mitra et al. (2018), that HEM has higher accuracy in observing heavy rainfall, while QPE and IMR products which correspond to the GPI and IMSRA approaches, respectively are better at observing light to moderate rainfall. A sample rainfall estimate using all three methods is shown in Fig. 21.6. However, the performance of all algorithms remains poor over the mountainous regions.

Considering the inconsistencies associated with the use of infrared channels for cloud and precipitation studies, the Tropical Rainfall Measurement Mission (TRMM) and its successor, the Global Precipitation Mission (GPM) were launched with a microwave imager and radar. The founding concept of GPM is the utilization of precipitation observations contributed by various organizations with the goal of generating a spatiotemporally consistent precipitation estimate. The GPM mission is divided into two aspects, one is the GPM core satellite (GPM-CO) and another is the GPM constellation of satellite. It is an advanced successor to the TRMM and consists to a dual frequency precipitation radar (DPR: 35.5 GHz and 13.6 GHz) and a 13 channel (10–183 GHz) microwave radiometer (GPM Microwave Imager: GMI). The DPR in GPM makes detailed 3D measurements of the clouds and the GMI retrieves heavy, moderate and light precipitation measurements. The DPR is used to provide 3D measurement of rain rate and drop size distribution, which is essential for classification of rainfall, for example into stratiform and convective rainfall (Gao et al. 2017). These informations of rainfall are essential for mountainous regions.

The GMI is used in the Integrated Multi-SatellitE Retrievals for GPM (IMERG) to inter-calibrate data from other microwave radiometers available in the GPM constellation (Huffman et al. 2015). The gaps in the microwave data are filled by morphing the data from Infrared estimates using Kalman Filter. The IMERG product is available at a spatial resolution of $0.1^\circ \times 0.1^\circ$ and at a temporal resolution of 30

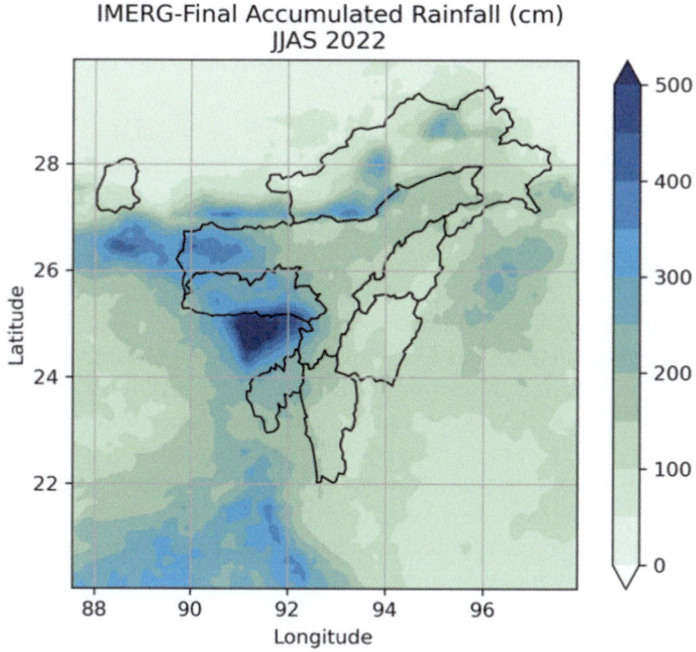

Fig. 21.7 Seasonal accumulated IMERG-final rainfall over the Eastern Himalayan region

min. The IMERG rainfall product consists of a microwave calibrated infrared rainfall (GIR), a merged inter-calibrated microwave product (GMW) and an IMERG infrared microwave merged rainfall product (GMS: MS stands from multi-sensor). The IMERG products are available at three stages of processing: early, late and final. As the IMERG final product is gauge calibrated from the Global Precipitation Climatology Centre (GPCC) data, it is observed to be the best among the three. Figure 21.7 shows accumulated rainfall over the Eastern Himalayan region during the monsoon season of 2022. Data from such space based sources are practically the only reliable and available data over most part of the Eastern Himalayan region.

Along with the GPM measurement of rainfall, for investigating the cloud properties, the CloudSat and CALIPSO has been used extensively. CloudSat is a satellite mission launched by NASA in 2006 and has been instrumental in advancing the knowledge on clouds and their roles in Earth's climate system. It is equipped with a 94 GHz Cloud Profiling Radar (CPR), and can penetrate most cloud layers, providing detailed vertical profiles of clouds, its thickness, layering, particle sizes and cloud phases. These informations on the clouds help in the characterization of clouds (Subrahmanyam and Kumar 2013). The Cloud-Aerosol Lidar and Infrared Pathfinder Satellite Observation (CALIPSO) provides high resolution cloud and aerosol information. CALIPSO's lidar can distinguishes between cloud layers, aerosol, and the surface, offering critical vertical profile of the cloud structure.

21.4.1 *Improving NWP Over Mountainous Regions Using Satellite Data*

Weather prediction is crucial for numerous aspects of human life and safety. In a global context, weather-related phenomena are responsible for a significant portion of natural disasters worldwide. According to a report by the United Nations Office for Disaster Risk Reduction (UNDRR), between 2005 and 2024, weather events have affected millions of people. During this period, floods impacted 41% of those affected, equating to 1.65 billion people, while droughts affected 35%, or 1.43 billion people. Additionally, storms impacted 18%, or 727 million people, with the remaining population affected by other disasters such as earthquakes, extreme temperatures, landslides, wildfires, and volcanic activities. While we cannot prevent natural disasters from occurring, we can mitigate their impact through the implementation of effective early warning systems. Fortunately, weather prediction, to a large extent, is possible today due to advancements in sophisticated weather prediction models. These models provide vital information that helps governments, communities, and industries prepare for and respond to impending weather-related threats, ultimately saving lives and reducing economic losses.

21.4.2 *Numerical Weather Prediction Models*

Numerical Weather Prediction (NWP) models form the foundation of weather forecasting by simulating the complex behavior of the atmosphere through mathematical equations. At the core of these models is the task of solving differential equations that govern atmospheric dynamics primarily based on principles of fluid dynamics, thermodynamics, and radiation. These equations describe how air masses move, how heat and moisture are exchanged, and how energy is distributed throughout the atmosphere. By starting with a set of initial conditions such as current temperature, pressure, humidity, and wind, NWP models attempt to predict how these variables will evolve over time. NWP models excel at providing weather predictions across different time scales, each serving specific forecasting needs. Short-term forecasts offer high precision for immediate weather events, medium-term forecasts guide preparations for weather systems evolving over days, and long-term forecasts help anticipate broader climate patterns. Together, they form a comprehensive system that allows meteorologists to track, analyze, and predict weather changes from hours to months ahead, enhancing our ability to plan and respond to varying weather conditions. However, it's important to note that as the forecast time frame lengthens, uncertainty increases. Small errors in the initial conditions of a forecast can amplify over time due to the chaotic nature of the atmosphere (a concept known as the "butterfly effect"). This is why short-term forecasts are generally more accurate and detailed, while long-term forecasts tend to be probabilistic in nature providing likely trends rather than exact predictions.

One of the key factors in improving the accuracy of NWP models is the process of data assimilation, which involves integrating real-time observational data from various sources into the model. Observations from weather stations, radiosondes (weather balloons), satellites, radars, aircraft, and even marine buoys contribute to providing a more accurate snapshot of current atmospheric conditions. This constant stream of fresh data helps correct and update the model's predictions, reducing the uncertainty that comes from relying solely on initial conditions. Additionally, high-resolution NWP models that operate on a finer spatial scale offer even more precise predictions by accounting for regional features that significantly influence local weather. For example, mountainous terrain can cause air to rise, creating orographic precipitation, while coastlines and large bodies of water affect humidity and temperature patterns. Similarly, urban landscapes create heat islands that can alter local temperature and wind dynamics. By incorporating such geographic and man-made features into their calculations, these high-resolution models are better equipped to predict localized weather events like thunderstorms, intense rainfall, fog, or coastal winds.

This section briefly introduces a globally used NWP model named as the Weather Research and Forecasting (WRF) model and its data assimilation module. This section also describes the data used for model initialization as well as data used for assimilation, sensitivity studies, primarily using the space based data. The studies shown here are focused over the NER of India only that is the most complex and dynamic region within the Eastern Himalayan region.

21.4.3 Weather Research and Forecasting Model

The WRF model is an advanced mesoscale, non-hydrostatic NWP system developed through the combined efforts of organizations including NOAA, the Naval Research Laboratory (NRL), the University of Oklahoma, the Air Force Weather Agency (AFWA), and the Federal Aviation Administration (FAA). This model mathematically represents the atmosphere using key atmospheric governing equations, enabling accurate simulations of weather and climate patterns. WRF serves as a flexible platform, supporting real-time weather forecasting, regional climate studies, idealized experiments, data assimilation, and other atmospheric research. Its strength lies in its ability to resolve meteorological systems across a wide range of spatial scales from small, local phenomena to large-scale systems that span thousands of kilometers. The model features two main dynamical cores: the Advanced Research WRF (ARW) and the Non-hydrostatic Mesoscale Model (NMM). The ARW core is managed by the Mesoscale and Microscale Meteorology (MMM) laboratory at NCAR, while the NMM core is maintained by the National Centers for Environmental Prediction (NCEP). Each core offers distinct advantages, allowing the WRF model to be widely used in various meteorological applications. This section focuses on the WRF-ARW core, with a schematic of the framework provided in Fig. 21.8 (Skamarock et al. 2008).

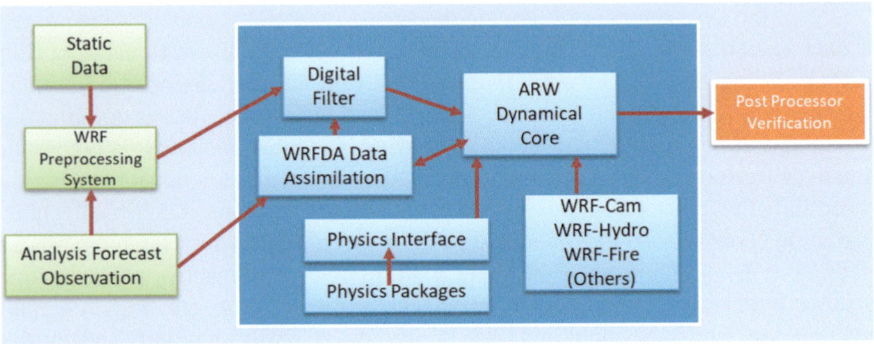

Fig. 21.8 WRF-ARW framework

21.4.3.1 WRF-ARW Core

The WRF Preprocessing System (WPS) prepares geographic and meteorological data, converting them into a format compatible with the model using its three subsystems: Geogrid, Ungrib, and Metgrid. These processed data are then utilized by the model's dynamical cores to solve the governing atmospheric equations. In the ARW solver, the dynamics are based on a fully compressible, non-hydrostatic system of equations in flux form, with an option for hydrostatic balance. This allows the simulation of both large- and small-scale atmospheric motions, making it particularly effective for mesoscale weather predictions, such as thunderstorms, hurricanes, and other severe weather events where vertical acceleration is critical.

The mathematical representations of the flux form equations used by ARW solver such as momentum equations, equation of continuity, equation of thermodynamics, the diagnostic relation for inverse density and equation of state are given in Eqs. 21.3–21.10:

$$\partial_t U + (\nabla.\mathbf{V}u) - \partial_x(p\phi_\eta) + \partial_\eta(p\phi_x) = F_U \tag{21.3}$$

$$\partial_t V + (\nabla.\mathbf{V}u) - \partial_y(p\phi_\eta) + \partial_\eta(p\phi_y) = F_V \tag{21.4}$$

$$\partial_t W + (\nabla.\mathbf{V}w) - g(\partial_\eta p - \mu) = F_W \tag{21.5}$$

$$\partial_t \Theta + (\nabla.\mathbf{V}\theta) = F_\Theta \tag{21.6}$$

$$\partial_t \mu + (\nabla.\mathbf{V}) = 0 \tag{21.7}$$

$$\partial_t \phi + \mu^{-1}\left[(\mathbf{V}.\nabla\phi) - gW\right] = 0 \tag{21.8}$$

$$\partial_\eta \phi = -\alpha \mu \tag{21.9}$$

$$p = p_0 (R_d \theta / p_0 \alpha)^\gamma \tag{21.10}$$

The details about the variables of the above differential equations are given in Table 21.2.

The ARW model uses the Arakawa C-grid staggering method for horizontal discretization and utilizes terrain-following vertical coordinates to define vertical grid spacing. For time integration, the model offers 2nd and 3rd order Runge–Kutta schemes to advance the equations numerically. The model's physics package captures sub-grid scale processes that are not directly resolved by the dynamical core, which operates on discrete grids. These processes are represented through various parameterization schemes, including microphysics, cumulus convection, planetary boundary layer (PBL) dynamics, land-surface interactions, and radiation. These schemes account for key atmospheric phenomena occurring at scales smaller than the model's grid resolution. Specifically, microphysics and cumulus parameterization are key for rainfall modeling. Microphysics parameterization (MP) schemes

Table 21.2 Description of variable used in the flux form equations of ARW solver

Variable	Description
$\eta = (p_d - p_t)/(p_s - p_t)$	Terrain-following hydrostatic-pressure vertical coordinate
p_d	Hydrostatic component of the pressure of dry air
p_s	Represents values of p_d along the surface of the model domain
p_t	Represents values of p_d along top boundaries of the model domain
$v = (u, v, w)$	The covariant velocities in the two horizontal and vertical direction
$\omega = \dot{\eta}$	The contravariant vertical velocity
$\phi = gz$	Geopotential
θ	Potential temperature
$\mathbf{V} = \mu v = (U, V, W), \Omega = \mu\dot{\eta}, \Theta = \mu\theta$	The flux form equations of the variables v, η and θ
$\mu(x, y)$	The mass per unit area within the atmospheric column in the model domain (x,y)
p	Pressure
F_U, F_V, F_W, F_Θ	The forcing variables occurring from model physics, turbulent mixing, spherical projections, and the Earth's rotation
R_d	The gas constant for dry air
p_0	The reference pressure
γ	The ratio of the heat capacity for dry air

handle grid-scale processes related to water vapor, cloud formation, and precipitation. Meanwhile, cumulus parameterization (CP) schemes address sub-grid scale cloud and precipitation processes, including vertical fluxes from unresolved updrafts and downdrafts, entrainment and detrainment, and compensatory motions outside clouds. The performance of the parameterization schemes in the WRF model is highly dependent on the region being simulated. As a result, it is recommended to conduct a sensitivity study to identify the most suitable schemes for accurately simulating weather events within the specific study area.

21.4.3.2 Data Assimilation Systems

NWP models rely heavily on initial conditions, making the accuracy of these conditions critical for model performance. Improving the initial conditions enhances the overall forecast quality. Data Assimilation (DA) is a key technique used to achieve this by integrating observational data with model forecasts, producing more accurate initial conditions, often referred to as the analysis. This process helps bridge the gap between real-world observations and model predictions, leading to better forecast outcomes (Kalnay 2003).

WRF Data Assimilation System (WRFDA) is the DA module of the WRF system that offers three DA techniques namely three-dimensional variational data assimilation (3DVAR), four-dimensional variational data assimilation (4DVAR) and hybrid Ensemble Transform Kalman Filter (ETKF)-3DVAR DA. Variational (VAR) and Ensemble are two distinct methods used for data assimilation in NWP model. VAR minimizes a cost function to optimize the initial conditions based on observational data, employing static error covariance matrices, which can be less responsive to changes in atmospheric conditions. It is particularly effective in handling strong nonlinearities, especially in its 4D-Var formulation, but it requires high computational resources compared to 3DVAR. In contrast, Ensemble method uses a probabilistic approach, updating an ensemble of forecasts to dynamically estimate flow-dependent error covariances. This ensemble-based method adapts better to evolving atmospheric states, although its performance might be influenced by the size of the ensemble members making it more computationally demanding. Overall, while VAR is suited for fitting observations over time and capturing complex dynamics in a cost-effective manner, Ensemble method offers flexibility and responsiveness to changing conditions, but with computational constraints. However, WRFDA provides a combined form of these two methods known as hybrid Ensemble Transform Kalman Filter (ETKF)-3DVAR DA (HYBRID hereafter). It combines the strengths of both VAR and Ensemble approaches, allowing the use of flow-dependent covariances within the variational framework, balancing accuracy and computational efficiency.

21.4.3.3 Data Used

The WRF model needs geographical as well as meteorological data for model initialization. The geographical data are taken from WRF download page whereas updated Land Use Land Cover (LULC) is obtained from ISRO. The meteorological data is obtained from the National Center for Environmental Prediction (NCEP) Global forecast system (GFS) data at $0.5° \times 0.5°$ gridded resolution at 6 hourly time intervals.

For data assimilation, global datasets from the Global Telecommunication System (GTS) are used, providing information on temperature, specific humidity, surface pressure, and wind, collected from various sources such as surface synoptic observations, radiosondes, ships, buoys, and satellite and aircraft wind reports. While global data forms the backbone for large-scale simulations, it is often beneficial to integrate region-specific local data to improve the simulation of weather phenomena that are locally driven. This section emphasizes the importance of assimilating regional data such as satellite radiances, satellite wind data, radar information and Automatic Weather Station (AWS) observations, to achieve more accurate and localized forecasts.

For model validation, GPM IMERG rainfall product at a 10 km resolution was used, along with AWS data from IMD and ISRO. Rainfall forecast verification is conducted using standard statistical skill scores, including the Bias score and Equitable Threat Score (ETS score) (Hamill and Juras 2006). Additionally, we employ three other statistical measures for spatial verification: Root Mean Square Error (RMSE), Mean Error (ME) also referred to as Bias and the Improvement Parameter (η) defined by Eq. 21.11.

$$IMP\_PAR(\eta)\left[1 - \frac{RMSE_{LocalData}}{RMSE_{GTS}}\right] \times 100 \qquad (21.11)$$

Here, the positive values of η represent percentage improvement due to local DA experiments concerning only global DA experiments.

21.4.3.4 Study Area

The study area covers the entire eastern Himalayan region but the analysis is shown only for the NER of India as shown in Fig. 21.9. The model outer domain covers the entire Indian sub-continent and the model is normally run in nested domain. By configuring the WRF model specifically for the unique topography of the NER of India, all efforts was made to improve weather predictions over this region.

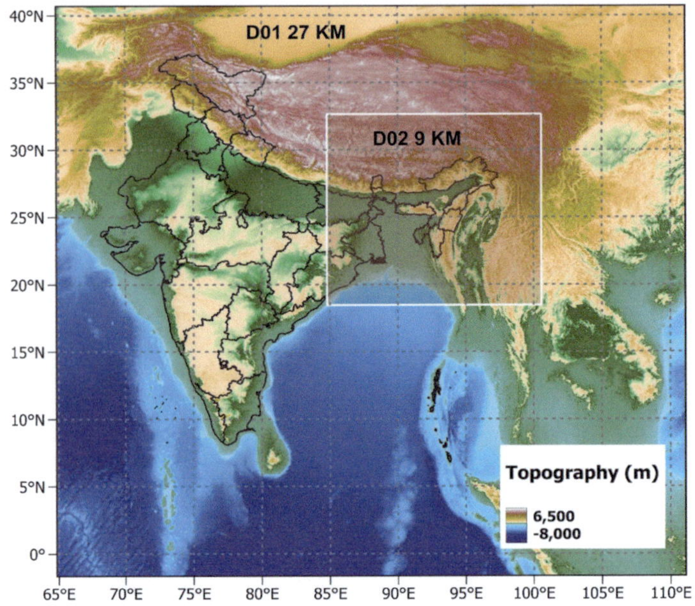

Fig. 21.9 Domain for running the WRF model and all assimilation experiments

21.4.4 Rainfall Forecast Over NER of India

The WRF model was run with the configuration shown in Table 21.3. Over the years, significant efforts have been made to improve the accuracy of these forecasts, recognizing that precise rainfall prediction is essential for managing and mitigating risk from flood, landslides, etc. The operational use of any NWP model requires extensive and systematic background research to configure the model specifically for the region of interest. This process is particularly complex for regions like the NER of India, which has challenging topography and diverse weather patterns.

NWP models rely on two primary datasets: geographical and meteorological data. The geographical data, which includes LULC and topography, plays a crucial role in how surface and atmospheric interactions are simulated within the model. These interactions significantly affect the formation and evolution of weather systems, such as storms, rainfall patterns, and wind dynamics. For example, mountainous terrain can induce orographic rainfall, while urban landscapes can influence surface heating and precipitation processes. To configure the model accurately for NER, a systematic sensitivity analysis was conducted. This involves studying how changes in the geographical datasets (such as different topography or LULC representations) impact the model's performance. Such analyses helped identify the most appropriate geographical datasets to use, ensuring that the model captures local weather phenomena more effectively. Similarly, the sensitivity analysis of the model to physical parameterization schemes was also conducted.

Table 21.3 The WRF model configuration

Model parameter	Setting
Domain size	2 domains; D01 extent (parent): 64.84°E–111.15°E, 1.66°S–40.7°N D02 extent (child): 84.79°E–100.57°E, 18.47°N–32.65°N
Map projection	Mercator
Spatial resolution	D01: 27 km, D02: 9 km
Vertical levels	36
Top boundary	50 hPa
Initialization	GFS 0.5 Degree 3 hourly
Forecast length	48 h with hourly interval

In terms of meteorological data, it has been discussed that NER of India is a data-sparse area. Compounding to this issue is the region's complex mountainous terrain, which adds another layer of difficulty in predicting rainfall. To address these challenges, the assimilation of local and high-resolution data is crucial. By assimilating such data into NWP models, the system can better simulate the interaction between terrain and atmospheric processes, leading to more accurate forecasts of localized weather phenomena, such as heavy rainfall. Experiments were conducted to evaluate the impact of assimilating INSAT-3D Atmospheric Motion Vectors (AMVs) and radiance data on rainfall simulations over the NER. Additionally, studies were performed to assess the effect of radar data assimilation on forecast accuracy. Both variational and ensemble-based data assimilation techniques were employed in these analyses. Figure 21.10 presents the seasonal validation of WRF model generated rainfall forecasts during the Indian summer monsoon season (June to September—JJAS, 2023) over the Eastern Himalayan region, providing context for the subsequent discussions.

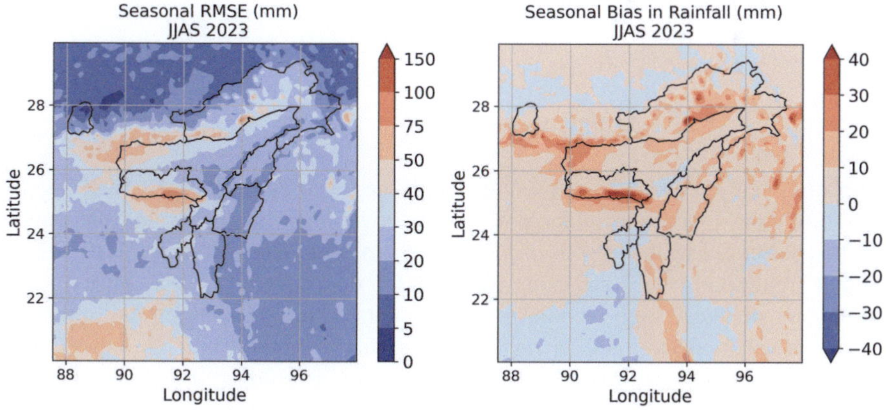

Fig. 21.10 Seasonal bias and RMSE of rainfall simulation compared to GPM data for JJAS 2023

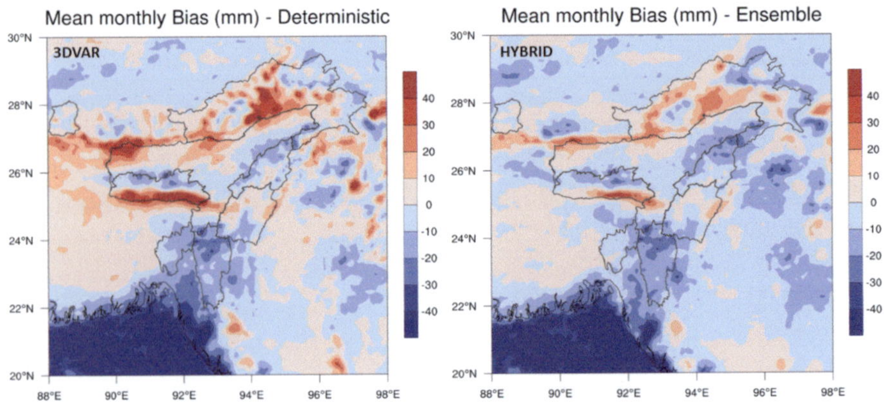

Fig. 21.11 Mean monthly bias for 3DVAR and HYBRID simulated mean rainfall over the Eastern Himalayan region

Significant bias and RMSE have been observed over western Assam, southern Meghalaya, and the hilly terrain of Arunachal Pradesh. To enhance the WRF model's performance in these regions, a more advanced data assimilation technique, the HYBRID method, was employed. In this method, the ensemble members were initialized at 0000UTC of every day for a whole month, by adding 50 random perturbations from WRF 3DVAR (Barker et al. 2004) and subsequently, advanced in time using ARW-WRF for a period of 72 h. To enhance clarity, the experiments are named according to the data assimilation (DA) method employed.

The results, presented in Fig. 21.11, demonstrate a notable improvement in HYBRID simulations compared to simulations done using 3DVAR in the same areas. However, HYBRID simulations are computationally more demanding. Further details on these experiments, as well as the outcomes of related sensitivity studies, are discussed in the subsequent sections.

21.4.5 Sensitivity Tests

A series of sensitivity study was conducted to see the sensitivity of the WRF model to LULC data, topography data and parameterization schemes so that the model can be configured well for NER. An example has been shown in Fig. 21.12, where rainfall simulations are done using three LULC data namely USGS, MODIS and ISRO. The advanced high resolution satellite data help in generating the improved LULC information over Indian region which was integrated into the model. The ISRO experiment shows lower bias in precipitation as compared to the rest of the experiments, mainly over central and the western Assam covering the Brahmaputra valley. Significant wet bias is observed when MODIS based LULC was used for simulations.

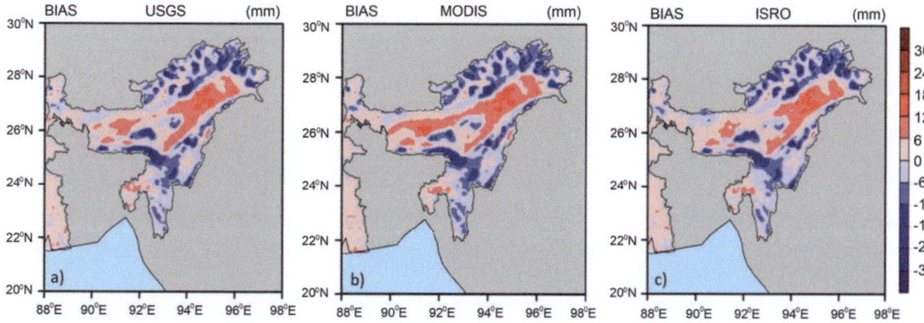

Fig. 21.12 Rainfall simulations using three different LULC data from different sources namely USGS, MODIS and ISRO

Similar experiments were conducted for the topographical data namely USGS, SRTM, and CartoDEM. However, none proved superior across all tests and so USGS data was used for all studies. To assess the sensitivity of WRF model to different parameterization schemes, 18 experiments are designed based on different combination of parameterization schemes available in WRF model given in the below Table 21.4.

The results show that as rainfall intensity increases, the ETS values which evaluates forecast accuracy by accounting for hits, misses, and false alarms; also improves across all parameterization combinations (as shown in Fig. 21.13). It was found that the combination of YSU (Yonsei University PBL), WSM6 microphysics, and Kain-Fritsch cumulus parameterization (1_6_1) is the most effective for simulating heavy rainfall over the NER of India. However, recognizing that parameterization schemes are region-specific, further analysis focusing solely on rainfall over the hilly terrain revealed that the Grell 3D scheme outperforms the Kain-Fritsch scheme, while the choice of microphysics (MP) schemes had little impact. Given that two-thirds of NER consists of mountainous regions, we opted to use Grell 3D instead of Kain-Fritsch for better accuracy. This highlights the importance of carefully selecting parameterization schemes based on the specific characteristics of the study domain for accurate operational forecasting.

Table 21.4 Parameterization schemes used for different simulation using the WRF model. The numerical values in the bracket represent the schemes as mentioned in the WRF document

PBL	MP	CP	Experiments
YSU (1)	Lin (2)	Kain-Fritsch (1)	1_2_1, 1_2_5, 1_2_93, 1_6_1, 1_6_5, 1_6_93, 1_8_1, 1_8_5, 1_8_93, 2_2_1, 2_2_5, 2_2_93, 2_6_1, 2_6_5, 2_6_93, 2_8_1, 2_8_5, 2_8_93
Mellor-Yamada-Janjic (2)	WSM6 (6)	Grell 3D (5)	
–	Thompson graupel (8)	Grell-Devenyi (93)	

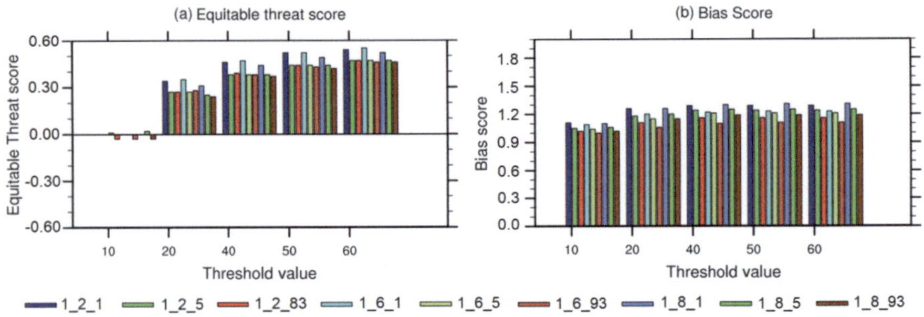

Fig. 21.13 Equitable threat score (ETS) and bias of simulated rainfall compared to GPM rainfall

21.4.6 Data Assimilation Tests

21.4.6.1 Assimilation of INSAT 3D AMVs

The INSAT 3D based AMVs (available in www.mosdac.gov.in) derived from low-level visible and mid-infrared (MIR) channels (600–950 hPa) and upper-level water vapor (WV) channels (100 hPa to 500 hPa) were assimilated in the model. Additionally, conventional in-situ observations and satellite-derived wind data from the GTS were also assimilated. In this study, 50 ensemble members were generated by adding random perturbations derived from WRF 3DVAR. To avoid spin-up issues typically encountered in NWP models, the ensembles are initialized at 0000 UTC, 24 h before the first analysis time. These ensemble members are then integrated over the next 24 h and updated using the Ensemble Transform Kalman Filter (ETKF). All DA experiments using 3DVAR and the HYBRID approach are initialized at 0000 UTC, with the ensemble mean. Observation error statistics for INSAT-3D AMVs are adapted from Kumar et al. (2017), while error statistics for GTS data are taken from NCEP. The impact of assimilating the INSAT 3D CMV improves the model performance which is further improved while using the HYBRID method, which is evident in Fig. 21.14, where the significant wet bias in the 3DVAR simulation is greatly reduced by the HYBRID method.

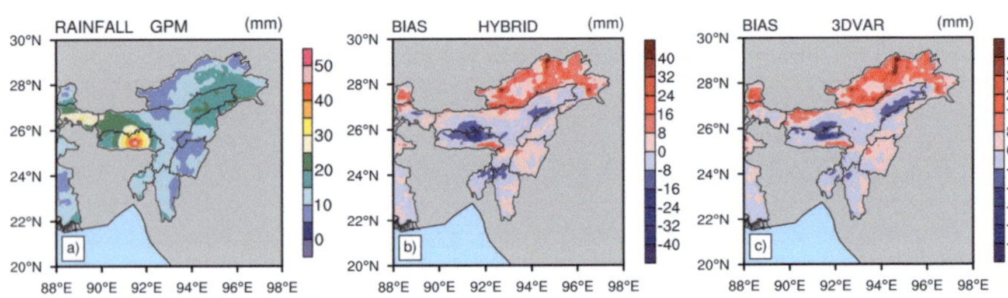

Fig. 21.14 GPM monthly average rainfall for (left), bias estimate for the model simulated monthly average 24 h forecasted rainfall for HYBRID (centre) and 3DVAR (right) using AMV DA

21.4.6.2 Assimilation INSAT-3D Radiance Data

AMV is one of the geophysical parameter retrieved from satellite data and it was observed that assimilation of AMV alone does not have any significant improvement in forecast. More experiments were conducted to assimilate the radiance data from INSAT-3D. Radiance data represents the raw measurements captured by satellite sensors in various spectral channels, and its direct assimilation can provide more accurate inputs for the model. The key advantage of using radiance data over processed geophysical products lies in its reduced error margin. Geophysical products, such as temperature or humidity profiles, are typically derived from radiance data through complex retrieval algorithms, which can introduce additional uncertainties. By assimilating radiance data directly, these uncertainties are minimized, allowing the model to make better use of the satellite observations for improved atmospheric simulations and weather predictions.

When assimilating data, it is crucial to apply error correction; otherwise, observational errors can lead to significant discrepancies between the forecast and actual conditions. The observed and RTTOV-calculated brightness temperatures (Tb) from the NWP background for INSAT-3D imager channels were compared to estimate warm/cold biases and then bias corrections were applied. The bias correction caused a substantial reduction in both RMSE and standard deviation (STD). The reduction in RMSE and STD could be of the order of 90% for channel 2, 87% for channel 3, and 92% for channel 4 of INSAT 3D radiance data.

Figure 21.15 shows the spatial distribution of the improvement parameter (IMP_PAR) over the NER of India which is a comparative assessment of the local DA experiments against only global DA experiments. The IMP_PAR indicates strong positive impact of INSAT-3D radiance data assimilation over almost entire NER of India compared to only GTS DA. The improvement is more significant over the hilly terrains such as Meghalaya and Arunachal Pradesh which was not observed so distinctly in the AMV DA.

21.4.6.3 Assimilation of Radar Data

In addition to satellite data assimilation, ground based Doppler Weather Radar (DWR) observations can also be assimilated to improve the NWP over the mountainous regions. DWR data could be of higher importance for forecasting convective processes due to their high spatiotemporal resolution. Two key variables from DWR, reflectivity and radial velocity, could be assimilated into NWP models. Reflectivity offers insights into hydrometeor distribution, while radial velocity captures atmospheric convective motions, both of which are vital for improving forecast accuracy. WRFDA enables both direct and indirect assimilation of radar reflectivity. In the direct method, radar data is assimilated as it is, while in the indirect method, rainwater and water vapor are derived from reflectivity and then assimilated into the model. In stable atmospheric conditions, direct radar data assimilation may be less effective because it primarily provides information on precipitation but lacks insights

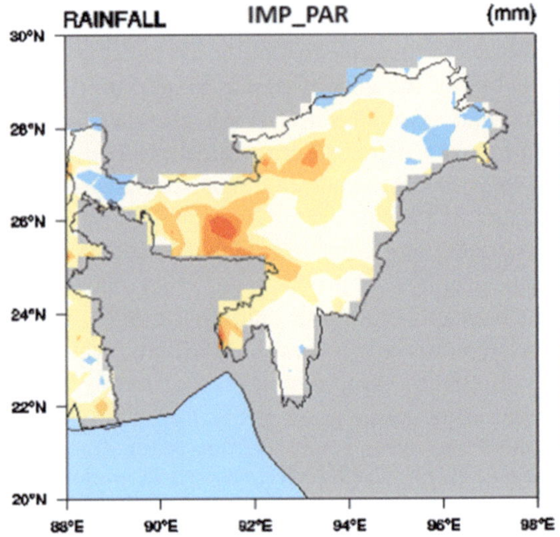

Fig. 21.15 Improvement parameter (IMP_PAR) depicting improvement of INSAT-3D radiance DA over only GTS DA

into the broader moisture and thermal structure of the atmosphere. In these scenarios, the indirect method, which adjusts moisture profiles, can be more useful for triggering convection and improving model performance. In one of the experiment to assimilate the DWR data, it was observed that such assimilation is able to simulate the moist warm core associated with a strong updraft, which normally enhances the instability needed to develop the thunderstorm. The experiment could identify the presence of tall clouds conducive to severe thunderstorm formation, which was absent in the CTRL experiment. Additionally, the severity of the thunderstorm was better captured when reflectivity is assimilated in the model indirectly, as mentioned above.

21.4.7 Prediction of Lightning and Severe Storms Over NE Region of India

Severe storm and lightning is one of the major forms of severe weather affecting the Eastern Himalayan region, particularly the NER of India. NER of India is one of the hotspot for the lightning (Albrecht et al. 2016), particularly during the pre-monsoon season. NER have a complex topography and the area is convergence zone of dry and cold westerly wind from Himalaya and warm and moist southerly from Bay of Bengal that intensifies the thunderstorms and lightning (Houze et al. 2007; Kumar and Kamra 2012; Qie et al. 2021) over the region. In NER, more than 100 thunderstorms days associated with more than 2000 lightning strokes reported in pre-monsoon to monsoon season. 10–15 thunderstorm cases are associated with squall line enhancing the severity. These thunderstorms and lightning are highest in pre-monsoon season with more number of events over the western part of the NER of

India. These events, in isolated cases produces hailstorm and cloud-burst and lead to loss of infrastructure and life (Tinmaker and Ali 2012; Singh and Singh 2015; Srivastava and Kundu 2020).

Early warning is one of the most efficient steps towards risk reduction from such severe weather. There are various methods of early warning including statistical, probabilistic, AI based, NWP, and physical modeling techniques. There are several mechanisms to warn the severe weather by studying the dynamical, electrical and microphysical features of the thunderstorms. Such properties of the atmosphere can be assessed using ground-based instruments (DWR and lightning detectors) or space based data (INSAT, GOES, etc.). Several algorithms like stand-alone methods (Srivastava et al. 2015; Bennett 2018), thunderstorms tracking (Betz et al. 2008; Dixon and Wiener 1993; Srivastava et al. 2022), as well as WRF model based system (Yair et al. 2010; Fierro et al. 2013; Srivastava et al. 2023) are capable to warn severe weather and lightning with several minutes to hours in advance. All of the above approaches utilize the electrical behaviours like atmospheric electric field, vertical electric field, Maxwell current, conductivity, and lightning etc., as important information for generating warning. The LIS on-board TRMM satellite and currently in ISS (International Space Station) and Lightning Mapping Imager (LMI) on-board the Fengyun-4A (FY-4A in geostationary orbit) satellite has been extremely useful to understand the spatial distribution of lightning and characterize the same over the Eastern Himalayan region. ISRO is planning to put lightning detectors in its upcoming 4th generation of meteorological satellites.

Ground based instruments provide continuous and additional information on lightning that is extremely useful to understand the micro-physical properties. These instruments work in network using VLF/LF frequency receivers to detect lightning, as these signals are radiated from the lightning which can travel several hundreds of km. There are large numbers of lightning location networks worldwide, like WWLLN, GLD360, ENTLN, EUCLID, ILLN, BLNET, KLDN, etc. The data from these networks that also covers the Eastern Himalayan region has been used to validate the satellite based data as well as calibrate the space based sensors.

21.4.8 WRF-Electricity (WRF-Elec) Module for Lightning Forecasting

The detail about the WRF model has been discussed in previous section. Lightning flash count can be estimated using WRF model by using parameterization scheme recommended by Price and Rind (1992). Later, lightning potential index (LPI) was introduced by Yair et al. (2010), in which charge separations regions depending on cloud microphysics was considered. Fierro et al. (2013) proposed to use cloud-resolving scale, explicit charging and discharging physics scheme to forecast lightning and developed an electricity module that was coupled in the WRF and known as WRF-Elec (Fierro et al. 2013; Gharaylou et al. 2020; Lu et al. 2022).

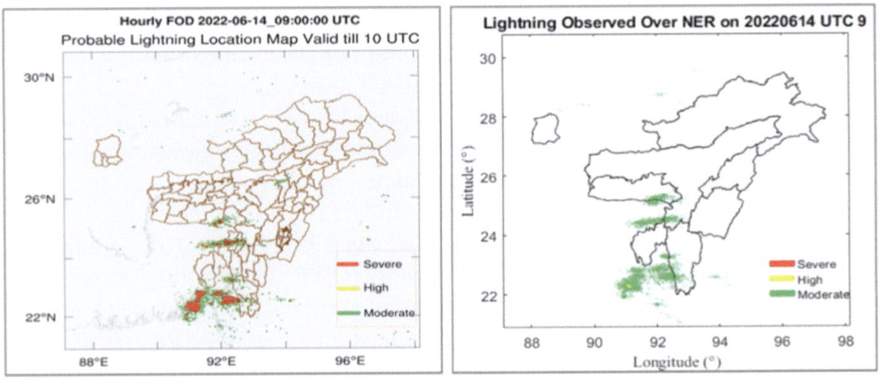

Fig. 21.16 Lightning nowcasting using the WRF-elec model (left) and observed lightning from ground based lightning detector network (right)

The ground based lightning data was assimilated in the WRF-Elec module to examine how effectively the model is able to forecast lightning over the NER of India. It was observed that in the absence of lightning data, the WRF-Elec is unable to nowcast lightning accurately. However, lightning data assimilation enables the model to identify the areas where lightning is likely to take place while identifying the areas that are unlikely to have lightning simultaneously and with equal accuracy. It was found that the model is capable to provide early warning upto 6 h with reasonable accuracy and it reaches above 80% for 3 h lead time. For the NER of India, such nowcasting is generated on a daily basis and made available through the https://www.nerdrr.gov.in/light/lightning.html. Figure 21.16 shows one such lightning early warning over the Eastern Himalayan region and its comparison with the actual observation. This depicts that the model was able to identify the zoned where lightning is likely as well as where lightning will not take place with very good accuracy. The lightning nowcast was little over-estimated in terms of severity.

21.4.9 Deep Learning Based Lightning Nowcasting Using INSAT-3D/3DR Data

As lightning is a very dynamic process and require analysis of large amount of data in near real time, the strength of deep learning based methodology could be handy to assess multi-platform data quickly and generate nowcasting services. Building on the innovative approach of ProbSevere LightningCast, a deep-learning model was developed to enhance lightning nowcasting capabilities using purely satellites imageries from INSAT-3D and INSAT-3DR datasets as inputs. This model leveraged high-temporal-resolution geostationary satellite data in conjunction with ground-based lightning detection data from the Indian Institute of Tropical Meteorology (IITM) to improve short-term lightning forecasts. At its core, the model is designed to

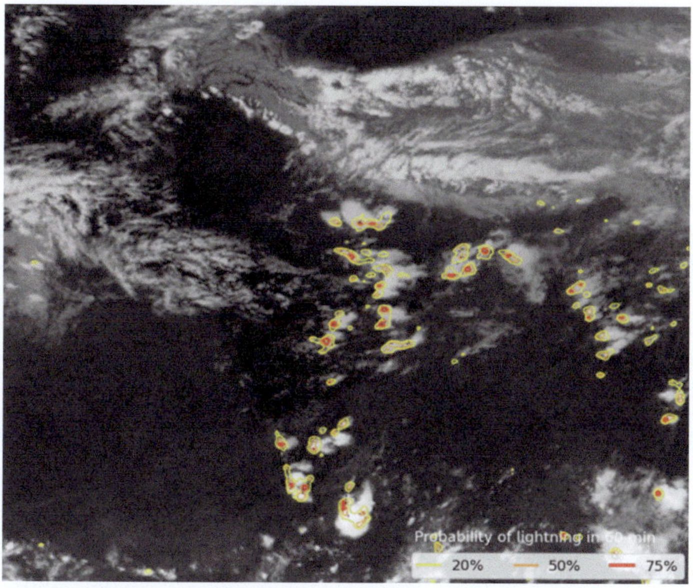

Fig. 21.17 Lightning areas identified and nowcast generated over the entire Indian region including the Eastern Himalayan region using ground based data and INSAT3DR data

process and analyze vast amounts of satellite imagery and meteorological data in real time. Using convolutional neural networks (CNNs), the model identifies spatial and temporal patterns that indicate favorable conditions for lightning formation. INSAT-3D and INSAT-3DR, with their frequent updates and multispectral observations, provide an ideal data source for capturing cloud-top properties, convective growth, and atmospheric instability over the Indian subcontinent and surrounding regions (Fig. 21.17).

This high-temporal-resolution data enhances the model's ability to detect convective initiation and forecast lightning activity up to 60 min in advance. A key aspect of this model is its integration with IITM's ground-based lightning detection network, which serves as a critical validation source. The ground-based lightning data will provide an extensive record of lightning occurrences, ensuring that the model's predictions are continuously refined for improved accuracy and reliability. The ongoing efforts focused on optimizing deep learning architectures, and fine-tuning model performance to make it more robust. Such a lightning nowcasting system, powered by space based lightning data, can be very effective for the complex Eastern Himalayan region in the days to come.

21.5 Space Based Geophysical Products for Severe Weather Assessment and Warning

Satellite-based remote sensing provides critical data over the Eastern Himalaya that helps in capturing the dynamic and thermodynamic state of the atmosphere. Among the key satellite-derived geophysical products aiding in severe weather prediction are relative vorticity, upper-level divergence, low-level convergence, wind shear, and other parameters derived from wind vector fields and atmospheric motion vectors (AMVs). These products, when assimilated into numerical weather prediction (NWP) models or used diagnostically, significantly improve the detection and forecasting of severe convective systems, including thunderstorms, hailstorms, and heavy precipitation events.

21.5.1 Brightness Temperature

Satellite-based brightness temperature (BT) products are crucial for monitoring cloud properties and identifying convective activity, especially over remote or mountainous regions. BT represents the temperature a blackbody would have to emit the observed radiance at a given wavelength, typically measured in the infrared (IR) and microwave (MW) bands. In meteorology, IR BT, particularly from thermal IR channels (e.g., 10.8 μm) is widely used to estimate cloud-top temperatures, where colder BT values (<233 K) often indicate deep convection and potential severe weather, including cloudbursts. Geostationary satellites like INSAT-3DR/3DS, Himawari-8, and GOES-R provide high-frequency BT data over Eastern Himalaya that support nowcasting of rapidly developing convective systems. BT trends, such as rapid cooling, are indicative of cloud growth and intensification, and are commonly used in nowcasting algorithms and convective initiation detection.

21.5.2 Relative Vorticity

Relative vorticity (RV) indicates the local spin or rotation of air parcels and is a critical parameter in diagnosing cyclonic circulation and storm development. It can be computed using the Eq. 21.12.

$$\textit{Relative vorticity} = \frac{\partial v}{\partial x} + \frac{\partial u}{\partial y} \tag{21.12}$$

where u and v are the zonal and meridional wind components, respectively. Positive vorticity generally indicates cyclonic rotation in northern hemisphere, a necessary condition for storm formation. Vorticity fields can be computed using wind products

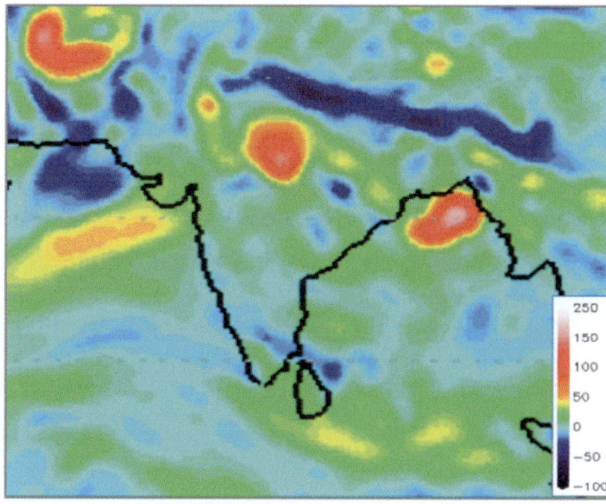

Fig. 21.18 RV at 700 mb pressure level over the Indian region

from satellite AMVs (like from INSAT-3DR / 3DS) or scatterometers (e.g., ASCAT on MetOp). Over mountainous areas, enhanced low to mid tropospheric vorticity along ridges can signal cyclogenesis or upslope-enhanced convection. Studies have shown enhanced vorticity over the eastern Himalayas linked to mesoscale convective systems and vice-versa. RV is a key diagnostic for assessing atmospheric dynamics across different pressure levels. At upper levels (200–300 hPa), it highlights jet streams and upper-level divergence zones favorable for deep convection. At mid-levels (around 500 hPa), it helps identify cyclonic systems and vorticity advection, critical for vertical motion and storm development. In the lower troposphere (850–700 hPa), high vorticity indicates low-level cyclonic circulations and convective potential, especially over complex terrain. Together, multi-level vorticity analysis aids in identifying dynamically active regions conducive to severe weather formation, particularly in mountainous areas. Figure 21.18 shows the RV at 700 mb pressure level with a negative RV over the Eastern Himalaya on 13th July, 2025 (summer monsoon season). Such a condition over the Eastern Himalaya is against development of any severe weather over the region and therefore this can be used a very effective tool for synoptic scale weather forecast.

21.5.3 Upper-Level Divergence

Divergence at upper levels (200–250 mb) is associated with upward motion and is often a precursor to severe convection. It contributes to mass evacuation aloft, thereby supporting the vertical growth of convective towers. Positive divergence at upper levels indicates favorable conditions for deep convection and vice versa. Geostationary satellites such as INSAT series, Himawari-8 and GOES-16 provide

AMV-derived wind fields from cloud tracking, enabling estimation of divergence fields.

21.5.4 Low-Level Convergence

Low-level convergence, often resulting from terrain-induced channeling and synoptic-scale flow interaction, is crucial for initiating convective updrafts. A region of convergence ($\nabla \cdot V < 0$) at low levels overlain by divergence aloft creates a favorable vertical motion profile. Convergence fields can be computed from scatterometer surface winds or low-level AMVs.

21.5.5 Vertical Wind Shear (VWS)

Vertical wind shear signifies the change of wind speed and direction with height, which is a vital condition required for the organization and longevity of severe convective storms. VWS estimates over remote mountainous terrain can be derived using multi-layer AMVs. When a tropical cyclone (TC) is present, the analysis within the storm region is adjusted by removing the storm's circulation within a defined radius. This approach provides a clearer representation of the environmental wind shear influencing the storm. Wind shear plays a crucial role in either inhibiting or supporting TC development. Low vertical wind shear creates favorable conditions for cyclogenesis, whereas high shear can prevent a developing TC from intensifying or even weaken a mature cyclone. Typically, shear values below 20 knots (10 m/s) are considered conducive for tropical cyclone formation. Figure 21.19 shows the VWS over the Indian region and highlights how effectively this can be used to focus on the areas that are likely to see development of severe weather.

21.5.6 Convective Available Potential Energy (CAPE) and Lifted Index (LI)

Satellite-derived data play a vital role in assessing atmospheric instability through parameters like Convective Available Potential Energy (CAPE) and the Lifted Index (LI), especially over regions lacking dense observational networks such as mountainous terrain. CAPE quantifies the buoyant energy available for convection, with higher values (>1500 J/kg) indicating a greater potential for strong updrafts and severe storms. The Lifted Index, derived from comparing the temperature of a lifted air parcel to the environment at 500 hPa, provides a quick estimate of atmospheric

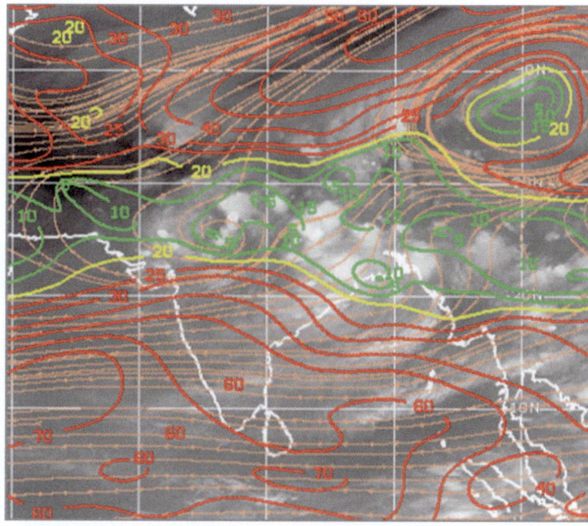

Fig. 21.19 Vertical wind shear over the Indian region highlighting the regions that are favorable (green contour), neutral (yellow) and unfavorable (red) areas for severe weather genesis

stability. Negative LI values (< –4) signify increasing instability. Geostationary satellites (e.g., INSAT-3DR, GOES) equipped with infrared and sounder instruments enable continuous retrieval of temperature and humidity profiles, from which CAPE and LI can be calculated in near real time. These products are instrumental in convective nowcasting and early warning of severe weather events.

21.5.7 Integrated Use of Satellite Based Geophysical Parameters

The products mentioned above, which are generated using the near real time data from meteorological satellites, can be effectively used to assess the atmospheric condition over a region and forecast the potential of severe weather development at synoptic to meso scale level. With greater experience about the region of study, one can use these data to even forecast at local scale with 6–12 h of lead time. This method has been very successfully used for forecasting rainfall over the NER of India under the Flood Early Warning System (FLEWS) project of NESAC. The weather advisory generated has been very useful to complement the limitations of NWP based forecast over the NER of India on the occasions, when NWP could not perform as expected (Fig. 21.20).

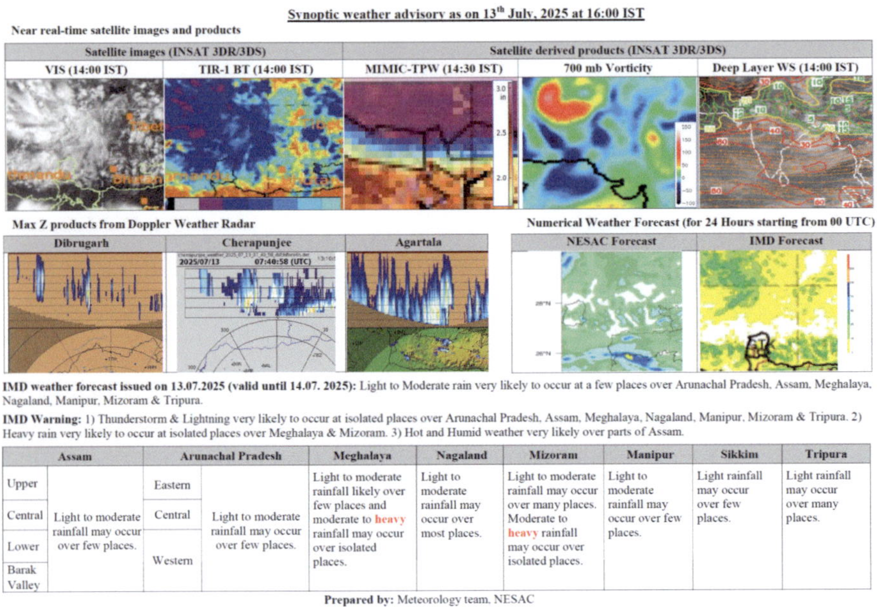

Fig. 21.20 Synoptic scale rainfall forecast over NER of India using the satellite based geophysical products

21.6 Forecasting of Cloud Bursts Over Eastern Himalaya

A cloudburst is typically defined as an extreme rainfall event with a rate exceeding 100 mm per hour over a localized area, often associated with strong convection and orographic uplift. Such sudden, intense rainfall events can cause devastating flash floods and landslides. While traditionally associated with western Himalayas like Uttarakhand and Jammu & Kashmir, recent studies report increasing frequency in the Eastern Himalaya, including parts of Sikkim, Darjeeling, and Arunachal Pradesh (Houze et al. 2017). These events, often localized and short-lived, are difficult to predict using conventional weather models and sparse ground observations alone. The region's complex topography, rapid orographic lifting, and high moisture availability further complicate forecasting efforts.

As cloud-burst is a very localized event, high frequency data is required to detect and forecast such events using satellite platform. The geostationary platform does not provide the required spatial resolution and vertical profiling, while the low earth orbit satellites does not provide the required temporal resolution. The Dual-frequency Precipitation Radar (DPR) on-board the GPM-Core satellite provides information on internal structure of convective clouds, helping identify intense updrafts and precipitation cores that are key indicators of impending cloudbursts (Hou et al. 2014). However, the limited pass over any area is a constraint towards using GPM as an operational tool for forecasting cloud burst. The INSAT-3DR satellite, positioned over the Indian subcontinent, provides half-hourly images from its multispectral imager and sounder, which can detect cold cloud tops ($\leq$ 233 K) associated with deep convection. This is particularly useful for identifying mesoscale convective systems (MCS) that may result in cloudbursts over mountainous regions.

Satellites such as CloudSat and EarthCARE carry radar and lidar payloads capable of vertical profiling of clouds, revealing details such as hydrometeor size, phase (ice/liquid), and vertical extent. Cloudburst-producing clouds typically have tall convective cores (>12 km), high ice water content, and embedded supercooled water droplets, which can be inferred from satellite retrievals.

Constellation of space based microwave radar and radiometers are planned by private agencies with a promise to provide space based radar observation every 30 min. Such a constellation could be very effective to monitor clouds with higher frequency and assimilate the data in NWP models to improve their nowcasting capability. Such constellations, while fully operational could help in better forecasting of cloud burst over the Himalayan region. IMD has been augmenting the ground based X-band DWR network to have very high frequency observation on cloud growth to facilitate improved forecasting of cloud bursts. However, covering the entire Himalayan region with DWRs could be challenging both in terms of setting up the network and their operation and maintenance.

21.7 Forecasting of Hailstorm Over Eastern Himalaya

Hailstorm are another form of severe weather affecting Eastern Himalaya and primarily occurs in Assam, Meghalaya and Tripura states of NER of India that causes damage to crops, structures, and automobile damage. Space based data can't be used as of now to directly detect precipitating hails, however, the favorable conditions for hailstorm can be assessed using the satellite data and products discussed in the previous sections. The DWR data also comes handy to detect the clouds that are likely to produce hails. The lightning data can also be used to assess the presence of hails in a cloud. The sudden increase in flash rate, known as Lightning jump (LJ), is leveraged to assess the potential of hailstorm in a system. The sudden fluctuations in the lightning flash rate are due to drastic changes in the hydrometeors insight the thunderstorm and such algorithms are well tested in USA and China (Chronis et al. 2015; Schultz et al. 2017; Tian et al. 2019). These processes are due to the alteration in microphysical and charging process of the thunderstorms. Most of LJ algorithms are based on the changes in flash rate using various mechanisms like change in mean, standard deviation etc. in the PCG (positive cloud to ground) or CG flashes. In this case study, change in standard deviation is considered in terms of PCG flashes and CG flashes. Here two layer of threshold is fixed for PCG and CG and when both are hit by the LJ, a hail storm is likely to happen. The algorithm was successfully tested to identify the Hailstorm over Riangdo, West Khasi Hills, Meghalaya and some other places in Meghalaya where ground verification of hailstorm could be done. With the availability of space based lightning data and radar data, the detection and forecasting of hailstorm using the space data alone could be possible in the days to come.

21.8 Challenges to Improve Weather Prediction Over Mountainous Regions

Weather prediction in mountainous regions presents significant challenges due to the complex interactions between the atmosphere and rugged terrain. The steep topography disrupts airflow, creating highly localized weather patterns over short distances, which are difficult for models to resolve accurately, especially when using coarser resolution. Orographic effects, such as forced lifting of air over mountains, lead to localized heavy precipitation, while sharp elevation changes cause variability in temperature, moisture, and wind patterns that models struggle to capture. Sudden, unpredictable weather changes, such as storms, fog, or snowfall, add to the difficulty of short-term forecasting. Compounding this issue is the limited observational data, as the scarcity of weather stations in remote mountainous areas hampers accurate model initialization and validation, reducing forecast reliability. In addition, common parameterization schemes used for processes like cloud formation and convection often fail to adequately represent the microphysical processes in mountainous regions, leading to forecast biases, particularly at higher altitudes. Local wind systems, such as valley breezes and downslope winds, add another layer of complexity. Addressing these challenges requires higher-resolution models, better data assimilation, and more localized observations to improve accuracy.

While observations are critical, resolution plays an equally important role, especially when forecasting orographic rainfall. Guwahati city in Assam, India experienced intense localized rainfall, with the AWS stations recording 81 mm and 113 mm of rainfall. The operational forecast issued for the day failed to capture the event. However, after assimilating AWS data, the model was able to predict the rainfall (as seen in Fig. 21.21), though it was slightly displaced to the south and underestimated in intensity. Notably, when the model resolution was increased, it predicted both the rainfall's location and intensity more accurately, demonstrating how critical high resolution is for capturing localized weather events effectively.

Besides model resolution and observed data, the data assimilation (DA) framework is another crucial factor. Our earlier discussion shows that the ensemble-based

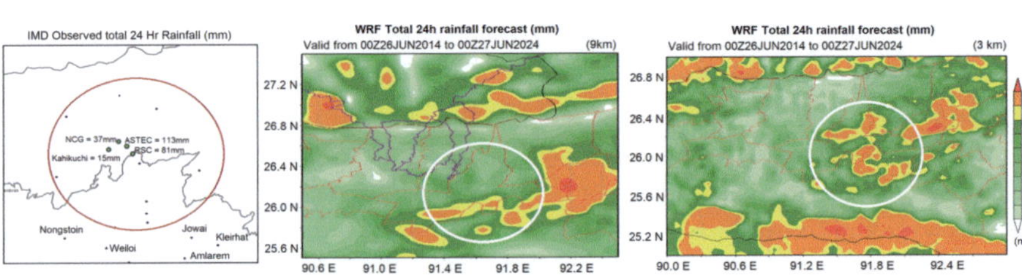

Fig. 21.21 Simulation of as localized heavy rainfall event over Guwahati using GTS and AWS DA at 9 km resolution (**b**) and at 3 km resolution (**c**). The observed rainfall is seen in (**a**)

DA system helps reduce the bias seen in the 3DVAR DA experiment. The HYBRID DA system, by incorporating flow-dependent ensemble background error covariance, produces optimal analysis through increments that align with the background flow and adapt to changes in the observing system. As a result, the impact of the observing system can vary depending on the DA system used, as evidenced by some of the results. However, to carry out such operational forecast at a very high resolution, computational constrain remains a challenge.

The Bharat Forecast System (BFS), with its high-resolution (6 km, can be downscaled further) modeling and advanced data assimilation, is set to significantly enhance weather forecasting over the NER of India. By better capturing the region's complex terrain, localized convection, and orographic rainfall, BFS is expected to improve the accuracy of short-range and extreme weather forecasts. Integration of satellite and radar data will fill observational gaps, while ensemble forecasting will support probabilistic, impact-based warnings. This will be especially valuable for predicting cloudbursts, flash floods, and thunderstorms, strengthening early warning systems and disaster preparedness across the vulnerable Himalayan and hill regions of the NER.

21.9 Conclusions and the Way Forward

Predicting severe weather in the Indian Himalayas and other mountainous regions is fraught with challenges due to complex topography, rapidly changing weather patterns, and the limited availability of ground-based observations. However, advancements in space technology, particularly satellite-based remote sensing, have provided new tools to improve weather forecasting in these regions. By integrating satellite data with ground-based observations and enhancing weather models, scientists continue to better predict severe weather events and reduce the impact of natural disasters on vulnerable communities in India's mountainous regions. Localized weather models that account for the unique topography and climatic conditions of the Indian Himalayas are crucial for improving forecast accuracy. Machine learning algorithms and artificial intelligence are being used to analyze historical data and develop predictive models tailored to the region's specific weather patterns. In addition to technological advancements, building community resilience is critical for mitigating the impacts of severe weather events. Early warning systems, based on satellite data, must be accompanied by public awareness campaigns and disaster preparedness programs to ensure that local communities are well-prepared to respond to extreme weather.

To improve weather predictions in mountainous regions, it is essential to increase the number of ground-based observation stations. The installation of additional automated weather stations (AWS), high-altitude observatories, and Doppler weather radars in the Himalayan region would provide more accurate data for weather models. This will enhance the ability to predict severe weather events like cloudbursts and flash floods. Future satellite missions, such as the proposed hyperspectral imaging satellites by ISRO, promise to offer higher spatial and temporal resolution data. These advancements will enable better monitoring of localized weather phenomena and improve the accuracy of weather forecasts in mountainous regions.

Satellite-based geophysical products such as relative vorticity, upper-level divergence, low-level convergence, and vertical wind shear offer indispensable insights into the atmospheric processes governing severe weather, especially over mountainous regions. As satellite capabilities and retrieval algorithms improve, their role in high-impact weather prediction will continue to grow.

In summary, space technology's ability to provide comprehensive, real-time data and feed it into advanced predictive models is crucial for overcoming the unique challenges of weather forecasting in mountainous regions like the Eastern Himalayas. The application of remote sensing and numerical modeling holds promise for better preparedness and disaster resilience in these vulnerable areas.

Acknowledgements The authors thank the ISRO/DOS for funding the atmospheric science facility at NESAC that allowed research towards improving the weather forecast over NER and Eastern Himalaya. The authors also thank IMD, MoES, IITM, NASA EarthData (GIOVANNI), MOSDAC, and NOAA for providing valuable data and modeling framework that facilitated the research that are mentioned in the chapter. The authors also extend their sincere gratitude to NARL, Tirupati, for providing the High-Performance Computing (HPC) facility, which was instrumental in carrying out the computationally intensive ensemble-based simulations.

References

R.I. Albrecht, S.J. Goodman, D.E. Buechler, R.J. Blakeslee, H.J. Christian, Where are the lightning hotspots on earth? Bull. Am. Met. Soc. **97**(11), 2051–2068 (2016). https://doi.org/10.1175/BAMS-D-14-00193.1

P.A. Arkin, The relationship between fractional coverage of high cloud and rainfall accumulations during GATE over the B-scale array. Mon. Weather. Rev. **107**(10), 1382–1387 (1979). https://doi.org/10.1175/1520-0493(1979)107%3C1382:TRBFCO%3E2.0.CO;2

M.B. Ba, A. Gruber, GOES multispectral rainfall algorithm (GMSRA). J. Appl. Meteorol. Climatol. **40**(8), 1500–1514 (2001). https://doi.org/10.1175/1520-0450(2001)040%3C1500:GMRAG%3E2.0.CO;2

S.R. Bajracharya, P.K. Mool, Glacial lake outburst floods in the Nepal Himalaya: hazard assessment and mitigation. Mt. Res. Dev. **29**(1), 29–35 (2009). https://doi.org/10.1659/mrd.0001

R.M. Banta, G. Berri, W. Blumen, D.J. Carruthers, G.A. Dalu, D. Durran, J. Egger, J.R. Garratt, S.R. Hanna, J.C.R. Hunt, R.N. Meroney, The role of mountain flows in making clouds. Atmospheric processes over complex terrain, pp.229–283. https://doi.org/10.1007/978-1-935704-25-6_9

D.M. Barker, W. Huang, Y.R. Guo, Q.N. Xiao, A three-dimensional (3DVAR) data assimilation system for use with MM5: implementation and initial results. Mon. Weather Rev. **132**, 897–914 (2004). https://doi.org/10.1175/1520-0493(2004)132%3c0897:atvdas%3e2.0.co;2

A.P. Barros, T.J. Lang, Monitoring the Monsoon in the Himalayas: observations in Central Nepal, June 2001. Mon. Weather Rev. **131**(7), 1408–1427 (2003). https://doi.org/10.1175/1520-0493(2003)131%3c1408:MTMITH%3e2.0.CO;2

A.P. Barros, D.P. Lettenmaier, Dynamic modeling of orographically induced precipitation. Rev. Geophys. **32**(3), 265–284 (1994). https://doi.org/10.1029/94RG00625

A.P. Barros, A.M. Wilson, J. Tao, D. Miller, Quantifying the role of orography on rainfall intensity and duration in the eastern Himalayas. J. Hydrometeorol. **19**(4), 833–850 (2018). https://doi.org/10.1175/JHM-D-17-0084.1

A.P. Barros, et al., Monsoon weather and climate of the Himalayas: climate dynamics and remote sensing. J. Hydrometeorology (2006)

P. Bauer, A. Thorpe, G. Brunet, The quiet revolution of numerical weather prediction. Nature **525**(7567), 47–55 (2015). https://doi.org/10.1038/nature14956

A.J. Bennett, Warning of imminent lightning using single-site meteorological observations. Weather **73**(6), 187–193 (2018). https://doi.org/10.1002/wea.2782

H.D. Betz, K. Schmidt, W.P. Oettinger, B. Montag, Cell-tracking with lightning data from LINET. Adv. Geosci. **17**, 55–61 (2008). https://doi.org/10.5194/adgeo-17-55-2008

M.R. Bhutiyani, V.S. Kale, N.J. Pawar, Climate change and the precipitation variations in the northwestern Himalaya: 1866–2006. Int. J. Climatol. **30**, 535–548 (2010). https://doi.org/10.1002/joc.1920

B. Bookhagen, D.W. Burbank, Toward a complete Himalayan hydrological budget: Spatiotemporal distribution of snowmelt and rainfall. J. Geophys. Res. (2010)

R.T. Bruintjes, T.L. Clark, W.D. Hall, Interactions between topographic airflow and cloud/precipitation development during the passage of a winter storm in Arizona. J. Atmos. Sci. **51**(1), 48–67 (1994). https://doi.org/10.1175/1520-0469(1994)051%3c0048:IBTAAC%3e2.0.CO;2

T. Chronis, L.D. Carey, C.J. Schultz, E.V. Schultz, K.M. Calhoun, S.J. Goodman, Exploring lightning jump characteristics. Weather Forecast. **30**(1), 23–37 (2015). https://doi.org/10.1175/WAF-D-14-00064.1

S. Das, A. Hazra, Thunderstorm climatology over northeastern India: a severe weather perspective. J. Earth Syst. Sci. **128**(5), 1–15 (2019)

S. Das, P. Acharya, P. Kumar, Satellite-based monitoring and prediction of cloudbursts and heavy rainfall events in the Himalayas. Int. J. Remote Sens. **40**(7), 2725–2738 (2019). https://doi.org/10.1080/01431161.2018.1547452

M. Desbois, M. Capderou, L. Eymard, R. Roca, N. Viltard, M. Viollier et al., Megha-Tropiques: un satellite hydrométéorologique franco-indien. La Météorologie **57**, 19–27 (2007). https://doi.org/10.4267/2042/18185

K.R. Dikshit, J.K. Dikhsit, North-East India: land, people and economy (2014). ISBN: 978-94-007-7054-6. https://doi.org/10.1007/978-94-007-7055-3

A.P. Dimri, D. Niyogi, A.P. Barros, J. Ridley, U.C. Mohanty, T. Yasunari, D.R. Sikka, Western disturbances: a review. Rev. Geophys. **53**, 225–246 (2015). https://doi.org/10.1002/2014RG000460

A.P. Dimri, T. Yasunari, B.S. Kotlia, U.C. Mohanty, D.R. Sikka, Indian winter monsoon: present and past. Earth Sci. Rev. **163**, 297–322 (2016). https://doi.org/10.1016/j.earscirev.2016.10.008

M. Dixon, G. Wiener, TITAN: thunderstorm identification, tracking, analysis, and nowcasting-a radar-based methodology. J. Atmos. Oceanic Technol. **10**(6), 785–797 (1993). https://doi.org/10.1175/1520-0426(1993)010%3C0785:TTITAA%3E2.0.CO;2

A.O. Fierro, E.R. Mansell, D.R. MacGorman, C.L. Ziegler, The implementation of an explicit charging and discharge lightning scheme within the WRF-ARW model: benchmark simulations of a continental squall line, a tropical cyclone, and a winter storm. Mon. Weather Rev. **141**(7), 2390–2415 (2013). https://doi.org/10.1175/MWR-D-12-00278.1

H. Frey, F. Paul, T. Strozzi, Compilation of a glacier inventory for the western Himalayas from satellite data: methods, challenges, and results. Remote Sens. Environ. **124**, 832–843 (2012). https://doi.org/10.1016/j.rse.2012.06.020

J. Gao, G. Tang, Y. Hong, Similarities and improvements of GPM dual-frequency precipitation radar (DPR) upon TRMM precipitation radar (PR) in global precipitation rate estimation, type classification and vertical profiling. Remote Sens. **9**(11), 1142 (2017). https://doi.org/10.3390/rs9111142

M. Gharaylou, M.M. Farahani, A. Mahmoudian, M. Hosseini, Prediction of lightning activity using WRF-ELEC model: impact of initial and boundary conditions. J. Atmosp. Solar-Terrestrial Phys. **210**, 105438 (2020). https://doi.org/10.1016/j.jastp.2020.105438

T.M. Hamill, J. Juras, Measuring forecast skill: is it real skill or is it the varying climatology? Q. J. R. Meteorol. Soc. **132**, 2905–2923 (2006). https://doi.org/10.1256/qj.06.25

K.B.R.R. Hari Prasad, V.S. Prasad, M. Sateesh, K. Amar Jyothi, The impact of Indian radar and lightning data assimilation on the short-range forecasts of heavy rainfall events. Q. J. R. Meteorol. Soc. **150**(759), 796(759), 796–819 (2024). https://doi.org/10.1002/qj.4623

A.Y. Hou, R.K. Kakar, S. Neeck, A.A. Azarbarzin, C.D. Kummerow, M. Kojima, R. Oki, K. Nakamura, T. Iguchi, The global precipitation measurement mission. Bull. Am. Meteor. Soc. **95**(5), 701–722 (2014). https://doi.org/10.1175/BAMS-D-13-00164.1

R.A. Houze, D.C. Wilton, B.F. Smull, Monsoon convection in the Himalayan region as seen by the TRMM precipitation Radar. Q. J. R. Meteorol. Soc. **133**, 1389–1411 (2007). https://doi.org/10.1002/qj.106

R.A. Houze, L.A. McMurdie, K.L. Rasmussen, A. Kumar, M.M. Chaplin. Multiscale aspects of the storm producing the June 2013 Flooding in Uttarakhand, India. Mon. Weather. Rev. **145**(11), 4447–4466 (2017). https://doi.org/10.1175/MWR-D-17-0004.1

R.A. Houze, Orographic effects on precipitating clouds. Rev. Geophys. **50**, RG1001 (2012). https://doi.org/10.1029/2011RG000365

G.J. Huffman, D.T. Bolvin D. Braithwaite, K. Hsu, R. Joyce, C. Kidd, E.J. Nelkin, P. Xie, NASA global precipitation measurement (GPM) integrated multi-satellite retrievals for GPM (IMERG). Algorithm Theoretical Basis Doc., version 4.5, 26 (2015). http://pmm.nasa.gov/sites/default/files/document_files/IMERG_ATBD_V4.5.pdf

W.W. Immerzeel, L. Petersen, S. Ragettli, F. Pellicciotti, The importance of observed gradients of air temperature and precipitation for modeling runoff in a glacierized catchment in the eastern Himalayas. Water Resour. Res. **50**(3), 2212–2226 (2014). https://doi.org/10.1002/2013WR014506

E. Kalnay, *Atmospheric Modeling* (Cambridge University Press, Data Assimilation and Predictability, 2003)

A. Karpatne, G. Atluri, J.H. Faghmous, M. Steinbach, A. Banerjee, A. Ganguly, Theory-guided data science: a new paradigm for scientific discovery from data. IEEE Trans. Knowl. Data Eng. **29**(10), 2318–2331 (2017). https://doi.org/10.1109/TKDE.2017.2720168

S.Q. Kidder, T.H. Vonder Haar, Satellite meteorology. Academic Press (1995). ISBN 9780124064300. https://doi.org/10.1016/B978-0-08-057200-0.50002-X

P.R. Kumar, A.K. Kamra, The spatiotemporal variability of lightning activity in the Himalayan foothills. J. Geophys. Res. Atmos. **117**, D24201 (2012). https://doi.org/10.1029/2012JD018246

P. Kumar, S.K. Deb, C. Kishtawal, P.J.T. Pal, Impact of assimilation of INSAT-3D retrieved atmospheric motion vectors on short-range forecast of summer monsoon 2014 over the south Asian region. Theor. Appl. Climatol. **128**, 575–586 (2017). https://doi.org/10.1007/s00704-015-1722-5

J. Lu, X. Qie, X. Xiao, R. Jiang, E.R. Mansell, A.O. Fierro, et al., Effects of convective mergers on the evolution of microphysical and electrical activity in a severe squall line simulated by WRF coupled with explicit electrification scheme. J. Geophys. Res.: Atmos. **127**, e2021JD036398 (2022). https://doi.org/10.1029/2021JD036398

A.K. Mitra, N. Kaushik, A.K. Singh, S. Parihar, S.C. Bhan, Evaluation of INSAT-3D satellite derived precipitation estimates for heavy rainfall events and its validation with gridded GPM (IMERG) rainfall dataset over the Indian region. Remote. Sens. Appl.: Soc. Environ. **9**, 91–99 (2018). https://doi.org/10.1016/j.rsase.2017.12.006

T. Phulpin, D. Renaut, H. Roquet, C. Camy-Peyret, B. Bell, J.-N. Thépaut, J Eyre The assimilation of satellite data in numerical weather prediction systems, in *Satellites for Atmospheric Sciences 2* ed. by T. Phulpin, D. Renaut, H. Roquet, C. Camy-Peyret (2023). https://doi.org/10.1002/9781394264780.ch4

C. Price, D. Rind, A simple lightning parameterization for calculating global lightning distributions. J. Geophys. Res.-Atmos. **97**, 9919–9933 (1992). https://doi.org/10.1029/92JD00719

K. Qie, X.S. Qie, W.S. Tian, Increasing trend of lightning activity in the South Asia region. Sci. Bull. **66**, 78–84 (2021). https://doi.org/10.1016/j.scib.2020.08.033

S. Ravela, K. Emanuel, D. McLaughlin, Data assimilation by field alignment. Physica D **230**(1–2), 127–145 (2007). https://doi.org/10.1016/j.physd.2006.09.035

D. Rosenfeld, G. Gutman, Retrieving microphysical properties near the tops of potential rain clouds by multispectral analysis of AVHRR data. Atmos. Res. **34**(1–4), 259–283 (1994). https://doi.org/10.1016/0169-8095(94)90096-5

C.J. Schultz, L.D. Carey, E.V. Schultz, R.J. Blakeslee, Kinematic and microphysical significance of lightning jumps versus nonjump increases in total flash rate. Weather Forecast. **32**(1), 275–288 (2017). https://doi.org/10.1175/WAF-D-15-0175.1

R.A. Scofield, The NESDIS operational convective precipitation-estimation technique. Mon. Weather Rev. **115**(8), 1773–1793 (1987). https://doi.org/10.1175/1520-0493(1987)1153C1773:TNOCPE3E2.0.CO;2

D. Shrestha, R. Deshar, K. Nakamura, Characteristics of summer precipitation around the Western Ghats and the Myanmar West Coast. Int. J. Atmos. Sci. 206016 (2015). https://doi.org/10.1155/2015/206016

M. Shrestha, S. Sharma, R.P. Shrestha, Landslides in the Himalayas: a comprehensive review of hazards, impacts, and adaptive strategies. Rural. Reg.Nal Dev. **3**, 10002 (2025). https://doi.org/10.70322/rrd.2025.10002

O. Singh, J. Singh, Lightning fatalities over India: 1979–2011. Meteorol. Appl. **22**, 770–778 (2015). https://doi.org/10.1002/met.1520

C. Skamarock, B. Klemp, J. Dudhia, O. Gill, D. Barker, G. Duda, X. Huang, W. Wang, G. Powers, A description of the advanced research WRF version 3 (2008). https://doi.org/10.5065/D68S4MVH.

G. Skofronick-Jackson, W.A. Petersen, W. Berg, C. Kidd, E.F. Stocker, D.B. Kirschbaum, R. Kakar, S.A. Braun, G.J. Huffman, T. Iguchi, P.E. Kirstetter, C. Kummerow, R. Meneghini, R. Oki, W.S. Olson, Y.N. Takayabu, K. Furukawa, T. Wilheit, The global precipitation measurement (GPM) mission for science and society. Bull. Am. Meteor. Soc. **98**(8), 1679–1695 (2017). https://doi.org/10.1175/BAMS-D-15-00306.1

R.B. Smith, The influence of mountains on the atmosphere. Adv. Geophys. **21**, 87–230 (Elsevier, 1979). https://doi.org/10.1016/S0065-2687(08)60262-9

P.K. Smolarkiewicz, R. Rotunno, Low froude number flow past three-dimensional obstacles. Part I: Baroclinically generated lee vortices. J. Atmos. Sci. **46**(8), 1154–1164 (1989). https://doi.org/10.1175/1520-0469(1989)046<1154:LFNFPT>2.0.CO;2

S. Sorooshian, A. AghaKouchak, P. Arkin, Global precipitation estimation from satellite imagery. Adv. Remote. Sens. Precip. 3–27 (2020). https://doi.org/10.1016/B978-0-12-814922-5.00001-4

A. Srivastava, M. Mishra, M. Kumar, Lightning alarm system using stochastic modelling. Nat. Hazards **75**(1), 1–11 (2015). https://doi.org/10.1007/s11069-014-1247-8

A. Srivastava, D. Liu, C. Xu, S. Yuan, D. Wang, B. Ogunsua, Z. Sun, Z. Chen, H. Zhang, Lightning nowcasting with an algorithm of thunderstorm tracking based on lightning location data over the Beijing area. Adv. Atmos. Sci. **39**(1), 178–188 (2022). https://doi.org/10.1007/s00376-021-0398-2

A. Srivastava, S.S. Kundu, Lightning characteristics in various phases of thunderstorm over north-eastern India associated with dual-polarization radar—a preliminary case study, in *Weather and Climate Services Over the Mountain Region Tropmet-2020*, Shillong, Meghalaya, December 14–17 (2020)

A. Srivastava, S.S. Kundu, S.D. Pawar, R.B. Gogoi, A. Chakravorty, A. Chhari, V. Gopalkrishnan, S.P. Aggarwal, Evaluation of WRF-ELEC model to forecast lightning over the North Eastern region of India. Meteorol. Atmos. Phys. **135**, 39 (2023). https://doi.org/10.1007/s00703-023-00977-y

K.V. Subrahmanyam, K.K. Kumar, CloudSat observations of cloud-type distribution over the Indian summer monsoon region. Annales geophysicae (Göttingen, Germany: Copernicus Publications, 2013) **31**(7), 1155–1162. https://doi.org/10.5194/angeo-31-1155-2013

R.J. Thayyen, J.T. Gergan, Role of glaciers in watershed hydrology: an emerging threat of deglaciation on a Himalayan basin water regime. Hydrol. Process. **24**(20), 2948–2954 (2010). https://doi.org/10.1002/hyp.7706

Y. Tian, X. Qie, Y. Sun, D. Wang, S. Yuan, Z. Sun, G. Lu, L. Yu, H. Sun, L. Li, C. Du, Total lightning signatures of thunderstorms and lightning jumps in hailfall nowcasting in the Beijing area. Atmos. Res. **230**, 104646 (2019). https://doi.org/10.1016/j.atmosres.2019.104646

M.I.R. Tinmaker, K. Ali, Space-time variation of lightning activity over northeast India. Meteorol Zeitschrift **21**, 135–143 (2012). https://doi.org/10.1127/0941-2948/2012/0227

M.L. Weisman, C. Davis, W. Wang, K.W. Manning, J.B. Klemp, Experiences with 0–36-h explicit convective forecasts with the WRF-ARW model. Weather Forecast. **23**(3), 407–437 (2008). https://doi.org/10.1175/2007WAF2007005.1

C.D. Whiteman, Mountain meteorology: fundamentals and applications. Oxford University Press (2000)

Y. Yair, B. Lynn, C. Price, V. Kotroni, K. Lagouvardos, E. Morin, A. Mugnai, M.D.C. Llasat, Predicting the potential for lightning activity in Mediterranean storms based on the weather research and forecasting (WRF) model dynamic and microphysical fields. J. Geophys. Res. **115**, D04205 (2010). https://doi.org/10.1029/2008JD010868

A. Yatagai, K. Kamiguchi, O. Arakawa, A. Hamada, N. Yasutomi, A. Kitoh, APHRODITE: constructing a long-term daily gridded precipitation dataset for Asia based on a dense network of rain gauges. Bull. Am. Meteor. Soc. **93**(9), 1401–1415 (2012). https://doi.org/10.1175/BAMS-D-11-00122.1

Y. Zhang, Z. Chen, X. Xiao, X. Qie, M. Chen, J. Lu, D. Wang, S. Yuan, H. Lyu, J. Feng, S. Fan, D. Liu, Combined assimilation of radar and lightning data for the short-term forecast of severe convection system. Atmos. Res. **283**, 106562 (2023). https://doi.org/10.1016/j.atmosres.2022.106562

Shyam S Kundu did his M.Sc. in Physics from University of North Bengal and Ph.D. in Physics from Gauhati University. He has more than 20 years of professional experience at NESAC and currently working as the Head of the Space and Atmospheric Sciences Division at the centre. He was instrumental in setting up the Space and Atmospheric Science research facility at NESAC with several instruments, a S-band Doppler Weather Radar, largest network of weather stations in NER covering difficult terrains, three aerosol observatories including two in Himalayan region, network of lightning detectors, etc. His research interest include extreme weather forecasting in NER, lightning physics, aerosol-cloud-precipitation interaction studies and its climatic impact, climate change assessment, etc. He took lead role to develop an operational lightning early warning system for the NE region of India. He has more than 50 publications in peer reviewed journals and books. He is making significant contribution to professional bodies like ISRS as its Joint Secretary and ISPRS as the Secretary of the WG-4 under TC V. He also represents ISRO in the Committee on Earth Observation Satellites for Precipitation constellations.

Rekha B Gogoi joined NESAC as Scientist/Engineer 'SC' on February 27, 2009, and has since led key projects applying space technology to disaster management in Northeast India. She was instrumental in operationalizing Numerical Weather Prediction (NWP) and ensemble-based data assimilation to enhance the region's Flood Early Warning System (FLEWS). She also led seismic hazard assessment under the HRVA project for Assam. As Focal Scientist for the NER-DRR program, she coordinates core activities, ensures smooth implementation, and works closely with national and state disaster management agencies during emergencies. Her efforts have strengthened both flood and seismic preparedness in the region through advanced modeling and inter-agency collaboration.

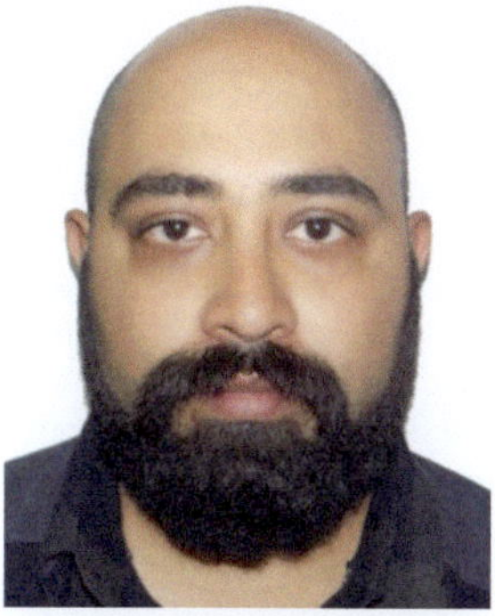

Aniket Chakravorty is a Scientist in the Space and Atmospheric Sciences Division at North Eastern Space Applications Centre, Department of Space, Government of India. He has a doctorate in land atmosphere interactions from the Indian Institute of Technology Delhi. He is working on understanding the changes in the climate of North East India with a special emphasis on extreme precipitation events. He has expertise in uncertainty quantification, time series analysis, and climate downscaling.

Abhay Srivastava is a Scientist of Space and Atmospheric Science division, North Eastern Space Application Centre (NESAC), Department of Space, Govt. of India. His research focuses are multidisciplinary direction on lightning physics, lightning detection, lightning associate events, thunderstorms, severe weather and global electric circuit. He received prestigious fellowships, DST-INSPIRE from Department of Science and Technology, India, CSIR-SRF-EXT (Interdisciplinary) from Council of Scientific and Industrial Research, India and prestigious Post-Doctoral Research Award President's International Fellowship (PIFI-Fellow) from Chinese Academy of Sciences, China. He authored/co-authored more than 25 research articles published in reputed scientific journals and delivered lectures in various conferences in National and International on his research theme. He has also severed as guest editor of Frontiers in Astronomy and Space Sciences and reviewer of important journals.

Arundhati Kundu is currently working as a Research Scientist at NESAC under the NER-DRR program. She was born and brought up in Maldah district, West Bengal, India. She completed her schooling at native place and obtained her B.Sc. in Physics from the University of Calcutta, India. She has completed her masters in Atmospheric Science from the same university and currently persuing a Ph. D. in Physics from Cotton University, India. Her research interests include aerosol cloud interaction and data assimilation for the prediction of extreme weather events. Outside her academic and research work, she loves reading psychological thrillers, watching anime, cooking, stitching and playing badminton.

Dr. Arup Borgohain is currently serving as a Senior Scientist in the Space and Atmospheric Science Division at the North Eastern Space Applications Centre (NESAC), under the Department of Space, ISRO. He holds an M.Sc. in Physics, an M.Tech. in Space Science, and a Ph.D. in Physics with a specialization in upper atmospheric studies. His main areas of research include aerosol–trace gas dynamics, atmospheric boundary layer processes, mass–energy exchange mechanisms, and climate science. He has conducted several field campaign experiments across the North Eastern Region of India to investigate aerosol–trace gas interactions and boundary layer dynamics. Dr. Borgohain has authored more than 50 research papers in reputed international journals. In addition to his scientific contributions, he is deeply committed to science communication and regularly writes popular science articles in leading Assamese magazines and newspapers. He is also actively involved in guiding Ph.D. scholars from reputed institutions such as the Indian Institute of Space Science and Technology (IIST), Dibrugarh University, Gauhati University, and Assam Science and Technology University.

Rosly Boy Lyngdoh is a Scientist in the Space and Atmospheric Sciences Division, North Eastern Space Applications Centre, Department of Space, Government of India. His research focuses on the application of Artificial Intelligence approaches for nowcasting and short-range prediction of features of weather, like lightning, rainfall, etc. using satellite remote sensing.

Chapter 22
Shifting Climate Patterns in North East India: Past Trends and Future Projections

Aniket Chakravorty, Shyam S. Kundu, Rosly B. Lyngdoh, Arup Borgohain, and Shiv P. Aggarwal

Abstract North East India has been undergoing marked climate change, with consistent warming observed from 1971 to 2020. The annual mean daily average air temperature has risen by 0.019 °C/year, with both maximum and minimum temperatures increasing. Winter nights are warming faster than days, likely due to Bay of Bengal warming and higher atmospheric moisture. Projections suggest further intensification of warming, especially under high-emission scenarios, with minimum temperatures rising more rapidly. The region's precipitation patterns have also changed significantly. Since the 1970s, total annual rainfall (PRCPTOT) has declined by approximately 9.81 mm/year, mainly due to fewer light and moderate rainfall events, longer dry spells, and shorter wet spells. In contrast, extremely rare, high intensity events have become more frequent, with frequency contributing more than intensity to the change. However, future climate projections indicate a reversal in total rainfall trends, with increases in both overall precipitation and extreme rainfall events under intermediate and high-emission pathways. For instance, SSP5-8.5 projects a 7.68 mm/year increase in PRCPTOT. Thus if these projections came to pass, North East India might become wetter than the present. The present trend of climate on the other hand suggests a shift towards higher risk of flash floods, heat waves, and droughts.

A. Chakravorty (✉) · S. S. Kundu · R. B. Lyngdoh · A. Borgohain · S. P. Aggarwal
North Eastern Space Applications Centre (NESAC), Umiam, Shillong, Meghalaya, India
e-mail: aniket.chakravorty@nesac.gov.in

S. S. Kundu
e-mail: ss.kundu@nesac.gov.in

R. B. Lyngdoh
e-mail: rosly.lyngdoh@nesac.gov.in

A. Borgohain
e-mail: arup.borgohain@nesac.gov.in

S. P. Aggarwal
e-mail: spaggarwal2010@gmail.com

S. P. Aggarwal et al. (eds.), *Remote Sensing of the North Eastern Himalayan Ecosystem*,
https://doi.org/10.1007/978-981-95-3221-6_22

Keywords North East India · Climate change · Future projections · Regional warming · Extreme events

22.1 Introduction

Climate is a complex system involving interactions between the atmosphere, land, oceans, cryosphere, and human society across multiple scales. It is defined as the average weather over a long period (typically 30 years), leading to the saying: "Climate is what you expect, weather is what you get." Climate evolves over time due to internal variability or external forcings—natural (e.g., volcanic eruptions, solar changes) or anthropogenic (e.g., greenhouse gases, aerosols). The system is driven by solar radiation, and its state is influenced by changes in:

- The intensity of incoming solar radiation,
- The fraction of radiation reflected by Earth, and
- The energy retained in the system, through the global greenhouse effect.

22.1.1 Global Greenhouse Effect

The greenhouse effect refers to the process where Earth's absorbed shortwave solar radiation is re-emitted as longwave thermal radiation, which is then absorbed by atmospheric gases—mainly triatomic molecules like H_2O and CO_2. These gases, known as greenhouse gases (GHGs), have vibrational and rotational modes that absorb and re-emit this thermal radiation, warming the atmosphere. To understand its significance, let us examine a simple greenhouse model.

Earth intercepts about 1367 W/m^2 of incoming solar energy flux, this is colloquially called Solar Constant, S_0. The solar energy is intercepted by Earth over a cross-sectional area of πa^2, where a is the radius of Earth. Thus,

$$\text{Total power of solar radiation incident on earth} = S_0 \pi a^2 \tag{22.1}$$

Earth reflects a significant portion of the incident solar radiation back to space due to the presence of various reflective surfaces, like, clouds, fresh snow, sand, deserts, sea ice, etc. The ratio of the reflected to incident solar radiation is called *albedo*, α. The average *planetary albedo* (α_p) of Earth is ~ 0.30 or 30%. Thus,

$$\text{Solar power absorbed by the Earth} = \left(1 - \alpha_p\right) S_0 \pi a^2 \tag{22.2}$$

Considering an Earth, without an atmosphere, as a isotropic black body in energy balance, the emitted energy is given according to the Stefan-Boltzmann Law as,

$$\text{Emitted radiation per unit area} = \sigma T_e^4 \tag{22.3}$$

where, σ is the Stefan-Boltzmann constant (5.67×10^{-8} Wm^{-2} K^{-4}) and T_e is the temperature in Kelvin at which the black body is emitting, also called *emission temperature* or *effective planetary temperature.* Since, this black body radiation is being emitted equally throughout the surface of the earth in energy balance, the *emission temperature* (T_e) can be derived as:

$$T_e = \left[\frac{S_0(1-\alpha_p)}{4\sigma}\right]^{1/4} \tag{22.4}$$

By putting in the numbers and solving it, we find that T_e, which is same as the *surface temperature* (T_s) of Earth without an atmosphere, to be 255 K (–18 °C). Which is significantly lower than the actual average T_s of 288 K. Life would not have been possible on Earth, if we were at 255 K. The difference between T_s and T_e is because of radiation being absorbed within the atmosphere because of the *greenhouse effect* and fluid motion, atmospheric circulation, which carry heat throughout the atmosphere.

For understanding the influence of atmosphere on the surface temperature of Earth, a simple green house model with a single layer opaque atmosphere which absorbs all the terrestrial radiation being emitted by the Earth (Fig. 22.1) can be analysed. In case of an opaque atmosphere, longwave radiation emitted from the Earth is absorbed completely by the atmosphere and re-emitted in all directions equally. Thus considering energy balance at the Earth's surface as well as at the top of the atmosphere, a relation between T_s and T_e can be derived as:

$$T_s = 2^{1/4} T_e \tag{22.5}$$

Thus considering a T_e of 255 K and for an atmosphere that can absorb all the longwave radiation emitted from the Earth's surface T_s would rise by a factor of $2^{1/4}$ = 1.189 to 303 K. However, Earth's surface is not at 303 K but at 288 K. This is because Earth's atmosphere is not an opaque atmosphere, absorbing all the longwave

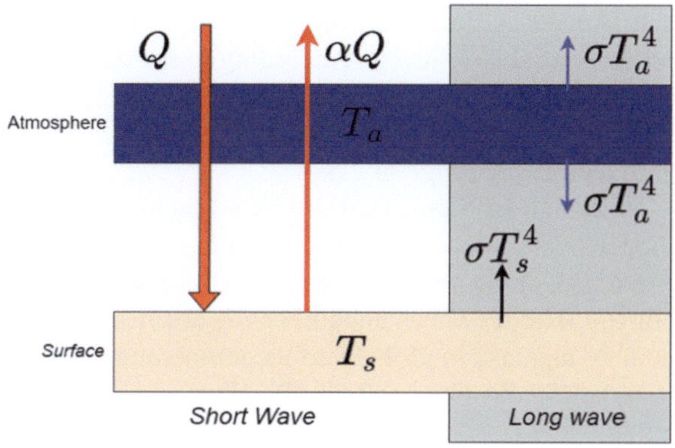

Fig. 22.1 Schematic representation of a simple green house model with a single layer opaque atmosphere

Table 22.1 The change of average surface temperature (K) with a change in absorptivity of atmosphere for an emission temperature of 255 K

Absorptivity	Surface temperature (K)
0	255.0
0.2	261.8
0.4	269.6
0.6	278.8
0.8	289.7
1.0	303.2

radiation emitted by Earth, but a leaky atmosphere. How leaky is an atmosphere, is determined by the absorptivity of the atmosphere (ϵ). Invoking the Kirchhoff's law, which states that the emissivity of a body is equal to its absorptivity. The absorptivity (ϵ) ranges from 0 to 1.0, with 0 corresponding to no atmospheric absorption and 1.0 corresponding to full absorption of the terrestrial thermal radiation by the atmosphere (opaque atmosphere). In case of a leaky atmosphere, Eq. 22.5 looks like:

$$T_s = \left(\frac{2}{2-\in}\right)^{1/4} T_e \tag{22.6}$$

Table 22.1 shows T_s corresponding to a constant T_e of 255 K for an increasing ϵ. The ϵ of the atmosphere increases as the concentration of greenhouse gases increase.

22.1.2 *Climate Sensitivity*

Earth is not warming exactly as shown in Table 22.1 because its atmosphere is a layered system with varying ϵ and temperature, and energy transfer involves both radiation and atmospheric circulation. Similarly, the surface shows spatial variability due to oceans and diverse land use. The expected rise in surface temperature from increased CO_2 is captured by metrics like Equilibrium Climate Sensitivity (ECS) and Transient Climate Response (TCR), which estimate warming from a doubling of pre-industrial CO_2 levels. According to the recent 6th Assessment Report (AR6) of the Inter-governmental Panel on Climate Change (IPCC), the best estimate of TCR is 1.8 °C with a probable range of 1.2–2.4 °C, and the best estimate of ECS is 3 °C with a probable range of 2.5–4 °C.

Although ECS and TCR are based on model simulations, this is no longer hypothetical—CO_2 concentrations reached ~ 422 ppm in 2024, up from ~280 ppm. This sharp rise, along with other greenhouse gases, has driven significant changes in the climate. The IPCC has unequivocally attributed these to human activity, including a notable increase in global surface temperature (1.09 °C) over the past four decades (Land: 1.59 °C; Ocean: 0.88 °C). Warming has accelerated since 1970, faster than any 50-year period in the past 2000 years (IPCC Synthesis Report 2023).

22.1.3 Climate Scenarios

This warming trend of the global climate and its effects on the climate system might continue in the future too, if no changes are made to how humans interact with the Earth system. For governments to plan the appropriate pathway for the future, climate scientist along with policy experts have come up with probable future scenarios that the world might take or can take. These scenarios are not a prediction of the future but are a glimpse into the futures that are expected if humans make the choices as stipulated in the framework of these scenarios. The climate change scenarios incorporate a range of plausible pathways that include changes in governmental policies, socio-economic dynamics, advances in energy efficient technologies, changes in emissions and concentrations of aerosols, etc. The development of these climate change scenarios has come a long way from the Special Report on Emission Scenarios (SRES) used in IPCC AR3 to the combination of Shared Socio-economic Pathways (SSP) and Representative Concentration Pathways (RCP) used in AR6.

The RCPs are climate scenarios based on different levels of radiative forcing by 2100 relative to pre-industrial times. Developed for IPCC's Fifth Assessment Report (AR5), they include GHG concentrations, aerosols, chemically active gases, and land use changes. The four original RCPs—RCP2.6, RCP4.5, RCP6.0, and RCP8.5—represent increasing radiative forcing levels from 2.6 to 8.5 W/m^2. RCP2.6, 4.5, and 6.0 are climate policy scenarios, while RCP8.5 assumes high emissions with a slowdown after 2050. Each was developed using different Integrated Assessment Models (IAMs) with distinct baseline assumptions (van Vuuren et al. 2011). Importantly, no probabilities were assigned to any RCP. Recognizing the need for more nuanced scenarios, additional pathways—RCP1.9, RCP3.4, and RCP7.0—were later introduced. Independent from the development of RCPs, IAM modelers explored a range of different technological, socio-economic and policy strategies that lead to specific concentration pathways (Moss et al. 2010). Five SSPs were designed, each representing various levels of challenges to mitigation and adaptation (O'Neill et al. 2014):

- SSP1: Sustainability (low challenge to both)
- SSP2: Middle of the Road (medium challenge)
- SSP3: Regional Rivalry (high challenge to both)
- SSP4: Inequality (low mitigation, high adaptation challenge)
- SSP5: Fossil-fueled Development (high mitigation, low adaptation challenge)

Each of the SSPs do have their own baseline scenarios, however these baselines (no-climate policy baselines) do not include any mitigation target which lead to higher warming compared to pre-industrial levels. The SSP-RCP pairs include these mitigation target and a good spread of warming is obtained, for example, SSP1-RCP1.9 (SSP1-1.9) refers to a scenario that limits warming to below 1.5 °C (Riahi et al. 2017).

These climate scenarios have been used to describe the probable changes in the global as well as the regional climates in various studies. According to IPCC AR6, the global climate keeps on warming, for all future scenarios, at least till the mid-century. For the very low GHG emission scenario of SSP1-1.9, the range of warming is projected to be 1–1.8 °C (best estimate: 1.4 °C) and for the very high GHG emission scenario of SSP5-8.5, it is projected to be 3.3–5.7 °C (best estimate: 4.4 °C) by 2081–2100, compared to the 1850–1900 period. This warming of the climate is projected to further intensify the water cycle, which will eventually lead to higher variability, stronger monsoons and increased severity of wet and dry extremes.

22.1.4 Climate Change: Indian Perspective

One of the major takeaways from IPCC AR6 is that without large scale reduction in emissions, it will not be possible to reduce the rise in temperature to below 1.5–2.0 °C. This does not bode well for the Indian sub-continent. India is already experiencing significant warming, with studies reporting rising air temperatures both annually and seasonally since the early twentieth century. The warming is non-uniform, both in time as well as in space, with the pre-monsoon season and northern India warming faster, accompanied by increases in humidity (Kothawale and Singh 2017; Mukhopadhyay et al. 2017; Srivastava et al. 2019; Vinnarasi et al. 2017). Temperature extremes have also intensified, with a greater number of warm days and nights, and a decrease in cold extremes, leading frequent heatwaves (Rohini et al. 2016; Sharma and Majumdar 2017). Kothawale and Singh (2017) also found that the tropospheric air temperature from surface till 500 hPa is also rising, further amplifying the extremes.

Global and regional climate model projections show continued warming across India under various future emission scenarios (Sanjay et al. 2020; Sarkar and Maity 2024). High emission scenarios like RCP8.5, showed that India could see warming beyond 4 °C, while low-emission scenario (RCP2.6) projects warming to be restricted between 2 °C and 1.5 °C, by the end of the century (Sanjay et al. 2020). Studies have revealed that the warming is strongest during the post-monsoon and winter seasons (Salunke et al. 2023; Sarkar and Maity 2024). The projected future warming over India will also exacerbate the hot and dry extremes as well as compound hot dry extremes by increasing their frequency and spatial extent (Kumar and Goyal 2024; Norgate et al. 2024).

While rising temperatures can enhance precipitation, but unlike temperature, precipitation processes are relatively more complex with multiple feedbacks. Studies analyzing the Indian Summer Monsoon have observed a relatively stable monsoon with no significant change in precipitation over the century (Goswami et al. 2006; Kulkarni 2012). However, a decline in Monsoon precipitation has been observed in the decades post 1970 (Kulkarni 2012). The decrease in precipitation over the Indian sub-continent has been attributed to various factors, namely, changes in anthropogenic aerosol concentration (Ramanathan et al. 2005), equatorial Indian Ocean warming (Kulkarni 2012) and land use/land cover change (Krishnan et al. 2016). The decline in annual precipitation over India is accompanied by an increase in its variability (Goswami et al. 2006). The increased variability has led to an increase in the frequency and magnitude of extreme rainfall events as well as an increase in frequency of dry spells (Singh et al. 2014). Vinnarasi and Dhanya (2016) have also indicated a trend of wet regions getting drier and dry regions getting wetter over India.

Contrary to the observed drying trend of Indian monsoon, all future warming scenarios showed a significant increasing trend of monsoon precipitation (Kushwaha et al. 2024). Katzenberger et al. (2021) examined 32 models and found that all models showed a significant increase in monsoon rainfall for all the warming scenarios: SSP5–8.5 (+24.3%), SSP3–7.0 (+18.6%), SSP2–4.5 (+11.9%), and SSP1–2.6 (+9.7%). Along with the annual precipitation, studies have also shown an increase in the frequency and intensity of high impact precipitation events (90–99.99th percentile) over India, especially for high emission scenarios (Myhre et al. 2019; Shahi et al. 2023).

Although all the studies of climate change emphasized the noticeable change in climate over India, they noted the significant observed and projected change over the North East region of India. Out of the eight most vulnerable states in India, three are northeastern states, namely, Mizoram, Assam, and Arunachal Pradesh, and the rest have been categorized as moderately vulnerable except for Sikkim which has the lowest vulnerability out of the eight northeastern states. This chapter will do a deep dive into these changes that has been reported and observed through analysis of climate data.

22.2 Climatology of North East India

Northeast India is a geographically diverse region, featuring a range of landscapes from the lush Brahmaputra Valley in Assam to the mountainous terrains of Arunachal Pradesh and Sikkim and bordered by Bhutan, China, Myanmar, and Bangladesh. The region is characterized by its dense forests, the mighty Brahmaputra River basin, and numerous biodiversity hotspots. The Eastern Himalayas run through parts of Arunachal Pradesh and Sikkim, while Meghalaya is famous for its rolling hills and highest rainfall areas like Sohra (Cherrapunjee) and Mawsynram. This varied geography supports rich flora and fauna unique to the region along with its diverse climatology. The North East India can be categorized into four meteorological subdivisions, namely, Arunachal Pradesh; Assam and Meghalaya; Nagaland-Manipur-Mizoram-Tripura (NMMT) and the Sub-Himalayan West Bengal (Fig. 22.2). The subdivision of Sub-Himalayan West Bengal consists of Sikkim and parts of northern West Bengal.

The climatology of rainfall and temperature of North East India is a characteristic of this region, with sub-zero temperatures in the northern reaches along the foothills of Himalaya and an average of 25 °C in the plains of Assam (Fig. 22.3a). The climatology of annual accumulated rainfall over the region shows a spatial mean rainfall of approx. 2000 mm and a maximum annual rainfall of approx. 9000 mm

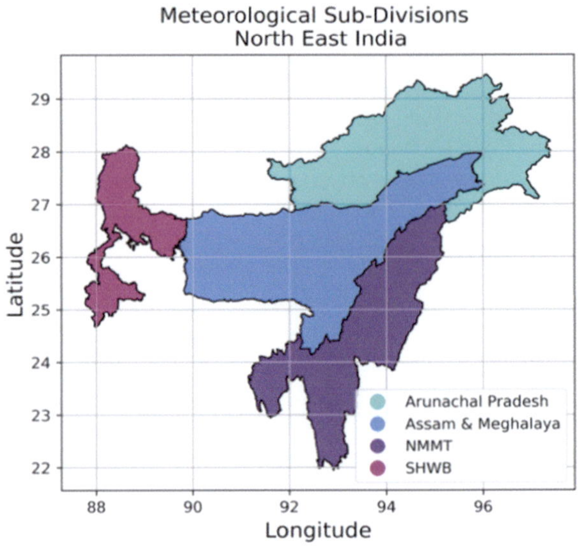

Fig. 22.2 The four meteorological Sub-divisions that make up the North East India: Arunachal Pradesh, Assam and Meghalaya, Nagaland-Manipur-Mizoram-Tripura (NMMT), and Sub-Himalayan West Bengal (SHWB)

(Fig. 22.3b). The maximum annual accumulated rainfall corresponds to the wettest state of North East India, Meghalaya. The orography of the region plays a significant role in the region's unique rainfall climatology with high convective activity resulting in significant number of extreme rainfall events (Goswami et al. 2010). Unlike the rest of India where the summer monsoon is for four months, the North East India experiences a summer monsoon of five months from June-October. The extension of monsoon beyond September over this region has been attributed to recycled precipitation during September. The North East region has a high recycling ratio of 25% due to the presence of high vegetation cover which contributes to higher evapotranspiration from this region (Pathak et al. 2014).

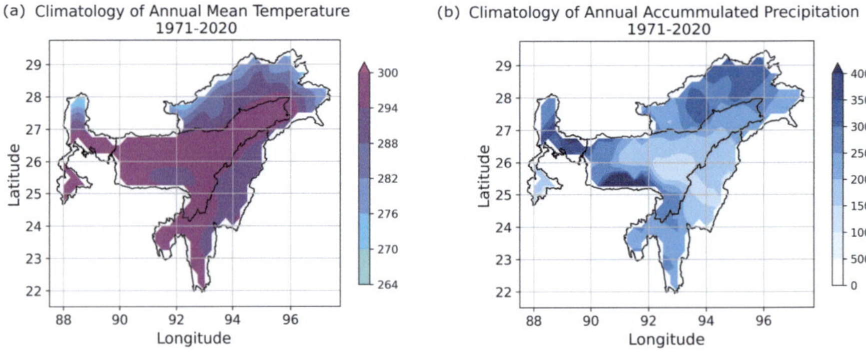

Fig. 22.3 Climatology of **a** annual mean daily average air temperature and **b** annual accumulated rainfall from 1971–2020 over North East India. *Data Source* ERA5

22.3 Data and Methodology for Climatological Trend Analysis

Climatological analysis relies on reliable, spatiotemporally consistent data. While in-situ observations reflect actual conditions, their limited coverage poses a challenge. To address this, reanalysis and multi-source merged products are widely used. Reanalysis datasets, such as ERA5 and IMDAA, combine models with historical observations using data assimilation. ERA5 offers global hourly data from 1940 at 0.25° resolution (Hersbach et al. 2020), while IMDAA provides regional hourly data over India from 1948 at 12.5 km resolution (Rani et al. 2021). These datasets include key atmospheric, oceanic, and land variables critical for understanding Earth system interactions.

Along with reanalysis products various multi-source merged products are also currently available especially of precipitation and soil moisture. NASA's Integrated Multi-satellitE Retrievals for GPM (IMERG: Huffman et al. 2015) and JAXA's Global Satellite Mapping of Precipitation (GSMaP: Kubuta et al. 2007) provide reliable long-term precipitation information at a high resolution of 0.1 ° from 2000 onwards at a half hourly temporal resolution. The India Meteorological Department, recognizing the need for long-term rainfall and temperature analysis developed a century long daily rainfall dataset at a spatial resolution of 0.25 ° from 1901 to present and simultaneously a daily average temperature dataset from 1950 onwards at a spatial resolution of 1° (Pai et al. 2014). The IMD gridded rainfall product is developed from the distribution of 6955 rain gauges and automatic weather stations installed across India.

Reanalysis and multi-source merged products help identify climate system changes due to global warming, while future trends under various warming scenarios are explored using CMIP simulations (Eyring et al. 2016). IPCC assessments rely heavily on CMIP outputs, making them key resources in climate research. The current CMIP6 phase provides simulations at coarse resolutions (~100–200 km). To improve spatial detail, the CORDEX initiative offers downscaled regional climate simulations, with South Asia covering the Indian subcontinent (Giorgi et al. 2009). This chapter uses ERA5 and IMD rainfall data to assess present climate over Northeast India, and an ensemble of 10 CMIP6 models to examine future precipitation trends.

The trend analysis performed in this chapter uses the Mann–Kendall Trend test and Sen's Slope. The Mann–Kendall (MK) test (Kendall 1975; Mann 1945) is a statistical method used to determine whether a variable shows a consistent increasing or decreasing trend over time. This trend, known as monotonic, indicates a steady increase or decrease, but it does not require the trend to follow a linear pattern. The MK test statistic S is given as:

$$S = \sum_{i=1}^{n-1} \sum_{j=i+1}^{n} sgn\left(x_j - x_i\right) \tag{22.7}$$

$$sgn(x_j - x_i) = \begin{Bmatrix} +1, if\,(x_j - x_i) > 0 \\ 0, if\,(x_j - x_i) = 0 \\ -1, if\,(x_j - x_i) < 0 \end{Bmatrix} \tag{22.8}$$

The MK trend test is applied to time series x_i, ranked from i = *1 to n–1*, and x_j that is ranked from $j = i + 1$ *to n*. The S statistics has a zero mean and variance (σ), which is used to reject the null hypothesis of no-trend. The magnitude of the trend is estimated using the Sen's slope as it is robust against the presence of outliers (Sen 1968).

$$\beta = median\left(\frac{x_i - x_j}{i - j}\right), j < i, 1 < j < i < n \tag{22.9}$$

A positive β indicate an increasing trend and a negative β indicate a decreasing trend. The MK trend test and Sen's Slope is applied onto a time series that represent the non-linear slow moving component of the raw time series. The slow moving component of a time series is decomposed using the Singular Spectrum Decomposition (SSD). The detailed description of SSD is available in Golyandina et al. (2001).

22.4 Status of Climate of North East India

22.4.1 Historical Status of Temperature

Studies in the past focusing on the North East region of India have examined the changing pattern of temperature: mean as well as the extremes (Maximum Temperature: *TX* and Minimum Temperature: *TN*). Dash et al. (2012) using IMD gridded temperature noted a significant rise of 0.04 °C/*year* in annual mean temperature over the region during the 1971–2005 period. They also observed a significant rise in minimum temperature in all the four seasons. This rise in daily minimum temperature is reflected in the increase in the frequency of summer warm nights over the region during 1971–2005. As is evident in Fig. 22.4a, the annual mean of daily average temperature shows a constant increase with a 5% significant trend of 0.019 °C/*year*. The rise is consistent for daily maximum (0.023 °C/*year*) as well as daily minimum (0.022/*year*) temperatures averaged over the North East region (Fig. 22.4b, c). Rising minimum temperature, over a region is of crucial importance not just from the perspective of heat waves due to their cascading effect on next day temperatures but also on agricultural yield (Rao et al. 2014). Studies have shown that although higher day-time as well as night-time temperatures, which correspond to daily maximum and daily minimum temperatures respectively, can have disastrous effects on the ecosystem of a region, the impact is more severe for higher night-time temperatures than for day-time temperatures (Sadok et al. 2020; Xia et al.

2010). Since, North East India hosts several ecological hotspots, the rise in minimum temperature can have disastrous effects on the flora and fauna of the region.

Figure 22.5 shows the spatial spread of the Sen's slope of long-term trend of daily mean, maximum and minimum temperatures. The examination of Fig. 22.5, shows that the rise in both maximum and minimum temperatures is significant all throughout the region. Furthermore, the rise in minimum temperatures is higher in the southern parts and the rise in maximum temperatures is higher in central and southern parts compared to rest of the parts. The same can be observed for rising trend of daily mean temperature (Fig. 22.5a). The higher rate of increase in temperatures down south can be attributed to the warming of Bay of Bengal (Goswami et al. 2022). It is evident from Fig. 22.5 that the temperatures all throughout the North East India are rising significantly. Similar results were observed by Jain et al. (2012), where they examined 30 IMD stations spread over the four meteorological sub-divisions and found a rising trend of daily mean, maximum and minimum temperatures over the region.

An analysis of the spatial spread of Sen's slope for the long-term trend of annual maximum and annual minimum of daily maximum and daily minimum temperatures across North East India show an interesting pattern (Fig. 22.6). The annual maximum of daily maximum refers to summer-time hottest days and annual minimum of daily maximum refer to winter-time hottest days, similarly, the annual maximum of daily minimum refers to summer-time coldest night and annual minimum of daily minimum refer to winter-time coldest night. Figure 22.6b and d indicate a significant winter warming in both day as well as night over the region. However, analysis of summer temperatures reveals a significant warming of day-time temperatures over the region, but the night-time warming is substantial only in the southern parts (Fig. 22.6a and c). Another facet of warm extremes is the number of summer days (days with maximum temperature > 25° C) over a region. The North East India has traditionally been cooler than the rest of the country, however, one of the effects of rising air temperatures has been the increase in the number of summer days (Fig. 22.7).

It is evident from Fig. 22.7b that the number of summer days has been steadily increasing over the region (from 182 days in 1971 to 203 days in 2020). This increase is especially evident in the NMMT meteorological sub-division and parts of Meghalaya (Fig. 22.7b). Furthermore, these are the regions which had climatologically lower number of summer days (Fig. 22.7a). These patterns of mean and extremes of temperature indicate a shift towards warmer conditions across North East India with noticeable warming of winter nights.

The warming up of the atmosphere can have a direct effect on the amount of moisture it can hold. This increase in moisture in the atmosphere can be measured by examining the changing pattern of dew point temperature. That is, higher the dew point, greater the amount of moisture in the air. Figure 22.8 shows spatial distribution of dew point temperature climatology and the slope of its trend. It is evident that, similar to the rise in daily mean temperatures, the amount of moisture in the atmosphere is also rising over the region and the rise is especially significant in the central part of the region covering parts of eastern Assam and all of Meghalaya. As water vapor is also a potent greenhouse gas, this rise in moisture content of the

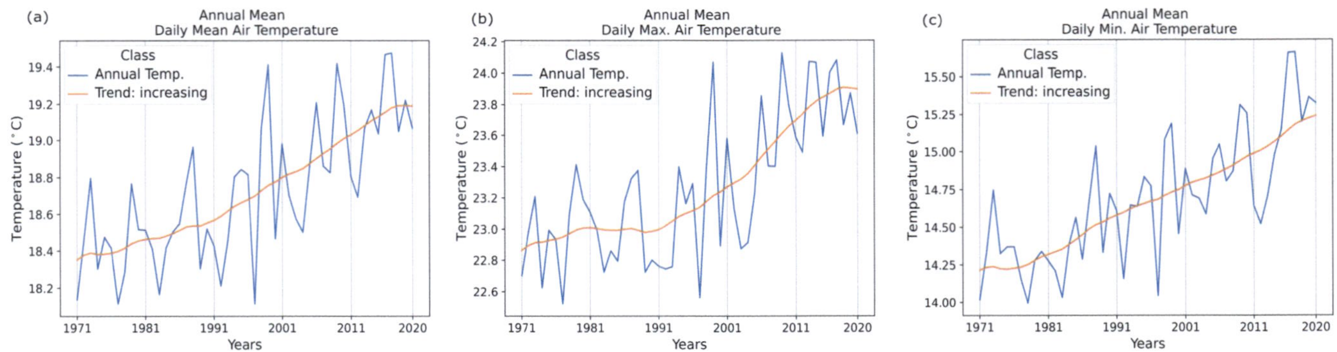

Fig. 22.4 Annual mean of **a** daily average air temperature (Sen's Slope: +0.019 °C/*year*), **b** daily maximum air temperature (Sen's Slope: +0.023 °C/*year*), **c** daily minimum air temperature (Sen's Slope: +0.022 °C/*year*) averaged over North East India from 1971–2020. The trend and Sen's slope are statistically significant ($p < 0.05$). The blue line shows the inter-annual behavior of temperature, and the orange line shows the non-linear trend calculated using Singular Spectrum Analysis. Slope is estimated using Mann–Kendall Trend test. *Data Source* ERA5

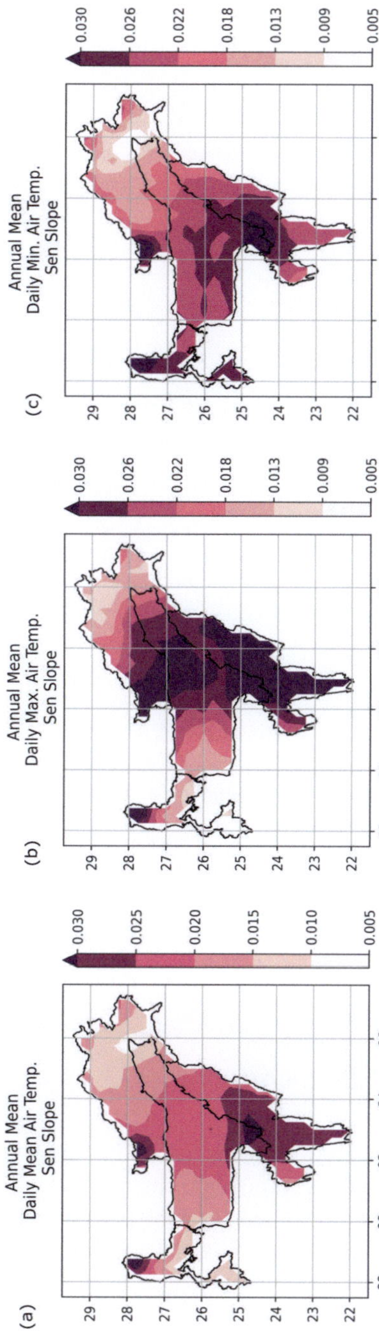

Fig. 22.5 Spatial map of Sen's slope of annual mean of **a** daily average, **b** daily maximum, and **c** daily minimum air temperature from 1971–2020. The Sen's Slopes are statistically significant ($p < 0.05$) and calculated using Mann–Kendall Trend test. *Data Source* ERA5

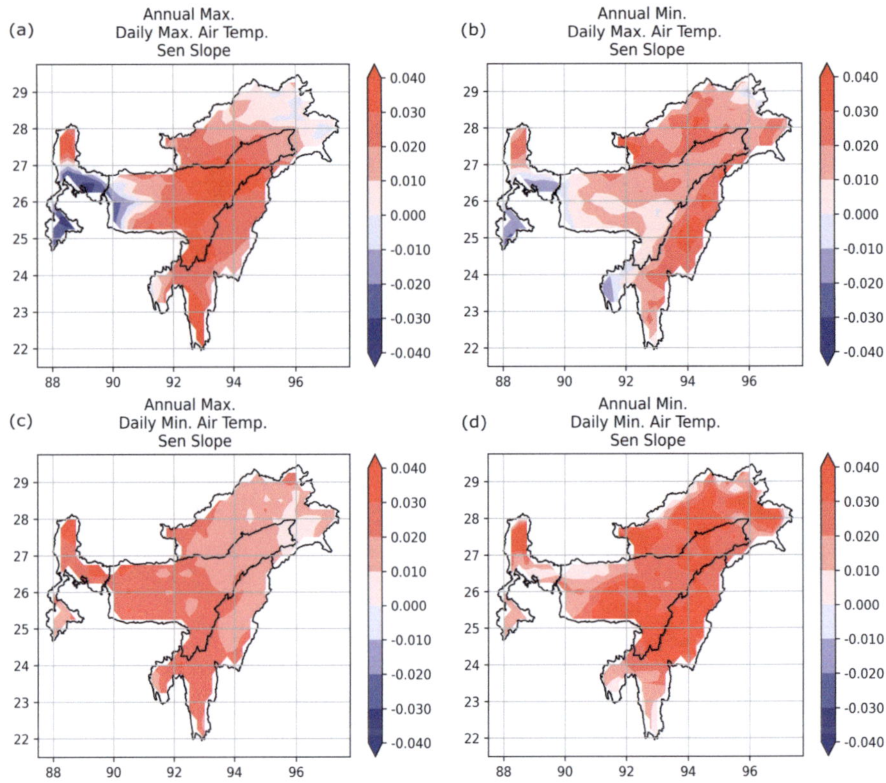

Fig. 22.6 Spatial map of Sen's slope for **a** annual maximum of daily maximum, **b** annual minimum of daily maximum, **c** annual maximum of daily minimum, **d** annual minimum of daily minimum. The Sen's Slope is calculated by using the Mann–Kendall Trend test and only the statistically significant slopes ($p < 0.05$) are presented. *Data Source* ERA5

atmosphere will have positive feedback to the day as well as night temperatures over this region.

22.4.2 Future Status of Temperature

The changing patterns of temperature in the retrospective analysis shows the current effects of global warming and the anthropogenic changes in land use and land cover over North East India. The future of temperature patterns in different warming scenarios shows a similar behavior. In a study by Paul and Maity (2023), they analyzed the behavior of 12 indices of temperature extremes over the contiguous region of North East India and Bangladesh and noted a shift towards warmer conditions, which is especially stronger in the far-future for all warming scenarios. It was observed that,

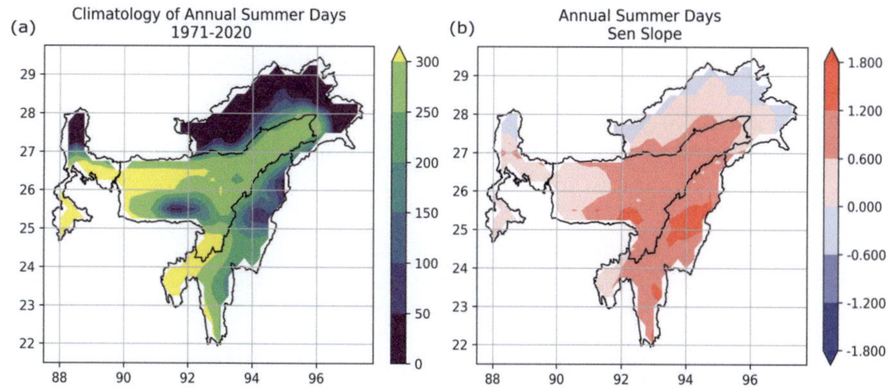

Fig. 22.7 **a** Annual climatology of summer days, **b** Spatial map of Sen's slope values of annual summer days, over North East India from 1971–2020. The Sen's slope is calculated using the Mann–Kendall Trend test and only the statistically significant ($p < 0.05$) trends and slopes are presented. *Data Source* ERA5

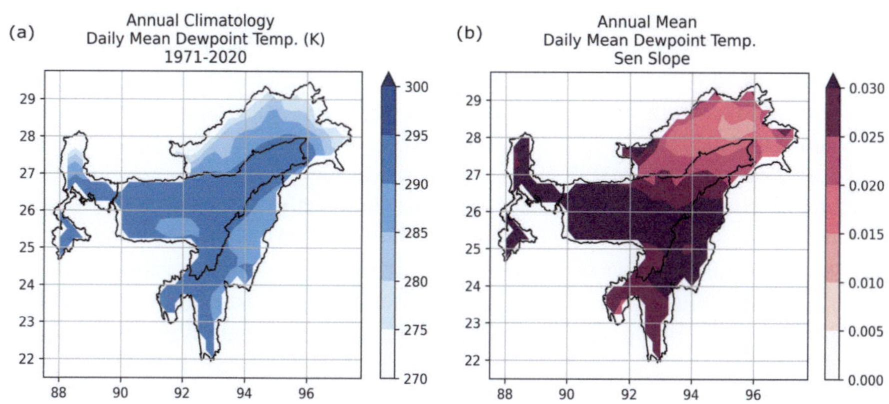

Fig. 22.8 **a** Annual climatology of daily mean dew point temperature (K), **b** Spatial map of Sen's slope of daily mean dew point temperature, over North East India from 1971–2020. The Sen's slope is calculated using the Mann–Kendall Trend test and only the statistically significant ($p < 0.05$) trends and slopes are presented. *Data Source* ERA5

the positive changes are substantially higher for SSP5-8.5 compared SSP1–2.6. A 4 °C rise in maximum and a 5.5 °C rise in minimum temperatures is noted by the end of twenty-first century for the high emission scenario of SSP5-8.5. Similar to the retrospective analysis from 1971–2020, the analysis of the future projections also indicated a higher rate of increase in minimum temperatures compared to maximum temperatures. This is inferred from negative change in the diurnal temperature range ($DTR = TX\text{-}TN$) in SSP1-2.6 (–0.27 °C) and SSP5-8.5 (–1.5 °C). Interestingly, SSP3-7.0 showed higher negative change in DTR compared to SSP5–8.5. The rise in temperature over North East India led to an increase in number of summer days too. Like the retrospective analysis (Fig. 22.7), the Paul and Maity (2023) study also found an increase in the number of summer days over this region with a substantial increase over NMMT and Meghalaya in the future warming scenarios. Additionally, due to the increase in maximum and minimum temperatures, an increase in warm spells and a decrease in cold spells has also been noted by studies. Thus, the shift towards a warmer North East India that has been observed in the retrospective analysis using reanalysis data from 1971–2020 continues in the future emission scenarios with higher warming projected in high emission scenarios (SSP3–7.0, SSP5–8.5) and relatively lower rate of warming in low emission scenarios (SSP1–2.6, SSP2–4.5).

22.4.3 Historical Status of Precipitation

The observed and projected changes in temperature under the influence of global warming has a direct correlation to the changes in precipitation over a region. To investigate this, eight climate change indicators of precipitation (Table 22.2) suggested by the Expert Team on Climate Change Detection Indices (ETCCDI) are utilized. Figure 22.9 shows the spatiotemporal changes in PRCPTOT over the North East region of India from 1901–2020. The long-term trend of PRCPTOT over the region showed a slight increase(+0.48 mm/*year*), however, a closer inspection of the trend reveals a shift in pattern around 1970s. The analysis showed a change in the secular trend of PRCPTOT around 1970 when the increasing trend (Fig. 22.9a: +5.25 mm/*year*) shifts to a significant decreasing trend (Fig. 22.9b: +9.81 mm/*year*). This shift from a positive to a negative trend of total rainfall is consistent over North East India, with northern parts of Arunachal Pradesh and parts of Meghalaya showing the most change (Fig. 22.9c and d). Analysis of the seasonal contribution of rainfall to PRCPTOT reveal a significant decreasing trend in all the seasons (Winter: –2.96 mm/*year*, Pre-Monsoon: –3.42 mm/*year*, Monsoon: –4.67 mm/*year*, Post-Monsoon: –5.27 mm/*year*). Post-monsoon showed the most rapid decrease in total precipitation over the region followed by the monsoon season. This decrease in monsoon rainfall can have far reaching consequences for agricultural yield of the region, especially of Tea and Rice, as most of the agriculture is rain-fed.

Interestingly, the decreasing trend of total precipitation shown in Fig. 22.9 is also observed in the behavior of annual maximum one-day rainfall (Rx1day) too. Figure 22.10 shows the spatiotemporal behavior of Sen's slope of Rx1day rainfall over the region. It also reiterates the sentiment of PRCPTOT, i.e., shifting from an overall positive trend pre-1970 (+0.23 mm/*year*) to a decreasing trend post-1970 (+0.21 mm/*year*). The spatial distribution of long-term trend of Rx1day showed a negative trend over a significant part of North East India, with a staggering decline observed for only a few pockets mostly in northern Arunachal Pradesh, Meghalaya and Sikkim (Fig. 22.10b). Although a positive trend is observed in a few pixels.

Similar to PRCPTOT and Rx1day, frequency of rainfall events of varying magnitude over the different political states of the region also showed a significant decline

Table 22.2 List of ETCCDI precipitation indicators

ETCCDI Indicator	Description
PRCPTOT	Annual total precipitation in wet days
Rx1 day	Annual maximum one-day precipitation
R10 mm	Annual count of days when precipitation ≥ 10 mm
R20 mm	Annual count of days when precipitation ≥ 20 mm
Rnn mm	Annual count of days when precipitation ≥ nn mm, nn = 67.5
CDD	Maximum number of consecutive dry days with rainfall < 1 mm annually
CWD	Maximum number of consecutive wet days with rainfall > 1 mm annually
RxxTOT	Annual total rainfall when rainfall > xx *percentile*, xx = 95, 99, 99.7, 99.9

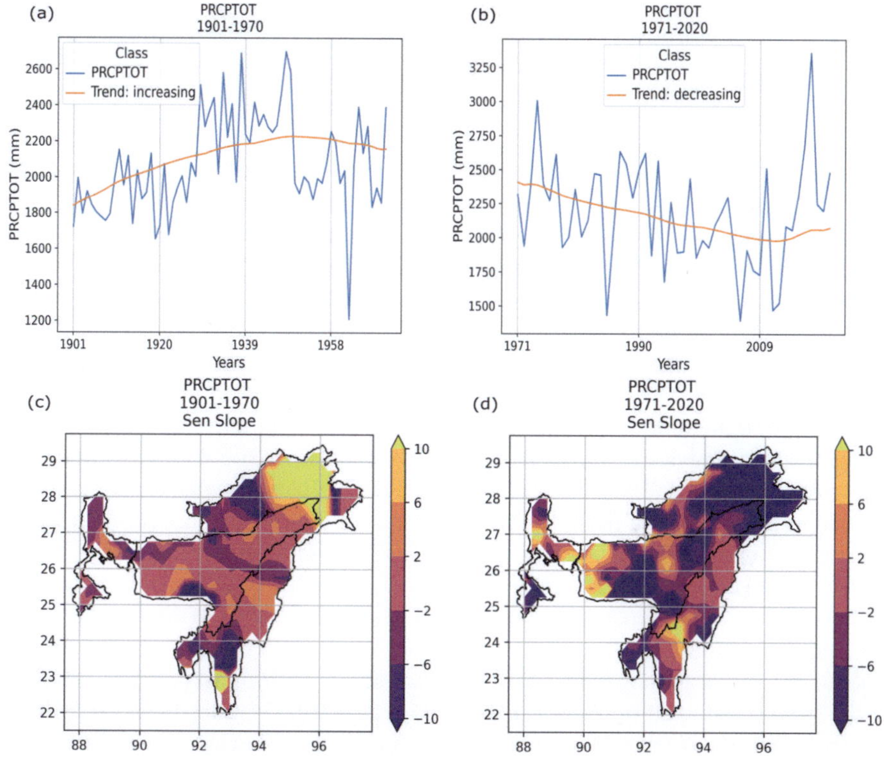

Fig. 22.9 Annual time series and trend of PRCPTOT from **a** 1901–1970, **b** 1971–2020 averaged over North East India. Spatial map of Sen's slope of PRCPTOT for the period **c** 1901–1970 and **d** 1971–2020. In **a** and **b**, the blue line shows the inter-annual behavior, and the orange line shows the non-linear trend calculated using Singular Spectrum Analysis. The Sen's slope is calculated using the Mann–Kendall Trend test and only the statistically significant ($p < 0.05$) trends and slopes are presented. *Data Source* IMD gridded rainfall

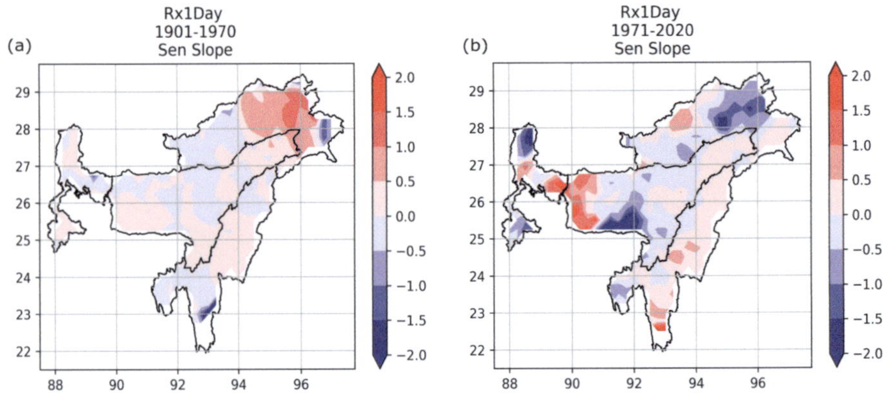

Fig. 22.10 Spatial map of Sen's slope of annual Rx1day for the period **a** 1901–1970, and **b** 1971–2020. The Sen's slope is calculated using the Mann–Kendall Trend test and only the statistically significant ($p < 0.05$) trends and slopes are presented. *Data Source* IMD gridded rainfall

post-1970 (Fig. 22.11). Analysis of the trend of area average timeseries of R10mm (–0.36 days/*years*) and R20mm (–0.19 days/*year*), which correspond to the frequency of light to moderate rainfall events, respectively, show significant decrease in their numbers. Interestingly, the decrease is also observed in heavy rainfall events of magnitude more than 67.5 mm (–0.04 days/*year*). However, the decrease is higher for light rainfall events compared to heavy rainfall events. The state-wise distribution reveals a decreasing trend in most states for all three intensity events, especially, Arunachal Pradesh, Meghalaya, Assam, Tripura. Manipur, however, showed a negative trend for light rainfall and a positive trend for heavy rainfall, and Sikkim showed a significant negative trend of heavy rainfall events and a slight positive trend for light rain (Fig. 22.11). Thus, the decrease in PRCPTOT can be explained by a secular decline in the number of light, moderate and heavy rainfall events over the region. The decline in PRCPTOT is further aggravated by a decline in the duration of wet spells (CWD) and an increase in the duration of dry spells (CDD).

Figure 22.12a, b and c show the spatiotemporal characteristics of dry spell duration over North East India. The rise in duration of dry spell events is evident in Fig. 22.12a and this increase is homogeneous for all the seasons, with the most increase observed in the Winter followed by Post-Monsoon (Fig. 22.12b). Contrary to an increase in dry spell duration over the region, the duration of wet spell events has significantly decreased and the most decline is observed in the Monsoon season (Fig. 22.12d, e). A significant decrease in the duration of wet spells is observed in all the states of North East India, except for Sikkim which showed an increase in the duration of wet spells and a decrease in dry spells (Fig. 22.12c and f).

The drying of North East India, evident from the analysis above, contradicts the increasing moisture content in the atmosphere above the region (Fig. 22.8). The question of where this significant moisture content present in the atmosphere goes, is amply answered by the analysis of extreme rainfall events over the region.

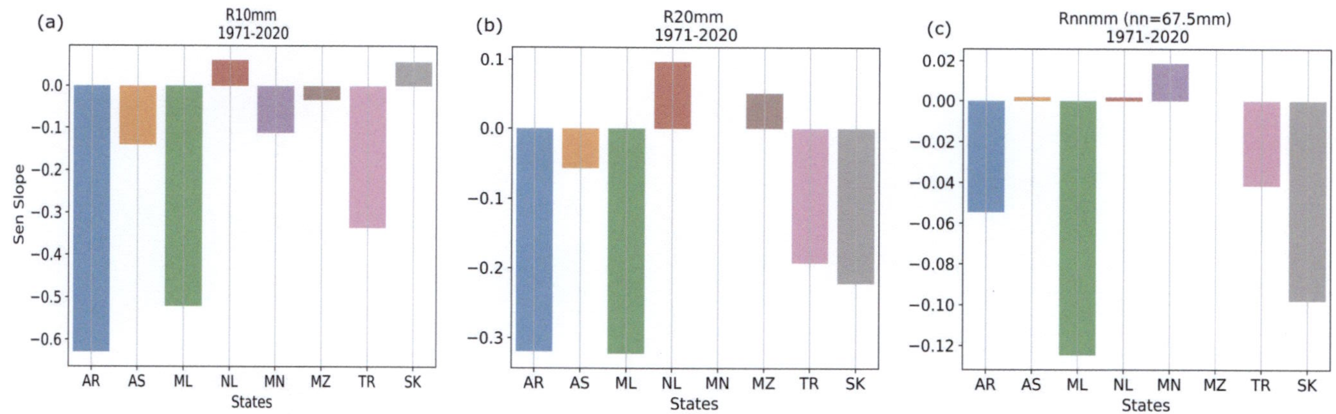

Fig. 22.11 State-wise (Arunachal Pradesh: AR, Assam: AS, Meghalaya: ML, Nagaland: NL, Manipur: MN, Mizoram: MZ, Tripura: TR, Sikkim: SK) Sen's slope of time series between 1971–2020 for **a** R10mm, **b** R20mm, **c** Rnnmm (nn = 67.5). The Sen's slope is calculated using the Mann–Kendall Trend test and only the statistically significant ($p < 0.05$) trends and slopes are presented. *Data Source* IMD gridded rainfall

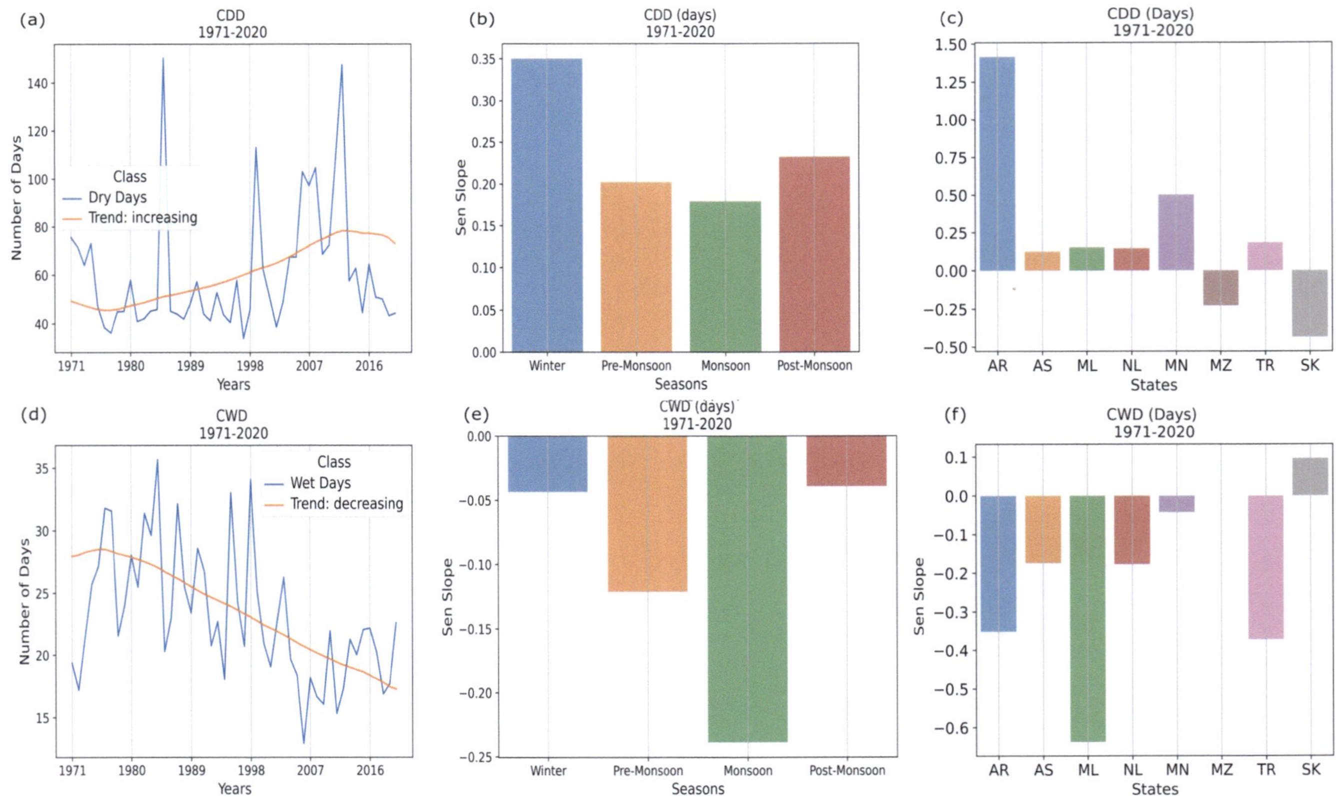

Fig. 22.12 Annual time series for the period 1971–2020 of **a** CDD, **d** CWD Sen's slope of seasonal time series of **b** CDD, **e** CWD. State-wise (Arunachal Pradesh: AR, Assam: AS, Meghalaya: ML, Nagaland: NL, Manipur: MN, Mizoram: MZ, Tripura: TR, Sikkim: SK) Sen's slope of time series between 1971–2020 for **c** CDD, **f** CWD. In Fig. 22.12a and d, the blue line shows the inter-annual behavior, and the orange line shows the non-linear trend calculated using Singular Spectrum Analysis. The Sen's slope is calculated using the Mann–Kendall Trend test and only the statistically significant ($p < 0.05$) trends and slopes are presented. *Data Source* IMD gridded rainfall

Figure 22.13 shows the magnitude and frequency of extreme rainfall events categorized based on their percentiles. Higher percentile events correspond to high impact rare events. The increasing trend of 99, 99.7, and 99.9th percentile events indicate an increasing magnitude of the rarer events (Fig. 22.13a). An analysis of the frequency of these events reveals the significant rise in the frequency of all rarer events, especially of the 99.9th percentile events over the region (Fig. 22.13b). The decreasing trend observed for the 95th percentile events matched the decreasing trend of Rnnmm (nn = 67.5 mm) events (Fig. 22.11c). Recognizing the contradictory behaviour of 95th percentile events compared to the rest, this chapter focuses on the higher percentile events of 99th, 99.7th, and 99.9th percentile. Interestingly, studies by Myhre et al. (2019) and Sarkar and Maity (2022) showed that, the contribution of frequency is substantially higher than the contribution of intensity to the total change in magnitude of these higher percentile events.

Following the approach used by Myhre et al. (2019) and Sarkar and Maity (2022), for identifying the relative contribution of frequency and intensity, revealed a similar finding of higher contribution of frequency to total change in magnitude. Figure 22.14a shows the total change in magnitude of extreme rainfall events, post-1970 (1971–2020) compared to the pre-1970 (1921–1970), with increasing rareness (99, 99.7, and 99.9th percentile). The significant change of more than 300% in the 99.9th percentile event post-1970 is evident. The analysis of Fig. 22.14b, c suggest a higher contribution of changes in frequency to the total change rather than intensity. A divide in terms of positive and negative change is clearly visible with the western parts showing a significant positive change whereas the eastern parts showing a slight negative change, suggesting higher occurrence of extreme to very extreme rainfall events in the West, especially over Meghalaya post 1970 (Fig. 22.14). The reason

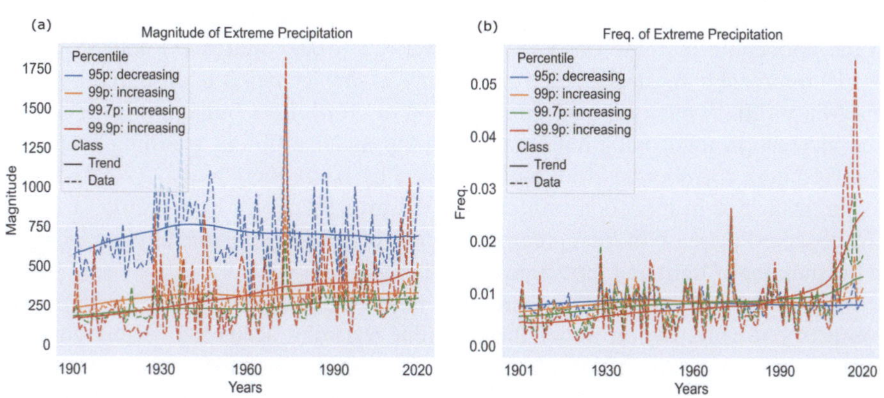

Fig. 22.13 Annual time series of **a** magnitude and **b** frequency of extreme rainfall above a threshold identified at 95th, 99th, 99.7th and 99.9th percentile during the period 1961–1990. The blue line shows the inter-annual behavior, and the orange line shows the non-linear trend calculated using Singular Spectrum Analysis. The Sen's slope is calculated using the Mann–Kendall Trend test and only the statistically significant ($p < 0.05$) trends and slopes are presented. *Data Source* IMD gridded rainfall

for this spatial divide is not entirely resolved, however the comparative roles of thermodynamic and dynamic factors can be attributed for this divide. The retrospective analysis of changing rainfall patterns indicates an overall drying of the region with increased frequency of highly intense rainfall events. This pattern of rainfall can lead to various kinds of disasters in this region, namely, flash floods, landslides due to cloud bursts, and even droughts. The pairing of hot and dry extremes as observed above can even lead to the occurrence of compound hot and dry extremes over this region.

22.4.4 Future Status of Precipitation

The retrospective analysis of rainfall does paint a grim picture. However, it is important to also investigate the pattern of these climate change indicators for the different future warming scenarios. Out of all the scenarios present, this chapter explores three emission scenarios, (SSP1 –2.6, SSP2 –4.5, and SSP5 –8.5) representing three warming scenarios, namely, 2 °C, 3 °C, and > 4 °C, for examining the projected behavior of rainfall over North East India. 10 CMIP6 models of varying climate sensitivities are selected, and the multi-model ensemble mean calculated (Table 22.3). This multi-model ensemble mean is used to calculate the ETCCDI indicators as noted in Table 22.2. Figure 22.15 shows the Sen's slope estimate of these indicators for the historical (1971–2014) and three emission scenarios from 2015 to 2100. Figure 22.15 show the consistent increasing trend of PRCPTOT for the three emission scenarios, with the highest emission scenario showing a staggering rise (+7.68 mm/*year*). Along with that, the three scenarios also showed an increase in Rx1day over the region. However, the increase is marginal for SSP1-2.6 (+0.004 mm/*year*) compared to the noticeable increase observed in SSP2–4.5 (+0.07 mm/*year*) and SSP5-8.5 (+0.13 mm/*year*). A similar trend of Rx1day in the scenarios is observed by Paul and Maity (2023) over the contiguous region of North East India and Bangladesh. The increase in total precipitation and Rx1day under different warming scenarios observed here corroborates well with the study by Katzenberger et al. (2021) where a significant rise in monsoon rainfall is noted over the North East region of India. The observed increase in total precipitation can be further explained by the increase in the number of light and moderate rainfall days.

Figure 22.15 also shows the increasing number of R10mm and R20mm events averaged over this region under the different warming scenarios. Although both light and moderate rainfall events are increasing in the low emission scenario, but the increase is significantly higher for light rainfall events compared to moderate rainfall events. The duration of dry spells (CDD) over this region showed only a slight decrease under the influence of warming, however, the wet spell duration (CWD) showed a significant increase in intermediate (+0.1 days/*years*) and high emission (+0.3 days/*year*) scenarios. Compared to that, the low emission scenario showed a slight decrease in the duration of wet spells (–0.007 days/*year*). Interestingly, the historical trend of Rx1day, R10mm, R20mm, CDD, and CWD estimated from the

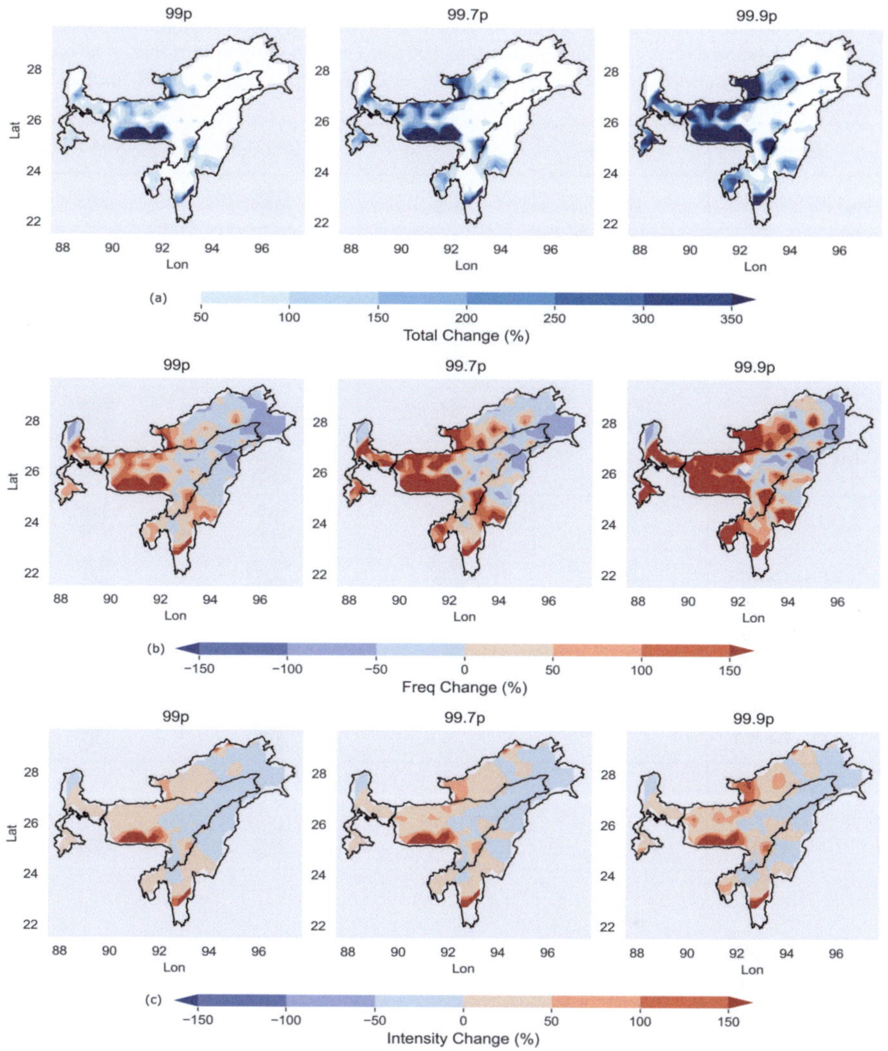

Fig. 22.14 Spatial map of percentage of **a** total change in magnitude, **b** change in frequency, and **c** change in intensity of extreme rainfall over the threshold of 99th, 99.7th and 99.9th percentile (estimated for the period of 1921–1970) between pre-1970 (1921–1970) and post-1970 (1971–2020). (*Data Source* IMD gridded rainfall)

multi-model ensemble mean, all behave the same way as that of the observation between 1971–2020 period. However, the magnitude of the trend of the models is weaker than that found in the observations. The behavior of total precipitation from the ensemble mean is positive in contrast to that of the observation which is significantly negative.

Table 22.3 List of CMIP6 global climate models used for the analysis of rainfall climatology. All models are downloaded at a spatial resolution of 100 km and at a temporal resolution of one-day under CMIP6 ScenarioMIP *Data Source* Eyring et al. 2016. https://esgf-metagrid.cloud.dkrz.de/search/cmip6-dkrz/

Models	Historical	SSP1-2.6	SSP2-4.5	SSP5-8.5
BCC-CSM2-MR	X	X	X	X
CESM2	X	X	X	X
CMCC-CM2-SR5	X	X	X	X
EC-Earth3	X	X	X	X
HadGEM3-GC31-MM	X	X		X
MPI-ESM1-2-HR	X	X	X	X
MRI-ESM2-0	X	X	X	X
NorESM2-MM	X	X	X	X
IPSL-CM6A-LR		X		
MIROC6		X		
GFDL-ESM4			X	X
INM-CM4-8			X	
TaiESM1			X	X

The behavior of extreme rainfall events beyond the thresholds of 99th, 99.7th, and 99.9th percentile in the multi-model ensemble mean showed a positive change in total magnitude (Total Change (%)) for the intermediate and high emission scenarios but a negative change for the low emission scenario (Fig. 22.16). Interestingly, the negative change in magnitude observed for SSP1–2.6 increases in the mid-future and decreases in the far-future and this behavior remains consistent for increasing rareness of events. Furthermore, for extremely rare events the total change in magnitude for the low emission scenario almost vanishes. This signifies a future behavior, of rarer extremes, that is similar to the historical period, if a low emission future happens to come to pass. Similar to the observational analysis in Fig. 22.14, the contribution of frequency is significantly higher than that of intensity in the total change in magnitude for future emission scenarios. The examination of the future scenarios thus shows a significant increase in event magnitude with an increase in event rareness, especially for the high emission scenario, with frequency contributing significantly to that change rather than intensity. The changes are estimated with respect to the historical period

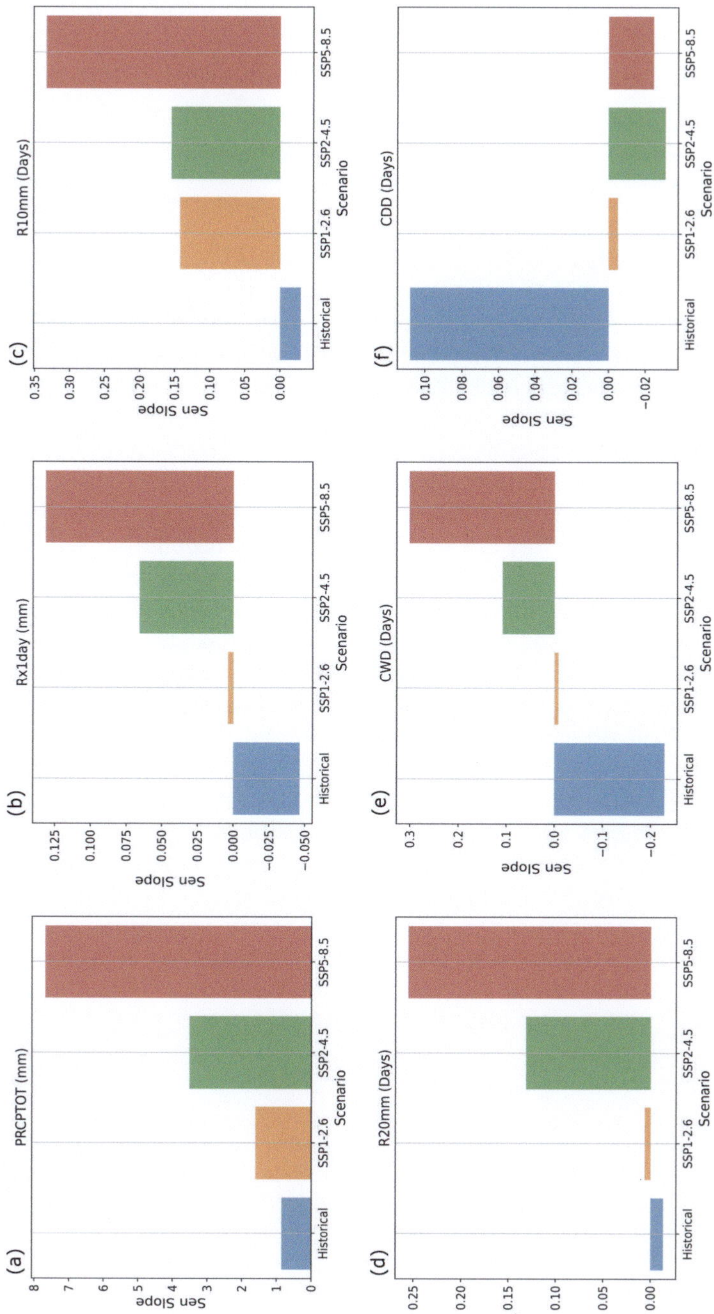

Fig. 22.15 Barplot of Sen's slope of PRCPTOT, Rx1day, R10mm, R20mm CDD, and CWD for SSP1–2.6, SSP2–4.5 and SSP5–8.5 time series averaged over North East India. The Sen's slope is calculated using the Mann–Kendall Trend test and only the statistically significant ($p < 0.05$) trends and slopes are presented. *Data Source* CMIP6 ScenarioMIP

of 1990–2010. A similar finding is noted in the study by Sarkar and Maity (2022). Thus, based on the examination of Figs. 22.15 and 22.16, its can be argued that the average effect of warming of the climate over North East India is to make the region wetter.

22.5 Summary

In summary, North East India is experiencing clear and significant climate change. Temperatures are rising, with winters warming faster than summers. States like Tripura, Mizoram, and Manipur show notable increases in daily mean temperatures. Both day and night-time summer temperatures have risen, with winter night-time temperatures rising even faster—likely due to Bay of Bengal warming and increased atmospheric moisture. These warming patterns are projected to further intensify under future high-emission scenarios.

Rising temperatures have influenced rainfall patterns. Contrary the positive temperature-precipitation relationship, North East India has seen a significant decline in total rainfall since the 1970s, across all seasons. This decline is reflected in fewer light, and moderate rainfall events annually, longer dry spells, and shorter wet spells—especially during the monsoon. The drying is most severe in parts of Assam, Arunachal Pradesh, Meghalaya, and Tripura. At the same time, the frequency and magnitude of extremely rare high impact rainfall events have increased, particularly in western parts like Meghalaya and western Assam. In contrast, eastern areas show declining trends, where the increase in rarer events is mainly due to frequency rather than intensity.

The changes observed or projected over the region all suggest a significant change in the climate of North East, which can have severe consequences for the region with the introduction of higher flash floods, heat waves, and even droughts. Keeping this thing in mind, extensive research on high resolution bias corrected simulations of the future scenarios and accurate understanding of the processes causing these extremes is required not just at the regional scale but also at local scale for effective planning.

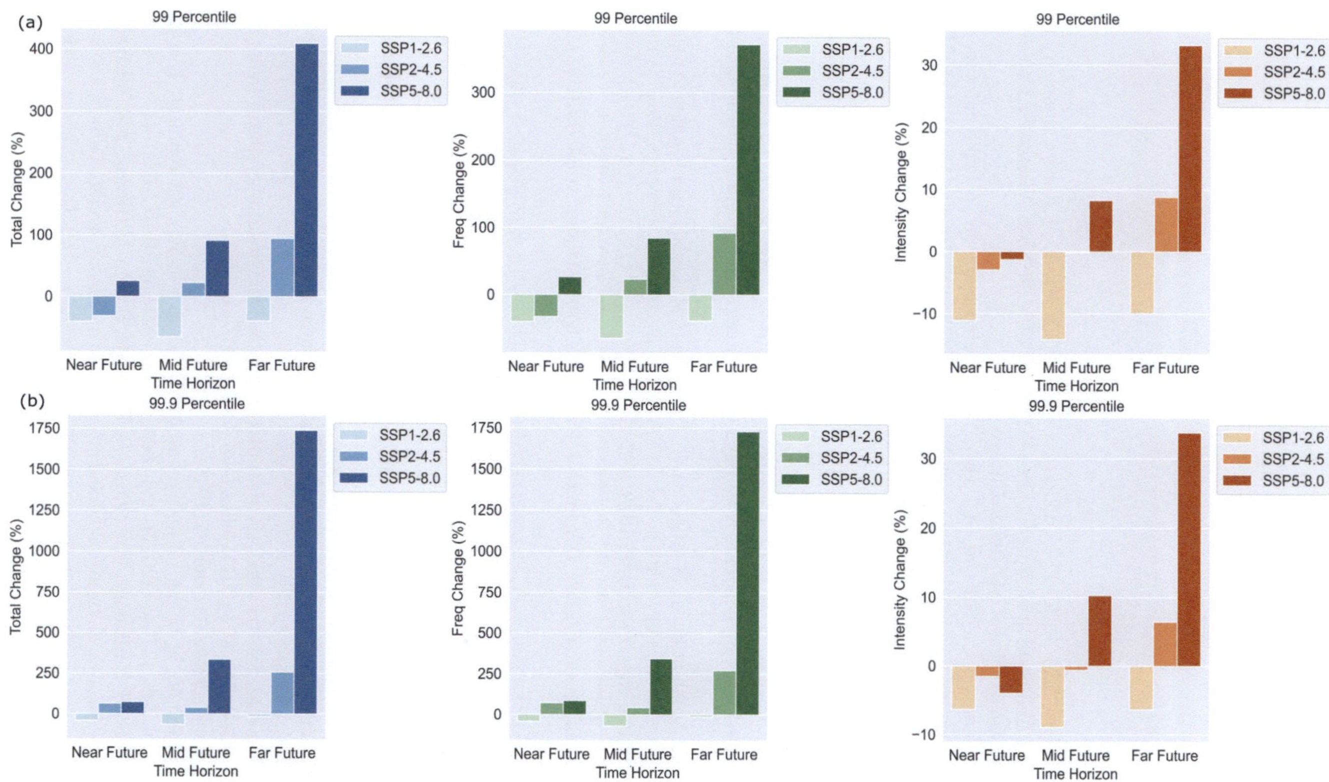

Fig. 22.16 Percentage change in total magnitude blue bars, frequency green bars, and intensity orange barsbetween historical and future warming scenarios (SSP1–2.6, SSP2–4.5, and SSP5–8.5) of extreme rainfall identified using the thresholds of **a** 99th, and **b** 99.9th percentile. *Data Source* CMIP6 ScenarioMIP

Acknowledgements The authors would like to thank North Eastern Space Applications Centre for facilitating this study. The authors are also grateful to India Meteorological Department (IMD) for making available the IMD gridded rainfall product and the European Centre for Medium range Weather Forecasting (ECMWF) for making available their Fifth generation of ECMWF Reanalysis (ERA5).

References

Core Writing Team, 2023. IPCC, 2023, Climate Change 2023: Synthesis Report, Summary for Policymakers. Contribution of Working Groups I, II and III to the Sixth Assessment Report of the Intergovernmental Panel on Climate Change (2023)

S.K. Dash, N. Sharma, K.C. Pattnayak, X.J. Gao, Y. Shi, Temperature and precipitation changes in the north-east India and their future projections. Global Planet. Change **98**, 31–44 (2012). https://doi.org/10.1016/j.gloplacha.2012.07.006

V. Eyring, S. Bony, G.A. Meehl, C.A. Senior, B. Stevens, R.J. Stouffer, K.E. Taylor, Overview of the coupled model intercomparison project phase 6 (CMIP6) experimental design and organization. Geosci. Model. Dev. **9**(5), 1937–1958 (2016). https://doi.org/10.5194/gmd-9-1937-2016

F. Giorgi, C. Jones, G.R. Asrar, Addressing climate information needs at the regional level: the CORDEX framework. World Meteorol. Organ. (WMO) Bulletin **58**(3), 175 (2009)

B.N. Goswami, V. Venugopal, D. Sengupta, M.S. Madhusoodanan, P.K. Xavier, Increasing trend of extreme rain events over India in a warming environment. Science **314**(5804), 1442–1445 (2006). https://doi.org/10.1126/science.1132027

B.B. Goswami, P. Mukhopadhyay, R. Mahanta, B.N. Goswami, Multiscale interaction with topography and extreme rainfall events in the northeast Indian region. J. Geophys. Res.: Atmos, **115**(D12) (2010). https://doi.org/10.1029/2009JD012275

B.B. Goswami, R. Murtugudde, An, S.I., Role of the Bay of Bengal warming in the Indian summer monsoon rainfall trend. Climate Dynamics 1–19 (2022). https://doi.org/10.1007/s00382-021-06068-1

N. Golyandina, V. Nekrutkin, A.A. Zhigljavsky, Analysis of time series structure: SSA and related techniques. CRC Press (2001). https://doi.org/10.1201/9780367801687

H. Hersbach, B. Bell, P. Berrisford, S. Hirahara, A. Horányi, J. Muñoz-Sabater, J. Nicolas, C. Peubey, R. Radu, D. Schepers, A. Simmons, The ERA5 global reanalysis. Q. J. r. Meteorol. Soc. **146**(730), 1999–2049 (2020)

G.J. Huffman, D.T. Bolvin, D. Braithwaite, K. Hsu, R. Joyce, P. Xie, S.H. Yoo, NASA global precipitation measurement (GPM) integrated multi-satellite retrievals for GPM (IMERG). Algorithm Theor. Basis Doc. (ATBD) version **4**(26), 30 (2015)

A.K. Jaswal, Changes in total cloud cover over India based upon 1961–2007 surface observations. Mausam **61**(4), 455–468 (2010)

S.K. Jain, V. Kumar, M. Saharia, Analysis of rainfall and temperature trends in northeast India. Int. J. Climatol. **33**(4), 968–978 (2013). https://doi.org/10.1002/joc.3483

M.G. Kendall, Rank correlation methods (1948)

A. Kulkarni, Weakening of Indian summer monsoon rainfall in warming environment. Theoret. Appl. Climatol. **109**, 447–459 (2012). https://doi.org/10.1007/s00704-012-0591-4

T. Kubota, S. Shige, H. Hashizume, K. Aonashi, N. Takahashi, S. Seto, M. Hirose, Y.N. Takayabu, T. Ushio, K. Nakagawa, K. Iwanami, Global precipitation map using satellite-borne microwave radiometers by the GSMaP project: production and validation. IEEE Trans. Geosci. Remote Sens. **45**(7), 2259–2275 (2007). https://doi.org/10.1109/TGRS.2007.895337

R. Krishnan, T.P. Sabin, R. Vellore, M. Mujumdar, J. Sanjay, B.N. Goswami, F. Hourdin, J.L. Dufresne, P. Terray, Deciphering the desiccation trend of the South Asian monsoon hydroclimate in a warming world. Clim. Dyn. **47**, 1007–1027 (2016). https://doi.org/10.1007/s00382-015-2886-5

R. Kumar, J. Kuttippurath, Long-term changes in tropospheric temperature in India: insights from radiosonde measurements, reanalysis data and CMIP6 model projections. Atmos. Res. **309**, 107583 (2024). https://doi.org/10.1016/j.atmosres.2024.107583

N. Kumar, M.K. Goyal, Projected changes in monsoonal compound dry-hot extremes in India. Atmos. Res. **310**, 107605 (2024). https://doi.org/10.1016/j.atmosres.2024.107605

P. Kushwaha, V.K. Pandey, P. Kumar, D. Sardana, A. Yadav, Projection of mean and extreme precipitation and air temperature over India: a CMIP6 analysis. J. Water Clim. Chang. p.jwc2024523 (2024). https://doi.org/10.2166/wcc.2024.523

D.R. Kothawale, H.N. Singh, Recent trends in tropospheric temperature over India during the period 1971–2015. Earth Space Sci. **4**(5), 240–246 (2017). https://doi.org/10.1002/2016EA000246

A. Katzenberger, J. Schewe, J. Pongratz, A. Levermann, Robust increase of Indian monsoon rainfall and its variability under future warming in CMIP6 models. Earth Syst. Dyn. **12**(2), 367–386 (2021). https://doi.org/10.5194/esd-12-367-2021

H.B. Mann, Nonparametric tests against trend. Econometrica:J. Econ. Soc.245–259 (1945)

R.H. Moss, J.A. Edmonds, K.A. Hibbard, M.R. Manning, S.K. Rose, D.P. Van Vuuren, T.R. Carter, S. Emori, M. Kainuma, T. Kram, G.A. Meehl, The next generation of scenarios for climate change research and assessment. Nature **463**(7282), 747–756 (2010). https://doi.org/10.1038/nature08823

G. Myhre, K. Alterskjær, C.W. Stjern, Ø. Hodnebrog, L. Marelle, B.H. Samset, J. Sillmann, N. Schaller, E. Fischer, M. Schulz, A. Stohl, Frequency of extreme precipitation increases extensively with event rareness under global warming. Sci. Rep. **9**(1), 16063 (2019). https://doi.org/10.1038/s41598-019-52277-4

P. Mukhopadhyay, A.K. Jaswal, M. Deshpande, Variability and trends of atmospheric moisture over the Indian region. Observed climate variability and change over the Indian region, pp.129–144. https://doi.org/10.1007/978-981-10-2531-0_8

M. Norgate, P.R. Tiwari, S. Das, D. Kumar, On the heat waves over India and their future projections under different SSP scenarios from CMIP6 models: climate change, heat wave, health. Int. J. Climatol (2024). https://doi.org/10.1002/joc.8367

B.C. O'Neill, E. Kriegler, K. Riahi, K.L. Ebi, S. Hallegatte, T.R. Carter, R. Mathur, D.P. Van Vuuren, A new scenario framework for climate change research: the concept of shared socioeconomic pathways. Clim. Change **122**, 387–400 (2014). https://doi.org/10.1007/s10584-013-0905-2

D.S. Pai, M. Rajeevan, O.P. Sreejith, B. Mukhopadhyay, N.S. Satbha, Development of a new high spatial resolution (0.25× 0.25) long period (1901–2010) daily gridded rainfall data set over India and its comparison with existing data sets over the region. Mausam **65**(1), 1–18 (2014)

P. Sabin, T. Krishnan, R. Ghattas, J. Denvil, S. Dufresne, J.L. Hourdin, F.T. Pascal, High resolution simulation of the South Asian monsoon using a variable resolution global climate model. Climate dynamics **41**, 173–194 (2013). https://doi.org/10.1007/s00382-012-1658-8

A.R. Paul, R. Maity, Future projection of climate extremes across contiguous northeast India and Bangladesh. Sci. Rep. **13**(1), 15616 (2023). https://doi.org/10.1038/s41598-023-42360-2

A. Pathak, S. Ghosh, P. Kumar, Precipitation recycling in the Indian subcontinent during summer monsoon. J. Hydrometeorol. **15**(5), 2050–2066 (2014). https://doi.org/10.1175/JHM-D-13-0172.1

S.I. Rani, T. Arulalan, J.P. George, E.N. Rajagopal, R. Renshaw, A. Maycock, D.M. Barker, M. Rajeevan, IMDAA: High-resolution satellite-era reanalysis for the Indian monsoon region. J. Clim. **34**(12), 5109–5133 (2021)

K. Riahi,D.P. Van Vuuren, E. Kriegler, J. Edmonds, B.C. O'neill, S. Fujimori, N. Bauer, K. Calvin, R. Dellink, O. Fricko, W. Lutz, The shared socioeconomic pathways and their energy, land use, and greenhouse gas emissions implications: an overview. Glob. Environ. Chang. **42**, pp. 153–168 (2017). https://doi.org/10.1016/j.gloenvcha.2016.05.009

V. Ramanathan, C. Chung, D. Kim, T. Bettge, L. Buja, J.T. Kiehl, W.M. Washington, Q. Fu, D.R. Sikka, M. Wild, Atmospheric brown clouds: Impacts on South Asian climate and hydrological cycle. Proc. Natl. Acad. Sci. **102**(15), 5326–5333 (2005). https://doi.org/10.1073/pnas.0500656102

P. Rohini, M. Rajeevan, A.K. Srivastava, On the variability and increasing trends of heat waves over India. Sci. Rep. **6**(1), 1–9 (2016). https://doi.org/10.1038/srep26153

B.B. Rao, P.S. Chowdary, V.M. Sandeep, V.U.M. Rao, B. Venkateswarlu, Rising minimum temperature trends over India in recent decades: implications for agricultural production. Global Planet. Change **117**, 1–8 (2014). https://doi.org/10.1016/j.gloplacha.2014.03.001

P.K. Sen, Estimates of the regression coefficient based on Kendall's tau. J. Am. Stat. Assoc. **63**(324), 1379–1389 (1968)

S. Sharma, P. Mujumdar, Increasing frequency and spatial extent of concurrent meteorological droughts and heatwaves in India. Sci. Rep. **7**(1), 15582 (2017). https://doi.org/10.1038/s41598-017-15896-3

J. Sanjay, J.V. Revadekar, M.V.S. Ramarao, H. Borgaonkar, S. Sengupta, D.R. Kothawale, J. Patel, R. Mahesh, S. Ingle, K. AchutaRao, A.K. Srivastava, Temperature changes in India. Assessment of climate change over the Indian region: a report of the Ministry of Earth Sciences (MoES), Government of India, pp. 21–45 (2020). https://doi.org/10.1007/978-981-15-4327-2_2

S. Sarkar, R. Maity, Unveiling climate change-induced temperature-based hotspots across India through multimodel future analysis from CMIP6. Int. J. Climatol. **44**(2), 627–646 (2024). https://doi.org/10.1002/joc.8348

P. Salunke, N.P. Keshri, S.K. Mishra, S.K. Dash, Future projections of seasonal temperature and precipitation for India. Frontiers in Climate **5**, 1069994 (2023). https://doi.org/10.3389/fclim.2023.1069994

A.K. Srivastava, J.V. Revadekar, M. Rajeevan, South Asia in "State of the climate in 2018." Bull Amer Meteor Soc **100**(9), S236–S240 (2019). https://doi.org/10.1175/2019BAMSStateoftheClimate.1

D. Singh, M. Tsiang, B. Rajaratnam, N.S. Diffenbaugh, Observed changes in extreme wet and dry spells during the South Asian summer monsoon season. Nat. Clim. Chang. **4**(6), 456–461 (2014). https://doi.org/10.1038/nclimate2208

N.K. Shahi, S. Rai, S. Verma, R. Bhatla, Assessment of future changes in high-impact precipitation events for India using CMIP6 models. Theoret. Appl. Climatol. **151**(1), 843–857 (2023). https://doi.org/10.1007/s00704-022-04309-9

S. Sarkar,R. Maity, R, Future characteristics of extreme precipitation indicate the dominance of frequency over intensity: a multi-model assessment from CMIP6 across India. J. Geophys: Res.: Atmos. **127**(16), p.e2021JD035539. https://doi.org/10.1029/2021JD035539

W. Sadok, S.K. Jagadish, The hidden costs of nighttime warming on yields. Trends Plant Sci. **25**(7), 644–651 (2020). https://doi.org/10.1016/j.tplants.2020.02.003

D.P. Van Vuuren, J. Edmonds, M. Kainuma, K. Riahi, A. Thomson, K. Hibbard, G.C. Hurtt, T. Kram, V. Krey, J.F. Lamarque, T. Masui, The representative concentration pathways: an overview. Clim. Change **109**, 5–31 (2011). https://doi.org/10.1007/s10584-011-0148-z

R. Vinnarasi, C.T. Dhanya, A. Chakravorty, A. AghaKouchak, Unravelling diurnal asymmetry of surface temperature in different climate zones. Sci. Rep. **7**(1), 7350 (2017). https://doi.org/10.1038/s41598-017-07627-5

R. Vinnarasi, C.T. Dhanya, Changing characteristics of extreme wet and dry spells of Indian monsoon rainfall. J. Geophys. Res.: Atmos. **121**(5), 2146–2160 (2016). https://doi.org/10.1002/2015JD024310

M. Wild, S. Wacker, S. Yang, A. Sanchez-Lorenzo, Evidence for clear-sky dimming and brightening in central Europe. Geophys. Res. Lett, **48**(6), p.e2020GL092216 (2021). https://doi.org/10.1029/2020GL092216

J. Xia, S. Chen, S. Wan, Impacts of day versus night warming on soil microclimate: Results from a semiarid temperate steppe. Sci. Total Environ. **408**(14), 2807–2816 (2010). https://doi.org/10.1016/j.scitotenv.2010.03.016

Z. Zuo, S. Yang, A. Kumar, R. Zhang, Y. Xue, B. Jha, Role of thermal condition over Asia in the weakening Asian summer monsoon under global warming background. J. Clim. **25**(9), 3431–3436 (2012). https://doi.org/10.1175/JCLI-D-11-00742.1

Aniket Chakravorty is a Scientist in the Space and Atmospheric Sciences Division at North Eastern Space Applications Centre, Department of Space, Government of India. He has a doctorate in land atmosphere interactions from the Indian Institute of Technology Delhi. He is working on understanding the changes in the climate of North East India with a special emphasis on extreme precipitation events. He has expertise in uncertainty quantification, time series analysis, and climate downscaling.

Shyam Sundar Kundu did his M.Sc. in Physics from University of North Bengal and Ph.D. in Physics from Gauhati University. He has more than 20 years of professional experience at NESAC and currently working as the Head of the Space and Atmospheric Sciences Division at the centre. He was instrumental in setting up the Space and Atmospheric Science research facility at NESAC with several instruments, a S-band Doppler Weather Radar, largest network of weather stations in NER covering difficult terrains, three aerosol observatories including two in Himalayan region, network of lightning detectors, etc. His research interest include extreme weather forecasting in NER, lightning physics, aerosol-cloud-precipitation interaction studies and its climatic impact, climate change assessment, etc. He took lead role to develop an operational lightning early warning system for the NE region of India. He has more than 50 publications in peer reviewed journals and books. He is making significant contribution to professional bodies like ISRS as its Joint Secretary and ISPRS as the Secretary of the WG-4 under TC V. He also represents ISRO in the Committee on Earth Observation Satellites for Precipitation constellations.

Rosly B. Lyngdoh is a Scientist in the Space and Atmospheric Sciences Division, North Eastern Space Applications Centre, Department of Space, Government of India. His research focuses on the application of Artificial Intelligence approaches for nowcasting and short-range prediction of features of weather, like lightning, rainfall, etc. using satellite remote sensing.

Arup Borgohain is a Scientist in the Space and Atmospheric Science Division, North Eastern Space Applications Centre, Department of Space, Government of India. His research focuses on aerosol and trace gas dynamics, atmospheric boundary layer physics, and mass-energy exchange processes.

Shiv Prasad Aggarwal is the Director of the North Eastern Space Applications Centre (NESAC), Dept. of Space, Govt. of India, Shillong. He holds a B. Tech. degree from Allahabad University, a Postgraduate and Ph.D. degree from the Indian Agricultural Research Institute (IARI), New Delhi. He did his postdoctoral research from ITC/IHE, The Netherlands. A distinguished space applications scientist and visionary leader, Dr. Aggarwal has been pivotal in spearheading transformative initiatives that harness space technology for sustainable development across the North Eastern Region of India. Under his leadership, NESAC has adopted cutting-edge technologies (AI, ML, UAVs), enhanced stakeholder engagement through NeUIM, and earned national accolades for excellence in geospatial applications. His contributions span a wide spectrum of applications, including natural resource management, climate-resilient infrastructure planning, disaster risk reduction, weather forecasting, satellite-based communication, and geo-governance. Earlier, he served as Group Head, Water Resources Department, at the Indian Institute of Remote Sensing, ISRO, Dehradun, where he played a crucial role in advancing remote sensing and GIS-based solutions for water resources management, flood forecasting, disaster management, and climate change impact assessment. He also held the position of Program Coordinator of

CSSTEAP, a United Nations-affiliated centre, where he significantly contributed to capacity-building in space science and technology education across the Asia-Pacific region. He has published over 200 scientific articles and book chapters, and has delivered more than 100 invited lectures at reputed national and international forums. He has supervised over 30 M. Tech. and Ph. D. theses, mentoring the next generation of scientists and engineers. In recognition of his exemplary contributions to the field of space applications and geospatial sciences, he has received numerous prestigious awards, including the National Geospatial Award and the National e-Governance Award. He is a life member and honorary fellow of several national and international scientific societies. Currently, he serves as the President of the Indian Society of Remote Sensing (ISRS), and has previously held key editorial and leadership roles, including Executive Editor of the Journal of ISRS.

Chapter 23
SATCOM Technology and Its Utilizations for Societal Applications in North Eastern Region of India

Ramani Kumar Das, Anjan Debnath, and Himangshu Jyoti Das

Abstract Satellite communication is an enabling technology that uses satellite as a relaying device to connect two points far apart geographically through radio frequency bandwidth. The radio frequency is used to carry information between the points. Such uses of satellite technology for societal development can provide significant benefit in terms of geographic coverage, reliability and availability. Such technology has been used for delivery of educational material over a network of connected schools for augmenting curriculum-based education under the Tele-Education Program. Similarly, under the Tele-Medicine program, health services have been made available to people of remote and inaccessible areas. ISRO has implemented such programs throughout the country for a long time which has benefitted lots of students and patients. This chapter focuses on such programs implemented by ISRO in collaboration with NESAC and various regional institutions/organizations/ Government Departments. The technologies for such programs, coverage, methodology for implementation and challenges to such programs have been highlighted in this chapter.

Keywords SATCOM · Tele-education · Tele-medicine · EDUSAT · SIT · ROT · VSAT · CME

R. K. Das (✉) · A. Debnath · H. J. Das
North Eastern Space Applications Centre (NESAC), Umiam, Shillong, Meghalaya, India
e-mail: rkdas@nesac.gov.in

A. Debnath
e-mail: anjan.debnath@nesac.gov.in

H. J. Das
e-mail: himangshu.das@nesac.gov.in

S. P. Aggarwal et al. (eds.), *Remote Sensing of the North Eastern Himalayan Ecosystem*,
https://doi.org/10.1007/978-981-95-3221-6_23

23.1 Introduction

Satellite communications (SATCOM) have revolutionized the way information is transmitted over long distances, providing global connectivity for telecommunications, navigation, broadcasting, and Earth observation. SATCOM technology relies on a constellation of components that work in harmony, from the satellite bus that houses the system to the payload dedicated to mission-specific tasks, and the transponders that receive and retransmit signals. This section offers an accessible yet comprehensive overview of SATCOM technology components for a general audience. We examine the key technical elements, describe their operational roles, and discuss the significant challenges and future trends that continue to shape this dynamic field.

Satellite systems are self-contained platforms, outlining the satellite bus's crucial functions in power management, propulsion, and telemetry. Each satellite carries a payload, which determines the satellite's functionality across various domains—whether for communication, navigation, or Earth observation. For a communication satellite, the heart of the system blocks is what is called transponders, indispensable devices that receive, amplify, and re-transmit signals, ensuring smooth communication between the satellite and ground stations. Finally, the chapter delves into the operational challenges and future trends that promise to further enhance SATCOM performance, such as advanced modulation techniques, high-throughput satellites (HTS), and the integration of optical and radio frequency (RF) systems.

This section aims to give a brief introduction to the underlying technologies of SATCOM. Readers will gain an understanding of how these systems operate, the interplay between different components, and the innovations that are driving the future of satellite communications.

23.1.1 Key Components of SATCOM Technology

Satellite Communication systems are complex integrations of hardware & software, designed to function reliably in the harsh conditions of space. This section outlines the preliminary components that enable SATCOM from the physical satellite bus to the operational transponders and mission-specific payload (Fig. 23.1).

23.1.1.1 Overview of Satellite Systems and the Bus

At the heart of SATCOM lies the satellite, a self-contained system engineered to perform specific communication tasks. A satellite is much more than just a relay station, it is a complete ecosystem comprising several subsystems that work together to ensure uninterrupted and reliable service. The satellite bus, often referred to as the "platform," supports various payloads and includes critical subsystems such as

Fig. 23.1 Components of SATCOM technology

power, propulsion, thermal management, and tracking telemetry and control (TT&C). By integrating all its components, the satellite bus serves as the foundation upon which mission-specific payloads are built. Each satellite is designed to endure the demanding environment of space while delivering high-reliability communication services.

23.1.1.2 The Role and Design of the Payload

The payload is the mission-specific component of a satellite, determining the particular function or service the satellite is intended to perform. In SATCOM, the payload is critical as it handles the reception, processing, and transmission of data signals. The payload consists of a suite of equipment that can be tailored for different applications such as telecommunications, navigation, and Earth observation. Key considerations in payload design include:

- **Frequency Bands**: The payload operates on designated frequency bands such as the C-band, Ku-band, and Ka-band, chosen according to the specific requirements of the mission. These frequency bands directly impact signal coverage, bandwidth available for service and data throughput.
- **Antenna Design**: The design and configuration of antennas within the payload determine signal directionality and gain. This aspect of the payload design is crucial for ensuring that transmitted signals reach their intended destinations with minimal degradation.
- **Signal Processing Capabilities**: Modern payloads incorporate sophisticated signal processing techniques that are responsible for amplification, filtering,

and modulation. Effective signal processing is necessary to counter potential challenges such as interference, noise, and signal degradation.

There are several types of payloads used in SATCOM, each tailored to its function:

23.1.1.3 Transponders: The Heart of Signal Relay

Transponders play a pivotal role in satellite communications by serving as the intermediary that receives incoming signals, processes them, and transmits them back to the intended destination. Essentially, a transponder is a combined transmitter and responder—a device that automatically responds to an incoming signal with a corresponding output. The fundamental functions of a transponder include Signal Reception, Signal Amplification and Frequency Translation and Signal Processing. There are two primary types of transponders used in SATCOM:

- **Bent Pipe Transponders (Repeaters)**: These transponders simply amplify the incoming signal and re-transmit it without altering its data content. They are compatible with both analog and digital signals and are commonly used in traditional satellite systems.
- **Regenerative Transponders**: These devices offer advanced functionality by performing demodulation and signal regeneration. By doing so, they improve the quality of the transmitted signal and provide enhanced flexibility in managing interference and noise

23.1.1.4 Additional Elements: Ground Stations, Antenna Systems, and Orbits

A comprehensive SATCOM system extends far beyond the satellite in orbit; it includes a network of ground segments that are critical to the overall communication infrastructure. Key additional elements include:

- **Ground Stations and Teleports**: These facilities serve as the terrestrial interface for satellite communications. Ground stations are responsible for sending uplink commands to satellites, receiving downlink data, and providing network connectivity with terrestrial infrastructure. Modern ground stations, sometimes known as teleports, are equipped with advanced transceivers, directional antennas, and encryption systems to ensure secure and reliable data transmission.
- **Antenna Systems**: Both spaceborne and ground-based antennas play crucial roles by directing radio frequency (RF) signals toward specific regions. The efficiency of the antenna system is a major determinant of the quality and reliability of signal transmission. Antennas are designed to optimize gain, directivity, and support various frequency bands depending on mission requirements. Mostly antennas used for SATCOM are large dish antennas technically called Very Small Aperture Terminals (VSAT). The Ground Station Antennas are generally quite big in size

like 11 m diameter while those installed at user premises can be as small as 0.3 m in diameter. The size of a VSAT antenna is related to the signal gain it can produce.

- **Orbital Considerations**: Satellites operate in different orbital regimes, each with unique advantages and constraints. Common orbits include:
  - **Low Earth Orbit (LEO)**: Characterized by shorter orbital periods and closer proximity to Earth, LEO satellites are ideal for applications that require low latency, such as real-time communications.
  - **Medium Earth Orbit (MEO)**: Frequently used in navigation systems like GPS, MEO offers a balance between coverage area and latency.
  - **Geostationary Orbit (GEO)**: Positioned at approximately 35,786 km above Earth, GEO satellites provide continuous coverage over specific areas, making them ideal for broadcasting and weather monitoring

23.1.2 Operational Environment and Challenges in SATCOM

Operating a SATCOM system is an intricate endeavour, with numerous environmental and technical challenges that impact performance, reliability, and service quality. This section examines these challenges and the strategies employed to overcome them.

23.1.2.1 Environmental Conditions in Space

Space is a harsh and unpredictable environment. Satellites must be designed to withstand extreme temperature variations, high levels of radiation, and micro-meteoroid impacts. Thermal management systems, such as radiators and heat shields, are essential for maintaining operational temperatures for sensitive electronics. Additionally, electromagnetic interference and solar flares can pose risks to communication integrity and overall satellite health, necessitating robust shielding and fault tolerance in system design.

23.1.2.2 Signal Interference and Degradation

Signal quality is paramount for SATCOM systems, yet it is vulnerable to various forms of interference and degradation. The challenges include:

- **Atmospheric Effects**: Atmospheric phenomena, such as rain fade and ionospheric disturbances, can reduce signal strength, particularly in higher frequency bands like Ka or V bands.
- **Interference from Other Signals**: The crowded radio frequency spectrum means that signals can interfere with one another if not properly isolated by frequency

translation and filtering. Advanced signal processing techniques, including error correction and redundancy, are used to mitigate these effects.
- **Noise and Distortion**: Both natural and man-made noise can degrade the quality of received signals. In such cases, regenerative transponders provide a significant advantage by demodulating and regenerating the transmitted signal, thus improving the overall signal-to-noise ratio (SNR).

23.1.2.3 Ground Segment Technical Challenges

The ground segment—comprising teleports and ground stations—faces its unique set of challenges.

- **Antenna Alignment and Stability**: Precise steering and alignment of ground-based antennas are critical to maintain a stable link with fast-moving LEO satellites. Automation and feedback control systems are essential to dynamically adjust antenna positioning.
- **Encryption and Security**: With the increasing importance of secure communications, both the onboard transponders and ground stations incorporate encryption modules to safeguard data. This adds complexity to the system and demands robust cyber security measures.
- **Integration with Terrestrial Networks**: SATCOM systems often serve as a complement to or substitute for terrestrial networks. Developing hybrid solutions that seamlessly integrate satellite and land-based communication paths requires advanced routing, signal switching, and protocol management capabilities.

23.1.2.4 Technical and Financial Constraints

Developing, launching, and operating satellites are capital-intensive endeavours. The need for high reliability in orbit, demands extensive testing on the ground, advanced materials, and sometimes even reusable launch vehicles. As technology evolves, efforts are made to reduce costs while enhancing performance through innovations such as software-defined payloads, high-throughput architectures, and miniaturized satellites.

23.1.3 Future Trends and Technological Advancements in SATCOM

Advancements in technology are continuously reshaping the SATCOM landscape. Several key trends are expected to drive the evolution of satellite communications in the coming years.

23.1.3.1 High-Throughput Satellites (HTS)

High-throughput satellites represent a significant leap forward in SATCOM capability. HTS systems are designed to deliver very high-capacity data links by using narrow spot beams and frequency reuse techniques. They often operate in the Ka and V bands and are a critical solution to the ever-growing demand for data bandwidth. HTS systems leverage advanced modulation schemes and artificial intelligence to dynamically manage multiple carriers, ensuring efficient spectrum usage even under variable network conditions.

23.1.3.2 Advanced Signal Modulation and Processing

The ability to transmit more data in smaller spectral footprints is essential as demand for broadband connectivity grows. Recent innovations in modulation techniques, such as quadrature amplitude modulation (QAM) and orthogonal frequency-division multiplexing (OFDM), allow satellites to support higher data rates while maintaining low latency. These advanced methods, while mathematically more complex, directly benefit users by increasing the robustness of communications against interference and nonlinear amplifier effects.

23.1.3.3 Optical Communication and Lasercom

Optical communications, including lasercom terminals, have emerged as a complementary technology to traditional RF systems. Optical communication offers several advantages, such as high data throughput, reduced susceptibility to interference, and the potential for inter-satellite links that form mesh networks in constellations. Such systems face their own unique challenges—chief among them being precise pointing and atmospheric effects—but continue to mature as technologies for both space-to-space and space-to-ground links.

23.1.3.4 Miniaturization and Constellations of Small Satellites

The rapid expansion of small satellites, including CubeSats and microsatellites, is fundamentally altering the SATCOM industry. Miniaturization has not only reduced launch costs but has also enabled the development of satellite constellations that provide global coverage with low latency. Constellations of small satellites are being deployed to deliver broadband internet services even in remote areas. Their reduced size imposes unique design challenges on payload and transponder integration, yet their economic advantages continue to drive significant research and development in this sector (Fig. 23.2).

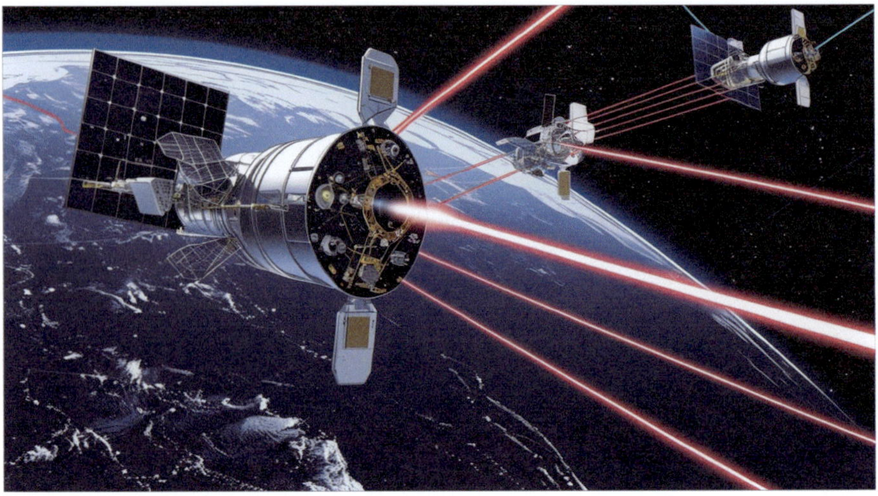

Fig. 23.2 An illustration of satellite based free space optical communication

23.1.3.5 Integration with 5G/6G and Terrestrial Networks

The future of SATCOM is also closely tied to advancements in mobile connectivity. Satellite-based non-terrestrial networks aim to expand 5G/6G coverage to areas where terrestrial infrastructure is sparse or absent—from rural communities to maritime platforms. These systems will rely on advanced SATCOM components and innovative integration strategies to ensure seamless operation with existing terrestrial networks. The convergence of satellite and terrestrial communications holds the promise of ubiquitous, resilient connectivity on a global scale (Fig. 23.3).

23.1.3.6 Research and Development in Advanced RF Engineering

Next-generation SATCOM systems rely on sophisticated RF engineering to meet the demands of high data throughput and robust performance. Companies such as Keysight are at the forefront of developing tools and techniques to simulate, test, and measure advanced SATCOM systems. Innovations in RF design, including beam steering, adaptive modulation, and advanced vector signal analysis, are critical to achieving the performance metrics required by modern satellite networks. These developments are expected to lead to SATCOM systems with unprecedented levels of capacity, reliability, and flexibility.

Fig. 23.3 An illustration of integration of SATCOM with 5G/6G & other terrestrial communication

23.2 Tele-Education: ISRO's Initiative for Societal Development Through Utilization of SATCOM

The first satellite developed in India specifically for the educational sector is GSAT-3 or EDUSAT. Its major goal is to satisfy the nation's need for an interactive remote learning system based on SATCOM networks. The network was basically a closed user group with a Hub in a central locations and Satellite Interactive Terminal (SIT) at the remote locations. Through the Telemedicine program, GSAT or INSAT satellites are used to connect specialist hospitals in urban regions with hospitals and healthcare facilities in rural and faraway areas. This is effectively used for the rural and distant population providing a better health care facility. Under the initiative of Village Resource Center (VRC) program, ISRO had provided the satellite bandwidth and network connectivity for natural resource management and various training through its interactive programs. ISRO had established a Virtual Private Node (VPN) network using GSAT 12 satellite in Ext. C band under Disaster management Support (DMS) program in 2008. A number of nodes comprising of monitoring, primary user and state nodes along with a hub had been established under ISRO DMS network using satellite based VSAT connectivity for effective disaster management. During disasters in Jammu & Kashmir and Uttarakhand, this network was successfully used.

23.3 Satellite Based Tele-Education

Education is the foundation of societal progress, enabling individuals to acquire knowledge, develop skills, and foster critical thinking. It is instrumental in shaping the economic, social, and technological advancements of a nation. However, ensuring equitable access to quality education remains a global challenge, particularly in geographically remote areas, underserved communities, and regions with inadequate infrastructure. Through the use of radio, television, and eventually online streaming technologies, the second generation of distance learning process have been evolved and introduced in the developing nations (Terry Anderson et al. 2020).

Tele-education, or satellite-based distance learning, has emerged as a transformative solution to bridge this gap. By utilizing modern communication technologies, tele-education enables real-time dissemination of educational content to distant learners, ensuring that geographical barriers do not hinder access to quality learning. This approach is particularly crucial for regions with limited access to skilled educators, educational resources, and structured learning environments. Tele-education facilitates interactive learning experiences, connects students with expert educators, and enhances the reach of educational institutions, thereby democratizing education at an unprecedented scale. After the Television technology started, efforts have been given to introduce it in the education sector as well to improve the quality of education with audio-visual aids prior to which only chalk-blackboard and Overhead projectors (OHP) were the primary means of doing the learning activities. The role of distance and self-learning over face to face or campus-based learning to extend the educational reach and to improve the quality of education were understood with the advent of ICT enabled education. Many government entities had been formed to support the activity. Following the launch of INSAT (Indian National Satellite) in eighties of twentieth century, national television, or Doordarshan, assumed production and broadcasting of educational TV programs for school children and teachers.

Based on the requirements for remote education with high coverage, ISRO launched its first fully thematic and dedicated educational satellite, GSAT-3 in 2004 which was named as EDUSAT. As the stakeholders for implementing various educational policies in the country are diverse, the implementation of the EDUSAT based distance education depended on extensive coordination among all stakeholders. ISRO initiated its satellite-based Tele-Education (TE) program in the year 2003 with a pilot project in three States (Madhya Pradesh, Maharashtra & Karnataka). In the later phase of the program, TE networks were also established in all the North-Eastern states, except Manipur, around the year 2007. EDUSAT was launched specifically for space-based management of educational content delivery over large distances using SATCOM bandwidth. This could create a network of virtual classrooms at remote area schools where educational content can be broadcast from a central location. In a state-wide network, this created a localized 'EDUSAT network' where teachers can sit at a central location and take classes from a multimedia enabled studio cum Teaching End. With the help of SATCOM Infrastructure, the class content in digital media form can be uploaded to the Satellite which then broadcasts the same which

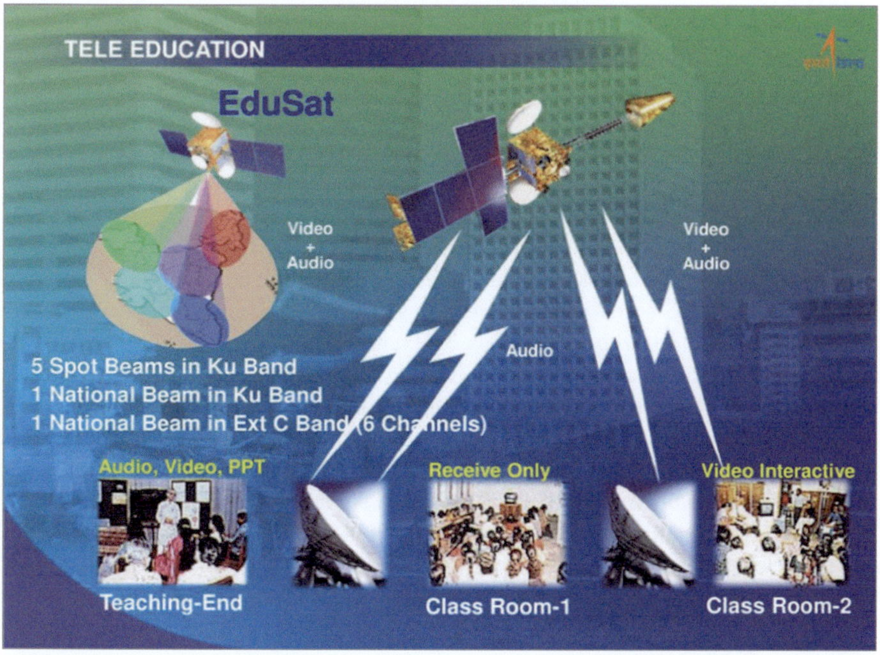

Fig. 23.4 Schematic of EDUSAT based Tele-Education *Source* space technology for health & education, L.S. Satyamurthy, Coordinator, Telemedicine Programs, ISRO HQ, Bangalore, October 21, 2008

can be received by the receiving nodes at remote end schools. Thus the satellite acts like a distribution point for educational content. The receiving end can be a simple Receive-Only Terminal (ROT) which only receives the multimedia educational content but the students sitting at a ROT cannot interact with the resource person, and it can also be a Satellite Interactive Terminal (SIT), where there is a Satellite return link for students to interact with the resource person. This element of interactivity makes the concept even more attractive to the students who no longer feel left away and this helps in their learning. EDUSAT facility has paved the way for national education organizations to start distance learning programs at all levels of education (Fig. 23.4).

23.3.1 Technical Description of EDUSAT (GSAT-3) Satellite

EDUSAT, also known as GSAT-3, was India's first dedicated educational satellite, launched in September 2004 by ISRO. Its primary purpose was to provide satellite-based, interactive distance education to various educational institutions across the country, from schools to higher education levels. The satellite was located at 74 deg

East longitude. EDUSAT was a pioneer in many technology developments & in-orbit testing compared to its INSAT predecessors. Use of multiple spot beams was the definitive technology which separated EDUSAT from other contemporary satellites. The Ku band spot beams generated by EDUSAT were directed by a 1.2 m reflector precisely towards ground what would become the footprint or the service area. It had many structural, mechanical & operational novelties for orbit and orientation maintenance. Advanced signal processing was its other strength thanks to integration of state-of-the-art electronics (www.isro.gov.in).

EDUSAT being a multi spot beam satellite, its coverage can be understood as the total coverage of its individual spot beams. The spot beams are divided into two national beams and five regional beams. The national beams are of wide beamwidth and cover almost the whole of Indian Mainland while the regional beams are narrower in beamwidth and focus on specific regions. The main objective of the satellite is to relay educational programs generated at a production cum teaching facility to student nodes, usually at schools. The satellite had flexibility in terms of frequency of service so that Ku-Band or Extended C Band signals can be used (Fig. 23.5).

The national beam was utilized by IGNOU, AICTE, ICAR, NCERT, UGC and many others academic and Research institution. The 5 regional Beams of EDUSAT were utilized by several institutions and universities to supplement the course curriculum. The North-Eastern Beam was utilized by prominent academic institutions of NER States.

Fig. 23.5 An artistic view of Multiple beams of GSAT-3 illuminating different regions of India

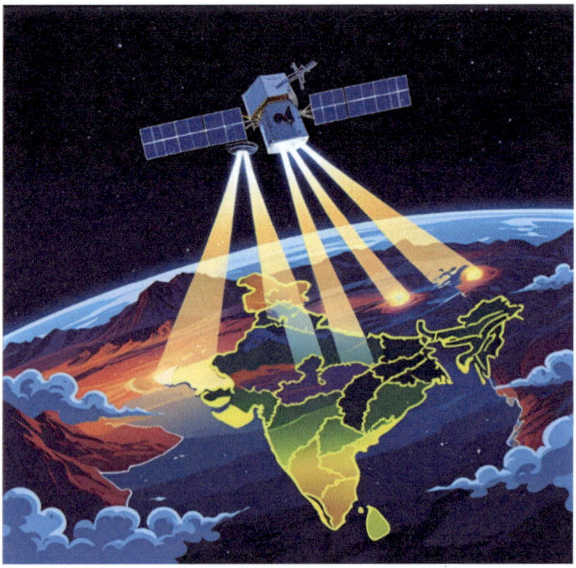

23.3.2 EDUSAT Utilization Program (EUP)

The first proposal for launching a dedicated satellite to cater to distance education was first put forward by ISRO/DoS in 2002. In a meeting with the Space Commission & Ministry of Human Resources Development (MHRD) The following factors inspired the recommendation:

- Inadequacy of teachers at the school level and technical education.

  Student drop out at the school level.
  Alternative mode of education for the vast majority of the country where conventional education was not sufficient.
  Augment curriculum-based teaching, impart effective teacher training, facilitate community participation and enable interaction between scholars and researchers.
  Alleviate the geographical divide for access to education and improving overall reception of technical education.

Based on the recommendation, Rs 85 crores budget was sanctioned for EDUSAT development. But actual infrastructure at ground level was also required for taking leverage of the satellite. Through technical discussions, a structure of the ground network was finalized. A distributed structure was conceived consisting of a hub and studio facility at state capital/premise of the state specific nodal agency for production of educational content with enlisted teachers in vernacular language and uplinking to satellite for broadcasting, for receiving such broadcast from satellite, receiving nodes were required at schools/beneficiary institute premise, called either Satellite Interactive Terminals (SITs) or Receive Only Terminals (ROTs). Additional budget was necessary for establishing all this ground hardware. Accordingly, Space Commission initiated the EDUSAT Utilization Program (EUP), a comprehensive program for coordination among all potential stakeholders of EDUSAT for utilization of the satellite resources. Both national and regional users were selected. The national ground network was set up by ISRO along with one hub and ten terminals at each State/Union Territory free of cost. Students of Secondary & Higher Secondary were basically the target audience for the program. GSAT-3(EDUSAT) continued its services till September-2010 and after EDUSAT stopped operation, Tele-education networks were served by other ISRO satellites like INSAT-4CR, GSAT-12, GSAT-18 etc. ensuring continuity of service at ground level (Pranab Das 2013).

The major objectives of EUP were as follows.

- Utilization of satellite resources for distributed education through low-cost ground segment
- Provide access to quality education throughout the country using satellite networks
- Provide sustainable distance education service and support formal and non-formal education in India

The program was organized in three phases, namely the pilot phase, semi-operational phase, and operational phase. Pilot phase was operated during 2004 primarily to validate the satellite communication based technology with the help of other Indian satellites. During semi-operational phase, EDUSAT program was expanded to most of the states and major national agencies and national and regional networks were set up. In this phase, the operation of tele-education networks commenced under the leadership of various educational agencies & major stakeholders of EUP. The networks were further expanded under operational phase with funding by respective state governments/user agencies.

23.4 Tele-Education Program in North Eastern States

ISRO's Tele-Education program has been running in NER to provide Satellite based connectivity to augment classroom-based teaching with digital classroom technology. Under the program, ISRO has established 330 SITs in all eight states of NER. Various State Government line departments and associated agencies had been identified as the nodal agencies to utilize the Tele-education networks for training and distance education in the respective north eastern states.

23.4.1 EDUSAT Network Configuration & Operation in NER

These networks consist of one Ku-Band Hub, Teaching End, around 50 Satellite Interactive Terminals (SITs) in each state. The networks operating in Ku band were initially supported through EDUSAT and later supported through INSAT-4CR. The NE networks are conFig.d with interactive terminals and utilized for various distance education & capacity building program in the remote areas of NE states.

23.4.1.1 Description of a Typical TE Network

The North Eastern States'Tele-Education program consists of Tele-Education networks in NE States. These networks have Space segment and Ground segment. The Space segment consists of transponder capacity in the satellite designated for Tele-Education. The ground segment consists of a Hub cum Teaching Studio, Baseband systems, Servers and Satellite Interactive Terminals (SITs). Presently, a Tele-Education network typically consists of the following:

- Satellite Capacity
- Satellite communication network associated with Teaching-End (Studio), Hub and Satellite link, SIT, and ROT

Tele education networks typically follow Star topology where SIT/ROTs are connected to a central Hub location from where all uplink and downlink communication with satellite is coordinated. Hub uplinks educational program or teacher material to satellite and satellite broadcasts this material to all SIT/ROT in the network. These SITs can, in turn, uplink their queries (student questions) which come to Hub through the return channel and are again uplinked to the satellite for rebroadcast. The teacher sitting at Hub answers the question which is again uplinked to the satellite for broadcasting. Multiple SITs and access the satellite through a digital access scheme called Time Division Multiplex/Time Division Multiple Access. In this, the SITs are given fixed time slots in which they communicate and all such individual streams are multiplexed together and uplinked to satellite as a mixed stream. This technology efficiently uses the costly satellite spectrum. For facilitating the radio frequency communication, both Hub & SIT/ROT are equipped with customized RF equipment like VSAT antenna, LNA & other components which form the Out Door Unit (ODU) which are finally connected to the digital signal processing chain inside called In Door Unit (IDU). The signal once received can be seen with the help of peripheral devices like PC, TV, Microphone, Camera etc. & other accessories. The can also run specific software like Learning Management Software (LMS) for coordinating all the distance education activities.

23.4.1.2 Components of the Tele-Education Network

There are three major components of Tele-education network i.e. Hub with RF equipment at teacher end, Satellite interactive terminal (SIT) and Learning management system (LMS). The TE hubs of Assam, Sikkim, Tripura, Arunachal & Mizoram are already operational. However, for non-functional Hubs some hardware items need to be repaired/replaced to make them operational. The spares for these hubs will be arranged by ISRO either by procurement or by using existing equipment installed in VRC (Village Resource Centres) hub located at ISTRAC, Bengaluru. In case any of the state hub could not be operationalized for some reason, the nearby state hub is used in sharing basis to provide the hub functionalities. The VSAT (Very Small Aperture Terminal) based satellite connectivity is being provided to all the remote nodes also termed as Satellite interactive terminal (SIT). However, there will be requirement of repair/replacement at non-functional SIT locations for network reactivation. The requirement of RF hardware replacement may arise at few remote nodes. To meet the requirements of hardware replacements and also to maintain spares for repairs/maintenance, it was planned to dismantle few existing and non-operational Village Resource Centre (VRC) nodes having similar RF hardware and relocate them at required NE hubs & remote locations. It is planned to procure and replace the old peripherals & accessories (Computer, handy camera, microphone, projector, etc.) with the new items. The third component of the network is the Learning Management System (LMS). TrainNet was used as LMS during its initial phase of inception. However, the support for the existing LMS software 'TrainNet' is discontinued by the software developer, M/s Robogroup, Israel. Finally, an LMS was developed by DECU-ISRO and was successfully introduced into the network.

23.4.2 Utilization of TE Network

Tele- education network is utilized to provide access to quality resource personnel, imparting quality teachers training along with improving the curriculum-based education to the under served section of the society. Tele-Education program supported higher education, especially development communication, by providing satellite connectivity to schools, colleges and other academic institutions through thousands of live and interactive programs.

23.4.3 Challenges and Achievements

Installation and maintenance of hardware required to set up such SATCOM based nodes in tough geographic location, power availability in the SITs were quite challenging in those early days. Unavailability of RF equipment required for rectification of technical issues is also causing difficulties to some extent for maintaining its operational condition. Moreover, getting skilled technical manpower to operate the satellite interactive terminal systems and minor maintenance are also challenging. Out of all these challenges, all the TE networks in the region have shown to be competent and successful in terms of reaching the unreached since its inception. Especially the TE networks in the states of Assam, Tripura and Meghalaya have shown an excellent example of successfully running the network and benefitting over lakhs of students through its large number of live and interactive as well as recorded sessions.

23.5 Telemedicine

One of the mandates of the government of the day is to provide basic minimum healthcare services to every individual across the width and breadth of the country. While this ideally means uniform distribution of medical infrastructure throughout the geography of the country, in actuality, it is seen to be clustered around the urban centres while sparsely available in the rural areas. With the revolution in Information Communication Technology (ICT), especially the combination of SATCOM and ICT, means for extending the benefits of medical infrastructure to remote and inaccessible areas has been a reality. ISRO has successfully implemented a number of societal development projects. Telemedicine/Tele-health is one such program which is of great social relevance to the country for taking advanced healthcare to the remote, rural and marginalized population.

23.5.1 Tele Medicine Applications

Tele-medicine is used to provide services to rural areas and to those who previously had limited access. Tele-education will be a part of tele-medicine as doctors take tele-courses to remain up-to-date on new technologies and methods of treatment.

Tele-medicine has many different applications:

- Tele-education/tele-training
- Tele-consultation
- Tele-diagnosis
- Tele-surgery
- Tele-presence

Tele-education for teaching/upgrading/training of primary care physicians, residents, medics, and nurses, tele-training—where medical staff/students are taught about the subject or are shown an operation, and tele-support of medical equipment operation and maintenance. It can also be used to inform people about health, hygiene, family planning or any other topic related with health. Tele-consultation provides remote access to a specialist physician which is not possible due to geographic distance. Tele-diagnosis means diagnosis of a patient by physician sitting far away using the help of technology. With Tele monitoring services, a patients' physiological values can be remotely monitored. In tele-monitoring the doctor can see all the parameters of a patient admitted in a hospital or at home and can suggest change in medicines/course of action. This kind of service is limited to chronic and/or high-risk patients. It also provides remote assistance in therapy to patients. These services are frequently used to monitor elderly people living in their own homes, by cable TV, POTS etc. This is also known as Tele-care, Homecare or Tele-nursing. Tele-presence is used by sick students who can attend the school from their home or hospital bed when they are convalescing. It consists of simple videoconferencing system. Tele-surgery means conducting a surgery from a remote location. This is done through controlling a robotic arm by a qualified doctor. Remote database access or Online-health system is a large database which can be accessed by medical fraternity like, patient, doctor, specialist or a research scholar. It will have all the information regarding- Health maintenance, hygiene, laws and directives, hospitals, medicines, pharmaceutical companies, ingredients of various medicines, side effects, contraindications, Medical equipments, physicians, specialists, doctors, medical education, medical jobs, Health clubs, medical books & journals, online expert, online consultation etc. (Rajashekhar and Ayyangar 2012).

23.5.2 ISRO Telemedicine Program

ISRO started its Telemedicine initiatives as a pilot mode from South India. In the year 2001, ISRO linked Apollo Hospital at Chennai to the Apollo Rural Hospital at

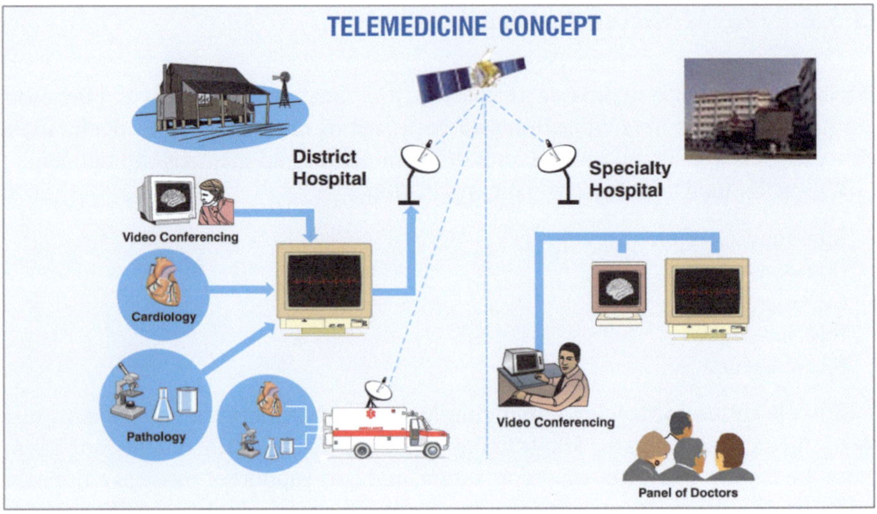

Fig. 23.6 Concept of ISRO tele-medicine *Source* telemedicine: healing touch through space, enabling specialty healthcare to the rural and remote population of India, 2005)

Aragonda village. Telemedicine combines SATCOM with ICT where the ICT system is built around a medical Electronic Medical Record (EMR) Software. This software is integrated with a computer which in turn interfaces with several medical diagnostic instruments. This ICT setup is linked to the SATCOM In-Door-Unit (IDU) which connects to the SATCOM Out-Door-Unit (ODU). The ODU basically means the extended C-band Very Small Aperture Terminal (VSAT) antenna at each location. The EMR Software follows the store-and-forward mechanism for transferring patient data while the actual doctor-patient interaction happens through the video-conferencing facility. The same facility is also used for conducting Continuing Medical Education (CME) through which doctors at entry level can be trained by experienced doctors. The responsibility of ISRO is to do all technical management, operations of Hub, user monitoring & maintenance of hardware & software and continuous user follow-up to ensure optimum utilization (Fig. 23.6).

23.5.2.1 History of ISRO's Tele Medicine

ISRO's Telemedicine initiatives started from Karnataka, where its headquarters is located. The Karnataka Telemedicine project linked Specialty & Super specialty hospitals in Bengaluru with the various district hospitals. Motivated by the success of the pilot phase and seeing the potential of the technology, ISRO went on to cover various state run hospitals & medical colleges or private/NGO run medical institutions to reach the maximum population. Thus, ISRO has introduced SATCOM based Telemedicine technology throughout the country to ensure that the hospitals get sufficient training and experience in running the facility so that the states can subsequently introduce telemedicine in an independent manner.

23.5.2.2 Technology of ISRO's Tele Medicine

Telemedicine brings together SATCOM & ICT, Biomedical Engineering and Medical Science. Telemedicine utilizes specialized hardware and software at Patient and doctor ends along with medical diagnostic equipment provided at the patient end. ISRO's Telemedicine technology is a pioneering initiative that leverages space technology to provide healthcare services to remote and underserved areas. By utilizing satellite communication, ISRO has enabled the delivery of medical consultations, diagnostics, and health education to regions with limited access to medical facilities. The system connects specialist doctors with patients in rural and remote locations, ensuring timely and effective healthcare intervention. This technology is especially valuable in India's diverse landscape, helping to bridge the gap between urban healthcare infrastructure and remote communities. ISRO's Telemedicine program exemplifies how space-based solutions can address critical healthcare challenges, improving health outcomes and expanding medical reach across the country.

23.5.2.3 Telemedicine Technology—Evolution

There are basically three types of telemedicine networks. Through the Point-to-Point System, one patient node is connected to one specialized hospital. The Point to Multi-point System allows one patient's end to link to many specialized doctor's nodes within the hospital at a time. The Multi Point to Multi Point System allows many patients' ends to connect to various doctors' ends at different hospitals in multiple locations simultaneously (Fig. 23.7).

23.5.2.4 Tele Medicine Initiatives

Building on ISRO's SATCOM-based ICT connectivity in its Telemedicine program, Continuing Medical Education (CME) initiatives have been launched to help doctors at rural healthcare centers enhance their medical knowledge and skills. These efforts enable rural practitioners to interact with specialists at major hospitals, thereby strengthening their capabilities. Additionally, the CME program has been integrated with the Tele-education initiative by establishing links between select Medical Institutions, Specialty Hospitals, and Research Centers, further promoting capacity building in rural healthcare.

The Telemedicine concept has potential to benefit a large number of underserved population if it can be freed from a permanent establishment. With this idea, a mobile Telemedicine unit has been conceived. A mobile Telemedicine Unit packs the whole Telemedicine setup including SATCOM, ICT & Biomedeical components

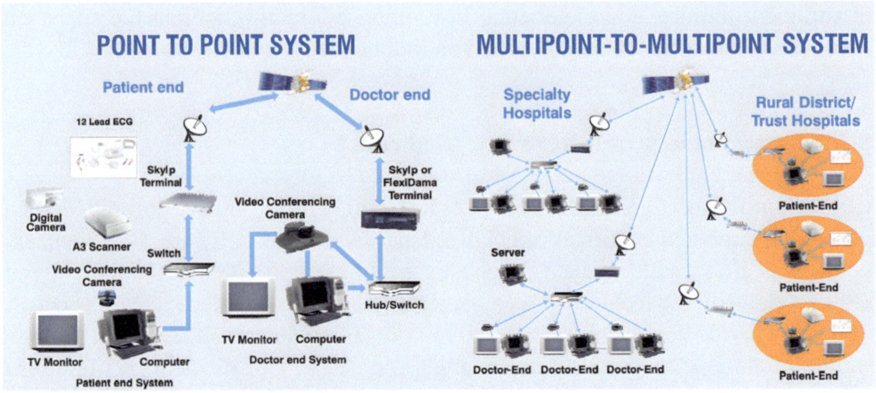

Fig. 23.7 ISRO's TeleMedicine technology *Source* telemedicine: healing touch through space, enabling specialty healthcare to the rural and remote population of India, 2005

in a Bus/Van that can establish a mobile Telemedicine centre at any place. A mobile telemedicine unit is an innovative healthcare delivery system that brings medical services directly to underserved and remote communities. Equipped with advanced diagnostic tools, telecommunication devices, and portable medical equipment, these units enable healthcare providers to conduct consultations, monitor vital signs, and perform preliminary assessments on-site. The utility of mobile telemedicine units lies in their ability to improve access to healthcare, especially in difficult-to-reach areas where permanent medical facilities are unavailable or limited. They facilitate timely diagnosis and treatment, reduce the need for patients to travel long distances, and help deliver specialized medical advice through telecommunication links. Overall, mobile telemedicine units play a crucial role in enhancing healthcare outreach, promoting early intervention, and supporting public health initiatives across diverse regions (Krishna Murthy et al. 2014).

Telemedicine connectivity has been provided at various popular Pilgrimage sites where lots of people visit for religious purposes and a medical emergency can happen virtually anytime.

23.5.3 ISRO-NEC Joint Tele Medicine Project

In July 2004, ISRO collaborated with the North Eastern Council (NEC) to initiate a joint telemedicine project aimed at establishing a SATCOM-based telemedicine network across the North Eastern Region (NER). The objective was to set up telemedicine nodes in all districts of the northeastern states, including Sikkim, totaling 72 nodes given the political and geographical landscape at that time. The primary goal was to connect district hospitals with regional and outside specialist and super-specialist hospitals, thereby improving healthcare access and quality at minimal cost through modern technology.

The project saw collaboration among ISRO, NEC, and the state governments of the region. The North Eastern Space Applications Centre (NESAC) played a crucial role in local technical coordination, handling installation, commissioning, manpower training, usage monitoring, and troubleshooting. By its completion, 25 telemedicine centers had been established. The operational management and technology usage were overseen by ISRO's Developmental Education & Communication Unit (DECU) and ISRO Headquarters. This project concluded in 2019.

Furthermore, ISRO established an Army Telemedicine Network in the NER in collaboration with the Indian Army from March 2008 to serve strategic users. Since Indian Army facilities are often located in remote, infrastructure-limited areas, a SATCOM-based telemedicine setup proved ideal. Several telemedicine nodes were commissioned in various army hospitals throughout the region. In addition, ISRO set up a few other telemedicine nodes across the NER via different civilian networks, expanding healthcare access further.

ISRO's telemedicine centers continue to serve the health needs of the people in the NER. They facilitate numerous tele-consultations and Continuing Medical Education (CME) sessions, connecting regional hospitals such as the Regional Institute of Medical Sciences (RIMS) in Imphal, Guwahati Medical College & Hospital (GMCH) in Guwahati, and the North Eastern Indira Gandhi Regional Institute of Health & Medical Sciences (NEIGRIHMS) in Shillong, as well as institutions outside the region. These services have significantly benefitted both patients and healthcare providers across the region.

A total of 39 tele medicine nodes had been established in north eastern states comprising of central and state medical college hospitals, district level civil hospitals, super specialty hospitals and military hospitals.

23.5.3.1 Description of a Tele-Medicine Node

The following tools and equipment were provided at the 25 ISRO-NEC joint Telemedicine sites. Satellite communication hardware includes a 1.8-m or 3.8-m VSAT antenna along with associated devices, power supply equipment, satellite receiver and UPS.

One computer table, one small equipment table, and one computer chair are also provided for each node.

Information Technology and Medical Hardware/Software:

- CPU With Medical Software
- Monitor
- Video Conferencing Camera (IP based with PTZ functionality) and Microphone
- Scanner
- Printer
- ECG Machine

- Switch (Ethernet HUB)
- Power Supply Board

23.5.4 Utilization

The tele-medicine network was effectively utilized by a number of super specialty hospitals such as medical college hospitals for tele consultation to the patients of remote nodes of this region. Specifically, the military hospital nodes have been highly benefitted by the TM network.

23.5.5 Challenges

While space technology holds immense promise for societal betterment, its effective implementation is not without its challenges. Establishing and maintaining the necessary ground infrastructure, particularly in remote and underserved regions, presents a significant hurdle. Coupled with this is the need for skilled personnel to operate and manage these sophisticated systems, a resource often lacking in such areas. The inherent costs associated with space-based solutions, from satellite development and launch to ongoing maintenance, necessitate careful consideration of affordability and long-term sustainability. Connectivity issues, even with satellite coverage, can still impede the smooth functioning of applications like telemedicine and tele-education. Furthermore, the creation of engaging and relevant content, especially in local languages, is crucial for the success of tele-education programs. Maintaining equipment in remote locations and ensuring timely repairs also poses a logistical challenge. Frequently changing the location of SIT or TM nodes within the school or hospital is also challenging, considering the reinstallation of hardware equipments and clear visibility of the respective satellites. Looking beyond terrestrial concerns, the growing problem of space debris demands responsible space practices and a commitment to environmental sustainability. Finally, effective implementation hinges on seamless coordination and collaboration among diverse stakeholders. Addressing these multifaceted challenges through strategic investments, capacity building, public–private partnerships, and a focus on cost-effective and user-centric solutions is essential to unlocking the full potential of space technology and realizing its transformative impact on society (Krishna Murthy et al. 2014).

23.6 Conclusion

ISRO's space initiatives have played a pivotal role in addressing societal challenges, particularly in remote and underserved regions like the North East. Early programs such as SITE, KCP, and JDCP demonstrated the potential of satellite communication for education and community development. Building on these successes, the launch of EDUSAT, India's first dedicated education satellite, transformed distance learning by enabling interactive virtual classrooms. This initiative connected thousands of schools and colleges across the country, overcoming geographical barriers in NER and improving access to quality education.

Additionally, the Telemedicine program has extended specialized healthcare to rural areas of the region by linking remote healthcare centers with urban specialty hospitals. These satellite-based solutions have significantly enhanced the reach and quality of education and healthcare, particularly in challenging terrains.

Despite the success of these programs, there is a growing need to upgrade the infrastructure to meet increasing demands for data rates, bandwidth, and throughput. Transitioning from closed user group systems to more flexible internet-based solutions, either terrestrial or satellite-based, will be essential to meeting future goals and expanding these services.

While challenges related to infrastructure, maintenance, and technical expertise persist, the continued evolution of space technology, particularly through broadband internet solutions, holds the potential to further enhance these services. ISRO's focus on societal applications highlights how space technology can address critical developmental needs, contributing to inclusive and sustainable development across the country.

Acknowledgements The authors would like to acknowledge the role of ISRO/DoS in funding the project, DECU, ISRO for overall technical supervision, fund transfer, training & other coordination. The authors would also like to acknowledge the various State Government Departments & NEC for their support towards the programs. The authors would also like to extend their thanks to various leaders at ISRO/DoS & NESAC for continuation of the programs in the NE region.

References

T. Anderson, P. Rivera-Vargas, A critical look at educational technology from a distance education perspective, Digital Education Review–Number 37 (2020)

P. Das, North East, 'the power house of India': prospects and problems. IOSR J. Hum.Ities Soc. Sci. **18**, 36–48 (2013). https://doi.org/10.9790/0837-1833648

Y.V.N. Krishna Murthy, P.L.N. Raju, S.K. Srivastava, Harish Karnataka; reach the unreached, IIRS outreach program for enhanced learning to all The International Archives of the Photogrammetry, Remote Sensing and Spatial Information Sciences, Vol. XL-8 (2014)

S.L. Rajashekhar, G. Ayyangar, Satellite technology to reach the unreached (India–A case study), in *2012 IEEE Global Humanitarian Technology Conference*

Telemedicine, Healing touch through space, enabling specialty healthcare to the rural and remote population of India, Published by: Publications and Public Relations Unit, ISRO Headquarters, Bangalore, February (2005)

Ramani Kumar Das has obtained M.Sc in Physics with specialization in Electronics from Gauhati University, Assam in the year of 2000 with First Class 1st Position (Gold Medalist). He is currently serving as Head, SATCOM & UAV Division (SUD) and Civil and Maintenance Division (CMD) at North Eastern Space Applications Centre (NESAC), Department of Space, Government of India, Shillong. His area of interest is Satellite Communication (SATCOM) and Satellite Navigation (SATNAV) Applications for Societal applications like Education, Health, communication support in disaster management, location-based services etc. and development of UAV technology and its applications. He is having more than 22 years of experience in the field. He is recipient ISRO's Team Excellence award in 2007 and ISRO's Young Scientist Award in 2009.

Anjan Debnath is currently serving as Scientist/Engineer-SF at North Eastern Space Applications Centre (NESAC) Dept. of Space, Govt. of India. He has taken important initiative towards effective utilization of Space Science and Technology for various Satellite Communication based societal developmental activities for NER such as Tele-Education, Tele-Medicine, Emergency Communication, etc. He specializes in Satellite Navigation, Internet of Things, wireless Sensor Networks etc. He has published multiple papers in National Peer Reviewed Journal. He is currently working as Safety Officer and Budget Coordinator of NESAC as well. He has mentored a number of interns at NESAC and acted as ISRO Mentor for various ISRO funded projects at STIC and under RESPOND programs. His area of interest is Microwave & Antenna Engineering, Ionospheric Earthquake Precursory Studies, GNSS related applications etc.

Himangshu Jyoti Das is currently working as Scientist/Engineer-SD at the North Eastern Space Applications Centre (NESAC), Umiam, Meghalaya. He completed his M.Sc. in Physics from Gauhati University and M. Tech. in Electronics Design & Technology from Tezpur University. He provides technical support for ISRO's societal application programs such as tele-education and telemedicine in the northeastern states. He also supports the GSAT-29 gateway located in Guwahati, as directed by ISRO Headquarters and SAC Ahmedabad. Currently, he is involved in the design and development of RF antennas. His research interests include RF and microwave antenna design for various uses. Additionally, he serves as the engineer-in-charge for electrical systems at NESAC, providing technical support for electrical maintenance as part of CMD activities.

Chapter 24
Unmanned Aerial Vehicle: Technological Development and Its Applications

Chirag Gupta, Sanjay Pandit, Ramani Kumar Das, and Shashi Bhusan Nayak

Abstract Unmanned Aerial Vehicles (UAVs), commonly known as drones, have transformed remote sensing and data acquisition by offering high-resolution, cost-effective, and real-time observation capabilities. Initially developed for military reconnaissance, UAVs have evolved into indispensable tools across civilian, scientific, and commercial domains. This chapter explores the historical evolution, technological classifications, and diverse applications of UAVs, with a focus on their use in environmental monitoring, agriculture, infrastructure mapping, disaster response, and payload delivery. UAVs are categorized into fixed-wing, rotary-wing, and hybrid VTOL systems, each with unique strengths suited for specific operational needs. Advances in propulsion, autonomy, sensors, AI, and communication systems, especially the integration of 5G and AI-driven swarm technologies, are accelerating UAV adoption across industries. The chapter highlights NESAC's contributions in UAV development tailored for Northeast India, including in-house UAV assembly, mission planning, testing infrastructure, and data processing capabilities. Despite their potential, UAVs face technical, ethical, and regulatory challenges related to battery limitations, privacy, airspace safety, and public acceptance. The chapter concludes with insights into future innovations such as smart-city integration and sustainable UAV ecosystems. Overall, UAVs are poised to play a pivotal role in geospatial technology and socio-economic development, particularly in remote and underserved regions.

Keywords Unmanned Aerial Vehicles (UAVs) · Rotary-wing · Fixed-wing UAV · UAV remote sensing applications · Autonomous navigation · UAV mission planning · Disaster management

---

C. Gupta (✉) · S. Pandit · R. K. Das · S. B. Nayak
North Eastern Space Applications Centre (NESAC), Umiam, Shillong, Meghalaya, India
e-mail: chirag.gupta@nesac.gov.in

S. Pandit
e-mail: pandit.sanjay@nesac.gov.in

R. K. Das
e-mail: rkdas@nesac.gov.in

S. B. Nayak
e-mail: shashi.bhusan1995@nesac.gov.in

S. P. Aggarwal et al. (eds.), *Remote Sensing of the North Eastern Himalayan Ecosystem*,
https://doi.org/10.1007/978-981-95-3221-6_24

24.1 Introduction

The use of aerial platforms for observation dates back to the mid-nineteenth century, with balloons being among the earliest technologies employed for this purpose. In 1858, Gaspard-Félix Tournachon, known as "Nadar," utilized a hot-air balloon to capture aerial photographs of Paris, marking a pioneering moment in aerial imaging. Subsequently, in 1882, English meteorologist E.D. Archibald advanced the field further by employing kites for aerial photography. Among the more unconventional methods was the use of pigeons equipped with small cameras, an idea developed by J. Neubronner and implemented by the Bavarian Pigeon Corps, showcasing early ingenuity in airborne surveillance (Colomina et al. 2014). The twentieth century saw significant experimentation with rotorcraft and vertical flight. Etienne Oehmichen, a French engineer, conducted numerous rotorcraft experiments during the 1920s. His Helicopter No. 2, featuring four rotors and eight propellers driven by a single engine, represented an early multi-rotor aircraft design. Around the same period, Dr. George de Bothezat and Ivan Jerome developed a helicopter with six-bladed rotors mounted on an X-shaped frame, further advancing rotary-wing aircraft capabilities. Later, the Curtiss-Wright Corporation introduced the VZ-7, a VTOL (Vertical Take-Off and Landing) aircraft developed for the United States Army. The VZ-7's control system relied on differential thrust across its four propellers-an approach that foreshadowed the control mechanisms of modern multirotor drones (Rumerman 2003). In recent decades, UAV technology has become increasingly accessible and widespread. The emergence of small, lightweight, and cost-effective quadcopters-such as the DJI Phantom and Parrot AR Drone-has expanded UAV applications well beyond military and scientific use into the commercial and consumer markets. The twentieth century also demonstrated the critical role of military research in catalyzing technological advancements with far-reaching civilian applications. For instance, surveillance satellites developed for defence purposes later enabled the Global Positioning System (GPS), and data transmission protocols created for secure military communication laid the foundation for the modern Internet. Drones, or Unmanned Aerial Vehicles (UAVs), developed under similar circumstances. Initially designed for reconnaissance, early UAV development remained largely obscure to the public. Their operational deployment began shortly after World War II (1939–1945), with the Ryan Firebee-an unmanned jet aircraft-being among the first widely used drones. The Vietnam War marked a notable period of UAV application, as documented in historical analyses of early drone warfare. Since then, the integration of UAVs into military operations has grown exponentially. This dramatic rise led The New York Times to describe drone warfare as a transformative paradigm in modern combat (Dixit et al. 2016).

UAVs, commonly referred to as drones, have revolutionized the field of remote sensing by providing unprecedented access to real-time, high-resolution, and cost-effective data acquisition. Initially developed for military applications, UAVs have rapidly evolved into indispensable tools for a wide range of civil, scientific, and industrial purposes. Their versatility stems from their ability to carry diverse sensors, including optical, thermal, multispectral, hyperspectral, and LiDAR systems, enabling precise data collection across various domains.

In the context of remote sensing, UAVs bridge the gap between satellite-based and ground-based observation systems. Unlike satellites, UAVs offer the flexibility to operate on-demand, providing detailed spatial and temporal resolutions tailored to specific project requirements. They can fly at low altitudes, making them particularly suitable for mapping small-scale features, monitoring environmental changes, and assessing disaster impacts with remarkable accuracy. Furthermore, their ability to access remote, hazardous, or otherwise inaccessible areas makes them invaluable for critical applications such as forestry, agriculture, urban planning, and disaster management. The increasing affordability of UAV technology, coupled with advancements in autonomous navigation, sensor miniaturization, and data processing capabilities, has significantly broadened their adoption. UAVs are extensively employed in applications, including precision agriculture, environmental monitoring, geological surveys, and infrastructure management. These platforms not only enhance data accuracy but also reduce the logistical and financial constraints traditionally associated with remote sensing.

This chapter delves into the technical, operational, and analytical aspects of UAVs in remote sensing, highlighting their potential to transform the way spatial data is captured and analysed. The chapter also aims to provide a comprehensive understanding of UAVs as a transformative tool in the domain of geospatial technology.

24.2 Categorisation of UAV

UAVs are engineered and assembled to meet specific operational objectives and can broadly be categorized into three main types: (1) fixed-wing UAVs, (2) rotorcraft or multirotor (rotary-wing) UAVs, and (3) hybrid UAVs that combine vertical take-off and landing (VTOL) capabilities with fixed-wing configurations (Sabour et al. 2023). Fixed-wing UAVs are typically designed for long-range missions, offering greater endurance and the capacity to carry heavier payloads. However, they come with certain limitations, including complex aerodynamic structures, the requirement for long runways, and operational challenges in uneven or mountainous terrain. In contrast, rotorcraft or multirotor UAVs, which include quadcopters and other rotary-wing platforms, have the distinct advantage of vertical take-off and landing. This makes them highly suitable for operations in confined, rugged, or high-altitude environments where fixed-wing UAVs may not be feasible. Despite their maneuverability and ease of deployment, multirotor UAVs generally suffer from limited flight duration and lower payload capacity compared to fixed-wing models (Boon et al. 2017). Hybrid VTOL UAVs aim to bridge this gap by integrating the endurance benefits of fixed-wing systems with the take-off flexibility of rotary-wing designs, offering a balanced solution for complex operational environments. This is because rotary wing must constantly generate lift by spinning all their rotors to stay in the air.

Unlike fixed-wing UAVs, which use wings to generate lift through forward motion (aerodynamic lift), rotary wing burn power just to hover, let alone fly forward. Also, most of the fixed wing UAVs are powered by gasoline-based internal combustion engine and the specific energy of chemical-based energy source is way higher than an electric/electrochemical based energy source such as a Lithium-Ion battery or an alkaline battery which is predominant power source for rotary wing UAVs (Depcik et al. 2020). The hybrid VTOL fixed wing UAV combines the advantage of fixed wing and multirotor categories. A comparison on features of different categories of UAV is presented in Table 24.1.

Table 24.1 Features of different categories of UAV

Feature	Copter/ Rotary Wing	Fixed-Wing UAV	Hybrid VTOL fixed-wing
Take-off & landing	Vertical take-off and landing, no runway required	Requires a runway or catapult launch	Vertical take-off and landing, no runway required
Flight efficiency	Less efficient, consumes more power for hovering	More efficient, uses lift from wings for longer endurance	Efficient, consumes more power for hovering, but lift from wings during cruise flight
Endurance	Shorter flight time (typically 20–60 min)	Longer flight time (typically 1–24 + hours)	Longer flight time, close to but less than fixed wing UAV
Maneuverability	Can hover, take off vertically and fly in tight spaces	Requires forward motion, less maneuverable	Can hover for very short duration, less maneuverable in forward flight
Payload capacity	Lower payload capacity due to high energy demand	Higher payload capacity, suitable for heavy payloads	Medium payload capacity, payload limitation during take-off and landing
Speed	Slower (0–80 km/hr)	Faster (50–200 + km/hr)	Faster (60–250 + km/hr)
Weather resistance	More affected by wind and turbulence	A little less effected by wind and turbulence	A little less effected by wind and turbulence
Operational use	Ideal for close-range tasks like inspection, surveillance, mapping survey, etc	Suitable for long-range missions like surveying, reconnaissance, and cargo transport	Suitable for long-range missions like surveying, reconnaissance, and cargo transport
Deployment complexity	Simple to deploy and operate	Requires more space and expertise to operate	Simple to deploy, but requires expertise to operate

24.2.1 Rotary Wing UAV

Rotary Wing or a rotorcraft UAV (RUAV) is an aircraft which relies on a rotor to move wings which in turn generates lift allowing the aircraft to fly. An example of a rotary wing aircraft is the Helicopter. In case of a rotary wing aircraft, the wings connected to the rotor mast moves. This rotating movement of the propellers/rotors generates the necessary pressure difference above and below the airfoil to generate lift. Since the velocity over the wings is generated by moving the rotors, RUAVs are capable of VTOL (Vertical Take-Off and Landing). The VTOL capability of the RUAVs allow them to be used in areas where there is lack of space or where focus on a certain area is required for an extended period of time. The presence of the rotary wing model also ensures very high manoeuvrability and the ability to hover in a set point in space for a long time. However, unlike most fixed wing models of UAVs, the rotary wing models generally have a much shorter flight time. Most rotary-wing based UAVs are multirotor, with generally more than one rotor. Based on number of rotor arms, the multirotors can be further sub-categorized as tricopter, quadcopter (Fig. 24.1), hexacopter (Fig. 24.2), octacopter (Fig. 24.3) etc. (Arjomandi et at. 2006). Multi-rotor also comes in various forms and configurations based on the arrangement of rotor arms such as " + ", "X", Y6, and X8 etc. (Quan et al. 2017).

A quadcopter is a UAV with four rotors arranged in a square layout, offering stable flight, vertical take-off, and precise hovering. For quadcopters, a diagonally opposite propellers pair rotate in one direction and the other pair rotates in other direction, but opposite to the first pair. Quadcopters primarily use X-configuration or + (plus)-configuration, defining motor alignment relative to the aircraft's front. In X-configuration, the motors are offset at 45° angles from the nose, forming an "X" when viewed from above. This design enhances agility and smooth control, making it ideal for FPV racing, freestyle flying, and aerial photography, as diagonal symmetry improves roll/pitch responsiveness. In + (plus)-configuration, one motor aligns

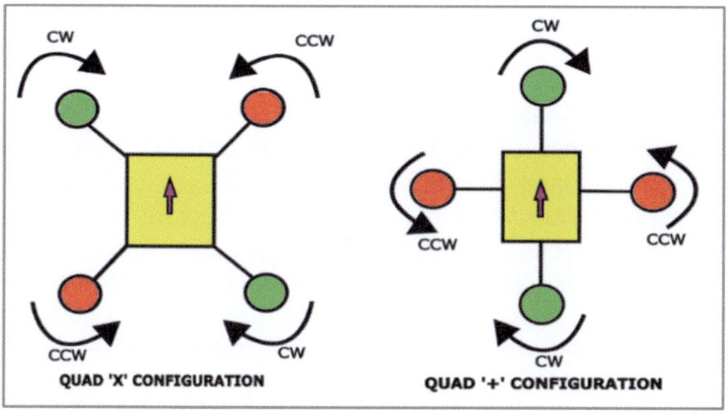

Fig. 24.1 Quadcopter and its 'X' and '+' configurations

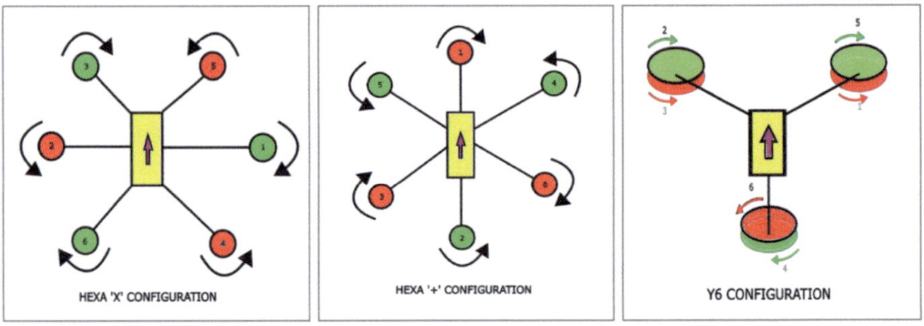

Fig. 24.2 Hexacopter and its 'X', '+' and Y6 configurations

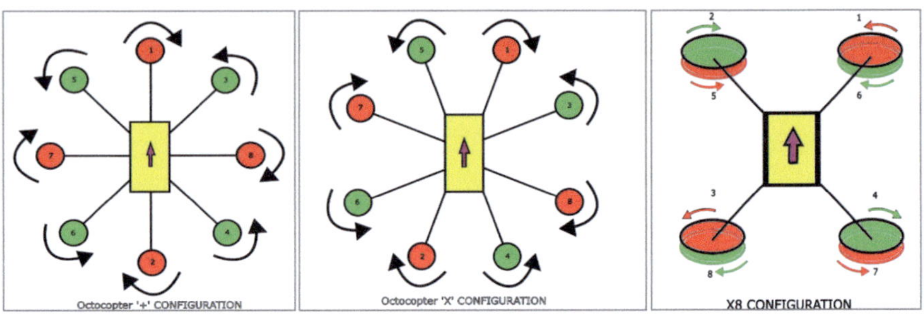

Fig. 24.3 Octocopter and its '+', 'X' and X8 configuration

directly forward, creating a " + " shape. This simpler layout offers intuitive orientation for beginners and stable flight for basic operations, often preferred in mapping and surveillance drones where forward-facing clarity matters. X-configuration dominates modern drones for versatility, while + -configuration remains niche for specific stability-focused applications. Both ensure stable lift and maneuverability through coordinated motor speed adjustments. All the two configurations of quadcopter are shown in Fig. 24.1.

A hexacopter is a UAV with six rotors arranged in a symmetrical layout, providing greater lift capacity, enhanced stability, and built-in redundancy compared to quadcopters. Like all even-numbered multirotors, its propellers follow a balanced rotation pattern—three spinning clockwise and three counterclockwise—to neutralize torque effects and maintain stable flight. Hexacopters typically use three configurations: Hexagon-X, Hexagon- +, and Y6 Coaxial. All the three configurations of hexacopter are shown in Fig. 24.2. In the Hexagon-X setup, motors are spaced 60° apart with the front arms angled outward, offering smoother forward flight and superior gimbal stability—making it ideal for professional cinematography and precision surveying. The Hexagon- + variant positions one motor directly forward, simplifying pilot orientation for industrial inspections. Alternatively, the Y6 coaxial design stacks pairs of motors vertically on three arms, creating a compact yet powerful drone capable of operating in tight spaces while maintaining redundancy—if one motor fails, the aircraft can still land safely. This configuration trades slight hover efficiency for agility and is favoured in search-and-rescue or confined-area missions. With greater payload capacity, wind resistance, and fault tolerance than quadcopters, hexacopters excel in heavy-lift applications like LiDAR mapping, film production, and critical infrastructure inspections where reliability is paramount. Their balanced motor rotation (3 CW, 3 CCW) ensures stable torque-neutral flight even under demanding conditions.

An Octocopter features eight rotors arranged in either flat or coaxial configurations to deliver unmatched lifting capacity, flight stability, and operational redundancy. In the standard flat configuration, motors are evenly spaced in either an X or + formation, with the X-layout (45° spacing) offering superior maneuverability for cinematic applications while the + -configuration provides simpler orientation for industrial use. The coaxial X8 design stacks pairs of motors vertically on four arms, combining compact dimensions with exceptional power density and fault tolerance—capable of maintaining controlled flight even with multiple motor failures. All the three configurations of octocopter are shown in Fig. 24.3. These robust systems excel in demanding professional applications, supporting heavy payloads up to 20 kg including full-frame cinema cameras, LiDAR systems, or specialized sensor packages. Their inherent redundancy makes them indispensable for critical missions like aerial firefighting, search-and-rescue operations, and infrastructure inspection in challenging environments. With precise torque-balanced control (4 CW/4 CCW motors) and advanced flight controllers, octocopters provide the ultimate combination of payload capacity, wind resistance, and operational safety for commercial and government applications where mission success cannot be compromised.

24.2.2 Fixed Wing UAV

A fixed-wing UAV is an aircraft that relies on rigid wings for lift, similar to traditional airplanes. Unlike rotary-wing UAVs (e.g., quadcopters), fixed-wing UAVs cannot hover in place but excel in long-duration, high-speed, and efficient flight over large areas. Lift in a fixed-wing aircraft is generated by the aerodynamic forces acting on the wings as air flows over and under them. The wings are designed with a specific shape, called an airfoil, which facilitates this process. Fixed-wing UAVs Requires forward motion to stay airborne, typically achieved through a propeller or jet engine. They offer longer endurance and range compared to rotary-wing UAVs. Fixed-wing UAVs come in various types, each designed for specific applications and environments. Some of the common types are conventional fixed-wing UAV, tail-sitter UAVs, Flying-wing UAVs (Elijah et al. 2021) as shown in Fig. 24.4.

Fixed-wing UAVs are Ideal for large-scale land surveys due to their ability to cover vast areas efficiently. They are commonly used for reconnaissance, border

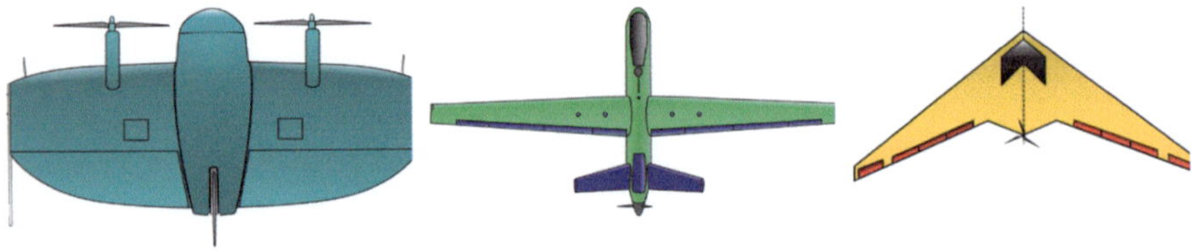

Fig. 24.4 4 Conventional fixed-wing, flying wing fixed-wing and Tail-sitter fixed-wing (left to right)

patrol, and intelligence gathering and are suitable for long-range delivery applications. Fixed Wing UAVs have greater flight range and endurance compared to rotary-wing UAVs. They also have higher cruising speeds and hence are energy-efficient for extended operations. Along with above advantages, fixed-wing UAVs also have certain limitation such as requirement of a runway, catapult, or launch system for takeoff and landing. They cannot hover or perform vertical takeoff and landing (VTOL), unless hybrid models are used.

24.2.3 Hybrid VTOL Fixed-Wing UAV

A hybrid VTOL (Vertical Take-Off and Landing) fixed-wing UAV combines the benefits of both fixed-wing and rotary-wing (typically multirotor) UAV designs. This type of UAV is designed to take off and land vertically, like a helicopter, while also having the long-range and high-speed capabilities of a fixed-wing aircraft once airborne (Zhou et al. 2020). VTOL feature allows the UAV to take off and land in confined spaces, eliminating the need for runways or large landing zones, once airborne, the UAV transitions to horizontal flight like a traditional fixed-wing aircraft, offering greater speed, endurance, and energy efficiency compared to rotary-wing UAVs. A Hybrid VTOL fixed-wing UAV is shown in Fig. 24.5. This is especially useful for operations in urban environments or rough terrain. Hybrid VTOL UAVs have sophisticated mechanisms (such as tilting motors, swivelling rotors, or separate propulsion systems) to seamlessly transition between vertical and horizontal flight modes. It offers major advantages in terms of versatility, increased efficiency, operational flexibility and reduced infrastructure requirements. However, this technology also has some challenges such as complexity of design, transition efficiency, power management, cost, etc.

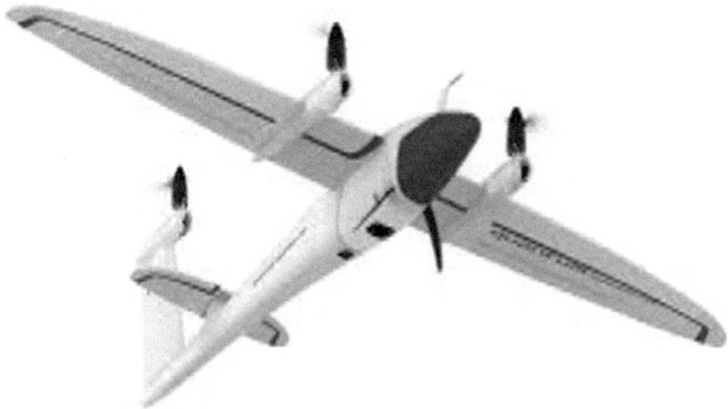

Fig. 24.5 Hybrid VTOL fixed-wing UAV

24.3 Applications of Unmanned Aerial Vehicles Across Various Sectors

UAVs are offering efficient, adaptable, and cost-effective solutions for various tasks. From aerial mapping and disaster response to agriculture, security, logistics, and environmental conservation, UAVs have enhanced data collection, monitoring and operational efficiency. Some of the prominent applications of UAVs are listed below:

24.3.1 Aerospace and Defence

UAVs have a wide range of applications in aerospace and defence sector. UAVs are used for intelligence gathering and monitoring activities in conflict zones or sensitive areas, providing real-time data without risking human lives (Hildmann et al. 2019). Armed drones can be deployed for targeted strikes, enabling military forces to engage threats from a distance with precision and reduced risk to personnel. UAVs can transport supplies, medical equipment, and ammunition to remote or inaccessible locations, enhancing operational efficiency and support. UAVs are used for monitoring borders and critical infrastructure, helping to detect illegal activities and ensuring national security. Some UAVs are equipped with capabilities to disrupt enemy communications and radar systems, playing a crucial role in modern warfare. UAVs collect vast amounts of data for analysis, which can be used for strategic planning and operational decision-making.

24.3.2 Agriculture

UAVs are revolutionizing Agriculture through precision farming and crop monitoring. Drones equipped with multispectral cameras can capture images in different wavelengths to assess plant health. This helps in identifying areas of stress due to disease, pests, or nutrient deficiencies. Normalized Difference Vegetation Index (NDVI) can be calculated from drone imagery to evaluate plant vigour and biomass, allowing farmers to target interventions effectively (Wahab et al. 2018). Drones can create detailed 3D maps of fields, helping in planning irrigation systems and optimizing field layouts. Aerial surveys can identify variations in soil properties, enabling targeted soil management practices.

As part of precision agriculture, UAVs can be used to apply fertilizers, pesticides, and herbicides with precision, reducing waste and minimizing environmental impact. Drones can monitor moisture levels across fields, assisting in efficient water usage and improving irrigation practices. Drones can be employed to count crops and estimate yield potential by analyzing crop density and health, which helps in better planning and resource allocation. Early detection of pests and diseases through aerial imagery allows for timely interventions, potentially saving crops and reducing pesticide use. UAVs can monitor livestock health and movement, allowing for better management and welfare practices. Some advanced drones are equipped to perform aerial seeding, especially in hard-to-reach areas or for reforestation efforts. Drones collect vast amounts of data that can be processed using AI and machine learning, providing insights for better decision-making and operational efficiency. UAVs can be used to assess the impact of agricultural practices on surrounding ecosystems, aiding in sustainable farming practices.

24.3.3 Environmental Monitoring

UAVs have a wide range of applications in environmental monitoring, particularly in wildlife conservation and habitat analysis. UAVs can capture high-resolution images and videos to estimate wildlife populations in hard-to-reach areas. Drones equipped with cameras can observe animal behaviour without disturbing them, providing insights into social interactions and feeding patterns. Drones can create detailed maps of vegetation types and distributions, helping to assess habitat quality and biodiversity. UAVs can monitor changes in habitats over time due to factors like climate change, human encroachment, or natural disasters (Ancin et al. 2020). UAVs can help track migratory patterns of birds and other animals by providing aerial perspectives over large areas. Drones can monitor how animals interact with their environment during migration, such as feeding and nesting behaviour. Drones can be equipped with sensors to assess water quality in lakes, rivers, and wetlands, identifying pollution sources or changes in aquatic habitats. UAVs can facilitate soil health assessments by providing data on vegetation cover and land use patterns. Drones can patrol protected areas to deter poaching activities and monitor illegal logging, providing real-time data to conservationists. UAVs can gather data on poaching hotspots, helping authorities to allocate resources effectively. Drones can evaluate the success of habitat restoration projects by assessing plant growth and wildlife return (Robinson, et al. 2022). Some innovative projects use drones for aerial seed dispersal in reforestation efforts, ensuring efficient coverage of hard-to-reach areas. UAVs can monitor the effects of climate change on ecosystems, such as shifts in species distributions and habitat alterations. Drones can assist researchers in collecting data necessary for modelling climate impacts on various species and habitats. Aerial footage can be used in campaigns to raise awareness about conservation efforts and the importance of protecting habitats. Drones can facilitate community-based monitoring programs, allowing local populations to participate in conservation efforts (Paneque-Galvez, et al. 2014).

24.3.4 Infrastructure Mapping and Monitoring

UAVs have revolutionized infrastructure inspection across various sectors. Drones can safely inspect hard-to-reach areas, such as under bridges, providing high-resolution imagery and data for structural assessments without the need for scaffolding or lane closures (Mandirola et al. 2022). UAVs equipped with thermal imaging cameras can detect hot spots and other anomalies in power lines, allowing for proactive maintenance and reducing the risk of outages. Drones can survey road conditions, assess traffic patterns, and identify maintenance needs, helping agencies prioritize repairs and improvements. For large commercial and residential structures, drones can inspect roofs, facades, and other hard-to-access areas, quickly identifying issues such as leaks or structural damage. Drones can perform visual and thermal inspections of wind turbines, identifying wear and tear on blades and other components, which helps in maintenance planning. UAVs can monitor railway tracks and surrounding areas for debris, vegetation overgrowth, and structural integrity, enhancing safety and operational efficiency. Drones can inspect oil and gas pipelines for leaks or damage, using both visual and thermal imaging, enabling faster responses to potential issues. UAVs can survey the structural integrity of dams and monitor water levels and sedimentation, ensuring safety and compliance with regulations.

24.3.5 Payload Delivery Services

Unmanned Aerial Vehicles (UAVs) play a crucial role in delivery services, especially in the realms of e-commerce and logistics. Drones can deliver packages quickly, reducing delivery times for last-mile logistics. They can bypass traffic and access hard-to-reach areas. Lower operational costs compared to traditional delivery methods, especially in remote or rural locations makes them a suitable futurist candidate for delivery system. Drones enable e-commerce companies to offer same-day or even instant delivery services, enhancing customer satisfaction and competitive advantage. Drones can assist in warehouse management by conducting inventory checks, helping to streamline logistics and reduce human error. Some restaurants and food delivery services are experimenting with drones to provide fast food delivery, improving service speed and efficiency. Drones can be equipped with tracking technology, allowing customers to receive real-time updates on their delivery status.

24.3.6 Emergency Response

UAVs have a wide range of applications in emergency response. UAVs can quickly survey disaster-affected areas, providing real-time imagery and data that help emergency responders assess damage, locate survivors, and plan rescue operations (Khan et al. 2022). Drones equipped with thermal imaging and high-resolution cameras can locate missing persons in hard-to-reach areas, especially in forests, mountains, or during urban disasters. Drones can transport medical supplies, food, and other essential items to remote or inaccessible locations, especially in situations where traditional transport methods are hindered. UAVs can inspect critical infrastructure such as bridges, roads, and power lines for damage after disasters, enabling faster repairs and maintenance. In the aftermath of disasters like floods or wildfires, UAVs can monitor environmental changes, assess the impact on ecosystems, and help in planning recovery efforts. Drones can monitor large gatherings or disaster sites to help manage crowds, ensuring safety and coordination among emergency services. It can gather data for predictive modelling and analysis, helping agencies prepare for future emergencies and improve response strategies. In situations where communication infrastructure is damaged, drones can act as temporary communication relays, facilitating coordination among emergency response teams. Drones can be used to monitor wildfires and provide real-time data on fire spread, helping teams strategize their response more effectively.

24.4 Basic Concepts of UAV Survey and Mission Planning

UAV mission planning is crucial for ensuring accurate, efficient, and reliable data acquisition across various applications such as mapping, surveillance, agriculture, and disaster response (Evers et al. 2011). It involves careful consideration of flight paths, altitude, sensor configurations, weather conditions, and regulatory constraints to optimize coverage and data quality. Pre-flight simulations, waypoint-based navigation and real time monitoring of the mission execution helps to enhance mission success while minimizing operational risks (Besada et al. 2018). A sample UAV mission plan for hilly area is shown in Fig. 24.6. With advancement in technology,

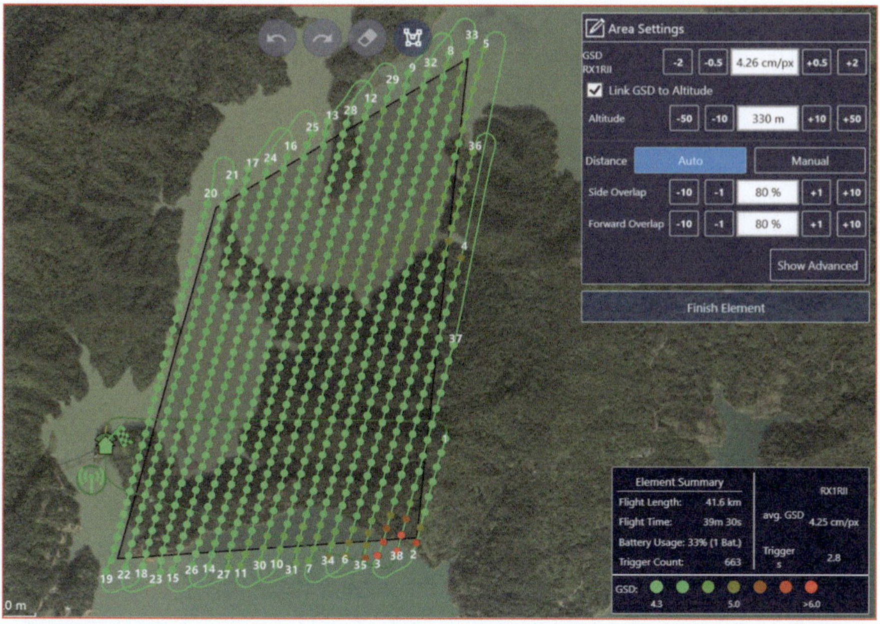

Fig. 24.6 UAV mission plan for mapping application

UAVs performs highly precise and autonomous data collection, enabling faster decision-making and improved analytical insights in diverse sectors. The different steps involved in mission planning and execution is given below.

24.4.1 Defining Area of Interest and Area to Survey

The area of interest is the area that the user provides in.kml/.kmz/.shp., etc. file format. The area of survey is created by generating an area with some buffer area. This is done in order to get proper overlap between images even at the edge of the area. These overlaps are very important to generate Orthomosic and DSM/DTM products even for the images at the corner.

24.4.2 Defining Overlap

Overlap is defined as the percentage of common area between two consecutive images. Based on the definition there are two types of overlaps, frontal overlap and side overlap. These two parameters are of paramount importance in aerial surveys as these dictates the quality of point cloud density, and detailing in DSM and orthomosic creation. As a general rule the side and frontal overlap in aerial UAV survey is kept > 60%. The flying area with dense vegetation or forest where very low number of automatic key points are generated, it is advisable to consider over > 75% overlap.

24.4.3 Defining Ground Sampling Distance (GSD)

GSD is defined as the least distance between two consecutive pixel centres measured on ground. So, in essence, lower the GSD better is the resolution of the Orthomosic and other data products. Also, low GSD means high data volume as the number of images increases. In general, the GSD from UAV data is less than 5 cm/pixel.

24.4.4 Considering the Local Conditions

In the area of interest, a number of challenges can emerge for example there can be a lot of variation in terrain elevation, there may be high-tension powerline, a designated no-fly zone, no suitable site for landing and take-off. All these challenges are considered during this planning step. A UAV survey mission is created taking care all these factors and the UAV is flown so that it remains at least 50-m AGL at any given point (safe flying height taking into account the SRTM and GPS errors in account). Also, it is ensured that the UAV mission is carried out in while being in Green Zone. Selection a suitable Take-off, landing and operations site is also considered at this stage. For considering a site suitable requires multiple factors such as availability of obstacle free area with 30 m radius, line of sight for control and command, and accessibility to the location by vehicle, etc.

24.4.5 Flight Plan Execution

Based on the weather conditions, lighting conditions, availability of take-off landing location, and line of sight requirement the flight plan is modified a little in the field. A total of nine flights of approximately 17 min each were executed. Considering the visual line of sight requirement, two take-off and landing sites were selected. The flight lines and location of images can be seen in the plan as shown in Fig. 24.7.

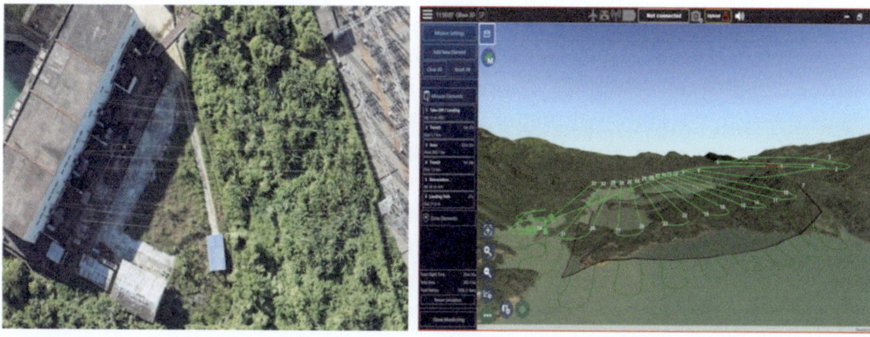

Fig. 24.7 Mission planning includes consideration such as high-tension power lines

24.5 Technological Advancements in UAVs

Technological advancements in UAVs have significantly transformed their capabilities, expanding their use across various industries. These advancements have not only improved their performance but also opened up new opportunities for automation, efficiency, and innovation (Mohsan et al. 2023). Here's an overview of some of the most important technological developments in UAVs:

24.5.1 Propulsion and Power Systems

UAVs are increasingly powered by electric motors that provide clean and quiet operation. The development of high-energy–density **lithium-ion** and **lithium-polymer batteries** has extended flight times and improved efficiency. Ongoing advancements in**solid-state batteries**and**fuel cell technologies**promise to offer higher energy density, faster charging times, and longer endurance (Janek et al. 2023).

Hybrid UAVs combine both electric and combustion engines, allowing for vertical takeoff and landing (VTOL) capabilities while providing extended flight range and endurance during horizontal flight. The integration of solar panels on UAVs, such as the **Solar Impulse** or **Aerospace Synergy**, enables perpetual flight without the need for recharging. This technology is particularly useful for long-duration flights, such as environmental monitoring or surveillance.

24.5.2 Autonomy and Artificial Intelligence (AI)

UAVs are increasingly equipped with autonomous flight systems that allow them to navigate and perform tasks without human intervention. These systems include **path planning algorithms**, **object recognition**, and **obstacle avoidance** systems.

AI-based systems enable UAVs to autonomously perform complex operations, such as identifying objects, following specific paths, or conducting inspections, using machine learning and computer vision algorithms (Rezwan et al. 2022).Swarm intelligence refers to the ability of multiple UAVs to work together, communicating and coordinating with each other. This is used in applications like surveillance, search-and-rescue, and military operations, where UAVs collaborate to cover larger areas or complete tasks more efficiently. AI-driven **computer vision** technologies enable UAVs to understand and interpret their environment, allowing for real-time decision-making in complex environments like crowded urban areas or disaster zones. The deep learning-based communication models are also being explored for future UAVs (Gupta et al. 2024 and Gupta et al. 2025).

24.5.3 Navigation and Communication Systems

Global Navigation Satellite Systems (GNSS) like GPS have been enhanced with technologies like **differential GPS (DGPS)** and **RTK (Real-Time Kinematic)** for centimeter-level accuracy, allowing UAVs to operate with higher precision in mission-critical tasks such as surveying, mapping, and agriculture. Visual Inertial odometry (VIO) is a navigation technology that uses cameras and inertial sensors (gyroscopes and accelerometers, Aqel et al. 2016) to allow UAVs to navigate in GPS-denied environments, such as indoor spaces or areas with weak satellite signals.

The advent of **5G networks** is expected to enhance UAV communication (Li et al. 2019) capabilities by providing ultra-low latency, high-speed data transfer, and improved connectivity. This can enable more real-time data sharing, particularly for applications in urban environments or large-scale operations. With satellite communication systems, UAVs can operate beyond the range of traditional ground-based communication networks, enabling global operations, such as long-range surveying, environmental monitoring, or military reconnaissance.

24.5.4 Sensors and Payloads

The integration of high-resolution **RGB cameras**, **thermal cameras**, and **infrared sensors** allows UAVs to capture detailed images for use in surveillance, search and rescue, mapping, and environmental monitoring. **LiDAR (Light Detection and Ranging)** technology enables UAVs to capture 3D topographical data, which is crucial for applications like forestry management, infrastructure inspection, and land surveying. **Hyperspectral sensors** can capture information across a wide range of wavelengths, enabling precise analysis of environmental conditions, vegetation health, and other factors. **Multispectral imaging** is used primarily in agriculture for crop health analysis, helping farmers monitor plant stress, irrigation needs, and pest infestation. UAVs equipped with gas sensors or chemical detection systems are increasingly used for environmental monitoring, industrial inspections, and emergency response, such as detecting hazardous gases or pollutants (Asadzadeh et al. 2022).

24.5.5 Stability and Control Systems

Modern flight control systems are more precise and capable of handling complex maneuvers, such as hovering, taking off vertically, and flying in turbulent conditions. **Gyroscopic stabilization** and **electronic stability control** have become standard features in UAVs, helping to maintain smooth flight even in challenging conditions. Redundant systems are becoming more common in high-end UAVs, particularly in commercial and industrial applications. These systems ensure that critical components like motors, flight controllers, and batteries have backups in place, enhancing safety during flight (Wanner et al. 2024).

24.5.6 Miniaturization and Durability

Advances in materials and engineering have led to the development of more compact and lightweight UAVs without sacrificing performance. This makes UAVs suitable for a wider range of applications, including indoor inspections, research in confined spaces, and low-cost consumer drones.UAVs are being designed to withstand harsher environments. **Weatherproofing** and **shock resistance** technologies allow UAVs (Anand et al. 2022) to operate in rain, snow, high winds, or extreme temperatures, expanding their use in industries like agriculture, disaster response, and military operations.

24.5.7 Regulatory and Safety Enhancements

UAVs are increasingly equipped with **geofencing capabilities** that automatically restrict them from entering restricted airspace or areas deemed unsafe. This is important for compliance with aviation regulations and for preventing accidents or interference with manned aircraft. To enhance safety, UAVs now come with **sense-and-avoid technology** that detects and avoids obstacles or other aircraft. This is critical for operating in shared airspace, especially for commercial and civilian UAVs (Grote et al. 2022).

24.6 Essentials for UAV Development in Northeast India

The development of UAV capabilities in the NER of India is critical for enhancing connectivity, security, disaster response, and socio-economic growth in this geographically diverse and often inaccessible area. To achieve this, a multi-pronged approach is required- investments in infrastructure such as drone corridors and testing zones, region-specific R&D for high-altitude and all-weather drones, and the establishment of training centres to build a skilled local workforce. Policy support tailored to the unique topography and strategic needs of the NER, along with collaboration between government, academia, and industry, will be essential in creating a sustainable UAV ecosystem. By focusing on localized applications like agricultural monitoring, border surveillance, and emergency logistics, UAV technology can become a transformative tool for the region's development.

NESAC has expanded the activities in the area of UAV technology and its applications for real time mapping and monitoring in the area of agriculture, infrastructure development, hydrology, forestry, disaster management support, security & surveillance etc. NESAC has been actively involved in region-specific R&D activities, collaborative works with academia and capacity building. It has developed UAV-based solutions to enhance data acquisition, analysis and decision-making for a range of applications. Some of the core contributions of NESAC for UAV development in NER are as following:

24.6.1 In-House Assembly and Tuning of Copter Type UAVs

NESAC has developed the capability to assemble and fine-tune copter-type UAVs using open source-platforms such as Ardupilot and PX4. The integration and tuning of UAVs have been carried out with a strong emphasis on autonomy, payload adaptability and, operations efficiency even under the difficult terrain and weather conditions of North East India. The centre also has expertise in integrating various sensors such as thermal, multispectral and hyperspectral sensors. The centre has also deployed inhouse assembled UAVs in other states of India for mapping surveys. This in-house development reduces dependency on commercial UAVs, offering cost-effective and customizable solutions tailored to specific project needs. NESAC has successfully developed and tuned quad-copters in small category, a heavy-duty hybrid hexacopter powered by a gasoline with 2 h flight time, a tethered UAV and a payload delivery mechanism.

24.6.2 Advanced Mission Planning Techniques

NESAC has integrated advanced mission planning techniques to ensure that UAV operations are conducted with precision, efficiency, and minimal risk, making the UAVs systems suitable for diverse applications in Remote Sensing and other missions such as disaster response. NESAC designs flight paths to maximize data collections while minimizing flight time and energy consumption, and conducts thorough risk evaluations, including terrain analysis and weather conditions assessments to enhance operations safety. NESAC has successfully carried out numerous missions in the challenging terrain of the unstable valley regions, including operations at altitudes as low as 500 m below take-off height.

24.6.3 State of the Art Lab Equipment

To support UAV development and validation, NESAC has established an advanced laboratory equipped with testing tools and fabrication facilities. These resources enable precise performance analysis, component fabrications and system optimization. Some of the key facilities include:

- **Propeller Thrust Stand**–Measures the efficiency of motors and propellers to optimize UAV propulsion. It records parameters such as current, voltage, RPM, motor and ESC temperatures, thrust, and torque. These measurements help define safe operating limits, ensuring the propulsion system operates efficiently and reliably. NESAC has setup a propeller thrust stand for drones up to 25 kgs all uplift weight Fig. 24.8.
- **Indoor UAV testing Machine**–Given Meghalaya's frequent rainfall, which hampers outdoor UAV testing and tuning, and to overcome the challenges associated with first take-off after assembling, NESAC has procured a state-of-the-art 3-axis (pitch, roll, yaw) UAV testing machine Fig. 24.9. This test bed allows controlled assessment and calibration of UAVs with a wheelbase of up to 1000 mm, ensuring precise performance evaluation in a controlled indoor environment.

- **CNC Machine and Spot-Welding machine**: NESAC is equipped with a 5-axis CNC machine capable of machining aluminium and alloy components with high precision Fig. 24.10. This enables the fabrication of complex UAV structures that require strength and durability. Additionally, NESAC has a spot welding (Fig. 24.11) used for in-house battery welding for assembling high-quality Li-ion battery packs, reducing dependency on external vendors while maintaining quality control for UAV power systems.

- **3D Laser Scanner and 3D printing machine**: To support rapid prototyping, structural repairs, and design modifications of UAV components, NESAC utilizes

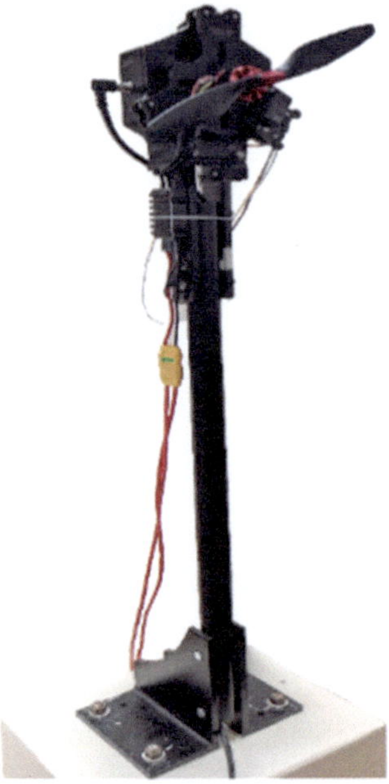

Fig. 24.8 Thrust testing stand

Fig. 24.9 3-axis UAV testing machine

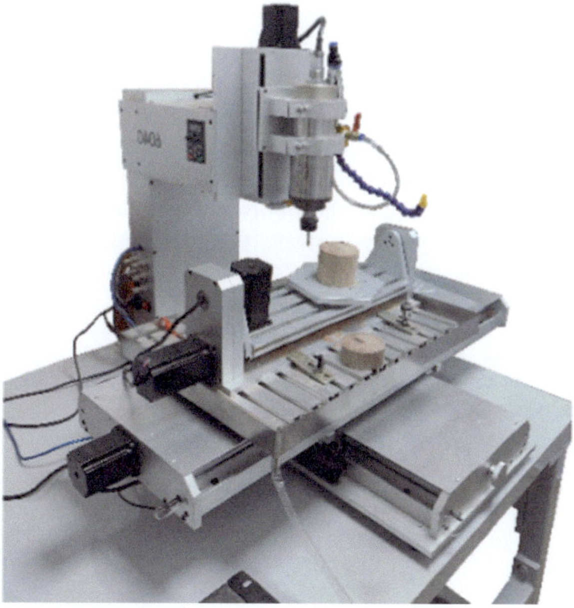

Fig. 24.10 5-axis CNC machine

Fig. 24.11 Spot welding machine

3D printer (Fig. 24.12) and 3D laser scanner (Fig. 24.13). The laser scanner facilitates high-precision measurements of existing UAV parts, while the 3D printer enables the fabrication of lightweight and optimized components, allowing for quick iterations and enhanced structural designs (Van et al. 2023).

- **Desing and Analysis Software**: NESAC is equipped with enterprise level design and analysis software to enhance UAV development. ANSYS Discovery and Fluent provide high-fidelity computational fluid dynamics (CFD) simulations for aerodynamic analysis, helping optimize UAV performance and efficiency. Siemens NX supports CAD based design, structural integrity assessment, and

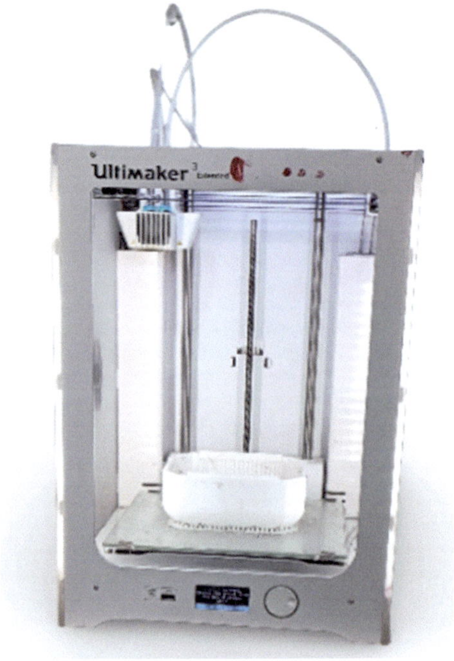

Fig. 24.12 FDM based 3D-printer

Fig. 24.13 3D laser scanner

components manufacturing processes. The integration of these tools enables precise design validation, minimizing prototyping iterations and ensuring UAVs meet operational and safety requirements.

24.6.4 High-End UAV Data Processing Facility

NESAC established a dedicated UAV data processing lab to transform raw UAV imagery into high-precision geospatial products. The lab is equipped with high-end computing systems and specialized software, including AgisoftMetashape, Pix4D, TerraScan, and Agisoft UAV Module, enabling efficient processing and analysis. Also, NESAC has developed field-ready workflows for UAV data analysis, enabling rapid assessment of survey results and gap-filling operations. Key data products generated include:

- **Ortho-mosaic**: A geometrically corrected, high-resolution aerial image stitched from multiple UAV-captured images. Unlike regular aerial photographs, an ortho-mosaic eliminates distortions caused by terrain variation, ensuring accurate spatial representation for land-use mapping, infrastructure monitoring, and disaster assessment. An illustration of an Orthomosic with GSD < 4 cm/pixel captured for Sohra region is given in (Fig. 24.14)

- **Digital Surface Model (DSM)**: Represents the Earth's surface, including natural and man-made features such as buildings, trees, and terrain variations. DSMs are useful for refining UAV mission planning, urban planning, 3D modelling, and line-of-sight analysis.
- **Digital Elevation Model (DEM)**: A bare-earth representation of terrain with vegetation and man-made structures removed. DEMs are crucial for hydrological modelling, flood risk assessment, and topographic analysis. An illustration of a DEM generated for a survey area in Sohra region is given in (Fig. 24.15)

- **Point Cloud**: A dense collection of georeferenced 3D points derived from UAV LiDAR or photogrammetry techniques. Point clouds provide detailed structural information and are used for volumetric analysis, contour generation, and precision mapping.
- **Vegetation Index Maps**: NESAC utilizes UAV-based multispectral imaging to capture reflectance data across different wavelengths, enabling detailed analysis of vegetation health, soil conditions, and water stress. Using specialized sensors, multispectral data is processed to generate Vegetation Index Maps, such as NDVI (Normalized Difference Vegetation Index) and EVI (Enhanced Vegetation Index). These indices help assess plant vigor, and detect disease outbreaks. The high spatial resolution of UAV multispectral imaging allows for monitoring of crop health, forestry studies, and environmental impact assessments, providing valuable insights for precision agriculture and sustainable land-use planning.

Fig. 24.14 Orthomosaic generated with UAV RGB images (GSD < 4 cm/pixel)

24.6.5 Collaboration with Other Users

NESAC has supported in setting up of UAV facility in each state remote sensing centre of NER. The facility includes a UAV, data processing software and training. NESAC is also providing continuous technical support to all the states of NER in maintaining their UAVs and assisting in data processing and field activities. NESAC has strong collaborations with premier academic institutes such as Anna University, Indian Institute of Science (IISC), IITs, NITs and other engineering collages and working on many joint projects related to UAV technology and its applications.

24.6.6 Capacity Building

NESAC has been conducting regular capacity building programs related to UAV technology and applications. These training courses ranges from 1–2 weeks and aimed at user specific needs. NESAC has also supported many B.Tech./M.Tech.

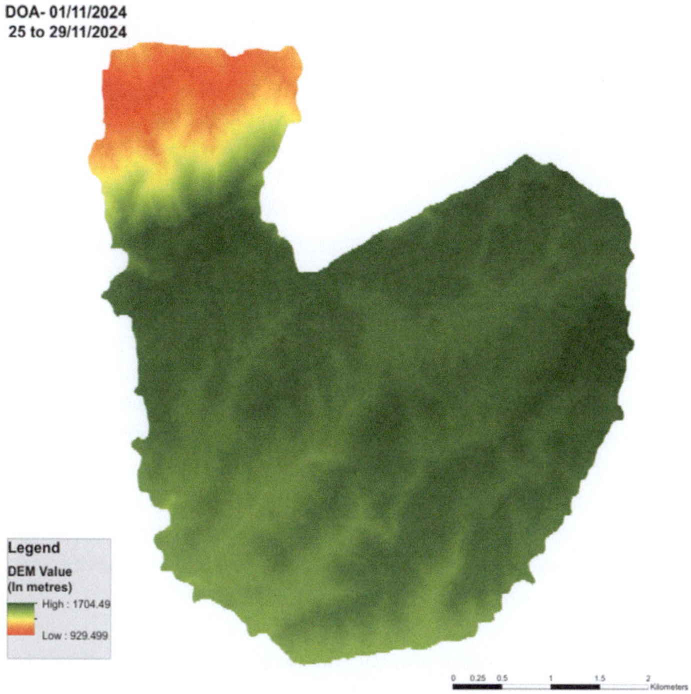

Fig. 24.15 DEM generated with UAV RGB images

students to work at UAV lab as interns and given them exposure of the cutting-edge technology.

24.7 Future Trends and Innovation in UAV Technology

As UAV technology and its applications are evolving day by day. There is lot of scope for future innovations related to software, communication system and applications. Some of the key areas are as given below.

24.7.1 Innovation in Software and Control System

Drone technology has evolved rapidly over the past decade, transforming from a niche gadget into an essential tool across various industries. As we look to the future, it's clear that drones will continue to innovate and expand their capabilities, influencing sectors from agriculture to entertainment.

Improved AI and machine learning will enable UAVs to operate autonomously with minimal human intervention (Challita et al. 2019). This includes advanced obstacle detection, navigation, and decision-making capabilities. Utilizing multiple UAVs that can communicate and coordinate with each other for tasks like surveillance, search and rescue, and agricultural monitoring. Swarm technology can enhance efficiency and coverage. Improved data analytics capabilities will allow UAVs to process and analyse data in real time, making them invaluable for applications in agriculture, infrastructure inspection, and disaster response. Innovations in vertical take-off and landing (VTOL) capabilities will enhance the versatility of UAVs, making them suitable for diverse environments and applications. Development of advanced sensors for environmental monitoring, surveillance, and other applications, enabling UAVs to gather more detailed data. The capabilities of drones have grown exponentially, driven by advancements in technology. Modern drones are equipped with high-definition cameras, advanced sensors, and powerful processors, enabling them to perform complex tasks with precision. One of the significant trends is the integration of artificial intelligence (AI) into drone systems. AI enhances autonomous flying, obstacle detection, and data analysis, making drones more efficient and versatile. Commercial applications of drones are on the rise, with businesses across various industries recognizing their value.

In agriculture, drones are used for precision farming, including crop monitoring and soil analysis. These drones provide farmers with detailed insights, enabling them to make informed decisions and improve crop yields. In-logistics, drones are revolutionizing delivery services by enabling faster and more efficient parcel delivery. Companies are experimenting with drone delivery systems to reduce delivery times and costs. As regulations evolve and technology improves, we can expect to see more widespread adoption of drones in commercial sectors.

Drone design is also undergoing significant changes, focusing on improving performance, durability, and user experience. Innovations such as foldable drones, which are compact and portable, are making it easier for enthusiasts and professionals to transport and deploy drones. Enhanced battery life and energy-efficient propulsion systems are extending flight times and enabling drones to cover larger areas.As drone technology advances, so does the need for comprehensive regulations to ensure safety and privacy. Governments around the world are working on developing and implementing regulations to address issues such as airspace management, privacy concerns, and the integration of drones into national airspace systems.Future regulations will likely focus on balancing the benefits of drone technology with safety and privacy considerations. This includes establishing clear guidelines for drone operations, integrating drones into existing air traffic control systems, and addressing concerns related to data security and privacy.

24.7.2 Potential Impacts of 5G on UAV Technology

5G technology is poised to significantly enhance UAV (drone) capabilities, offering faster data transmission, lower latency, and improved connectivity. Here are the key impacts:

Low latency and high-speed connectivity allow UAVs to process data in real-time, enabling advanced autonomous functions like obstacle avoidance and collaborative operations (e.g., drone swarms). 5G ensures reliable remote control for UAVs, even in remote areas, with minimal delay, making it ideal for critical applications like search and rescuewith 5G, UAVs can be networked for real-time communication and tracking, improving safety and enabling better management of crowded airspace.5G supports UAVs in sectors like delivery services, telemedicine, smart cities, and industrial inspections, enabling real-time data sharing and advanced analytics.

24.7.3 Role of UAVs in Smart Cities

Unmanned Aerial Vehicles (UAVs) play a crucial role in the development and operation of smart cities, enhancing efficiency, safety, and sustainability. Key applications include: monitoring traffic, providing aerial surveillance for public safety, and assisting in disaster response with real-time data. It can inspect buildings, bridges, power lines, and roads, enabling proactive maintenance and reducing downtime. UAVs measure air quality, detect pollution, and map urban heat islands, aiding in environmental protection and sustainability efforts. It can support last-mile deliveries, including medical supplies, reducing traffic and speeding up service. It monitors urban agriculture and public parks, optimizing resource use and maintaining green spaces. Drones assist in search-and-rescue operations, assess environmental risks, and provide data for disaster preparedness. It collects real-time traffic data, optimizing traffic flow and supporting autonomous vehicle systems. UAVs gather data for 3D mapping, urban modeling, and smart city analytics, supporting informed decision-making. Drones can inspect solar panels and energy grids, ensuring efficient energy management in urban environments.

24.8 Challenges and Limitations

Unmanned Aerial Vehicles (UAVs) have become increasingly popular across a variety of industries, including agriculture, military, logistics, and environmental monitoring. However, despite their widespread adoption, UAVs face several challenges and limitations that can affect their performance, safety, and utility.

24.8.1 Technical Challenges

UAVs, especially those used for commercial or military purposes, often have limited flight times (from 20 min to a few hours). The battery life can be a significant limitation, particularly for tasks that require long-duration flights, surveillance, or transportation.

The range of a UAV is constrained by battery life, environmental factors (e.g., wind, weather), and communication signal strength. Beyond a certain distance, UAVs may lose connection with their operators, which can result in crashes or mission failure. UAVs can only carry a limited payload, which restricts their ability to carry heavier equipment, multiple sensors, or large cargo. The capacity often limits their use in tasks like cargo delivery or industrial inspections. UAVs are susceptible to environmental conditions such as strong winds, rain, and extreme temperatures. Inclement weather can cause malfunctions, crashes, or the need for mission cancellations.

24.8.2 Ethical and Privacy Concerns

UAVs equipped with cameras and sensors raise concerns about surveillance and the unauthorized collection of personal data. In both commercial and military use, UAVs can potentially track individuals, gather sensitive data, or conduct surveillance without consent.

UAVs collect vast amounts of data during operations—whether it's video footage, sensor data, or real-time location information. If this data is not securely stored or transmitted, it could be intercepted, hacked, or misused by malicious actors. UAVs have been widely used in military operations, raising ethical concerns regarding their role in warfare, particularly around drone strikes and the targeting of individuals. The use of UAVs for targeted killings (such as in counterterrorism efforts) often faces criticism due to the potential for civilian casualties and lack of accountability. Autonomous UAVs used in both military and civilian settings raise concerns about moral and ethical decision-making, particularly in scenarios where the UAV must make life-or-death decisions without human intervention (e.g., autonomous combat drones or self-piloting delivery drones).The increasing number of UAVs in the skies presents ethical challenges related to airspace management, congestion, and coordination with manned aircraft. Improperly managed airspace could lead to dangerous collisions or unintended disruptions.

24.8.3 Public Perception and Acceptance

UAVs are seen as potentially unsafe due to the risks of technical failure, signal loss, or crashes. High-profile incidents, such as UAVs crashing in public spaces, airports, or even into manned aircraft, can create public fear and mistrust. People often feel uncomfortable with UAVs flying overhead, particularly when they are used for surveillance. Concerns over privacy violations (e.g., UAVs capturing images or videos without consent) contribute to public resistance. The legal framework for UAV operations is still evolving. Inconsistent rules and regulations across regions or countries can confuse operators and the general public. The lack of clear and unified guidelines for UAV use—especially in urban environments—can hinder public acceptance. UAVs, particularly larger ones, can generate noise that disrupts local communities, especially when flying at low altitudes for extended periods. Noise pollution from UAVs is a significant concern in residential areas, particularly in the case of commercial delivery drones or surveillance drones operating at all hours. As UAVs become more capable in performing tasks traditionally handled by humans (e.g., surveillance, inspections, deliveries), there are concerns about job losses in industries like transportation, logistics, and agriculture. UAVs can be weaponized or used for malicious purposes (e.g., smuggling, terrorism, or industrial espionage). This potential misuse fosters public anxiety and a perception that UAVs could pose a threat to security and safety.

24.8.4 Regulatory Framework

The regulatory framework for Unmanned Aerial Vehicles (UAVs), or drones, varies significantly across the globe, reflecting differing priorities related to safety, privacy, and innovation. Also, these frameworks are dynamic, changing per demand and technology. Drone technology has rapidly evolved, bringing significant opportunities across various sectors such as agriculture, logistics, surveillance, and public services. To ensure the safe and responsible use of drones, the Government of India, through the Directorate General of Civil Aviation (DGCA), has established comprehensive regulations. (Srivastava et al. 2020).

24.9 Conclusion

This chapter provided a brief history of UAV technology development along with different types of UAVs and applications. Some important concepts of UAV survey and mission planning are also introduced along with present technological advancement. The chapter also highlights some important work done by NESAC related to UAV technology development. The chapter also gives some insights on future trends and innovations in UAV technology along with related challenges and limitations.

Acknowledgements The authors gratefully acknowledge the support provided by the North Eastern Space Applications Centre (NESAC), Department of Space, Government of India, for facilitating the research, development, and deployment of UAV technologies in the North Eastern

Region. We sincerely thank Dr. S.P. Aggarwal, Director, NESAC, for his visionary leadership. The authors also acknowledge the valuable contributions of dedicated staff of the UAV Lab at NESAC, whose consistent efforts made this work possible.

References

L.C.C. Agisoft AgisoftPhotoScan. http://www.agisoft.com

S. Anand, A.K. Mishra, High-performance materials used for UAV manufacturing: Classified review. Int. J. All Res. Educ. Sci. Methods **10**(7), 2811–2819 (2022)

F.J. Ancin-Murguzur, L. Munoz, C. Monz, C.V.H. Hausner, Drones as a tool to monitor human impacts and vegetation changes in parks and protected areas. Remote Sens. Ecol. Conserv. **6**, 105–113 (2020). https://doi.org/10.1002/rse2.127

M. Aqel, M.H. Marhaban, M.I. Saripan, N. Ismail, Review of visual odometry: Types, approaches, challenges, and applications. SpringerPlus. **5**, (2016). https://doi.org/10.1186/s40064-016-3573-7

M. Arjomandi, S. Agostino, M. Mammone, M. Nelson, T. Zhou, Classification of unmanned aerial vehicles. *Report for Mechanical Engineering class, University of Adelaide, Adelaide, Australia*, pp. 1–48

S. Asadzadeh, W.J. de Oliveira, C.R. de Souza Filho, UAV-based remote sensing for the petroleum industry and environmental monitoring: State-of-the-art and perspectives. J. Pet. Sci. Eng. **208**, 109633, (2022). ISSN 0920-4105. https://doi.org/10.1016/j.petrol.2021.109633

M.A. Boon, A.P. Drijfhout, S. Tesfamichael, Comparison of a fixed-wing and multi-rotor UAV for environmental mapping applications: a case study. Int. Arch. Photogramm. Remote. Sens. Spat. Inf. Sci. **42**, 47–54 (2017)

J.A. Besada, L. Bergesio, I. Campaña, D. Vaquero-Melchor, J. López-Araquistain, A.M. Bernardos, J.R. Casar, Drone mission definition and implementation for automated infrastructure inspection using airborne sensors. Sensors **18**, 1170 (2018)

Bryan Phillip Fitzpatrick, *Unmanned aerial system for surveying and mapping: cost comparison of UAS versus traditional methods of data acquisition* (Thesis, University of Southern California, MSc, 2016)

U. Challita, A. Ferdowsi, M. Chen, W. Saad, Machine learning for wireless connectivity and security of cellular-connected UAVs. IEEE Wirel. Commun. **26**(1), 28–35 (2019) https://doi.org/10.1109/MWC.2018.1800155

F. Chiabrando, E. Donadio, F. Rinaudo, SFM for orthophoto generation: a winning approach for cultural heritage knowledge, in *ISPRS–International Archives of the Photogrammetry, Remote Sensing and Spatial Information Sciences*, XL-5/W7(September), pp. 91–98 (2015)

I. Colomina, P. Molina, Unmaned aerial systems for photogrammetry and remote sensing: a review–ISPRS. J. Photogram. Remote Sens. **92**, 79–97 (2014)

H. Das, N.D. Choudhury, C. Gupta, An estimation of long endurance power supply system for a rotary wing unmanned aerial vehicle. AJET, ISSN: 2348-7305, vol. 9, no. 2, 009021153(8PP) (2020)

N. Delavarpour et al., A technical study on UAV characteristics for precision agriculture applications and associated practical challenges. Remote Sens. **13**(6), 1204 (2021). https://doi.org/10.3390/rs13061204

C. Depcik, T. Cassady, B. Collicott, S.P. Burugupally, X. Li, S.S. Alam, J. Hobeck, Comparison of lithium ion Batteries, hydrogen fueled combustion Engines, and a hydrogen fuel cell in powering a small unmanned aerial vehicle. Energy Convers. Manage. **207**, 112514 (2020)

A. Dixit, A. Awasthi, V. Akram, H. Sharma, D. Ranawat, M. Kumar, *Rajiv Gandhi Proudyogiki Vishwavidyalaya, Bhopal (M.P.), Fabrication of Drone, Mahakal Institute of Technology and Management*, Ujjain, Bhopal, pp. 10–13 (2016)

H. Eisenbeiß, UAV photogrammetry. University of Technology Dresden (2009)
G. Elert, Author, Illustrator, Webmaster Chemical Potential Energy-The Physics Hypertextbook. Physics. info.N.p (2016). Web. Accessed 13 Sep 2016
O. Elijah, T.A. Rahman, I. Orikumhi, C.Y. Leow, M.N. Hindia, A comprehensive survey of UAV communication channel modeling. IEEE Access. **9**, 32141–32173 (2021). https://doi.org/10.1109/ACCESS.2021.3060851
J. Everaerts, The use of unmanned aerial vehicles (UAVs) for remote sensing and mapping, in *The International Archives of the Photogrammetry, Remote Sensing and Spatial Information Sciences*, Beijing Vol. XXXVII. Part B1, pp. 1187–1191 (2008)
L. Evers, T. Dollevoet, A. Barros, H. Monsuur, Robust UAV mission planning. Ann. Oper. Res. **222** (2011). https://doi.org/10.1007/s10479-012-1261-8
Grc.nasa.gov. N.p, The Lift Equation, Web 3 June (2016)
M. Grote, A. Pilko, J. Scanlan, T. Cherrett, J. Dickinson, A. Smith, A. Oakey, G. Marsden, Sharing airspace with Uncrewed Aerial Vehicles (UAVs): Views of the general aviation (GA) community, J. Air Transp. Manage. **102**, 102218, ISSN 0969-6997 (2022) https://doi.org/10.1016/j.jairtraman.2022.102218.
G. Grenzdörffer, A. Engel, B. Teichert, The photogrammetric potential of low-cost UAVs in forestry and agriculture. Int. Arch. Photogram. Remote Sens. Spat. Inform. Sci. **31**, 1207–1214 (2008)
C. Gupta et al., GANCE: Generative adversarial network assisted channel estimation for unmanned aerial vehicles empowered 5g and beyond wireless networks. in *IEEE Access*, **13**, 198–213 (2024). https://doi.org/10.1109/ACCESS.2024.3522847
C. Gupta et al, Deep learning based channel estimation for UAVs: a modified U-Net approach. Adv. Electr. Comput. Eng. (AECE) **25**(1), 61–70 (2025). https://doi.org/10.4316/AECE.2025.01007
T. Henderson, inertia and mass. Physicsclassroom.com.N.p, (2016). Web Accessed 13 Sept 2016
R. Higgons, Qi3 insight: unmanned aerial vehicles-growing markets in a changing worlds (2014). http://www.qi3.co.uk/wp-content/uploads/2014/02/Qi3-Insights-White-Paper-UAVs-Growing-Markets-in-a-Changing-World-2014021903.pdf
H. Hildmann, E. Kovacs, Using unmanned aerial vehicles (UAVs) as mobile sensing platforms (MSPs) for disaster response, civil security and public safety. Drones **3**(3), 59 (2019)
Hung, C., Xu, Zhe., and Sukkarieh, S., 2014. Feature learning based approach for weed classification using high resolution aerial images from a digital camera mounted on a UAV. Remote Sensing, **6**, 12037–12054. https://doi.org/10.3390/rs61212037
L. Jin et al., E-agriculture in action-DRONES FOR AGRICULTURE, *2017. The Food and Agriculture Organization of the United Nations and International Telecommunication Union*, Bangkok (2017)
J. Janek, W. Zeier, Challenges in speeding up solid-state battery development. Nat. Energy **8**, 1-11 (2023) https://doi.org/10.1038/s41560-023-01208-9
J. He, J.Y. Li, K. Zhang, Research of UAV flight planning parameters. Positioning **3**, 41–43 (2012)
B. Li, H. Zhang, UAV communications for 5G and beyond. arXiv preprint arXiv:1901.06637 (2019) https://arxiv.org/pdf/1901.06637.pdf
X. Liu et al., Accuracy assessment of a UAV direct georeferencing method and impact of the configuration of ground control points. Drones **6**(2), 30 (2022). https://doi.org/10.3390/drones6020030
A. Khan, S. Gupta, S.K. Gupta, Emerging UAV technology for disaster detection, mitigation, response, and preparedness. J. Field Robot. **39**, 905–955 (2022). https://doi.org/10.1002/rob.22075
A. Matese et al, Intercomparison of UAV, aircraft and satellite remote sensing platforms for precision viticulture. Remote Sens, **7**(3), 2971–2990 (2015)
M. Mandirola, C. Casarotti, S. Peloso, I. Lanese, E. Brunesi, I. Senaldi, Use of UAS for damage inspection and assessment of bridge infrastructures. Int. J. Disaster Risk Reduction **72,** 102824 (2022) ISSN 2212-4209. https://doi.org/10.1016/j.ijdrr.2022.102824. https://www.sciencedirect.com/science/article/pii/S2212420922000437

S.A.H. Mohsan, M.A. Khan, Y.Y. Ghadi, Editorial on the advances, innovations and applications of UAV technology for remote sensing. Remote Sens. **15**(21), 5087 (2023). https://doi.org/10.3390/rs15

J. Paneque-Gálvez, M.K. McCall, B.M. Napoletano, S.A. Wich, L.P. Koh, Small drones for community-based forest monitoring: An assessment of their feasibility and potential in tropical areas. Forests **5**(6), 1481–1507 (2014). https://doi.org/10.3390/f5061481

Pix4D tutorial Accessed online at www. pix4d.com

Q. Quan, *Introduction to multicopter design and control*, Vol. 10, pp. 978–981. Singapore: Springer (2017)

S. Rezwan, W. Choi, Artificial intelligence approaches for UAV navigation: Recent advances and future challenges. IEEE Access, **10**, 26320–26339 (2022) https://doi.org/10.1109/ACCESS.2022.3157626

J.M. Robinson, P. Harrison, S. Mavoa, M. Breed, Existing and emerging uses of drones in restoration ecology. Methods Ecol. Evol. 1–13 (2022). https://doi.org/10.1111/2041-210X.13912

J. Rumerman, U.S. Army Aircraft 1947–1960, in *NASA SP-2003-4531: NASA Historical Data Book, Volume II — NASA Launch Systems, Space Transportation, Human Spaceflight, and Aeronautics, NASA History Office,* ed. by J. Wegg (2003)

Saadatseresht M., et al., Uav photogrammetry: a practical solution for challenginging mapping projects, in *International Conference on Sensors & Models in Remote Sensing & Photogrammetry*, 23–25 Nov, Kish Island, Iran (2015)

M.H. Sabour, P. Jafary, S. Nematiyan, Applications and classifications of unmanned aerial vehicles: a literature review with focus on multi-rotors. The Aeronautical J **127**(1309), 466–490 (2023)

P. Sivanandi, C. Gupta, H. Durai, CFD and XFLR5 analysis of an active aeroelastic wing, Springer Open Accessed 22 Nov 2022(2022). https://doi.org/10.21203/rs.3.rs-2263166/v1

P. Sivanandi, C. Gupta, H. Durai, A review on evolution of aeroelastic assisted wing. Int. J. Aeronaut. Space Sci. Published online on 14th April, 2023 (2023). https://doi.org/10.1007/s42405-023-00583-7

S. Srivastava, S. Gupta, O. Dikshit, S. Nair, *A Review of UAV Regulations and Policies in India.* (2020) https://doi.org/10.1007/978-3-030-37393-1_27

S. Suresh Kumar, T. Indiran, G. Vadakkekara Itty, C. Gupta, Simulation of nonlinear proportional integral derivative controller for a nonlinear model of a HexSoon EDU 450 quadrotor aerial robot, in *2022, 3rd International Conference on Intelligent Computing Instrumentation and Control Technologies (ICICICT)*, pp. 1724–1728 (2022). (https://doi.org/10.1109/ICICICT54557.2022.9917674)

TCPO, MoHUA, GOI, Formulation of GIS based master plans for small and medium towns: design & standards for application of Drone/UAV Technology (2020). http://tcpo.gov.in/sites/default/files/newfile/dron-min.pdf

D. Wanner, H.A. Hashim, S. Srivastava, A. Steinhauer, UAV Avionics safety, certification, accidents, redundancy, integrity and reliability: A comprehensive review and future trends. (2024). https://doi.org/10.36227/techrxiv.171328078.80258471/v1.

I. Wahab, O. Hall, M. Jirström, Remote sensing of yields: Application of UAV imagery-derived NDVI for estimating maize vigor and yields in complex farming systems in sub-saharan africa. Drones **2**(3), 28 (2018). https://doi.org/10.3390/drones2030028

J. Titus, Careful designers get the most from brushless DC Motors. Electronic Component News. N.p, Web Accessed 13 Sept 2016

M. Ustuner, F.B. Sanli, S. Abdikan, M.T. Esetlili, Y. Kurucu, Crop type classification using vegetation indices of Rapideye imagery. *The International Archives of the Photogrammetry, Remote Sensing and Spatial Information Sciences*, XL-7:194–198 (2014)

L. van Brügge, K.M Çetin, S.J. Koeberle, et al, Application of 3D-scanning for structural and geometric assessment of aerospace structures. CEAS Aeronaut. J. **14**, 455–467 (2023).https://doi.org/10.1007/s13272-023-00654-1

M. Zhou, Z. Zhou, L. Liu, J. Huang, Z. Lyu, Review of vertical take-off and landing fixed-wing UAV and its application prospect in precision agriculture. Int. J. Precis. Agric. Aviat. **3**(4), 8–17 (2020)

Chirag Gupta is working as Scientist/Engineer 'SE' at NESAC, Umiam, Meghalaya. He completed Bachelor of Engineering in Electronics and Communication in 2012 from SIT, Tumkur, Karnataka and pursuing PhD in Advance Communication Systems from National Institute of Technology (NIT), Meghalaya. He worked at Tata Consultancy Services, Mumbai for a year before joining Indian Space Research Organisation (ISRO) in 2014, since than he is posted at NESAC, Umiam. At NESAC he is leading the UAV Technology Development, Operations & Maintenance team and also working as In-charge UAV Lab/Facility. He is having an experience of more than 10 years in the field of UAV systems design, development, operations and its applications (e.g. Agriculture, Forestry, Disaster Management, and Town Planning etc.), 3D printing technology, capacity building and training. He is also a DGCA certified UAV pilot and Trainer under SMALL class, Rotorcraft category drones. He has contributed to more than 15 papers/research articles in the field of UAV systems and applications. He is also helping B.Tech/M.Tech students to carry out their project work at NESAC UAV Lab.

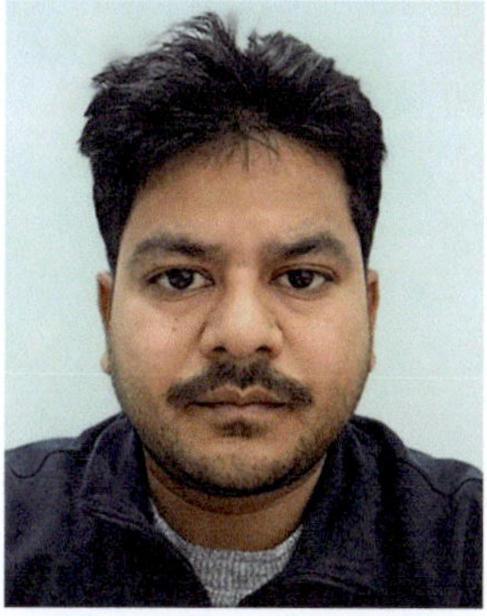

Sanjay Pandit is working as Scientist/Engineer 'SD' at North Eastern Space Applications Centre (NESAC), Umiam, Meghalaya. He has completed his Bachelor of Technology in Aerospace Engineering in 2017 from Indian Institute of Space Science and Technology (IIST), Trivendrum, Kerala. At NESAC, he is working as a team member in UAV Technology Development, Maintenance and deployment for various applications. He is having an experience of 8 years working in SATCOM and UAV division. He has contributed significantly in UAV system design, development, payload integrations, and UAV applications. He also has expertise in FDM based 3D printing technology, and has contributed in capacity building programs and trainings. He is a DGCA certified UAV pilot and Trainer under SMALL class, Rotorcraft category UAVs. He has also guided 12+ BTech student to carry out their internship project at NESAC UAV Lab.

Ramani Kumar Das has obtained M.Sc in Physics with specialization in Electronics from Gauhati University, Assam in the year of 2000 with First Class 1st Position (Gold Medalist). He is currently serving as Head, SATCOM & UAV Division (SUD) and Civil and Maintenance Division (CMD) at North Eastern Space Applications Centre (NESAC), Department of Space, Government of India, Shillong. His area of interest is Satellite Communication (SATCOM) and Satellite Navigation (SATNAV) Applications for Societal applications like Education, Health, communication support in disaster management, location-based services etc. and development of UAV technology and its applications. He is having more than 22 years of experience in the field. He is recipient ISRO's Team Excellence award in 2007 and ISRO's Young Scientist Award in 2009.

Shashi Bhusan Nayak is a Senior Technical Assistant 'A' at the North Eastern Space Applications Centre (NESAC), Umiam, Meghalaya. He holds a Diploma in Electronics and Telecommunication Engineering (2013) from IGIT, Sarang and a B.Tech in the same field (2017) from VSSUT, Burla. Prior to joining ISRO in 2020, he worked as a Junior Research Fellow at DRDO, Chandipur. At NESAC, he is involved in UAV operations, maintenance, and development. With over 4.5 years of experience, his expertise includes UAV systems design, sensor integration, testing, tuning of open-source UAVs, and remote sensing applications. He is also skilled in 3D printing, capacity building, and training. Shri Nayak is a DGCA-certified UAV pilot and trainer (Small class, Rotorcraft category) and actively supports B.Tech students with their projects at the NESAC UAV Lab.

Chapter 25
North Eastern Regional Node for Disaster Risk Reduction: A Single Window Platform

Rekha B. Gogoi, Diganta Barman, Kuntala Bhusan, Gopal Sharma, Chirag Gupta, Puyam S. Singh, Victor Saikhom, Ranjit Das, Jonali Goswami, Kasturi Chakraborty, Jenita M. Nongkynrih, M. Somorjit Singh, Dibyajyoti Chutia, Ramani Kumar Das, Shyam S. Kundu, Bijoy K. Handique, K. K. Sarma, and Shiv P. Aggarwal

Abstract The North Eastern Region of India is one of the most disaster-prone areas in the country, regularly affected by floods, landslides, thunderstorms, forest fires, and high seismic activity. The risk of a major earthquake comparable to those in 1897 and 1950 remains a pressing concern due to rising population and infrastructure vulnerability. Effective disaster preparedness and response in such a region demand

R. B. Gogoi (✉) · D. Barman · K. Bhusan · G. Sharma · C. Gupta · P. S. Singh · V. Saikhom · R. Das · J. Goswami · K. Chakraborty · J. M. Nongkynrih · M. S. Singh · D. Chutia · R. K. Das · S. S. Kundu · B. K. Handique · K. K. Sarma · S. P. Aggarwal
North Eastern Space Applications Centre (NESAC), Umiam, Shillong, Meghalaya, India
e-mail: rekha.bharali@nesac.gov.in

D. Barman
e-mail: diganta.barman@nesac.gov.in

K. Bhusan
e-mail: kuntala.bhusan@nesac.gov.in

G. Sharma
e-mail: gopal.sharma.nesac@nesac.gov.in

C. Gupta
e-mail: chirag.gupta@nesac.gov.in

P. S. Singh
e-mail: ss.puyam@nesac.gov.in

V. Saikhom
e-mail: victor.saikhom@nesac.gov.in

R. Das
e-mail: ranjit.das@nesac.gov.in

J. Goswami
e-mail: jonali.goswami@nesac.gov.in

K. Chakraborty
e-mail: kasturi.c@nesac.gov.in

S. P. Aggarwal et al. (eds.), *Remote Sensing of the North Eastern Himalayan Ecosystem*,
https://doi.org/10.1007/978-981-95-3221-6_25

timely access to accurate geospatial data on infrastructure and hazard-prone zones, which traditional methods cannot efficiently provide. To address this, the Department of Space (DoS) has established the North Eastern Regional Node for Disaster Risk Reduction (NER-DRR) at NESAC. The node hosts a dynamic GIS-based geoportal with interoperable OGC data services, offering multi-hazard monitoring, alert dissemination, and statistical analysis tools. Organized into four major services hydrological, meteorological, geological, and forest fire, the platform supports real-time and seasonal data updates. In 2024 alone, it recorded 223 landslides, 428 earthquakes, 24 floods, and over 37,000 fire incidents. NER-DRR also integrates UAV-based monitoring to complement satellite observations, delivering high-resolution data for response planning. By combining remote sensing, UAV data, and geospatial analysis, the portal enables informed decision-making and strengthens regional disaster resilience through timely alerts, vulnerability assessment, and emergency response planning.

Keywords Disaster risk reduction · Space technology · Geoportal · Hazard · Vulnerability · Early warning

25.1 Introduction

Spanning approximately 1200 km, the Eastern Himalaya forms a distinct section of the Himalayan orogenic belt. It stretches longitudinally from eastern Nepal to Arunachal Pradesh and latitudinally from the foredeep plains of Assam in the south to the elevated Tibetan plateau in the north (Bansal et al. 2022; Shrestha et al. 2025). Geologically, it is characterized by tectonic instability resulting from the ongoing

J. M. Nongkynrih
e-mail: jm.nongkynrih@nesac.gov.in

M. S. Singh
e-mail: m.somorjit@nesac.gov.in

D. Chutia
e-mail: d.chutia@nesac.gov.in

R. K. Das
e-mail: rkdas@nesac.gov.in

S. S. Kundu
e-mail: ss.kundu@nesac.gov.in

B. K. Handique
e-mail: bk.handique@nesac.gov.in

K. K. Sarma
e-mail: kk.sarma@nesac.gov.in

S. P. Aggarwal
e-mail: spaggarwal2010@gmail.com

collision between the Indian and Eurasian plates, while climatically, it receives some of the heaviest rainfall anywhere on Earth. The region's rugged topography, combined with its active geophysical setting, makes it exceptionally prone to a range of natural disasters, including earthquakes, landslides, floods, lightning, and glacial lake outburst floods (GLOFs).

Among the most defining natural hazards in the region are earthquakes. This region lies within the Himalayan Seismic Belt and are classified as a high-risk zone for seismic activity. The Eastern Himalayas are traversed by major tectonic structures such as the Main Frontal Thrust (MFT), Main Boundary Thrust (MBT), and Main Central Thrust (MCT), which are known to be sources of significant seismic events. Several significant earthquakes have marked the history of this region, including the 1897 Great Assam earthquake, the 1934 Bihar–Nepal earthquake, the 1950 Assam–Tibet earthquake, the strongest recorded on land in India and more recent events such as the 2011 Sikkim-Nepal border earthquake and 2015 Nepal earthquake. These seismic events not only cause direct damage but also often trigger secondary disasters like landslides and river damming, amplifying their impact across borders. Studies suggest the region is overdue for another major Himalayan event, with potentially catastrophic cascading effects (Bilham and Ambraseys 2005; Bilham 2019; Sharma et al. 2020).

Landslides are another persistent hazard in the Eastern Himalayas. Steep slopes, fragile geology, and intense monsoonal rains create ideal conditions for slope failures. These landslides can be sudden and devastating, as evidenced by the 1968 Sikkim–Darjeeling event that claimed over a thousand lives and caused extensive infrastructural damage. Apart from the catastrophic 1968 event, several other significant historical landslides have occurred in the region, notably in 1976, 1991, 1993, 2003, 2004, 2005, 2006, 2009, 2011, and 2015 (Pal et al. 2016; Bansal et al. 2022). Landslides frequently occur in Bhutan, the Darjeeling hills, the north eastern hilly states of India, with risks further compounded by infrastructure development, such as road construction and hydropower projects, on unstable slopes.

Flooding in the Eastern Himalayas includes both large-scale riverine floods and sudden flash floods. Major rivers like the Teesta, Yarlung Tsangpo/Brahmaputra, and their tributaries are prone to annual flooding during the monsoon. These floods are often intensified when upstream landslide-dammed lakes breach, as occurred during the 2000 Siang flood, which originated from a landslide in Tibet and caused widespread destruction in Arunachal Pradesh.

GLOFs are rapidly emerging as one of the most dangerous hazards in the Eastern Himalaya, driven by climate change and accelerated glacial retreat. Studies indicate that the Eastern Himalayas constitute a GLOF hotspot, with risk levels estimated to be nearly three times higher than in other parts of the Himalayan arc (Veh et al. 2020). Notable events such as the 1994 Lugge Tso outburst in Bhutan and the October 2023 South Lhonak Lake GLOF in Sikkim underscore the catastrophic potential of glacial lake outburst floods, as evidenced by the scale of damage they inflicted.

In parallel, lightning has also emerged as a recurrent and deadly threat, although it is often overshadowed by larger and more visibly destructive hazards. The Eastern Himalaya's rugged terrain and intense convective activity during the pre-monsoon and monsoon seasons make it highly susceptible to frequent lightning strikes. The NER, India alone accounts for nearly 19% of lightning-related deaths in India (Mishra et al. 2024), disproportionately affecting rural and highland populations who often lack access to early warning systems or lightning-safe infrastructure. The issue is not confined to India, as mentioned by NDRRMA officials, Nepal reported 237 deaths and over 900 injuries due to lightning strikes over a period ranging from July 2021 to May 2025 (The Himalayan 2025). Despite the rising toll, lightning remains underreported and insufficiently addressed within formal disaster risk management frameworks.

These hazards GLOFs, lightning, flash floods and others are symptomatic of a broader shift in the region's hazard profile, increasingly influenced by climate change. Warming across the Himalayan arc and the Tibetan Plateau has accelerated at approximately 0.2 °C per decade between 1951 and 2014 (Sabin et al. 2020). This rise in temperature is not only driving glacial melt but also permafrost thaw, which destabilizes mountain slopes and moraine dams, heightening the likelihood of GLOFs and landslides. Simultaneously, the monsoonal regime, once marked by steady rainfall, is now shifting toward short-duration, high-intensity precipitation events, which significantly increase the risk of flash floods and debris flows. These changes are contributing to a rise in compound disasters, where hazards such as landslides, lightning and storm-triggered floods occur in rapid succession or as cascading events, thereby amplifying their overall impacts and complicating response efforts.

Given the Eastern Himalaya's complex terrain and exposure to multiple, often interrelated hazards, effective disaster risk reduction increasingly depends on integrated, multi-hazard early warning systems, supported by real-time data and advanced modeling. In this context, space-based technologies have become indispensable. Earth observation satellites provide continuous monitoring of key environmental variables such as precipitation, snow cover, glacier dynamics, land surface deformation, and vegetation health, which are critical for anticipating floods, landslides, and GLOFs. When integrated with geographic information systems (GIS) and digital elevation models (DEMs), this data enables the creation of hazard zonation maps, vulnerability assessments, and predictive risk models. Moreover, communication satellites play a crucial role in ensuring connectivity during disasters, particularly in remote mountainous areas where terrestrial networks are frequently disrupted. They support the real-time dissemination of alerts, enable coordination among response agencies, and facilitate public communication when it is most needed. This chapter explores the application of geospatial technology in disaster risk reduction, with particular emphasis on its relevance and implementation in the Northeastern Region of India.

25.1.1 Overview of Disaster Risk Reduction (DRR)

Disaster, as defined by the UNISDR Terminology on Disaster Risk Reduction 2009, is "a *serious disruption of the functioning of a community or a society involving widespread human, material, economic or environmental losses and impacts, which exceeds the ability of the affected community or society to cope using its own resources.* These events threaten the stability of communities and can easily overwhelm local response capabilities. Disasters stem from a range of causes, including natural hazards such as earthquakes, landslides, floods, cyclones, and tsunamis; human-induced events like industrial accidents and conflicts; and public health crises, such as pandemics. However, the severity of their impact is not solely determined by the event itself but also by a community's exposure and vulnerability. Factors such as socio-economic disparities, fragile infrastructure, environmental degradation, governance quality, and access to early warning systems play a critical role in shaping disaster outcomes. Given the evolving global disaster landscape and the increasing frequency of hazards, building community resilience through proactive disaster risk reduction measures is more urgent than ever. Prioritizing preparedness, strengthening infrastructure, and addressing socio-economic vulnerabilities are essential to mitigating the devastating impacts of future crises.

The World Disasters Report 2022 highlights a notable rise in the frequency of natural hazard-triggered disasters since 1970, despite annual fluctuations. This pattern is effectively illustrated in Fig. 25.1, which is based on EM-DAT disaster occurrence data spanning the period 1970–2023. A significant trend observed is the dominance of climate-related hazards, which accounted for 90.3% of all recorded disasters in 2020 and 2021, compared to only 7.3% from geological hazards such as earthquakes and volcanic activity. This dominance is also evident in the number of people affected by disasters. Over the past 12 years, between 97 and 98% of all those affected by disasters experienced a climate- or weather-related event, with an annual average of about 100 million people impacted. Disasters linked to geological hazards constitute the next largest proportion of affected populations. These trends depict the intensifying effects of climate change and highlight the urgent need for strengthened disaster preparedness, response strategies, and risk reduction measures to protect vulnerable populations and reduce disaster-related risks. Traditional disaster response, reacting only after disasters strike, has proven to be insufficient in the face of increasingly frequent and severe hazards. To build resilience and reduce the long-term impacts of disasters, proactive and preventive approaches are essential.

DRR takes a proactive approach to identifying, assessing, and mitigating disaster risks to minimize their impact. Rather than focusing solely on disaster response, it addresses the root causes of risk, thereby enhancing community resilience to future hazards (NDMA Report, 2022–23). Reflecting these principles, India's Disaster Management Act, 2005 (DMA) was enacted during the era of the Hyogo Framework for Action 2005–2015, the first global framework to provide a structured approach to disaster risk reduction (UNISDR 2005).

The Hyogo Framework laid the groundwork for strengthening resilience through improved early warning systems, enhanced risk governance, and capacity building at all levels of society. While the Hyogo Framework drove significant progress, it also revealed critical gaps, particularly in addressing underlying risk factors, defining clear priorities, fostering resilience across all levels, and securing adequate resources

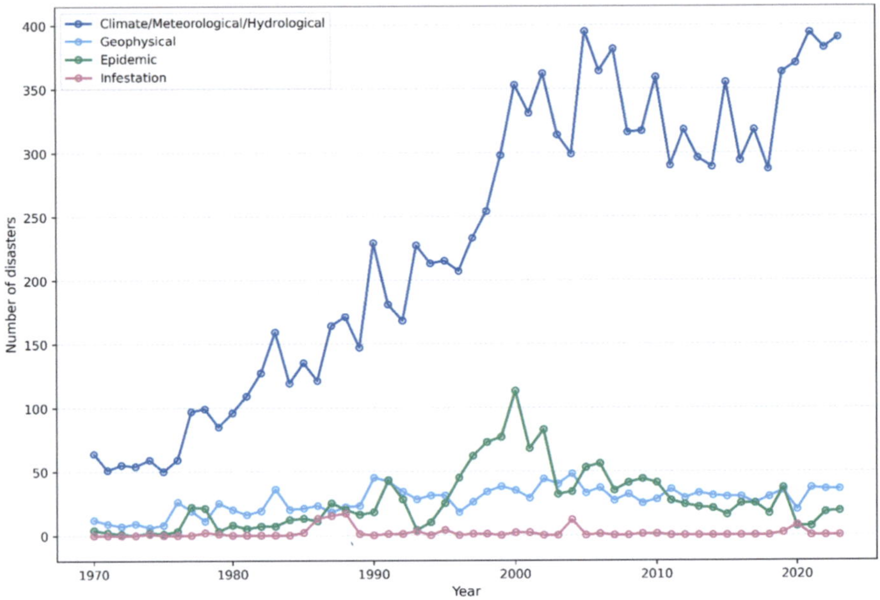

Fig. 25.1 Number of Disaster triggered by natural hazards globally per year, 1970–2023 based on EM-DAT data

for implementation. These challenges underscored the need for a more dynamic and action-oriented framework that would enable governments and stakeholders to identify key disaster risks, streamline investments, and enhance resilience-building efforts.

In response to these lessons, the Sendai Framework for Disaster Risk Reduction 2015–2030 (UNISDR 2015) introduced a more comprehensive and forward-thinking approach. It emphasizes four key priorities:

i. Understanding disaster risk through comprehensive risk assessments and data-driven insights.
ii. Strengthening disaster risk governance to enable coordinated, effective decision-making.
iii. Investing in disaster resilience to reduce vulnerabilities and safeguard development gains.
iv. Enhancing disaster preparedness and recovery efforts with a focus on "building back better.

India has aligned its evolving disaster management strategies with the principles of the Sendai Framework, enhancing institutional mechanisms to reduce risk and strengthen resilience. The national disaster management policy has laid the foundation for critical institutional structures responsible for formulating, implementing, and monitoring disaster management plans. It emphasizes proactive measures by government agencies to prevent and mitigate disasters, while ensuring a swift and coordinated response when crises occur (Biju et al. 2019). Over the years, these initiatives have contributed to reducing the human toll of disasters.

Historically, the absence of a coordinated disaster management system led to significant human losses. For instance, the 1993 flood affected eight northern states and claimed 530 lives, followed the 2001 Gujarat earthquake, which resulted in over 20,000 deaths. In 2004, severe flooding in Bihar caused 885 fatalities, while the 2013 Uttarakhand floods led to the loss of more than 6,000 lives. The 2014 Kashmir floods further added to this toll, with 277 casualties reported. These events underscore the need for robust preparedness, risk governance, and early warning systems.

In recent years, however, advancements in disaster risk reduction, particularly in cyclone forecasting, early warning dissemination, and community evacuation protocols, have significantly improved outcomes and demonstrated India's growing capacity to protect lives during extreme events. A notable illustration of this progress is the marked difference in outcomes between the 1999 Odisha Super Cyclone and Cyclone Phailin in 2013. The 1999 event, which occurred prior to the establishment of formal disaster management institutions, resulted in nearly 10,000 fatalities, largely due to inadequate early warnings and limited preparedness. By contrast, in 2013, the coordinated efforts of the India Meteorological Department (IMD), supported by real-time satellite data from Indian Space Research Organization (ISRO), enabled accurate cyclone forecasting and timely alerts. The National Disaster Management Authority (NDMA), in collaboration with the Odisha State Disaster Management Authority (OSDMA), executed the mass evacuation of over one million people, limiting casualties to fewer than 50 deaths (Singh et al. 2013).

This success highlights the vital role of accurate forecasting, geospatial technologies, and coordinated response efforts in disaster management. While challenges persist, especially with high-impact disasters like earthquakes, flash flood India's progress in DRR has significantly enhanced its resilience. By continuously refining its disaster management strategies, the nation is not only reducing the risk to human lives and infrastructure but also setting a benchmark for disaster preparedness and response on a global scale.

25.1.2 Significance of DRR in the Northeast Region (NER)

The NER of India is a unique area in terms of topography and culture that consists of eight states: Arunachal Pradesh, Assam, Manipur, Meghalaya, Mizoram, Nagaland, Sikkim, and Tripura. This region covers approximately 8% of India's total geographical area and is inhabited by around 4% of the nation's population. The NER can be broadly divided into three geographical zones: the Surma Valley, the North Eastern Hill Basin, and the Brahmaputra Valley. Each of these zones showcases unique characteristics, including variations in natural resource availability, population density, habitation patterns, and climatic conditions. The plains, especially in the Brahmaputra Valley, are known for their agricultural richness and higher population density, while the hilly regions are characterized by lower population density and are often more remote and less accessible.

Situated within the geologically dynamic Eastern Himalaya, the NER of India is highly prone to recurrent natural disasters that vary in intensity and frequency across states and seasons. The region is particularly vulnerable to a wide spectrum of hydro-meteorological hazards, including floods, riverbank erosion, hailstorms, lightning, thunderstorms, cloudbursts, cyclones, and GLOFs largely due to its complex terrain, high rainfall, and climatic variability. Its seismic location further exposes the region to frequent earthquakes, landslides, mudflows, and potential dam failures, especially in areas with aging or geologically unstable infrastructure. Although less frequent, chemical and industrial disasters pose emerging risks in urban and industrial pockets, where safety regulations may not be robust. The NER is also susceptible to accident-related disasters, such as forest fires, urban fires, mine flooding, major building collapses, and boat capsizing, often exacerbated by challenging terrain, unplanned development, and aging infrastructure. Moreover, the region has witnessed biological disasters, including epidemics and pest infestations, which threaten public health, food security, and ecological balance.

25.1.2.1 Floods

All eight states of NER are vulnerable to flooding, though the frequency, intensity, and underlying causes vary across the region. Among them, Assam experiences the most frequent and severe flooding, particularly during the monsoon season. The Brahmaputra and Barak rivers, along with their tributaries, flow through the state with considerable force, swelling rapidly during intense rains. This combination of high river discharge, sediment deposition, and topographical characteristics makes Assam particularly susceptible to seasonal inundation. Approximately 40% of the state's land area is flood-prone, placing millions of people at risk each year. The economic and human costs are staggering, Assam incurs an estimated ₹200 crore in annual losses due to flood-related damages, including widespread agricultural destruction and damage to homes and infrastructure. The 2022 flood season alone displaced over 4 million residents and caused damages exceeding ₹1,000 crore, underscoring the urgent need for robust flood management and disaster preparedness strategies (Bhuyan et al. 2023).

While Assam faces the brunt of floods in the region, other northeastern states are also significantly affected. Arunachal Pradesh, with its steep terrain and fast-flowing rivers such as the Siang and Subansiri, is prone to flash floods and landslides. A catastrophic flood in June 2000, caused by a dam-breach upstream in Tibet, raised the Siang River by over 30 m, submerging parts of four districts, killing around 7 people. More recently, incessant monsoon rains in 2025 affected over 61,000 people statewide, as landslides cut off roads and a heavy rainfall inundated part of the capital Itanagar. In Changlang district alone, more than 2,000 residents were left homeless by these floods. Meghalaya faces serious flood threats in its low-lying regions along the Bangladesh border. In 2022, extreme rainfall impacted over 600,000 people and resulted in 34 deaths, especially in the Garo Hills and border districts (https://morungexpress.com/over-6-lakh-affected-by-inclement-weather-in-meghalaya). Similarly, Tripura suffers from frequent urban and rural flooding, with the 2018 floods alone causing damages worth ₹1,300 crore. In 2024, more than 1.7 million residents were affected and at least 23 lives lost due to river overflows and poor drainage. Manipur's flood vulnerability is concentrated in the Imphal Valley, where seasonal inundation is exacerbated by encroachments and weak drainage systems. In 2023, over 165,000 people were affected, and more than 35,000 houses were damaged in one of the worst flood events in recent memory (https://www.indiatodayne.in/opinion/story/). Mizoram is frequently hit by flash floods and landslides, especially in valley settlements during heavy monsoons. In 2018, more than 1,000 residents were evacuated and multiple deaths reported as rain-induced disasters swept through Aizawl and Lunglei districts. Similarly, Nagaland's mountainous terrain makes it prone to both landslides and flash floods. The 2018 monsoon affected over 50,000 families in 538 villages, prompting the state to explore parametric insurance schemes to mitigate future losses. Sikkim, although less frequently flooded, faces high risks from GLOFs in addition to seasonal river flooding. These diverse but recurrent flood events across NER demonstrate that while Assam remains the epicenter, the entire region faces multi-hazard flood risks shaped by terrain, river systems, and changing climatic patterns. A region-wide strategy integrating early warning systems, embankment strengthening, slope stabilization, and glacial monitoring is vital for building long-term flood resilience in Northeast India.

25.1.2.2 Severe Storm and Lightning

The NE region of India along with the eastern India and Bangladesh is one of the hot spots of severe weather events including severe storms with gusting winds and lightning. Lightning is the natural disaster that causes highest number of life loss per year by any disaster in India (annual reports from NCRB). The data from the ground-based lightning detectors managed by the Indian Institute of Tropical Meteorology (IITM) and the National Remote Sensing Centre (NRSC) of ISRO, has been helping to have a detailed study on the spatial distribution of lightning events including classification of inter-cloud/intra-cloud (IC) lightnings and cloud to ground (CG) lightning. The Annual Lightning Report for 2023 by CROPC reported a significant 53% increase in Cloud-to-Ground (CG) lightning flashes between 2019 and 2023, while an increase in number of lightning detectors across India over the same period of time is helping in detecting more lightning events, the possibility of increase in number of severe weather events in recent years could also be a reason of increase in lightning events. The intense lightning activity in NE region of India typically occurs during the pre-monsoon season covering the months of March, April, and May. This phenomenon results in severe consequences, including substantial loss of life and damage to property. On an average, approximately 50 lightning induced fatalities are reported across the NE region of India every year.

25.1.2.3 Earthquakes

NER, India lies in Seismic Zone V as per the Bureau of Indian Standards code IS 1893 (2002) and is considered one of the most seismically active zones in the world (Longbir et al. 2017). This region is situated within a complex tectonic framework, shaped by the ongoing convergence of the Indian and Eurasian plates, which results in crustal deformation, active faulting, and frequent seismic activity. Earthquake data from 1900 to June 2025 from the National Center for Seismology (NCS), covering NER and its surrounding (Fig. 25.2) indicated a total of 4,091 earthquakes in the region during this125 years. These earthquakes include 113 events of magnitude Mw 0–2, 1,449 events of magnitude Mw 2–4 range, 2,468 events of magnitude Mw 4–6 range, and 61 events of magnitude Mw greater than 6. This distribution shows that the majority of recorded earthquakes were of moderate strength, although several high-magnitude events have also occurred.

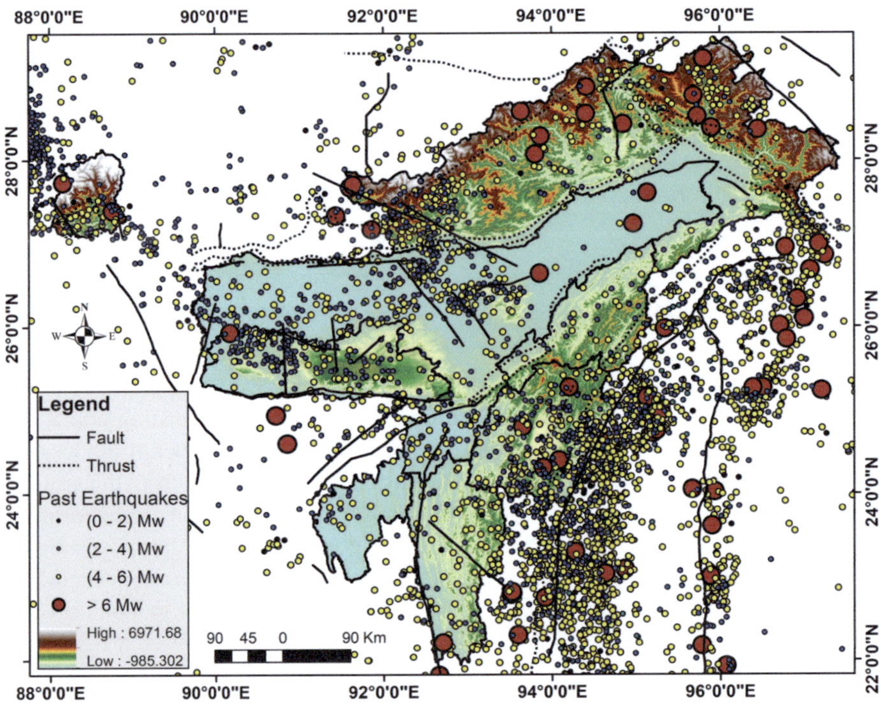

Fig. 25.2 Map illustrating the spatial distribution and magnitude of past seismic events that occurred across the Eastern Himalayan region from Dec 1908 to June 2025 *Source* NCS, USGS

25.1.2.4 Landslides

The region's hilly terrain and unstable soil conditions make it highly prone to landslides, especially during heavy rainfall or seismic events. To mention some of the recent devastating landslide, in 2019, Tamenglong District in Manipur experienced a series of landslides that caused widespread destruction and displaced numerous families. A year later, in 2022, intense rainfall in Dima Hasao, Assam, triggered landslides that disrupted rail and road connectivity, inflicted extensive damage to infrastructure, and affected thousands of residents. In 2023, a catastrophic landslide in Noney District, Manipur, claimed over 50 lives and left many injured. More recently, in late August 2024, Tripura experienced its worst monsoon in over 40 years, leading to in excess of 2,000 landslides displacing more than 3 lakh residents. These recurring incidents highlight the urgent need for strengthened disaster preparedness and risk reduction strategies to reduce the impacts of such natural hazards in the region.

25.1.2.5 GLOF Outbreak

GLOFs are sudden, catastrophic floods that occur when water is abruptly released from glacial lakes, often causing extensive destruction. These events pose a significant threat to life, infrastructure, and communities, especially in the NER region, where Sikkim and Arunachal Pradesh are particularly vulnerable due to the presence of large Himalayan glaciers. A recent GLOF in Sikkim illustrates the risks of these phenomena. The South Lhonak glacial lake in Sikkim burst on October 4, 2023, leading to flash floods and landslides that dramatically swelled the Teesta River. The surge destroyed multiple bridges, damaged parts of National Highway 10, and impacted the Chungthang Dam. Small towns along the upper Teesta Valley suffered heavy damage. According to reports from the **National Emergency Response Centre**, as of October 28, 2023, a total of 104 lives were lost, 77 individuals remained missing, and 5,665 people were evacuated. The floods affected approximately 88,400 residents and damaged 2,002 homes, with 1,423 completely destroyed. Several bridges were washed away, disrupting access and impacting the people across all four districts. This event underscores the critical need for targeted disaster preparedness in the region.

25.1.2.6 Droughts

Although not as common as other types of natural disasters, droughts can still impact Northeast India, significantly affecting agriculture and water supply. As reported by State Disaster Management Authority (SDMA), Nagaland, the year 2009 marks the first officially recorded drought-like situation in the state, affecting three districts: Peren, Dimapur, and Mon. The Jalukie Valley in Peren District, often referred to as the "Rice Bowl of Nagaland," has been particularly hard-hit by this drought, with large areas of paddy fields left uncultivated due to severe water shortages.

In addition to natural challenges, the cultural diversity of the NER shapes how communities manage disaster risks. Traditional knowledge and indigenous practices are integral to disaster preparedness and resilience. For instance, farmers in the hilly states of Northeast India practice Jhum cultivation, a traditional shifting agricultural method that, when managed sustainably, can help reduce soil erosion and maintain fertility, thereby mitigating landslide risks. In Assam, grassroots early warning systems based on weather patterns and river behavior help communities prepare for floods (ASDMA Report). The Mishing tribe constructs elevated stilt houses to protect against seasonal flooding and cultivates flood-resistant crops to ensure food security. Similarly, indigenous forest management practices in Arunachal Pradesh stabilize soil and reduce erosion. While these traditional practices strengthen community resilience, they are not sufficient to meet the increasing complexities of disaster risks today. Therefore, building trust with local communities and incorporating their knowledge into modern DRR strategies is essential. Involving local people in efforts like early warning systems, comprehensive planning, and geospatial technologies can bridge the gap between traditional wisdom and modern advancements.

A single-window geospatial platform plays a vital role in the integration of satellite data, ground observations, hazard assessments, and traditional knowledge into a unified system for disaster preparedness and response. Real-time hazard monitoring and faster dissemination of alerts enhance early warning systems and empower communities to respond promptly. Additionally, geospatial analytics help decision-makers assess risks, identify vulnerable areas, and prioritize mitigation measures more effectively. By integrating traditional knowledge with cutting-edge technologies, disaster management authorities in Northeast India can adopt a more holistic and resilient approach.

25.2 Disaster Risk Reduction Systems

25.2.1 National Agenda for Disaster Risk Reduction

As part of Sendai Framework, India announced the 10-Point Agenda for DRR in recognition of the country's high vulnerability to natural disasters (Press Information, Ministry of Home Affairs). The 10-Point Agenda aims to address the root causes of disaster vulnerability and build resilience across all sectors of society. It recognizes that DRR is not just a reactive measure to respond to disasters but a proactive approach to prevent and mitigate disaster risks. The agenda emphasizes the importance of mainstreaming DRR into public expenditure, investing in risk coverage for all, promoting greater involvement and leadership of women, investing in mapping disaster risk, promoting early warning systems, investing in climate resilient infrastructure, protecting ecosystems for DRR, enhancing international cooperation and partnerships, empowering local communities and Panchayati Raj Institutions, and promoting education and awareness for DRR. The Prime Minister's 10-Point Agenda for DRR emphasizes a comprehensive approach to enhancing resilience and preparedness against disasters. Key elements include:

- All development sectors must imbibe the principles of disaster risk management.
- Risk coverage must include all, starting from poor households to SMEs to multinational corporations to nation states.
- Women's leadership and greater involvement should be central to disaster risk management.
- Invest in risk mapping globally to improve global understanding of nature and disaster risks.
- Leverage technology to enhance the efficiency of disaster risk management efforts.
- Develop a network of universities to work on disaster-related issues.
- Utilize the opportunities provided by social media and mobile technologies for disaster risk reduction.
- Build on local capacity and initiative to enhance disaster risk reduction.
- Make use of every opportunity to learn from disasters and, to achieve that, there must be studies on the lessons after every disaster.
- Bring about greater cohesion in international response to disasters.

25.2.2 Disaster Risk Reduction System for NER

The disaster risk reduction system in Northeast India operates within a multi-tiered institutional framework that is part of India's overall disaster management structure (Fig. 25.3). It encompasses various stakeholders at the national, state, and district levels. At the apex of this system is the NDMA, which is responsible for formulating and overseeing the national disaster management strategy. The NDMA ensures that policies and procedures are effectively implemented across the country, coordinating with state and local authorities to enhance disaster preparedness and response efforts.

Beneath the NDMA, the State and District Disaster Management Authorities (SDMAs/DDMAs) play crucial roles in coordinating disaster preparedness, response, and recovery activities specific to their regions. SDMAs are tasked with managing disaster risks at the state level, while DDMAs focus on localized strategies and immediate response to incidents within their respective districts. This hierarchical structure allows for a streamlined approach to disaster management, ensuring that resources and information flow efficiently from national to local levels.

A vital component of this disaster management framework is the geospatial support provided by the Department of Space (DOS). The organization has developed a robust and versatile program dedicated to delivering space-based inputs that enhance the effectiveness of disaster management efforts across India. Over the years, DOS's contributions have spanned all phases of disaster management, focusing on disaster risk reduction, preparedness, response, and recovery. The Decision Support Centre established at the NRSC, Hyderabad plays a central role in offering these

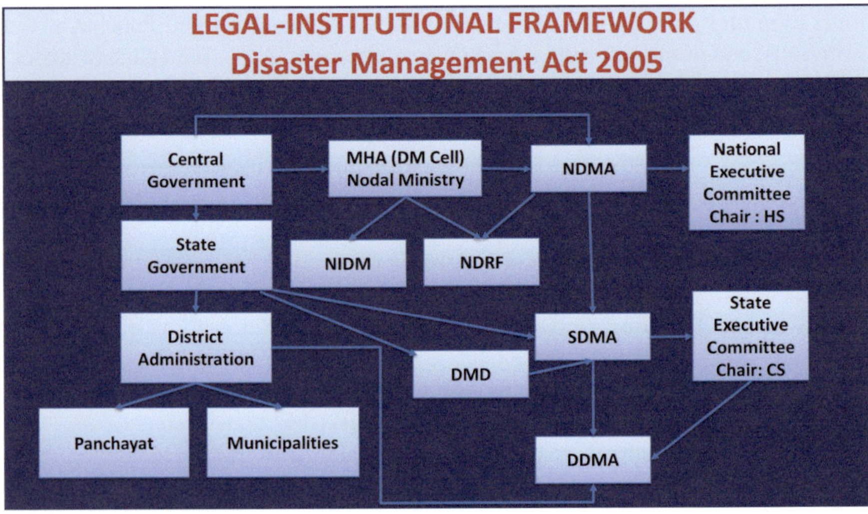

Fig. 25.3 Institutional framework for disaster management *Source* disaster management in India, ministry of home affairs publication

services across the country, utilizing satellite data and remote sensing technology to inform decision-making during disasters.

Recognizing that Northeast India is one of the most disaster-prone regions in the country, DOS established the North Eastern Regional node for Disaster Risk Reduction (NER-DRR) at the North Eastern Space Applications Centre (NESAC) in 2012. This dedicated node aims to provide a single-window delivery system for all space-based support related to disaster management in the region. Since its inception, the NER-DRR has been instrumental in offering timely and relevant geospatial information, facilitating more effective management of major disasters affecting NER.

The activities for disaster management support at NESAC under NER-DRR has been centered around the following points:

- Build a comprehensive geospatial database for different disaster vulnerable areas within the region and making them available in a central server along with the available database thus meeting the major objectives of disaster preparedness and mitigation plans.
- Development of support tools (including for simulation studies) for decision making exclusively for the NER using the available geospatial data sets in centralized data server.
- Establish satellite-based connectivity through VPN/ Space net with the National and State Emergency Operations Centre and with the district authorities for faster and more effective dissemination of space enabled services during and post disaster.

- Develop actionable products and services in tune with the requirement of the region in partnership with all stakeholders in the region, and service the region in an operational manner.
- To ensure most advanced and reliable mode of communication with redundancy.

DRR is inherently a collaborative effort, requiring strong institutional linkages across various sectors. NESAC plays a pivotal role in this framework, partnering with disaster management authorities at both the state (SDMAs) and district levels (DDMAs), as well as key agencies responsible for disaster management, including the Central Water Commission (CWC) and Water Resources Department (WRD) and other central bodies like IMD, Geological Survey of India (GSI), Survey of India (SOI), NRSC, etc. In addition, NESAC collaborates with State Remote Sensing Application Centres (SRSAC), and prominent research and academic institutions such as NEIST, Jorhat, Central Universities of NER, IITs, NITs.

The NER-DRR initiative encompasses a comprehensive range of disaster-related services, including operational support for flood, lightning and forest fire management, monitoring of landslides, embankment breaches, and earthquakes, and conducting crop damage assessments for both floods and droughts. It also maintains detailed inventories of landslides, forest fire and earthquakes and offers specialized high-resolution Hazard, Vulnerability, and Risk assessments upon request. Furthermore, based on user demand the initiative extends its capabilities to assess industrial and health-related hazards, ensuring a holistic approach to disaster risk reduction. NER-DRR also provides UAV and satellite-based services during emergency situation to manage disaster. Finally, all the database that has been created under NER-DRR are brought to a common platform called NER-DRR geoportal (www.nerdrr.gov.in) for better visualization and decision making.

25.3 Geospatial Techniques in Disaster Risk Reduction

Geospatial technology, including Remote Sensing (RS), GIS and Global Positioning System (GPS), provides critical tools and data for enhancing disaster preparedness, response, and recovery operations. Remote Sensing offers real-time imagery for situational awareness, GIS integrates multiple data layers to identify risk zones, and GPS supports precise location tracking for rescue coordination. These technologies collectively enable risk assessment, hazard mapping, and real-time monitoring, strengthening global disaster mitigation and emergency management strategies. The following sections discuss the various segments of geospatial technology for DRR.

25.3.1 Space Based Support for DRR

The space-based support includes both the earth observation data support and communication support. From India's perspective, DOS has established a comprehensive satellite-based monitoring framework that delivers crucial data for early warning, damage assessment, and disaster preparedness, significantly strengthening the nation's disaster risk management capabilities. An overview of Indian Satellites for DRR are given in Table 25.1. India's space policy has undergone significant changes, with ISRO/DOS making satellite data of 5-m resolution or above available for free through platforms like Bhoonidhi. This policy aims to democratize access to geospatial data, allowing broader use by Government agencies, private industries and academic institutions. Similarly, platforms like NDEM (National Database for Emergency Management) and MOSDAC (Meteorological and Oceanographic Satellite Data Archival Centre) play a key role in supporting disaster-related decision-making.

Table 25.1 Indian Satellites and their applications for DRR

Satellite	Purpose	Applications in DRR
INSAT Series (INSAT-3D/3DR/3DS)	Primarily meteorological satellites; INSAT-3D and its successors provide real-time weather monitoring, cyclone detection, and severe storm monitoring	Monitoring of tropical cyclones, heavy rainfall, and droughts. The data is crucial for issuing timely warnings and evacuations
Oceansat	It provides data on sea surface temperature, wind speed, and chlorophyll concentration	It is particularly beneficial for cyclone tracking, tsunami warnings, and fishermen safety by giving real-time data on ocean conditions
RISAT Series (RISAT-1, RISAT-2)	These radar-based imaging satellites can penetrate through clouds and darkness, offering continuous day-night surveillance	Helps in monitoring floods, landslides, forest fires etc.
Cartosat Series	Designed for high-resolution cartographic applications, It offer detailed imagery for land-use mapping, infrastructure planning, and disaster management	Assists in evaluating damaged buildings, roads, and other infrastructure
ResourceSat Series	Delivers multi-spectral images useful for environmental monitoring	Helps in vegetation monitoring, urban planning, and soil conservation, crucial for disaster mitigation in flood-prone and drought-affected regions

ISRO collaborates with numerous international organizations and agencies to facilitate the exchange of satellite data. Through global initiatives such as Sentinel Asia and the International Charter "Space and Major Disasters," ISRO plays a crucial role in providing satellite data for disaster management (Navalgund et al. 2010; Jayaraman 2013). This international cooperation enables the sharing of satellite data, contributing to the global effort to mitigate the impact of disasters.

25.3.2 Role of Geospatial Tools in DRR

Geospatial technologies encompass a broad range of tools and methods that allow us to collect, analyze, and visualize data related to geographic locations (Chang 2016), and have revolutionized disaster management by playing an indispensable role across all phases: preparedness, mitigation, response, and recovery. In preparedness, advanced hazard mapping and real-time early warning systems pinpoint vulnerable areas and enable timely alerts, while simulations and scenario modeling enhance readiness and decision-making. Mitigation efforts are strengthened through GIS-driven risk assessments and land-use planning that help avoid hazard-prone zones and reduce vulnerabilities. During response, real-time situational awareness through satellite imagery and GPS ensures rapid, precise resource allocation and prioritization of rescue efforts. In recovery, geospatial tools accelerate damage assessments, guide efficient reconstruction planning, and track long-term progress, ensuring resilient rebuilding. By integrating cutting-edge spatial data with strategic decision-making, geospatial technologies empower disaster managers to act faster, more efficiently, and with greater precision, ultimately saving lives and reducing disaster impacts (Van Westen 2002; Joyce et al. 2009; Roy et al. 2010; Elliot et al. 2020).

25.3.2.1 Hazard Mapping

A hazard refers to a potential threat that can cause loss of life, injury, or damage to property and the environment. Hazard mapping plays a crucial role in identifying and visualizing the geographic distribution of these threats within a specific area. This process is crucial for visualizing where hazards are likely to occur and understanding their potential severity. Geospatial data plays a pivotal role in hazard mapping by enabling the creation of detailed and accurate maps that depict the locations and extents of various hazards. For example, satellite imagery and aerial photography can provide up-to-date visuals of hazard-prone areas, while GIS can layer multiple data sources to create comprehensive hazard maps. Additionally, historical geospatial data can reveal patterns in the frequency and intensity of past disasters, helping to identify areas at higher risk. By integrating real-time data such as weather patterns or seismic activity into hazard maps, authorities can enhance their predictive capabilities during emergencies. Effective hazard mapping not only aids in risk assessment but also improves communication among stakeholders, facilitating public awareness and education efforts about the potential dangers in their regions.

25.3.2.2 Vulnerability and Exposure Analysis

Vulnerability and exposure analysis is a critical component of disaster risk management, as it assesses the susceptibility of populations, properties, and infrastructure to the effects of hazards. Vulnerability analysis identifies the factors that increase a community's risk, while exposure analysis determines the extent to which people and assets are at risk of being impacted by hazards. Geospatial data is essential in this context, as it allows for the integration of demographic and socioeconomic information with hazard maps, highlighting which communities are most at risk. For instance, GIS can be used to map critical infrastructure such as hospitals, schools, and transportation networks to identify assets that are particularly exposed to hazards. Furthermore, understanding land use patterns through geospatial data helps to identify areas where development may increase vulnerability, guiding better planning and zoning regulations. Accessibility analysis can also be conducted using geospatial tools to assess how well vulnerable populations can reach resources and emergency services, ensuring that responses are effectively tailored to those in need.

25.3.2.3 Risk Modeling and Prediction

Risk modeling and prediction are vital for understanding the potential impacts of hazards on vulnerable populations and assets. Risk modeling involves assessing the likelihood and severity of adverse outcomes, often incorporating statistical methods and simulations to forecast potential disaster scenarios. Geospatial data significantly enhances these models by providing the spatial context necessary for thorough risk assessments. For example, combining hazard data with socioeconomic information enables a more nuanced understanding of risk factors, allowing stakeholders to identify high-risk areas effectively. Geospatial modeling tools can also simulate various disaster scenarios, such as flooding under different rainfall conditions, earthquake scenario helping stakeholders explore potential impacts and develop appropriate mitigation strategies. Moreover, geospatial data supports multi-hazard assessments, allowing for the evaluation of multiple threats simultaneously such as the combined risks of flooding and landslides which provides a comprehensive view of risks and informs integrated risk management approaches. By incorporating real-time data into risk models, authorities can conduct dynamic assessments that adapt to changing conditions, thereby improving emergency preparedness and response efforts.

25.3.2.4 Real-Time Monitoring Using Space Technology

Satellites equipped with high-resolution sensors can capture data on weather patterns, natural events, and environmental changes as they happen. This allows agencies to track the progress of hazards like cyclones, floods, wildfires, and landslides in near real-time, offering critical information on their speed, direction, and intensity. Real-time satellite data enables early warnings and immediate response measures, helping authorities to issue timely evacuation orders, mobilize resources, and protect vulnerable communities. Moreover, advancements in geostationary and polar-orbiting satellites, along with data processing systems, have made it possible to visualize and assess unfolding events from a global perspective, thus supporting coordinated efforts in disaster management across regions. By providing up-to-the-minute information, space technology enhances the ability to mitigate impacts and save lives during emergencies.

25.3.2.5 Post-Event Damage Evaluation Using Space Technology

Space technology enables comparison between pre- and post-disaster scenarios, which helps identify the specific locations and severity of impacts. This information is valuable for agencies to document losses, support humanitarian aid distribution, and guide the rebuilding process. Satellite data also plays a role in evaluating long-term environmental damage, such as changes to coastlines or deforestation, which may impact future disaster vulnerability. Overall, space technology streamlines post-event evaluation, making it faster, more accurate, and more effective.

25.4 Application of Geospatial Techniques Under NER-DRR

25.4.1 Flood Management Using Geospatial Technology

NER of India experience severe flooding, particularly during the monsoon season, which spans June to September. The significant sediment transport from upstream areas, combined with the rivers' diminished carrying capacity, exacerbates flooding during peak monsoon months, resulting in enormous damage to life and property. To reduce the impact of such disaster, NESAC leverages advanced geospatial technologies tailored to each stage of the disaster management cycle. The efforts span from preparedness and early warning through real-time response and post-event damage assessment, forming a comprehensive and resilient framework that saves lives and informs decision-makers.

25.4.1.1 Preparedness

Geospatial technology plays a crucial role in disaster preparedness, particularly in monitoring long-term hazards such as river bank erosion and strengthening flood resilience. One of the major challenges faced by the northeastern states, especially Assam, is river bank erosion caused by the Brahmaputra, Barak, and their tributaries. This phenomenon, driven by high inflows, sediment overload, and shifting river channels, results in the loss of fertile agricultural land and causes extensive damage to infrastructure. Economic losses due to erosion run into several hundred crores annually, with the Brahmaputra widening up to 15 km in some stretches. To quantify and monitor these changes, Landsat 5 and Sentinel-2 satellite data over a 15-year period (2005–2020) has been used, mapping the loss of approximately 33,860.68 ha of agricultural land across 21 riverine districts representing 1.64% of their combined agricultural area and 1.16% of the total agricultural land in Assam. Barpeta and Morigaon were among the most affected, losing 4,294.56 ha and 4,000.96 ha respectively (Goswami et al., 2024). The analysis also captured sediment deposition zones that have evolved into new landforms, supporting informed land-use planning.

To further enhance flood preparedness, NESAC has laid the groundwork for comprehensive resilience planning through detailed river atlas development and hazard zonation mapping. Initiated in 2017, the Centre prepared a high-resolution (1:5,000 scale) atlas of Assam's rivers, embankments, sluice gates, and critical infrastructure. This geo-referenced dataset supports floodplain zoning, evacuation route planning, and risk-informed land-use strategies, thereby strengthening preparedness at both community and administrative levels. Recognizing the importance of regional scalability, NESAC is expanding this initiative across entire Northeast India including Arunachal Pradesh, Manipur, Meghalaya, Mizoram, Nagaland, Tripura, and Sikkim ensuring these states also benefit from advanced geospatial tools for disaster preparedness and resilient development planning.

In addition to these long-term strategic interventions, geospatial technology also enables real-time monitoring of embankment breaches, which is essential for operational flood preparedness and response. Embankments are the primary flood control structures, and their failure can lead to sudden and devastating inundation. Rapid identification of breach locations facilitates timely emergency response, prioritization of repair work, and safer evacuation planning. To address this, NER-DRR has operationalized embankment breach monitoring using satellite-based geospatial analysis (Fig. 25.4). By comparing pre-flood and post-flood imagery, breach locations are accurately identified and validated. Multi-temporal dual-polarized (VV and VH) Sentinel-1 synthetic-aperture radar (SAR) data, along with LISS-IV optical data acquired in May, August, and October, are utilized for detection. This analysis is further supported by embankment alignments, river networks, and road infrastructure digitized from high-resolution optical images. As a result, 40 vulnerable reaches showing continuous erosion from 2016 to 2022 have been identified along the Brahmaputra. During the 2023 flood season, 16 embankment breaches were detected across 8 districts of Assam, providing critical inputs for infrastructure strengthening and disaster preparedness planning.

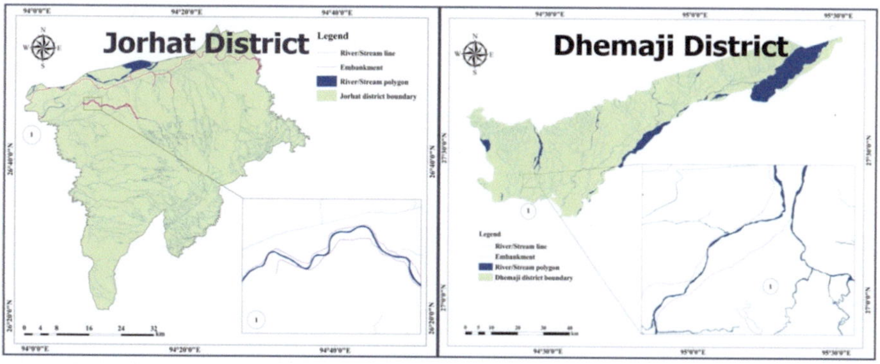

Fig. 25.4 Mapping and monitoring of embankment breach locations in two districts of Assam & identification of vulnerable locations

25.4.1.2 Early Warning

Flood Early Warning System (FLEWS) was introduced in Lakhimpur District of Assam in 2009 as a pilot project and has been progressively implemented across all districts from 2012 to 2024. Early warning systems designed for flood-prone rivers are mostly dependent on hydrological models for runoff prediction. In this FLEWS project, the focus is placed on estimating discharge at the outlets of tributaries within the Brahmaputra and Barak basins using the HEC-HMS model (Guduru et al., 2022). Various geospatial datasets are incorporated, including a 30 m resolution DEM from the Shuttle Radar Topography Mission (SRTM), land use and land cover data at a 1:10,000 scale generated through supervised classification, and soil data at a 1:250,000 scale obtained from the National Bureau of Soil Survey and Land Use Planning (NBSS-LP). Gridded meteorological data from the Global Precipitation Model (GPM) are utilized for simulations, along with flow data sourced from the CWC and the WRD, Assam. Using these geospatial data, HEC-HMS models have been developed for the major tributaries of the Brahmaputra and Barak basins.

Apart from hydrological models, the flood alert generation is also dependent on meteorological models for the rainfall information. This framework uses Weather Research Forecasting (WRF) (Skamarock et al. 2008) model for rainfall forecast. Since WRF is a global model, various sensitivity studies such as model sensitivity to LULC data, elevation data, parameterization schemes have been performed to configure the model for NER. Also, for increasing the forecast accuracy, data assimilation (Kalnay et al. 1997) has been performed using data from INSAT series of satellites, AWS, radar along with the global data. The final output from the WRF model is a 48-h rainfall forecast, which serves as input to all HEC-HMS models developed for flood-prone tributaries in the Brahmaputra and Barak basins. These HEC-HMS models are then run to produce forecasted discharge hydrographs at the outlets of respective tributaries. However, significant emphasis is placed on real-time

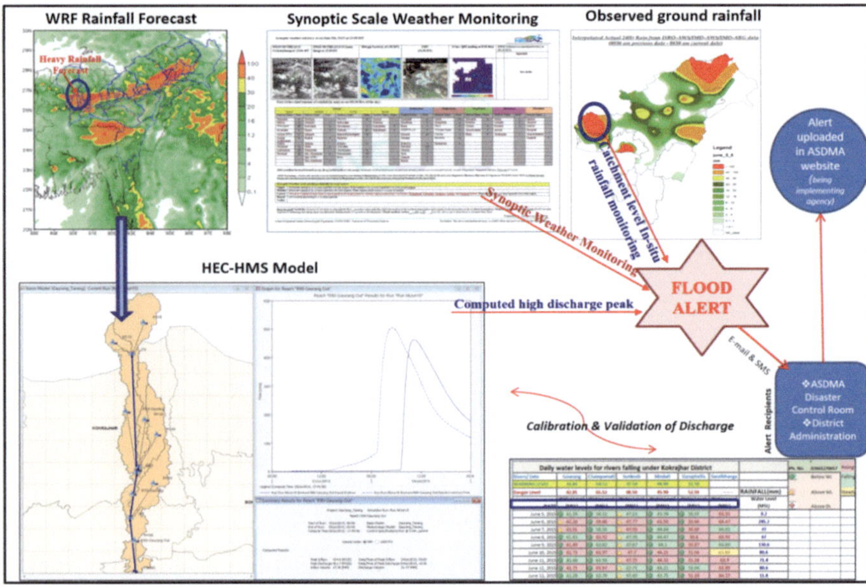

Fig. 25.5 FLEWS work flow

weather monitoring and observed rainfall data to reduce the false alarm rate associated with flood alerts derived from numerical weather prediction (NWP) model outputs (Fig. 25.5).

Currently, the FLEWS project has expanded its coverage to encompass all flood-prone districts in Assam, delivering actionable flood alerts to the SDMA, Assam at the revenue circle level. Since its launch, the system has achieved an impressive average success rate of 88% in alert generation year after year, with lead times for alerts ranging from 12 to 48 h. During the 2024 monsoon season, 55 alerts were issued to districts throughout Assam. Flood alerts are issued via detailed messages, accompanied by relevant maps, and sent by email to the concerned groups of SDMA, Assam. These alerts are also integrated into the NER-DRR Geoportal, where an interactive dashboard enables real-time visualization of flood-affected areas, enhancing situational awareness and facilitating data-driven decision-making during emergencies (Fig. 25.6).

Building on the effective implementation of FLEWS in Assam, NESAC has extended these services on an experimental basis to other flood-prone districts across the NER. Hydrological models have been built for Meghalaya, Arunachal, Manipur, Mizoram, Nagaland, Tripura & Sikkim and as a part of this, FLEWS alert has been disseminated and number of alerts for 2024 are: Meghalaya –10, Arunachal Pradesh –10, Tripura –19 and Mizoram – 01 alert. Some of the FLEWS basins for NER states are shown in Fig. 25.7.

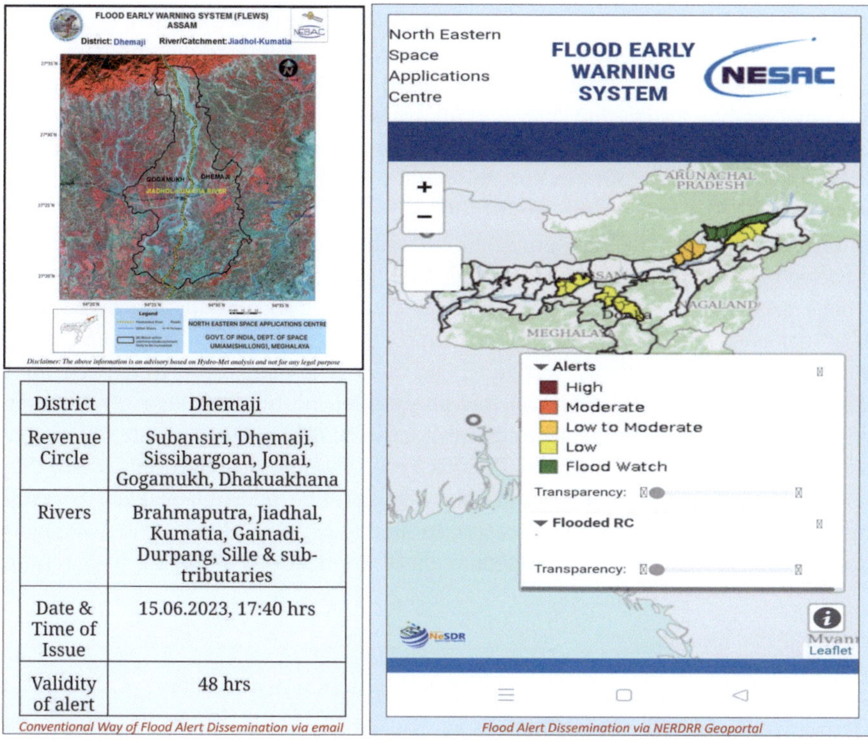

Fig. 25.6 Flood alert issued by NESAC

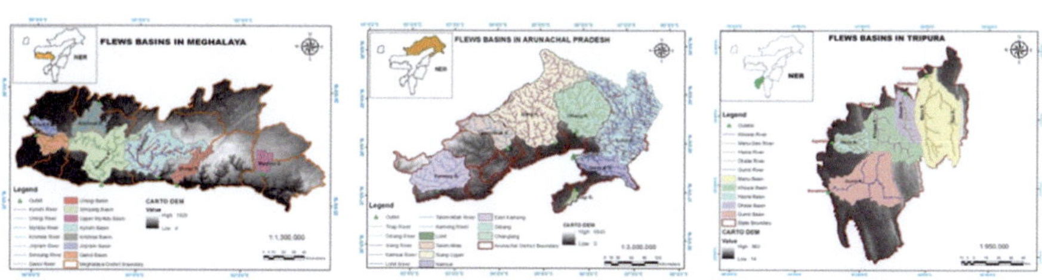

Fig. 25.7 FLEWS basins for some of the NER states: Meghalaya, Arunachal Pradesh and Tripura respectively (Left to Right)

25.4.1.3 Flood Response and Post-Event Damage Assessment

Flood response and post-event damage assessment are two critical phases in the disaster management cycle, both significantly enhanced by the use of geospatial technology. During an active flood event, rapid inundation mapping provides real-time visualization of flood extent, helping authorities prioritize evacuations, identify safe zones, and allocate rescue resources efficiently. SAR satellites such as Sentinel-1 and

EOS-4 enable uninterrupted monitoring under cloud cover and nighttime conditions, making them ideal for monsoon-season surveillance. Leveraging this capability, the NER-DRR platform integrates SAR-based flood mapping into its operational workflow, enabling state and district authorities to make timely, data-driven decisions such as rerouting rescue efforts, closing vulnerable routes, and pre-positioning relief supplies. Figure 25.8 presents the flood inundation map provided to SDMA, Tripura by NESAC during the August 2024 flood event, serving as a critical input for mitigation planning and response coordination.

Following the flood event, geospatial tools play a critical role in post-disaster damage assessment. By integrating high-resolution flood inundation layers with LULC datasets, it becomes possible to quantify the extent of inundation across agricultural fields, built-up areas, wetlands, and other land categories. In collaboration with the NRSC, NESAC conducts this analysis at the block/village level, generating detailed reports to share with the respective SDMAs. Furthermore, all analytical outputs are uploaded to the NER-DRR Flood Inundation geoportal, where they are visualized through interactive dashboards enabling decision-makers to assess impacts in near real-time, adjust response strategies, and plan targeted recovery efforts with greater precision. Together, these applications illustrate how remote

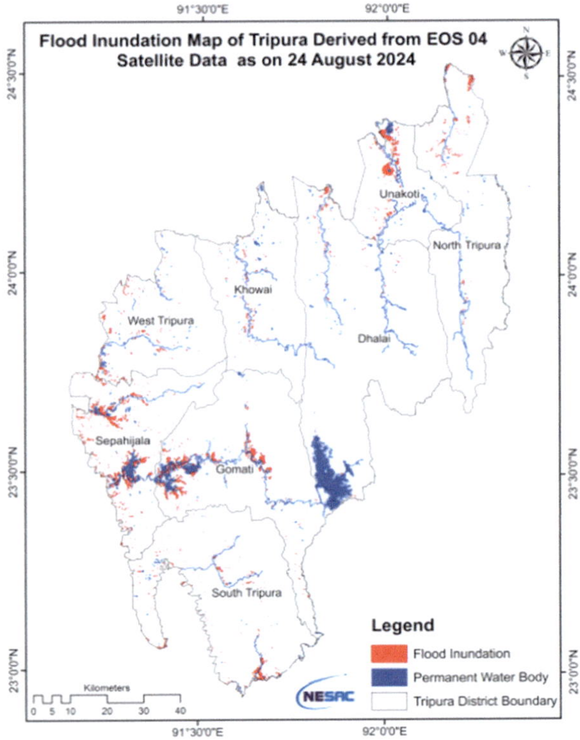

Fig. 25.8 Flood inundation map of Tripura (EOS 04 Satellite Data, 24 August 2024)

sensing and geospatial intelligence enable both real-time response and long-term recovery, making flood management more efficient, data-driven, and life-saving.

25.4.2 *Crop Damage Assessment and Monitoring Using Geospatial Tools*

Crop damage assessment using geospatial tools is crucial for timely, accurate, and large-scale evaluation of agricultural losses due to disasters such as floods, droughts, hailstorms, pests, or diseases. These tools integrate satellite imagery, remote sensing techniques, and GIS to map affected areas, quantify crop loss, and support evidence-based decision-making. High-resolution satellite data and time-series analysis help detect vegetation stress, monitor changes in crop health, and differentiate between damaged and healthy zones. This enables rapid response by authorities, ensures targeted compensation to farmers, guides insurance claims, and aids in planning recovery strategies. In disaster-prone regions like Northeast India, such assessments are vital for safeguarding livelihoods and ensuring food security. To showcase how geospatial technology supports accurate crop damage assessment, a few representative studies are outlined below.

25.4.2.1 Agricultural Drought in 2009

Unlike most natural hazards, drought lacks a clearly marked beginning or end, yet its repercussions can persist for long periods. Researchers have devised many methods to pinpoint when meteorological or agricultural drought starts and to measure its impact on crops. Rainfall is the primary driver on its own, or combined with other weather and hydrologic indicators, it remains the chief criterion for identifying drought onset and cessation. For agriculture, however, the timing and spatial pattern of rainfall matter more than the season's total amount. Agricultural drought occurs whenever soil moisture declines below the level required to support healthy crop growth. Effective management of this disaster requires vigilant monitoring of drought indicators and the establishment of early warning systems. Agricultural assessment using space technology has been operational in India since 1989, through a project "National Agricultural Assessment and Monitoring System–NADAMS", developed by NRSC, provides near real-time information on prevalence, severity level and persistence of agricultural drought at state/district/sub-district level. Mahalanobis National Crop Forecast Centre (MNCFC) carries out the assessment using rainfall, Remote Sensing Vegetation Index and Moisture adequacy Index during the Kharif season. The states that are covered by the program are primarily agricultural and vulnerable to drought. However, the eight states of NER are not covered under the NADAMS, primarily because this region is not prone to drought.

NER experiences significant rainfall during the monsoon season (June to September), which annually replenishes its rivers. However, in recent decades, the NER has witnessed an increasing frequency of deficit rainfall during kharif crop season. This reduction in precipitation has led to drought-like conditions, adversely impacting agricultural production in the region. Agricultural drought is a situation with a weekly rainfall of 5 cm or less during the period from mid of May to mid of October (the kharif season) when 80 percent of the country's total crop is planted (Ray et al. 2015). Again, there should be a monthly rainfall of 200 mm to grow lowland rice.

Figure 25.9 depicts the affected districts in each of the eight NER states in the year 2009 which was one of the worst years in Indian history for summer monsoon rainfall deficits, with 23% less than the Long Period Average (LPA). The NER also received 27% less rainfall than LPA. Since rice is the predominant crop in NER during Kharif season, assessment was done based on number of consecutive weeks (CW) with rainfall less than the water requirements of rice crop during various growth stages. Within the affected districts, severity of agricultural drought was assessed based on number of CW with rainfall less than the water requirements of rice crop, in terms of low (3CWs), moderate (4CWs) and severe (>4CWs).

Further it was noted that as per the IMD reports, NER recorded much lower rainfall than normal, affecting the Rabi seasonal crops (https://imdpune.gov.in). Agricultural practice in the Northeastern states is mostly rainfed, especially in the upland regions. Therefore, in order to reduce the impact of drought on agriculture, precise assessment and appropriate planning are required. A variety of indices and parameters, including rainfall-based indicators like dry spell, rainfall deviation, SPI (Standardized Precipitation Index), indices related to vegetation conditions like NDVI (Normalized Difference Vegetation Index), NDWI (Normalized Difference Water Index), VCI (Vegetation Condition Index), etc., and crop situation-related indicators like LSWI (Land Surface Water Index), SMI (Soil Moisture Index), and SASI (Shortwave Angle Slope Index), are necessary for monitoring agricultural drought

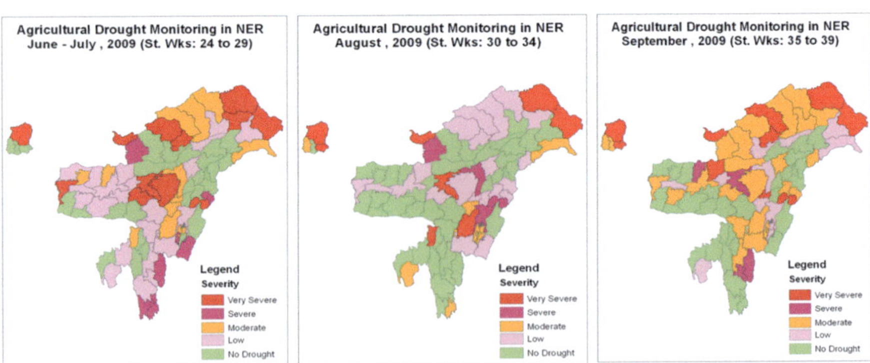

Fig. 25.9 Agricultural draught severity mapping in NER of India using Surface rainfall observation and TRMM multi satellite rainfall estimate

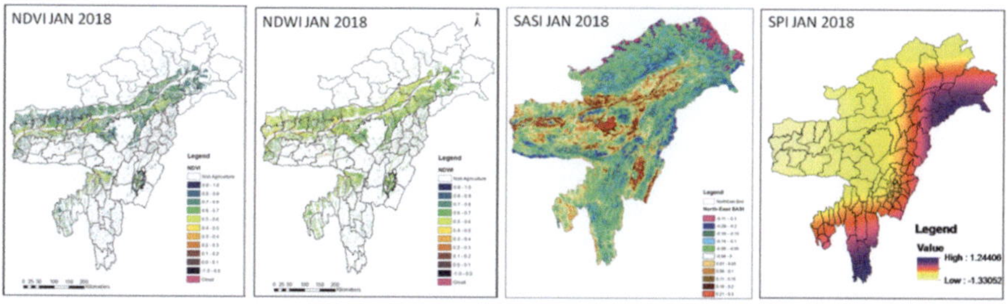

Fig. 25.10 Indices for drought monitoring in NER

and assessing its severity (Fig. 25.10). The impact of agricultural drought on crop varies from district to district depending on the type of crops grown and land use practices.

25.4.2.2 Severe Flood and Cyclone in 2022

Flooding in Assam was particularly intense in 2022, beginning unusually early in mid-May and persisting through the month. A second wave followed in mid-June, continuing until mid-July, affecting people across all districts. Prolonged and intense rainfall led to river overflows that inundated large tracts of agricultural land, causing significant damage to Kharif rice seedlings in many areas. To assess the extent of crop damage during the Kharif season, evaluations were carried out at both the district and revenue circle levels. The assessment process involved mapping rice transplanting dates and monitoring crop growth stages across the state using Sentinel-1A SAR data in a systematic manner. Damage estimation was based on the inundation status as recorded by the end of October. While Kharif rice transplanting typically takes place from mid-June to mid-July, in 2022, early flooding in May followed by subsequent events in June and July led to delayed sowing in many regions, pushing transplanting activities into August.

Based on flood impact severity, the affected areas were categorized into severe, moderate, and low damage zones. The analysis revealed that approximately 18,536.58 ha (0.83% of the total Kharif rice area) suffered severe flood-induced damage, while an additional 89,401.95 ha experienced moderate to low impact, with potential for crop recovery. Furthermore, damage assessment due to Cyclone Sitrang at the end of October in Tripura indicated that 0.16% of the aman rice area suffered severe damage. This analysis was also conducted using Sentinel-1A SAR data, underscoring the value of satellite-based geospatial tools for timely and accurate crop damage evaluation.

25.4.3 Severe Weather Forecast

Extreme weather events, such as Thunderstorms are often accompanied by heavy rainfall, strong wind, and intense lightning, posing significant challenges to the NER of India, due to its high frequency and intensity over this region. This vulnerability is amplified by the region's complex topography and unique climatic conditions, which make it a hotspot for convective weather systems (Das et al. 2019). Reducing lightning risk over a region requires a combination of activities that should include establishment of monitoring system, development of early warning systems, infrastructure improvements, public awareness, as well as policy measures.

The lightning detection started by the mid of twentieth century when ground-based electromagnetic sensors were first used to detect radio waves produced by lightning strikes. By the 1990s, satellite-based systems like the Lightning Imaging Sensor (LIS) on-board the TRMM satellite enabled the global monitoring of lightning for the first time. In India, lightning detection gained momentum with the setting up of national lightning location network by IITM, Pune and subsequently another network by NRSC, Hyderabad. There were some other agencies like Indian Air Force, who set up similar detectors, but the near real time data on lightning was made available to the research communities by the IITM and NRSC network. The NE region of India, being highly prone to cloud-to-ground lightning strikes due to its complex topography and convective weather patterns, has seen sincere efforts to improve the early warning services in recent years. The Indian meteorological satellites like the INSAT 3DR / 3DS have been providing very important information on the deep convective clouds that has been used by the researchers to understand the physical processes and conditions favorable for cloud electrification and lightning. With advancements in ground-based stations, remote sensing, and numerical modeling systems, lightning detection and forecasting have become significantly more precise and timelier, enabling early warnings with greater confidence and shorter lead times.

Several activities in the field of lightning risk reduction like, development of location-based lightning early warning system, lightning hazard mapping, augmenting the national lightning detectors network over the NE region, etc. have been taken up by NESAC. Thunderstorm Potential Maps (TPM) is prepared during the pre-monsoon and early monsoon season using the WRF model which identifies the potential region where chances of occurrence of thunderstorms are high. NESAC has developed a numerical weather prediction (WRF-ELEC) model-based lightning early warning system to nowcast lightning over the NER of India during the pre-monsoon seasons.

Additionally, NESAC has developed a severe weather tracking system, operational under the NER-DRR Lightning Tracking Decision Support System (DSS), which serves as a notable example of applied geospatial technology. This system is designed to nowcast lightning by monitoring sudden changes in lightning flash rates an important precursor to severe convective weather. By leveraging real-time geospatial data, it can identify specific areas likely to be affected by thunderstorms with a lead time of up to 45 min. The system makes extensive use of near real-time

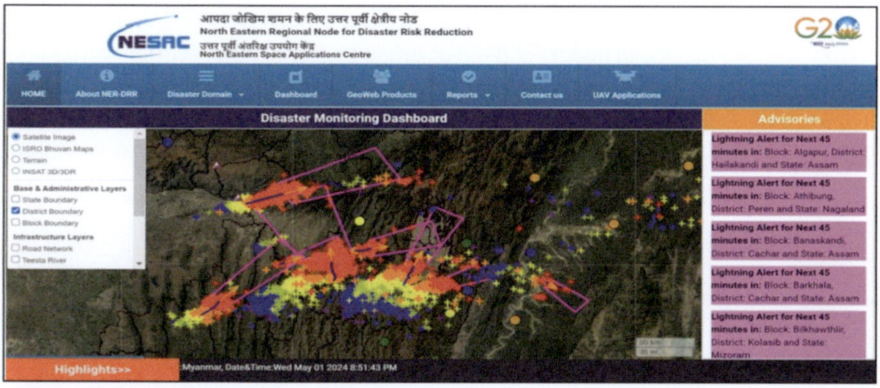

Fig. 25.11 The severe weather tracking system

total lightning data provided by IITM, Pune, and integrates it with spatial algorithms to track storm development and intensity across the Northeast Region. The outputs are visualized geospatially within the DSS environment, allowing decision-makers to interpret the lightning activity in both spatial and temporal dimensions (Fig. 25.11). A Daily Lightning Report is also generated as part of the system, offering a region-wide summary of lightning events.

25.4.4 Geospatial Approaches for Landslide Risk Mitigation

Landslide early warning systems are a critical component of disaster risk reduction, particularly in highly susceptible and topographically complex regions such as NER of India. A foundational element of such systems is the development and maintenance of seasonal landslide inventories, which document the spatial and temporal distribution of landslide occurrences. The inventories capture seasonal trends, enabling precise hazard assessments, refined predictive models, and targeted resource allocation during high-risk periods such as the monsoon. They also support resilient infrastructure planning and effective mitigation strategies while aiding climate adaptation by identifying shifts in landslide patterns. Recognizing the need for systematic landslide monitoring in the NER, the NER-DRR node at NESAC has been cataloguing landslide incidents annually, thereby developing a comprehensive geospatial database and an accompanying analytical report. This database maps landslide-prone zones in relation to critical infrastructure such as settlements, roadways, and transport corridors providing an invaluable resource for assessing localized risk, analyzing event frequency and magnitude, and enhancing regional resilience. The landslide inventory draws from two primary data sources: satellite imagery and news/media reports, including online publications. For the year 2024, a total of 1,137 landslide incidents were documented across various states of the Northeast, with the majority attributed to intense rainfall events. To operationalize this data, the "Landslide Information System" geoportal under NER-DRR offers near real-time, geo-tagged updates on newly reported landslides. By integrating spatial location data with base layers such as roads, settlements, health centres, and other critical infrastructure, the system supports rapid impact assessment and facilitates timely, informed decision-making during geological hazard events.

25.4.5 Earthquake Cataloguing Using Geospatial Platform

An earthquake catalogue, also known as a seismic catalogue or earthquake database, is a systematic record of earthquakes that have occurred in a particular region or globally (Weatherill 2016). It contains information about the location and depth of epicenter, magnitude, and other characteristics of each earthquake. Earthquake catalogues are essential documents for seismologists, earthquake engineers, and other professionals involved in studying and mitigating the impact of earthquakes. The combination of geological complexity, steep terrain, and population centers located in vulnerable valleys makes seismic risk in the Eastern Himalayas particularly high. Therefore, the creation and maintenance of a comprehensive earthquake catalogue are critical for understanding regional seismicity and improving disaster preparedness and mitigation strategies.

NER-DRR has been actively engaged in earthquake cataloguing by utilizing seismic data from the NCS, India, and the United States Geological Survey (USGS). The process begins with the systematic collection of earthquake event data across NER and it's adjoining seismically active zones, including the Eastern Himalayas. The acquired data is organized based on key seismic parameters such as date and time of occurrence, geographical coordinates, focal depth, magnitude, and reported location. This curated dataset is then imported into a GIS environment to enable spatial visualization and pattern analysis of earthquake epicenters across the NER and Eastern Himalayan belt. For enhanced regional insight, the data is further disaggregated at the state level, allowing more localized mapping and interpretation. In parallel, tectonic discontinuities and fault lines, sourced from Bhukosh (GSI), are overlaid in the GIS platform to assess their spatial correlation with seismic activity. This integrated geospatial analysis supports a deeper understanding of the region's seismotectonic framework and contributes to improved seismic risk assessment and preparedness planning. As part of NER-DRR geospatial services, an "Earthquake Information System" has been developed to visualize geotagged earthquake epicentres alongside relevant thematic maps (Fig. 25.12).

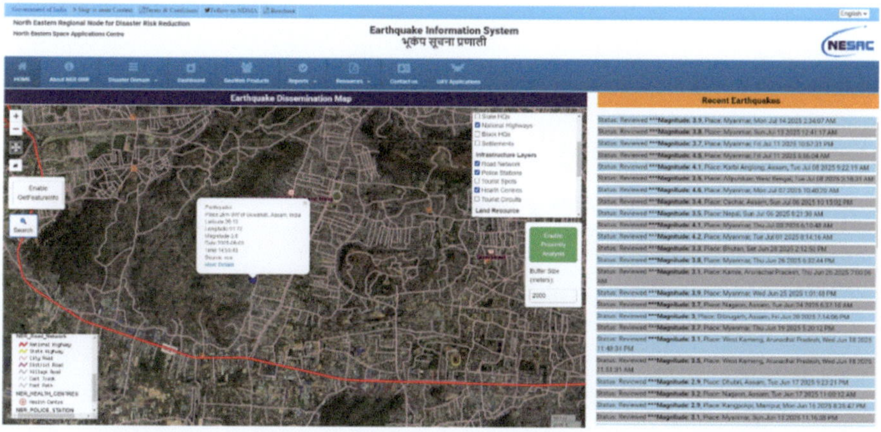

Fig. 25.12 Earthquake information system

25.4.6 *Forest Fires and Geospatial Solutions*

Forest fires are a major environmental problem in the forest ecosystems of northeast India, with large forest areas being affected every season. The causes of forest fire may be many but a high proportion of these fires are anthropogenic in nature either deliberately or accidentally. The spatial and temporal distribution of forest fire varies every year. Therefore, it is important to assess the number of forest fire events of this region. Under NER-DRR, the historical annual archived data of forest fire incident of each year (from 2003) has been obtained from Forest Survey of India (FSI) and a geospatial database has been prepared for better visualization and decision making (Fig. 25.13). In addition, value added forest fire alerts are being disseminated to all the concerned forest departments of NER since 2016 on an operational mode.

Daily fire alerts information is disseminated with information on latitude and longitude of fire location, fire occurrence date, state and district name, type of LULC, type of road connectivity, slope and aspect, presence of settlements/water bodies and identified into different categories viz., very high, high, moderate, low, and very low, based on likelihood of fire spread. This is done district wise, and is being sent to respective state and district forest departments and disaster management authorities through email and uploaded to "Forest Fire Services" geoportal of the NER-DRR geoportal for effective decision making.

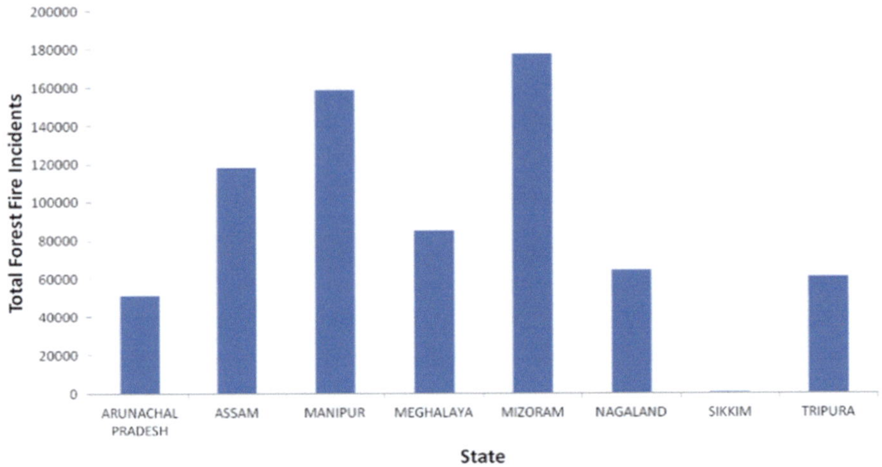

Fig. 25.13 State wise forest fire incidents over North Eastern Spates (2003–2024)

25.4.7 Remote Sensing and GIS Based Input for Hazard, Vulnerability and Risk Assessment

Hazard, Vulnerability, and Risk Assessment (HVRA) is essential for effective disaster management as it systematically identifies potential hazards, evaluates community vulnerabilities, and quantifies associated risks. This assessment provides critical data that informs policymakers and disaster management agencies, allowing them to formulate targeted disaster management plans and prioritize resource allocation to the most vulnerable areas. Geospatial technology serves as an integral component of HVRA by enabling spatial mapping of hazard zones, analyzing exposure of population and infrastructure, and overlaying multi-layered datasets such as land use, topography, and socio-economic indicators. Remote sensing, GIS, and ground-based inputs help in visualizing risk patterns at multiple scales and in near real-time, thus supporting data-driven decision-making. By enhancing community preparedness and guiding mitigation efforts, HVRA supported by geospatial insights, strengthens resilience and ensures effective disaster response and recovery at the community level.

In this context, a HVRA exercise was proposed for selected urban areas in the NER, focusing on landslides, floods, and lightning/thunderstorms. As of now it has covered urban areas of Meghalaya and Tripura. To gain a comprehensive understanding of the risks faced by the selected urban areas, a detailed map has been created by integrating a geodatabase with socio-economic survey and census data using geospatial techniques. This mapping approach allows for the identification of potential hazards, vulnerabilities, and risk areas at a scale of 1:4000. The work aimed to create Hazard Zonation Maps and evaluate Vulnerability and Risk for floods, landslides, and thunderstorms/lightning hazards. Various modeling techniques were used to generate the designed outflow which includes RS & GIS system for landslide analysis, a hydraulic model for flood hazard mapping, and lightning data were gathered from the Earth Network Total Lightning Network (ENTLN) for lightning hazard mapping. A dashboard has been created and the maps pertaining to Meghalaya HVRA is uploaded for visualization in the NER-DRR geoportal which has 69 geospatial layers pertaining to different base layer, hazard layer and susceptibility classes.

25.5 NER-DRR Geoportal for Disaster Risk Reduction Services

This section provides a consolidated summary of all activities discussed throughout the paper under the NER-DRR Geoportal initiative. This geoportal is a comprehensive online platform for visualization of disaster specific multi-scale geospatial database coupled with operational DSS designed and developed for addressing various disasters affecting NER, India. The updated and relevant geospatial database along with geospatial analytical tools, should assist disaster officials in the states for disaster/emergency management. All geospatial data have been generated as interoperable data services for ease of access into various applications developed for disaster risk reduction. Further more than 600 geospatial interoperable data services have been published for north eastern region from NDEM at https://www.nerdrr.gov.in/geoserver. These analysis ready datasets may be consumed into different applications. The NER-DRR geoportal https://www.nerdrr.gov.in now serves as a complete geoportal gateway for offering application services for disaster risk reduction in NER (Fig. 25.14).

The geoportal has nine independent sub-portals under four major operational services: (a) Hydrological Services; (b) Meteorological Services; (c) Geological Services; (d) Forest Fire Services (Fig. 25.15). All the portals have dynamic geospatial datasets which are fed during its respective seasons. The "Hydrological Services" include flood inundation monitoring and early warning systems, helping predict and manage flood risks. "Meteorological Services" offer weather forecasting, thunderstorm and lightning tracking, providing real-time updates for weather related hazards. The "Geological Services" cover earthquake and landslide information, assisting in preparedness and response to these events for systematic inventory and its analysis. The geoportal namely "Forest-Fire Services" monitor forest fires, offering early detection to mitigate fire risks by providing various value-added services. All these disaster related alerts for floods and lightning as well as value-added alerts for forest fire are disseminated in operational mode, while the current and historical occurrences of landslides, earthquakes, embankment breach locations and forest fires are systematically categorized within the NER-DRR portal. Together, all these modules enable comprehensive disaster monitoring, supporting timely alerts and informed decision-making for the region. For instance, when a landslide occurs in the NER

Fig. 25.14 Home page of NER-DRR portal showing different modules on alerts and decision support systems

and is reported in the news, the NER-DRR portal is quickly updated with the relevant information. This real-time update provides critical location data and allows users to overlay various data layers, such as district boundaries, road networks, and settlements. By integrating this information, users can assess potential impacts, identify damaged roads, find alternate routes, and determine the nearest settlements for emergency response. This capability significantly enhances situational awareness and supports informed decision making during geological disasters.

For example, in 2024 DSS system of NER-DRR recorded a total of 223 landslide events, 428 earthquakes, 37,436 fire incidents while the dashboard was updated with 24 flood occurrences based on the NRSC near-real time flood inundation data during the monsoon season in Assam. The geoportal is equipped with interactive features such as a measurement tool, opacity tool, layer identification information, and legend information. Further, it has a common geospatial dashboard having data on different disaster themes along with geo-analytical tools for visualizing, analyzing and running custom geospatial operations at regional scale.

In additional, a dedicated state level forest fire system and a mobile app has been developed for Meghalaya Govt as per their request. The dashboard and mobile app are open to all the citizens and have been designed as a single point source of information for near real-time and past forest fire events, forest fire vulnerability and prone zone,

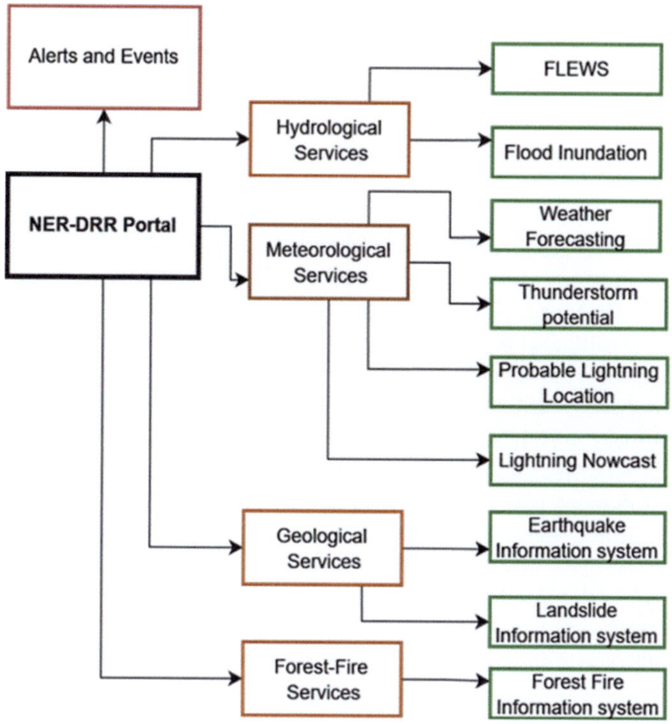

Fig. 25.15 NER-DRR operational disaster risk reduction services

and historical burnt areas for the entire Meghalaya which will help decision makers in mitigating forest fires and contributing to the sustainable management of the state's forest resources.

25.6 Role of Geospatial Technology in Emergency Support

During emergencies, timely access to accurate spatial data can mean the difference between a swift, coordinated response and a delayed, fragmented effort. Tools such as satellite imagery, GIS, UAVs, and real-time geospatial dashboards offer critical support for decision-making, logistics, damage assessment, and recovery planning.

A comparison of the 2011 Sikkim earthquake and the 2015 Nepal earthquake clearly demonstrates how the integration or absence of geospatial tools can significantly influence emergency outcomes. In Sikkim, despite the swift deployment of central agencies, the state's limited preparedness, poor situational awareness, and lack of real-time data severely hampered the efficiency of rescue and relief operations. Blocked roads caused by landslides, communication failures, and untrained personnel further compounded these challenges, while damage assessment remained largely manual and time-consuming (Khanna et al., 2015). In stark contrast, during the Nepal earthquake, geospatial technologies played a transformative role. Within 48 h, high-resolution satellite imagery enabled rapid mapping of affected zones, identification of collapsed structures, and detection of blocked transport corridors. UAVs were used to capture aerial imagery of hard-to-reach areas, while real-time updates from global volunteers via platforms like OpenStreetMap helped coordinate humanitarian response. These tools allowed rescue teams to prioritize resources, guide search operations, and restore access routes far more effectively than traditional methods alone (Clark 2015; Vukic 2017). Had similar technologies been institutionalized in Sikkim, early identification of blocked routes, dynamic resource coordination, and faster damage assessment could have significantly improved outcomes.

Recognizing these critical gaps, NER-DRR has emerged as a key geospatial support institution for emergency response across the Northeast Region of India. Leveraging its expertise in satellite remote sensing, NESAC generates near-real-time flood inundation maps using data from satellites like Sentinel-1, RISAT, and EOS-04. These maps, integrated with land use/land cover and administrative layers, help identify inundated agriculture, settlements, and critical infrastructure. The processed maps are shared with SDMAs and uploaded to the NER-DRR Geoportal, aiding rapid and informed decision-making.

In addition to space-based support, NESAC has developed a dedicated UAV capability to address limitations posed by cloud cover, spatial resolution, and terrain in satellite imagery. These UAVs are deployed during floods, landslides, and urban fire incidents to collect high-resolution aerial data over affected zones. The resulting imagery is processed into orthomosaics, DEMs, and 3D terrain visualizations, which are instrumental in planning rescue routes, assessing damage, and supporting post-disaster recovery.

For instance, in May 2022, a series of landslides struck Dima Hasao district in Assam due to continuous early monsoon rainfall. One of the worst-affected locations was New Haflong railway station, which was completely devastated by debris flows. NESAC rapidly deployed UAV teams to capture aerial imagery of the affected site (Fig. 25.16). The data was analyzed to detect the extent of damage to the station, surrounding houses, and submerged railway infrastructure, and was promptly shared with the SDMA, Assam. Similarly, on October 14, 2023, a major fire broke out in the Janata Furniture building in Shillong's densely populated Police Bazaar. NESAC's UAV teams conducted rapid aerial surveys over the site, capturing high-resolution imagery that revealed the spread of the fire, smoke plume direction, and spatial layout of nearby multistorey structures and narrow access lanes. This data was critical for fire and emergency responders working under constrained urban conditions.

These examples underscore how geospatial and UAV support fills critical information gaps during emergencies, enabling faster, more targeted, and coordinated action. Institutionalizing such capabilities particularly in remote, hazard-prone states will be key to enhancing India's resilience in the face of increasing natural and human-induced disasters. operating amidst challenging narrow roads and dense construction.

Fig. 25.16 UAV view of landslides in New Haflong railway station

25.7 Challenges and Opportunities

25.7.1 Challenges in Implementing Geospatial Techniques

Implementing geospatial techniques presents several challenges that can impede their effective application and integration. The NE region of India is characterized by a complex and varied topography, including hills, valleys, and dense forests, which complicates data collection efforts. Traditional survey methods may struggle in such rugged terrain, and environmental factors such as landslides, erosion, and flooding further complicate the gathering and interpretation of data. In this context, satellite data plays a crucial role in monitoring changes and managing resources. However, the revisit times of satellites can significantly hinder the ability to track dynamic events such as landslides, flooding. For example, if a satellite revisits a location only every few days or weeks, critical data may be missed during essential periods, potentially delaying response and mitigation efforts. Furthermore, the temporal resolution of satellite data varies based on the satellite's orbit and operational capabilities. In emergency situations, such as floods or landslides, timely access to data is vital for making informed decisions. If satellites are unable to provide frequent updates, relying solely on their data can create significant gaps in situational awareness, compromising the effectiveness of disaster response efforts. Therefore, it is essential to explore complementary data sources such as UAVs. However, flying UAVs during disasters is challenging due to regulatory, security, and operational constraints. Airspace restrictions, no-fly zones, and the need for flight permits can delay deployment. Security concerns, such as data privacy and potential threats, further complicate their use. Additionally, coordination with ground teams, limited battery life, and adverse weather conditions can hinder operations. To make UAVs more accessible during emergencies, policies should streamline flight authorizations, improve coordination with security agencies, and invest in the necessary infrastructure and communication systems to support real-time operations. Some other challenges are mentioned below:

Data Availability and Quality: Data availability and quality pose significant challenges to the effective implementation of geospatial techniques for NER. The region's frequent cloud cover, particularly during the monsoon season, often hampers the acquisition of clear satellite imagery, making it difficult to obtain reliable data for monitoring land use, environmental changes, and disaster assessments. Traditional optical satellite data can be obscured by clouds, resulting in gaps that can hinder timely decision-making and analysis. To mitigate the limitations of cloud cover on satellite imagery, the use of SAR data becomes increasingly important. SAR technology can penetrate clouds and capture images under various weather conditions, making it especially valuable for monitoring terrain changes and disaster impacts in regions like Eastern Himalaya. However, the effective utilization of SAR data comes with its own challenges, primarily the need for skilled manpower to process and analyze the complex data it generates.

Infrastructure and Technology Limitations: The region's infrastructure is often underdeveloped, with inadequate internet connectivity and limited access to reliable power sources, especially in remote areas. This hinders the use of advanced geospatial technologies that require stable internet for data transfer and processing. Moreover, the lack of proper storage facilities for data can also impact the effectiveness of geospatial applications.

Capacity and Skill Gaps: There is a notable shortage of trained professionals in geospatial technologies within local institutions in NER. Many organizations lack the necessary expertise to effectively utilize GIS, remote sensing, and other geospatial tools. This skills gap can lead to underutilization of available technologies and hinder the successful implementation of geospatial projects. To circumvent the issues, NER-DRR provide regular training to officials from different disaster management authorities on application of geospatial technology for disaster risk management. Training on UAV technology is also provided to user departments as per their demand.

Interagency Coordination: Effective implementation of geospatial techniques often requires collaboration among various stakeholders, including government agencies, non-governmental organizations (NGOs), and local communities. However, bureaucratic differing priorities, and lack of communication can complicate coordination efforts. Establishing a cohesive framework for collaboration is essential but can be difficult to achieve.

Cultural and Socioeconomic Factors: NER is home to diverse ethnic groups, each with its unique cultural practices and socioeconomic conditions. Engaging local communities and ensuring that geospatial applications align with their needs and knowledge systems is crucial for acceptance and success. Resistance to new technologies or methodologies can impede effective implementation if community concerns are not addressed.

Funding and Resource Constraints: Limited financial resources can restrict the acquisition of necessary technologies, training programs, and data collection efforts. Securing adequate funding for geospatial initiatives is a persistent challenge, and without sufficient investment, projects may struggle to meet their objectives or sustain long-term impact.

Regulatory and Policy Framework: The absence of clear policies and regulations governing the use of geospatial data can create uncertainty for stakeholders and hinder investment in geospatial initiatives. Establishing a supportive regulatory environment is essential for fostering innovation and ensuring that geospatial technologies are effectively integrated into development planning and disaster management.

25.7.2 Opportunities for Future Research and Application

Emerging technologies, particularly artificial intelligence (AI) and machine learning (ML), can play a transformative role in enhancing geospatial techniques. These technologies significantly improve data processing and analysis, allowing for the efficient handling of vast amounts of geospatial data, which is crucial for timely decision-making during disasters. AI and ML algorithms excel in recognizing patterns and trends in complex datasets, enabling the prediction of future occurrences of environmental events such as landslides and floods, thus enhancing preparedness and response strategies.

25.8 Conclusion

NER is a highly disaster-prone area, with frequent occurrences of floods, landslides, earthquakes, forest fires, and lightning. Its complex and remote terrain makes relief and rescue operations especially challenging during disasters, creating a "dangerous combination" of high vulnerability and difficult access. In this context, disaster preparedness and planning are critical to reducing disaster risk and safeguarding communities in the region. The NER-DRR node of NESAC plays a pivotal role in supporting these efforts by providing space-based data and timely alerts to assist the eight states in NER.

NER-DRR's contributions are multifaceted and operationally essential. Currently, the node provides real-time flood alerts, forest fire alerts, and lightning nowcasting to user agencies, equipping local governments and responders with the information needed to act quickly. Flood alerts help prepare communities in advance, minimizing loss of life and property, while forest fire alerts protect vital forest ecosystems and nearby settlements. Lightning nowcasting helps reduce fatalities in a region where lightning strikes are a significant risk. Additionally, NESAC supports weather forecasting, which helps communities and disaster management agencies plan effectively, especially during monsoon season when extreme weather events are frequent. The NER-DRR node also undertakes seasonal landslide inventory preparation and earthquake cataloguing, which are invaluable for long-term planning and risk assessment. By maintaining a landslide inventory, NESAC aids in identifying high-risk areas, allowing for preventive measures and infrastructure planning to avoid future damage. Earthquake cataloguing supports seismic risk assessment, contributing to safer building practices and informing policies that improve structural resilience in this seismically active region. Forest fire data archiving and embankment breach location mapping are other regular activities under NER-DRR, aiding in resource management and the protection of both natural landscapes and human settlements. Together, these services and data products enable disaster preparedness and risk reduction across NER, creating a robust framework for addressing the region's unique and complex challenges.

The NER-DRR geoportal (www.nerdrr.gov.in) serves as a centralized platform that brings together all geospatial databases developed for Disaster Risk Reduction in the Northeast Region of India. Through this comprehensive geoportal, users have access to an integrated collection of resources, including four dedicated decision support systems that provide services across hydrological, meteorological, geological, and forest fire domains. The geoportal hosts both near-real-time and historical datasets, covering critical hazards such as landslides, forest fires, and earthquakes specifically for the NER. This wealth of data allows users to monitor ongoing risks and analyze past events to support planning and preparedness. In addition to hazard-specific information, the platform offers various base layers, including district boundaries, road networks, and settlement locations, which users can overlay with hazard data to better visualize affected areas. This feature facilitates detailed situational awareness, enabling informed and timely decision-making in response to natural hazards. For better utilization of the geospatial database, regular trainings are provided to officials from different user departments such as SDMAs, DDMAs, Forest Departments, NDRF etc. Nevertheless, miles to go to achieve a fully space technology based DRR system for NER.

Acknowledgements The authors express their sincere gratitude to ISRO/DOS for financing the NER-DRR node at NESAC, in recognition of the critical role of geospatial technology in strengthening disaster risk reduction efforts in the disaster-prone region of Northeast India. Effective disaster risk reduction is inherently collaborative, and the authors gratefully acknowledge the vital support extended by the State and District Disaster Management Authorities (SDMAs and DDMAs), as well as key partner agencies including CWC, SRSAC, WRD, IMD, GSI, SOI, and NRSC.

References

Agenda on Disaster Risk Reduction, Posted On: 19 DEC 2023 5:23PM by PIB Delhi, Ministry of Home Affairs (2023)

Annual Lightning Report. Climate resilient observing systems promotion council (CROPC) (2023)

B.K. Bansal, M. Verma, A.K. Gupta, R.A. Prasath, On mitigation of earthquake and landslide hazards in the eastern Himalayan region. Nat. Hazards **114**(2), 1079–1102 (2022)

S. Bhuyan, Economic impact of flash floods on the people of Assam. An Anal. Int. J. Adv. Eng. Manag. (IJAEM) **5**(3), 534–537 (2023)

M.R. Biju, M.A. Padmanabha, Disaster management in India: Paradigm Shift in public policy. Int. J. Polit. Sci. **5**(1), 5–15 (2019)

R. Bilham, N. Ambraseys, Apparent Himalayan slip deficit from the summation of seismic moments for Himalayan earthquakes, 1500–2000. Current science, 1658–1663 (2005)

R. Bilham, Himalayan earthquakes: a review of historical seismicity and early 21st century slip potential. Geological Society, London, Special Publications **483**(1), 423–482 (2019)

Chang, K.-T, Introduction to geographic information systems (8th ed.). McGraw-Hill Education (2016)

L. Clark, How Nepal's earthquake was mapped in 48 hours. Wired UK, 2015–04 (2015)

Community Based Disaster Preparedness, Assam state disaster management authority & Doctors For You (2013)

S. Das, A. Hazra, Thunderstorm climatology over northeastern India: a severe weather perspective. J. Earth Syst. Sci. **128**(5), 1–15 (2019)

Disaster Management Act 2005, Ministry of Home Affairs, Govt of India (2005). http://www.nidm.net/disastermanagement act

Disaster Management in India. https://ndmindia.mha.gov.in/ndmi/publication

J.R. Elliott, Earth observation for the assessment of earthquake hazard, risk and disaster management. Surv. Geophys. **41**(6), 1323–1354 (2020)

J. Goswmai, S. Velliyangiri, K.K. Sarma, S.P. Aggarwal, Assessment of agricultural land area lost due to Brahmaputra riverbank line erosion in Assam. Curr. Sci. **127**(1), 20 (2024)

EM-DAT, CRED / UCLouvain—with major processing by Our World in Data (2024)

J.U. Guduru, N.B. Jilo, Z.A. Rabba, W.G. Namara, Rainfall-runoff modeling using HEC-HMS model for Meki River watershed, rift valley basin. Ethiopia. J African Earth Sci **197**, 104743 (2023)

International federation of red cross and red crescent societies, World Disasters Report (2022)

V. Jayaraman, Space-based disaster risk reduction: Indian framework & perspectives. Executives development programme (EDP) Space Applications Centre, Ahmedabad (2013)

K.E. Joyce, S.E. Belliss, S.V. Samsonov, S.J. McNeill, P.J. Glassey, A review of the status of satellite remote sensing and image processing techniques for mapping natural hazards and disasters. Prog. Phys. Geogr. **33**(2), 183–207 (2009)

E. Kalnay, A.J. Busalacchi, S.E. Cohn, Data assimilation in the Ocean and in the atmosphere. J. Meteorol. Soc. Jpn **75**(1B), 489–496 (1997)

N. Khanna, J. Verma, B.K. Khanna, Sikkim earthquake: perils of poor preparedness. Journal of Defence Studies **6**(1), 77–90 (2012)

S. Longbir, S. Arjun, Assessment of liquefaction potential of Guwahati city based on geotechnical standard penetration test (SPT) data (2017)

M. Mishra, T. Acharyya, R. Guria, N.R. Rout, C.A.G. Santos, R.M. da Silva, A.K. Pradhan, Lightning-related fatalities in India (1967–2020): a detailed overview of patterns and trends. Environ, Dev. Sustain. 1–30 (2024)

National disaster management authority (NDMA) Annual Report (2022–2023)

R. Navalgund, V.A. Menshikov, J. Akinyede (eds.), Space-based disaster management: the need for international cooperation. International Academy of Astronautics (2010)

R. Pal, S.S. Biswas, B. Mondal, M.K. Pramanik, Landslides and floods in the Tista Basin (Darjeeling and Jalpaiguri Districts): historical evidence, causes and consequences. J Ind Geophys Union **20**(2), 66–72 (2016)
P.S. Roy, R.S. Dwivedi, Remote sensing applications. NRSC (National Remote Sensing Centre), ISRO (2010)
T.P. Sabin, R. Krishnan, R. Vellore, P. Priya, H.P. Borgaonkar, B.B. Singh, A. Sagar, Climate change over the Himalayas. Assessment of climate change over the Indian region: a report of the Ministry of earth sciences (MoES), Government of India, 207–222 (2020)
Y. Sharma, S. Pasari, K.E. Ching, O. Dikshit, T. Kato, J.N. Malik, J.Y. Yen, Spatial distribution of earthquake potential along the Himalayan arc. Tectonophysics **791**, 228556 (2020)
M. Shrestha, S. Sharma, R. Pradhan Shrestha, Landslides in the Himalayas: a comprehensive review of hazards, impacts, and adaptive strategies. Rural. Reg.Nal Dev. **3**(1), 10002 (2025)
D. Singh, A. Jeffries, A, Cyclone Phailin in Odisha, October 2013: Rapid damage and needs assessment report (2013)
Situation report regarding Flood/Heavy rainfall in the country as on 28.10.2023 at1800 hrs, Ministry of Home Affairs, Disaster Management Division, National Emergency Response Center
W.C. Skamarock, J.B. Klemp, J. Dudhia, D.O. Gill, D.M. Barker, M.G. Duda, J.G. Powers, A description of the advanced research WRF version 3. NCAR Technical Note **475**(125), 10–5065 (2008)
The Himalayan, Published: 04:35 pm. May 20, 2025 (2025). https://thehimalayantimes.com/nepal/lightning-kills-237-injures-908-across-the-country-in-four-years
UN International Strategy for Disaster Reduction (UNISDR), Hyogo Framework for 2005–2015: Building the Resilience of Nations and Communities to Disasters (2005). https://www.unisdr.org/2005/wcdr/intergover/official-doc/L-docs/Hyogo-framework-for-action-english.pdf
UN International Strategy for Disaster Reduction (UNISDR), Terminology on disaster risk reduction (2009). https://www.undrr.org/publication/2009-unisdr-terminology-disaster-risk-reduction
UN International Strategy for Disaster Reduction (UNISDR), Sendai Framework for Disaster Risk Reduction 2015–2030 (p. 32). United Nations Office for Disaster Risk Reduction (UNISDR) (2015). https://www.unisdr.org/we/inform/publications/43291
C.J. Van Westen, Remote sensing and geographic information systems for natural disaster management. Environ. Model. GIS Remote. Sens. **6**, 200–222 (2002)
G. Veh, O. Korup, A. Walz, Hazard from Himalayan glacier lake outburst floods. Proc. Natl. Acad. Sci. **117**(2), 907–912 (2020)
L. Vukic, The value of rapid disaster response with UAVs (Master's thesis, University of Twente) (2017)
G.A. Weatherill, M. Pagani, J. Garcia, Exploring earthquake databases for the creation of magnitude-homogeneous catalogues: tools for application on a regional and global scale. Geophys. J. Int. **206**(3), 1652–1676 (2016)

Rekha B Gogoi joined NESAC as Scientist/Engineer 'SC' on February 27, 2009, and has since led key projects applying space technology to disaster management in Northeast India. She was instrumental in operationalizing Numerical Weather Prediction (NWP) and ensemble-based data assimilation to enhance the region's Flood Early Warning System (FLEWS). She also led seismic hazard assessment under the HRVA project for Assam. As Focal Scientist for the NER-DRR program, she coordinates core activities, ensures smooth implementation, and works closely with national and state disaster management agencies during emergencies. Her efforts have strengthened both flood and seismic preparedness in the region through advanced modeling and inter-agency collaboration. She holds a PhD in Numerical Weather Modeling from IIST, Trivandrum, M.Tech degree in Computational Sesimology from Tezpur University and M.Sc. degree in Mathematics from University of Delhi. She has been awarded ISRO Young Scientist Merit Award for 2016.

Diganta Barman a senior scientist with over 22 years of experience, heads the Water Resources Division at NESAC. He holds a PhD in Civil Engineering (Water Resources & Hydrology) from IIT and specializes in hydro-modelling, watershed management, flood forecasting, and reservoir operations. He led the development of India's first operational Hydro-met Flood Early Warning System (FLEWS) for the Brahmaputra and Barak basins, a system he has steered since 2009. Dr. Barman has also led major projects like the River Atlas (Assam & NER), SWaRA, NEEPCO's Hydro-met monitoring, and flood hazard zonation for Assam. He currently mentors the HIU Team under the World Bank-funded AIRBMP project. A member of various expert committees on water and flood management in NER, he has authored 20+ peer-reviewed papers and supervised multiple PhD and postgraduate theses. He collaborates with national and international agencies including CWC, IMD, IRADe, Columbia Water Center, and ICIMOD.

Kuntala Bhusan is working as a Scientist in North Eastern Space Applications Centre (NESAC), Govt of India, Dept of Space, Umiam, Meghalaya since 2003. She did her B.Sc. (1st Class 1st position) from Assam University and M.Sc. (1st Class 3rd position) in Geology and Ph. D. from Gauhati University, Assam. She has coordinated several National, User funded, and R&D projects taken up by NESAC related to geoscience and landslide hazard. Currently, she is also contributing as the Project Manager under North Eastern Regional Node for Disaster Risk Reduction (NER-DRR) program of NESAC for Landslide Hazard. She has a couple of publications in national and international journals, conference proceedings and several technical reports in the related subject.

Gopal Sharma has obtained a PhD in Crustal deformation study from IIT Dhanbad. Presently, he is a Senior Scientist at the North Eastern Space Applications Centre (NESAC), Shillong, Department of Space, Government of India. His research interests include Earthquake precursors, Crustal Deformation, and strain accumulation measurements. His ongoing research focuses on large-network GNSS

Chirag Gupta is working as Scientist/Engineer 'SE' at NESAC, Umiam, Meghalaya. He completed Bachelor of Engineering in Electronics and Communication in 2012 from SIT, Tumkur, Karnataka and pursuing PhD in Advance Communication Systems from National Institute of Technology (NIT), Meghalaya. He worked at Tata Consultancy Services, Mumbai for a year before joining Indian Space Research Organisation (ISRO) in 2014, since than he is posted at NESAC, Umiam. At NESAC he is leading the UAV Technology Development, Operations & Maintenance team and also working as In-charge UAV Lab/Facility. He is having an experience of more than 10 years in the field of UAV systems design, development, operations and its applications (e.g. Agriculture, Forestry, Disaster Management, and Town Planning etc.), 3D printing technology, capacity building and training. He is also a DGCA certified UAV pilot and Trainer under SMALL class, Rotorcraft category drones. He has contributed to more than 15 papers/research articles in the field of UAV systems and applications. He is also helping B.Tech/M.Tech students to carry out their project work at NESAC UAV Lab.

Puyam S. Singh currently works as Scientist/Engineer–SF at the North Eastern Space Applications Centre (NESAC), specializing in Geo-spatial Intelligence and Computer Vision. He holds an M.E. in Computer Science from PEC University of Technology, Chandigarh. Previously, he worked as a research fellow at DRDO's Snow and Avalanche Study Establishment (SASE), contributing to avalanche forecasting models. At NESAC, he works on the development of space-based geospatial applications for societal benefit and e-Governance, integrating satellite and aerial data. His research focuses on computer vision algorithms for automated feature extraction from remote sensing imagery. He has mentored over 54 postgraduate and undergraduate students and has published 43+ papers in reputed journals and conferences. His innovative contributions have earned National and regional e-Governance Awards in 2015, 2018, and 2023. Puyam continues to advance AI-driven remote sensing solutions for operational and research applications in governance and development sectors.

Victor Saikhom is currently serving as Scientist/Engineer-SF at the Geoinformatics & IT Division, NESAC, Department of Space, with a Ph.D. in Remote Sensing, M.Tech in Remote Sensing & GIS, and B.E. in Electrical & Electronics Engineering. With over two decades of combined experience in industry and research, he specializes in Photogrammetry, 3D GIS, UAV Remote Sensing, and Geoinformatics. He has led key geospatial projects in land use mapping, infrastructure planning, terrain modeling, and precision agriculture using satellite, UAV, DGPS, and TLS data. His expertise includes DEM generation, orthomosaics, point cloud modeling, and multispectral analysis for disaster response and environmental monitoring. Dr. Saikhom also mentors students and professionals and serves as Course Director for UAV Remote Sensing at NESAC. He has published widely and received the National e-Governance Award (2018) and the 8th eNorthEast Award (2023) for his and his team's contributions to geospatial solutions for governance and regional development.

Ranjit Das is basically a civil engineer specialization in water resources. He joined North Eastern Space Applications Centre as scientist in the year 2004. He has been involved large number of studies carried out at NESAC. In 2009, he received an ISRO Team Excellence Award for the project "Assessment of Irrigation Potential Created in AIBP Funded Irrigation Projects in India Using Cartosat Data."This project was carried out under the auspices of NESAC. Presently, he involves as principal investigator in couple of regional, national projects carried out at NESAC. He published few research papers as a 1st or co-author in national/international conferences and Journals.

Jonali Goswami currently working as the Scientist-SF and Head Outreach & Capacity Building at North Eastern Space Applications Centre under Department of Space, Government of India, has more than 22 years of experience in space technology applications and research. Her research expertise spans Geospatial technology applications in agriculture and allied sectors, disaster management and epidemiological studies. She has developed expertise in hyperspectral and Unmanned Aerial Vehicle (UAV) data analysis and their applications in different fields of research and applications towards precision and climate resilience agriculture. She has published more than 50 research papers in national and international journals and conference proceedings. She has also many book chapters publications and guided a number of M.Sc./M.Tech and Ph.D. students.

Kasturi Chakraborty completed her M.Sc. and M.Phil in Ecology and completed her Ph.D. on 'Land use/land cover and forest change using Remote sensing and GIS in Barak basin'. She has led project works and contributed as team member for successful completion of important projects at NESAC. Few of the important are in the field of forest fire, forest working plan, forest change analysis at regional scale, land use land cover mapping, wetland conservation and management. She has published research articles on forest and land use change analysis using geospatial techniques. She is a recipient of ISRO, young scientist award (2009) and ISRO project team award (2019).

Jenita Mary Nongkynrih is currently Head Urban and Regional Planning Division at North Eastern Space Applications Centre under Department of Space, Government of India. She has more than 21 years of experience in space technology applications for urban & regional planning, natural resource management and disaster management. She has been leading mainly user specific projects relating to Risk-informed Master plan preparations, city disaster management plans and hazard risk vulnerability assessment for urban centers in North Eastern India. She has published research papers in national and international journals. She has guided a number of M.Sc./ M.Tech and Ph.D. students. She has been awarded ISRO Young Scientist Merit Award for 2011.

M. Somorjit Singh Heading the Geoscience Division at the North Eastern Space Applications Centre (NESAC), Shillong, Department of Space, Government of India. In 20 years of service at NESAC, he has contributed to various operational and R&D activities/projects. He has been honored 'ISRO Team Excellence Award for his significant contribution as a team member in 'National Geomorphology and Lineament Mapping (NGLM) project' which was initiated by National Remote Sensing Centre (NRSC), Hyderabad in collaboration with Geological Survey of India (GSI) under Natural Resources Census program. His research interests includes Geomorphology & Lineament Studies/Mapping, Infrastructural Planning (Site Suitability/feasibility Studies, road & railway alignment/realignment, Natural/Geo-hazard, Manmade Hazard (Industrial Hazard), Environmental Studies/Pollution (Heavy Metals) etc.

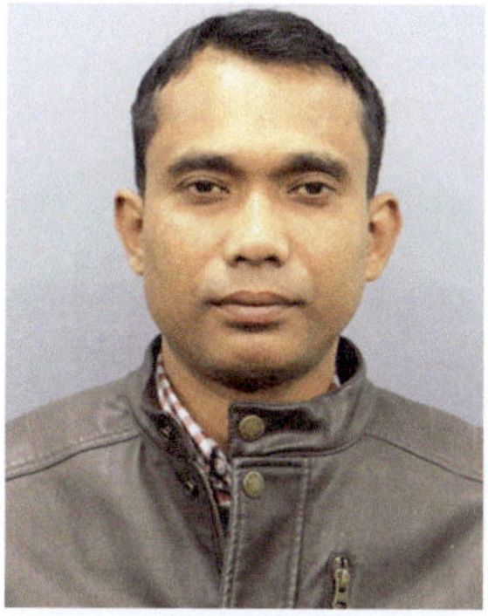

Dibyajyoti Chutia received the M. Tech. degree in information technology and the Ph.D. degree in computer science & engineering with specialization in soft computing technique for classification of geospatial data from Tezpur Central University, Tezpur, India, in 2000 and 2015, respectively. He is currently serving as head of Geoinformatics and IT division at North Eastern Space Applications Centre (NESAC), Department of Space, Government of India. His research interests include soft computing, machine learning, and deep learning applications for remotely sensed data. He has authored or coauthored more than 65 research articles in peer-reviewed journals, conferences, book chapters, etc. He has received the National e-Governance Award (2018, 2022), National Geospatial award (2020) and the eNorthEast Award (2020, 2023) for his and his team's contributions to geospatial innovations in governance and regional development.

Ramani Kumar Das has obtained M.Sc in Physics with specialization in Electronics from Gauhati University, Assam in the year of 2000 with First Class 1st Position (Gold Medalist). He is currently serving as Head, SATCOM & UAV Division (SUD) and Civil and Maintenance Division (CMD) at North Eastern Space Applications Centre (NESAC), Department of Space, Government of India, Shillong. His area of interest is Satellite Communication (SATCOM) and Satellite Navigation (SATNAV) Applications for Societal applications like Education, Health, communication support in disaster management, location-based services etc. and development of UAV technology and its applications. He is having more than 22 years of experience in the field. He is recipient ISRO's Team Excellence award in 2007 and ISRO's Young Scientist Award in 2009.

Shyam S Kundu did his M.Sc. in Physics from University of North Bengal and Ph.D. in Physics from Gauhati University. He has more than 20 years of professional experience at NESAC and currently working as the Head of the Space and Atmospheric Sciences Division at the centre. He was instrumental in setting up the Space and Atmospheric Science research facility at NESAC with several instruments, a S-band Doppler Weather Radar, largest network of weather stations in NER covering difficult terrains, three aerosol observatories including two in Himalayan region, network of lightning detectors, etc. His research interest include extreme weather forecasting in NER, lightning physics, aerosol-cloud-precipitation interaction studies and its climatic impact, climate change assessment, etc. He took lead role to develop an operational lightning early warning system for the NE region of India. He has more than 50 publications in peer reviewed journals and books. He is making significant contribution to professional bodies like ISRS as its Joint Secretary and ISPRS as the Secretary of the WG-4 under TC V. He also represents ISRO in the Committee on Earth Observation Satellites for Precipitation constellations.

Bijoy K. Handique currently working as the Head, Agriculture and Soil Division and Head, Program Planning and Evaluation Group at North Eastern Space Applications Centre under Department of Space, Government of India has more than 22 years of experience in space technology applications and research. He has been leading a number of research and applications programs at NESAC particularly in space technology applications in agriculture and allied areas. He is also leading a team at NESAC in using unmanned aerial vehicles for various agricultural applications. He has published about 60 research papers in national and international journals and conference proceedings. He has also published 12 book chapters and guided a number of M.Sc./M.Tech and Ph.D. students. He has been awarded ISRS-SPECK Award for 2005, ISRO Young Scientist Merit Award for 2008 and ISRO-ASI Young Scientist Award for 2014.

K. K. Sarma Ph. D, leads the Remote Sensing Applications Group (RSAG) at the North Eastern Space Applications Centre (NESAC), where he has been a key figure since 2002. Specializing in remote sensing for forestry and biodiversity, his work leverages space technology for critical applications in Biodiversity, Ecosystems, Forestry, and Natural Resource Management. Dr. Sarma has spearheaded over 60 diverse geospatial projects, contributing significantly to natural resource mapping and monitoring across the North Eastern Region (NER). Currently, he coordinates 110 "Plan of Action" projects, focused on using space technology to improve livelihoods in the NER. He also served as the Principal Investigator and NER Coordinator for all ISRO natural resource management projects, including comprehensive Land Use Land Cover Mapping and National Wasteland Updating, Monitoring, and Change Analysis. With over 75 research publications, Dr. Sarma has successfully supervised six Ph.D. students and actively mentors many others across NER Universities.

S. P. Aggarwal is the Director of the North Eastern Space Applications Centre (NESAC), Dept. of Space, Govt. of India, Shillong. He holds a B. Tech. degree from Allahabad University, a Postgraduate and Ph.D. degree from the Indian Agricultural Research Institute (IARI), New Delhi. He did his post-doctoral research from ITC/IHE, The Netherlands. A distinguished space applications scientist and visionary leader, Dr. Aggarwal has been pivotal in spearheading transformative initiatives that harness space technology for sustainable development across the North Eastern Region of India. Under his leadership, NESAC has adopted cutting-edge technologies (AI, ML, UAVs), enhanced stakeholder engagement through NeUIM, and earned national accolades for excellence in geospatial applications. His contributions span a wide spectrum of applications, including natural resource management, climate-resilient infrastructure planning, disaster risk reduction, weather forecasting, satellite-based communication, and geo-governance. Earlier, he served as Group Head, Water Resources Department, at the Indian Institute of Remote Sensing, ISRO, Dehradun, where he played a crucial role in advancing remote sensing and GIS-based solutions for water resources management, flood forecasting, disaster management, and climate change impact assessment. He also held the position of Program Coordinator of CSSTEAP, a United Nations-affiliated centre, where he significantly contributed to capacity-building in space science and technology education across the Asia-Pacific region. He has published over 200 scientific articles and book chapters, and has delivered more than 100 invited lectures at reputed national and international forums. He has supervised over 30 M. Tech. and Ph. D. theses, mentoring the next generation of scientists and engineers. In recognition of his exemplary contributions to the field of space applications and geospatial sciences, he has received numerous prestigious awards, including the National Geospatial Award and the National e-Governance Award. He is a life member and honorary fellow of several national and international scientific societies. Currently, he serves as the President of the Indian Society of Remote Sensing (ISRS), and has previously held key editorial and leadership roles, including Executive Editor of the Journal of ISRS.

Publisher Correction to: Advancement in Drone Data Processing Analytics and Applications

Victor Saikhom, Puyam S. Singh, Nilay Nishant, and Dibyajyoti Chutia

Publisher Correction to: Chapter 19: S. P. Aggarwal et al. (eds.), *Remote Sensing of the North Eastern Himalayan Ecosystem*, https://doi.org/10.1007/978-981-95-3221-6_19

The original version of the title was inadvertently published. Belated corrections received for Chapter 19, and the same have been carried out. Accordingly, pagination has been revised from page 579 onwards.

The updated version of this chapter can be found at
https://doi.org/10.1007/978-981-95-3221-6_19

S. P. Aggarwal et al. (eds.), *Remote Sensing of the North Eastern Himalayan Ecosystem*,
https://doi.org/10.1007/978-981-95-3221-6_26

Glossary

3DVAR 3 Dimensional Variational Data Assimilation
4DVAR 4 Dimensional Variational Data Assimilation
AASMAN Agromet Advisory Services and Management of Farm Resources
ABL Atmospheric Boundary Layer
ABLH ABL Height
ACI Aerosol-Cloud Interaction
ADI Amplitude Dispersion Index/Area Diversity Index
AFWA Air Force Weather Agency
AGB Above Ground Biomass
AGRI Advanced Geosynchronous Radiation Imager
AHI Advanced Himawari Imager
AHP Analytical Hierarchy Process
AI Artificial Intelligence
AICTE All India Council of Technical Education
AIRS Atmospheric Infrared Sounder
ALADIN Atmospheric Laser Doppler Instrument
ALERT Automated Local Evaluation in Real Time
ALOS Advanced Land Observing Satellite
AMRUT Atal Mission for Rejuvenation and Urban Transformation
AMSR-E Advanced Microwave Scanning Radiometer for EOS
AMV Atmospheric Motion Vectors
ANN Artificial Neural Network
AOD Aerosol Optical Depth
AP Afternoon Peak
API Application Programming Interface
ARG Automatic Rain Gauge
ARI Aerosol-Radiation Interaction
ARW Advanced Research WRF
ASCAT Advanced Scatterometer
ASDMA Assam State Disaster Management Authority

S. P. Aggarwal et al. (eds.), *Remote Sensing of the North Eastern Himalayan Ecosystem*,
https://doi.org/10.1007/978-981-95-3221-6

ASI Italian Space Agency
ASIS Agricultural Stress Index System
ASSAC Assam State Space Application Centre
ASTER Advanced Spaceborne Thermal Emission and Reflection Radiometer
ATLID Atmospheric Lidar
ATMS Advanced Technology Microwave Sounder
ATSR Along Track Scanning Radiometer
AVHRR Advanced Very High-Resolution Radiometer
AWC Available Water Capacity
AWD Alternate Wetting and Drying
AWiFS Advanced Wide Field Sensor
AWS Automatic Weather Station
BAI Burned Area Index
BAIM Improved Burned Area Index
BBR Broad Band Radiometer
BC Black Carbon
BFS Bharat Forecast System
BIS Bureau of Indian Standards
BLNET Beijing Lightning Network
BPH Brown Plant Hopper
BrC Brown Carbon
BREB Bowen Ratio Energy Balance
BRV Brahmaputra River Valley
BSZ Barapani Shear Zone
BVLOS Beyond Visual Line of Sight
C&RD Community and Rural Development
CAIPEEX Cloud Aerosol Interaction and Precipitation Enhancement Experiment
CALIOP Cloud-Aerosol Lidar with Orthogonal Polarization
CALIPSO Cloud-Aerosol Lidar and Infrared Pathfinder Satellite Observations
CAPE Convective Available Potential Energy
CartoDEM Cartosat-1 Digital Elevation Model
CBRI Central Building Research Institute
CCDC Continuous Change Detection and Classification
CCE Competitive Complex Evolution
CCN Cloud Condensation Nuclei
CDD Consecutive Dry Days
CD-ROM Compact Disc-Read Only Memory
CERES Clouds and The Earth's Radiant Energy System
CG Cloud to Ground
CHAMAN Coordinated Horticulture Assessment and Management using Geo-Informatics
CHM Canopy Height Model
CI Chlorophyll Index/Consistency Index
CLUI Cultivated Land Utility Index
CMA China Meteorological Administration

CME Continuing Medical Education
CMF Churachandpur Mao Fault
CMIP6 Coupled Model Intercomparison Project Phase-6
CMORPH Climate Prediction Centre Morphing Technique
CMV Cloud Motion Vector
CN Curve Number
CNC Computer Numerical Control
CNES Centre National D'études Spatiales/ National Centre for Space Studies
CNN Convolutional Neural Network
CODE Centre of Orbit Determination in Eu
CORS Continuously Operating Reference Stations
COVID19 COronaVIrus Disease of 2019
CP Cumulus Physics
CPC Climate Prediction Centre
CPR Cloud Profiling Radar
CR Consistency Ratio
CRA Climate-Resilient Agriculture
CRED Centre for Research on the Epidemiology of Disasters
CRF Community Reserve Forest
CRIS Cross-Track Infrared Sounder
CSB Central Silk Board
CSRM Crop Surface Reflectance Modeling
CT Canyon Thrust
CTGT Carbon Dioxide Temperature Gradient Tunnel
CV Coefficient of Variation/Computer Vision
CW Consecutive Weeks
CWC Central Water Commission
CWD Consecutive Wet Days
CWT Concentrated Weighted Trajectory
DA Data Assimilation
DBH Diameter at Breast Height
DBMS Database Management System
DDI Developmental Disability Index
DDMA District Disaster Management Authority
DECU Developmental Education and Communication Unit
DEM Digital Elevation Model
DF Dauki Fault
DGPS Differential Global Positioning System
DGPS Differential GPS
DHaBReT Drone Haka Buit Rep Thymai
DHI Drought Hazard Index
DInSAR Differential Interferometric Synthetic Aperture Radar
DL Deep Learning
DLR German Aerospace Center (Deutsches Zentrum für Luft- und Raumfahrt)
DMA Disaster Management Act

DMS Disaster Management Support
dNBR Delta Normalized Burn Ratio
DoS Department of Space
DPR Dual-Frequency Precipitation Radar
DPRs Detailed Project Reports
DR Double Rotation
DRF Direct Radiative Forcing
DRR Disaster Risk Reduction
DSAS Digital Shoreline Analysis System
DSC Decision Support Centre
DSM Digital Surface Model
DSS Decision Support System
DSSAT Decision Support System for Agrotechnology Transfer
DT Dapsi Thrust
DTC Decision Tree Classification
DTM Digital Terrain Models
DTR Diurnal Temperature Range
DTRL Defense Terrain Research Laboratory
DWLR Doppler Water Level Recorder
DWR Doppler Weather Radar
EAL Elevated Aerosol Layer
EarthCARE Earth, Cloud, Aerosol and Radiation Explorer
EC Elemental Carbon
ECG Electrocardiogram
ECMWF European Centre for Medium-Range Weather Forecasts
ECS Equilibrium Climate Sensitivity
EFFIS European Forest Fire Information System
EFFS European Flood Forecasting System
EFM Electric Field Mill
EH Eastern Himalayan
EHR Eastern Himalayan region
EHS Eastern Himalaya Syntaxis
EM-DAT Emergency Events Database
EMR Electronic Medical Record
EMS Electromagnetic Spectrum
ENTLN Earth Network Total Lightning Network
EO Earth Observation
EOS Earth Observation Satellite
EPR End Point Rate
ERA5 European Centre for Medium Range Weather Forecasts Version-5
ERFS Extended Range Forecasts
EROS Earth Remote Observation Satellite
ESA European Space Agency
ET Evapotranspiration
ETCCDI Expert Team on Climate Change Detection Indices

ETKF Ensemble Transform Kalman Filter
ETM+ Enhanced Thematic Mapper Plus
ETS Equitable Threat Score
EUCLID European Cooperation for Lightning Detection
EUMETSAT European Organization for The Exploitation of Meteorological Satellites
EVI Enhanced Vegetation Index
EVM Electronic Voting Machine
EWS Early Warning System
FAA Federal Aviation Administration
FAO Food and Agriculture Organization
FASA Forecasting Agricultural Output Using Space, Agro-Meteorology and Land-Based Observation
FCC False Color Composite
FFG Flash Flood Guidance
FHZ Flood Hazard Zonation
FIMMA Fire Identification, Mapping and Monitoring Algorithm
FIRMS Fire Information for Resource Management System
FLDAS Famine Early Warning Systems Network Land Data Assimilation System
FLEWS Flood Early Warning System
FPV First-Person View
FRP Fire Radiative Power
FSI Forest Survey of India
FY-4A Fengyun 4A
GAN Generative Adversarial Network
GBA2000 Global Burnt Area 2000
GCP Ground Control Point
GDP Gross Domestic Product
GDPR General Data Protection Regulation
GEE Google Earth Engine
GEO Geostationary Earth Orbit/Group on Earth Observations
GeoAI Geospatial Artificial Intelligence
GEOS ADAS Goddard Earth Observing System Atmospheric Data Assimilation System
GFCS Global Framework for Climate Services
GFED4S Global Fire Emissions Database, Version 4 with Small Fires
GFFG Gridded Flash Flood Guidance
GFMC Global Fire Monitoring Center
GFS Global Forecast System
GHADC Garo Hills Autonomous District Council
GHG Green House Gases
GIIRS Geostationary Interferometric Infrared Sounder
GIR GPM Infrared
GIS Geographic Information System
GISAT GEO Imaging Satellite

GKMS Gramin Krishi Mausam Sewa
GLD360 Global Lightning Detection Network 360
GLM Geostationary Lightning Mapper
GLOBSCAR Global Burn Scars
GLOF Glacial Lake Outburst Flood
GMI GPM Microwave Imager
GMS GPM Multi-Sensor
GMW GPM Microwave
GNDVI Green Normalized Difference Vegetation Index
GNN Graph Neural Network
GNSS Global Navigation Satellite Systems
GOES Geostationary Operational Environmental Satellite
GOFC/GOLD Global Observations of Forest and Land Cover Dynamics
GoI Govt of India
GPCC Global Precipitation Climatology Centre
GPF General Planar Fit
GPI GOES Precipitation Index
GPM Global Precipitation Measurement
GPS Global Positioning System
GRAS Global Navigation Satellite System Receiver for Atmospheric Sounding
GRD Ground Range Detected
GSAT Geostationary Satellite
GSD Ground Sampling Distance
GSI Geological Survey of India/Global Statistical Interpolation
GSMaP Global Satellite Mapping of Precipitation
GTS Global Telecommunication System
GWIS Global Wildfire Information System
HBV Hydrological Bureau of Vatten
HEC RAS Hydrologic Engineering Center-River Analysis System
HEC-HMS Hydrologic Engineering Center-Hydrologic Modeling System
HEM Hydro-Estimator Method
HEP Hydro-Electric Project
HFL High Flood Level
HFT Himalayan Frontal Thrust
HIPAA Health Insurance Portability and Accountability Act
HIS Hyper-Spectral Imaging HSI
HPC High Performance Computing
HTS High Throughput Satellite
HVRA Hazard, Vulnerability, and Risk Assessment
HYBRID Hybrid ETKF-3DVAR DA
Hyperion Hyperspectral Imager
HySIS Hyperspectral Imaging Satellite
HYSPLIT Hybrid Single Particle Lagrangian Integrated Trajectory
IAM Integrated Assessment Model
IASI Infrared Atmospheric Sounding Interferometer

IC Inter-Cloud/Intra-Cloud
ICAR Indian Council of Agricultural Research
ICESat-2 Ice, Cloud, and Land Elevation Satellite-2
ICIMOD International Centre for Integrated Mountain Development
ICR Itanagar Capital Region
ICT Information and Communication Technology
IDU In Door Unit
IGNOU Indira Gandhi National Open University
IGP Indo-Gangetic Plains
IGRF12 International Geomagnetic Reference Field
IGS International GNSS Service
IIoT Industrial Internet of Things
IIT Indian Institute of Technology
IITM Indian Institute of Tropical Meteorology
IIZ Imaging Infrared Radiometer
ILLN Indian Lightning Location Network
IMD India Meteorological Department
IMDAA Indian Monsoon Data Assimilation and Analysis
IMERG Integrated Multi-satellitE Retrievals for GPM
IMR INSAT Multi-Spectral Rainfall
IMSRA INSAT Multi-Spectral Rainfall Algorithm Technique
INFFRAS Indian Forest Fire Response and Assessment System
InSAR Interferometric Synthetic Aperture Radar
INSAT Indian National Satellite System
IoT Internet of Things
IPCC Intergovernmental Panel on Climate Change
IRGASON Integrated CO_2 and H_2O Open-Path Gas Analyzer and 3D Sonic
IRS Indian Remote Sensing Satellite
ISFR India State of Forest Report
ISRO Indian Space Research Organisation
ISS International Space Station
ITSZ Indus Tsangpo Suture Zone
JAXA Japan Aerospace Exploration Agency
JJAS June July August September
JMA Japan Meteorological Agency
JPSS Joint Polar Satellite System
KaPR Ka-Band Precipitation Radar
KISAN Pradhan Mantri Kisan Samman Nidhi
KLDN Korea Meteorological Administration Lightning Detection Network
KNN k-Nearest Neighbour
KOMPSAT Korea Multipurpose Satellite
KuPR Ku-Band Precipitation Radar
KVKs Krishi Vigyan Kendras
LAI Leaf Area Index
LCC Land Capability Classification

LEO Low Earth Orbit
LEWS Landslide Early Warning System
LF Low Frequency
LGP Length of the Growing Period
LHEF Landslip Hazard Evaluation Factor
LHZ Landslide Hazard Zonation
LI Lifted Index
LiDAR Light Detection and Ranging
LISS Linear Imaging Self-Scanning Sensor
LLM Large Language Model
LMI Lightning Mapping Imager
LMS Learning Management System
LOD Level-of-Detail
LOS Line of Sight
LPA Long Period Average
LPG Liquefied petroleum gas
LPI Lightning Potential Index
LPM Laser Precipitation Monitor
LR Linear Regression
LRF Long-Range Forecasts
LSM Landslide Susceptibility Mapping
LST Land Surface Temperature
LSTM Long Short-Term Memory
LSU Linear Spectral Unmixing
LSWI Land Surface Water Index
LT Lohit Thrust
LULC Land Use Land Cover
LUP Land-Use Planning
MADRAS Microwave Analysis and Detection of Rain and Atmospheric Structures
MAE Mean Absolute Error
MAM March–April–May
MBDA Meghalaya Basin Development Agency
MBT Main Boundary Thrust
MCD45A1 MODIS Burned Area Product (Collection 5.1)
MCDA Multi-Criteria Decision Analysis
MCDM Multi-Criteria Decision-Making
MCI Multiple Cropping Index
MCS Mesoscale Convective System
MCT Main Central Thrust
MDoNER Ministry of Development of North Eastern Region
ME Mean Error
MeFFIS Meghalaya Forest Fire Information System
MEO Medium Earth Orbit
MERRA Modern Era-Retrospective Analysis for Research and Applications
Metop Meteorological Operational Satellite

MF Mat Fault/Matched Filtering
MFT Main Frontal Thrust
MI Moisture Index
MIR Mid Infrared
MIRAS Microwave Imaging Radiometer Using Aperture Synthesis
MK Mann-Kendall (Trend Test)
ML Machine Learning
MLC Maximum Likelihood Classification
MLR Multiple Linear Regression
MMCFS Monsoon Mission Climate Forecast System
MMF Morgan-Morgan-Finney model
MMM Mesoscale and Microscale Meteorology
MNCFC Mahalanobis National Crop Forecast Centre
MNF Minimum Noise Fraction
MoAFW Ministry of Agriculture and Farmers Welfare
MODIS ET Moderate Resolution Imaging Spectroradiometer Evapotranspiration
MoEFCC Ministry of Environment, Forest and Climate Change
MoES Ministry of Earth Sciences
MoM Ministry of Mines
MOSDAC Meteorological and Oceanographic Satellite Data Archival Centre
MSI Multi Spectral Instrument/Multi-Spectral Imager
MSS Multispectral Scanner
MT Mishmi Trust
NADAMS National Agricultural Drought Assessment and Monitoring System
NAPCC National Action Plan on Climate Change
NASA National Aeronautical and Space Administration
NBR Normalized Burn Ratio
NBSS&LUP National Bureau of Soil Survey and Land Use Planning
NCAR National Center for Atmospheric Research
NCEP National Centers for Environmental Prediction
NCERT National Council of Educational Research and Technology
NCMRWF National Centre for Medium Range Weather Forecasts
NCRB National Crime Report Bureau
NCS National Centre for Seismology
NDEM National Database for Emergency Management
NDMA National Disaster Management Authority
NDRE Normalized Difference Red Edge
NDRRMA Nepal National Disaster Risk Reduction and Management Authority, Nepal
NDSI Normalized Difference Snow Index
NDVI Normalized Difference Vegetation Index
NDWI Normalized Difference Water Index
NEC North Eastern Council
NEEPCO North Eastern Electric Power Corporation Limited
NEH Northeastern Himalayas

NEHR North East Himalayan Region
NEIST North East Institute of Science and Technology
NER North Eastern Region
NERDRR North Eastern Regional Node for Disaster Risk Reduction
NERTPS North East Region Textile Promotion Scheme
NESAC North Eastern Space Applications Centre
NeSDR North Eastern Spatial Data Repository
NESIDS Northeast Special Infrastructure Development Scheme
NFI National Forest Inventories
NGO Non-Government Organization
NHB National Horticultural Board
NICRA National Innovations in Climate Resilient Agriculture
NIR Near Infrared
NISAR NASA ISRO Synthetic Aperture Radar
NIT National Institute of Technology
NLP Natural Language Processing
NMMT Meteorological Sub-Division of Nagaland, Manipur, Mizoram, and Tripura
NMSA National Mission for Sustainable Agriculture
NOAA National Oceanic and Atmospheric Administration
NPSTA National Programme on the Use of Space Technology for Agriculture
NRI Normalize Ratio Index
NRL Naval Research Laboratory
NRSC National Remote Sensing Centre
NSE Nash-Sutcliffe Efficiency
NSIL New Space India Limited
NSMP National Soil Mapping Programme
NSSO National Sample Survey Office
NT Naga Thrust
NWP Numerical Weather Prediction
NWPC National Working Plan Code
OBIA Object-Based Image Analysis
OC Organic Carbon
ODU Out Door Unit
OFDM Orthogonal Frequency Division Multiplexing
OlF Oldham Fault
OLI Operational Land Imager
OMI Ozone Monitoring Instrument
OMPS Ozone Mapping and Profiler Suite
OSCAT Oceansat Scatterometer
OSDMA Odisha State Disaster Management Authority
PAC Precipitation Accumulation
PAH Polycyclic Aromatic Hydrocarbons
PAN Panchromatic
PBIAS Percent Bias
PBL Planetary Boundary Layer

PCA Principal Component Analysis
PERSIANN Precipitation Estimation from Remotely Sensed Information Using Artificial Neural Networks
PET Potential Evapotranspiration
PF Peak Flow
PM Particulate Matter
PMKSY Pradhan Mantri Krishi Sinchayee Yojana
PMN Post Monsoon
POA Plan of Action
PolSAR Polarimetric Synthetic Aperture Radar
POTS Plain Old Telephone Service
PPI Plan Position Indicator
PPS Post-Processed Kinematic
PRCPTOT Annual Total Precipitation in Wet Days
PRISMA Precursore IperSpettrale della Missione Applicativa
PS Persistent Scatterers
PS-InSAR Persistent Scatterer Interferometry
PSSPa Pigmnt Specific Simple Ratio
QAM Quadrature Amplitude Modulation
QPE Qualitative Precipitation Estimation
QPF Quantitative Precipitation Forecasts
QSM Quantitative Structure Model
RADAR Radio Detection and Ranging
RAID Redundant Array of Independent Disks
RBAC Role-Based Access Control
R-CNN Region-Based Convolutional Neural Network
RCP Representative Concentration Pathways
REDD+ Reducing Emissions from Deforestation and Forest Degradation
RF Radio Frequency/Random Forest
RGB Red Green Blue
RH Relative Humidity
RI Random Index
RIMS Regional Institute of Medical Sciences
RMSE Root Mean Square Error
ROI Region of Interest
ROT Receive Only Terminal
RS Remote Sensing
RSOD Remote Sensing Object Detection
RTK Real-Time Kinematic
RTK-GNSS Real-Time Kinematic Global Navigation Satellite System
RTTOV Radiative Transfer for TOVS (TIROS Operational Vertical Sounder)
RUAV Rotary-Wing Unmanned Aerial Vehicle
RV Relative Vorticity
RVI Radar Vegetation Index
SAM Spectral Angle Mapper

SAPHIR Sounder for Probing Vertical Profiles of Humidity
SAR Synthetic Aperture Radar
SASI Shortwave Angle Slope Index
SATCOM Satellite Communication
SAVI Soil Adjusted Vegetation Index
SBAS Satellite-Based Augmentation Systems/Small Baseline Subset
ScaRaB Scanner for Radiation Budget
SCATSAT-1 Scatterometer Satellite-1
S-CCD Satellite Charge-Coupled Device
SCE Shuffled Complex Evolution
SCI System of Crop Intensification
SCS-CN Soil Conservation Service-Curve Number
SDGs Sustainable Development Goals
SDI Spatial Data Infrastructure
SDMA State Disaster Management Authority
SDRF State Disaster Response Force
SDSS Spatial Decision Support System
SEBAL Surface Energy Balance Algorithm for Land
SEOC State Emergency Operations Centres
SEP Space Environment Package
SEVIRI Spinning Enhanced Visible and Infrared Imager
SF Safety Factor
SF6 Sulfur Hexafluoride
SFF Spectral Feature Fitting
SfM Structure from Motion
SHC Soil Health Card
SHF Sensible Heat Flux
SHWB Sub-Himalayan West Bengal
SILAAS Satellite Integrated Landslide Assessment and Alert System
SIT Satellite Interactive Terminal
SMA Soil Moisture Accounting/Spectral Mixture Analysis
SMAP Soil Moisture Active Passive
SMI Soil Moisture Index
SMOS Soil Moisture and Ocean Salinity
SNPP Suomi National Polar-Orbiting Partnership
SNR Signal-to-Noise Ratio
SOA Secondary Organic Aerosols
SODAR Sound Detection and Ranging
SOI Survey of India
SPA Shillong Planning Area
SPF Sector-Wise Planar Fit
SPI Standardized Precipitation Index
SRES Special Report on Emission Scenarios
SRI Surface Rainfall Intensity/System of Rice Intensification
SRSAC State Remote Sensing Application Centre

SRTM Shuttle Radar Topography Mission
SSA Single Scattering Albedo
SSD Singular Spectrum Decomposition
SSP Shared-Socioeconomic Pathways
SSTM Sea Surface Temperature Monitor
STD Standard Deviation
STEC Slant Total Electron Content
STEM Sulfur Transport dEposition Model
Suomi-NPP Suomi National Polar-Orbiting Partnership
SVD Singular Value Decomposition
SVM Support Vector Machine
SWAT Soil and Water Assessment Tool
SWIR Shortwave Infrared
Tandem-X Terrasar-X Add-On for Digital Elevation Measurement
Tb/BT Brightness Temperature
TCN Tropospheric Columnar Nitrogen dioxide
TCR Transient Climate Response
TDR Time Domain Reflectometry
TE Tele-Education
TEHD Total Estimated Hazard
TI Thermal Index
TIR Thermal Infrared
TIROS-1 Television and Infrared Observation Satellite
TIRS Thermal Infrared Sensor
TKE Turbulent Kinetic Energy
TLS Terrestrial Laser Scanner
TM Thematic Mapper
TMI TRMM Microwave Imager
TMPA TRMM Multi-Satellite Precipitation Analysis
TOA Top-of-Atmosphere
TOPMODEL Topography-Based Hydrological Model
TPM Thunderstorm Potential Maps
TPS Town Planning Schemes
TR Triple Rotation
TRMM Tropical Rainfall Measuring Mission
TROPOMI Tropospheric Monitoring Instrument
TSEB Two-Source Energy Balance
TT&C Telemetry, Tracking and Command
UAV Unmanned Aerial Vehicle
UGC University Grants Commission
UKMO UKMet Office
UM Unified Model
UN/ISDR United Nations International Strategy for Disaster Reduction
UNCED United Nations Conference on Environment and Development
UNDRR United Nations Office for Disaster Risk Reduction

UNEP United Nations Environment Programme
UNFCCC United Nations Framework Convention on Climate Change
UNISDR United Nations International Strategy for Disaster Reduction
URDPFI Urban and Regional Development Plans Formulation and Implementation
USA United States of America
USACE United States Army Corps of Engineers
USDA United States Department of Agriculture
USGS United States Geological Survey
USLE Universal Soil Loss Equation
UTC Universal Time Mercator
VAR Variational
VCI Vegetation Condition Index
VHR Very High Resolution
VHRR Very High Resolution Radiometer
VIC LSM Variable Infiltration Capacity Land Surface Model
VIC Variable Infiltration Capacity
VIIRS Visible Infrared Imaging Radiometer Suite
VIO Visual Inertial Odometry
VIRS Visible and Infrared Scanner
VIS Visible Spectrum
ViT Vision Transformer
VLF Very Low Frequency
VMF1 Vienna Mapping Function
VNIR Visible and Near Infrared
VOC Volatile Organic Compound
VRC Village Resource Centre
VSAT Very Small Aperture Terminal
VTEC Vertical Total Electron Content
VTOL Vertical Take-Off and Landing
VVPAT Voter Verifiable Paper Audit Trail
VWS Vertical Wind Shear
WAPCOS Water and Power Consultancy Services (India) Limited
WHO World Health Organization
WMO World Meteorological Organization
WPS WRF Preprocessing System
WRD Water Resources Department
WRF Weather Research and Forecasting
WRFDA WRF Data Assimilation System
WRF-ELEC Weather Research and Forecasting Coupled with the Electrification (ELEC) Module
WSM6 WRF Single Moment 6 Class
WSN Wireless Sensor Network
WV Water Vapour
WWLLN World-Wide Lightning Location Network
YF Yemla Fault

YLF Yigonge Lulang Strike Slip Fault
YOLO You Only Look Once
YSU Yonsei University PBL